Natural Disasters

TENTH EDITION

Patrick L. Abbott
San Diego State University

Mc
Graw
Hill
Education

NATURAL DISASTERS, TENTH EDITION

Published by McGraw-Hill Education, 2 Penn Plaza, New York, NY 10121. Copyright © 2017 by
McGraw-Hill Education. All rights reserved. Printed in the United States of America. No part of this
publication may be reproduced or distributed in any form or by any means, or stored in a database or retrieval
system, without the prior written consent of McGraw-Hill Education, including, but not limited to, in any
network or other electronic storage or transmission, or broadcast for distance learning.

Some ancillaries, including electronic and print components, may not be available to customers outside the
United States.

This book is printed on acid-free paper.

1 2 3 4 5 6 7 8 9 0 RMN/RMN 1 0 9 8 7 6

ISBN 978-0-07-802298-2
MHID 0-07-802298-3

Senior Vice President, Products & Markets: *Kurt L. Strand*
Vice President, General Manager, Products & Markets: *Marty Lange*
Vice President, Content Design & Delivery: *Kimberly Meriwether David*
Managing Director: *Thomas Timp*
Brand Manager: *Michael R. Ivanov, PhD*
Director, Product Development: *Rose Koos*
Product Developer: *Jodi Rhomberg*
Executive Marketing Manager: *Tamara Hodge*
Director of Development: *Rose Koos*
Director, Content Design & Delivery: *Linda Avenarius*
Program Manager: *Lora Neyens*
Content Project Managers: *Sherry Kane/Tammy Juran*
Buyer: *Susan K. Culbertson*
Design: *Keith McPherson/Tara McDermott/Egzon Shaqiri*
Content Licensing Specialists: *DeAnna Dausener/Carrie K. Burger*
Cover Image: NASA's Goddard Space Flight Center and NASA Center for Climate Simulation
Video and images courtesy of NASA/GSFC/William Putman
Compositor: *SPi Global*
Printer: *R. R. Donnelley*

All credits appearing on page or at the end of the book are considered to be an extension of the copyright
page.

Library of Congress Cataloging-in-Publication Data
Abbott, Patrick L.
  Natural disasters / Patrick L. Abbott, San Diego State University. – Tenth edition.
    pages cm
  ISBN 978-0-07-802298-2 (alk. paper)
    1. Natural disasters. I. Title.
GB5014.A24 2017
904′.5–dc23

2015025069

The Internet addresses listed in the text were accurate at the time of publication. The inclusion of a website
does not indicate an endorsement by the authors or McGraw-Hill Education, and McGraw-Hill Education
does not guarantee the accuracy of the information presented at these sites.

www.mhhe.com

About the Author

Patrick L. Abbott Patrick Abbott is a native San Diegan. Pat earned his MA and PhD degrees in geology at the University of Texas at Austin. He benefited greatly from the depth and breadth of the faculty in the Department of Geological Sciences at Austin; this was extended by their requirement to take five additional graduate courses outside the department. Developing interests in many topics helped lead to writing this textbook.

Pat's research has concentrated on the Mesozoic and Cenozoic sedimentary rocks of the southwestern United States and northwestern Mexico. Studies have focused on reading the history stored within the rocks—depositional environments, provenance, paleoclimate, palinspastic reconstructions, and high-energy processes.

Pat has long been involved in presenting Earth knowledge to the public, primarily through TV news. He has produced award winning videos for TV broadcast. He was one of the main cast members in the TV series *The Real Gilligan's Island* on TBS, *Serial Killer Earth* on H2 (The History Channel 2), and *So You Think You'd Survive* on The Weather Channel. During part of each year, Pat works as a Smithsonian lecturer visiting all continents and oceans.

Brief Contents

*Available in expanded form as chapter 18 The Great Dyings

Contents

CHAPTER 4

Plate Tectonics and Earthquakes 77

CHAPTER 5

Earthquakes Throughout the United States and Canada 107

CHAPTER 14

Fire 378

CHAPTER 15

Mass Movements 407

Preface

Why Study Natural Disasters?

Natural disasters occur every day and affect the lives of millions of people each year. Many students have been affected by earthquakes or tornadoes or hurricanes or floods or landslides or wildfires or other events. They are interested in lectures that explain these processes, and lively discussions commonly ensue.

During decades of teaching courses at San Diego State University, I found that students have an innate curiosity about "death and destruction"; they want to know why natural disasters occur. Initiation of a Natural Disasters course led to skyrocketing enrollments that now exceed 5,000 students per year. Some of these experiences are described in a *Journal of Geoscience Education* article by Pat Abbott and Ernie Zebrowksi [v 46 (1998), pp. 471–75].

Themes and Approach

This textbook focuses on explaining how the normal processes of the Earth concentrate their energies and deal heavy blows to humans and their structures. The following themes are interwoven throughout the book:

- Energy sources underlying disasters
- Plate tectonics
- Climate change
- Earth processes operating in rock, water, and atmosphere
- Significance of geologic time
- Complexities of multiple variables operating simultaneously
- Detailed and interesting case histories

New to This Edition

- Many of the Tables and Figures have been updated and more than 60 new ones have been added.
- Chapter 1: Extensive updating of all disaster and demographic data.
- Chapter 2: Isostasy coverage expanded with new figure.

- Chapter 3: New *In Greater Depth* section compares seismic waves of earthquakes versus nuclear bomb blast waves.
- Chapter 4: Expanded text on 2011 Japan earthquake; added the 2015 Nepal earthquake; rewrite earthquakes in the Holy Land.
- Chapter 5: New section on earthquake early warning system; expanded section on human-triggered earthquakes, fracking and the U.S. economy; added *Side Note* explaining trial of Italian scientists over lack of warning before L'Aquila earthquake.
- Chapter 6: Major expansion of flood basalts to include their role in mass extinctions; major rewrite of *Side Note* How a Geyser Erupts with new data from Old Faithful, new photo and new figure.
- Chapter 7: Describe surprise eruption and deaths on Mount Ontake, Japan; Add new *In Greater Depth* explaining new understanding of the rapid assembly and rise of magma bodies.
- Chapter 8: Expand description of 2011 Japanese tsunami.
- Chapter 10: New images of tornadoes, hail, lightning.
- Chapter 11: Expands on Hurricane Sandy and transformation to a post-tropical cyclone. Adds *In Greater Depth* on How to Build a Home Near the Coastline.
- Chapter 12: Covers IPCC Assessment Report 5. Expanded discussion of Arctic Ocean sea ice. Added 21st-century sea-level rise, ocean acidification and fisheries.
- Chapter 13: Major rewrite of Red River of the North. Increased discussion of runoff reduction.
- Chapter 14: Adds information about houses as fuel.
- Chapter 15: Adds coverage of Oso, Washington landslide and debris flow. Expanded discussion of landslide mitigation: reshaping topography; strengthening slopes; draining water.
- Chapter 16: Adds deaths by shark bite. Building of massive structures to protect U.S. cities from sea-level rise versus Maldives protected by Mother Nature.
- Chapter 17: Cover Chelyabinsk meteor explosion. First landings on planets, their moons; asteroid; comet.
- Epilogue: Expansion of causes of mass extinctions. Added text on Australia and New Zealand.

Acknowledgments

I am deeply appreciative of the help given by others to make this book a reality. The photograph collection in the book is immeasurably improved by the aerial photographs generously given by the late John S. Shelton, the greatest geologist photographer of them all. Please see John's classic book *Geology Illustrated*.

The quality of the book was significantly improved by the insights provided by comments from the following reviewers:

Baylor University, John Dunbar
California State Polytechnic University–Pomona, Jon Nourse
California State University–Fullerton, Stephen I. Wareham
California State University–Los Angeles, Hassan Rezaie Boroon
California State University, Los Angeles, Richard W. Hurst
California State University, Sacramento, Lisa Hammersley
Chandler–Gilbert Community College, John Dassinger
Colby College, James Rodger Fleming
College of Southern Idaho, Shawn P. Willsey
Dartmouth College, Leslie Sonder
Erie Community College, Buffalo State College, Karen S. Wehn
Fairleigh Dickinson University, Edward Catanzaro
Fort Lewis College, David Gonzales
Grand Valley State University, Patrick Colgan
Illinois State University, Robert S. Nelson
Indiana University, Bingming Shen-Tu
Lindenwood University, Sandra Allen
Manchester Community College, Eszter Samodai
Minnesota State University–Mankato, Cecil S. Keen
North Hennepin Community College, John Dooley
Northeastern University, Jennifer Cole
Northeastern University, Langdon D. Clough
Penn State University–Altoona, Timothy J. Dolney
Pennsylvania State University, Kevin P. Furlong
Radford University, Jonathan Tso
Rio Hondo Junior College, Michael Forrest
Salisbury University, Brent R. Skeeter
San Diego State University, Jim Rickard
San Diego State University, Victor E. Camp
San Francisco State University, Bridget James
San Francisco State University, Mary Leech
San Francisco State University, Oswaldo Garcia
Sonoma State University, Terry Wright
Southeast Arkansas University, Steven Sumner
Southeast Missouri State University, Ernest L. Kern
St. Cloud State University, Alan Srock

SUNY–Stony Brook, Christiane Stidham
Texas State University, Philip Suckling
The Arizona Geological Survey, Michael Conway
The Ohio State University, Michael Barton
Tulane University, Stephen A. Nelson
University at Albany, Michael G. Landin
University of British Columbia, Roland Stull
University of California–Santa Barbara, Cathy Busby
University of California Santa Cruz, Thorne Lay
University of California–Davis, John F. Dewey
University of California–Riverside, Peter Sadler
University of California–San Diego, Gabi Laske
University of Colorado–Boulder, Charles R. Stern
University of Colorado–Colorado Springs, Paul K. Grogger
University of Colorado, Alan Lester
University of Illinois at Urbana–Champaign, Wang-Ping Chen
University of Kansas, David Braaten
University of Kansas, Don Steeples
University of Kentucky–Lexington, Kevin Henke
University of Michigan, Youxue Zhang
University of Nebraska at Kearney, Jeremy S. Dillon
University of Nebraska–Kearney, A. Steele Becker
University of Nebraska–Kearney, Jean Eichhorst
University of Nebraska–Kearney, Stanley Dart
University of Nebraska–Kearney, Vijendra Boken
University of North Carolina–Greensboro, John Hidore
University of Oklahoma, Judson Ahern
University of Portland, Robert Butler
University of Southern California, John P. Wilson
University of Wisconsin–LaCrosse, George Hupper
Utah State University–Logan, Sue Morgan
Washington University–St. Louis, Carol Prombo
Yale University, David Bercovici

Special thanks to the following individuals who wrote and/or reviewed learning goal-oriented content for **LearnSmart**.

California State University–Sacramento, Lisa Hammersley
Northern Arizona University, Sylvester Allred
Roane State Community College, Arthur C. Lee

I sincerely appreciate the talents and accomplishments of the McGraw-Hill professionals in Dubuque who took my manuscript and produced it into this book. For the shortcomings that remain in the book, I alone am responsible. I welcome all comments, pro and con, as well as suggested revisions.

Pat Abbott
professor_pat_abbott@yahoo.com

McGraw-Hill Connect®
Learn Without Limits

Connect is a teaching and learning platform that is proven to deliver better results for students and instructors.

Connect empowers students by continually adapting to deliver precisely what they need, when they need it and how they need it, so your class time is more engaging and effective.

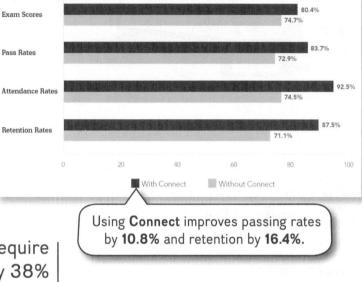

Course outcomes improve with Connect.

	With Connect	Without Connect
Exam Scores	80.4%	74.7%
Pass Rates	83.7%	72.9%
Attendance Rates	92.5%	74.5%
Retention Rates	87.5%	71.1%

■ With Connect ■ Without Connect

Using Connect improves passing rates by 10.8% and retention by 16.4%.

88% of instructors who use **Connect** require it; instructor satisfaction **increases** by 38% when **Connect** is required.

Analytics

Connect Insight®

Connect Insight is Connect's new one-of-a-kind visual analytics dashboard—now available for both instructors and students—that provides at-a-glance information regarding student performance, which is immediately actionable. By presenting assignment, assessment, and topical performance results together with a time metric that is easily visible for aggregate or individual results, Connect Insight gives the user the ability to take a just-in-time approach to teaching and learning, which was never before available. Connect Insight presents data that empowers students and helps instructors improve class performance in a way that is efficient and effective.

Connect helps students achieve better grades

	A	B	C	D
With Connect	36%	29.5%	22%	4.3%
Without Connect	27.2%	22.3%	25.6%	9.8%

Based on McGraw-Hill Education Connect Effectiveness Study 2013

Students can view their results for any **Connect** course.

Mobile

Connect's new, intuitive mobile interface gives students and instructors flexible and convenient, anytime–anywhere access to all components of the Connect platform.

Adaptive

THE FIRST AND ONLY **ADAPTIVE READING EXPERIENCE** DESIGNED TO TRANSFORM THE WAY STUDENTS READ

More students earn **A's** and **B's** when they use McGraw-Hill Education **Adaptive** products.

SmartBook®

Proven to help students improve grades and study more efficiently, SmartBook contains the same content within the print book, but actively tailors that content to the needs of the individual. SmartBook's adaptive technology provides precise, personalized instruction on what the student should do next, guiding the student to master and remember key concepts, targeting gaps in knowledge and offering customized feedback, driving the student toward comprehension and retention of the subject matter. Available on smartphones and tablets, SmartBook puts learning at the student's fingertips—anywhere, anytime.

Over **4 billion questions** have been answered making McGraw-Hill Education products more intelligent, reliable & precise.

STUDENTS WANT SMARTBOOK®

95% of students reported **SmartBook** to be a more effective way of reading material

100% of students want to use the Practice Quiz feature available within **SmartBook** to help them study

100% of students reported having reliable access to off-campus wifi

90% of students say they would purchase **SmartBook** over print alone

95% reported that **SmartBook** would impact their study skills in a positive way

*Findings based on a 2015 focus group survey at Pellissippi State Community College administered by McGraw-Hill Education

Prologue: Energy Flows

Earth, the Blue Marble as seen from Apollo 17 in 1972.
NASA.

LEARNING OUTCOMES

Earth is a planet with varied flows of energy that can cause problems for humans. After studying the Prologue you should

- know the main flows of energy on Earth.
- comprehend how internal energy creates land.
- understand how external energy destroys land.
- be familiar with the rock cycle.

Disasters occur where and when Earth's natural processes concentrate energy and then release it, killing life and causing destruction. Our interest is especially high when this energy deals heavy blows to humans. As the growth of the world's population accelerates, more and more people find themselves living in close proximity to Earth's most hazardous places. The news media increasingly present us with vivid images and stories of the great losses of human life and destruction of property caused by natural disasters. As the novelist Booth Tarkington remarked: "The history of catastrophe is the history of juxtaposition."*

To understand the natural processes that kill and maim unwary humans, we must know about the energy sources that fuel them. Earth is an active planet with varied flows of energy from: (1) Earth's interior, (2) the Sun, (3) **gravity,** and (4) impacts with **asteroids** and **comets.**

Internal energy flows unceasingly from Earth's interior toward the surface. The interior of the Earth holds a tremendous store of heat accumulated from the initial impacts that formed our planet and from the heat released by the ongoing decay of **radioactive isotopes.** Over short time spans, internal energy is released as eruptions from **volcanoes** and as **seismic waves** from **earthquakes.** Over longer intervals of geologic time, the flow of internal energy has produced our **continents,** oceans, and **atmosphere.** On a planetary scale, this outflow of internal energy causes continents to drift and collide, thus constructing mountain ranges and elevated plateaus.

External energy is delivered by the Sun. About a quarter of the Sun's energy that reaches Earth evaporates and lifts water into the atmosphere. At the same time, the constant pull of gravity helps bring atmospheric moisture down as snow and rain. On short timescales, these processes bring us **hail, lightning, tornadoes, hurricanes,** and floods. Solar energy is also stored in plant tissue to be released later as fire. On a long timescale, the Sun and gravity power the agents of **erosion—glaciers,** streams, underground waters, winds, ocean waves and currents—that wear away the continents and dump their broken pieces and dissolved remains into the seas. Solar radiation is the primary energy source because it evaporates and elevates water, but gravity is the immediate force that drives the agents of erosion.

Gravity is an attractional force between bodies. At equal distances, the greater the mass of a body, the greater its gravitational force. The relatively great mass of the Earth has powerful effects on smaller masses such as ice and rock, causing ice to flow as avalanches and hillsides to fail in landslides and **debris flows.**

An energy source for disasters arrives when visitors from outer space—asteroids and comets—impact Earth. Impacts were abundant early in Earth's history. In recent times, collisions with large bodies have become infrequent. However, asteroids and comets traveling at velocities in excess of 30,000 mph occasionally slam into Earth, and their deep impacts have global effects on life.

The sequence of chapters in this book is based on energy sources, in the following order: Earth's internal energy, external energy supplied by the Sun, gravity, and impacts with space objects.

Earth's internal energy fuels volcanism, as well as providing the energy for earthquakes. Here, lava flows from the Pu'u O'o-Kupaianaha eruption in Hawaii meet the ocean, 18 August 2010.

Michael Poland/U.S. Geological Survey.

External energy from the Sun fuels tornadoes, as well as hurricanes, floods, and wildfires. Here, a powerful tornado spins down from a supercell thunderstorm and travels along an Oklahoma road.

© 2010 Willoughby Owen/Getty Images RF.

The pull of gravity brings down hillsides. This earthquake-triggered debris flow destroyed homes and killed 585 people in Santa Tecla, El Salvador on 13 January 2001.

Ed Harp/U.S. Geological Survey.

High-velocity comets and asteroids can impact the Earth and kill life worldwide. Here the Comet Lovejoy nears Earth's horizon behind airglow in the night sky.

Photo by NASA astronaut Dan Burbank from the International Space Station on 22 December 2001.

Processes of Construction versus Destruction

Another way to look at energy flow on Earth is by understanding the rock cycle and the construction and destruction of land (continents). Energy flowing up from Earth's interior melts rock that rises as **magma** and then cools and crystallizes to form **igneous rocks;** they are **plutonic rocks** if they solidify at depth or **volcanic rocks** if they cool and harden at the surface. These newly formed rocks help create new land. Igneous-rock formation is part of the internal energy–fed **processes of construction** that create and elevate landmasses.

At the same time, the much greater flow of energy from the Sun, working with gravity, brings water that weathers the igneous rocks exposed at or near the surface and breaks them down into **sediments. Physical weathering** disintegrates rocks into **gravel** and **sand,** while **chemical weathering** decomposes rock into **clay minerals.** The sediments are eroded, transported mostly by water, and then deposited in topographically low areas, ultimately the ocean. These external, energy-fed **processes of destruction** work to erode the lands and dump the debris into the oceans.

These land-building and land-destroying processes result from Earth's energy flows that create, transform, and destroy rocks as part of the rock cycle. Think about the incredible amount of work done by the prodigious flows of energy operating over the great age of Earth. There is a long-term conflict raging between the internal-energy-powered processes of construction, which create and elevate landmasses, and the external-energy-powered processes of destruction, which erode the continents and dump the continental debris into the ocean basins. Visualize this: If the interior of Earth cooled and the flow of internal energy stopped, mountain building and uplift also would stop; then the ongoing solar-powered agents of erosion

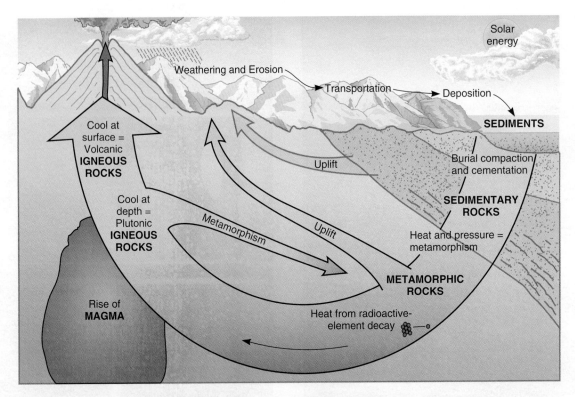

The rock cycle. Follow the cycle clockwise beginning in the lower left. Magma cools and solidifies to form igneous rocks. Rocks exposed at Earth's surface break down and decompose into sediments (e.g., gravel, sand, clay), which are transported, deposited, and hardened into sedimentary rock. With increasing burial depth, temperature and pressure increase, causing changes (or metamorphosis) of rocks into metamorphic rocks.

would reduce the continents to sea level in just 45 million years. There would be no more continents, only an ocean-covered planet.

Think about the timescales involved in eliminating the continents. At first reading, 45 million years of erosion may seem like an awfully long time, but the Earth is more than 4.5 billion years old. The great age of Earth indicates that erosion is powerful enough to have leveled the continents about 100 times. This shows the power of the internal processes of construction to keep elevating old continents and adding new landmasses. And woe to human and other life-forms that get too close to these processes of construction and destruction, for this is where natural disasters occur.

Terms to Remember

asteroid 1	igneous rock 3
atmosphere 1	lightning 1
chemical weathering 3	magma 3
clay minerals 3	physical weathering 3
comet 1	plutonic rocks 3
continent 1	processes of construction 3
debris flow 1	processes of destruction 3
earthquake 1	radioactive isotope 1
erosion 1	sand 3
glacier 1	sediment 3
gravel 3	seismic wave 1
gravity 1	tornado 1
hail 1	volcanic rocks 3
hurricane 1	volcano 1

CHAPTER 1

Natural Disasters and the Human Population

"Mankind was destined to live on the edge of perpetual disaster. We are mankind because we survive."

—JAMES A. MICHENER, 1978, *CHESAPEAKE.*, RANDOM HOUSE

The world population of humans continues to increase exponentially. Photo of shopping area in New Delhi, India.

© Dr. Parvinder Sethi

LEARNING OUTCOMES

The human population is growing rapidly. Natural disasters are causing great numbers of deaths and economic losses. After studying this chapter you should

- recognize the differences between a natural hazard, a natural disaster and a great natural disaster.
- be familiar with the processes that cause the deadliest natural disasters.
- understand the relationship between frequency and magnitude of natural disasters.
- know the size of the human population.
- understand the significance of exponential growth.
- recognize the demographic transition of human populations.
- be able to explain the concept of carrying capacity.

OUTLINE

- Great Natural Disasters
- Human Fatalities and Economic Losses in Natural Disasters
- Natural Hazards
- Overview of Human Population
- Future World Population
- Carrying Capacity

In 2013, there were 150 **natural disasters** that claimed 20 or more human lives. They were primarily caused by **earthquakes, hurricanes** (= **cyclones** = **typhoons**), floods, winter storms and heat waves; they killed more than 20,000 people. The 16 deadliest events are listed in table 1.1. As horrible as the 2013 death total is, it is markedly less than in 2010, when about 286,000 people were killed in two events alone (Haiti earthquake: 230,000; Russian heat wave: 56,000). All these disasters were the result of natural processes operating at high **energy** levels for brief times in restricted areas.

Great Natural Disasters

The Japan earthquake and tsunami in 2011, the Haiti earthquake in 2010 and the Myanmar cyclone and China earthquake in 2008 combined to kill almost 500,000 people. They are examples of **great natural disasters:** these events so overwhelm regions that international assistance is needed to rescue and care for people, clean up the destruction, and begin the process of reconstruction. Great natural disasters commonly kill thousands of people, leave hundreds of thousands homeless, and overwhelm the regional economy.

Today, in earthquake-active areas of the world, several hundred million people live in buildings that will collapse during a strong earthquake. An earthquake killing more than 100,000 people could happen any day in Teheran, Iran; in Istanbul, Turkey; or in other large cities. Today, people by the millions are moving to the ocean shores, where they can be hit by tsunami, hurricanes, and floods. We need to learn how to build disaster-resistant communities to lessen the human fatalities and economic losses resulting from natural disasters.

Human Fatalities and Economic Losses in Natural Disasters

The 40 deadliest disasters in the 44-year period from 1970 to 2013 are shown in table 1.2. The most frequent megakillers were earthquakes (25) and hurricanes (8). Notice that 27 of the 40 worst natural disasters occurred in a belt running from China and Bangladesh through India and Iran to Turkey. Nine happened in the Americas but none were in the United States or Canada.

What is the correlation between human population density and the number of natural-disaster deaths? The data of table 1.2 paint a clear picture: densely populated Asia dominates the list of fatalities. The Asian experience offers a sobering view of what may befall the global population of humans if we continue our rapid growth. Where humans

TABLE 1.1

The 16 Deadliest Natural Disasters in 2013

Fatalities	Date	Event	Country
7,345	11 Aug	Typhoon Haiyan	Philippines
5,748	14 Jun	Floods	India
760	6 Aug	Heat wave	United Kingdom
531	1 Apr	Heat wave	India
399	24 Sep	Earthquake	Pakistan
388	1 Jan	Cold wave	India, Bangladesh
275	24 Aug	Cold wave	Peru
246	17 Jan	Floods	Mozambique, Zimbabwe
234	1 Aug	Floods	Pakistan
230	15 Oct	Earthquake	Philippines
218	15 Sep	Floods	Cambodia, Vietnam
217	20 Apr	Earthquake	China
200	5 Jul	Floods	China
174	9 Jul	Floods	India
169	13 Sep	Hurricane Manuel	Mexico
162	8 Nov	Cyclone	Somalia
17,296 Total deaths			

Source: Data from Swiss Reinsurance Company (2014).

are concentrated, disasters can kill many more people during each high-energy event.

THE ROLE OF GOVERNMENT IN NATURAL-DISASTER DEATH TOTALS

As the global population of humans increases, the number of deaths by natural disasters is expected to rise, but the relationship has complexities. Analyses by Gregory van der Vink and students at Princeton University show that between 1964 and 1968, about 1 person in 10,000 was killed by a natural disaster. Between 2000 and 2004, even though the population of humans doubled, the death rate by natural disaster dropped to about 1 person in 100,000. Yet, great natural disasters still result in horrific death totals in some countries. What relationships, in addition to population size, explain the locations of great natural disasters? Van der Vink and students compared natural-disaster deaths to the levels of democracy and economic development within 133 nations with populations greater than 1 million that

TABLE 1.2
The 40 Deadliest Natural Disasters, 1970–2013

Fatalities	Date/Start	Event	Country
300,000	14 Nov 1970	Hurricane (Bhola)	Bangladesh
255,000	28 Jul 1976	Earthquake (Tangshan)	China
245,000	26 Dec 2004	Earthquake and tsunami	Indonesia, Sri Lanka, India, Thailand
230,000	12 Jan 2010	Earthquake	Haiti
140,000	2 May 2008	Hurricane Nargis	Myanmar
140,000	29 Apr 1991	Hurricane Gorky	Bangladesh
88,000	8 Oct 2005	Earthquake	Pakistan
87,500	12 May 2008	Earthquake	China
66,000	31 May 1970	Earthquake and debris flow (Nevados Huascaran)	Peru
55,630	15 Jun 2010	Heat wave and fire	Russia
50,000	21 Jun 1990	Earthquake (Gilan)	Iran
35,000	Aug 2003	Heat wave	Europe
27,000	26 Dec 2003	Earthquake (Bam)	Iran
25,000	7 Dec 1988	Earthquake	Armenia
25,000	16 Sep 1978	Earthquake (Tabas)	Iran
23,000	13 Nov 1985	Volcanic eruption and mudflows (Nevado del Ruiz)	Colombia
22,000	4 Feb 1976	Earthquake	Guatemala
20,103	26 Jan 2001	Earthquake (Gujarat)	India
19,184	11 Mar 2011	Earthquake and tsunami	Japan
19,118	17 Aug 1999	Earthquake (Izmit)	Turkey
18,000	15 Dec 1999	Flooding and debris flows	Venezuela
15,000	19 Sep 1985	Earthquake (Mexico City)	Mexico
15,000	1 Sep 1978	Floods (monsoon rains in north)	India
15,000	29 Oct 1999	Hurricane (Orissa)	India
11,000	22 Oct 1998	Hurricane Mitch	Honduras
11,000	25 May 1985	Hurricane	Bangladesh
10,800	31 Oct 1971	Floods	India
10,000	20 Nov 1977	Hurricane (Andhra Pradesh)	India
9,500	30 Sep 1993	Earthquake (Marashtra state)	India
8,000	16 Aug 1976	Earthquake (Mindanao)	Philippines
7,345	8 Nov 2013	Hurricane Haiyan	Philippines
6,425	17 Jan 1995	Earthquake (Kobe)	Japan
6,304	5 Nov 1991	Typhoons Thelma and Uring	Philippines
5,778	21 May 2006	Earthquake	Indonesia
5,748	14 Jun 2013	Floods	India
5,422	30 Jun 1976	Earthquake (West Irian)	Indonesia
5,374	10 Apr 1972	Earthquake (Fars)	Iran
5,300	28 Dec 1974	Earthquake	Pakistan
5,112	15 Nov 2001	Floods and debris flows	Brazil
5,000	23 Dec 1972	Earthquake (Managua)	Nicaragua

2,053,643 Total deaths

Source: Data from Swiss Reinsurance Company (2014).

experienced five or more natural disasters between 1964 and 2004. Democracy is assessed by the World Bank's Democracy Index, and economic development by gross domestic product (GDP).

The Princeton researchers state that more than 80% of deaths by natural disasters between 1964 and 2004 took place in 15 nations, including China, Bangladesh, and Indonesia. For these 15 countries, 87% are below the median democracy index and 73% are below the median GDP. The correlation between high GDP and low death totals shows exceptions in Iran and Venezuela, two oil-rich nations with significant GDP but low democracy indices. These exceptions suggest a greater importance for democracy than GDP: the stronger the democracy index, the lower the death totals from natural disasters. The mega-killer natural disasters of recent years fit this trend also: Pakistan earthquake in 2005 (88,000 dead), Myanmar cyclone in 2008 (140,000 dead), China earthquake in 2008 (87,500 dead), and Haiti earthquake in 2010 (230,000 dead).

In a thought-provoking paragraph in their conclusion, van der Vink and students state: "Deaths from natural disasters can no longer be dismissed as random acts of nature. They are a direct and inevitable consequence of high-risk land use and the failures of government to adapt or respond to such known risks."

HUMAN RESPONSES TO DISASTER

Decades of social science research help us understand how most human beings react to natural disasters, and the news is good. Our behavior in ordinary times changes following disasters. In day-to-day life, most people are primarily concerned with their own needs and those of their immediate families; other relationships tend to be more superficial. After a natural disaster, many people change from inward-directed concerns to outward-directed actions. After an initial response of shock and disbelief, our emotions of sympathy and empathy tend to dominate. Personal priorities may be set aside and humanitarian and community-oriented actions take over. People reach out to others; they give aid and comfort to strangers; they make great efforts to provide help. Following a natural disaster, people become better connected and cohesive; they experience a heightened and compelling desire to add to the common good.

ECONOMIC LOSSES FROM NATURAL DISASTERS

The deaths and injuries caused by natural disasters grab our attention and squeeze our emotions, but in addition, there are economic losses. The destruction and disabling of buildings, bridges, roads, power-generation plants, and transmission systems for electricity, natural gas, and water, plus all the other built works of our societies, add up to a huge dollar cost. But the economic losses are greater than just damaged structures; industries and businesses are knocked out of operation, causing losses in productivity and wages for employees left without places to work.

In 2013 there were 308 natural and human-caused disasters with losses greater than US$95 million. The total economic losses were around US$140 billion. This is well below the inflation-adjusted 10-year average of US$190 billion.

Insured Portion of Economic Losses

The 40 greatest disasters between 1970 and 2013 from the insurance company perspective of dollar losses are listed in table 1.3. Notice that 39 of the 40 most expensive disasters were due to natural processes. The list of most expensive events is dominated by weather events (32 of 40), whereas earthquakes contributed seven. Compare the events on the 40 deadliest disasters list (see table 1.2) with table 1.3.

The locations of the worst dollar-loss disasters for the insurance industry (table 1.3) are different from the worst locations for fatalities (see table 1.2). The highest insurance dollar losses occurred in the United States (24 of 40), Europe (7), and Japan (5). Wealthy countries are better insured and their people live in safer buildings.

The extent of economic and insured losses may take years to become known. For example, the insured losses from the January 1994 Northridge earthquake were listed at $2.8 billion in February 1994, but they grew to $10.4 billion in January 1995 and increased to $15.3 billion in April 1998.

Natural Hazards

Many sites on Earth have not had a natural disaster in recent time, but are hazardous nonetheless. **Natural hazards** may be assessed as the probability of a dangerous event occurring. For example, people migrate and build next to rivers that are likely to flood, on the shoreline of the sea awaiting a powerful storm, and on the slopes of volcanoes that will eventually erupt. Decades, or even centuries, may pass with no great disasters, but the hazard remains.

Sites with natural hazards must be studied and understood. Their risks must be evaluated. Then we can try to prevent natural hazards from causing natural disasters. Remember: *Natural hazards are inevitable, but natural disasters are not.*

In the process of **mitigation,** we make plans and take actions to eliminate or reduce the threat of future death and destruction when natural hazards suddenly become great threats. The mitigating actions taken to protect us may be engineering, physical, social, or political.

Another need for mitigation occurs after great disasters, because people around the world tend to reoccupy the same site after a disastrous event is done. Earthquakes knock cities down, and then the survivors may use the same bricks

TABLE 1.3

The 40 Costliest Insurance Disasters, 1970–2013

Losses in Millions of 2013 US$	Fatalities	Date/Start	Event	Country
80,373	1,836	29 Aug 2005	Hurricane Katrina	USA
37,665	19,184	11 Mar 2011	Earthquake and tsunami	Japan
36,890	237	24 Oct 2012	Hurricane Sandy	USA
27,594	43	24 Aug 1992	Hurricane Andrew	USA
25,664	2,982	11 Sep 2001	Terrorist attack	USA
22,857	61	17 Jan 1994	Earthquake (Northridge)	USA
22,751	136	6 Sep 2008	Hurricane Ike	USA
17,218	181	2 Sep 2004	Hurricane Ivan	USA
16,519	815	27 Jul 2011	Floods (monsoon)	Thailand
16,142	181	22 Feb 2011	Earthquake	New Zealand
15,570	35	16 Oct 2005	Hurricane Wilma	USA
12,510	34	20 Sep 2005	Hurricane Rita	USA
11,594	123	15 Jul 2012	Drought (corn belt)	USA
10,313	24	11 Aug 2004	Hurricane Charley	USA
10,031	51	27 Sep 1991	Typhoon Mireille	Japan
8,924	71	15 Sep 1989	Hurricane Hugo	USA
8,876	562	27 Feb 2010	Earthquake	Chile
8,648	95	25 Jan 1990	Winter Storm Daria	Europe
8,426	110	25 Dec 1999	Winter Storm Lothar	Europe
7,856	354	22 Apr 2011	Tornadoes (Alabama)	USA
7,587	155	20 May 2011	Tornadoes (Missouri)	USA
7,112	54	18 Jan 2007	Winter Storm Kyrill	Europe
6,602	22	15 Oct 1987	Storm	Europe
6,593	38	26 Aug 2004	Hurricane Frances	USA
6,400	63	17 Oct 1989	Earthquake (Loma Prieta)	USA
6,274	55	22 Aug 2011	Hurricane Irene	USA
5,909	64	26 Feb 1990	Winter Storm Vivian	Europe
5,869	26	22 Sep 1999	Typhoon Bart	Japan
5,548	—	4 Sep 2010	Earthquake	New Zealand
5,240	600	20 Sep 1998	Hurricane Georges	USA, Caribbean
4,925	41	5 Jun 2001	Tropical Storm Allison	USA
4,872	3,034	13 Sep 2004	Hurricane Jeanne	USA, Haiti
4,593	45	6 Sep 2004	Typhoon Songda	Japan
4,250	135	26 Aug 2008	Hurricane Gustav	USA
4,216	45	2 May 2003	Tornadoes	USA
4,134	25	27 July 2013	Floods	Europe
4,100	70	10 Sep 1999	Hurricane Floyd	USA, Bahamas
3,979	59	4 Oct 1995	Hurricane Opal	USA
3,926	6,425	17 Jan 1995	Earthquake (Kobe)	Japan
3,406	25	24 Jan 2009	Winter Storm Klaus	France, Spain
$512 Billion	38,096 Total deaths			

Source: Data after Swiss Reinsurance Company (2014).

and stones to rebuild on the same site. Floods and hurricanes inundate towns, but people return to refurbish and again inhabit the same buildings. Volcanic eruptions pour huge volumes of magma and rock debris onto the land, burying cities and killing thousands of people, yet survivors and new arrivals build new towns and cities on top of their buried ancestors. Why do people return to a devastated site and rebuild? What are their thoughts and plans for the future? For a case history of a natural hazard, let's visit Popocatépetl in Mexico.

POPOCATÉPETL VOLCANO, MEXICO

Popocatépetl is a 5,452 m (17,883 ft) high **volcano** that lies between the huge populations of Mexico City (largest city in Mexico) and Puebla (fourth largest city in Mexico) (figure 1.1). The volcano has had numerous small eruptions over thousands of years; thus its Nahuatl name, Popocatépetl, or Popo as it is affectionately called, means smoking mountain. But sometimes Popo blasts forth with huge eruptions that destroy cities and alter the course of

Figure 1.1. Popocatepetl in minor eruption. The cathedral was built by the Spanish on top of the great pyramid at Cholula, an important religious site in a large city that was mostly buried by an eruption around 822 CE.

© Florian Kopp/imagebroker/Corbis RF.

civilizations. Around the year 822 CE (common era), Popo's large eruptions buried significant cities. Even its smaller eruptions have affected the course of human affairs. In 1519, Popo was in an eruptive sequence as Hernán Cortéz and about 500 Spanish conquistadors marched westward toward Tenochtitlan, the Aztec capital city. The superstitious Aztec priest-king Montezuma interpreted the eruptions as omens, and they affected his thinking on how to deal with the invasion.

Popocatépetl has helped change the path of history, but what is the situation now? Today, about 100,000 people live at the base of the volcano; they have been attracted by the rich volcanic soil, lots of sunshine, and fairly reliable rains. Millions more people live in the danger zone extending 40 km (25 mi) away. The Nahuatl people consider Popo to be divine—a living, breathing being. In their ancient religion, God, rain, and volcano are intertwined. Most do not fear the volcano; rather, they believe that God decides events and that with faith, things will work out. Thus, good opportunities for farming, coupled with faith and fatalism, bring people back.

Volcanic activity on Popo resumed on 21 December 1994 with eruptions of ash and gases. The sequence of intermittent eruptions continues today. How do we evaluate this hazard? Is this just one of the common multiyear sequences of small eruptions that gave the volcano its name? Or are these little eruptions the forewarnings of a giant killing eruption that will soon blast forth? We cannot answer these questions for sure. How would you handle the situation? Would you order the evacuation of 100,000 people to protect them, and in so doing, have them abandon their homes, sell their livestock, and leave their independent way of life for an unknown length of time that could be several years? Or would you explain the consequences of an unlikely but possible large eruption and let them decide whether to stay or go? If they decide to stay and then die during a huge volcanic blast, would this be your fault?

It is relatively easy to identify natural hazards, but as the Popocatépetl case history shows, it is not easy to decide how to answer the questions presented by this volcanic hazard. We are faced with the same types of questions again and again, for earthquakes, landslides, tornadoes, hurricanes, floods, and fire.

MAGNITUDE, FREQUENCY, AND RETURN PERIOD

Earth is not a quiet and stable body. Our planet is dynamic, with major flows of energy. Every day, Earth experiences earthquakes, volcanic eruptions, landslides, storms, floods, fires, meteorite impacts, and extinctions. These energy-fueled events are common, but their **magnitudes** vary markedly over space and time.

Natural hazards and disasters are not spaced evenly about Earth. Some areas experience gigantic earthquakes and some areas are hit by powerful hurricanes; some are hit by both, while other areas receive neither.

During a period of several years or even several decades, a given area may experience no natural disasters. But given enough time, powerful, high-energy events will occur in every area. It is the concentrated pulses of energy that concern us here, for they are the cause of natural disasters—but how frequent are the big ones? In general, there is an inverse correlation between the **frequency** and the magnitude of a process. The frequent occurrences are low in magnitude, involving little energy in each event. As the magnitude of an event increases, its frequency of occurrence decreases. For all hazards, small-scale activity is common, but big events are rarer. For example, clouds and rain are common, hurricanes are uncommon; streams overflow frequently, large floods are infrequent.

Another way of understanding how frequently the truly large events occur is to match a given magnitude event with its **return period,** or recurrence interval, which is the number of years between same-sized events. In general, the larger and more energetic the event, the longer the return period.

A U.S. Geological Survey mathematical analysis of natural-disaster fatalities in the United States assesses the likeliness of killer events. Table 1.4 shows the probabilities of 10- and 1,000-fatality events for earthquakes, hurricanes, floods, and tornadoes for 1-, 10-, and 20-year intervals, and estimates the return times for these killer events. On a yearly basis, most low-fatality events are due to floods and tornadoes, and their return times are brief, less than one year. High-fatality events are dominantly hurricanes and earthquakes, and their return times for mega-killer events are much shorter than for floods and tornadoes.

Knowing the magnitude, frequency, and return period for a given event in a given area provides useful information, but it does not answer all our questions. There are still the cost-benefit ratios of economics to consider. For example, given an area with a natural hazard that puts forth a dangerous pulse of energy with a return period of about 600 years, how much money should you spend constructing a building that will be used about 50 years before being torn down and replaced? Will your building be affected by a once-in-600-year disastrous event during its 50 years? Should you spend the added money necessary to guarantee that your building will withstand the rare destructive event? Or do economic considerations suggest that your building be constructed to the same standards as similar buildings in nearby nonhazardous areas?

ROLE OF POPULATION GROWTH

The world experiences significant numbers of great natural disasters and increasing economic losses from these events. The losses of life and dollars are occurring at the same time the global population of humans is increasing (figure 1.2). Population growth places increasing numbers of people in hazardous settings. They live and farm on the slopes of active volcanoes, build homes and industries in the lowlands of river floodplains, and move to hurricane-prone coastlines. How have the numbers of people grown so large? The present situation can best be appreciated by examining the record of population history.

TABLE 1.4

Probability Estimates for 10- and 1,000-Death Natural Disasters in the United States

	Likeliness of a 10-Fatality Event			
	During 1 Year	During 10 Years	During 20 Years	Return Time (in years)
Earthquake	11%	67%	89%	9
Hurricane	39	99	>99	2
Flood	86	>99	>99	0.5
Tornado	96	>99	>99	0.3
	Likeliness of a 1,000-Fatality Event			
	During 1 Year	During 10 Years	During 20 Years	Return Time (in years)
Earthquake	1%	14%	26%	67
Hurricane	6	46	71	16
Flood	0.4	4	8	250
Tornado	0.6	6	11	167

Source: US Geological Survey Fact Sheet (unnumbered).

Figure 1.2 The number of people on Earth continues to grow rapidly.

Photo courtesy of Pat Abbott.

Overview of Human Population

The most difficult part of human history to assess is the beginning, because there are no historic documents and the fossil record is scanty. In 2003, modern human fossils discovered in Ethiopia were dated as 160,000 years old. Our species appears to have began in Africa about 200,000 years ago. The rate of population growth and the number of people alive early in human history were so small that they cannot be plotted accurately on the scale of figure 1.3. The growth from a few thousand people 160,000 years ago to more than 7.34 billion people in the year mid-2015 did not occur in a steadily increasing, linear fashion. The growth rate is exponential.

THE POWER OF AN EXPONENT ON GROWTH

The most stunning aspect of figure 1.3 is the peculiar shape of the human population curve; it is nearly flat for most of human time and then abruptly becomes nearly vertical. The marked upswing in the curve shows the result of **exponential growth** of the human population. Possibly the least appreciated concept of present times is what a growth-rate exponent does to the size of a population over time. Exponential growth moves continuously in ever-increasing increments; it leads to shockingly large numbers in surprisingly short times. Probably our most familiar example of exponential growth occurs when interest is paid on money.

It can be difficult to visualize the results of exponential growth when it is expressed only as a percentage over time, such as the very small growth rate of the human population in 160,000 years or as 7% interest on your money for 50 years. It is easier to think of exponential growth in terms of doubling time—the number of years required for a population to double in size given an annual percentage growth rate. A simple formula, commonly called the rule of 70, allows approximation of doubling times:

$$\text{Doubling time (in years)} = \frac{70}{\%\ \text{growth rate/year}}$$

Learning to visualize annual percentage growth rates in doubling times is useful whether you are growing your money in investments or spending it by paying interest on debts (especially at the high rates found with credit-card debt). Table 1.5 shows how interest rates affect how quickly your money will grow.

THE PAST 10,000 YEARS OF HUMAN HISTORY

The long, nearly flat portion of the population curve in figure 1.3 certainly masks a number of small-scale trends, both upward and downward. The fossil record is not rich

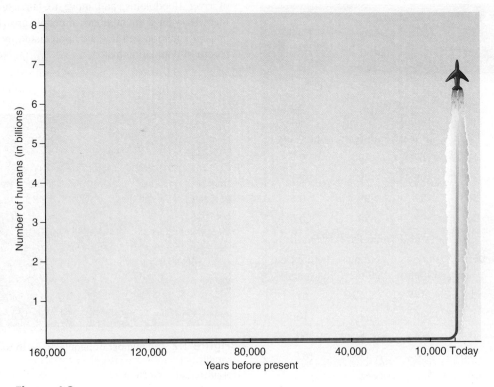

Figure 1.3 Human population growth since its start about 160,000 years ago.

TABLE 1.5

Doubling Times at Some Common Percentage Rates

Growth Rate (% per year)	Doubling Time (years)
0.02	3,500
0.5	140
1	70
1.2	58
2	35
5	14
7	10
10	7
17	4

enough to plot a detailed record, but surely at times when weather was pleasant and food from plants and animals was abundant, the human population must have risen (figure 1.4). Conversely, when weather was harsh, food was scarce, and diseases were rampant, the human population must have fallen.

The nearly flat population growth curve began to rise about 8,000 years ago, when agriculture became established and numerous species of animals were domesticated. The world population is estimated to have been about 8 million people by 10,000 years ago. After the development of agriculture and the taming of animals removed much of the hardship from human existence, the population growth rate is likely to have increased to 0.036% per year, yielding a net gain of 360 people per million per year. This increased rate of population growth probably caused the human population to reach 200 million people by 2,000 years ago.

As humans continued to improve their ability to modify the environment with better shelter and more reliable food and water supplies, the world population grew at faster rates.

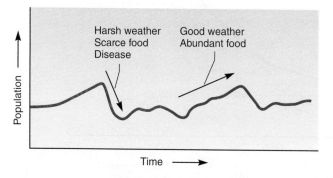

Figure 1.4 Good weather and plentiful food cause upsurges in population; bad weather, disease, and scarce food cause downswings in population.

From about 1 CE to 1750, world population grew to about 800 million. Growth occurred at an average rate of 0.056% per year, meaning that another 560 people were added per million per year.

Throughout the history of the human race, high rates of birth were required to offset high rates of infant mortality and thus maintain a viable-sized human population. The 18th century saw many of the intellectual advances that set the stage for the present phase of cultural change. At long last, the causes of many diseases were being recognized. The health necessities of clean water, sanitation, and nutrition led to the principles of public health being established. Advances in the medical world, including immunization, greatly improved the odds for the survival of individual humans through their reproductive years. No longer were many mothers and great numbers of children dying during childbirth and infancy.

The 18th century saw death rates drop dramatically, but birth rates remained high and population doubling times dropped dramatically; thus population size soared. About 1804, the human population reached 1 billion; by 1922, it had grown to 2 billion; in 1959, it reached 3 billion; by 1974, it was 4 billion; by early 1987, it was 5 billion; in 1999, it reached 6 billion; it passed 7 billion in October 2011 (figure 1.6). Notice the continuing decline in the number of years it takes for a net gain of another 1 billion people on Earth.

The 20th-century growth of the human population is unprecedented and breathtaking. The number of humans doubled twice—from about 1.5 billion to 3 billion and again to more than 6 billion. The increased population used 16 times more energy, increased industrial output 40 times, used 7 times more water, caught 35 times more fish, and expanded the cattle population to 1.4 billion. The effect of exponential growth is racing ahead. In his book *Wealth of Nations,* published in 1776, Adam Smith said, "Men, like all other animals, naturally multiply in proportion to the means of their subsistence."

THE HUMAN POPULATION TODAY

At present, the world population is growing at about 1.2% per year for a doubling time of 58 years (table 1.6). The 1.2% gain is a net figure derived by measuring the birth rate (**fertility** rate) and subtracting the death rate (**mortality** rate). Even after subtracting all the human lives lost each year to accidents, diseases, wars, and epidemics such as AIDS, the human population still grows by more than 80 million people per year. Each year, the world population increases by about the total population of Germany.

The net growth of the human population can be grasped by viewing it on short timescales (figure 1.7). There is a net addition of 2.6 people every second, a rate comparable to a full jetliner landing a load of new people every minute. The monthly net growth of people is greater than the population of Massachusetts.

Side Note

Interest Paid on Money: An Example of Exponential Growth

Compare the growth of money in different situations (figure 1.5). If $1,000 is stashed away and another $100 is added to it each year, a linear growth process is in operation. Many of the processes around us can be described as linear, such as the growth of our hair or fingernails.

If, in contrast, another $1,000 is stashed away but this time earns interest at 7% per year and the interest is allowed to accumulate, then an exponential growth-rate condition exists. Not only does the $1,000 earn interest, but the interest from prior years remains to earn its own interest in compound fashion.

Notice that an exponential growth curve has a pronounced upswing, or J shape. A comparison of the linear and exponential curves in figure 1.5 shows that they are fairly similar in their early years, but as time goes on, they become remarkably different. The personal lesson here is to *invest money now.* Smaller amounts of money invested during one's youth will become far more important than larger amounts of money invested later in life. Individuals who are disciplined enough to delay some gratification and invest money while they are young will be wealthy in their later years. Albert Einstein described compound interest, the exponential growth of money, as one of the most powerful forces in the world.

Here is a riddle that illustrates the incredible rate of exponential growth; it shows the significance of doubling times in the later stages of a system. Suppose you own a pond and add a beautiful water lily plant that doubles in size each day. If the lily is allowed to grow unchecked, it will cover the pond in 30 days and choke out all other life-forms. During the first several days, the lily plant seems small, so you decide not to worry about cutting it back until it covers half the pond. On what day will that be?

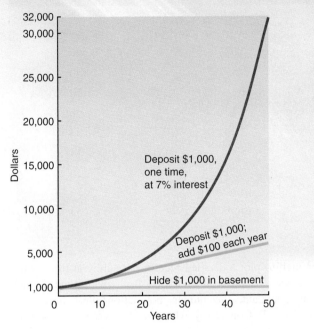

Figure 1.5 Amounts of money versus time. Compound interest (exponential growth) produces truly remarkable sums if given enough time.

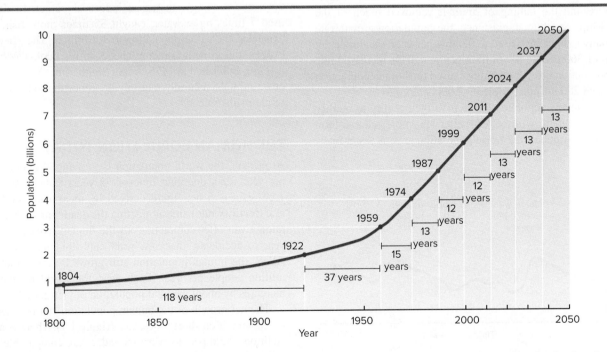

Figure 1.6 Growth of the world population of humans. Notice how the time to add another billion people has decreased to date but is projected to start increasing in the future.

Source: US Census Bureau.

TABLE 1.6

World Population Data, Mid-2015

	Population (millions)	Birth Rate (per 1,000)	Death Rate (per 1,000)	Yearly Growth %	Doubling Time (in years)	Projected Population in 2050 (millions)
World	7,337	20	8	1.2	58	9,804
More-developed countries	1,254	11	10	0.1	700	1,310
Less-developed countries	6,082	22	7	1.4	50	8,495
Least-developed countries*	938	34	9	2.4	29	1,887
Africa	1,171	36	10	2.5	28	2,473
Asia	4,397	18	7	1.1	64	5,324
Europe	742	11	11	0	—	728
Northern America	357	12	8	0.5	140	445
Latin America	630	18	6	1.3	54	776
Oceania	40	18	7	1.1	64	59

†Subset of less-developed countries

Source: World Population Data Sheet (2015).

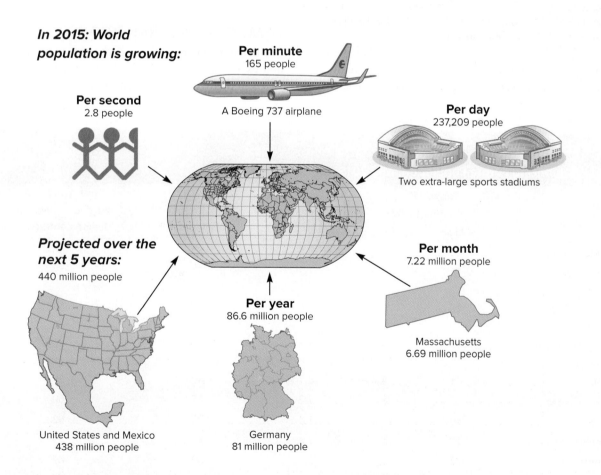

In 2015: World population is growing:

Per minute
165 people

A Boeing 737 airplane

Per second
2.8 people

Per day
237,209 people

Two extra-large sports stadiums

Projected over the next 5 years:
440 million people

Per month
7.22 million people

Massachusetts
6.69 million people

Per year
86.6 million people

United States and Mexico
438 million people

Germany
81 million people

Figure 1.7 Growth of world population over differing lengths of time.
Source: Modified from US Census Bureau.

Future World Population

Today, most of the more-developed countries have gone through **demographic transitions;** they have gone from high death rates and high birth rates to low death rates and low birth rates. But many less-developed countries have low to moderate death rates and high birth rates; will they go through demographic transitions? In demographic transition theory, both mortality and fertility decline from high to low levels because of economic and social development. Yet even without significant economic development, Population Reference Bureau estimates of the rates of world population growth are dropping: from 1.8% in 1990, to 1.6% in 1997, to 1.4% in 2000, and to 1.2% in 2015. What is causing this decrease in fertility? It appears to be due largely to urbanization and increased opportunities for women. At the beginning of the 20th century, less than 5% of people in less-developed countries lived in cities, but by the year 2015, about half of the people were living in urban areas (table 1.7). This is a change from farmer parents wanting many children to work in the fields and create surplus food, to city parents wanting fewer children to feed, clothe, and educate. Urban women have greater access to education, health care, higher incomes, and family-planning materials. When presented with choices, many women choose to have fewer children and to bear them later. Both of these choices lower the rate of population growth.

In the last 50 years of the 20th century, population grew from about 2.5 billion to over 6 billion, an increase of 3.5 billion people. Even with the recent decreases in fertility rates, the population explosion is not over. A growth rate of 1.2% per year will cause the world population of humans to approach 10 billion by the year 2050 (see table 1.6), an increase of another 3.5 billion people within 50 years. Population growth is not evenly distributed around the world. In general, wealthy countries have low or even negative rates of population growth. Many poor nations have high rates of population growth (figure 1.8).

An important factor in estimating future growth is the age distribution of the population (table 1.7). Nearly 30% of the population today is less than 15 years old, meaning their prime years for childbearing lie ahead. The century from 1950 to 2050 will see the world population grow from 2.5 billion to about 10 billion people.

The number of births per woman has a dramatic effect on human population growth. Starting in the year 2000 with a world population in excess of 6 billion people, look at three scenarios for population size in the year 2150 based on births per woman: (1) if women average 1.6 children, world population drops to 3.6 billion; (2) if women average 2 children, population grows to 10.8 billion; (3) if women average 2.6 children, population grows to 27 billion. The difference between a world population of 3.6 billion or 27 billion rests on a difference of only one child per woman.

DEMOGRAPHIC TRANSITION

The demographic transition model is based on the population experiences of economically wealthy countries in the past few centuries. Up through the 17th century, a woman

---

TABLE 1.7
Data Influencing Future Population, Mid-2015

	Percent of Population of Age		Average Number of Children Born per Woman	Percent Urban (cities >2,000 people)	Percent of Married Women Using Modern Contraception
	<15	65+			
World	26	8	2.5	53	56
More-developed countries	16	17	1.7	77	59
Less-developed countries	28	6	2.6	48	55
Least-developed countries	40	4	4.3	29	32
Africa	41	4	4.7	40	29
Asia	25	8	2.2	47	60
Europe	16	17	1.4	73	62
Northern America	19	15	1.8	81	73
Latin America	27	7	2.1	80	67
Oceania	24	12	2.5	70	58

Source: World Population Data Sheet (2015).

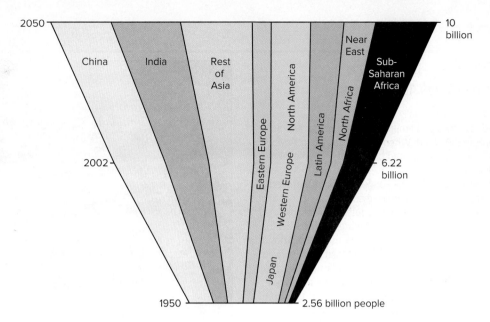

Figure 1.8 World population by region: 1950, 2002, 2050.

Source: US Census Bureau.

had to bear several children to have a few survive to adulthood and replace the prior generation. Births had to be numerous to compensate for the high rates of infant mortality. Beginning in the 18th century, discoveries in public health, medicine, and immunization caused the death rate to drop dramatically. During this time, birth rates stayed high, so overall population grew rapidly. As time passed and people realized that most of their children would survive to adulthood, birth rates dropped and population stabilized at a new and higher level.

The demographic transition takes place in phases:

1. Before the transition: high death rates are offset by high birth rates to maintain a population.
2. During the transition: low death rates coupled with continuing high birth rates cause population to soar
3. After the transition: low death rates combine with low birth rates to achieve a stable population at a significantly higher level.

Today the transition is taking place at different rates in different countries (figure 1.9). Most of the population growth is occurring in the poorest areas of the poorest countries. Some of the wealthiest countries now have more deaths than births each year.

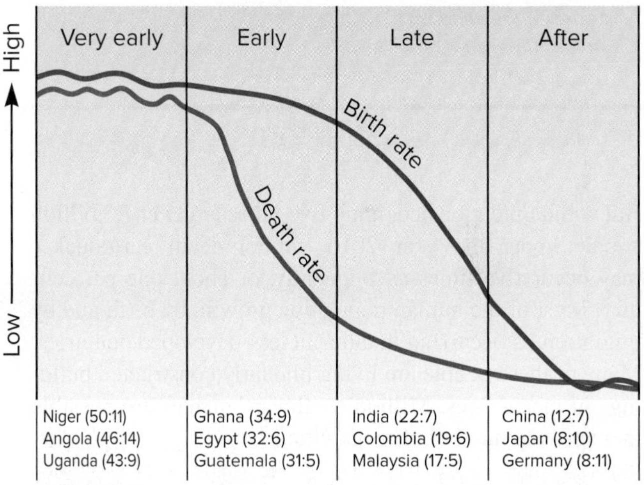

URBANIZATION AND EARTHQUAKE FATALITIES

During the past 500 years, global earthquakes killed about 5 million people. Average numbers of deaths were about 1 million per century, or 100,000 per decade. These simple averages are misleading because they hide the effects of the

Demographic transition

	Very early	Early	Late	After
Niger (50:11)	Ghana (34:9)	India (22:7)	China (12:7)
Angola (46:14)	Egypt (32:6)	Colombia (19:6)	Japan (8:10)
Uganda (43:9)	Guatemala (31:5)	Malaysia (17:5)	Germany (8:11)

Figure 1.9 Demographic transition. In mid-2014, the shifts in birth rates and death rates vary markedly between countries. Birth and death rates are both expressed in number of people per 1,000 each year. For example, Uganda has 43 births and 9 deaths per 1,000 people each year (43:9).

Data from Population Reference Bureau.

deadliest earthquakes, such as the 250,000 people killed by the Tangshan, China, event in 1976. The mega-killer earthquakes of the past 500 years occurred in China, Indonesia, Pakistan, Iran, Turkey, Italy, Japan, and Haiti—and they may occur there again.

An analysis of the past 500 years by Roger Bilham shows that, with an average population of about 1.5 billion people, there was one earthquake that killed nearly a million people.

A Classic Disaster: Influenza (FLU) Pandemic of 1918

In July 1914, a major war, eventually known as World War I, broke out in Europe. The countries and empires involved contained more than half the people in the world. When the war ceased in November 1918, almost 7 million soldiers had been killed in battle, along with about 1 million civilians.

As bad as 8 million war deaths sounds, a far more deadly natural disaster began during that time: the influenza pandemic of 1918–1919. The flu pandemic killed about 50 million people; this was 3% of the world's population. Estimates of total deaths range up to 100 million people. The influenza migrated around the world in waves. In the United Kingdom, the first wave arrived in the spring of 1918. In the fall of 1918, a longer-lasting, deadlier wave of flu swept the world, followed in 1919 by yet a third wave. Most flu victims were healthy young adults rather than the more typical elderly or juvenile victims of influenza (figure 1.10).

World War I did not cause the flu, but the global movements of millions of troops, weakened by stress and battle, increased the spread and deadly effects of the **virus**. Another 3 million soldiers died, not from World War I battles, but from influenza. In 1918, children skipped rope to this rhyme:

I had a little bird
Its name was Enza
I opened the window
And in-flu-enza.

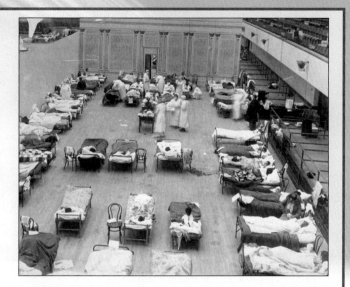

Figure 1.10 A typical scene during the 1918 flu pandemic. The Oakland Municipal Auditorium was used as a temporary hospital, allowing volunteer nurses to tend to the sick.

Photo by Edward A. "Doc" Rogers. From The Joseph R. Knowland Collection at the Oakland History Room, Courtesy Oakland Public Library

But with population becoming five times larger at 7.5 billion people about the year 2016, million-death earthquakes may occur five times as frequently, or about one per century. Most of the human population growth, by birth and by migration, is occurring in cities in less-developed countries. Many of these people are living in poorly constructed buildings in mega-cities. Million-death earthquakes are possible in a growing number of mega-cities.

DISEASE PANDEMICS

Throughout recorded history, deadly diseases have swept throughout the world, killing millions of people in **pandemics.** For example, the bacterium *Yersinia pestis*, transmitted to humans by fleas, caused the bubonic plague—the Black Death that killed about 75 million people in Europe in the 14th century.

Viruses have also caused pandemics via smallpox, HIV, polio, **influenza,** and other diseases. For example, in 1918–1919, the influenza virus A (H1N1) spread around the world, killing about 50 million people. With the human population now exceeding 7 billion people, with more than 50% of people now living in cities, and with the rapid movement of people worldwide via jet airplanes, the potential exists for a new pandemic disease.

Viruses

Viruses are life in the simplest form. They are genetic material (DNA or RNA) coated by fat and protein. A **virus** might have only 4 genes, whereas a bacterium might have 4,000 genes, and a human 24,000 genes. Viruses cannot reproduce by themselves; they must invade a host cell and cause the host to reproduce the virus.

Viruses infect many forms of life, including animals, plants, and even bacteria. The same viruses commonly exist in humans, pigs, and birds, and move easily between them. There are an estimated 1 billion pigs and 20 billion chickens in the world. Because humans commonly live and interact with birds and pigs, the transfer of viruses between them is especially likely. Other transfers of viruses to humans include HIV/AIDS from chimpanzees and Ebola from bats. When two different viruses enter a single cell, their genes can form new combinations, creating a new type of virus. On the surface of a virus are molecules shaped into unique configurations that might match a living cell and allow entry, much like a unique key will open a specific lock.

Influenza A Viruses Influenza A viruses cause recurrent **epidemics** and pandemics, as in 1918–1919. Type A viruses examined on the basis of their haemagglutinin (HA)

and neuraminidase (NA) molecules are divided into 16 HA subtypes (H1 to H16) and 9 NA subtypes (N1 to N9). In 2005, researchers reported the results of a study of the 1918–1919 influenza virus collected from samples preserved from World War I flu-victim soldiers and from historic individuals buried in Arctic permafrost (frozen soil). The 1918–1919 influenza was type A (H1N1), a subtype with an early history in birds.

Early in 2009, a flu epidemic broke out near La Gloria in the state of Veracruz, Mexico. By 23 April 2009, 23,000 cases had been reported. By 7 May 2009, the flu had spread to become a pandemic, with cases identified in 21 countries on five continents. Laboratory analyses showed that this new virus was type A (H1N1) and was made up of genes from four different flu viruses: from North American pigs (30.6%), Eurasian pigs (17.5%), North American birds (34.4%), and humans (17.5%). People were worried. Could this virus evolve into as big a killer as the one in 1918–1919?

Analysis of H1N1 deaths in 2009 from 214 countries showed 44,100 deaths—a significant total, but far less severe than in 1918. Like the 1918 influenza, though, most of the deaths occurred in young people; 73% of deaths were people 29 years old and younger. The death percentages by age groups include:

- 37% were 10 to 19 years old
- 22% were less than 9 years old
- 14% were 20 to 29 years old

In the reverse of a typical flu year, people 60 years old and older suffered only 3% of the deaths. If one views the 2009 H1N1 figures as deaths only, then it was not as bad a year as had been feared. But if one considers the number of years of life lost by the young victims, then the 2009 pandemic would be more equivalent to 250,000 deaths in a typical flu year.

Carrying Capacity

How many people can Earth support? At this time, the question is unanswerable. Nonetheless, many people worry about dangers resulting from the unprecedented growth of the human population, such as more and greater natural disasters, increased global warming, decreasing supplies of fresh water, depletion of fossil fuels, increased pollution, increased desertification, and the increased rate of extinction of species. Other people see no big problems and point out that humans have already increased the carrying capacity of Earth for us via agriculture, water storage and purification, and advances in public health; they feel that any upcoming problems will be solved just like others have been in the past.

In the natural world, biologists studying **carrying capacity** of the environment for individual species of mammals, birds, frogs, and other animals find that population size is regulated by the resources available. For example, when a resource such as available food increases, a feeding population grows in size. If that food resource decreases due to drought, competition, or other causes, the population dependent on that food dies back and decreases in size.

A fundamental principle of biology is that a population of animals cannot increase forever because they live in a finite ecosystem. Ultimately, population growth is controlled by negative feedbacks such as starvation, predation, and disease.

Ireland in the 1840s

Ireland in the 1840s provides a human example of carrying capacity. The European explorers of the 1500s brought the potato back from South America. The potato is a highly nutritious food. A diet of potatoes, milk (from animals fed potatoes), and greens constitute a nutritionally complete diet. An acre of potatoes could feed an Irish family of six for a year. In Ireland, the potato was the wonder crop that allowed a child-loving population to grow explosively. By 1841, Ireland's population had grown well past 8 million, with nearly half the people surviving wholly or mostly on potatoes. In 1845, heavy spring rains aided growth and spread of a fungal infestation, the potato blight, which caused potatoes to rot during storage. But when the potatoes rotted there was no substitute food. Malnourishment became common. Then the winter of 1846–1847 hit with unusual severity, causing weakened people to suffer even more. The toll was severe: a million people died from disease, and another 1.5 million people emigrated. During their travel to the United States and Canada, 1 in 7 emigrés died.

The carrying capacity of Irish land increased for humans when the potato arrived. Potato plants covered the lands, even extending into bogs and up steep mountain slopes. The human population fed by the increased food supply grew rapidly. But when the potato supply dropped suddenly, so did the human population.

Easter Island (Rapa Nui)

Easter Island (Rapa Nui) is a triangular-shaped, volcanic land with an area of about 165 km^2 (64 mi^2). It lies over 2,000 km (more than 1,200 mi) east of Pitcairn Island and over 3,700 km (more than 2,200 mi) west of Chile (figure 1.11). Easter Island is isolated; it has high temperatures and humidity, poorly drained and marginal soils, no permanent streams, no terrestrial mammals, about 30 native plant species including trees in locally dense growths, and few varieties of fish in the surrounding sea. Year-round water is available only in little lakes within the volcano caldera.

About 1,000 years ago, seafaring Polynesian people arrived on Rapa Nui with 25 to 50 settlers. They were part of the great Polynesian expansion outward from southeast Asia that led them to discover and inhabit islands from Hawaii in the north to New Zealand in the southwest and to

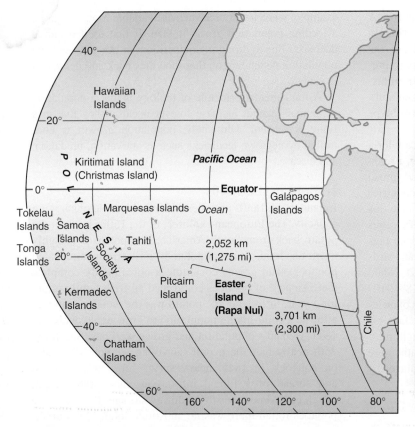

Figure 1.11 Easter Island (Rapa Nui) is an isolated outpost of Polynesian civilization nearly lost in the vast Pacific Ocean.

Figure 1.12 Rapa Nui inhabitants spent much of their energy creating giant statues (moai).
© Adalberto Rios Szalay/Sexto Sol/Getty Images RF

Rapa Nui in the southeast. The wide-ranging voyagers colonized islands over a Pacific Ocean area more than twice the size of the United States.

The colonizers of Rapa Nui brought chickens and rats, along with several of their food plants. The climate was too severe for most of their plants except the yam. Their resulting diet was based on easily grown chickens and yams, and housing was fashioned using wood from native trees; the people had lots of free time.

The islanders used their free time to develop a complex social system divided into clans that practiced elaborate rituals and ceremonies. Their customs included competition between the clans in shaping and erecting mammoth statues. The statues were carved out of volcanic rock using obsidian (volcanic glass) tools. The statues (moai) were more than 6 m (20 ft) high, weighed about 15 tons apiece, and were erected on ceremonial platforms (ahu) (figure 1.12).

The peak of the civilization occurred about 1550 CE, when the human population had risen to about 7,000; statues numbered more than 600, with half as many more being shaped in the quarries. But from its peak, the civilization declined rapidly and savagely, as first witnessed by the crew of a Dutch ship on Easter Sunday, 5 April 1722. The Europeans found about 2,000 people living in caves in a primitive society engaged in almost constant warfare and practicing

cannibalism. What caused this cultural collapse? It appears that human activities so overwhelmed the environment that it was no longer able to support the greatly enlarged human population. The customs of society dissolved in the fight of individuals and clans to survive.

Carving the giant statues had not been particularly difficult, but transporting them was physically and environmentally strenuous. Trees were cut down and placed under statues as rollers. Islanders pushed the heavy statues from the quarry and levered them onto their ceremonial platforms. The competition between clans to create the most statues helped destroy the forests. Without trees, houses could not be built, and people had to move to caves. There was little fuel for cooking or to ward off the chill of colder times. Soil erosion increased and agricultural production dropped. Without trees, there were no canoes, and so islanders caught fewer fish. Without canoes, there was no escape from the remote and isolated island. As food resources declined, the social system collapsed, and the statue-based religion disintegrated. Clans were reduced to warfare and cannibalism in the struggle for food and survival.

Competition between clans was so consuming that they did not consider the health of the environment and thus paid a price: the human population on the island collapsed. Easter Island is one of the most remote inhabited areas on Earth, a tiny island virtually lost in the vast Pacific Ocean. When problems set in faster than the Rapa Nui customs could solve them, there was no place to turn for help, no place to escape. The carrying capacity of the land had been exceeded, and the human population suffered terribly. What lesson does Easter Island have for the whole world? Earth is but a tiny island lost in the vast ocean of the universe (figure 1.13); there is no realistic chance of the human population escaping to another hospitable planet.

The Easter Island example raises interesting philosophical questions. If climate change decreases global food production, causing the human population to exceed Earth's carrying capacity, could human value systems change fast enough to solve the problem? If all the people on Earth had to face the Easter Island situation, how would we fare?

Figure 1.13 Earth is an isolated outpost nearly lost in the vast "ocean" of the universe.
Courtesy of Pat Abbott.

Summary

Great natural disasters killed almost 500,000 people in four recent events: 2011 Japan earthquake and tsunami, 2010 Haiti earthquake, 2008 hurricane in Myanmar and earthquake in China. Over time, the two deadliest events are tropical storms (hurricanes) and earthquakes. In 2011, the known economic losses from natural disasters were about US$375 billion. The long-term trend is for economic losses to increase.

Natural hazards exist in areas of obvious danger, such as cities built on the slopes of active volcanoes or on the floodplains of rivers. For these sites, it is only a matter of time before the hazard is realized as a disaster. At any one site, the greater the magnitude of a disaster, the less frequently it occurs. Large disasters have longer return periods.

The curve describing the history of human population growth is flat to gently inclined for 160,000 years, and then it rises rapidly in the last three centuries. In the past, women bore numerous children, but many died, so overall population growth was slow. With the arrival of the scientific-medical revolution and the implementation of the principles of public health, the human population has soared. Birth rates remain high in much of the world, even though death rates have plummeted. The population reached 1 billion in about 1804, 2 billion in 1922, 3 billion in 1959, 4 billion in 1974, 5 billion in 1987, 6 billion in 1999, and it passed 7 billion in 2011.

A steeply rising growth curve is exponential; in terms of population, more people beget ever more people. One way to visualize exponential growth is by using doubling time, the length of time needed for a population to double in size. Doubling times can be approximated by the rule of 70:

$$\text{Doubling time (in years)} = \frac{70}{\% \text{ growth rate/year}}$$

At present, after subtracting deaths from births, world population increases 1.2% per year for a doubling time of 58 years.

Much hope is placed in the demographic transition model, which holds that economic wealth, combined with knowing that one's children will survive, leads to dramatic drops in birth rates. This model holds for some more-developed countries. Now some less-developed countries are experiencing drops in birth rates, presumably due to urbanization and more choices for women. Even at lower rates of growth, human population is likely to reach 10 billion in 2050. The rapid growth in human population sets the stage for mega-death earthquakes and hurricanes.

New flu viruses are commonly created where people live closely with birds and pigs. These new viruses which have the potential to kill millions of people, can rapidly spread around the world.

Carrying capacity is an estimate of how many individuals of a species the environment can support. How many people can Earth support? The answer is not known, but it is the subject of much debate.

Terms to Remember

carrying capacity   19
CE   10
cyclone   6
demographic transition   16
earthquake   6
energy   6
epidemic   18
exponential growth   12

fertility   13
frequency   11
great natural disaster   6
hurricane   6
influenza   18
magnitude   10
mitigation   8
mortality   13

Questions for Review

1. What types of natural disasters killed the most people in the past 40 years? Where in the world are deaths from natural disasters the highest? Where in the world are insurance losses from natural disasters the highest?
2. What is a great natural disaster?
3. What is the difference between a natural disaster and a natural hazard? How do economic losses differ from insured losses?
4. For nations, what is the relationship between natural-disaster deaths, gross domestic product, and level of democracy?
5. What is the relationship between the magnitude of a given disaster and its frequency of occurrence?
6. Draw a curve showing the world population of humans in the past 100,000 years. Why has the curve changed shape so dramatically?
7. Explain the concept of exponential growth.
8. What is the size of the world population of humans today? Extrapolating the current growth rate, what will the population be in 100 years? In 200 years? Are these large numbers environmentally realistic?
9. What are the population doubling times given these annual growth rates: Africa, 2.4%; world, 1.2%?
10. For nations, what are demographic transitions?
11. How much time does it take for a flu pandemic to infect people all around the world?
12. What is the relationship between earthquake fatalities and cities?
13. Explain the concept of carrying capacity for a species. What processes might limit the numbers of a species?

Questions for Further Thought

1. Would we call a large earthquake or major volcanic eruption a natural disaster if no humans were killed or buildings destroyed?
2. Which single disaster could kill the most people—a flu pandemic, an earthquake, or a hurricane?
3. Could global building designs be made disaster-proof, thus reducing the large number of fatalities?
4. What is the carrying capacity of Earth for humans—that is, how many humans can Earth support? What factors are most likely to slow human population growth?
5. Compare the rate of change of human populations to the rate of change in religious and cultural institutions. Can religious and cultural institutions change fast enough to deal with world population growth?
6. Evaluate the suggestion that the overpopulation problem on Earth can be solved by colonizing other planets.
7. Is a nation's destiny determined by its demographics?

CHAPTER 2

Internal Energy and Plate Tectonics

Such superficial parts of the globe seemed to me unlikely to happen if the Earth were solid to the centre. I therefore imagined that the internal parts might be a fluid more dense, and of greater specific gravity than any of the solids we are acquainted with; which therefore might swim in or upon that fluid. Thus the surface of the globe would be a shell, capable of being broken and disordered by the violent movements of the fluid on which it rested.

—BENJAMIN FRANKLIN, 1780

Satellite view of Arabia moving northeast away from Africa.
NOAA.

LEARNING OUTCOMES

Internal energy has caused the Earth to differentiate into layers. Throughout the Earth, materials move vertically and horizontally. After studying this chapter you should

- know the layering of the Earth and how it formed.
- be familiar with the sources of energy inside the Earth.
- understand the behavior of materials.
- be able to explain how plate tectonics operates.
- comprehend Earth's magnetic field and the evidence it provides for plate tectonics.
- know the age of the Earth and how it is determined.
- appreciate the thought processes used to understand the Earth.

OUTLINE

- Origin of the Sun and Planets
- Earth History
- The Layered Earth
- Internal Sources of Energy
- Plate Tectonics
- The Grand Unifying Theory
- How We Understand Earth

At 3 a.m., on 12 April 1977, a 1.4 kg (3 lb) meteorite tore through the roof of a parked car in the city of Chambery in the French Alps. Friction created while speeding through the atmosphere made the exterior of the meteorite so hot that it set the car on fire. The car's owner was awakened by the impact and fire but refused to believe it was caused by a meteorite. He filed an arson complaint with the police. But sometimes things are exactly what they seem to be—the fire was caused by a meteorite.

Origin of the Sun and Planets

Impacts of material are not rare and insignificant events in the history of our Solar System; they probably were responsible for its formation. The most widely accepted hypothesis of the origin of the Solar System was stated by the German philosopher Immanuel Kant in 1755. He thought the Solar System had formed by growth of the Sun and planets through collisions of matter within a rotating cloud of gas and dust.

The early stage of growth began within a rotating spherical cloud of gas, ice, dust, and other solid debris (figure 2.1a). Gravity acting upon matter within the cloud attracted particles, bringing them closer together. Small particles stuck together and grew in size, resulting in greater gravitational attraction to nearby particles and thus more collisions. As matter drew inward and the size of the cloud decreased, the speed of rotation increased and the mass began flattening into a disk (figure 2.1b). The greatest accumulation of matter occurred in the center of the disk, building toward today's Sun (figure 2.1c). The two main constituents of the Sun are the lightweight elements hydrogen (H) and helium (He). As the central mass grew larger, its internal temperature increased to about 1,000,000 degrees **centigrade** (°C), or 1,800,000 degrees **Fahrenheit** (°F), and the process of **nuclear fusion** began. In nuclear fusion, the smaller hydrogen atoms combine (fuse) to form helium, with some mass converted to energy. We Earthlings feel this energy as **solar radiation** (sunshine).

The remaining rings of matter in the revolving Solar System formed into large bodies as particles continued colliding and fusing together to create the planets (figure 2.1d). Late-stage impacts between ever-larger objects would have been powerful enough to melt large volumes of rock, with some volatile elements escaping into space. The inner planets (Mercury, Venus, Earth, Mars) formed so close to the Sun that solar radiation drove away most of their volatile gases and easily vaporized liquids, leaving behind rocky planets. The next four planets outward (Jupiter, Saturn, Uranus, Neptune) are giant icy bodies of hydrogen, helium, and other frozen materials from the beginning of the Solar System.

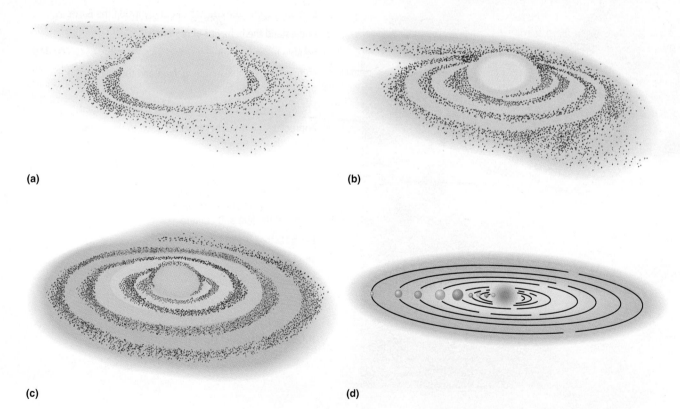

(a) (b)

(c) (d)

Figure 2.1 Hypothesis of the origin of the Solar System. (a) Initially, a huge, rotating spherical cloud of ice, gas, and other debris forms. (b) The spinning mass contracts into a flattened disk with most of its mass in the center. (c) Planets grow as masses collide and stick together. (d) The ignited Sun is surrounded by planets. Earth is the third planet from the Sun.

IMPACT ORIGIN OF THE MOON

Large impacts can generate enough heat to vaporize and melt rock; they can produce amazing results. For example, the dominant hypothesis on the origin of Earth's Moon involves an early impact of the young Earth with a Mars-size body, a mass about 10 times larger than the Moon. The resultant impact generated a massive cloud of dust and vapor, part of which condensed and accumulated to form the Moon. This theory suggests that the Moon is made mostly from Earth's rocky **mantle.** The theory accounts for the lesser abundance of iron on the Moon (iron on Earth is mostly in the central **core**) and the Moon's near lack of lightweight materials (such as gases and water), which would have been lost to space.

Earth History

To understand the origin and structure of Earth, we must know the flows of energy throughout the history of our planet. Studying early history is difficult because Earth is a dynamic planet; it recycles its rocks and thus removes much of the record of its early history. The older the rocks, the more time and opportunities there have been for their destruction. Nonetheless, the remaining early Earth rocks, along with our growing knowledge of the processes in Earth's interior and in the Solar System, allow us to build an increasingly sophisticated approximation of early Earth history.

Earth appears to have begun as an aggregating mass of particles and gases from a rotating cloud about 4.6 billion years ago. During a 30- to 100-million-year period, bits and pieces of metal-rich particles (similar to iron-rich meteorites), rocks (similar to stony meteorites), and ices (composed of water, carbon dioxide, and other compounds), accumulated to form Earth. As the ball of coalescing particles

enlarged, the gravitational force may have pulled more of the metallic pieces toward the center, while some of the lighter-weight materials may have concentrated near the exterior. Nevertheless, Earth in its infancy probably grew from random collisions of debris that formed a more or less homogeneous mixture of materials.

But Earth did not remain homogeneous. The very processes of planet formation (figure 2.2) created tremendous quantities of heat, which fundamentally changed the young planet. The heat that transformed Earth came primarily from (1) impact energy, (2) decay of radioactive isotopes, (3) gravitational energy, and (4) differentiation into layers (figure 2.2).

As the internal temperature of Earth rose beyond 1,000°C (1,800°F), it passed the melting points of iron at various depths below the surface. Iron forms about one-third of Earth's mass, and although it is much denser than ordinary rock, it melts at a much lower temperature. The buildup of heat caused immense masses of iron-rich meteorites to melt. The high-density liquid iron was pulled by gravity toward Earth's center. As these gigantic volumes of liquid iron moved inward to form Earth's core, they released a tremendous amount of gravitational energy that converted to heat and probably raised Earth's internal temperature by another 2,000°C (3,600°F). The release of this massive amount of heat would have produced widespread melting likely to have caused low-density materials to rise and form: (1) a primitive **crust** of low-density rocks at the surface of Earth; (2) large oceans; and (3) a denser atmosphere. The formation of the iron-rich core was a unique event in the history of Earth. The planet was changed from a somewhat homogeneous ball into a density-stratified mass with the denser materials in the center and progressively less-dense materials outward to the atmosphere.

The low-density materials (magmas, waters, and gases), freed by the melting, rose and accumulated on Earth's exterior

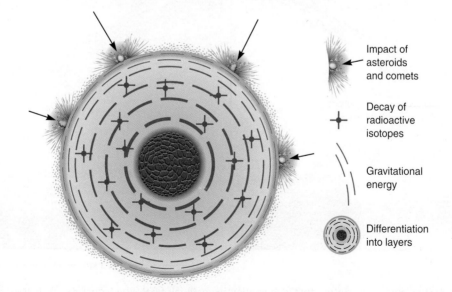

Impact of
asteroids
and comets

Decay of
radioactive
isotopes

Gravitational
energy

Differentiation
into layers

Figure 2.2 Heat-generating processes during the formative years of Earth include (1) impact energy, (2) decay of radioactive isotopes, and (3) gravitational energy. Increasing heat caused Earth to differentiate into layers.

as continents, oceans, and atmosphere. It seems that oceans and small continents existed by 4.4 billion years ago, life probably was present as photosynthetic bacteria 3.5 billion years ago, large continents were present at least 2.5 billion years ago, and the outer layers of Earth were active in the process of plate tectonics by at least 1.5 billion years ago.

The Layered Earth

Earth today is differentiated into layers of varying densities. As we have noted, much of the densest material was pulled toward the center, and some of the least dense substances escaped to the surface (figure 2.3). At the center of Earth is a dense, iron-rich core measuring about 7,000 km (4,350 mi) in diameter. The inner core is a solid mass 2,450 km (1,520 mi) in diameter with temperatures up to 4,300°C (7,770°F). The outer core is mostly **liquid,** and the **viscous** movements of convection currents within it are responsible for generating Earth's magnetic field. The entire iron-rich core is roughly analogous in composition to a melted mass of metallic meteorites.

Surrounding the core is a rocky mantle nearly 2,900 km (1,800 mi) thick, with a composition similar to that of stony meteorites. The mantle comprises 83% of Earth's volume and 67% of its mass. The rocks of the mantle can be approximated by melting a stony meteorite in the laboratory; this produces a separation in which an upper froth rich in low-density elements rises above a residue of denser

minerals/elements. The low-density material is similar to continental crust that by melting and separation has risen above the uppermost mantle. All the years of heat flow toward Earth's surface have "sweated out" many low-density elements to form a continental crust. Today, the continents make up only 0.1% of Earth's volume. Floating above the rocky layers of Earth are the oceans and the atmosphere.

Earth's layering can be described as based on either (1) different strengths or (2) different densities due to varying chemical and mineral compositions (figure 2.3). Both temperature and pressure increase continuously from Earth's surface to the core, yet their effects on materials are different. Increasing temperature causes rock to expand in volume and become less dense and more capable of flowing under pressure and in response to gravity. Increasing pressure causes rock to decrease in volume and become more dense and more rigid. Visualize tar at Earth's surface. On a cold day, it is solid and brittle, but on a hot day, it can flow as a viscous fluid. Similar sorts of changes in physical behavior mark different layers of Earth. In fact, from a perspective of geological disasters, the crust-mantle boundary is not as important as the boundary between the rigid **lithosphere** (from the Greek word *lithos,* meaning "rock") and the "soft plastic" **asthenosphere** (from the Greek word *asthenes,* meaning "weak") (figure 2.4). The **mesosphere,** the mantle below the asthenosphere (see figure 2.3), is solid; it is a "stiff plastic," but it is not brittle like the lithosphere. The differences in strength and mechanical behavior between solid, "plastic," and fluid states are partly responsible for earthquakes and volcanoes.

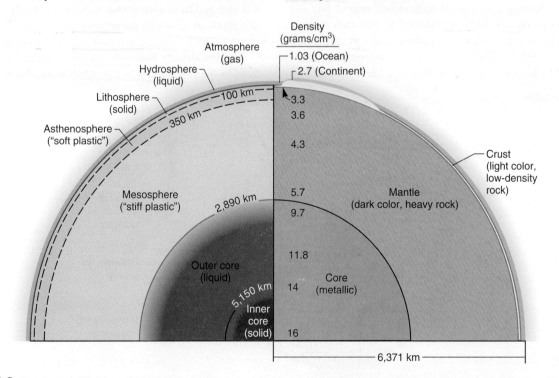

Figure 2.3 Density stratification within Earth—that is, lower-density materials float atop higher-density materials. Pressure and temperature both increase from the surface to the center of Earth. Layers illustrated on the left show the differences in physical properties and strengths. Layers on the right emphasize different mineral and chemical compositions.

Side Note

Mother Earth

The history of the 4.6-billion-year-old Earth has been metaphorically contrasted with the life history of a 46-year-old woman by Nigel Calder in his book *The Restless Earth*. In this metaphor, each of "Mother" Earth's years equals 100 million years of geologic time. The first seven of her years are mostly lost to the biographer. Like human memory, the early rock record on Earth is distorted; it emphasizes the more recent events in both number and clarity. Most of what we know of "Mother" Earth happened in the past six years of her life. Her continents had little life until she was 42. Flowering plants did not appear until her

45th year. Her pet dinosaurs died out eight months ago. In the middle of last week, some ancestors of present apes evolved into human ancestors. Yesterday, modern humans (*Homo sapiens*) evolved and began hunting other animals, and in the last hour, humans discovered agriculture and settled down. Fifteen minutes ago, Moses led his people to safety; five minutes later, Jesus was preaching along the same fault line; and after another minute, Muhammad taught in the same region. In the last minute, the Industrial Revolution began, and the number of humans increased enormously.

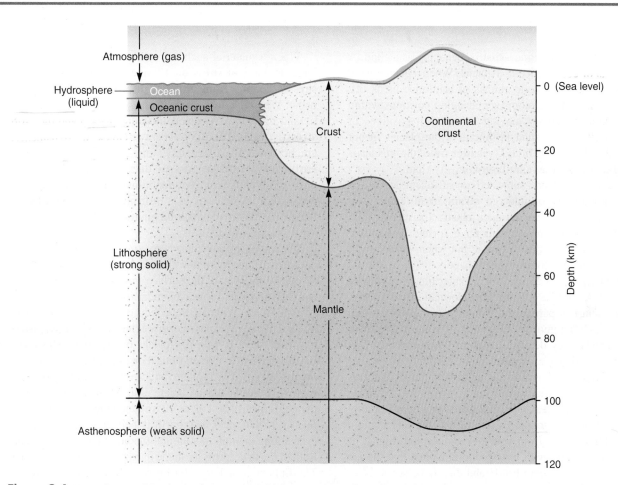

Figure 2.4 Upper layers of Earth may be recognized (1) compositionally, as lower-density crust separated from the underlying higher-density mantle, or (2) on the basis of strength, as rigid lithosphere riding atop "soft plastic" asthenosphere. Notice that the lithosphere includes both the crust and the uppermost mantle.

BEHAVIOR OF MATERIALS

The concepts of gas, liquid, and solid are familiar. Gases and liquids are both **fluids,** but a gas is capable of indefinite expansion, while a liquid is a substance that flows readily and has a definite volume but no definite shape. A solid is firm; it offers resistance to pressure and does not easily change shape. What is not stated but is implicit in these definitions is the effect of time. All of these definitions describe behavior at an instant in time—but how do the substances behave when viewed over a longer timescale? Specifically, some solids yield to long-term pressure such that at any given moment, they are solid, yet internally they

Side Note

Volcanoes and the Origin of the Ocean, Atmosphere, and Life

The **elements** in volcanic gases are predominantly hydrogen (H), oxygen (O), carbon (C), sulfur (S), chlorine (Cl), and nitrogen (N). These gaseous elements combine at Earth's surface to make water (H_2O), carbon dioxide (CO_2), sulfur dioxide (SO_2), hydrogen sulfide (H_2) with its rotten egg smell, carbon monoxide (CO), nitrogen (N_2), hydrogen (H_2), hydrochloric acid (HCl), methane (CH_4), and numerous other gases. The dominant volcanic gas is water vapor; it commonly makes up more than 90% of total gases.

The elements of volcanic gases (C, H, O, N, S, Cl) differ from the elements of volcanic rocks: oxygen (O), silicon (Si), aluminum (Al), iron (Fe), calcium (Ca), magnesium (Mg), sodium (Na), and potassium (K). The elements of volcanic gases make up the oceans, the atmosphere, and life on Earth, but they are rare in rocks. The 4.5 billion years of heat flow from Earth's interior have "sweated" out many lightweight elements and brought them to the surface via volcanism. Billions of years of volcanism on Earth go a long way toward explaining the origin of the continents, the oceans, the present atmosphere, and the surface concentration of the CHON elements (carbon, hydrogen, oxygen, nitrogen) of which all life on Earth is composed and on which it depends.

---

are deforming and flowing—that is, behaving as a fluid. A familiar example is the ice in a glacier. When a glacier is hit with a rock hammer, solid chunks of brittle ice break off. Yet, inside the glacier, atoms are changing positions within the ice and dominantly moving to downhill positions of lower stress. At no instant in time does the glacier fit our everyday concept of a liquid, yet over time, the glacier is flowing downhill as an ultra-high-viscosity fluid.

When materials are subjected to sufficient **stress, or force,** they deform or undergo **strain** in different ways (figure 2.5). Stress may produce **elastic** (or recoverable) deformation, as when you pull on a spring. The spring deforms while you pull or stress it, but when you let go, it recovers and returns to its original shape.

If greater stress is applied for a longer time or at higher temperatures, **ductile** (or plastic) deformation may occur, and the change is permanent. You can visualize this with a wad of chewing gum or Silly Putty. If you squeeze them in your hands, they deform. Set them down and they stay in the deformed shape; this is ductile deformation. Another example occurs deep within glaciers where the ice deforms and moves with ductile flow.

If stress is applied rapidly to a material, it may abruptly fracture or break into pieces, called **brittle** deformation. Take a chunk of ice from your refrigerator and drop it or hit it; it will shatter with brittle failure. Notice that the ice in a glacier exhibits both brittle and ductile behavior. Near the surface, there is little pressure on the rigid ice and it abruptly fractures when stressed. Deep within the glacier, where the weight of overlying ice creates a lot of pressure, the ice deforms and moves by ductile flow. The style of ice behavior depends on the amount of pressure confining it.

The type of mechanical behavior illustrated by ice deep within a glacier typifies that of the rock within the Earth's mantle. This rock is **plastic** in the sense used by William James in his 1890 *Principles of Psychology.* He defined *plastic* as "possession of a structure weak enough to yield to an influence, but strong enough not to yield all at once."

When a material such as rock is subjected to the same large amounts of stress on all sides, it compresses. When stresses coming from different directions vary, strain can occur. When the differences in stress are low, strain is elastic and reversible. As stress differences increase, the **yield stress** is reached and permanent strain occurs. Most rocks are brittle at the low temperatures and low pressures at Earth's surface. Most rocks are ductile at the high temperatures and high pressures at depth inside Earth. In the asthenosphere, rock deforms in a "soft plastic" fashion. Most of the deeper mantle rock is solid but not brittle; it behaves as a "stiff plastic"—it deforms.

The top of the asthenosphere comes to the surface at the ocean's volcanic mountain chains but lies more than 100 km (about 60 mi) below the surface in other areas. It has gradational upper and lower boundaries and is about 250 km (155 mi) thick. What are the effects of having this "soft plastic" ductile zone so near Earth's exterior? Within the asthenosphere, there is a lot of flowage of rock that helps cause Earth's surface to rise and fall. For example, Earth is commonly described as a sphere, but it is not. Earth may be more properly described as an oblate ellipsoid that is flattened at the poles (nearly 30 km, or 19 mi) and bulged at the equator (nearly 15 km, or 9 mi). Earth is neither solid enough nor even strong enough to spin and maintain a spherical shape. Rather, Earth deforms its shape in response to the spin force. The flattening of Earth during rotation is analogous to the flattening of the early Solar System from a sphere to a disk (see figure 2.1).

ISOSTASY

From a broad perspective, Earth is not a homogeneous, solid ball but rather a series of floating layers where less dense materials successively rest upon layers of more dense materials. The core, with densities up to 16 gm/cm^3, supports the mantle, with densities ranging from 5.7 to 3.3 gm/cm^3. Atop the denser mantle float the continents, with densities around

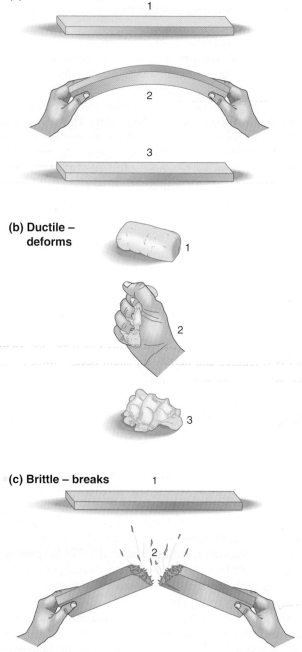

(a) Elastic – recovers

1

2

3

(b) Ductile – deforms

1

2

3

(c) Brittle – breaks

1

2

Figure 2.5 Behavior of materials. (a) Elastic: bend a thin board; let it go and the board recovers its original shape. (b) Ductile: squeeze a wad of bubblegum or Silly Putty; let it go and the mass stays in the deformed shape. (c) Brittle: bend a thin board sharply and it breaks.

2.7 gm/cm^3, which in turn support the salty oceans, with densities of about 1.03 gm/cm^3, and then the least dense layer of them all—the atmosphere. The concept of floating layers holds true on smaller scales as well. For example, the oceans are made of layered masses of water of differing densities. Very cold, dense Antarctic waters flow along the ocean bottoms and are overlain by cold Arctic water, which is overlain by extra-salty waters, which in turn are overlain

by warmer, less dense seawater. Earth is composed basically, from core through atmosphere, of density-stratified layers.

The concept of **isostasy** was developed in the 19th century. It applies a principle of **buoyancy** to explain how the low-density continents and mountain ranges literally float on the denser mantle below. Just as an iceberg juts up out of the ocean while most of its floating mass is beneath sea level, so does a floating continent jut upward at the same time it has a thick "root" beneath it (see figure 2.4). Visualize a boat floating in water: Add a load onto the boat and it sinks downward; remove the load and the boat rises upward. So it is with a continent. Add a load onto the land, such as a large glacial ice mass, and the land will sink downward as rock at depth flows outward within the asthenosphere; remove the load (the ice melts), and the land rises or rebounds upward as rock flows inward in the asthenosphere (figure 2.6). An example of this buoyancy effect, or isostatic equilibrium, was defined by carefully surveying the landscape before and after the construction of Hoover Dam across the Colorado River east of Las Vegas, Nevada. On 1 February 1935, the impoundment of Lake Mead began. By 1941, about 24 million **acre feet** of water had been detained, placing a weight of 40,000 million tons over an area of 232 square miles. Although this is an impressive reservoir on a human scale, what effect can you imagine it having on the whole Earth? In fact, during the 15 years from 1935 to

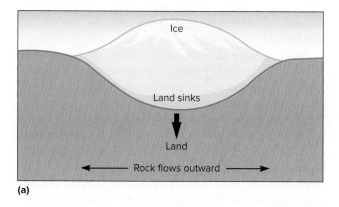

Ice

Land sinks

Land

Rock flows outward

(a)

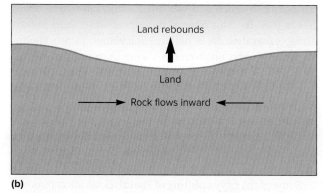

Land rebounds

Land

Rock flows inward

(b)

Figure 2.6 Isostatic equilibrium. (a) Land sinks as weight of ice causes rock at depth to flow outward. (b) Land rebounds as ice melts and removes weight, causing rock at depth to flow inward.

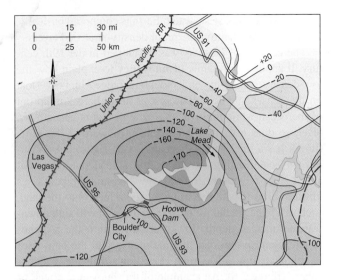

Figure 2.7 Isostatic downwarping caused by the weight of Lake Mead, from 1935 to 1950. Black circular lines (contour lines) define the depressed land surface. In the center is a –170 line where land sank 170 mm (7 in).

Source: Smith, W.O. et al. Comprehensive Survey of Sedimentation in Lake Mead, 1948-49, in *US Geological Survey Professional Paper 295,* 1960.

1950, the central region beneath the reservoir sank up to 175 mm (7 in) (figure 2.7). The relatively simple act of impounding water behind the dam triggered an isostatic adjustment as asthenosphere rock flowed away from the pressure of the overlying reservoir, causing the area to subside.

Just how solid and firm is the surface of the earth we live on? Larger-scale examples are provided by the great ice sheets of the recent geologic past. The continental glacier that buried the Finland-Sweden region was up to 3 km (2 mi) thick less than 20,000 years ago. The land was depressed beneath this great weight. By 10,000 years ago, the ice sheet had retreated and melted, and the water returned to the ocean. The long-depressed landmass, now freed from its heavy load, is rebounding upward via isostatic adjustment. In the past 10,000 years, northeastern coastal Sweden and western Finland have risen about 200 m (650 ft). This upward movement was vividly shown during excavation for a building foundation in Stockholm, Sweden. Workers uncovered a Viking ship that had sunk in the harbor and been buried with mud. The ship had been lifted above sea level, encased in its mud shroud, as the harbor area rose during the ongoing isostatic rebound. Gravity measurements of this region show a negative anomaly, indicating that another 200 m (650 ft) of isostatic uplift is yet to come. The uplift will add to the land of Sweden and Finland and reduce the size of the Gulf of Bothnia between them.

Some of the early uplifting of land after ice-sheet removal occurred in rapid movements that ruptured the ground surface, generating powerful earthquakes. In northern Sweden, there are ground ruptures up to 160 km (100 mi) long with parallel cliffs up to 15 m (50 ft) high. The rocks in the region are ancient and rigid, suggesting that ruptures may go 40 km (25 mi) deep and that they generated truly large earthquakes.

Vertical movements of the rigid lithosphere floating on the flexible asthenosphere are well documented. If we add a load on the surface of Earth, we can measure the downward movement. For example, Antarctica is buried beneath ice up to 4,470 m (2.8 mi) thick. A 100-meter-thick ice mass will cause the land to sink about 27.5 m (90 ft). Thus, Antarctica is depressed up to 1,230 m (4,000 ft), placing most of the continent below sea level. If the ice is removed, Antarctica will slowly rise up and become the fifth largest land mass on Earth. The surface of Earth clearly is in a delicate vertical balance. Do major adjustments and movements also occur horizontally? Yes, there are horizontal movements between lithosphere and asthenosphere, which will bring us into the realm of plate tectonics (described later in this chapter).

Internal Sources of Energy

The flow of energy from Earth's interior to its surface comes mainly from three sources: impact energy, gravitational energy, and the ongoing decay of radioactive isotopes.

IMPACT ENERGY AND GRAVITATIONAL ENERGY

The impact energy of masses colliding with the growing Earth produced heat. Tremendous numbers of large and small asteroids, meteorites, and comets hit the early Earth, their energy of motion being converted to heat on impact.

Gravitational energy was released as Earth pulled into an increasingly dense mass during its first tens of millions of years. The ever-deeper burial of material within the growing mass of Earth caused an increasingly greater gravitational pull that further compacted the interior. This gravitational energy was converted to heat.

The immense amount of heat generated during the formation of Earth did not readily escape because rock conducts heat very slowly. Some of this early heat is still flowing to the surface today.

RADIOACTIVE ISOTOPES

Energy is released from **radioactive isotopes** as they decay. Radioactive isotopes are unstable and must kick out subatomic particles to attain stability. As radioactive isotopes decay, heat is released.

In the beginning of Earth, there were abundant, short-lived radioactive isotopes, such as aluminum-26, that are now effectively extinct, as well as long-lived radioactive isotopes, many of which have now expended much of their energy (table 2.1). Young Earth had a much larger complement of radioactive isotopes and a much greater heat

TABLE 2.1
Some Radioactive Isotopes in Earth

Parent	Decay Product	Half-Life (billion years)
Aluminum-26	Magnesium-26	0.00072 (720,000 years)
Uranium-235	Lead-207	0.71
Potassium-40	Argon-40	1.3
Uranium-238	Lead-206	4.5
Thorium-232	Lead-208	14
Rubidium-87	Strontium-87	47
Samarium-147	Neodymium-147	106

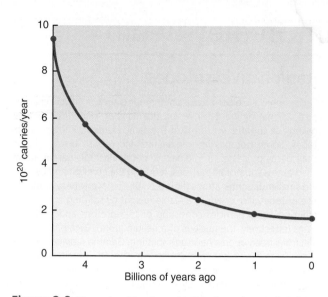

Figure 2.8 The rate of heat production from decay of radioactive atoms has declined throughout the history of Earth.

production from them than it does now (figure 2.8). With a declining output of radioactive heat inside the Earth, the flow of energy from Earth's interior is on a slow decline heading toward zero.

The radioactive-decay process is measured by the **half-life,** which is the length of time needed for half the present number of atoms of a radioactive isotope (parent) to disintegrate to a decay (daughter) product. As the curve in figure 2.9 shows, during the first half-life, one-half of the atoms of the radioactive isotopes decay. During the second half-life, one-half of the remaining radioactive atoms decay (equivalent to 25% of the original parent atoms). The third half-life witnesses the third halving of radioactive atoms

present (12.5% of the original parent atom population), and so forth. Half-lives plotted against time produce a negative exponential curve; this is the opposite direction of a positive exponential curve, such as interest being paid on money in a savings account.

The sum of the internal energy from impacts, gravity, and radioactive isotopes, plus additional energy produced by **tidal friction,** is very large. The greater abundance of radioactive isotopes at Earth's beginning combined with the early gravitational compaction and more frequent meteorite impacts to elevate Earth's internal temperature during its

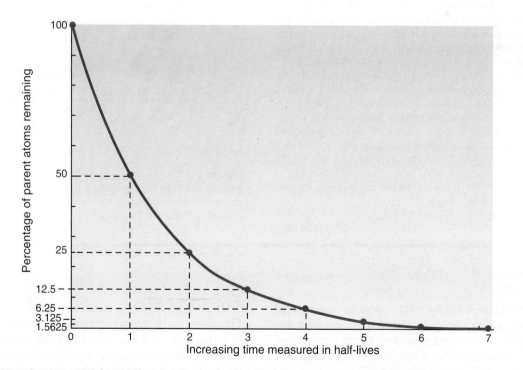

Figure 2.9 Negative exponential curve showing decay of radioactive parent atoms to stable daughter atoms over time. Each half-life witnesses the disintegration of half the remaining radioactive parent atoms.

In Greater Depth

Radioactive Isotopes

Each chemical element has a unique number of positively charged protons that define it. However, the number of neutrons varies, giving rise to different forms of the same element, known as isotopes. Some isotopes are radioactive and release energy during their decay processes. In radioactive decay, unstable parent atoms shed excess subatomic particles, reducing their weight and becoming smaller daughter atoms (figure 2.10). The overly heavy radioactive isotopes slim down to a stable weight by splitting apart, as in emitting alpha particles consisting of two protons and two neutrons (effectively, the nucleus of a helium atom). Beta particles are electrons freed upon a neutron's splitting. Gamma radiation, which is similar to X-rays but with shorter wavelength, is emitted, lowering the energy level of a nucleus. As the rapidly expelled particles are slowed and absorbed by surrounding matter, their energy of motion is transformed into heat.

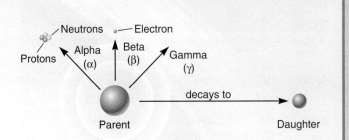

Figure 2.10 A radioactive parent atom decays to a smaller daughter atom by emitting alpha particles (such as the nucleus of a helium atom, i.e., two protons and two neutrons), beta particles (electrons), and gamma radiation (such as X-rays).

DATING THE EVENTS OF HISTORY

The same decaying radioactive isotopes producing heat inside Earth, Moon, and meteorites also may be read as clocks that date events in history. For example, uranium-238 decays to lead-206 through numerous steps involving different isotopes and new elements (figure 2.11). By emitting alpha and beta particles, 32 of the 238 subatomic particles in the U-238 nucleus are lost, leaving the 206 particles of the Pb-206 nucleus. Laboratory measurements of the rate of the decay process have given us the U-238-to-Pb-206 half-life of 4.5 billion years. These facts may be applied to quantifying history by reading the radiometric clocks preserved in some minerals. For example, some **igneous rocks** (crystallized from **magma**) can be crushed, and the very hard mineral zircon (from which zirconium, the diamond substitute in jewelry, is synthesized) separated from it. Zircon crystals contain uranium-238 that was locked into their atomic structure when they crystallized from magma, but they originally contained virtually no lead-206. Thus, the lead-206 present in the crystal must have come from decay of uranium-238.

The collected zircon crystals are crushed into a powder and dissolved with acid under ultraclean conditions. The sample is placed in a mass spectrometer to measure the amounts of parent uranium-238 and daughter lead-206 present. Then, with three known values—(1) the amount of U-238, (2) the amount of Pb-206, and (3) the half-life of 4.5 billion years for the decay process—it is easy to calculate how long the U-238 has been decaying into Pb-206 within the zircon crystal. In other words, the calculation tells us how long ago the zircon crystal formed and consequently the time of formation of the igneous rock.

See the Geologic Time Scale in the Epilogue.

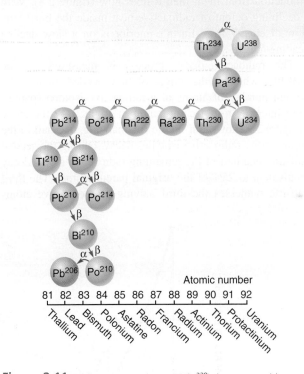

Figure 2.11 Radioactive uranium-238 (U^{238}) decays to stable lead-206 (Pb^{206}) by steps involving many intermediate radioactive atoms. The atomic number is the number of protons (positively charged particles) in the nucleus.

early history. It is noteworthy that this heat buildup reached a maximum early in Earth's history and has declined significantly since then. Nonetheless, the flow of internal heat toward Earth's surface today is still great enough to provide the energy for continents to drift, volcanoes to erupt, and earthquakes to shake.

AGE OF EARTH

The oldest Solar System materials are about 4.57 billion (4,570 million) years old. The 4.57-billion-year age has been measured using radioactive isotopes and their decay products collected from Moon rocks and meteorites. The oldest Earth rocks found to date are in northwest Canada, they are

In Greater Depth

Radioactivity Disasters

The term *radioactivity disasters* brings to mind the meltdown of the uranium-rich core of a nuclear-power plant, as happened at Chernobyl in Ukraine, part of the former Soviet Union, on 26 April 1986. This human-caused disaster occurred when the night-shift workers made a series of mistakes that unleashed a power surge so great that the resultant explosions knocked off the 1,000-ton lid atop the nuclear reactor core, blew out the building's side and roof, triggered a partial meltdown of the reactor core's radioactive fuel, and expelled several tons of uranium dioxide fuel and fission products, including cesium-137 and iodine-131, in a 5 km (3 mi) high plume. As many as 185 million **curies** of radioactivity were released. (The worst U.S. incident released 17 curies from the Three Mile Island nuclear-power plant in Pennsylvania in 1979.) After the 1:24 a.m. explosion, people near Chernobyl were at least fortunate that they were indoors and thus somewhat sheltered, there was no rain in the area, and the contaminant plume rose high instead of hugging the ground. The cloud of radioactive contaminants affected people, livestock, and agriculture from Scandinavia to Greece. In the Chernobyl power plant area, about 50 people died directly. Most of the deaths will come later from cancer and other diseases. The worst contaminant is radioactive iodine-131, which lodges in the thyroid. Cancer of the thyroid is expected to be common in the area; it is estimated that it will shorten the lives of about 8,000 people.

An earthquake may have helped trigger this disaster. The Chernobyl power-plant workers were having difficulties in the early morning hours of 26 April, and then a magnitude 3 earthquake occurred 12 km (7 mi) away. The panicked supervisor thought the shaking meant the power plant was losing control, and he quickly implemented emergency maneuvers, but they jammed the internal works of the reactor, leading to the fateful explosion 22 seconds after the earthquake.

Earthquakes and radioactivity disasters entered the news again on 11 March 2011 when a great magnitude 9 earthquake and resultant tsunami in Japan destroyed several nuclear-power plants. This event is described in Chapter 8 on tsunami.

Chernobyl was a human-caused disaster. What can happen under natural conditions? Today, on Earth and Moon, uranium is present mostly as the heavier U-238 isotope, which has a combined total of 238 protons and neutrons in each uranium atom nucleus. The lighter-weight uranium isotope, U-235, makes up only 0.7202% of all uranium atoms. In nuclear-power plants, the uranium ore fed to nuclear reactors is enriched to 2–4% U-235 to promote more potent reactions. Remember from table 2.1 that U-235 has a half-life of 0.71 billion years, whereas the half-life of U-238 is 4.5 billion years. Because U-235 decays more rapidly, it would have been relatively more abundant in the geologic past. In fact, at some past time, the U-235 natural percentage relative to U-238 would have been like the U-235 percentage added to U-238 and fed as ore to nuclear reactors today.

Have natural nuclear reactors operated in the geologic past? Yes. A well-documented example has been exposed in the Oklo uranium mine near Franceville in southeastern Gabon, a coastal country in equatorial West Africa. At Oklo, 2.1 billion years ago, sands and muds accumulated along with organic carbon from the remains of fossil bacteria. These carbon-bearing **sediments** were enriched in uranium; U-235 was then 3.16% of total uranium. The sand and mud sediments were buried to shallow depths, and at least 800 m³ (1,050 yd³) of uranium ore sustained **nuclear fission** reactions that generated temperatures of about 400°C (750°F) regionally and much higher temperatures locally. At Oklo, 17 sites started up as natural nuclear reactors about 1.85 billion years ago; they ran for at least 500,000 years (and maybe as long as 2 million years). Nine of the natural reactors that have been carefully studied are estimated to have produced at least 17,800 megawatt years of energy.

4.055 billion years old. These rocks are of crustal composition, implying that they were recycled and formed from even older rocks. The oldest ages obtained on Earth materials are 4.37 billion years, measured on sand grains of the mineral zircon collected from within a 3.1-billion-year-old sandstone in western Australia.

Our understanding of the age of Earth is improving rapidly as new technologies allow measurement of more types of radioactive isotopes. It now seems that Earth has existed as a coherent mass for about 4.54 billion years. Earth must be younger than the 4.57-billion-year-old materials that collided and clumped together to form the planet. The time it took to build Earth is possibly as short as 30 million years. The collision of Earth with the Mars-size body that formed our Moon seems to have occurred between 4.537 and 4.533 billion years ago, suggesting that Earth was already a large, coherent mass at that time. Coming from the other direction, Earth must be older than the

4.37-billion-year-old zircon grains collected from sandstone in Australia. In sum, our planet has existed for about 4.5 billion years.

The work to exactly determine the age and early history of Earth continues today. It is challenging to try and find the oldest minerals and rocks because Earth is such an energetic planet that surface rocks are continually being formed and destroyed. Because of these active earth processes, truly old materials are rarely preserved; there have been too many events over too many years that destroy rocks.

Plate Tectonics

The grand recycling of the upper few hundred kilometers of Earth is called the **tectonic cycle.** The Greek word *tekton* comes from architecture and means "to build"; it has been adapted by geologists as the term **tectonics,** which describes

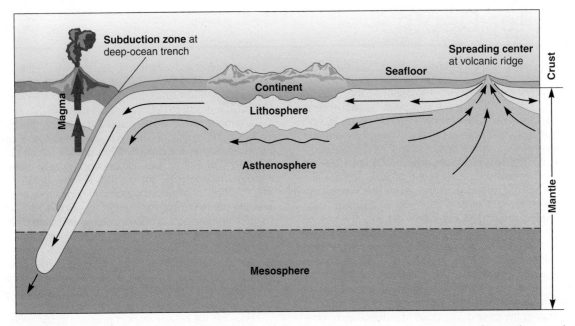

Figure 2.12 Schematic cross-section of the tectonic cycle. Magma rises from the asthenosphere to the surface at the oceanic volcanic ridges where it solidifies and adds to the plate edges. As the igneous rock cools, the plate subsides and gravity pulls the plates from their topographic highs. The plate continues to cool, grows thicker at its base, becomes denser, collides with a less-dense plate, and turns down into the mantle, where it is ultimately reassimilated.

the building of **topography** and the deformation and movement within Earth's outer layers.

Adding the horizontal components of movements on Earth allows us to understand the tectonic cycle. Ignoring complexities for the moment, the tectonic cycle can be simplified as follows (figure 2.12). First, melted asthenosphere flows upward as magma and cools to form new ocean floor/lithosphere. Second, the new lithosphere slowly moves laterally away from the zones of oceanic crust formation on top of the underlying asthenosphere; this phenomenon is known as **seafloor spreading.** Third, when the leading edge of a moving slab of oceanic lithosphere collides with another slab, the older, colder, denser slab turns downward and is pulled by gravity back into the asthenosphere, a process called **subduction,** while the less-dense, more buoyant slab overrides it. Last, the slab pulled into the mantle is reabsorbed. The time needed to complete this cycle is long, commonly in excess of 250 million years.

If we adopt the perspective of a geologist-astronaut in space and look down upon the tectonic cycle, we see that the lithosphere of Earth is broken into pieces called **plates** (figure 2.13). The study of the movements and interactions of the plates is known as **plate tectonics.** The gigantic pieces of lithosphere (plates) pull apart during seafloor spreading at **divergence zones,** slide past at **transform faults,** or collide at **convergence zones.** These plate-edge interactions are directly responsible for most of the earthquakes, volcanic eruptions, and mountains on Earth.

Another way that plate tectonics can be visualized is by using a hard-boiled egg as a metaphor for Earth. Consider the hard-boiled egg with its brittle shell as the lithosphere, the slippery inner lining of the shell as the asthenosphere, the

egg white (albumen) as the rest of the mantle, and the yolk as the core. Before eating a hard-boiled egg, we break its brittle shell into pieces that slip around as we try to pluck them off. This hand-held model of brittle pieces being moved atop a softer layer below is a small-scale analogue to the interactions between Earth's lithosphere and asthenosphere.

DEVELOPMENT OF THE PLATE TECTONICS CONCEPT

Our planet is so large and so old that the combined efforts of many geologists and philosophers over the past few hundred years have been required to amass enough observations to begin understanding how and why Earth changes as it does. The first glimpse of our modern understanding began after the European explorers of the late 1400s and 1500s made maps of the shapes and locations of the known continents and oceans. These early world maps raised intriguing possibilities. For example, in 1620, Francis Bacon of England noted the parallelism of the Atlantic coastlines of South America and Africa and suggested that these continents had once been joined. During the late 1800s, the Austrian geologist Eduard Suess presented abundant evidence in support of **Gondwanaland,** an ancient southern supercontinent composed of a united South America, Africa, Antarctica, Australia, India, and New Zealand, which later split apart. This process of the continents moving, splitting, and recombining is known as **continental drift.** The most famous and outspoken of the early proponents of continental drift was the German meteorologist Alfred Wegener. In his 1915 book, *The Origin of Continents and Oceans,* he collected all available evidence, such as similar

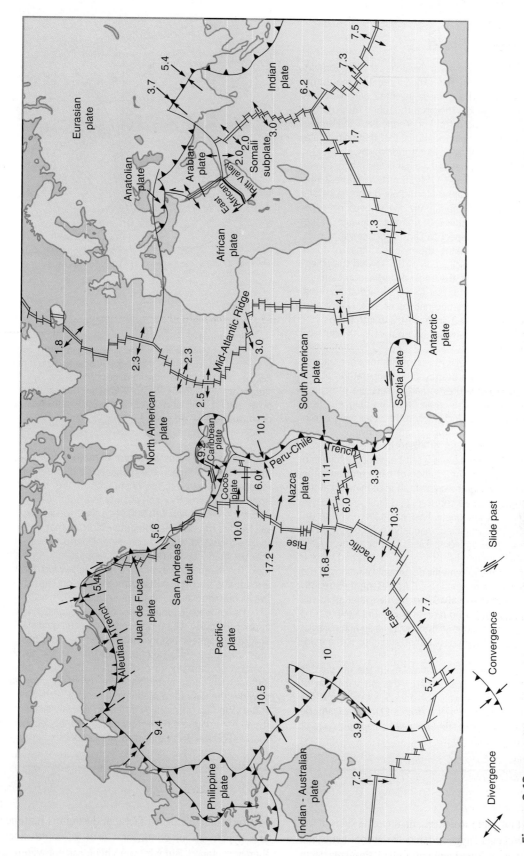

Figure 2.13 Map of the major tectonic plates with arrows showing directions of movement. Rates of movement are shown in centimeters per year.

In Greater Depth

Earth's Magnetic Field

Anyone who has ever held a compass and watched the free-turning needle point toward the north has experienced the **magnetic field** that surrounds Earth. The Chinese invented and were the first to use magnetic compasses. They in turn taught 14th-century European travelers, who brought this knowledge back to Europe, where it was developed into the navigational tool that helped late 15th-century explorers make their voyages of discovery.

Earth's magnetic field operates as if a gigantic bar magnet were located in the core of Earth inclined 11° from vertical (figure 2.14). The **magnetic pole** and geographic North Pole do not coincide, but the magnetic pole axis has apparently always been near the rotational pole axis. Notice in figure 2.14 that the inclination of the magnetic lines of force with respect to Earth's surface varies with **latitude.** At the magnetic equator, the magnetic lines of force are parallel to Earth's surface (inclination of 0°). Toward the poles, either northward or southward, the angle of inclination continuously increases until it is perpendicular to the surface at both the north and south magnetic poles (inclinations of 90°). Notice also that the lines of force are inclined downward and into Earth's surface near the North Pole and upward and out of Earth's surface near the South Pole.

In reality, the interior of Earth is much too hot for a bar magnet to exist. **Magnetism** in rocks is destroyed by temperatures above 550°C (1,020°F), and temperatures in Earth's core are estimated to reach 5,800°C (10,470°F). The origin of Earth's magnetic field involves movements of the iron-rich fluid in the outer core, which generate electric currents that in turn create the magnetic field. Fluid iron is an excellent conductor of electricity. The molten iron flowing around the solid inner core is a self-perpetuating dynamo deriving its energy both from the rotation of Earth and from the convection of heat released by the crystallization of minerals at the boundary of the inner and outer cores.

A closer look at Earth's magnetic field yields several problems awaiting resolution. The simplified magnetic field portrayed in figure 2.14 does not show the complexities that occur over years and centuries as the magnetic field's strength waxes and wanes. More than 400 years of measurements document variations in the strength and stability of the magnetic field. At present, the strength is 10% weaker than in the year 1845, but the field strength is still about twice as strong as the long-term average. The flow of fluid iron in the outer core has regions of turbulence, including motions as complex as whirlpools. Change is normal, and in turn, the magnetic field fluctuates.

In addition, the magnetic pole moves about the geographic North Pole region in an irregular pattern. The rapid rotation of Earth holds the magnetic pole near the pole of rotation, but the magnetic pole wanders enough that it crosses 5° to 10° of latitude each century. In recent decades, the magnetic pole has moved at rates of 10 to 40 km (6 to 25 mi) per year.

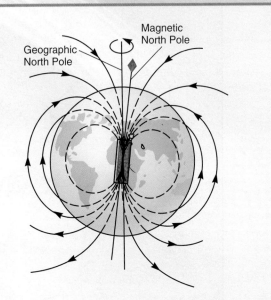

Figure 2.14 Schematic diagram of Earth's magnetic field. The bar magnet pictured does not exist, but it would create the same magnetic field achieved by the electrical currents in Earth's liquid, iron-rich outer core. Notice that (1) the magnetic pole and the rotational pole do not coincide, (2) the magnetic lines of force are parallel to Earth's surface at the magnetic equator and perpendicular at the magnetic poles, and (3) the lines of force go into Earth at the North Pole and out at the South Pole.

Every several thousand to tens of millions of years, a highly dramatic change occurs in the magnetic field: the magnetic polarity reverses. In a reversal, the orientation of the magnetic field flip-flops from a north (normal) polarity to a south (reverse) polarity or vice versa. It has been 780,000 years since the last long-term reversal. Models run on supercomputers indicate that reversals take less than a thousand years to complete. During a reversal, it appears that the magnetic field does not disappear; it just gets more complex. The magnetic lines of force become twisted and tangled, but a magnetic field still exists to protect life on Earth from much of the incoming solar and space radiation.

The change in orientation of the magnetic field leaves its imprint in rocks, where geologists (paleomagnetists) can read it. The paleomagnetic history contained in the rocks has provided the most important evidence of seafloor spreading; it also has allowed charting of the paths of continents as they have moved through different latitudes. In addition, the record of magnetic reversals provides the data for a magnetic timescale, a third geologic timescale. (The first timescale is based on the irreversible sequence of fossils occurring in sedimentary rocks, and the second timescale is founded on the decay of radioactive isotopes.)

rocks, fossils, and geologic structures, on opposite sides of the Atlantic Ocean. Wegener suggested that all the continents had once been united in a supercontinent called **Pangaea** (*pan* meaning "all" and *gaea* meaning "earth").

Much is made of the fact that during his lifetime, Wegener's hypothesis of continental drift garnered more ridicule than acceptance. But why were his ideas not widely accepted? Wegener presented an intriguing hypothesis well supported with observations and logic, but his mechanism was deemed impossible. Geologists and geophysicists could not visualize how a continent could break loose from the underlying rocks and plow a path over them. The breakthrough in understanding came when the ocean floors were studied and the data were best explained by the formation of new seafloor that spread apart and later was consumed by subduction. When scientists realized that the lithosphere decouples from the asthenosphere and moves laterally, they understood how the relatively small, low-density continents, set within the oceanic crust, could be carried along as incidental passengers (see figure 2.12).

In the mid-1960s, evidence abounded, mechanisms seemed plausible, and the plate tectonic theory was developed and widely accepted. Wegener was restored to an elevated status. Scientific understanding grew with the addition of new data, old hypotheses were modified, and new theories were created. Science is never static; it is a growing, evolving body of knowledge that creates ever-better understanding of how Earth works.

It is rare in science to find widespread agreement on a large-scale hypothesis such as plate tectonics. But when data from Earth's magnetic field locked inside seafloor rocks were widely understood, skeptics around the world became convinced that seafloor spreading occurs and that the concept of plate tectonics is valid. These paleomagnetic data are so powerful that we need to understand their story so that plate tectonics can be seen as real.

MAGNETIZATION OF VOLCANIC ROCKS

Lava is magma that erupts from a volcano, flows outward as a sheetlike mass, slows down, and stops. Then, as the lava cools, minerals begin to grow as crystals. Some of the earliest formed crystals incorporate iron into their structures. After the lava cools below the **Curie point,** about 550°C (1,020°F), atoms in iron-bearing minerals become magnetized in the direction of Earth's magnetic field at that time and place. The lined-up atoms in the iron-rich crystals behave like compass needles pointing toward the magnetic pole of their time (measured as declination or "compass bearing"); they also become inclined at the same angle as the lines of force of the magnetic field (measured as inclination or dip). Ancient magnetic fields have been measured in rocks as old as 3.5 billion years.

Lava flows pile up as sequences of stratified (layered) rock, and the magnetic polarity of each rock layer can be measured (figure 2.15). Many of the volcanic rocks also contain minerals with radioactive isotopes that allow us to

Figure 2.15 A stratified pile of former lava flows of the Columbia River Basalt exposed in the east wall of Grand Coulee, Washington. The oldest flow is on the bottom and is overlain by progressively younger flows.
University of Washington Libraries, Special Collections, John Shelton Collection, Shelton KC10288

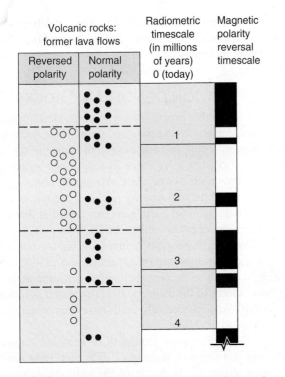

Volcanic rocks: former lava flows		Radiometric timescale (in millions of years) 0 (today)	Magnetic polarity reversal timescale
Reversed polarity	Normal polarity		

Figure 2.16 A portion of the magnetic polarity timescale. Magnetic polarity measurements in volcanic rocks combined with radiometric ages determined from the same rocks allow formation of a timescale based on magnetic polarity reversals. Notice the unique and nonrepetitive pattern of the polarity reversals.

Source: From P. J. Wyllie, *The Way the Earth Works*. Copyright © 1976 John Wiley & Sons, Inc., New York.

determine the age of the volcanic rock—that is, how long ago the lava flow solidified. When this information is plotted together in a vertical column, a timescale of magnetic polarities emerges (figure 2.16). It is interesting to note that the timing of polarity reversals appears to be random. There is no discernible pattern to the lengths of time the magnetic field was oriented either to the north or to the south. The processes that reverse the polarity of the magnetic field are likely related to changes in the flow of the iron-rich liquid in the outer core. The reversal-causing mechanism does not occur at any mathematically definable time interval.

Magnetization Patterns on the Seafloors

Since the late 1940s, oceanographic research vessels crisscrossing the Atlantic Ocean have towed magnetometers to measure the magnetization of the seafloor. As the number of voyages grew and more data were obtained, a striking pattern began to emerge (figure 2.17). The floor of the Atlantic Ocean is striped by parallel bands of magnetized rock that show alternating polarities. The pattern is symmetrical and parallel with the midocean volcanic **ridge** (spreading center). That is, each striped piece of seafloor has its twin on the other side of the oceanic mountain range.

A remarkable relationship exists between the time of reversals of magnetic polarity, as dated radiometrically from a sequence of solidified lava flows (see figure 2.16), and the widths of alternately polarized seafloor (figure 2.17)—they are comparable. How stunning it is that the widths of magnetized seafloor strips have the same ratios as the lengths

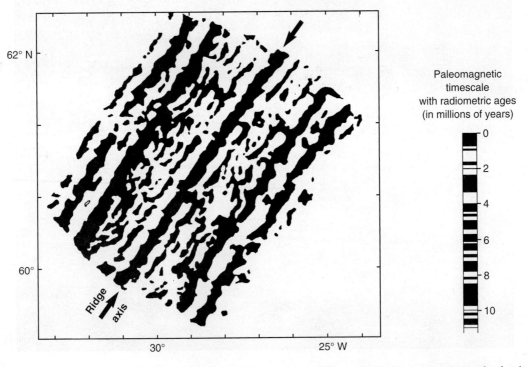

Figure 2.17 Map of the magnetically striped Atlantic Ocean floor southwest of Iceland. Black areas are magnetized pointing to a north pole and white areas to a south pole. Notice the near mirror images of the patterns on each side of the volcanic ridge (spreading center).

of time between successive reversals of Earth's magnetic field. This means that distance in kilometers is proportional to time in millions of years. Now, if Earth's magnetic field is reversing polarity in a known timescale and if that timescale reappears in distances, then the relationship must take the form of a velocity. That is, magma is injected into the oceanic ridges where it is imprinted by Earth's magnetic field as it cools to form new rock. Then the seafloor/ocean crust/lithosphere is physically pulled away from the oceanic ridges as if they were parts of two large conveyor belts going in opposite directions (figure 2.18).

The evidence provided by the paleomagnetic time-scale and the magnetically striped seafloors is compelling. These phenomena are convincing evidence that seafloor spreading occurs and that plate tectonics is valid.

Let's now consider other evidence supporting plate tectonics.

Earthquake Evidence

The map of earthquake **epicenters** (figure 2.19) can be viewed as a connect-the-dots puzzle. Each epicenter represents a place where one major section of rock has moved

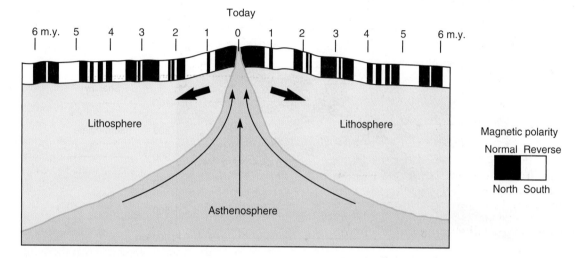

Figure 2.18 Cross-section of magnetically striped seafloor. Numbers above the seafloor are radiometrically determined ages in millions of years (m.y.). The near mirror-image magnetic pattern is like a tape recorder that documents "conveyor belt" movements away from volcanic ridges.

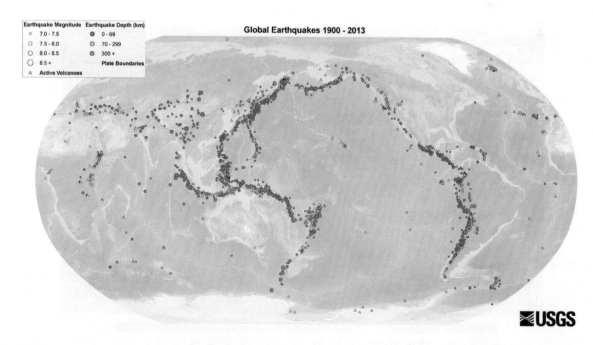

Figure 2.19 Map of earthquake epicenters, 1975–1995. Notice that epicenters are concentrated in linear belts. Color-coding of epicenters indicates depths of earthquakes; notice that depths increase toward continents.

U.S. Geological Survey.

past another section. Take your pen or pencil, connect the dots (epicenters), and you will outline and define the edges of the tectonic plates, the separately moving pieces of lithosphere. Remember that these plates are about 100 km (60+ mi) thick and can be thousands of kilometers across.

Earthquakes at depth commonly occur along inclined planes (figure 2.20) adjacent to deep-ocean trenches. These deep earthquakes define the subducting plates being pulled forcefully back into the mantle. In figure 2.19, the earthquake epicenters are color-coded according to depth. The increasing depths toward and beneath continents define the subducting oceanic plates.

Ages from the Ocean Basins

One of the most stunning facts discovered during the recent exploration of the oceans is the youthfulness of the ocean basins. The oldest rocks on the ocean floors are about 200 million years in age; this is less than 5% of the age of Earth (figure 2.21). Remember that some continental rocks are more than 4,000 million years old. Meteorites are more than 4,500 million years old. Some Moon rocks are more than 4,500 million years old, and none are younger than 3,100 million years. But the ocean basins (not the water in them) and their contained volcanic mountains, sediments, and fossils are all much, much younger. Why? Because the ocean

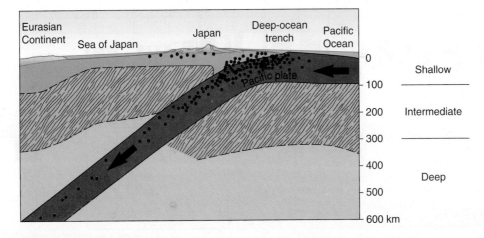

Figure 2.20 Cross-section showing earthquake (fault movement) locations at depth; notice the inclined plane defined by the earthquake sites (black dots). The earthquake locations define the subducting plate beneath Japan. At shallow depths, earthquakes are generated in brittle rocks in both subducting and overriding plates. At greater depths, only the interior of the subducting Pacific plate is cold enough to maintain the rigidity necessary to produce earthquakes. Striped areas are hot rocks defined by relatively lower-velocity seismic waves.

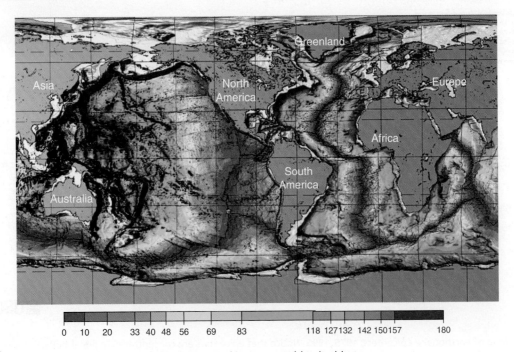

Figure 2.21 Age of the ocean floor in millions of years. Red is youngest; blue is oldest.

Dr. Peter Sloss/NOAA/NGDC.

basins are young features that are continuously being formed and destroyed.

Along the oceanic ridges, volcanism is active, and new seafloor/oceanic crust is forming (see figure 2.18). Moving away from the ridges, the seafloor volcanic rocks and islands become progressively older. The oldest seafloor rocks are found at the edges of the ocean basins.

At certain locations, deep-seated **hot spots** produce more heat, causing hotter rocks with lower density. These masses of buoyant hot rock rise upward as **plumes** through the mantle, begin to melt near the top of the overlying asthenosphere, and pass up through the lithosphere as magma. Hot spots have active volcanoes above them on Earth's surface. The volcanoes rest on moving plates that carry them away from their hot-spot source. This process forms lines of extinct volcanoes on the ocean floor, from youngest to oldest, pointing in the direction of plate movement (figure 2.22). The hot-spot–fed plume moves also and probably is responsible for the prominent bend in the island/seamount chain.

The blanket of sediment on the seafloor ranges from very thin to nonexistent at the volcanic ridges and thickens toward the ocean margins (figure 2.23). The older the seafloor, the more time it has had to accumulate a thick cover of sand, silt, clay, and **fossils.**

Oceanic Mountain Ranges and Deep Trenches

The greatest mountain ranges on Earth lie on the ocean bottoms and extend more than 65,000 km (40,000 mi). These long and continuous volcanic mountains are seen to form at **spreading centers** where plates pull apart and magma rises to fill the gaps.

The ocean bottom has an average depth of 3.7 km (2.3 mi), yet depths greater than 11 km (nearly 7 mi) exist in elongate, narrow **trenches** (see top of figure 2.22a). The long and deep trenches were known since the *Challenger* oceanographic expedition in the 1870s, but they were not understood until the 1960s, when geologists recognized that they are the tops of the subducting plates turning downward to reenter the mantle.

Systematic Increases in Seafloor Depth

Above the oceanic ridges, the ocean water depths are relatively shallow. However, moving progressively away from the ridges, the ocean water depths increase systematically with seafloor age (figure 2.23). This is due to the cooling and contraction of the oceanic crust with a resultant increase in density. Also, some isostatic downwarping occurs due to the weight of sediments deposited on the seafloor. The progressive deepening of the seafloor with increasing age also testifies to the existence of seafloor spreading.

The Fit of the Continents

If the continents have really drifted apart, then we should be able to take a map, cut out the continents to make puzzle pieces, and then reassemble them in their former

(a) Map

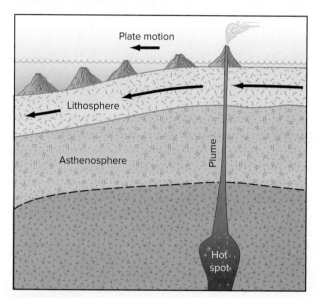

(b) Cross-section

Figure 2.22 A hot spot and its path. (a) Map shows the Hawaiian Islands–Emperor seamount chain of hot-spot–fed volcanoes with plots of their radiometric ages in millions of years. The map pattern of volcano ages testifies to movement of the Pacific plate through time. (Ma = mega-annum = 1 million years.) (b) Cross-section shows a hot spot at a depth where hot mantle rock rises up through the asthenosphere and passes through the lithosphere as a plume of magma supplying a volcano. Because the lithospheric plate keeps moving, new volcanoes are formed.

(a) NOAA/NGDC.

configuration. In fact, this can be done if we know where to cut the map. On two-dimensional world maps, the landmasses occupy about 29% of Earth's surface and the oceans the other nearly 71%. If we cut the puzzle pieces at the land-sea shoreline and then attempt to reassemble them, the fit will not be good. The problem here is that the significant boundary is not between land and water but instead at the

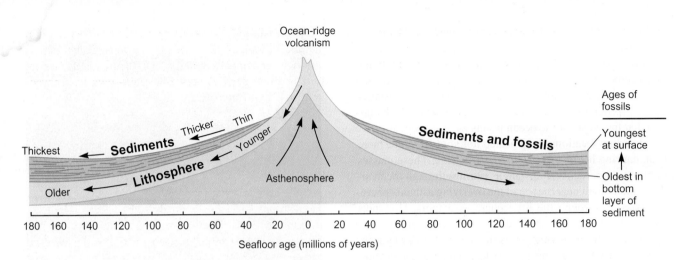

Figure 2.23 Schematic cross-section through oceanic lithosphere perpendicular to a volcanic ridge. Moving away from the ridge: (1) radiometric ages of oceanic lithosphere increase, (2) thicknesses of accumulated sediments increase, and (3) ages of fossils in the sediments increase. The systematic increases in water depth are due to cooling, shrinking, increasing density of the aging seafloor rocks.

real edge of the continent—the change from low-density continental rocks to higher-density oceanic rocks. This change occurs at about a 1,800 m (6,000 ft) water depth. If we remove the oceans, we find that the continental masses cover 40% of Earth's surface and the ocean basins the other 60%. If we cut the puzzle pieces at the 1,800-meter water-depth line, then the continental puzzle pieces fit together quite well. There are some overlaps and gaps, but these are reasonably explained by changes during the last 220 million years, since the last major split of the continents. Examples of the changes include deformation during the process of

rifting; growth of river deltas, volcanoes, and coral **reef** masses; erosion of the continents; and land movements.

Changing Positions of the Continents

Undoing the seafloor spreading of the last 220 million years restores the continents of today into the supercontinent Pangaea, which covered 40% of Earth (figure 2.24). Although the present continents had yet to form, this figure shows their relative positions within Pangaea before its breakup. The remaining 60% of the Earth's surface was a massive ocean called **Panthalassa** (meaning "all oceans").

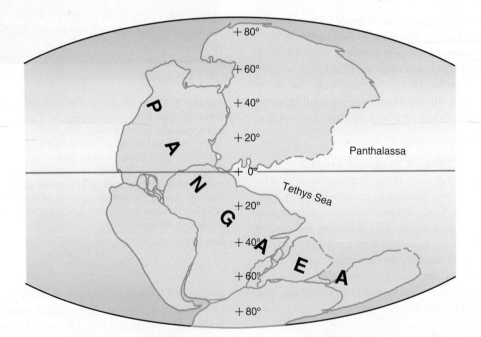

Figure 2.24 Pangaea, the supercontinent, 220 million years before present. The modern continents are drawn to be recognizable in this restoration. The superocean of the time (Panthalassa) exists today in shrunken form as the Pacific Ocean.

After R. S. Dietz and J. C. Holden, "Reconstruction of Pangaea: Breakup and Dispersion of Continents, Permian to Present" in *Journal of Geophysical Research* 75:4, 939–56, 1970. Copyright © 1970 American Geophysical Union.

Figure 2.25a shows the breakup of Pangaea at 180 million years before present. An equatorial spreading center separated the northern supercontinent **Laurasia** from the southern supercontinent Gondwanaland. Much of the sediment deposited in the Tethys Sea at that time has since been uplifted to form mountain ranges, from the Himalayas to the Alps. Another spreading center began opening the Indian Ocean and separating Africa-South America from Antarctica-Australia.

At 135 million years ago, seafloor spreading had begun opening the North Atlantic Ocean, India was moving toward Asia, and the South Atlantic Ocean was a narrow sea similar to the Red Sea today (figure 2.25b).

By 65 million years ago, seafloor spreading had opened the South Atlantic Ocean and connected it with the North Atlantic, and Africa came into contact with Europe, cutting off the western end of the Tethys Sea to begin the Mediterranean Sea (figure 2.25c). Although the modern world had become recognizable, note that North America and Eurasia were still connected and that Australia had not yet left Antarctica.

Nearly half of the present ocean floor was created during the last 65 million years (figure 2.25d). India has rammed into Asia, continued opening of the North Atlantic has split Eurasia from North America, and Australia has moved a long way from Antarctica.

The Grand Unifying Theory

Figure 2.26 shows how Earth's outer layers are operating today in plate-tectonic action. The following model explains how it happens. Rising hot rock in the mantle reaches the asthenosphere and begins to melt. The buildup of hot rock, magma, and heat causes expansion and topographic elevation of the overlying oceanic lithosphere or continent, which then fractures because of the uplift and begins to be pulled apart laterally by gravity. A continent can be split, forming a rift zone, a young divergent plate boundary such as the East African Rift Valley. In an ocean basin, the pulling apart of oceanic lithosphere reduces pressure on superheated

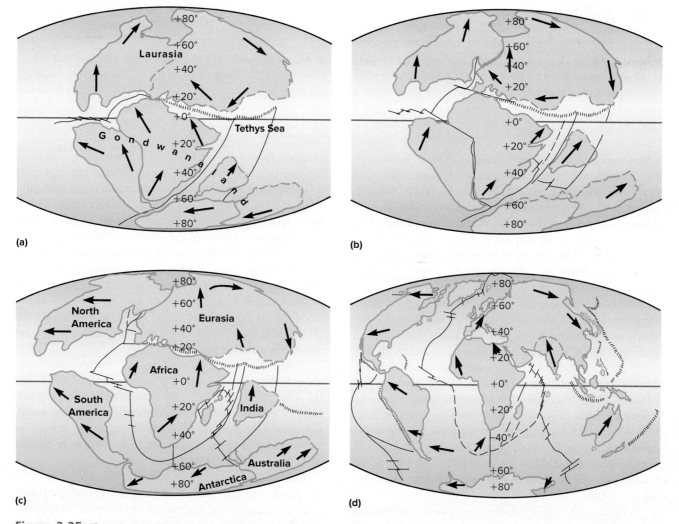

Figure 2.25 Changing positions of the continents. (a) 180 million years ago. (b) 135 million years ago. (c) 65 million years ago. (d) Today.

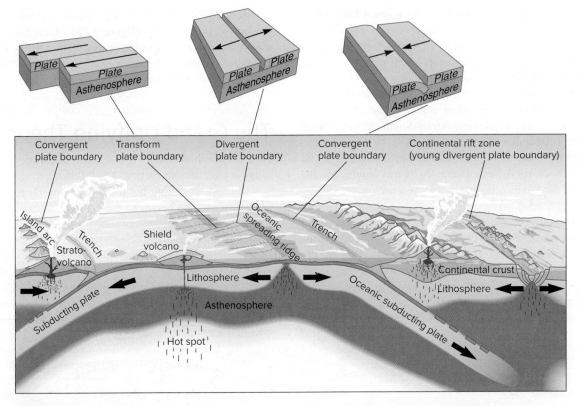

Figure 2.26 Three-dimensional schematic view of tectonic plates with divergent, convergent, and transform boundaries plus volcanoes above subducting plates and a hot spot.

From Kious and Tilling, *US Geological Survey.*

asthenosphere rock; some liquefies and rises upward to fill fractures and create new oceanic lithosphere via sea-floor spreading. The continuing elevation of the volcanic mountain chain (ridge) forms a setting for gravity to keep pulling material downward and outward (spreading). The lateral spreading is aided by the push of positive buoyancy at spreading ridges and may be aided by **convection** cells of mantle heat, which rise and move laterally beneath the litho-sphere before descending.

As the lithosphere spreads, cools, and becomes denser, it is pulled ever more strongly by gravity. When oceanic litho-sphere collides with another plate, the denser (older, colder) plate goes beneath the less-dense (younger, warmer) plate in the process of subduction. If an oceanic plate goes beneath another oceanic plate at a convergent plate boundary, an island arc of volcanoes next to a trench can form, such as the Aleutian Islands of Alaska. If the subducting oceanic plate is pulled beneath a continent-carrying plate, the top of the down-bending oceanic plate forms a trench, and a line of active volcanoes builds on the continent edge, such as the Cascade Ranges of northern California, Oregon, Washing-ton, and British Columbia.

As the leading edge of the negatively buoyant subducting plate turns downward, gravity exerts an even stronger pull on it, which helps tear the trailing edge of the plate away

from the spreading center. The combination of gravity pull-ing on elevated spreading-center mountains and especially on denser, down-going plates at subduction zones (slab pull) keeps the lithospheric plates moving. Thus, an ongoing tec-tonic cycle operates whereby each moving part stimulates and maintains motions of the others in a large-scale, long-term recycling operation. Subducted plates are reassimilated into the mantle as physical slabs that remain solid enough to be recognized by their effects on the travel velocities of seismic waves. Plate movements are now so well under-stood due to the magnetic record of seafloor rocks that the plates are not only outlined, but their rates of movement are defined as well (see figure 2.13).

Note also on figure 2.26 that oceanic spreading ridges are offset by faulting at transform plate boundaries. Deep below the realm of plate tectonics, hot spots send up plumes of hot, buoyant rock that turns into magma near the surface, build-ing shield volcanoes on the seafloor, such as in Hawaii, or explosive mega-volcanoes on continents, such as in Yellow-stone National Park, Wyoming.

Plate tectonics is a great scientific concept. It provides us with new perspectives about Earth that are quite different from those encountered in our life or historical experiences. Because Earth is so much older and so much larger than a human being, we must set aside our personal time and size scales. Our lives are measured in decades, and our personal measuring rods are our 5- to 6-feet-tall bodies; with these as reference guides, we can be only mystified by Earth. However, if we change our time perspective to millions and billions of years and our size scales to continents and plates, then—and only then—can we begin to understand Earth. An active plate may move 1 cm (0.4 in) in a year; this is only 75 cm (30 in) in a human lifetime. The rates of plate movement are comparable to those of human fingernail growth.

How can we explain the building of mountains or the formation of ocean basins? We must consider Earth over its own time span of 4,570 million years. Then there is plenty of time for small events to add up to big results. The plate moving 1 cm/yr travels 10 km (6+ mi) in just 1 million years. The 1 cm/yr process is fast enough to uplift a mountain in a small amount of geologic time.

How We Understand Earth

Thousands of years ago, human thought had already made great advances in topics such as philosophy, government, religion, drama, and engineering. But our understanding of Earth was insignificant until Earth's great age was realized. This recognition came late in human history; it started with James Hutton in the 1780s. Hutton carefully observed his Scottish landscape and thought deeply about it. For example, he saw rock walls built by the Romans that had stood for 15 centuries with only slight change. If 1,500 years was not long enough to break down a wall, Hutton wondered, how much time had been required to break down some of the hard rock masses of Scotland into the abundant pebbles and sand grains he saw? And how much more time had been necessary to lift the pebbly and sandy sedimentary rocks to form hills?

All the active processes Hutton observed worked slowly, so his answer to the questions was that great lengths of time were required. In 1788, Hutton described the history of Earth as follows: "The result, therefore, of our present inquiry is that we find no vestige of a beginning, no prospect of an end." And this was Hutton's great gift to human thought: time is long, and everyday changes on Earth add up to major results.

UNIFORMITARIANISM

Hutton's thought pattern, called **uniformitarianism,** has revolutionized our understanding of Earth. Uniformitarianism implies that natural laws are uniform through time and space. Physical and biological laws produce certain effects today, as they have in the past, and will in the future. If we can understand how Earth works today, we can use this knowledge to read the rock and fossil record to understand Earth's history. The present is the key to the past.

The term *uniformitarianism* has come under attack by some who assume it says that earth processes have always acted at a uniform and slow rate, but we all know that rates can vary. For example, seafloor spreading has operated at slower rates in the past, and it has also run at faster rates, but the laws governing how seafloor spreading operates do not change just because the rates vary. Some suggest using the term **actualism** instead of uniformitarianism, but the concept is basically the same. Actualism tells us to understand physical, chemical, and biological processes actually operating on, in, and outside Earth today, and to use these known and testable processes to interpret the past; it advises us *not* to invent undemonstrated and untestable supernatural causes to explain away problems.

How do we go about understanding Earth? We study the present to understand the past and then make probabilistic forecasts about the future.

Summary

Massive amounts of internal heat within the early Earth caused widespread melting. Gravity has pulled Earth into layers of differing density, beginning with a heavy metallic core and proceeding outward through layers of decreasing density: from the mantle to the continents, the ocean, and finally, the atmosphere. These layers exist in a state of flotational equilibrium known as isostasy. Up-and-down movements of the land due to isostatic adjustments are readily measurable.

The radioactive isotopes that help heat Earth's interior by their decay do so at measurable rates quantified by half-lives. Elements that radioactively decay act as clocks that can be used to date the events of Earth history. Earth is 4.57 billion years old.

The outer layers of Earth are involved in a grand recycling known as the tectonic cycle. Hot buoyant rock and magma rise up from the mantle, through the lithosphere, to build world-encircling mountain ranges of volcanoes (oceanic ridges). The injection of magma elevates ridges that are pulled apart by gravity (slab pull) in gigantic slabs (plates) to form ocean basins in the process known as seafloor spreading. When these moving lithospheric plates collide, if one plate is composed of denser rock, it will turn back down into the mantle, a process called subduction, to become melted and reabsorbed. The entire lithosphere is fractured into plates that pull apart (diverge), slide past, and collide (converge) with each other. The plate collisions cause mountains to rise, seafloors to bend down forming trenches that are elongate and deep, volcanoes to erupt, and earthquakes to be generated; this cyclic process is the topic of plate tectonics. Continents are composed of lower-density rock that rides on top of the denser rock of the moving plates.

The evidence for plate tectonics is overwhelming. Ancient magnetic fields locked into iron-bearing minerals in rocks point toward former south or north magnetic poles in patterns, indicating that seafloor spreading and continental drift occur. The ages of rocks, sediments, and fossils, as well as the depth of water, all increase away from the oceanic ridges, indicating that oceanic crust/lithosphere is continuously forming and spreading apart. The oldest rocks and fossils in the ocean basins are less than 5% of Earth's age, indicating that oceanic material is destroyed by recycling into the mantle.

Terms to Remember

acre foot	29	magnetic field	36
actualism	45	magnetic pole	36
asthenosphere	26	magnetism	36
brittle	28	mantle	25
buoyancy	29	mesosphere	26
centigrade	24	nuclear fission	36
continental drift	34	nuclear fusion	24
convection	44	Pangaea	36
convergence zone	34	Panthalassa	42
core	25	plastic	28
crust	25	plate	34
curie	33	plate tectonics	34
Curie point	37	plume	41
divergence zone	34	radioactive isotope	30
ductile	28	reef	42
elastic	28	ridge	38
element	28	seafloor spreading	34
epicenter	39	sediment	33
Fahrenheit	24	solar radiation	24
fluids	27	spreading center	41
fossil	41	strain	28
Gondwanaland	34	stress	28
half-life	31	subduction	34
hot spot	41	tectonic cycle	33
igneous rock	32	tectonics	33
isostasy	29	tidal friction	31
latitude	36	topography	34
Laurasia	43	transform fault	34
lava	37	trench	41
liquid	26	uniformitarianism	45
lithosphere	26	viscous	26
magma	32	yield stress	28

Questions for Review

1. Describe how Earth became segregated into layers of differing density.
2. How did Earth's continents, oceans, and atmosphere form?
3. Describe some examples of isostasy.
4. What energy sources caused the interior of the early Earth to heat up?
5. How does nuclear fusion differ from nuclear fission?
6. What is the age of Earth? How is this determined?
7. After freeing zircon crystals from an igneous rock, how could you determine when the rock formed (solidified)?
8. Where are the oldest known Earth rocks found? How old are they?
9. What are the differences between brittle, ductile, and elastic behavior?
10. Draw and label a cross-section that explains the tectonic cycle.
11. Explain some other evidence indicating that seafloors spread.
12. What is a tectonic plate?
13. What are the ages of the oldest (a) rocks on the continents, and (b) rocks making up the ocean floor?
14. Explain several lines of evidence indicating that the continents move about Earth.
15. Describe a deep-ocean trench. How does one form?
16. Why do deep earthquakes tend to occur within inclined bands?
17. Draw a cross-section that shows a hot spot and plume. How do they help determine the directions of plate motions?
18. How is Earth's magnetic field formed? Describe the reversals of magnetic polarity from north to south.
19. Explain the concept of uniformitarianism.
20. What is the Curie temperature? How is this related to magnetism?
21. Explain the paleomagnetic evidence for seafloor spreading.
22. How can the magnetic record inside a volcanic rock be used to determine the latitude at which the lava cooled?

Questions for Further Thought

1. Earth is commonly called *terra firma*. Does this make good geologic sense?
2. Are new natural nuclear reactors likely to spring into action on Earth?
3. Your lifetime will be what percentage of geologic time?
4. How much does the ground sink under a load or rise after removal of a load during isostatic adjustments?
5. Why does the polarity of Earth's magnetic field switch from north to south and back again?
6. How can the rate of motion of a plate be calculated?

CHAPTER 3

Earthquake Geology and Seismology

Diseased nature oftentimes breaks forth
In strange eruptions: oft the teeming earth
Is with a kind of colic pinch'd and vex'd
By the imprisoning of unruly wind
Within her womb; which, for enlargement striving,
Shakes the old beldam earth, and topples down
Steeples, and moss-grown towers.
—WILLIAM SHAKESPEARE, 1598, *KING HENRY IV*

Houses built on vertical posts in Bosques de las Lomas, Mexico City, have precious little shear strength to respond to seismic waves.
Courtesy of Pat Abbott.

LEARNING OUTCOMES

Earthquakes are shaking most commonly caused by earth movements along faults. Energy from movements is carried long distances by seismic waves. After studying this chapter, you should:

- be able to describe the types of faults.
- know the types of seismic waves.
- understand the different ways of calculating earthquake magnitude.
- be familiar with the variables that determine earthquake intensity, as in the Mercalli intensity scale.
- comprehend the relationships between periods and frequencies of seismic waves, buildings, and geologic foundations.
- recognize the types of buildings and building materials that fail during earthquakes.
- understand how to construct buildings that do not fail during earthquakes.

OUTLINE

A Classic Disaster

The Lisbon Earthquake of 1755

Portugal in the 18th century, and especially its capital city of Lisbon, was rich with the wealth its explorers brought from the New World. Portugal's decline probably began with a set of earthquakes. On the morning of 1 November 1755—All Saints Day—Lisbon rocked under the force of closely spaced earthquakes originating offshore under the Atlantic Ocean. On this day of religious observance, the churches were full of worshippers. About 9:40 a.m., a thunderous underground sound began, followed by violent ground shaking. The severe ground movement lasted two to three minutes, causing widespread damage to the buildings in this city of more than 250,000 people. Most of Lisbon's churches were built of masonry; they collapsed into the narrow streets, killing thousands of trapped and fleeing people. Tapestries fell onto candles and lamps—all lit on this holy day—and started fires that burned unchecked for six days.

Before an hour had passed, crippled Lisbon was rocked by a second earthquake, more violent but shorter-lived than the first. In the panic, many of the frightened survivors of the first earthquake had rushed to the shore for safety, only to be swept away by quake-caused sea waves up to 10 m (33 ft) high. These walls of water spilled onto the land, carrying boats and cargo more than 0.5 km inland. As the seawater withdrew, it dragged people and debris from the earthquake-shattered structures back to the ocean.

The two earthquakes killed almost 70,000 people and destroyed or seriously damaged about 90% of the buildings in Lisbon (figure 3.1). At the time, the city was rich in bullion, jewels, and merchandise, and it had great commercial and cultural importance. The destruction of this famous city by earthquakes and their resulting sea waves and fires was a shock to Western civilization. Not only were the losses of lives and buildings staggering, but the fires also incinerated irreplaceable libraries, maps and charts of the

Figure 3.1 The Lisbon earthquake.

Portuguese voyages of discovery, and paintings by such masters as Titian, Correggio, and Rubens. The Lisbon earthquakes did more than devastate a city; they changed the prevailing philosophies of the era. All was not well in the world after all.

The earth beneath our feet moves, releasing energy that shifts the ground and sometimes topples cities. Some earthquakes are so immense that their energy is equivalent to thousands of atomic bombs exploded simultaneously. The power of earthquakes to destroy human works, to kill vast numbers of people, and to alter the very shape of our land has left an indelible mark on many civilizations.

Earthquake unpredictability instills an uneasy respect and fear in humankind that, through the millennia, have helped shape thought about life and our place in it. Ancient accounts of earthquakes tend to be quite incomplete. Instead of providing rigorous descriptions of Earth behavior, they emphasize interpretations. For more than 2,000 years, based on Aristotle's ideas, many explanations of earthquakes were based on winds rushing beneath Earth's surface. Even Leonardo da Vinci wrote in his *Notebooks,* about 1500 CE, that:

When mountains fall headlong over hollow places they shut in the air within their caverns, and this air, in order to escape, breaks through the Earth, and so produces earthquakes.

Despite the profound effects that earthquakes have had on civilizations for so many centuries, scientific observations did not begin until the early 19th century, when good descriptions were made of earthquake effects on the land. Today, less than two centuries later, our knowledge of earthquakes has increased enormously. We have a fairly comprehensive understanding of what earthquakes are, why and where they happen, and how big and how often they occur at a given site. Our scientific data and theories allow us to understand phenomena that even the greatest minds of the past could not have glimpsed. Such are the rewards from the pyramidal building of knowledge we call science.

Understanding Earthquakes

The word *earthquake* is effectively a self-defining term—the Earth quakes, the Earth shakes, and we feel the vibrations. Earthquakes, or **seisms,** may be created by volcanic activity, meteorite impacts, undersea landslides, explosions of nuclear bombs, and more; but most commonly, they are caused by sudden earth movements along faults. A **fault** is a **fracture** surface in the Earth across which the two sides move past each other (figure 3.2). Stresses build up in rocks, but **friction** along fault surfaces holds the rocks together. When **stress** builds high enough, the rocks along the fault snap and move suddenly, releasing energy in waves we feel as the shaking of an earthquake.

Figure 3.2 Offset of tilled farmland by 1979 movement of the Imperial fault, southernmost California. View is to the east; the west side of the fault (closest to you) has moved northward (to your left).

© Kerry Sieh.

To visualize this fault movement, snap your fingers. As you prepare your finger snap, you push your thumb and finger together and sideways, but friction resists their moving past each other. When stress builds high enough, your thumb and finger slip rapidly, releasing energy as sound waves. Both a fault rupture in the earth and your finger snap feature the same sudden slips that release energy in waves.

FAULTS AND GEOLOGIC MAPPING

The 19th-century recognition that fault movements cause earthquakes was a fundamental advance that triggered a whole new wave of understanding. With this relationship in mind, geologists go into the field to map active faults, which in turn identifies earthquake-hazard belts. Because a fault moves formerly continuous rock layers apart, the careful mapping of different rock masses can define sharp lines that separate offset segments of single rock masses. Fault surfaces can be vertical, horizontal, or at any angle to Earth's surface. Some faults rupture the ground, some do not.

The principles that help us understand faults begin with some of the earliest recognized relationships about rocks, which are still useful today. In 1669, the Danish physician Niels Steensen, working in Italy and known by his Latinized name of Steno, set forth several laws that are fundamental in interpreting geologic history. His **law of original horizontality** explains that sediments (sands, gravels, and muds) are originally deposited or settled out of water in horizontal layers. This is important because some older sedimentary rock layers are found at angles ranging from horizontal to vertical. But since we know they started out as horizontal layers (figure 3.3), their postdepositional history of deformation can be unraveled by mentally returning their orientations back to horizontal (figure 3.4).

Figure 3.3 North wall of the upper Grand Canyon. At the canyon bottom, the once horizontal sedimentary rock layers have been tilted to the east. Their uptilted ends have been eroded and buried by horizontal younger rock layers.

University of Washington Libraries, Special Collections, John Shelton Collection, Shelton 1081.

In the **law of superposition,** Steno stated that in an unde-formed sequence of sedimentary rock layers, each successive layer is deposited on top of a previously formed, and hence older, layer. Thus, each sedimentary rock layer is younger than the bed beneath it but older than the bed above it (figures 3.3 and 3.4).

Figure 3.4 These sedimentary rocks were deposited in horizontal layers, but have since been compressed into contorted layers by movements of the San Andreas fault.

University of Washington Libraries, Special Collections, John Shelton Collection, Shelton KC13887.

Steno's **law of original continuity** states that sediment layers are continuous, ending only by butting up against a topographic high, such as a hill or a cliff, by pinching out due to lack of sediment, or by gradational change from one sediment type to another. This relationship allows us to appreciate the incongruity of a sedimentary rock layer that abruptly terminates. Something must have happened to terminate it. For example, a stream may have eroded through it, or a fault may have truncated it. Geologists spend a lot of time locating and identifying offsets of formerly continuous rock layers. In this way, we can determine the lengths of faults and estimate the magnitude of earthquakes they produce. Longer lengths of fault rupture create bigger earthquakes.

On a much broader scale, we can find large offsets on long-acting, major faults. figure 3.5 shows a pronounced line cutting across the land in a northeast-southwest trend; this is the Alpine fault on the South Island of New Zealand. The west (left) side has been moved 480 km (300 mi) toward the north. In Otago province in the southern part of the South Island, gold was discovered in 1861 in stream gravels. This set off a gold rush that brought in prospectors and miners from all over the world. The gold fever that had attracted so many fortune seekers to California in 1849 now moved to New Zealand. Prospectors panned the streams and worked their way upstream into bedrock hills to find the source of the gold. Yet much of the wealth lay 480 km to the

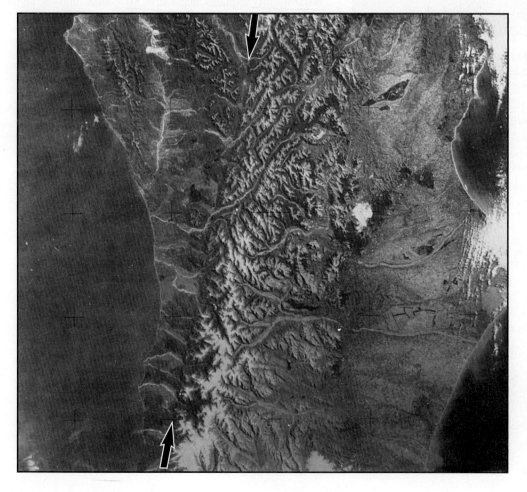

Figure 3.5 Aerial photo of part of South Island, New Zealand (see figure 3.6 for location). The Alpine fault cuts a prominent slash from near the lower left (southwest) corner of the photo to the top center (northeast). Arrowheads line up with the fault.

Courtesy of Pat Abbott.

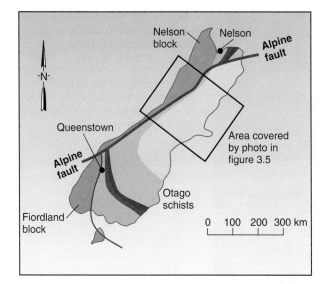

Figure 3.6 Generalized geologic map of South Island, New Zealand. Each map color records a different type of rock. Locate the Alpine fault, and then match up the rock patterns across the fault. The gold-bearing rocks near Queenstown have been offset 480 km (300 mi) to near Nelson.

northeast in Nelson province, where the same gold-bearing rock masses had been offset along the Alpine fault by more than 23 million years of fault movements (figure 3.6). As this example shows, fault studies also can have tremendous implications for locating mineral wealth.

Types of Faults

As tectonic plates move, mountains are elevated and basins are warped downward. The brittle rocks of the lithosphere respond by fracturing (also called **jointing** or cracking). When regional forces create a large enough stress differential in rocks on either side of a fracture, then movement occurs and the fracture becomes a fault. Accumulated movements of rocks along faults range from millimeters to hundreds of kilometers. These movements can cause originally horizontal sedimentary rock layers to be tilted and folded into a wide variety of orientations (figure 3.7a). To describe the location in three-dimensional (3-D) space of a deformed rock layer, a fault surface, or any other planar feature, geologists make measurements known as **dip** and **strike.** Dip is seen in the two-dimensional (2-D) vertical view (**cross-section**) as the angle of inclination from the horizontal of the tilted rock layer (figure 3.7b). It is also important to note the compass direction of the dip in the horizontal plane—for example, toward the northeast. Strike is viewed in the 2-D horizontal view (**map**) as the compass bearing of the rock layer where it pierces a horizontal plane.

DIP-SLIP FAULTS

The classification of faults uses some terminology of early miners. Many ore veins were formed in ancient fault zones.

(a)

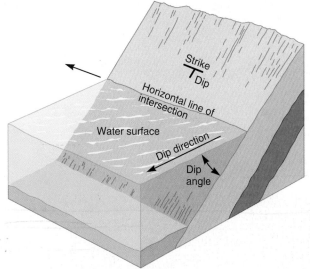

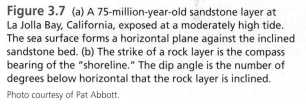

(b)

Figure 3.7 (a) A 75-million-year-old sandstone layer at La Jolla Bay, California, exposed at a moderately high tide. The sea surface forms a horizontal plane against the inclined sandstone bed. (b) The strike of a rock layer is the compass bearing of the "shoreline." The dip angle is the number of degrees below horizontal that the rock layer is inclined.

Photo courtesy of Pat Abbott.

Thus, many mines consist of adits (passages) dug along old, inactive faults. Ores are common along faults because when one block of rocks moves past another in a fault zone, the tremendous friction tends to shatter and pulverize the rocks in the fault zone. The broken rock creates an avenue of **permeability** through which water can flow. If the underground water carries a concentration of dissolved metals, they may precipitate as valuable elements or minerals within the fault zone. Early miners working in excavated fault zones called the floor beneath their feet the **footwall** and the rocks above their heads the **hangingwall** (figure 3.8). This terminology is used to define the types of faults dominated by vertical movements, called **dip-slip faults.** Faults with the major amounts of their offset in the dip or vertical direction are caused by either a pulling (**tension**) or a pushing (**compression**) force.

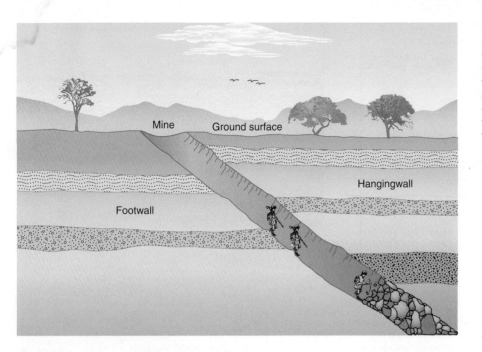

Figure 3.8 Schematic cross-section of miners excavating ore that precipitated in broken rock within an old fault zone. Notice that the rock layers in the footwall and hangingwall are no longer continuous; this gives evidence of the movements that occurred along the fault in the past.

Mine   Ground surface

Hangingwall

Footwall

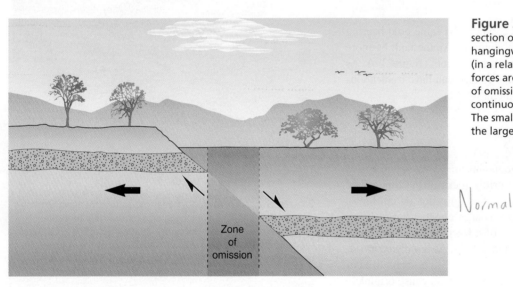

Figure 3.9 Schematic cross-section of a normal fault; that is, the hangingwall has moved downward (in a relative sense). Extensional forces are documented by the zone of omission, where the originally continuous rock layers are missing. The small arrows indicate movement; the larger arrows show force.

Normal

Zone of omission

There are two major types of dip-slip faults: normal faults and reverse faults. A **normal fault** occurs when the hangingwall moves down relative to the footwall. The dominant force is extensional, as recognized by the separation of the pulled-apart rock layers in a zone of omission (figure 3.9). The word *normal* as a name for this type of fault is unfortunate because it carries a connotation of normalcy, as if this were the standard or regular mode of fault movement; such is not the case. Extensional or normal-style faults are typical of the faults at seafloor spreading centers and in regions of continents being pulled apart.

If the dominant force that creates a fault movement is compressional, then the rock layers are pushed together, or repeated, when viewed in cross-section (figure 3.10). With compressional forces, the hangingwall moves upward relative to the footwall; this type of fault is referred to as a

reverse fault. The compressional motions of reverse faults are commonly found at areas of plate convergence where subduction or continental collision occurs.

The extensional versus compressional origins of movement can have enormous economic implications. Look again at figures 3.9 and 3.10. Visualize the emphasized (dotted) rock layer in each figure as being an oil reservoir. Now imagine yourself to be the landowner above either the zone of omission or the zone of repetition. In one case, it could mean poverty; in the other, great wealth.

STRIKE-SLIP FAULTS

When stress produces **shear** and causes most of the movement along a fault to be horizontal (parallel to the strike direction), the fault is referred to as a **strike-slip fault.**

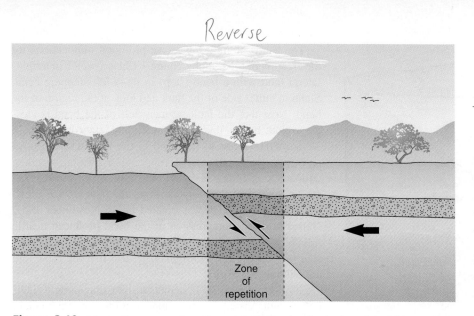

Reverse

Figure 3.10 Schematic cross-section of a reverse fault; that is, the hangingwall has moved upward (in a relative sense). Compressional forces are documented by the zone of repetition, where the originally continuous rock layers have been split, shoved together, and stacked above each other.

These fault offsets are seen in map view as though from a balloon or airplane looking down on Earth's surface. Strike-slip faults are further classified on the basis of the relative movement directions of the fault blocks. If you straddle a fault and the block on your right-hand side has moved relatively toward you, then it is called a **right-lateral,** or dextral, **fault** (figure 3.11). Notice that this convention for naming the fault works no matter which way you are straddling the

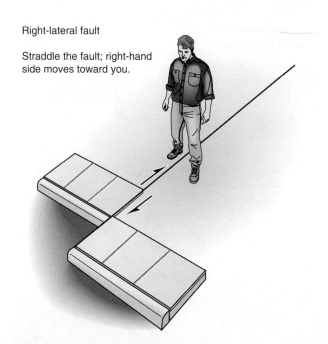

Right-lateral fault

Straddle the fault; right-hand side moves toward you.

Figure 3.11 Map of a right-lateral, strike-slip fault. As the man straddles the fault, the right-hand side of the fault has moved relatively closer to him. If he turns around, will the right-hand side of the fault still have moved closer to him?

fault; try it facing both directions with figure 3.11. Similarly, if features on the left-hand side of the fault have moved closer to you, then it is a **left-lateral,** or sinistral, **fault.**

We have looked at a large strike-slip fault in New Zealand, the Alpine fault, but the most famous strike-slip fault in the world is the San Andreas in California. This right-lateral fault is more than 1,300 km (800 mi) long. On 18 April 1906, a 430 km (265 mi) long segment of the San Andreas fault ruptured and moved horizontally as much as 6.5 m (20 ft) in 60 seconds. The great burst of energy generated by the fault movement was actually the release of elastic energy that had built up and been stored in the rocks for many decades.

Faults are not simple planar surfaces that glide readily when subjected to stress. Instead, faults are complex zones of breakage where rough and interlocking rocks are held together over an irregular surface that extends many miles below the ground. Stress must build up over many years before enough potential energy is stored to allow a rupture on a fault. The initial break occurs at a weak point on the fault and then propagates rapidly along the fault surface. Much of the energy stored in the rocks is released as radiating seismic waves that humans call an earthquake. The point where the fault first ruptures is known as the **hypocenter,** or focus. The point on Earth's surface directly above the hypocenter is called the *epicenter* (figure 3.12).

A fault rupture is not a simple, one-time movement that produces "the earthquake." In fact, we never have just one earthquake. The stresses that build up in the rocks in an area are released by a series of movements along the fault, or several faults, that continue for weeks to months to years. Each fault movement generates an earthquake.

Steps in Strike-Slip Faults

Strike-slip faults do not simply split the surface of Earth along perfectly straight lines. The rupturing fault tears apart the rocks along its path in numerous subparallel breaks that stop and start, bend left, and bend right. For analogy, visualize a sheet cake or pan of moist mud. Put your right hand on the upper right corner and your left hand on the lower left corner. Now pull toward you with your right hand and push away with your left. Do you visualize the cake ripping along one straight line? Or along several breaks that stop and start, bend left and bend right? So it is with Earth when it ruptures during an earthquake-generating fault movement. Normal and reverse faults also have bends; we just don't see them as easily on the surface.

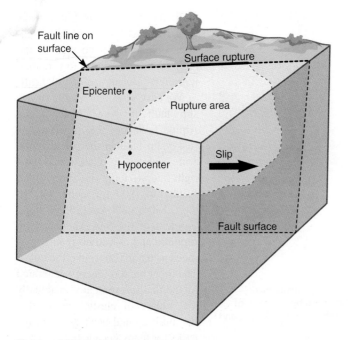

Figure 3.12 Block diagram of a fault surface. The hypocenter *(focus)* is the point on the fault surface where the rupture began; the epicenter is the point on Earth's surface directly above the hypocenter. Notice that because the fault surface is inclined (it dips), the epicenter does not plot on the trace of the fault at the surface.

Source: J. Ziony, ed., "Earthquakes in the Los Angeles Region." *US Geological Survey.*

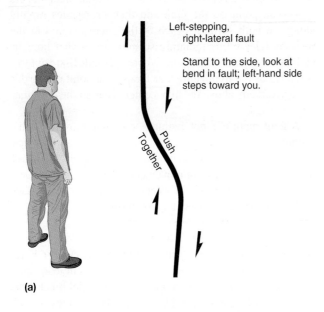

Figure 3.13 (a) Left step in right-lateral fault. Notice that the land is pushed together at the fault bend whenever the fault moves. Movements will create a hill, which could grow to a mountain if the fault remains active for a long enough time. (b) Land offset along the Superstition Hills right-lateral fault during its 16 November 1987 earthquake. See the left step and the uplift at the bend. (Black arrows indicate directions of land movement.)

Photo courtesy of Pat Abbott.

The bends along a fault have profound implications for the creation of topography. figure 3.13a is a sketch of a right-lateral fault with a bend (step) in it—a left-stepping bend. Stand to either side of the fault and look at the region of the bend. Note that the fault segment left of the bend is closest to you; hence, this is a left-stepping, right-lateral fault. Notice what occurs at the bend in the fault when the two sides slide past each other—compression, pushing together, collision, constraint. The photo in figure 3.13b shows a left step in the right-lateral Superstition Hills fault west of Brawley, California, which was created on 16 November 1987. Notice how the compression at the bend produced a little hill. What size could this hill attain if movements at this left step were to occur for millions of years? It could grow into a mountain.

Similarly, figure 3.14a depicts a right step along a right-lateral fault. Visualize what happens at the bend in the fault. In this case, the two sides pull apart from each other, extend, diverge, release. The photo in figure 3.14b is from the same earthquake, along a different length of the same fault, as in figure 3.13b. At this right step, the two sides pulled apart and created a down-dropped area—a wide crack or a little basin.

TRANSFORM FAULTS

Transform faults are a special type of horizontal-movement fault first recognized by the Canadian geologist J. Tuzo Wilson in 1965. Figure 3.15 depicts how a transform fault

(b)

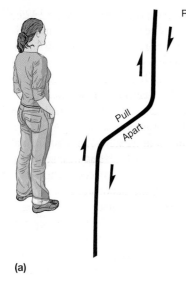

Right-stepping,
right-lateral fault

Stand to the side, look at
bend in fault; right-hand side
steps toward you.

Pull
Apart

(a)

(b)

Figure 3.14 (a) Right step in right-lateral fault. Notice that the land is pulled apart at the fault bend whenever the fault moves. Movements will create a hole, which could become a basin if the fault stays active for a geologically long time. (b) Land offset along the Superstition Hills right-lateral fault during its 1987 rupture. See the right step and the pull apart at the bend. (Black arrows indicate directions of land movement.)

Photo courtesy of Pat Abbott.

Spreading
center

Deep-ocean
trench

Transform fault

Fracture zone

Spreading
center

Subduction zone

Continent

Lithosphere

Magma

Asthenosphere

Figure 3.15 Plate-tectonic model of a transform fault. Notice that the transform fault connects the two separated spreading centers; the seafloor moves in opposite directions here. Beyond the spreading centers, the two plates move in the same direction and are separated by a fracture zone; there is no transform fault here.

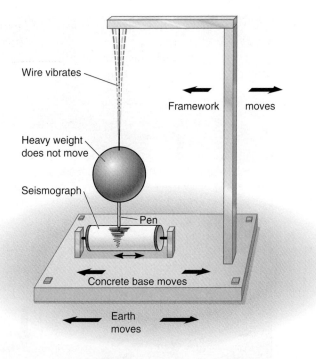

Wire vibrates

Framework moves

Heavy weight does not move

Seismograph

Pen

Concrete base moves

Earth moves

Figure 3.16 A basic seismograph. Earth moves, the seismograph framework moves, and the hanging wire vibrates, but the suspended heavy mass and pen beneath it remain relatively steady. Ideally, the pen holds still while Earth moves beneath the pen to produce an inked line. Three seismometers sensing vibrations in orthogonal directions of ground shaking are required to record the full 3-D shaking at a point.

forms. Seafloor crust forms at oceanic volcanic ridges and is pulled apart by gravity and slab pull of subducting plates. When plates collide, the denser plate subducts. But what happens along the sides of the plates? They slide past each other at transform faults. Visualize this process in three dimensions. The spreading plates are rigid slabs of oceanic rock, tens of kilometers thick, that are being wrapped around a near-spherical Earth. How does a rigid plate move about a curved surface? The plates must fracture, and these fractures are transform faults. In fact, transform faults must link spreading centers or connect spreading centers with subduction zones.

In figure 3.15, notice that in the region between the two spreading centers, the relative motions of the two plates are in opposite directions in typical strike-slip fault fashion. However, passing both to the right and left of the spreading centers, notice that the two slabs are moving in the same direction and there they are called fracture zones there is no active offset across a fracture zone.

Development of Seismology

The study of earthquakes is known as **seismology** (after *seism*, meaning "earthquake"). The earliest earthquake-indicating device known was invented in China in 132 CE by Chang Heng. The modern era of seismologic instrumentation began about 1880. Instrumentation continues to evolve through many different styles, but a basic need is to record the 3-D movement of earthquake waves. This is achieved by having instruments detect Earth motions (**seismometers**) and record them (**seismographs**) as north-south horizontal movements, east-west horizontal movements, and vertical movements. To accurately record the passage of seismic waves, a seismometer must have a part that remains as stationary as possible while the whole Earth beneath it vibrates. One way to accomplish this is by building a frame that suspends a heavy mass (figure 3.16). The support frame rests on Earth and moves as Earth does, but the mass suspended by a wire must have its **inertia** overcome before it moves. The principle of inertia explains that a stationary object—for example, the suspended mass—tends to remain stationary. The differences between motions of the frame and the hanging mass are recorded on paper by pen and ink or, increasingly, as digital data. Visualize the process this way: hold an ink pen steady in your hand and then vibrate the entire Earth beneath your pen to make an inked line.

Other important pieces of information to record include the arrival times and the durations of the various seismic waves. This is accomplished by having time embedded in the seismographic record either as inked tick marks on the paper graph or within the digital data. Time is standardized in the United States by the national clock in Boulder, Colorado.

First-order analysis of the seismic records allows seismologists to identify the different kinds of seismic waves generated by the fault movement, to estimate the amount of energy released (magnitude), and to locate the epicenter/ hypocenter (where the rock hit the water, so to speak).

WAVES

Throw a rock into a pond, play a musical instrument, or experience a fault movement, and the water, the air, or the Earth will transmit waves of energy that travel away from the initial disturbance. All these waves have the following similarities: **amplitude,** the height of the wave above the starting point (figure 3.17); **wavelength,** the distance between successive waves; **period,** the time between waves measured in seconds; and **frequency,** the number of waves passing a given point during 1 second. Frequencies are measured in **hertz (Hz),** where 1 Hz equals one cycle

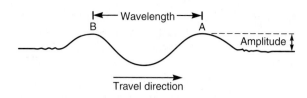

Figure 3.17 Wave motion. Amplitude is the height of the wave above the starting point. Wavelength is the distance between wave crests B and A. Period is the amount of time in seconds for wave crest B to travel to site A.

per second. Note that period and frequency are inversely related:

$$\text{Period} = \frac{1}{\text{frequency (in hertz)}}$$

For example, if five waves passed a given point in 1 second, then the frequency is 5 Hz and the period of time between each wave is 0.2 second.

Seismic Waves

When a fault slips, or an explosion occurs, it releases energy in **seismic waves** that pass through the whole body of the planet (**body waves**) and others that move near the surface only (**surface waves**).

BODY WAVES

Body waves are the fastest and are referred to as either primary or secondary waves. Body waves ranging from about 0.02 Hz to tens of Hz produce measurable ground shaking. These high-frequency, short-period waves are most energetic for short distances close to the hypocenter/epicenter.

Primary Waves

The **primary (P) wave** is the fastest and thus the first to reach a recording station. P waves move in a push-pull fashion, alternating pulses of compression (push) and extension (pull); this motion is probably best visualized using a Slinky toy (figure 3.18a). P waves radiate outward from their source in an ever-expanding sphere, like a rapidly inflating balloon. They travel through any material, be it solid, liquid, or gas. Their speed depends on the density and compressibility of the materials through which they pass. The greater the resistance to compression, the greater the speed of the seismic waves passing through packed atomic lattices. Representative velocities for P waves in hard rocks (e.g., **granite**) are about 5.1 to 5.5 km/sec (about 11,400 to 12,300 mph). P waves in water slow to 1.4 km/sec (about 3,100 mph). Because P waves and sound waves are both compressional waves, they can travel through air. P waves may emerge from the ground, and if you are near the epicenter, you may be

able to hear those P waves pulsing at around 15 cycles per second as low, thunderous noises. The arrival of P waves at your home or office is similar to a sonic boom, including the rattling of windows.

Secondary Waves

The **secondary (S) wave** is the second wave to reach a recording station. S waves are transverse waves that propagate by shearing or shaking particles in their path at right angles to the direction of advance. This motion is probably most easily visualized by considering how a jump rope moves when you shake one end up and down (figure 3.18b). S waves travel only through solids. S waves do not propagate through fluids. On reaching fluid or gas, the S wave energy is reflected back into rock or is converted to another form. The velocity of an S wave depends on the density and resistance to shearing of materials. Fluids and gases do not have shear strength and thus cannot transmit S waves. Representative velocities for S waves in dense rocks (e.g., granite) are about 3 km/sec (about 6,700 mph). With their up-and-down and side-to-side motions, S waves shake the ground surface and can do severe damage to buildings.

SEISMIC WAVES AND EARTH'S INTERIOR

Large earthquakes generate body waves energetic enough to be recorded on seismographs all around the world. These P waves and S waves do not follow simple paths as they pass through Earth; they speed up, slow down, and change direction, and S waves even disappear. Analysis of the travel paths of the seismic waves gives us our models of Earth's interior (figure 3.19). Earth is not homogeneous. Following the paths of P and S waves from Earth's surface inward, there is an initial increase in velocity, but then a marked slowing occurs at about 100 km (62 mi) depth; this is the top of the asthenosphere. Passing farther down through the mantle, the velocities vary but generally increase until about 2,900 km (1,800 mi) depth; there, the P waves slow markedly and the S waves disappear. This is the mantle-core boundary zone. The disappearance of S waves at the mantle-core boundary, due to their reflection or conversion to P waves, indicates that the outer core is mostly liquid. Moving into the core, P wave velocities gradually increase until a jump is reached at about 5,150 km (3,200 mi) depth, suggesting that the inner core is solid.

SURFACE WAVES

Surface waves are created by body waves disturbing the surface. They are of two main types—Love waves and Rayleigh waves. Both Love and Rayleigh waves are referred to as L waves (long waves) because they take longer periods of time to complete one cycle of motion and are the slowest moving. The frequencies of surface waves are low—less than one cycle per second. The low-frequency, long-period waves carry significant amounts of energy for much greater distances away from the epicenter.

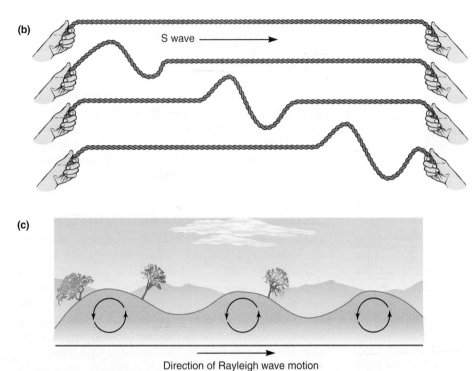

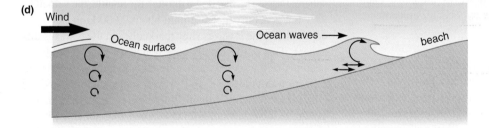

Figure 3.18 Types of seismic waves. (a) P waves exhibit the push-pull motion of a Slinky toy. (b) S waves move up and down perpendicular to the direction of advance, like a shaken jump rope. (c) Rayleigh waves advance in a backward-rotating motion, as opposed to (d) wind-blown ocean waves, which cause water to move in forward-rotating circles.

advance; to understand this, visualize the jump rope in figure 3.18b lying on the ground. Love waves generally travel faster than Rayleigh waves. Like S waves, they do not move through water or air.

Rayleigh Waves

Rayleigh waves were predicted to exist by Lord Rayleigh 20 years before they were actually recognized. They advance in a backward-rotating, elliptical motion (figure 3.18c) similar to the orbiting paths of water molecules in wind-blown waves of water, except that waves in water are forward-rotating (figure 3.18d). The shaking produced by Rayleigh waves causes both vertical and horizontal movement. The shallower the hypocenter, the more P and S wave energy will hit the surface, thus putting more energy into Rayleigh waves. The rolling waves pass through both ground and water. The often-heard report that an earthquake feels like being rocked in a boat at sea well describes the passage of Rayleigh waves. These waves have long periods, and once started, they go a long way.

SOUND WAVES AND SEISMIC WAVES

Waves are fundamental to both music and seismology. Musicians use instruments to produce the sound waves we hear as music. For example, a trombone player controls the amount of sound with his breath, and changes the frequencies of the sound waves by extending and retracting the slide on the trombone. Earthquakes generate body and surface waves; seismologists record and analyze the seismic wave frequencies to understand the earthquake.

Love Waves

Love waves were recognized and first explained by the British mathematician A. E. H. Love. Their motion is similar to that of S waves, except it is from side-to-side in a horizontal plane roughly parallel to Earth's surface. As with S waves, their shearing motion is at right angles to the direction of

Music is a common part of our lives and we are familiar with hearing sound waves. Sound waves and seismic waves can be presented in the same visual form. Waveforms for a trombone and a moderate-size earthquake are shown in figure 3.20. Both a trombone and an earthquake have more

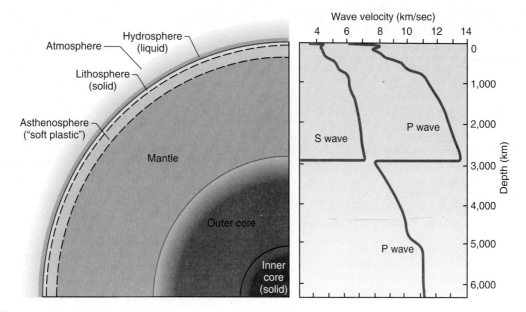

Figure 3.19 Varying velocities of P waves and S waves help define the internal structure of Earth.

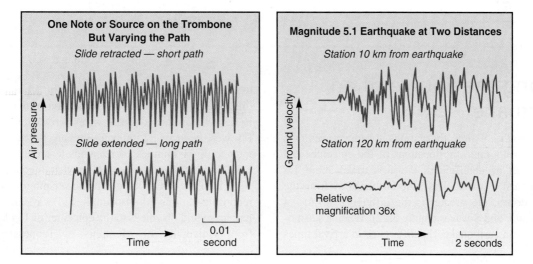

Figure 3.20 Comparison of wave patterns for a trombone and an earthquake for short and long-distance travel paths.

Source : A. Michael, S. Ross, and D. Schaff, "The Music of Earthquakes; Waveforms of Sound and Seismology" originally presented at Sigma Xi conference on Science and Art. USGS.

higher-frequency waves if a shorter path is traveled—that is, the trombone is retracted and has a short length, and the fault-rupture length is short. As the travel paths become longer for both trombone (extended) and earthquake (longer fault rupture), the number of low-frequency waves increases. Musically, as the path through the trombone lengthens, the vibrations per second decrease, the frequencies are lower, and the tone is lower. Seismically, a rupturing fault sends off high-frequency seismic waves, but as the fault rupture grows longer, more low-frequency seismic waves are generated. The ranges of some common frequencies are listed in table 3.1.

TABLE 3.1

Some Common Frequencies (in hertz)

Sound Waves

30,000 Hz—heard by dogs

15–20 Hz to 15,000–20,000 Hz—range of human hearing

15–20 Hz—P waves in air heard by humans near epicenter

Seismic Waves

0.02–30 Hz—body waves

0.002–0.1 Hz—surface waves

In Greater Depth

SEISMIC WAVES FROM NUCLEAR BOMB BLASTS VERSUS EARTHQUAKES

North Koreans buried an atom bomb and detonated it on 12 February 2013, releasing energy equivalent to a magnitude 5.1 earthquake. The seismic wave pattern recorded at the IRIS/USGS Global Seismic Network Station in Mudanjiang, China (figure 3.21a) shows an explosion of compressional energy yielding an abundance of P waves, with lesser shearing, S wave energy.

Compare the bomb-blast seismic record with that of a magnitude 5.0 earthquake recorded at the same seismic station in China (figure 3.21b). The natural earthquake has less compressional energy as shown by lesser P waves. The earthquake has much greater shear wave energy as shown by the prominent S wave development.

The different P and S wave patterns are useful for distinguishing between human-caused and natural events.

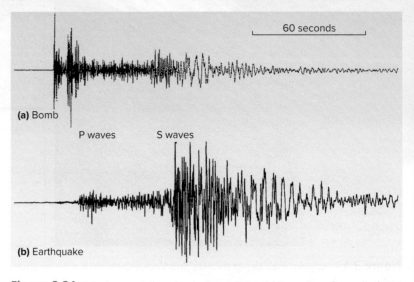

Figure 3.21 Seismic records from Mudanjiang, China. (a) Recording of magnitude 5.1 bomb blast set off in North Korea on 12 February 2013. Note the prominent development of the early arriving P waves. (b) Recording of magnitude 5.0 earthquake. Note the lesser P waves and prominent development of later arriving S waves.

Locating the Source of an Earthquake

Using the lengths of time the various seismic waves take to reach a seismograph, the locations of the epicenter and hypocenter can be determined. P waves travel about 1.7 times faster than S waves. Thus, the farther away from the earthquake origin, the greater is the difference in arrival times between P and S waves (figure 3.22). When a seismograph records an earthquake, the difference in arrival times of P and S waves is determined by subtracting the P arrival time from the S time (S–P). Inspection of the **seismogram** in figure 3.23 shows that S waves arrived 11 minutes after P waves. Figure 3.22 indicates that an S–P arrival time difference of 11 minutes corresponds to an earthquake about 8,800 km (5,400 mi) away. But in what direction?

Epicenters can be located using seismograms from three recording stations. As an example, S–P wave arrival time differences yield distances to the epicenter of 164 km (102 mi) from University of Memphis in Tennessee, 236 km

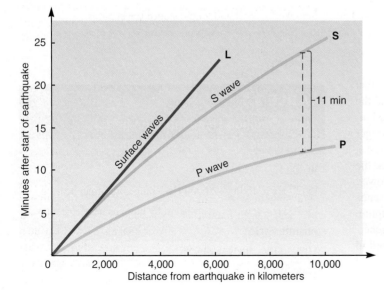

Figure 3.22 Plot of travel time versus distance from earthquake for seismic waves. Note that the arrival time difference for P and S waves of 11 minutes in figure 3.23 corresponds to a distance of about 8,800 km (5,400 mi).

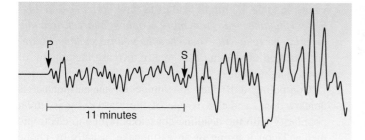

Figure 3.23 Seismogram recorded in Finland of the Sumatran earthquake on 26 December 2004. Notice that the difference in arrival times of P and S waves is 11 minutes. See figure 3.22 to read the distance traveled by the seismic waves.

(146 mi) from St. Louis University in Missouri, and 664 km (412 mi) from Ohio State University in Columbus. If the distance from each station is plotted as the radius of a circle, the three circles will intersect at one unique point—an epicenter at New Madrid, Missouri (figure 3.24). Computers usually make the calculations to determine epicenter locations; however, a better mental picture of the process is gained via the hand-drawn circles.

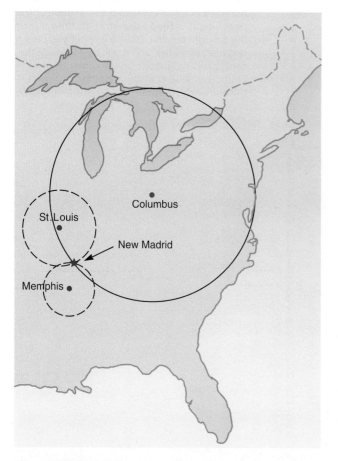

Figure 3.24 Location of an earthquake epicenter. S–P arrival time difference calculations gave a radius of 164 km from Memphis, 236 km from St. Louis, and 664 km from Columbus. The circles plotted with these values intersect uniquely at New Madrid, Missouri—the epicenter.

The difference in arrival times of P and S waves (S–P) actually measures the distance from the recording station to the hypocenter (or focus) of the earthquake, the site of initial fault movement (see figure 3.12). If the hypocenter is on Earth's surface, then the hypocenter and epicenter are the same. However, if the hypocenter is deep below the surface, it will affect the arrival time of surface (L) waves because L waves do not begin until P waves strike the Earth's surface. The depth to a hypocenter is best determined where an array of seismometers is nearby, thus allowing careful analysis of P wave arrival times.

Magnitude of Earthquakes

Magnitude is an estimate of the relative size or energy release of an earthquake. The magnitude is proportional to the area of the fault surface that moves or **slips** and how much it slips. It is commonly measured from the seismic wave traces on a seismogram.

RICHTER SCALE

In 1935, Charles Richter of the California Institute of Technology devised a quantitative scheme to describe the magnitude of California earthquakes, specifically events with shallow hypocenters located near (less than 300 mi from) the seismometers. Richter based his scale on the idea that the bigger the earthquake, the greater the shaking of Earth and thus the greater the amplitude (swing) of the lines made on the seismogram. To standardize this relationship, he defined magnitude as:

> *the logarithm to the base ten of the maximum seismic wave amplitude (in thousandths of a millimeter) recorded on a standard seismograph at a distance of 100 kilometers from the earthquake center.*

Because not all seismometers will be sitting 100 km from the epicenter, corrections are made for distance. Richter assigned simple, whole numbers to describe magnitudes; for every 10-fold increase in the amplitude of the recorded seismic wave, the Richter magnitude increases one number—for example, from 4 to 5. The energy released by earthquakes increases even more rapidly than the 10-fold increase in amplitude of the seismic wave trace. For example, if the amplitude of the seismic waves increased 10,000 times (10 × 10 × 10 × 10), the Richter magnitude would move up from a 4 to an 8. However, the energy release from 4 to 8 increases by 2,800,000 times (table 3.2).

What does this increase mean in everyday terms? If you feel a magnitude 4 earthquake while sitting at your dinner table, and then a magnitude 8 comes along while you are still at the table, would you really be shaken 2,800,000 times as hard? No. The greater energy of the magnitude 8 earthquake would be spread out over a much larger area, and over a time interval about 20 times longer (e.g., 60 seconds as

TABLE 3.2
Energy of Richter Scale Earthquakes

Richter Magnitude	Energy Increase		Energy Compared to Magnitude 4
4			1
5	=	48 Mag 4 EQs	48
6	=	43 Mag 5 EQs	2,050
7	=	39 Mag 6 EQs	80,500
8	=	35 Mag 7 EQs	2,800,000

opposed to 3 seconds). At any one location, the felt shaking in earthquakes above magnitude 6 does not increase very much more (maybe three times more for each step up in magnitude); it certainly does not increase as much as the

values in table 3.2 might lead us to think. *In effect, the bigger earthquake means that more people in a larger area and for a longer time will experience the intense shaking.* A longer duration of shaking can greatly increase the amount of damage to buildings.

Computing a Richter magnitude for an earthquake is quickly done, and this is one of the reasons for its great popularity with the deadline-conscious print and electronic media. Upon learning of an earthquake, usually by phone calls from reporters, one can rapidly measure (1) the amplitude of the seismic waves and (2) the difference in arrival times of P and S waves. Figure 3.25 has reduced Richter's equation to a nomograph, which allows easy determination of magnitude. Take a couple of minutes to figure out the magnitude of the earthquake whose seismogram is printed above the nomograph.

Each year, Earth is shaken by millions of quakes that are recorded on seismometers. Most are too small to be felt by

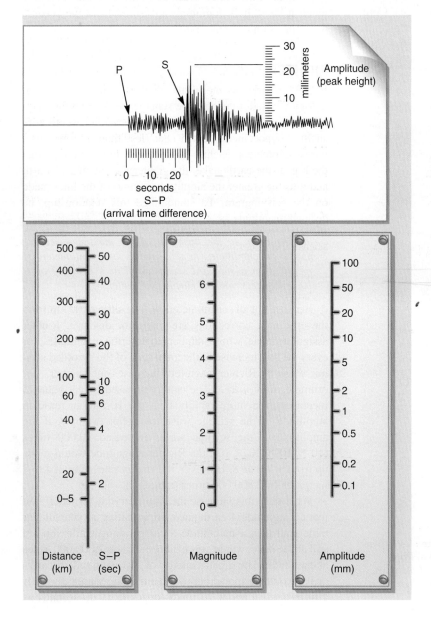

Figure 3.25 Nomograph of the Richter scale allowing earthquake magnitudes to be estimated. On the seismogram, read the difference in arrival times of P and S waves in seconds and plot the value on the left column of the nomograph. Next read the amplitude of the peak height of the S wave and plot this value on the far right column. Draw a line between the two marked values, and it will pass through the earthquake magnitude on the center column. Check your answer in Questions for Review at the end of the chapter.

humans. Notice the distinctive "pyramidal" distribution of earthquakes by size—the smaller the earthquake magnitude, the greater their numbers (table 3.3). Yet the fewer than 20 major and great earthquakes (magnitudes of 7 and higher) each year account for more than 90% of the energy released by earthquakes. At the upper end of the magnitude scale, the energy increases are so great that more energy is released going from magnitude 8.9 to 9 than from magnitude 1 to 8. These facts underscore the logarithmic nature of the Richter scale; each step up the scale has major significance.

OTHER MEASURES OF EARTHQUAKE SIZE

An earthquake is a complex event, and more than one number is needed to assess its magnitude. Although the Richter scale is useful for assessing moderate-size earthquakes that occur nearby, the 0.1- to 2-second-period waves it uses do not work well for distant or truly large earthquakes. The short-period waves do not become more intense as an earthquake becomes larger. For example, the Richter scale assesses both the 1906 San Francisco earthquake and the 1964 Alaska earthquake as magnitude 8.3. However, using other scales, the San Francisco earthquake is a magnitude 7.8 and the Alaska seism is a 9.2. The Alaska earthquake was at least 100 times bigger in terms of energy.

The Richter scale is now restricted to measuring only local earthquakes with moderate magnitudes (noted as M_L). Because earthquakes generate both body waves that travel through Earth and surface waves that follow Earth's uppermost layers, two other magnitude scales have long been used: m_b and M_s. The body-wave (m_b) scale uses amplitudes of P waves with 1- to 10-second periods, whereas the surface-wave scale (M_s) uses Rayleigh waves with 18- to 22-second periods. Early on, all magnitude scales were considered equivalent, but now we know that earthquakes generate different proportions of energy at different periods. For example, larger earthquakes with their larger fault-rupture surfaces radiate more of their energy in longer-period seismic waves. Thus, for great and major earthquakes, body-wave magnitudes (m_b) will significantly underestimate the actual size of the earthquake. Even a composite of these three methods of determining earthquake magnitude (M_L, m_b, and M_s) does not necessarily yield the true size of an earthquake.

Moment Magnitude Scale

Seismologists have moved on to other measures to more accurately determine earthquake size. The **seismic moment** (M_o) relies on the amount of movement along the fault that generated the earthquake; that is, M_o equals the shear strength of the rocks times the rupture area of the fault times the average displacement (slip) on the fault. Moment is the most reliable measure of earthquake size; it measures the amount of strain energy released by the movement along the whole rupture surface. Seismic moment has been incorporated into a new earthquake magnitude scale by Thomas Hanks and Hiroo Kanamori, the moment magnitude scale (M_w), where:

$$M_w = 2/3 \log_{10}(M_o) - 10.7$$

The moment magnitude scale is used for big earthquakes. It is more accurate because it is tied directly to physical parameters such as fault-rupture area, fault slip, and energy release. For great earthquakes, it commonly takes weeks or months to determine M_w because time is required for the aftershocks to define the area of the rupture zone.

Some of the largest moment magnitudes calculated to date are the 1960 Chile earthquake (M_s of 8.5; M_w of 9.5), the 1964 Alaska earthquake (M_s of 8.3; M_w of 9.2), the 2004 Sumatra event (M_w of 9.1), and the 2011 Japan seism (M_w of 9.0). These gigantic earthquakes occurred at subduction zones. A variety of energetic events are placed on a logarithmic scale for comparison in figure 3.26. Each step or increment up the scale is a 10-fold increase in magnitude.

FORESHOCKS, MAINSHOCK, AND AFTERSHOCKS

Large earthquakes do not occur alone; they are part of a series of movements on a fault that can go on for years. The biggest earthquake in a series is the **mainshock.** Smaller earthquakes that precede the mainshock are **foreshocks,** and those that follow are **aftershocks.** Realistically, there are no differences between these earthquakes other than size; they are all part of the same series of stress release on the fault.

TABLE 3.3
Earthquakes in the World Each Year

Magnitude	Number of Quakes per Year	Description
8.5 and up	0.3	
8–8.4	1	Great
7.5–7.9	3	
7–7.4	15	Major
6.6–6.9	56	
6–6.5	210	Strong (destructive)
5–5.9	800	Moderate (damaging)
4–4.9	6,200	Light
3–3.9	49,000	Minor
2–2.9	350,000	Very minor
0–1.9	3,000,000	

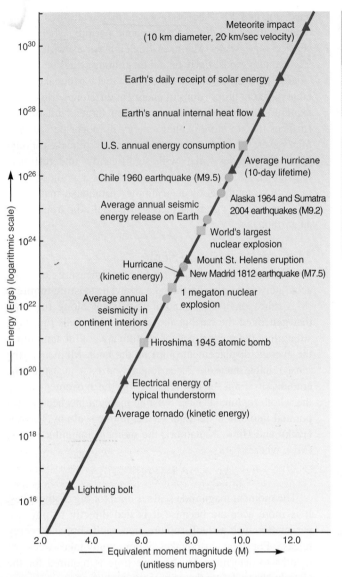

Figure 3.26 Equivalent moment magnitude of a variety of seismic (green dots), human-made (yellow squares), and other phenomena (red triangles).

Source: A. C. Johnston, "An earthquake strength scale for the media and the public" in *Earthquakes and Volcanoes* 22 (no. 5): 214–16. *US Geological Survey.*

A large-scale fault movement increases the stress on adjacent sections of a fault, helping trigger the additional fault movements that we feel as aftershocks. The danger of large aftershocks is greatest in the three days following the mainshock. Sometimes a big earthquake is followed by an even bigger earthquake, and then the first earthquake is reclassified as a foreshock.

MAGNITUDE, FAULT-RUPTURE LENGTH, AND SEISMIC-WAVE FREQUENCIES

Fault-rupture length greatly influences earthquake magnitude. As approximations, these fault-rupture lengths yield the following earthquake magnitudes:

TABLE 3.4
Rupture Length and Duration

	Magnitude	Rupture Length (km)	Duration (seconds)
1964 Alaska	9.2	1,000	420
1906 San Francisco, CA	7.8	400	110
1992 Landers, CA	7.3	70	24
1983 Borah Peak, ID	7.0	34	9
2001 Nisqually, WA	6.8	20	6
1933 Long Beach, CA	6.4	15	5
2001 Yountville, CA	5.2	4	2

- 100 m (328 ft) rupture ≈ magnitude 4
- 1 km (0.62 mi) rupture ≈ magnitude 5
- 10 km (6.2 mi) rupture ≈ magnitude 6
- 40 km (25 mi) rupture ≈ magnitude 7
- 400 km (250 mi) rupture ≈ magnitude 8
- 1,000 km (620 mi) rupture ≈ magnitude 9

A rupture along a fault during an earthquake typically moves 2 to 4 km/sec. A lengthier rupture gives a lengthier duration of movement (table 3.4).

Fault-rupture lengths and durations in seconds also affect the frequencies of seismic waves produced during earthquakes. Faults that move for short distances and short amounts of time generate mostly high-frequency seismic waves. Faults that rupture for longer distances and longer times produce increasingly greater amounts of low-frequency seismic waves.

Seismic waves die off with distance traveled. High-frequency seismic waves die out first—at shorter distances from the hypocenter. Low-frequency seismic waves carry significant amounts of energy farther—through longer distances. High-frequency seismic waves cause much damage at short distances from the epicenter. But at longer distances, it is the low-frequency seismic waves that do most of the damage.

Ground Motion During Earthquakes

Seismic waves radiate outward from a fault movement. The interactions among the various seismic waves move the ground both vertically and horizontally. Buildings usually are designed to handle the large vertical forces caused by the weight of the building and its contents. They are designed with such large factors of safety that the additional vertical forces imparted by earthquakes are typically not a problem.

In Greater Depth

Figure 3.27 This inadequately braced house failed due to horizontal acceleration during the 1971 San Fernando earthquake.

Courtesy Al Boost.

Usually, the biggest concern in designing buildings to withstand large earthquakes is the sideways push from the horizontal components of movement (figure 3.27).

ACCELERATION

Building design in earthquake areas must account for **acceleration.** As seismic waves move the ground and buildings up and down, and back and forth, the rate of change of velocity is measured as acceleration. As an analogy, when your car is moving at a velocity of 25 mph on a smooth road, you feel no force on your body. But if you stomp on the car's accelerator and rapidly speed up to 55 mph, you feel a force pushing you back against the car's seat. Following the same thought, if you hit the brakes and decelerate rapidly, you feel yourself being thrown forward. This same type of accelerative force is imparted to buildings when the ground beneath them moves during an earthquake.

The usual measure of acceleration is that of a free-falling body pulled by gravity; it is the same for all objects, regardless of their weight. The acceleration due to gravity is 9.8 m/sec^2 (32 ft/sec^2), which is referred to as 1.0 g and is used as a comparative unit of measure. Weak buildings begin to suffer damage at horizontal accelerations of about 0.1 g. At accelerations between 0.1 to 0.2 g, people have trouble keeping their footing, similar to being in the corridor of a fast-moving train or on a small boat in high seas. A problem for building designers is that earthquake accelerations have locally been in excess of 1 g. For example, in the hills above Tarzana, California, the 1994 Northridge earthquake generated phenomenal accelerations—1.2 g vertically and 1.8 g horizontally.

PERIODS OF BUILDINGS AND RESPONSES OF FOUNDATIONS

The concepts of period and frequency also apply to buildings. Visualize the shaking or vibration of a 1-story house and a 30-story office building. Do they take the same amount of time to complete one cycle of movement, to shake back and forth one time? No. Typical periods of swaying for buildings are about 0.1 second per story of height. The 1-story house shakes back and forth quickly at about 0.1 second per cycle. The 30-story building sways much slower, with a period of about 3 seconds per cycle.

The periods of buildings are also affected by their construction materials. A building of a given height and design will have a longer period if it is made of flexible materials such as wood or steel; its period will be shorter if it is built with stiff materials such as brick or concrete.

The velocity of a seismic wave depends on the type of rock the wave is traveling through. Seismic waves move faster through hard rocks and slower through softer rocks and loose sediments. Seismic waves are modified by the rocks they pass through; they become distorted. When

In Greater Depth

seismic waves pass from harder rocks into softer rocks, they slow down and thus must increase their amplitude to carry the same amount of energy. Shaking tends to be stronger at sites with softer sediments because seismic waves move more slowly but with greater amplitude.

When seismic waves of a certain period carry a lot of energy and their period matches the period of a building, the shaking is amplified and **resonance** can occur. The resonance created by shared periods for seismic waves and buildings is a common cause of the catastrophic failure of buildings during earthquakes.

Understanding the concept of shared periods and resonance may be advanced by visualizing a tall flagpole with a heavy metal eagle on top. First, if you shake this pole, you will quickly learn that the pole has a strong tendency to move back and forth only at a certain rate or period. If the flagpole swings a complete cycle in 2 seconds, it has a period of 2 seconds. Second, if seismic waves of a 2-second period begin to shake the ground, the amount of movement of the flagpole starts to increase. The pole is now resonating, the forces it must withstand have increased, and the greater forces created by the combined periods may cause destruction.

Earthquake Intensity— What We Feel During an Earthquake

During the tens of seconds that a large earthquake lasts, we feel ourselves rocked up and down and shaken from side to side. It is an emotional experience, and the drama of our personal accounts varies according to our location during the shaking and our personalities. But for personal narratives to have meaning that can be passed on to succeeding generations, common threads are needed to bind the accounts together. In the late 1800s, descriptive schemes appeared that were based on the intensity of effects experienced by people and buildings. The most widely used scale came from the Italian professor Giuseppi Mercalli in 1902; it was modified by Charles Richter in 1956. The Mercalli Intensity Scale has 12 divisions of increasing intensity labeled by Roman numerals (table 3.5).

Earthquake magnitude scales are used to assess the energy released during an earthquake; earthquake intensity scales assess the effects on people and buildings (table 3.6). The difference between magnitude and intensity can be illuminated by comparison to a lightbulb. The wattage of a lightbulb is analogous to the magnitude of an earthquake. Wattage is a measure of the power of a lightbulb, and magnitude is a measure of the energy released during an earthquake.

A lightbulb shining in the corner of a room provides high-intensity light nearby, but the intensity of light decreases toward the far side of the room. The intensity of shaking caused by a fault movement is great near the epicenter, but in general, it decreases with distance from the epicenter. (This generalization is offset to varying degrees by variations in geologic foundations and building styles.)

Mercalli intensities also are crucial for assessing magnitudes of historical events before there were instrumented records, thus allowing us to assess recurrence intervals between major earthquakes.

TABLE 3.5
Modified Mercalli Scale of Earthquake Intensity

I. Not felt except by a very few people under especially favorable circumstances.

II. Felt by only a few people at rest, especially those on upper floors of buildings or those with a very sensitive nature. Delicately suspended objects may swing.

III. Felt quite noticeably indoors, especially on upper floors, but many people do not recognize it as an earthquake. Vibrations are like those from the passing of light trucks. Standing automobiles may rock slightly. Duration of shaking may be estimated.

IV. Felt indoors during the day by many people, outdoors by few. Light sleepers may be awakened. Vibrations are like those from a passing heavy truck or a heavy object striking a building. Standing automobiles rock. Windows, dishes, and doors rattle; glassware and crockery clink and clash. In the upper range of IV, wooden walls and frames creak.

V. Felt indoors by nearly everyone, outdoors by many or most. Awakens many. Frightens many; some run outdoors. Some broken dishes, glassware, and windows. Minor cracking of plaster. Moves small objects, spills liquids, rings small bells, and sways tall objects. Pendulum clocks misbehave.

VI. Felt by all; many frightened and run outdoors. Excitement is general. Dishes, glassware, and windows break in considerable quantities. Knickknacks, books, and pictures fall. Furniture moves or overturns. Weak plaster walls and some brick walls crack. Damage is slight.

VII. Frightens all; difficult to stand. Noticed by drivers of automobiles. Large bells ring. Damage negligible in buildings of good design and construction, slight to moderate in well-built ordinary buildings, considerable in badly designed or poorly built buildings, adobe houses, and old walls. Numerous windows and some chimneys break. Small landslides and caving of sand and gravel banks occur. Waves appear on ponds; water becomes turbid.

VIII. Fright is general and alarm approaches panic. Disturbs drivers of automobiles. Heavy furniture overturns. Damage slight in specially designed structures; considerable in ordinary substantial buildings, including partial collapses. Frame houses move off foundations if not bolted down. Most walls, chimneys, towers, and monuments fall. Spring flow and well-water levels change. Cracks appear in wet ground and on slopes.

IX. General panic. Damage considerable in masonry structures, even those built to withstand earthquakes. Well-built frame houses thrown out of plumb. Ground cracks conspicuously. Underground pipes break. In soft sediment areas, sand and mud are ejected from ground in fountains and leave craters.

X. Most masonry structures are destroyed. Some well-built wooden structures and bridges fail. Ground cracks badly with serious damage to dams and embankments. Large landslides occur on river banks and steep slopes. Railroad tracks bend slightly.

XI. Few, if any, masonry structures remain standing. Great damage to dams and embankments, commonly over great distances. Supporting piers of large bridges fail. Broad fissures, earth slumps, and slips occur in soft and wet ground. Underground pipelines completely out of service. Railroad tracks bend greatly.

XII. Damage nearly total. Ground surfaces seen to move in waves. Lines of sight and level distort. Objects thrown up in air.

TABLE 3.6
Comparison of Magnitude, Intensity, and Acceleration

Magnitude		Mercalli Intensity	Acceleration (% g)
2 and less	I–II	Usually not felt by people	Less than 0.1–0.19
3	III	Felt indoors by some people	0.2–0.49
4	IV–V	Felt by most people	0.5–1.9
5	VI–VII	Felt by all; building damage	2–9.9
6	VII–VIII	People scared; moderate damage	10–19.9
7	IX–X	Major damage	20–99.9
8 and up	XI–XII	Damage nearly total	More than 100 = more than 1 g

MERCALLI SCALE VARIABLES

The Mercalli intensity value at a given location for an earthquake depends on several variables: (1) earthquake magnitude; (2) distance from the hypocenter/epicenter; (3) type of rock or sediment making up the ground surface; (4) building style—design, kind of building materials, height; and (5) duration of the shaking. These factors must be considered in assessing the earthquake threat to any region and even to each specific building.

1. *Earthquake Magnitude:* The relation between magnitude and intensity is obvious—the bigger the earthquake (the more energy released), the higher the odds are for death and damage.

2. *Distance from Hypocenter/Epicenter:* The relation between distance and damage also seems obvious; the closer to the hypocenter/epicenter, the greater the damage. But this is not always the case, as will be seen in chapter 4 with the 1989 World Series (Loma Prieta) and 1985 Mexico City earthquakes.

3. *Foundation Materials:* The types of rock or sediment foundation are important. For example, hard rock foundations can vibrate at high frequencies and be excited by energetic P and S waves near an epicenter; the shaking of soft or water-saturated sediments can be amplified by surface (L) waves from distant earthquakes; and steep slopes often fail as landslides when severely shaken.

4. *Building Style:* Building style is of vital importance. What causes the deaths during earthquakes? Not the shaking of the earth, but the buildings, bridges, and other structures that collapse and fall on us. *Earthquakes don't kill, buildings do.* Buildings have frequencies of vibration in the same ranges as seismic waves. The vibrations of high-frequency P and S waves are amplified by (1) rigid construction materials, such as brick or stone, and (2) short buildings. If this type of building is near the epicenter, beware!

   The movements of low-frequency surface waves are increased in tall buildings with low frequencies of vibration. If these tall buildings also lie on soft, water-saturated sand or mud and are distant from the epicenter, disaster may strike.

5. *Duration of the Shaking:* The duration of the shaking is underappreciated as a significant factor in damages suffered and lives lost. Consider the ranges of shaking times in table 3.7. For example, if a magnitude 7 earthquake shakes vigorously for 50 seconds, rather than 20, the increase in damages and lives lost can be enormous.

TABLE 3.7
Magnitude versus Duration of Shaking

Richter Magnitude	Duration of Strong Ground Shaking in Seconds
8–8.9	30 to 180
7–7.9	20 to 130
6–6.9	10 to 30
5–5.9	2 to 15
4–4.9	0 to 5

A Case History of Mercalli Variables: The San Fernando Valley, California, Earthquake of 1971

The San Fernando Valley (Sylmar) earthquake of 9 February 1971 occurred within the northwestern part of the Los Angeles megalopolis at 6:01 a.m., causing 67 deaths (including nine heart attacks). One of the most critical factors in determining life loss from earthquakes is the time of day of the event. In California, the best time for an earthquake for most people is when they are at home; their typical one- and two-story woodframe houses are usually the safest buildings to occupy.

1. *Earthquake Magnitude:* The magnitude was 6.6, with 35 aftershocks of magnitude 4.0 or higher occurring in the first 7 minutes after the main shock. This is a lot of energy to release within an urban area.

2. *Distance from Epicenter:* The distance from the epicenter was a fairly consistent variable in this event. A rather regular bull's-eye pattern resulted from contouring the damages reported in Mercalli numerals (figure 3.28).

3. *Foundation Materials:* The types of foundation materials were not a major factor in this event.

4. *Building Style:* Poorly designed buildings, bridges, and dams were the major problem. Three people died at the Olive View Hospital with the collapse of its "soft" first story featuring large plate-glass windows. "Soft" first-story buildings support the heavy weight of upper floors without adequate shear walls or braced frames to withstand horizontal accelerations (figure 3.29a). Many of these buildings still exist, despite their known high odds of failure during earthquakes (figure 3.29b).

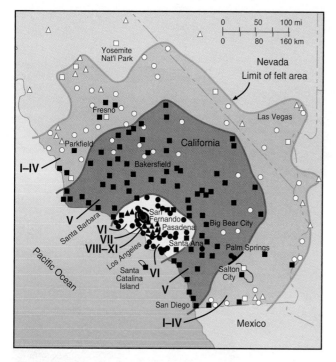

Figure 3.28 Contour map of Mercalli intensities from the San Fernando earthquake of 9 February 1971 shows an overall decrease in intensity away from the epicenter.

Another hospital failure was responsible for 47 deaths. Some of the pre-1933 buildings at the Veterans Administration Hospital used hollow, clay-tile bricks to build walls designed to carry only a vertical load. Many of the hollow-core clay bricks shattered under horizontal accelerations that measured up to 1.25 times gravity.

Freeway bridges collapsed and took three lives. A freeway bridge is a heavy horizontal mass (roadbed) suspended high atop vertical columns. Swaying of these top-heavy masses, which have poor connections between their horizontal and vertical elements, resulted in collapse as support columns moved out from under elevated roadbeds. The lessons learned from these 1971 failures had not been acted upon by 1989, when the Interstate 880 elevated roadway collapsed, killing 42 people in Oakland during the World Series earthquake. Failure happened again in Los

Figure 3.29 Buildings with "soft" first stories. (a) Bracing is inadequate on the first floor, and there are no shear walls to transmit seismic loads to the ground. Thus, seismic stresses are concentrated at the join between the first and second floors. When the ground accelerates to the right, the building lags behind and the first story flattens. (b) This eight-story medical-office building atop a "soft" first story is located in a California city near active faults.

Sources: (a) "Improving Seismic Safety of New Buildings," 1986, Federal Emergency Management Agency; (b) Photo by Pat Abbott

Figure 3.30 This freeway collapsed in Los Angeles during the 1994 Northridge earthquake. Vertical supports and horizontal roadbeds move at different periods. If not bound together securely, they separate and fall when shaken.

NOAA/NGDC, Mehmet Celebi, U.S. Geological Survey.

(a)

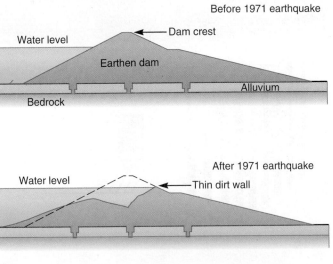

(b)

Figure 3.31 Failure of the Lower Van Norman Dam. (a) A few more seconds of strong shaking would have unleashed the deadly force of 11,000 acre-feet of water on San Fernando Valley residents below the dam. (b) Landsliding lowered the dam by 30 feet.

(a) E.V. Leyendecker/U.S. Geological Survey.

Angeles in 1994 during the Northridge earthquake (figure 3.30).

5. *Duration of Shaking:* The strong ground shaking lasted 12 seconds. Earthquakes in the magnitude 6 range typically shake from 10 to 30 seconds (see table 3.7). The significance of the relatively short time of strong shaking in the San Fernando Valley earthquake is enormous. The Lower Van Norman Reservoir held 11,000 acre-feet of water at the time of the quake. Its dam was begun in 1912 as a hydraulic-fill structure where sediment and water were poured into a frame to create a large mass; this is not the way to build a strong dam. During the earthquake, the dam began failing by landsliding and had lost 30 ft of its height (800,000 cubic yards of its mass) and stood only 4 ft above the water level when

the shaking stopped (figure 3.31). If the strong shaking had lasted another 5 seconds, the dam would have failed and released the water onto a 12-square-mile area below the dam where 80,000 people were at home.

LEARNING FROM THE PAST

The 1971 San Fernando Valley earthquake unequivocally demonstrated the hazard in this region. It has been eloquently stated that *"past is prologue"* and that *"those who do not learn the lessons of history are doomed to repeat them."* How well were the lessons of 1971 learned? Another test was painfully administered on 17 January 1994, when the magnitude 6.7 Northridge earthquake struck the immediately adjacent area. This time, 57 people died and damages escalated to $30 billion. The same types of buildings again failed, and freeway bridges again fell down. Not all the lessons from 1971 were learned.

Building in Earthquake Country

One of the problems in designing buildings for earthquake country is the need to eliminate the occurrence of resonance. This can be done in several ways: (1) Change the height of the building; (2) move most of the weight to the lower floors; (3) change the shape of the building; (4) change the type of building materials; and (5) change the degree of attachment of the building to its foundation. For example, if the earth foundation is hard rock that efficiently transmits short-period (high-frequency) vibrations, then build a flexible, taller building. Or if the earth foundation is a thick mass of soft sediment with long-period shaking (low frequency), then build a stiffer, shorter building. For building materials, wood is flexible and lightweight, has small mass, and is able to handle large accelerations. Concrete has great compressional strength but suffers brittle failure all too easily under tensional stress. Steel has ductility and great tensional strength, but steel columns fail under compressive stress.

Ground motion during an earthquake is horizontal, vertical, and diagonal—all at the same time. The building components that must handle ground motion are basic. In the horizontal plane are floors and roofs. In the vertical plane are walls and frames. An important component in building resistance is how securely the floors and roofs are tied or fastened to the walls so they do not separate and fail.

SHEAR WALLS AND BRACING

Walls designed to take horizontal forces from floors and roofs and transmit them to the ground are called *shear walls.* In a building, shear walls must be strong themselves, as well as securely connected to each other and to roofs and floors.

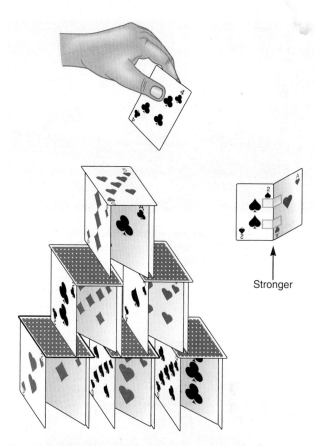

Figure 3.32 A "house of cards" is a structure with walls and floors but no strength. Earthquake resistance is greatly increased by tying the walls and floors together with tape.

Source : "Improving Seismic Safety of New Buildings: 1986, Federal Emergency Management Agency

In a simple building, seismic energy moves the ground, producing inertial forces that move the roofs and floors. These movements are resisted by the shear walls, and the forces are transmitted back to the ground.

Even a "house of cards" is a shear-wall structure, although each "wall" does not have much strength. The walls must be at right angles and preferably in a simple pattern (figure 3.32). The house of cards is made enormously stronger if horizontal and vertical elements are all securely fastened—for example, by taping them together.

A structure commonly built with insufficient shear walls is the multistory parking garage. Builders do not want the added expense of more walls, which eliminate parking spaces and block the view of traffic inside the structure. These buildings are common casualties during earthquakes (figure 3.33).

Bracing is another way to impart seismic resistance to a structure. Bracing gives strength to a building and offers resistance to the up, down, and sideways movements of the ground (figure 3.34). The bracing should be made of ductile materials that have the ability to deform without rupturing.

Figure 3.33 This three-story parking structure for automobiles at the Northridge Fashion Center collapsed during the 17 January 1994 earthquake.

E.V. Leyendecker/U.S. Geological Survey.

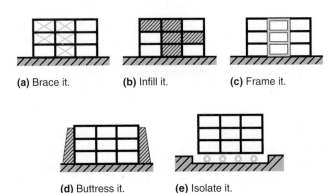

(a) Brace it. **(b)** Infill it. **(c)** Frame it.

(d) Buttress it. **(e)** Isolate it.

Figure 3.35 How to strengthen buildings. (a) Add braces. (b) Infill walls. (c) Add frames to exterior or interior. (d) Add buttresses. (e) Isolate building from the ground.

Source: After AIA/ACSA Council on Architectural Research.

Figure 3.34 A six-story building with a braced frame incorporated in its design.

Photo courtesy of Pat Abbott.

RETROFIT BUILDINGS, BRIDGES, AND HOUSE CONSTRUCTION

The process of reinforcing existing buildings to increase their resistance to seismic shaking is known as **retrofitting.** Figure 3.35 shows how some common designs in building retrofits give seismic strength to a building.

Highway bridges and elevated roadways commonly collapse during major earthquakes. Part of the problem comes from the different frequencies of movement of vertical supports and horizontal roadbeds, but part comes from the behaviors of different construction materials. Bridge builders combine steel (for its ductility) with concrete (for its strength). During the 1994 Northridge earthquake, support-column failures occurred as concrete cracked and steel deformed (figure 3.36a). The rebuilding process employs additional alternating layers of concrete and steel to avoid future failures (figure 3.36b).

Modern one- and two-story woodframe houses perform well during seismic shaking. Houses must be able to move up, down, and sideways without failing. The ability to withstand earth movements is given by building shear walls and by using bracing and other elements that tie the walls, foundation, and roof together (figure 3.37).

For retrofitting, older houses must have these same resisting elements added to the foundation walls that hold the house above the ground. Additionally, much of the damage, injury, and even death during an earthquake occurs inside homes as personal items are thrown about—items such as unsecured water heaters, ceiling fans, cabinets, bookshelves, and electronic equipment. Bolt down or secure with Velcro your personal items so they don't become airborne missiles inside your home during an earthquake.

BASE ISOLATION

When the earth shakes, the energy is transferred to buildings. How can buildings be saved from this destructive energy? One approach is to build structures so huge and strong that an earthquake cannot knock them down. But earthquakes *can* knock them down. For an example, see the failure of the massive support column in the 6.9 M_W event in Kobe, Japan (figure 3.38). If buildings cannot stand up against the most powerful seismic waves, then we need to learn to roll with them. Modern designs employ **base isolation** whereby devices are placed on the ground or within the structure to absorb part of the earthquake energy. For example, visualize yourself standing on Rollerblades during an earthquake. Would you move as much as the earth? Base isolation uses wheels, ball bearings, shock absorbers, "rubber doughnuts," rubber and steel

(a)

(b)

Figure 3.36 Support columns on Freeway 118 in Simi Valley, California. (a) Problem: This column failed during the 1994 earthquake when brittle concrete cracked and ductile steel rebar buckled. (b) Solution: New columns have vertical steel rebar wrapped by circular rebar, and both are encased in concrete. In addition, columns are confined by bolted steel jackets that will be encased in concrete.

Courtesy Peter W. Weigand, CSU Northridge.

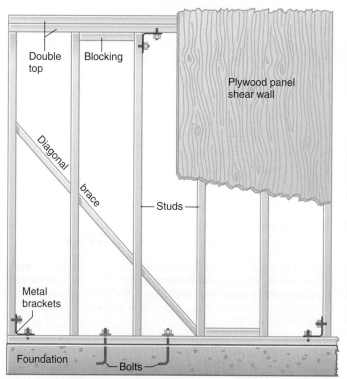

Double top

Blocking

Plywood panel shear wall

Diagonal brace

Studs

Metal brackets

Foundation

Bolts

Figure 3.37 How can a house be built to resist seismic waves?

Bolt it.

Bracket it.

Brace it.

Block it.

Panel it.

FIGURE 3.38 Despite their huge size, the stiff and massive beams (note car for scale) supporting the elevated expressway in Kobe, Japan failed in the 17 January 1995 earthquake of 6.9 M_w.
Dr. Richard Hutchison/NOAA.

sandwiches, and other creative designs to isolate a building from the worst of the ground shaking (figure 3.39). The goal is to make the building react to shaking much like your body adjusts to accelerations and decelerations when you are standing in a moving train or bus. This concept has recently been used in building San Francisco's new airport terminal. The 115-million-pound building rests on 267 stainless steel sliders that rest in big concave dishes. When the earth shakes, the terminal will roll up to 20 inches in any direction.

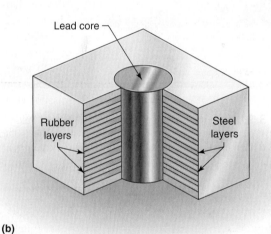

(b)

(a)

Figure 3.39 (a) The Office of Disaster Preparedness in San Diego County is housed in a two-story, 7,000 ft^2 building sitting on top of 20 lead-impregnated rubber supports (base isolators) that each weigh 1 ton. (b) An example of a base isolator. Cutaway view into a 1 m wide by 1 m tall sandwich shows alternating layers of rubber (each 15 mm thick) and steel (each 3 mm thick) with a central core of lead. During an earthquake, the rubber and steel flex and the lead absorbs energy.
Photo courtesy of Pat Abbott.

Summary

Earthquakes are shaking ground caused most often by sudden movements along cracks in the Earth called faults. Some major faults acting for millions of years have offset rock layers by hundreds of kilometers. Sedimentary rock layers originally are continuous, horizontal, and in superpositional order (oldest on bottom, youngest on top); however, fault movements cut rocks into discontinuous masses, and in places, fault deformation has tilted rock layers and even overturned the superpositional sequence. Geologists measure the 3-D orientation of rock layers via dip (angle and direction of inclination) and strike (compass bearing of rock cutting a horizontal plane).

Dip-slip fault types have dominantly vertical movements. Normal faults are due to extensional (pull-apart) forces. Reverse faults are due to compressional (push-together) forces. Strike-slip fault types have dominantly horizontal (shear) offsets. Straddling the fault, if the right-hand side moves toward you, it is a right-lateral fault; if the left-hand side moves toward you, it is a left-lateral fault. Bends (steps) in strike-slip faults cause the land to either uplift or downdrop. Another type of fault, called a transform fault, connects offset spreading-center segments.

Earthquakes, also called seisms, disperse their energy in seismic waves that radiate away from the hypocenter or point of fault rupture. The point on the surface above the fault rupture is the epicenter. Some seismic waves pass through the body of Earth; these are the P waves (primary waves with a push-pull motion) and the S waves (secondary waves with a shearing motion). Other seismic waves travel along the surface (Love and Rayleigh waves).

Earthquake energy is assessed by its magnitude. Different estimates of magnitude are derived from different methods, based on local shaking (Richter scale), body waves (m_b), surface waves (M_s), or seismic moment (M_w). Earth has more than a million earthquakes each year, but more than 90% of the energy is released by the 12 to 18 largest events.

Seismic waves have different periods (time between cycles) and frequencies (number of cycles per second):

$$\text{Period} = \frac{1}{\text{frequency}}$$

P waves commonly have from 1 to 20 cycles per second; surface waves commonly have 1 cycle every 1 to 20 seconds. Where the frequencies of seismic waves match the vibration frequencies of foundations and buildings, destruction may be great.

Earthquake effects on structures and people are assessed via the Mercalli Intensity Scale. Its variables are earthquake magnitude, distance from the hypocenter/epicenter, type of rock or sediment foundation, building style, and duration of shaking. Mercalli intensities are of more than just scientific interest because earthquakes don't kill, buildings do.

Building components that must stand up to seismic shaking are horizontal (floors, roofs) and vertical (walls, frames). But horizontal and vertical components move at different frequencies. For buildings to stand up to earthquakes, the horizontal and vertical components must be securely tied together using bolts, brackets, braces, and such. New designs of large buildings utilize energy-absorbing base isolation devices placed between the building and the ground.

Terms to Remember

acceleration 65	mainshock 63
aftershock 63	map 51
amplitude 56	normal fault 52
base isolation 72	period 56
body waves 57	permeability 51
compression 51	primary (P) wave 57
cross-section 51	resonance 66
dip 51	retrofit 72
dip-slip fault 51	reverse fault 52
fault 49, 53	right-lateral fault 53
footwall 51	secondary (S) wave 57
foreshock 63	seism 49
fracture 49	seismic moment 63
frequency 56	seismic wave 57
friction 49	seismogram 60
granite 57	seismograph 56
hangingwall 51	seismology 56
hertz (Hz) 56	seismometer 56
hypocenter 53	shear 52
inertia 56	slip 61
joint 51	stress 49
law of original	strike 51
continuity 50	strike-slip fault 52
law of original	surface waves 57
horizontality 49	tension 51
law of superposition 50	transform fault 54
left-lateral fault 53	wavelength 56
magnitude 61	

Questions for Review

Ans. In figure 3.25, the earthquake magnitude is close to 5.
1. Draw a cross-section of a sequence of sedimentary rock layers. Label and explain the laws of original horizontality, superposition, and original continuity.
2. Draw cross-sections of a normal fault and a reverse fault. What are the differing forces that determine which one forms? Which one involves tension? Compression?

3. Draw a map of a left-stepping, right-lateral fault. Explain what happens to the land at the step (bend) in the fault.
4. Draw a cross-section showing an inclined fault with a hypocenter at 15 km (9 mi) depth. Does the epicenter plot on the surface trace of the fault?
5. Sketch a map of a strike-slip and a transform fault. Explain their similarities and differences.
6. What do P and S seismic waves tell us about the nature of Earth's interior?
7. How can arrival times of P and S waves be used to determine distance to the epicenter?
8. How are foreshocks distinguished from aftershocks?
9. What are typical P wave velocities in hard rock? Water? Air?
10. What are typical S wave velocities in hard rock?
11. How damaging to buildings are P waves? S waves? Rayleigh waves?
12. What is the frequency of a seismic wave with a period of 1 second? ¼ second? ¹⁄₁₀ second?
13. What are typical frequencies for 1-story buildings? 10-story? 30-story?
14. Will a tall building be affected more by high- or low-frequency seismic waves? Why?
15. Is resonance more likely for a 20-story building when shaken by P waves or Rayleigh waves?
16. Building designers must account for acceleration. What does this statement mean?
17. What are the differences between earthquake magnitude and earthquake intensity?
18. List five main variables affecting Mercalli intensities.
19. How does the Richter magnitude scale for earthquakes differ from moment magnitude?
20. Explain how base isolation systems can reduce the shaking of buildings during an earthquake.
21. How do recorded P and S wave patterns differ from a bomb blast versus an earthquake?

Questions for Further Thought

1. Immediately after the start of a big earthquake, how can the greater velocity of P waves be utilized to provide some protection for hospitals, computer systems, and trains?
2. What is the quake potential of the Moon (moonquakes)? Does the Moon have similar numbers and magnitudes of quakes as Earth? Why?
3. If you are in an airplane over the epicenter of a great earthquake, what will you experience?
4. How earthquake safe is your home or office? What are the nearest faults? What kind of earth materials is your home or office built upon? How will your building size, shape, and materials react to shaking? What nearby features could affect your home? What hazards exist inside your home?
5. Make a list of the similarities between snapping your fingers and the movement of a fault.

Disaster Simulation Game

Your challenge is to protect a city from earthquake disaster by constructing new buildings and retrofitting old ones. You are given a budget. Then you have real choices to make.

The city you must protect has a specified population of people. You are provided with a map of the town and charged with protecting as many people, buildings, and livelihoods as possible. You must build a hospital and two schools plus retrofit 10 old buildings. Are you ready for the challenge? Go to http://www.stopdisastersgame.org. Click on Play Game. On the next page, click on Play Game again. Select the Earthquake scenario, then choose your preferred difficulty level: Easy (small map); Medium (medium-size map); or Large (large map).

Good luck! Save as many people as you can.

CHAPTER 4

Plate Tectonics and Earthquakes

A bad earthquake at once destroys our oldest associations: the earth, the very emblem of solidity, has moved beneath our feet like a thin crust over a fluid;—one second of time has created in the mind a strange idea of insecurity, which hours of reflection would not have produced.

—CHARLES DARWIN, 1835, NOTES FOR *THE VOYAGE OF THE BEAGLE*

LEARNING OUTCOMES

Movements along tectonic-plate edges are responsible for many large earthquakes. After studying this chapter, you should:

- be able to describe the types of movements along tectonic-plate edges and the resultant earthquake magnitudes.

- be able to explain why subduction-zone earthquakes have the greatest magnitudes.

- understand the seismic-gap method of forecasting earthquakes.

- recognize the relationship between buildings and earthquake fatalities.

OUTLINE

- Tectonic-Plate Edges and Earthquakes
- Spreading-Center Earthquakes
- Convergent Zones and Earthquakes
- Subduction-Zone Earthquakes
- Continent-Continent Collision Earthquakes
- The Arabian Plate
- Transform-Fault Earthquakes

During the Northridge earthquake, the ground moved rapidly to the north and pulled out from under this elevated apartment building, causing it to fall back onto its parking lot in Canoga Park, California, 17 January 1994.

Courtesy Peter W. Weigand, CSU Northridge.

TABLE 4.1

Mega-Killer Earthquakes, 2003–2015

Year	Place	Magnitude	Deaths	Tectonic Setting
2015	Nepal	7.8	~9,000	continent collision
2011	Japan	9.0	~20,000	subduction
2010	Haiti	7.0	~230,000	transform fault
2008	China	7.9	87,500	continent collision
2005	Pakistan	7.6	88,000	continent collision
2004	Indonesia	9.1	~245,000	subduction
2003	Iran	6.6	31,000	continent collision

Recent years have brought a staggering number of mega-killer earthquakes (table 4.1). The causes of these earthquakes are best understood using their plate-tectonic settings.

Tectonic-Plate Edges and Earthquakes

Most earthquakes are explainable based on plate-tectonics theory. The lithosphere is broken into rigid plates that move away from, past, and into other rigid plates. These global-scale processes are seen on the ground as individual **faults** where Earth ruptures and the two sides move past each other in earthquake-generating events.

Figure 4.1 shows an idealized tectonic plate and assesses the varying earthquake hazards that are concentrated at plate edges:

1. The divergent or pull-apart motion at spreading centers causes rocks to fail in tension. Rocks rupture relatively easily when subjected to tension. Also, much of the rock here is at a high temperature, causing early failures. Thus, the spreading process yields mainly smaller earthquakes that do not pose an especially great threat to humans.
2. The slide-past motion occurs as the rigid plates fracture and move around the curved Earth. The plates shear and slide past each other in the dominantly horizontal movements of transform faults. This process creates large earthquakes as the irregular plate boundaries retard movement because of irregularities along the faults. It takes a lot of stored energy to overcome the rough surfaces, nonslippery rocks, and bends in faults. When these impediments are finally overcome, a large amount of seismic energy is released.
3. The convergent or push-together motions at subduction zones and in continent-continent collisions cause rocks to fail in compression. These settings store immense amounts of energy that are released in Earth's largest tectonic earthquakes. The very processes of pulling a 70 to 100 km (45 to 60 mi) thick oceanic plate back into the mantle via a subduction zone or of pushing continents

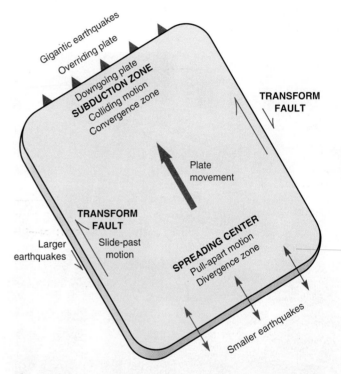

Figure 4.1 Map view of an idealized plate and the earthquake potential along its edges.

together—such as India slamming into Asia to uplift the Himalayas—involve incredible amounts of energy. This results in Earth's greatest earthquakes.

Moving from an idealized plate, let's examine an actual plate—the Pacific plate. Figure 4.2 shows the same type of plate-edge processes and expected earthquakes. The Pacific plate is created at the spreading centers along its eastern and southern edges. The action there produces smaller earthquakes that also happen to be located away from major human populations.

The slide-past motions of long transform faults occur: (1) in the northeastern Pacific as the Queen Charlotte fault, located near a sparsely populated region of Canada; (2) along the San Andreas fault in California with its famous earthquakes; and (3) at the southwestern edge of the Pacific Ocean where

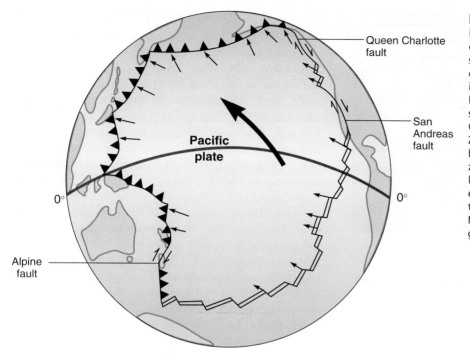

Figure 4.2 The Pacific plate is the largest in the world; it underlies part of the Pacific Ocean. Its eastern and southern edges are mostly spreading centers characterized by small- to intermediate-size earthquakes. Three long transform faults exist along its sides in Canada (Queen Charlotte), California (San Andreas), and New Zealand (Alpine); all are characterized by large earthquakes. Subduction zones (shown by black triangles) lie along the northern and western edges, from Alaska to Russia to Japan to the Philippines to Indonesia to New Zealand; all are characterized by gigantic earthquakes.

the Alpine fault cuts across the South Island of New Zealand (see figures 3.5 and 3.6).

The Pacific plate subducts along its northern and western edges and creates enormous earthquakes, such as the 2011 Japan seism, the 1964 Alaska event, and the 1931 Napier quake on the North Island of New Zealand.

Our main emphasis here is to understand plate-edge effects as a means of forecasting where earthquakes are likely to occur and what their relative sizes may be.

Spreading-Center Earthquakes

A look at earthquake epicenter locations around the world (see figure 2.19) reveals that earthquakes are not as common in the vicinity of spreading centers or divergence zones as they are at transform faults and at subduction/collision zones. The expanded volumes of warm rock in the oceanic ridge systems have a higher heat content and a resultant decrease in rigidity. These heat-weakened rocks do not build up and store the huge stresses necessary to create great earthquakes.

ICELAND

The style of spreading-center earthquakes can be appreciated by looking at the earthquake history of Iceland, a nation that exists solely on a hot-spot–fed volcanic island portion of the mid-Atlantic ridge spreading center (figures 4.3 and 4.4). The Icelandic geologist R. Stefansson reported on catastrophic earthquakes in Iceland and stated that in the portions

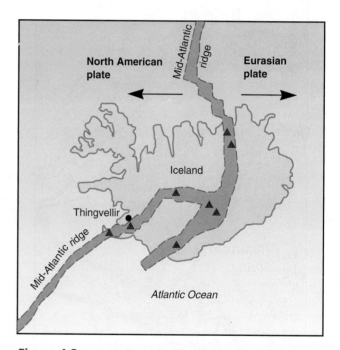

Figure 4.3 Iceland sits on top of a hot spot and is being pulled apart by the spreading center in the Atlantic Ocean. Triangles mark sites of some active volcanoes.

of the country underlain by north-south-oriented spreading centers, stresses build up to cause earthquakes too small to destroy buildings or kill people. These moderate-size earthquakes tend to occur in swarms, as is typical of volcanic areas where magma is on the move. Iceland does have large earthquakes, but they are associated with east-west-oriented transform faults between the spreading-center segments.

Figure 4.4 Looking south along the fissure at Thingvellir, Iceland. This is the rift valley being pulled apart in an east-west direction by the continuing spreading of the Atlantic Ocean basin.

University of Washington Libraries, Special Collections, John Shelton Collection [no number].

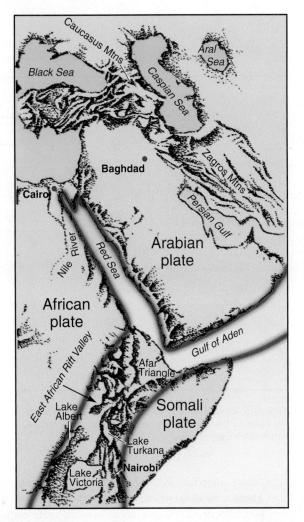

Figure 4.5 Topography in northeastern Africa and Arabia. Northeastern Africa is being torn apart by three spreading centers: Red Sea, Gulf of Aden, and East African Rift Valley. The spreading centers meet at the triple junction in the Afar Triangle.

RED SEA AND GULF OF ADEN

Iceland has been built on a mature spreading center that has been opening the North Atlantic Ocean basin for about 180 million years. What would a young spreading center and new ocean basin look like? Long and narrow. In today's world, long and narrow ocean basins exist in northeast Africa as the Red Sea and the Gulf of Aden (figure 4.5). Following is a model explaining how spreading began: the northeastern portion of Africa sits above an extra-hot area in the upper mantle. The heat contained within this mantle hot zone is partially trapped by the blanketing effect of the overlying African plate and its embedded continent (figure 4.6a). The hot rock expands in volume, and some liquefies to magma. This volume expansion causes doming of the overlying rocks, with resultant uplift of the surface to form topography (figure 4.6b). The doming uplift sets the stage for gravity to pull the raised landmasses downward and apart, thus creating pull-apart faults with centrally located, down-dropped **rift** valleys, also described as pull-apart basins (figure 4.6c). As the fracturing/faulting progresses, magma rises up through the cracks to build volcanoes. As rifting and volcanism continue, seafloor spreading processes take over, the down-dropped linear rift valley becomes filled by the ocean, and a new sea is born (figure 4.6d).

Figure 4.5 reveals another interesting geometric feature. Three linear pull-apart basins meet at the south end of the Red Sea; this point where three plate edges touch is called a **triple junction.** Three rifts joining at a point may concentrate mantle heat, or a concentration of heat in the upper mantle may begin the process of creating this triple junction. Earth's surface may bulge upward into a dome, causing the elevated rocks to fracture into a radial pattern (figure 4.7). Gravity can then pull the dome apart, allowing magma to well up and fill three major fracture zones, and the spreading process is initiated.

The triple junction in northeast Africa is geologically young, having begun about 25 million years ago. To date, spreading in the Red Sea and Gulf of Aden has been enough to split off northeast Africa and create an Arabian plate and to allow seawater to flood between them. But the East African Rift Valley has not yet been pulled far enough apart for the sea to fill it (see figure 4.5). The East African Rift Valley is a truly impressive physiographic feature. It is 5,600 km (3,500 mi) long and has steep escarpments and dramatic valleys. Beginning at the Afar triangle at its northern end and moving southwest are the domed and stretched highlands of Ethiopia, beyond which the Rift Valley divides into two major branches. The western rift is markedly curved and has many deep lakes, including the world's second deepest lake, Lake Tanganyika. The eastern rift is straighter and holds shallow, alkaline lakes and volcanic peaks, such as Mount Kilimanjaro, Africa's highest mountain. The Rift Valley holds the oldest humanoid fossils found to date and is the probable homeland of the first human beings. Will the spreading continue far enough to split a Somali plate from Africa? It is simply too early to tell.

Figure 4.6 A model of the stages in the formation of an ocean basin. (a) Stage 1, _Centering:_ Moving lithosphere centers over an especially hot region of the mantle. (b) Stage 2, _Doming:_ Mantle heat causes melting, and the overlying lithosphere/continent extends. The increase in heat causes surface doming through uplifting, stretching, and fracturing. (c) Stage 3, _Rifting:_ Volume expansion causes gravity to pull the uplifted area apart; fractures fail and form faults. Fractures/ faults provide escape for magma; volcanism is common. Then, the dome's central area sags downward, forming a valley such as the present East African Rift Valley. (d) Stage 4, _Spreading:_ Pulling apart has advanced, forming a new seafloor. Most magmatic activity is seafloor spreading, as in the Red Sea and the Gulf of Aden.

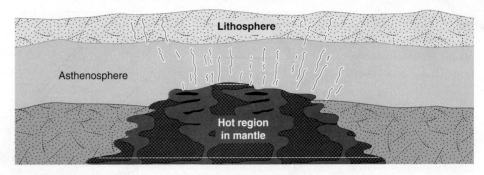

(a) Stage 1, Centering

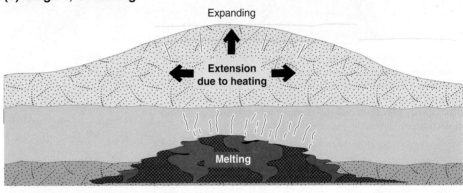

(b) Stage 2, Doming

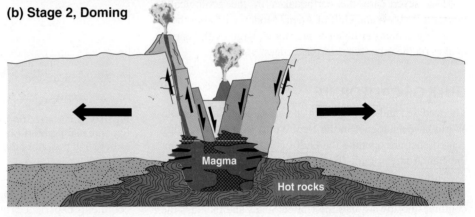

(c) Stage 3, Rifting

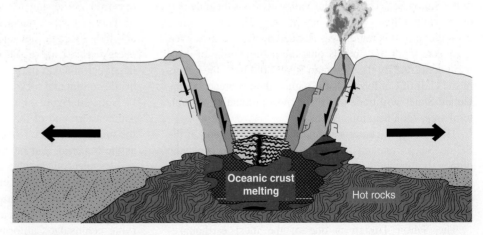

(d) Stage 4, Spreading

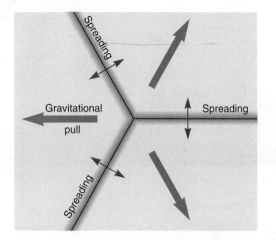

Figure 4.7 Schematic map of a triple junction formed by three young spreading centers. Heat may concentrate in the mantle and rise in a magma plume, doming the overlying lithosphere and causing fracturing into a radial set with three rifts. Gravity may then pull the dome apart, initiating spreading in each rift.

How severe are the earthquakes in the geologically youthful Red Sea and Gulf of Aden? Moderately—but these spreading-center earthquakes are not as large as the earthquakes on the other types of plate edges.

GULF OF CALIFORNIA

The Red Sea and Gulf of Aden spreading centers of the Old World have analogues in the New World with the spreading centers that are opening the Gulf of California and moving the San Andreas fault (figure 4.8). Geologically, the Gulf of California basin does not stop where the sea does at the northern shoreline within Mexico. The opening ocean basin continues northward into the United States and includes the Salton Sea and the Imperial and Coachella valleys at the ends of the Salton Sea. The Imperial Valley region is the only part of the United States that sits on opening ocean floor. In the geologic past, this region was flooded by the sea. However, at present, fault movements plus the huge volume of sediment deposited by the Colorado River hold back the waters of the Gulf of California. If the natural dam is breached, the United States will trade one of its most productive agricultural areas for a new inland sea.

The spreading-center segment at the southern end of the Salton Sea is marked by high heat flow, glassy volcanic domes, boiling mud pots, major geothermal energy reservoirs (subsurface water heated to nearly 400°C (750°F) by the magma below the surface), and swarms of earthquakes associated with moving magma (figure 4.8).

The Salton Trough is one of the most earthquake-active areas in the United States. There are seisms caused by the splitting and rifting of continental rock and swarms of earthquakes caused by forcefully moving magma. The

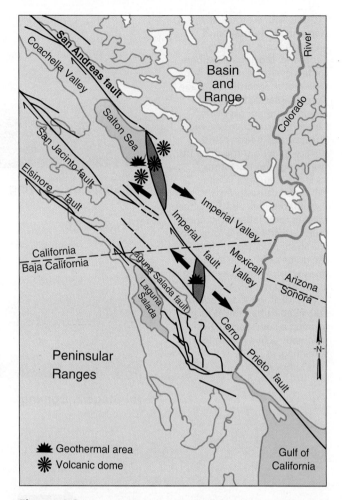

Figure 4.8 Map of northernmost Gulf of California. Note the two spreading centers (shown in red and by large, diverging arrows) and the right-lateral (transform) faults associated with them.

Brawley seismic zone at the southern end of the Salton Sea commonly experiences hundreds of earthquakes in a several-day period. For example, in four days in January 1975, there occurred 339 seisms with magnitudes (M_L) greater than 1.5; of these, 75 were greater than $3M_L$, with the largest tremor at $4.7M_L$. Because hot rock does not store stress effectively, energy release takes place via many smaller quakes. The larger earthquakes in the valley are generated by ruptures in brittle continental rocks.

Notice in figure 4.8 that the San Andreas fault ends at the southeastern end of the Salton Sea at the northern limit of the spreading center. Notice also that other major faults, such as the Imperial, San Jacinto system, Cerro Prieto, Elsinore, and Laguna Salada, also appear to be transform faults that line up with spreading-center segments. From a broad perspective, all these subparallel, right-lateral, transform faults are part of the San Andreas plate boundary fault system carrying peninsular California to the northwest. Large earthquakes on these faults in recent years in the area covered by figure 4.8 include a $6.9M_w$ quake on the Imperial fault in 1940, a $6.6M_w$ event in the San Jacinto system in 1942, a

6.4M$_w$ quake on the Imperial fault in 1979, a 6.6M$_w$ quake in the San Jacinto system in 1987, and a 7.25M$_w$ seism in the Laguna Salada fault system in 2010. These are large earthquakes, but deaths and damages for each event typically were not high because the region is sparsely inhabited, most buildings are low, and the frequent shakes weed out inferior buildings.

Convergent Zones and Earthquakes

The greatest earthquakes in the world occur where plates collide. Three basic classes of collisions are (1) oceanic plate versus oceanic plate, (2) oceanic plate versus continent, and (3) continent versus continent. These collisions result in either subduction or continental upheaval. If oceanic plates are involved, subduction occurs. The younger, warmer, less-dense plate edge overrides the older, colder, denser plate, which then bends downward and is pulled back into the mantle. If a continent is involved, it cannot subduct because its huge volume of low-density, high-buoyancy rocks simply cannot sink to great depth and cannot be pulled into the denser mantle rocks below. The fate of oceanic plates is destruction via subduction and reassimilation within the mantle, whereas continents float about on the asthenosphere in perpetuity. Continents are ripped asunder and then reassembled into new configurations via collisions, but they are not destroyed by subduction.

Subduction-Zone Earthquakes

Subduction zones are the sites of great earthquakes. Imagine pulling a 100 km (62 mi) thick rigid plate into the weaker, deformable rocks of the mantle that resist the plate's intrusion. This process creates tremendous stores of energy, which are released periodically as great earthquakes. Most of the really large earthquakes in the world are due to subduction (table 4.2). Subduction occurs on a massive scale. At the present rates of subduction, oceanic plates with an area equivalent to the entire surface area of Earth will be pulled into the mantle in only 180 million years.

A descending slab of oceanic lithosphere is defined by an inclined plane of deep earthquakes or fault-rupture locations (see figure 2.20). Earthquakes at subduction zones result from different types of fault movements in shallow versus deeper realms. At shallow depths (less than 100 km, or 62 mi), the two rigid lithospheric plates are pushing against each other. Earthquakes result from compressive movements where the overriding plate moves upward and the subducting plate moves downward. Pull-apart fault movements also occur near the surface within the subducting plate as it is bent downward and snaps in tensional failure and as the overriding plate is lifted up from below. Notice in

TABLE 4.2
Earth's Largest Earthquakes, 1904–2014

Rank	Location	Year	M$_w$	Cause
1.	Chile	1960	9.5	Subduction—Nazca plate
2.	Alaska	1964	9.2	Subduction—Pacific plate
3.	Indonesia	2004	9.1	Subduction—Indian plate
4.	Japan	2011	9.0	Subduction—Pacific plate
5.	Kamchatka	1952	9.0	Subduction—Pacific plate
6.	Chile	2010	8.8	Subduction—Nazca plate
7.	Ecuador	1906	8.8	Subduction—Nazca plate
8.	Alaska	1965	8.7	Subduction—Pacific plate
9.	Indonesia	2005	8.6	Subduction—Indian plate
10.	Assam	1950	8.6	Collision—India into Asia
11.	Alaska	1957	8.6	Subduction—Pacific plate
12.	Indonesia	2007	8.5	Subduction—Indian plate
13.	Kuril Islands	1963	8.5	Subduction—Pacific plate
14.	Banda Sea	1938	8.5	Subduction—Pacific/Indian plate
15.	Russia	1923	8.5	Subduction—Pacific plate
16.	Chile	1922	8.5	Subduction—Nazca plate

figure 2.20 that the shallow earthquakes occur (1) in the upper portion of the down-going plate, (2) at the bend in the subducting plate, and (3) in the overriding plate.

Compare the locations of the shallow earthquake sites to those of intermediate and deep earthquakes (see figure 2.19). At depths below 100 km, earthquakes occur almost exclusively in the interior of the colder oceanic lithosphere, the heart of the subducting slab. The high temperatures of rock in the upper mantle cause it to yield more readily to stresses and thus not build up the stored energy necessary for gigantic earthquakes. At depth, the upper and lower surfaces of the subducting slabs are too warm to generate large earthquakes. Thus, the earthquakes occur in the cooler interior area of rigid rock, where stress is stored as gravity pulls against the mantle resistance to slab penetration. In the areas of most rapid subduction, the down-going slab may remain rigid enough to spawn large earthquakes to depths in excess of 700 km (435 mi). A great earthquake that occurs deep below the surface has much of its seismic energy dissipated while traveling to the surface. Thus, the biggest disasters are from the great earthquakes that occur at shallow depths and concentrate their energy on the surface.

Most of the subduction-zone earthquakes of today occur around the rim of the Pacific Ocean or the northeastern Indian Ocean. This is shown by the presence of most of the deep-ocean trenches (see figure 2.13) and by the dense concentrations of earthquake epicenters (see figure 2.19).

When will the next large earthquake occur in the northwestern Pacific Ocean region? A popular way of forecasting the locations of future earthquakes is the **seismic-gap method.** If segments of one fault have moved recently, then it seems reasonable to expect that the unmoved portions will move next and thus fill the gaps. Looking at figure 4.9, where would you forecast large earthquakes to occur? It is easy to see the gaps in earthquake locations, and although seismic-gap analysis is logical, it yields only expectations, not guarantees. One segment of a fault can move two or more times before an adjoining segment moves once. In 2011, a seismic gap was filled in Japan by a 9.0M_W earthquake (figure 4.9).

JAPAN, 2011: STUCK SEGMENTS OF SUBDUCTING PLATE

At first glance, the plate-tectonic setting of eastern Japan looks fairly simple. The Pacific plate moves northwest 8.3 cm (3.3 in) per year and dives under Japan (figure 4.9). But in detail, the tectonic setting of the Japan region has complexities as the huge Pacific, North American, and Eurasian plates meet here, along with smaller plates such as the Philippine, Okhotsk, and Amur. The subduction process causes strain energy to build until there is enough to rupture individual fault segments, resulting in earthquakes of 7 to 8 M. This had been the pattern of the last several centuries. Some scientists felt that the highly segmented plate and fault pattern limited the size of future fault ruptures and thus limited the magnitude of future seisms. But on 11 March 2011, five adjacent segments ruptured together over an area with length greater than 600 km (375 mi). Two central segments had enormous slips up to 50 m (165 ft), contibuting to the 9.0M_w event, the 4th largest earthquake in the past 100+ years (table 4.2).

The 9.0M_w mainshock was preceded two days earlier by a 7.2M_w event 40 km (25 mi) away and three other seisms greater than 6M. But we don't know how to recognize that an earthquake is a foreshock until after the mainshock occurs; thus, no warnings were given. The maximum acceleration offshore is calculated as 3 g, that is, three times stronger than the pull of gravity. Acceleration of 2.7 g was measured onshore in Miyagi Prefecture.

Before the earthquake, 15 years of **global positioning system (GPS)** data showed that the eastern edge of Japan was being dragged downward. Apparently the down-going Pacific plate was stuck to the overlying continental plate and was warping it downward. The strain of this movement accumulated during many centuries. When the stuck zone ruptured as the Pacific plate moved downward, the overlying plate carrying Japan sprang upward and released elastic strain.

The Japanese have kept the best historic records of earthquakes and tsunami in the world. A look farther back in their records shows that a similar-size earthquake and tsunami hit the same area of northeast Japan on 13 July 869 CE. The 2011 event surprised many experts who had disregarded the older historic record. But 1,142 years is a short time in geologic history.

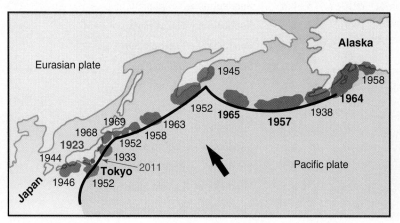

Figure 4.9 Brown patterns show severely shaken areas, with dates, from recent earthquakes caused by Pacific plate subduction. The 1957, 1964, and 1965 Alaska earthquakes are three of the largest in the 20th century. They were part of an earthquake cluster. Using the seismic-gap method, where are the next earthquakes most likely to occur? See the seismic gap filled in 2011.

A Classic Disaster

The Tokyo Earthquake of 1923

Early on Saturday morning, 1 September 1923, the cities of Tokyo and Yokohama were shattered by a deadly series of earthquakes. The principal shock occurred as the floor of Sagami Bay dropped markedly and sent 11 m (36 ft) high seismic sea waves (tsunami) crashing against the shore. The waves washed away hundreds of homes. Yet fishermen spending their day at sea were unaware of the monster waves. At day's end, as they sailed toward home through Sagami Bay, they were sickened to find the floating wreckage of their houses and the bodies of their families. Devastation on land was great; houses were destroyed, bridges fell, tunnels collapsed, and landslides destroyed hills. The shaking caused the collapse of flammable house materials onto cooking fires, and the flames, once liberated, quickly raced out of control. Little could be done to stop their spread because the earthquake had broken the water mains. Shifting winds pushed the fires for two and a half days, destroying 71% of Tokyo and 100% of Yokohama.

Possibly the most tragic event in this disaster occurred when 40,000 people, clutching their personal belongings, attempted to escape the flames by crowding into a 250-acre garden on the edge of the Sumida River. People packed themselves into this open space so densely that they were barely able to move. At about 4 p.m., several hours after the earthquake, the roaring fires approached on all three landward sides of the crowd. Suddenly the fire-heated winds spawned a tornado that carried flames onto the huddled masses and their combustible belongings. After the flames had died, 38,000 people lay dead, either burned or asphyxiated. The usual instinct to seek open ground during a disaster was shockingly wrong this time.

The combined forces of earthquakes, tsunami, and fires killed 99,331 people and left another 43,476 missing and presumed dead. Yet, despite this immense catastrophe, the morale of the Japanese people remained high. They learned from the disaster. They have rebuilt their cities with wider streets, more open space, and less use of combustible construction materials.

The historic record of earthquakes in the region is thought provoking. The region 80 km (50 mi) southwest of Tokyo has been rocked by five very strong earthquakes in the last 400 years. The seisms have occurred roughly every 73 years, the most recent in 1923.

A concern now is that the March 2011 earthquake transferred stress southward, closer to Tokyo, which was hit hard in 1923. In November 2011, an official Japan earthquake assessment committee forecast a 30% chance of a 9M event closer to Tokyo in the next 30 years. The fault there is closer to shore, meaning more intense shaking for buildings and less time before tsunami come ashore.

The death and destruction caused by the 2011 earthquake are not known because many of its effects were erased by the tsunami that followed. We will pick up this story again in chapter 8 on tsunami.

INDONESIA, 2004: ONE EARTHQUAKE TRIGGERS OTHERS

The Indian-Australian plate moves obliquely toward western Indonesia at 5.3 to 5.9 cm/yr (2 to 2.3 in/yr). The enormous, ongoing collision results in subduction-caused earthquakes that are frequent and huge (figure 4.10). Many of these earthquakes send off tsunami. On 26 December 2004, a 1,500 km (930 mi) long fault rupture began as a 100 km (62 mi) long portion of the plate-tectonic boundary ruptured and slipped during 1 minute. The rupture then moved northward at 3 km/sec (6,700 mph) for 4 minutes, then slowed to 2.5 km/sec (5,600 mph) during the next 6 minutes in a $9.1M_w$ event. At the northern end of the rupture, the fault movement slowed drastically and only traveled tens of meters during the next half hour.

On 28 March 2005, the subduction zone broke again and this time ruptured 400 km (250 mi) southward from the southern end of the 2004 rupture in an $8.6M_w$ event (figure 4.10). Was this second rupture event, just 92 days later, a continuation of the earlier earthquake? It appears that a bend or scissors-like tear in the subducting plate may have delayed the full rupture in December 2004. The history of the region suggested there were more big earthquakes to come.

And come they did. An **earthquake cluster** is under way. On 12 September 2007, there were two earthquakes, an $8.4M_w$ followed 12 hours later by a $7.9M_w$; on 20 February 2008 there was a $7.4M_w$; on 30 September 2009 there was a $7.6M_w$; in 2010 there was a $7.8M_w$ on 6 April and a $7.7M_w$ on 25 October; and on 11 April 2012 there was an $8.6M_w$ and an $8.2M_w$ (figure 4.10). And there are more seismic gaps to fill.

Another earthquake cluster consisting of a $9.2M_w$ and several magnitude 8s occurred in the mid-20th century at the Pacific plate subduction zone along Alaska, Russia, and northern Japan (see figure 4.9 and table 4.2).

MEXICO CITY, 1985: LONG-DISTANCE DESTRUCTION

On Thursday morning, 19 September 1985, most of the 18 million residents of Mexico City were at home, having their morning meals. At 7:17 a.m., a monstrous earthquake broke loose some 350 km (220 mi) away. Seismic waves traveled far to deal destructive blows to many of the 6- to 16-story buildings that are heavily occupied during the working day (figure 4.11). Building collapses killed more than 9,000 people.

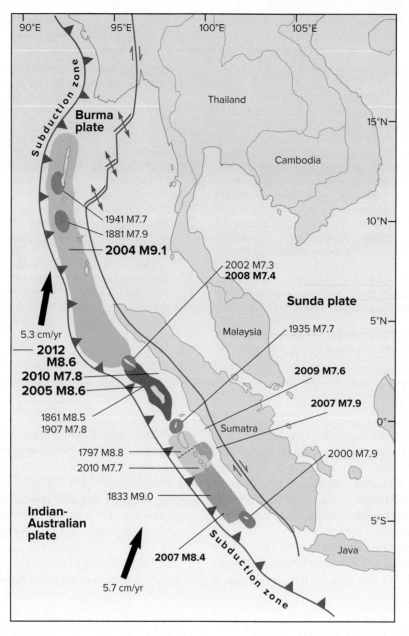

Figure 4.10 Subduction of the Indian-Australian plate beneath Indonesia was the cause of the huge earthquakes in 2004, 2005, 2007, 2008, 2009, 2010, and 2012. The region has a long history of large earthquakes, and more will occur.

What caused this earthquake? The Cocos plate made one of its all-too-frequent movements. This time, a 200 km (125 mi) long front, inclined 18° east, thrust downward and eastward about 2.3 m (7.5 ft) in two distinct jerks about 26 seconds apart (figure 4.12). The mainshock had a surface wave magnitude (M_s) of 8.1. It was followed on 21 September by a 7.5M_s aftershock and by another on 25 October of 7.3M_s. The earthquakes were not a surprise to seismologists. Before these seisms occurred, the area was called the Michoacan seismic gap, and many instruments had been deployed in the region to measure the expected big event.

As figure 4.12 shows, another large seismic gap waits to be filled by a major movement of the Cocos plate. The Guerrero seismic gap lies near Acapulco and is closer to Mexico City than the Michoacan epicenter.

Resonance Matters

Many of the coastal towns near the epicenter received relatively small amounts of damage. Yet in Mexico City, more than 5,700 buildings were severely damaged, with 15% of them collapsing catastrophically. Why did so many buildings collapse and kill so many people when Mexico City lies 350 km (220 mi) from the epicenter? It was largely due to resonance between seismic waves, soft lake-sediment foundations, and improperly designed buildings. The duration of shaking was increased due to seismic energy trapped within the soft sediments.

Mexico City is built atop the former Aztec capital of Tenochtitlan. The Aztecs built where they saw the favorable omen—an eagle sitting on a cactus and holding a writhing snake in its mouth. The site was Lake Texcoco, a broad lake surrounded by hard volcanic rock. Over time, the lake basin was partially filled with soft, water-saturated clays. Portions of Lake Texcoco have been drained, and large buildings have been constructed on the weak lake-floor sediments.

Building damages were the greatest and the number of deaths the highest where three factors combined and created resonance: (1) the earthquakes sent a tremendous amount of energy in seismic waves in the 1- to 2-second frequency band; (2) the thick, soft muds (clays) amplified the seismic waves (figure 4.13); and (3) buildings of 6 to 16 stories vibrated in the 1- to 2-second frequency band. Where all three factors were in phase, disaster struck.

There were design flaws in the failed buildings (figure 4.14), including soft first stories, poorly joined building wings, odd-shaped buildings prone to twist on their foundations, and buildings of different heights and vibration frequencies that sat close together and bumped into each other during the earthquake (figures 4.14c and 4.15).

CHILE, 1960: THE BIGGEST ONE

Earth's biggest measured earthquake is a 9.5M$_w$ event that sprang forth on Sunday afternoon, 22 May 1960, in southern Chile. Here the Nazca plate converges with the South American plate at 8 m/century (see figure 2.13). In 1960, the Nazca

Figure 4.11 This 15-story building collapsed during the 1985 Mexico City earthquake, crushing all its occupants as its concrete floors pancaked.

Photo by M. Celebi, *US Geological Survey.*

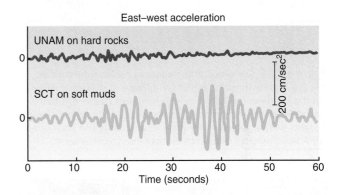

Figure 4.13 Some east–west accelerations recorded in Mexico City in 1985. The Universidad Nacional Autonoma de Mexico (UNAM) sits on a hard-rock hill and received small accelerations. The Secretaria de Comunicaciones y Transportes site (SCT) sits on soft lake sediments that amplified the seismic waves.

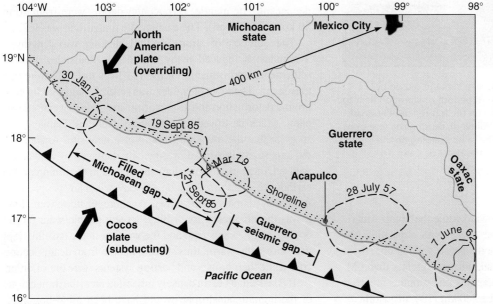

Figure 4.12 Map of coastal Mexico showing dates of earthquakes and fault areas moved (dashed lines) during Cocos plate subduction events. The Michoacan seismic gap was filled by the 1985 seisms. The Guerrero seismic gap is overdue for a major movement.

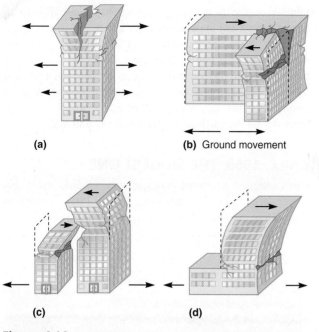

(a)

(b) Ground movement

(c)

(d)

Figure 4.14 Some building-response problems during the Mexico City earthquake. (a) The amplitude of shaking increases up the building. (b) Buildings with long axes perpendicular to ground motion suffer more shaking. (c) Buildings with different heights sway at different frequencies and bang into each other. (d) A building with different heights tends to break apart.

Figure 4.15 Mexico City earthquake damage caused by constructing buildings with different periods of vibration next to each other. The four-story building on the left repeatedly struck the taller Hotel de Carlo (middle building), causing collapse of its middle floors (see figure 4.14c). The taller building on the right was damaged by hammering from the Hotel de Carlo.

© Chris Arnold.

plate moved eastward and downward while the South American plate moved westward and above with slips of 20 to 30 m (65 to 100 ft). In the 33 hours before the big one, there were foreshocks of 8.1M$_w$ and six others of greater than 6M. Over a period of days, the subduction-zone ruptures involved a 1,000 km (620 mi) length and a 300 km (185 mi) width.

Chile, 1835 and 2010: Emptying, Filling, and Emptying Elastic Strain in a Seismic Gap

Giant earthquakes are common phenomena in Chile. During his epic voyage on the HMS *Beagle,* Charles Darwin was resting on his back in the woods near Valdivia, Chile, on 20 February 1835 when a huge earthquake struck. His well-written description of large areas of land being uplifted, giant sea waves crossing the shoreline, and two volcanoes being shaken into eruptions are instructive even today. The modern estimate is that the elastic strain released in Darwin's earthquake yielded an 8.5M$_w$ event.

During the 175 years following the 1835 seism, tectonic-plate convergence in the region was about 14 m (46 ft), but few earthquakes were being recorded. A seismic gap was recognized and the region was heavily instrumented. GPS measurements showed that the area was not moving; it was locked. Frictional resistance was causing elastic strain to accumulate in the rocks; the seismic gap was filling with strain. A big earthquake was expected, and on 27 February 2010, one happened: the 8.8M$_w$ event is the 6th biggest earthquake ever measured. A 500 km (310 mi) long subduction-zone interface ruptured with bilateral movement at 3.1 km/sec (6,930 mph). Some areas of fault surface moved up to 15 m (50 ft), whereas other areas had low to no slip. South America, from Chile to the Argentina coast, had all moved westward.

ALASKA, 1964: SECOND BIGGEST ONE

Saint Matthew's account of the first Good Friday included: "And, behold . . . the earth did quake, and the rocks rent." His words applied again, more than 1,900 years later, on Good Friday, 27 March 1964. At 5:36 p.m., in the wilderness at the head of Prince William Sound, a major subduction movement created a gigantic earthquake. This was followed in sequence by other downward thrusts at 9, 19, 28, 29, 44, and 72 seconds later as a nearly 1,000 km (more than 600 mi) long slab of 400 km (250 mi) width lurched its way deeper into the mantle. Hypocenter depths were from 20 to 50 km (12 to 30 mi). The earthquake magnitude was 9.2M$_w$.

The duration of strong ground shaking was lengthy—7 minutes; it induced many avalanches, landslides, ground settlements, and tsunami. Of the 131 lives lost, 122 were due to tsunami. The town of Valdez was severely damaged by both ground deformation and a submarine landslide that caused tsunami, which destroyed the waterfront facilities. Damage was so great that the town was rebuilt at a new site. Anchorage, the largest city in Alaska, was heavily damaged by landslides. And yet, there were some elements of luck in the timing of this earthquake. It occurred late on a Friday, when few people were in heavily damaged downtown Anchorage; tides were low; it was the off-season for fishing, so few people were on the docks or in the canneries; and the weather in the ensuing days was seasonally warm, thus sparing people from death-dealing cold while their homes and heating systems were out of order.

If Alaska had been a densely inhabited area, the dimensions of the human catastrophe would have been mind-boggling.

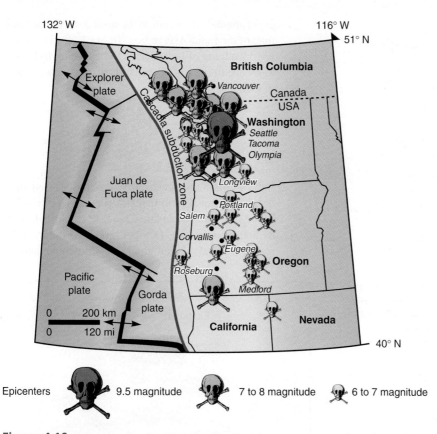

132° W 116° W

51° N

British Columbia

Explorer plate

Vancouver

Canada
USA

Cascadia subduction zone

Washington
Seattle
Tacoma
Olympia

Juan de Fuca plate

Longview

Portland

Salem

Corvallis

Eugene

Roseburg

Oregon

Pacific plate

Gorda plate

Medford

0   200 km

0   120 mi

California   Nevada

40° N

Epicenters   9.5 magnitude   7 to 8 magnitude   6 to 7 magnitude

Figure 4.16 Epicenters for the 1960 Chile earthquake sequence are plotted over the Cascadia subduction zone. One earthquake had a magnitude of 9.5; nine had magnitudes of 7 to 8; and 28 had magnitudes of 6 to 7.

buoyant and is best subducted when overridden by continental lithosphere. The North American continent is moving southwest at 2.5 cm/yr (1 in/yr) and colliding with the oceanic plate, which is subducting along a N 68° E path at 3.5 cm/yr (1.4 in/yr). Thus, it seems certain that the subduction zone is storing energy in elastic strain.

The Cascadia subduction zone is 1,100 km (680 mi) long. Its characteristics of youthful oceanic plate and strong coupling with the overriding plate are similar to situations in southwestern Japan and southern Chile. Events of Chilean magnitude could unlock the entire Cascadia subduction zone. Figure 4.16 is a plot of the epicenters of the 1960 Chile mainshock, foreshocks, and aftershocks over a map of the Pacific Northwest to give an idea of what could happen in British Columbia, Washington, and Oregon. Could the Pacific Northwest experience a magnitude 9 earthquake? Yes—in fact, it already has.

In 2004, essentially the same-size earthquake and tsunami occurred in Indonesia, killing at least 245,000 people in the region, compared to 131 in the Alaska event.

In the United States, California is commonly called "earthquake country," but it is clear from table 4.2 that Alaska is more deserving of this title. Over the past 5 million years, about 290 km (180 mi) of Pacific plate have been pulled under southern Alaska in the vicinity of Anchorage.

PACIFIC NORTHWEST: THE UPCOMING EARTHQUAKE

The 1985 Mexico City earthquake was caused by eastward subduction of a small plate beneath the North American plate. Other small plates are subducting beneath North America at the Cascadia subduction zone (figure 4.16). No gigantic seisms have occurred in the Pacific Northwest in the 200 or so years since Europeans settled there. Will this area remain free of giant earthquakes? Could the Cascadia subduction zone be plugged up like a clogged drain, meaning that subduction has stopped? No. The active volcanoes above the subducting plates testify that subduction is still occurring. Could the subduction be taking place smoothly and thus eliminating the need for giant earthquakes? Probably not. The oceanic lithosphere being subducted is young, only about 10 million years old. Young lithosphere is more

Earthquake in 1700: The Trees Tell the Story

Recent work by Brian Atwater has shown that the last major earthquake in the Pacific Northwest occurred about 9 p.m. on 26 January 1700 and was about magnitude 9. This is indicated by two converging lines of evidence: (1) Counting the annual growth rings in trees of drowned forests along the Oregon–Washington–British Columbia coast shows that the dead trees have no rings after 1699. Apparently the ground dropped during an earthquake, and seawater got to the tree roots, killing them between August 1699 and May 1700, the end of one growing season and the beginning of the next (figure 4.17). (2) The Japanese maintain detailed records of tsunami occurrences and sizes that they correlate to earthquake magnitudes and locations around the Pacific Ocean. Tsunami of 2 m (7 ft) height that hit Japan from midnight to dawn point to a 9 p.m. earthquake along the Washington–Oregon coast on 26 January 1700.

What will the British Columbia–Washington–Oregon region experience during a magnitude 9 earthquake? Three to five minutes of violent ground shaking will be followed by tsunami 10 m (33 ft) high surging onshore 15 to 40 minutes after the earthquake. Energy will be concentrated in long-period seismic waves, presenting challenges for tall buildings and long bridges.

What will the next magnitude 9 earthquake, along with its major aftershocks, do to cities such as Portland, Tacoma, Seattle, Vancouver, and Victoria? When will the next magnitude 9 earthquake occur in the Pacific Northwest?

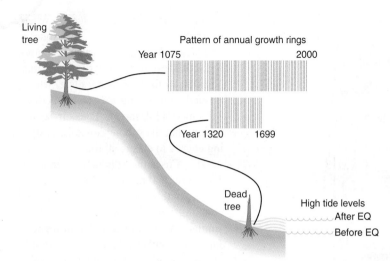

Living tree

Pattern of annual growth rings

Year 1075   2000

Year 1320   1699

Dead tree

High tide levels
After EQ
Before EQ

Figure 4.17 Annual growth rings in drowned trees along the Oregon–Washington–British Columbia coast tell of their deaths after the 1699 growing season. Seawater flooding occurred as land dropped during a magnitude 9 earthquake (EQ).

Continent-Continent Collision Earthquakes

The grandest continental pushing match in the modern world is the ongoing ramming of Asia by India. When Gondwanaland began its breakup, India moved northward toward Asia. The 5,000 km (3,000 mi) of seafloor (oceanic plate) that lay in front of India's northward path had all subducted beneath Asia by about 40 million years ago. Then, with no seafloor left to separate them, India punched into the exposed underbelly of Asia (figure 4.18). Since the initial contact, the assault has remained continuous. India has moved another 2,000 km (1,250 mi) farther north, causing complex accommodations within the two plates as they shove into, under, and through each other accompanied by folding, overriding, and stacking of the two continents into the huge mass of the Himalayas and the Tibetan Plateau. The precollision crusts of India and Asia were each about 35 km (22 mi) thick. Now, after the collision, the combined crust has been thickened to 70 km (44 mi) to create the highest-standing continental area on Earth. The Tibetan Plateau dwarfs all other high landmasses. In an area the size of France, the average elevation exceeds 5,000 m (16,400 ft). But what does all of this have to do with earthquakes? Each year, India continues to move about 5 cm (2 in) into Asia along a 2,000 km (1,250 mi) front. This ongoing collision jars a gigantic area with great earthquakes. The affected area includes India, Pakistan, Afghanistan, the Tibetan Plateau, much of eastern Russia, Mongolia, and most of China.

A relatively simple experiment shows how earthquake-generating faults may be caused by continental collision (figure 4.19). The experiment uses a horizontal jack to push into a pile of plasticine, deforming it under the force. The

experimental deformation is similar to the tectonic map of the India-Asia region (figure 4.20). The northward wedging of India seems to be forcing Indochina to escape to the southeast and is driving a large block of China to the east.

CHINA, PAKISTAN, AND INDIA, 2008, 2005, AND 2001: CONTINENT COLLISION KILLS

India's continuing push into Asia has caused three deadly earthquakes in the 21st century (figure 4.20). The May 2008 Sichuan, China, event killed about 87,500 people; the October 2005 Kashmir, Pakistan, earthquake killed about 88,000 people; and the

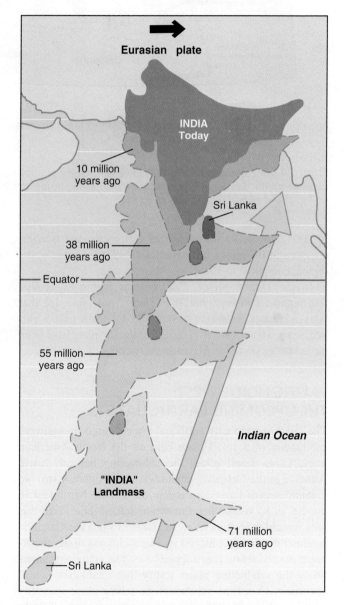

Figure 4.18 Map showing the movement of India during the past 71 million years. India continues to shove into Eurasia, creating great earthquakes all the way through China.

Figure 4.19 Simulated collision of India into Asia. A wedge is slowly jacked into layered plasticine confined on its left side but free to move to the right. From top to bottom of figure, notice the major faults that form and the masses that are compelled to move to the right. Compare this pattern to the tectonic map of India and Asia in figure 4.20.

After P. Tapponier, et al. (1982). *Geology,* 10, 611–16.

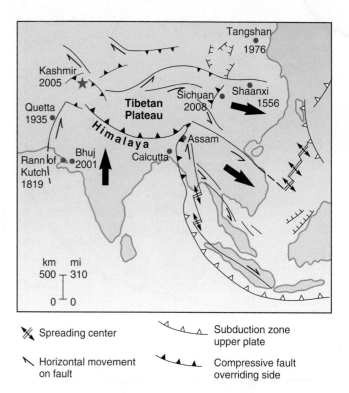

Symbol	Description	Symbol	Description
✗	Spreading center	⊥⊥⊥	Subduction zone upper plate
⌐	Horizontal movement on fault	▲▲▲	Compressive fault overriding side

Figure 4.20 Tectonic map showing India pushing into Asia. The ongoing collision causes devastating earthquakes, each killing tens or hundreds of thousands of people. The list includes two in the Indian state of Gujarat (in 1819 at Rann of Kutch and in 2001 near Bhuj), three in China (in 1556 at Shaanxi, in 1976 at Tangshan, and in 2008 at Sichuan), and two in Pakistan (in Quetta in 1935 and in Kashmir in 2005).

January 2001 Gujarat, India, event killed more than 20,000. These earthquakes were close together in space and time. Is this just a coincidence, or will they be part of a cluster of ongoing killer events? We don't know. But there are seismic gaps waiting to be filled in this region, and some are large, including a 600 km (375 mi) long section in the central Himalaya front that has not moved since 1505. The rapid population growth in these countries has resulted in the construction of millions of new buildings. Many buildings were and are being built without seismic-safety inspections to guide their construction. Even where codes exist, poor construction practices have led to catastrophic failures of many buildings during shaking. It has been suggested that many of these buildings can be viewed as weapons of mass destruction.

The earthquakes of recent years have been deadly, but none of them have been a direct hit on the mega-cities of the region. Some Indian plate–caused earthquakes in China show how disheartening the death totals can be.

CHINA, 1556: THE DEADLIEST EARTHQUAKE

In Shaanxi Province, the deadliest earthquake in history occurred in 1556 when about 830,000 Chinese were killed in and near Xi'an on the banks of the mighty Huang River (once known as the Yellow River). The region has numerous hills composed of deposits of windblown silt and fine sand that have very little **cohesion** (ability to stick together). Because of the ease of digging in these loose sediments, a tremendous number of the homes in the region were caves dug by the inhabitants. Most of the residents were in their cave homes at 5 a.m. on the wintry morning of 23 January, when the seismic waves rolled in from the great earthquake. The severe shaking caused many of the soft silt and sand sediments of the region to vibrate apart and literally behave like fluids. Most of the cave-home dwellers were entombed when the once-solid walls of their homes liquefied and collapsed.

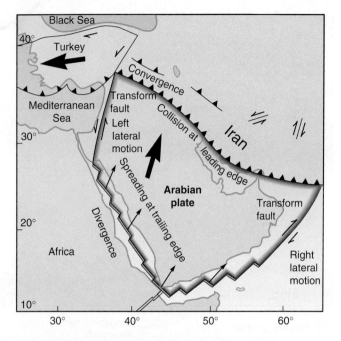

Figure 4.21 The Arabian plate pulls away from Africa, pushes into Eurasia, slices through the Holy Land with a transform fault, and squeezes Turkey westward.

TABLE 4.3

Earthquake Fatalities in Iran, 1962–2005

Fatalities	Date	Location
612	22 Feb 2005	Zarand
41,000	26 Dec 2003	Bam
50,000	21 Jun 1990	Rudbar-Tarom
3,000	28 Jul 1981	Kerman
3,000	11 Jun 1981	Golbas
25,000	16 Sep 1978	Tabas-e-Golshan
5,000	24 Nov 1976	Northwest
5,044	10 Apr 1972	Fars
12,000	31 Aug 1968	Khorasan
12,225	1 Sep 1962	Buyin-Zara
156,881 Total deaths		

The Arabian Plate

The emergence of the geologically young spreading centers in the Red Sea and Gulf of Aden has cut off the northeast tip of the African continent (figure 4.21; see figure 4.5) and created the Arabian plate. Analysis of the movement of the Arabian plate gives us good insight into different earthquake types.

CONTINENT-CONTINENT COLLISION EARTHQUAKES

The Red Sea and Gulf of Aden areas may not have many large earthquakes, but their spreading centers are responsible for shoving the Arabian plate into Eurasia, causing numerous devastating earthquakes there. The rigid continental rocks of the Arabian plate are driven like a wedge into the stiff underbelly of Eurasia. The force of this collision uplifts mountain ranges (e.g., Caucasus and Zagros in figure 4.5) and moves many faults that create the killer earthquakes typical of this part of the world.

Iran, 1962–2005: Mud-Block Buildings Kill

Catastrophic earthquakes occur along the entire length of Iran. For example, during a 44-year interval, 10 earthquakes killed more than 150,000 people (table 4.3). The deadliest natural disaster in 2003 was the earthquake that shook loose 8 km (5 mi) below the city of Bam at 5:27 a.m. on Friday, 26 December. Traditional construction methods there consist of sun-dried, mud-block walls topped by heavy roofs. When the earthquake shaking began, the walls crumbled and the heavy roofs crashed down. Collapsing homes killed 41,000 people.

TRANSFORM-FAULT EARTHQUAKES

The Arabian plate is moving away from Africa and is pushing into Eurasia, but what is happening along the sides of the Arabian plate? The slide-past movements of transform faults. On the eastern side, the plate-boundary fault occurs beneath the Indian Ocean and has scant effect on humans. But look where the slide-past fault movements occur along the western side of the Arabian plate (figures 4.21, 4.22, and 4.23).

Dead Sea Fault Zone: Holy Land Disasters

The Dead Sea fault zone is an Eastern Hemisphere analogue of the San Andreas fault in California. It not only runs right through the Holy Land but has also created much of the area's well-known topography. Notice in figure 4.22 that there are four prominent overlaps or steps in the Dead Sea fault zone. Fault movements on both sides of these steps have created pull-apart basins that are filled by historically famous water bodies, such as the Dead Sea and the Sea of Galilee. The Dead Sea fault zone has been operating for as long as the Red Sea has been opening. During that time, there has been 105 km (65 mi) of offset, and 40 km (25 mi) of this movement has happened in the past 4.5 million years. This computes to an average slip (movement) rate of more than 5 mm/yr over the longer time frame or 9 mm/yr over the more recent time span. However, the rough, frictionally resistant faults do not easily glide along at several millimeters each year. The rocks along the fault tend to store stress until they can't hold any more, and then they rupture in an earthquake-producing fault movement. How often do these earthquakes occur? Table 4.4 is a partial list from Amos Nur of Stanford University.

It is interesting to ponder the effects of these earthquakes on the thinking of the religious leaders in this region, the birthplace of Judaism and Christianity and an important area

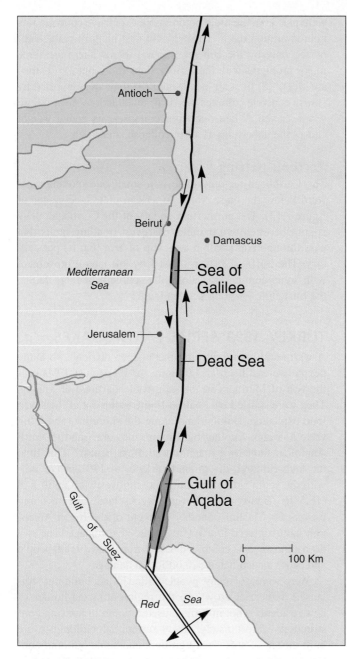

Figure 4.22 Map of the Dead Sea fault zone. Notice that the subparallel faults have pull-apart basins in the steps between faults. The Dead Sea basin is deep; it has a 7 km (greater than 4 mi) thick infill of sediments below its water. On 21 November 1995, a magnitude 7.2 earthquake in the Gulf of Aqaba killed people as far away as Cairo, Egypt.

Figure 4.23 Space shuttle view of the northern Red Sea. The Nile River is in the center left; the Suez Canal, in the middle; and the Gulf of Aqaba, pointing toward the Dead Sea in the upper right.

NASA/JSC

TABLE 4.4			
Some Earthquakes in the Holy Land			
Year	**Magnitude**	**Year**	**Magnitude**
1927	6.5	1068	6.6
1834	6.6	1033	7.0
1759	6.5	749	6.7
1546	6.7	658	6.2
1293	6.4	363	7.0
1202	7.2	31 BCE	6.3
		759 BCE	7.3

Source: "When the Walls Came Tumbling Down" (1991). [Video] Amos Nur, Stanford University.

Transform-Fault Earthquakes

The transform faults forming the sides of some tectonic plates have dominantly horizontal movements that cause major earthquakes. Examples include the Alpine fault of New Zealand, the San Andreas fault in California, the North Anatolian fault in Turkey, and the Enriquillo-Plantain Garden fault in Haiti.

HAITI, 2010: EARTHQUAKES DON'T KILL, BUILDINGS DO

Year 2010 began with a horrifying event. The earth shook in a $7.0M_w$ event in the Republic of Haiti and much of its capital city of Port-au-Prince collapsed, killing an estimated 230,000 people, seriously injuring another 300,000, and displacing 1.1 million more people. What lies behind this tragedy?

to Islam. These great early leaders lived in stiff, mud-block and stone buildings along one of the world's major strike-slip faults. How would they explain and interpret these powerful earthquakes that destroyed entire cities and killed many thousands of people?

Recent historical and archaeological investigations in the Holy Land have shown that many of the destroyed buildings and cities of the past did not meet their ends by time or humans alone; many fell to earthquakes.

Figure 4.24 Houses collapsed as poorly made concrete floors, support beams, and walls all failed; Port-au-Prince, Haiti, January 2010.

U.S. Air Force photo by Tech. Sgt. James L. Harper Jr.

In 1751, most of the buildings in Port-au-Prince were destroyed in an earthquake. In 1770, another earthquake demolished most of the reconstructed city. In response to this double destruction, the French authorities required that buildings be constructed with wood, and they banned the use of construction relying on concrete (masonry).

Haiti gained independence from France in 1804. During its two centuries of independence, the Haitian population grew to 9.35 million people, most of whom suffer poverty, low rates of literacy, and life in poorly constructed, concrete buildings. The lessons of their past about building construction were forgotten.

In 2008, four tropical storms hit Haiti, killing 800 people, displacing 10% of the population, and reducing economic output by 15%. But for every bad, there is a worse—and life became much worse at 4:53 pm on Tuesday, 12 January 2010, when the powerful earthquake occurred. About 250,000 houses collapsed and another 30,000 commercial buildings fell (figure 4.24). The buildings that collapsed were great and small, rich and poor. Destruction ranged from shacks in shantytowns to the National Palace, the National Assembly building, the United Nations headquarters, the main Catholic cathedral, the upscale Hotel Montana, the Citibank building, schools, hospitals, both fire stations, the main prison, and more.

The death total from this earthquake is more than double that from any previous 7M earthquake anywhere in the world. The deaths were due to bad buildings, all of which were doomed during construction. Building construction was not supervised, allowing bad materials to be used and bad construction practices to be employed. Bad materials include brittle steel; not enough cement; and too much dirty/salty sand in the concrete mixture. Bad construction practices include stopping vertical rods of reinforcement steel at horizontal floors just where they needed to be strongest and to be connected. But the major cause of deaths was widespread use of a concrete-blocks-as-filler style of slab construction. Weak, cheap concrete blocks containing large open spaces were placed in the mold before more expensive concrete was poured around them. These blocks just hang in place, supported only by the concrete around them. There are three major problems with these floor slabs (horizontal): (1) they are weak, (2) they are heavy, and (3) they are supported by flimsy concrete columns (vertical) that undergo shear failure when shaken. When one floor collapses, its added weight causes the underlying floor to collapse, and so on.

Tectonic Setting

The Caribbean tectonic plate is a small one moving eastward below the huge North American plate (see center of figure 2.13). The northern boundary of the Caribbean plate is split between two parallel, east-west trending faults that pass through Haiti with a total slip of 2 m (6.6 ft) per century. The 2010 seism was caused by left-lateral movement with some compression along the southern of these faults, the Enriquillo-Plantain Garden fault.

TURKEY, 1999: SERIAL EARTHQUAKES

A warm and humid evening made sleep difficult, so many people were still up at 3:01 a.m. on 17 August 1999 near the Sea of Marmara in the industrial heartland of Turkey. They were startled by a ball of flame rising out of the sea, a loud explosion, sinking land along the shoreline, and a big wave of water. Another big rupture moved along the North Anatolian fault as a magnitude 7.4 earthquake. This time the fault ruptured the ground surface for 120 km (75 mi), with the south side of the fault moving westward up to 5 m (16.5 ft) (figure 4.25). Several weeks later, after evening prayers for Muslims, another segment of the North Anatolian fault ruptured in a 7.1 magnitude earthquake. The two devastating events combined to kill more than 19,000 people and cause an estimated $20 billion in damages.

Why were so many people killed? Bad buildings collapsed. Industrial growth in the region attracted hordes of new residents who, in turn, caused a boom in housing construction. Unfortunately, many residential buildings were built on top of soft, shaky ground, and some building contractors cut costs by increasing the percentage of sand in their concrete, causing it to crumble as the ground shook.

The North Anatolian fault is not on the Arabian plate, but it is caused by that plate (see figure 4.21). As the Arabian plate pushes farther into Eurasia, Turkey is forced to move westward and slowly rotate counterclockwise in **escape tectonics.** Bounded by the North Anatolian fault in the north and by the east Anatolian fault in the southeast, Turkey is squeezed westward like a watermelon seed from between your fingers.

The North Anatolian fault is a 1,400 km (870 mi) long fault zone made of numerous subparallel faults that split and combine, bend and straighten. A remarkable series of earthquakes began in 1939 near the eastern end of the fault with the magnitude 7.9 Erzincan earthquake, which killed 30,000 people. Since 1939, 11 earthquakes with magnitudes greater than 6.7 have occurred as the fault ruptures westward in a semiregular pattern that is unique in the world (figure 4.25). At intervals ranging from 3 months to 32 years, more than 1,000 km (620 mi) of the fault has moved in big jumps.

What is likely to happen next? There is every reason to expect the fault rupture to keep moving to the west, ever closer to Istanbul. The Sea of Marmara fills a basin partly created by movements along subparallel strands of the North Anatolian fault. The region had big earthquakes in 1063, 1509, and 1766. The next big earthquake will likely occur near Istanbul, a city of 13 million people and about 1 million buildings. A study in 2011 forecast that a 7.25M$_w$ earthquake on the main fault under the Sea of Marmara could kill more than 50,000 people and destroy or severely damage up to 800,000 housing units.

SAN ANDREAS FAULT TECTONICS AND EARTHQUAKES

The plate-tectonic history of western North America explains why earthquakes occur. As the Atlantic Ocean basin widens further, both North and South America move westward into the Pacific Ocean basin, helping reduce its size (see figure 2.13). At 30 million years ago, most of the northern portion of the Farallon plate had subducted eastward beneath North America (figure 4.26). At about 28 million years ago, the first segment of the Pacific spreading center collided with North America at about the site of Los Angeles today. The spreading centers to the north and south still operated as before. What connected the northern and southern spreading centers? A transform fault, specifically the ancestor of the San Andreas fault.

In the last 5.5 million years, the Gulf of California has opened about 300 km (190 mi). This rifting action has torn Baja California and California west of the San Andreas fault (including San Diego, Los Angeles, and Santa Cruz) from the North American plate and piggybacked them onto the Pacific plate (figure 4.26). The Gulf of California continues to open and is carrying the western Californias on a Pacific plate ride at about 56 mm/yr (2.2 in/yr).

The San Andreas fault is part of a complex system of subparallel faults (figure 4.27). The San Andreas fault proper is a 1,200 km (750 mi) long, right-lateral fault. In 1906, the northernmost section of the fault broke loose just offshore of the city of San Francisco, rupturing northward and southward simultaneously (figure 4.28). When it stopped shifting, the ground between Cape Mendocino and San Juan Bautista had been ruptured; this is a distance of 400 km (250 mi). The earthquake had a moment magnitude estimated at 7.8 resulting from 110 seconds of fault movement. When movement stopped, the western side had shifted northward a maximum of 6 m (20 ft) horizontally. In the peninsula south of San Francisco, fault movements have formed elongate topographic low areas now filled by lakes, and some of the land offset by the 1906 movements was smoothed out and built upon (figure 4.29). The amount of fault movement in 1906 died out to zero at the northern and southern ends of the rupture.

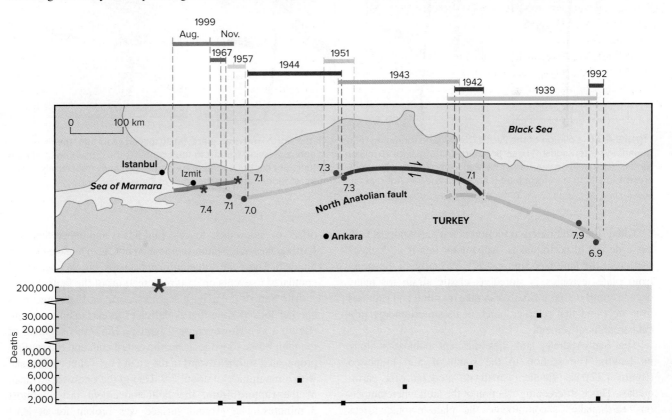

Figure 4.25 The North Anatolian fault accommodates the movement of Turkey westward into the Mediterranean basin. Note the time sequence of the fault ruptures from east to west. What does the near future hold for Istanbul?

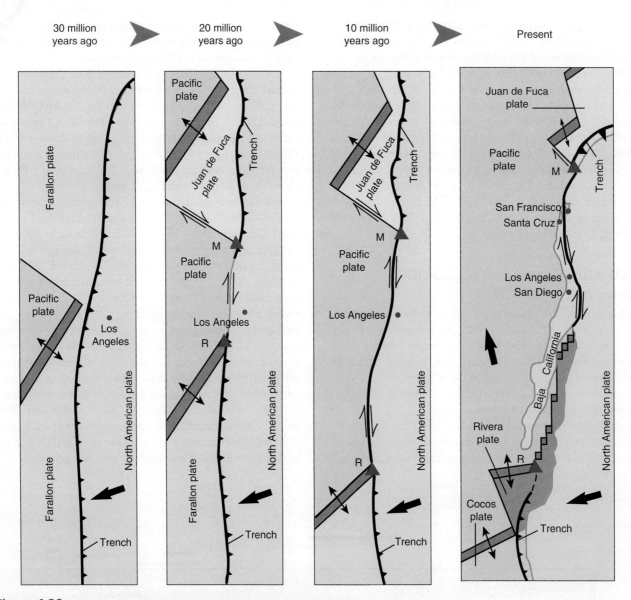

Figure 4.26 Collision of the Pacific Ocean basin spreading center with the North American plate: (a) 30 million years ago, the first spreading-center segment nears Southern California; (b) 20 million years ago, a growing transform fault connects the remaining spreading centers; (c) 10 million years ago, the Mendocino (M) and Rivera (R) triple junctions continue to migrate north and south respectively; (d) at present, the long transform fault is known as the San Andreas fault. (Interpretations based on the work of Tanya Atwater.)

Source: Kious, W. J., and Tilling, R. I., *This Dynamic Earth. US Geological Survey*, p. 77.

Today, the San Francisco section of the San Andreas fault has a deficit of earthquakes. Apparently this is a "locked" section of the fault (see figure 4.27). Virtually all the stress from plate tectonics is stored as elastic strain for many decades until the fault finally can take no more and ruptures in a big event that releases much of its stored energy in a catastrophic movement.

The San Andreas fault has different behaviors along its length. The section to the south of San Francisco (figure 4.27) has frequent small- to moderate-size earthquakes. This is a "creeping" section of the fault where numerous earthquakes accommodate the plate-tectonic forces before they build to high levels. The creeping movements of the fault are shown by the millimeters per year of ongoing offset of sidewalks, fences, buildings, and other features. Earthquakes in this fault segment do not seem to exceed magnitude 6. These are still significant seisms, but they are small compared to events on adjoining sections of the fault.

The San Andreas fault segment north of Los Angeles is another locked zone that is deficient in earthquake activity (figure 4.27). However, on 9 January 1857, this segment of the fault broke loose at its northwestern end, and the rupture propagated southeastward in the great Fort Tejon earthquake with a magnitude of about 7.9. Due to the one-way advance of the rupture front, the fault movement lasted almost 3 minutes. The ground surface was broken for at least 360 km (225 mi), and the maximum offsets in the Carrizo Plain (figure 4.30) were a staggering 9.5 m (31 ft). One of

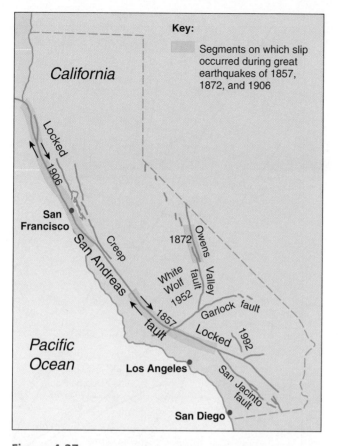

Figure 4.27 Historic behavior of some California faults. The northern "locked" section of the San Andreas fault ruptured for 250 mi in 1906 (magnitude 7.8). The central "creeping" section has frequent smaller earthquakes. The south-central "locked" section ruptured for 225 mi in 1857 (magnitude 7.9). The southernmost San Andreas awaits a major earthquake. The Owens Valley fault ruptured for 70 mi in 1872 (magnitude 7.3). A magnitude 7.5 seism occurred on White Wolf fault in 1952, and a magnitude 7.3 seism happened in the Mojave Desert in 1992.

Source: "The San Andreas Fault," *US Geological Survey.*

Figure 4.28 Looking south-southeast down the San Andreas fault. View is over Bodega Head and Tomales Bay toward the epicenter of the 1906 San Francisco earthquake.

University of Washington Libraries, Special Collections, John Shelton Collection, Shelton 980.

Figure 4.29 Looking southeast along the trace of the San Andreas fault. The San Francisco airport and part of San Francisco Bay are in left center. Linear lakes in right center (e.g., Crystal Springs Reservoir) are in the fault zone. In bottom center, the land offset by the 1906 fault movement was bulldozed and covered with houses!

University of Washington Libraries, Special Collections, John Shelton Collection, Shelton 6742.

Figure 4.30 The San Andreas fault slashes across the Carrizo Plain. Notice the ridges and basins caused by local squeezing and pulling apart.

Courtesy of Pat Abbott.

the offset features was a circular corral for livestock that was split and shifted to an S-shape by the fault movement. In 1857, the region was sparsely settled, so the death and damage totals were small. The next time a great earthquake occurs here, the effects may be disastrous.

The southernmost segment of the San Andreas fault, from San Bernardino to its southern end at the Salton Sea, has not generated a truly large earthquake in California's recorded history. But we can extend our knowledge of earthquakes into the prehistoric past by measuring offsets in sedimentary rock layers. For example, we've learned that the last truly big earthquake on the southern San Andreas fault occurred about the year 1690. These techniques will be discussed in chapter 5.

World Series (Loma Prieta) Earthquake, 1989

In 1989, the World Series of baseball was a Bay Area affair. It pitted the American League champion Oakland Athletics against the National League champion San Francisco Giants. Game 3 was scheduled in San Francisco's Candlestick Park, where the Giants hoped the home field advantage would help them win their first game. It was Tuesday, 17 October, and both teams had finished batting practice, which was watched by 60,000 fans at the park, along with a television crowd of another 60 million fans in the United States and millions more around the world. At 5:04 p.m., 21 minutes before the game was scheduled to start, a distant rumble was heard, and a soft thunder rolled in from the southwest, shaking up the fans and stopping the game from being played. San Francisco was experiencing another big earthquake, and this time, it shared it with television viewers. After the earthquake, the San Franciscans at Candlestick Park broke into a cheer, while many out-of-staters were seen heading for home.

What caused this earthquake? An 83-year-long pushing match between the Pacific and North American plates resulted in a 42 km (26 mi) long rupture within the San Andreas fault system. The southernmost section of the fault zone that moved in 1906 had broken free and moved again. There were several different aspects to the 1989 earthquake: (1) The fault rupture took place at depth; (2) the fault movement did not offset the ground surface; (3) there was significant vertical movement; and (4) the fault rupturing lasted only 7 seconds, an unusually short time for a magnitude 6.9 event.

Movement occurred in a gently left-stepping constraining bend of the San Andreas fault zone (figure 4.31). Long-term compressive pressures along this left step have uplifted the Santa Cruz Mountains. This step in the San Andreas fault is near where the Calaveras and Hayward faults split off and run up the east side of San Francisco Bay. The epicenter of the 1989 seism was near Loma Prieta, the highest peak in the Santa Cruz Mountains. Loma Prieta is the official name of this earthquake; it follows the rule of taking the name

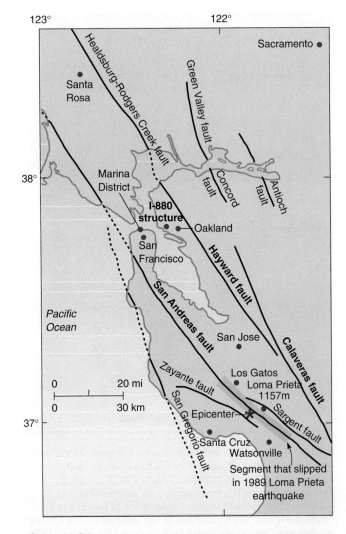

Figure 4.31 Map showing the epicenter of the World Series (Loma Prieta) earthquake. The San Andreas fault takes an 8° to 10° left step in the ruptured section. The left step also is where the Calaveras and Hayward faults split off from the main San Andreas trend.

Source: "Lessons Learned from the Loma Prieta Earthquake of October 17, 1989" in *US Geological Survey Circular 1045*, 1989.

from the most prominent geographic feature near the epicenter. Nonetheless, this event remains known to many people as the World Series earthquake.

It is difficult for a fault to move around a left-stepping bend. Constraining bends commonly "lock up"; thus, movements at a bend tend to be infrequent and large. This left step in the San Andreas zone also causes the fault plane to be inclined 70° to the southwest (figure 4.32). The fault movement began at 18.5 km (11.5 mi) depth and slipped for 2.3 m (7.5 ft). The motion can be resolved into 1.9 m (6.2 ft) of horizontal movement (strike slip) and 1.3 m (4.3 ft) of vertical movement (reverse slip). Stated differently, the western or Pacific plate side moved 6.2 ft to the northwest, and a portion of the Santa Cruz Mountains was uplifted 36 cm (14 in). Although the fault

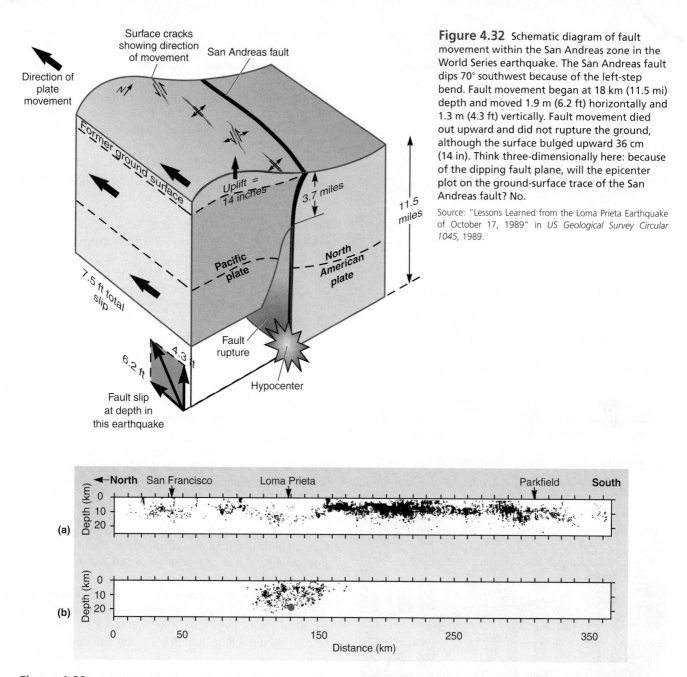

Figure 4.32 Schematic diagram of fault movement within the San Andreas zone in the World Series earthquake. The San Andreas fault dips 70° southwest because of the left-step bend. Fault movement began at 18 km (11.5 mi) depth and moved 1.9 m (6.2 ft) horizontally and 1.3 m (4.3 ft) vertically. Fault movement died out upward and did not rupture the ground, although the surface bulged upward 36 cm (14 in). Think three-dimensionally here: because of the dipping fault plane, will the epicenter plot on the ground-surface trace of the San Andreas fault? No.

Source: "Lessons Learned from the Loma Prieta Earthquake of October 17, 1989" in *US Geological Survey Circular 1045*, 1989.

Figure 4.33 Cross-sections of seismicity along the San Andreas fault, 1969 to early 1989. (a) Notice the dense concentrations of hypocenters in the central creeping section of the fault from south of Loma Prieta to Parkfield, as well as the "seismic gap" in the Loma Prieta area. (b) Notice the deep hypocenter (in red) of the 1989 mainshock plus the numerous aftershocks. Putting the two cross-sections together fills the seismic gap. Are there other seismic gaps in cross-section (a)? Yes, south of San Francisco in the Crystal Springs Reservoir area (see figure 4.29), just west of the densely populated midpeninsula area. When will this seismic gap be filled?

Source: "Lessons Learned from the Loma Prieta Earthquake of October 17, 1989" in *US Geological Survey Circular 1045*, 1989.

did not rupture the surface, the uplifted area was 5 km (3 mi) wide and had numerous fractures in the uplifted and stretched zone. Many of the cracked areas became the sites of landslides.

The mainshock had a surface-wave magnitude (M_s) of 7.1 and a moment magnitude (M_w) of 6.9; numerous aftershocks followed, as is typical for large earthquakes. The Loma Prieta area had been a relatively quiet zone for earthquakes since

the 1906 fault movement (figure 4.33); before 1989, the Loma Prieta region had been a *seismic gap*. As the numerous epicenters in figure 4.33a show, the San Andreas fault section to the south moves frequently, generating numerous small earthquakes. But the same plate-tectonic stresses affecting the creep zone also affect the locked or seismic-gap zone. How does a locked zone catch up with a creep zone? By infrequent but large fault movements. Notice in

Magnified **Magnified**

(a)

(b)

Figure 4.34 (a) Water-saturated sediment usually rests quietly (left). However, when seismic waves shake, sand grains and water can form a slurry and flow as a liquid (right). When earth materials liquefy, building foundations may split and buildings may fail. (b) A typical Marina District building collapse. Three residential stories sat above a soft first story used for car parking; now, the four-story building is three stories tall.

figure 4.33b how the World Series (Loma Prieta) mainshock and aftershocks filled in the seismic gap in cross-section (a). This demonstrates some merit for the seismic-gap method as a forecasting tool. Figure 4.33 also shows another seismic gap, south of San Francisco in the heavily populated midpeninsula area (this is the area of elongate lakes shown in figure 4.29). The 1989 fault movement has increased the odds by another 10% for a large earthquake in the Crystal Springs Reservoir area in the next 30 years.

In the World Series earthquake, the fault ruptured at greater than 2 km/sec in all directions simultaneously, upward for 13 km (8 mi), and both northward and southward for over 20 km (13 mi) each. Table 3.7 indicates that earthquakes with magnitudes of 7 usually rupture for about 20 seconds; this radially spreading, 6.9-magnitude rupture lasted only 7 seconds. Had it lasted the expected 20 seconds, numerous other large buildings and the double-decker Embarcadero Freeway in San Francisco would have failed catastrophically. As it was, the event left 67 people dead or dying, 3,757 injured, and more than 12,000 homeless; caused numerous landslides; disrupted transportation, utilities, and communications; and caused about $6 billion in damages.

Building Damages

In the epicentral region, serious damage was dealt to many older buildings. The short-period P and S waves wreaked their full effects on low buildings built of rigid materials. Common reasons for failure included poor connections of houses to their foundations, buildings made of unreinforced masonry (URM) or brick-facade construction, and two-to-five-story buildings deficient in shear-bearing internal walls and supports. In Santa Cruz, four people died, and the Pacific Garden Mall, the old city center of historic brick and stone buildings that had been preserved and transformed into a tourist mecca, was virtually destroyed.

From the epicentral region, the seismic waves raced outward at more than 3 mi/sec. Some longer-period shear waves remained potent even after traveling 100 km (more than 60 mi). Upon reaching the soft muds and artificial-fill foundations around San Francisco Bay, these seismic waves had their vibrations amplified. Ground motion at some of these soft-sediment foundation sites was 10 times stronger than at nearby sites on rock.

Marina District

The Marina District is one of the most beautiful areas in San Francisco. It sits on the northern shore of the city next to parks, the Golden Gate Bridge, and the bay itself. In this desirable and expensive district, five residents died, building collapses were extensive, and numerous building-eating fires broke out due to: (1) amplified shaking, (2) deformation and liquefaction of artificial-fill foundations, and (3) soft first-story construction, which led to building collapses (figure 4.34).

Much of the Marina District is built on artificial fill dumped onto the wetlands of the bay to create more land

Figure 4.35 The Cypress double-decker section of Interstate 880 in Oakland was completed in 1957. It failed in the 1989 seism and dropped 1.25 mi of upper roadbed onto the lower roadbed, crushing many vehicles and people.

Courtesy of AECOM.

for development. Ironically, much of the artificial fill was the debris from the San Francisco buildings ruined by the 1906 earthquake. Seismic waves in 1989 were amplified in this artificial fill. Some fill underwent permanent deformation and settling, and some formed **slurries** as underground water and loose sediment flowed as fluids in the process of **liquefaction** (figure 4.34a). Liquefaction in the Marina District in 1989 brought to the surface pieces of glass, tar paper, redwood, and other debris from 1906 San Francisco.

The central cause of building failure was flawed design. Because the Marina District is home to many affluent people, they need places to park their cars. But where? The streets are already overcrowded, and basement parking garages would be below sea level and thus flooded. A common solution has been to clear obstructions from the first stories of buildings to make space for car parking. That means removing the internal walls, lateral supports, and bracing needed to support the upper one to four stories. This creates a "soft" first story, so that in an earthquake, buildings simply

Figure 4.37 The support columns of the Interstate 880 structure failed at the joints. There were 20 #18 bars of steel in each column, but they were discontinuous at the joints and failed there.

Courtesy of AECOM.

pancake and become one story shorter (figure 4.34b). It is estimated that there are 2,800 blocks of soft first-story residences in San Francisco today and another 1,500 blocks in Oakland.

Interstate 880

The most stunning tragedy associated with the World Series earthquake was the crushing of 42 people during the collapse of a double-decker portion of Interstate 880 in Oakland (figure 4.35). The elevated roadway was designed in 1951 and completed in 1957. A 2 km (1.25 mi) long section

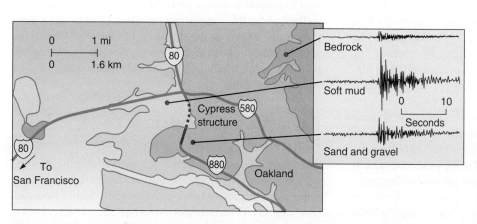

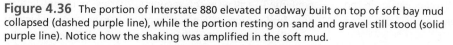

Figure 4.36 The portion of Interstate 880 elevated roadway built on top of soft bay mud collapsed (dashed purple line), while the portion resting on sand and gravel still stood (solid purple line). Notice how the shaking was amplified in the soft mud.

A Classic Disaster

The San Francisco Earthquake of 1906

Early in the 20th century, San Francisco was home to about 400,000 people who enjoyed a cosmopolitan city that had grown during the economic boom times of the late 19th century. During the evening of 17 April 1906, many thrilled to the special appearance of Enrico Caruso, the world's greatest tenor, singing with the Metropolitan Opera Company in Bizet's *Carmen*. But several hours later, at 5:12 a.m., the initial shock waves of a mammoth earthquake arrived to begin the destruction of the city. One early riser told of seeing the earthquake approach as the street before him literally rose and fell like a series of ocean swells moving toward shore.

During a noisy minute, the violently pitching Earth emitted dull booming sounds joined by the crash of human-made structures. When the ground finally quieted, people went outside and gazed through a great cloud of dust to view the destruction. Unreinforced masonry buildings lay collapsed in heaps, but steel-frame buildings and wooden structures fared much better. Another factor in the building failures was the nature of the ground they were built on. Destruction was immense in those parts of the city that were built on artificial fill that had been dumped onto former bay wetlands or into stream-carved ravines.

As repeated aftershocks startled and frightened the survivors, another great danger began to grow. Smoke arose from many sites as fires fed on the wood-filled rubble. Unfortunately, the same earthquake waves that wracked the buildings also broke most of the water lines, thus hindering attempts to stop the growing fires. From the business district and near the waterfront, fires began their relentless intrusion into the rest of the city. Desperate people tried dynamiting buildings to stop the fire's spread, but they only provided more rubble to feed the flames or even blew flaming debris as far as a block away, where it started more fires.

The fires did about 10 times as much damage as the earthquake itself; fire destroyed buildings covering 490 city blocks. More than half the population lost their homes. Death and destruction were concentrated in San Francisco, where 315 people died, but the affected area was much larger. About 700 deaths occurred in a 430 km (265 mi) long belt of land running near the San Andreas fault. Towns within the high-intensity zone, such as San Jose and Santa Rosa, were heavily damaged, yet other cities to the east of the narrow zone, such as Berkeley and Sacramento, were spared significant damage. Problems continued in the months that followed as epidemics of filth-borne diseases sickened Californians; more than 150 cases of bubonic plague were reported. When all the fatalities from earthquake injuries and disease are included, the death total from the earthquake may have been as high as 5,000.

Total financial losses in the event were almost 2% of the US gross national product in 1906; for comparison, Hurricane Katrina economic losses were much less than 1%. Politicians and the press in their desire to restore the city called the disaster a fire-related event and listed the death total at about 10% of actual life loss. In their desire to rebuild, the emphasis was on quickness, not on increasing safety. This problem haunts us today because much of the early rebuilding was done badly and is likely to fail in the next big earthquake.

One of the intriguing aspects of disasters is their energizing effects on many survivors. Hard times shared with others bring out the best in many people. Shortly after this earthquake, the resilient San Franciscans were planning the Panama-Pacific International Exposition that was to impress the world and leave behind many of the beautiful buildings that tourists flock to see today. You can't keep a good city down.

---

collapsed: 44 slabs of concrete roadbed, each weighing 600 tons, fell onto the lower roadbed and crushed some vehicles to less than 30 cm (1 ft) high. The section that collapsed was built on young, soft San Francisco Bay mud. The elevated freeway structure had a natural resonance of two to four cycles per second; the bay-mud foundation produced a five to eight-fold amplification of shaking in that range. The seismic waves excited the mud (figure 4.36), causing the heavy structure to sway sharply. The portion of I-880 elevated roadway built on firmer sand and gravel stood intact; the portion standing on soft mud collapsed catastrophically.

The weak foundation was compounded by a flawed structural design. The joints where roadbeds (horizontal) were connected to concrete support columns (vertical) were not reinforced properly. Cracks initiated at the joints caused failure of supporting columns, which slid off the crushed areas of the joints and dropped the upper roadbed onto the lower level (figure 4.37). Was this bridge failure a surprise? Not really. The lessons had been learned 18 years earlier in the 1971 San Fernando earthquake, but no one had corrected this disaster-in-waiting.

An ironic and deadly footnote to this disaster lay in the mode of failure. There was a delay between the initial shock and the final collapse, which allowed some people a brief time to plan. Some maneuvered their vehicles under beams next to support columns, and others got out of their cars and walked under the same supports, thinking that these were the strongest parts of the structure; but the steel bars in the support columns were discontinuous. Tragically, these were the weak spots, where failure was most catastrophic, and no one survived there.

BAY AREA EARTHQUAKES— PAST AND FUTURE

The historic record of California earthquakes is accurate only back to about 1850, and thus is shorter than the recurrence times for major movements on most faults. Nonetheless, the San Francisco Bay Area has enough information contained in newspaper accounts, diaries, personal letters, and similar sources to piece together a fairly accurate history of 19th-century earthquakes, and it is quite different from the

20th-century record. During the 19th century, earthquakes with magnitudes greater than 6 were much more common (figure 4.38). There were seven destructive seisms in the 70 years before the 1906 San Francisco earthquake, averaging a large earthquake every decade. Then came the monstrous movement of the San Andreas fault in 1906. This 250-mile-long rupture removed so much of the plate-tectonic stress stored in the rocks that several decades of the 20th century were effectively free of large earthquakes (figure 4.39). But large earthquakes returned to the southern part of the Bay Area beginning in the 1970s (figure 4.38). We can identify three patterns in these data.

Pattern 1

Common Large Earthquakes versus Rare Giant Shakes The movement of the Pacific plate past the North American plate in the Bay Area seems to be satisfied by either a magnitude 6 to 7 earthquake roughly every decade (19th century) or a magnitude 8 earthquake every century (20th century). Which pattern is preferable for this heavily developed and populated region (not that we have any choice)? Which pattern causes the least amount of death, damage, and psychological distress? Will the 21st century be like the 19th or the 20th?

Pattern 2

Pairings of Earthquakes In 1836, the Monterey Bay area experienced a quake of about magnitude 6.5; this was followed two years later on the San Francisco Peninsula with a seism of about magnitude 6.8 (see figure 4.38). In the southern Bay Area, a large shallow earthquake near Santa Cruz in 1865 was followed three years later by a shake of about magnitude 6.9 near Hayward. Will this pattern of paired earthquakes reoccur?

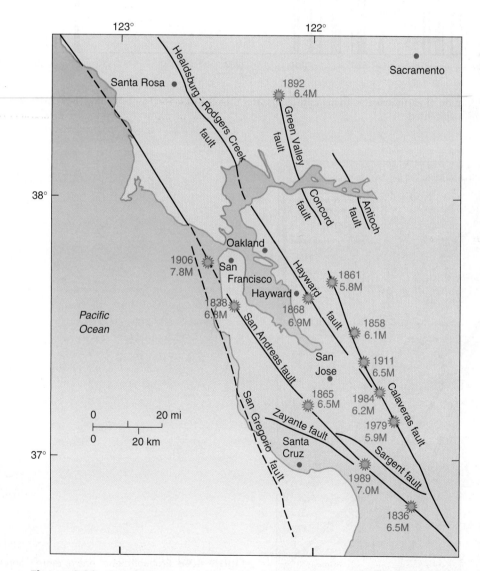

Figure 4.38 Locations and approximate sizes of some larger Bay Area earthquakes.
Source: *US Geological Survey.*

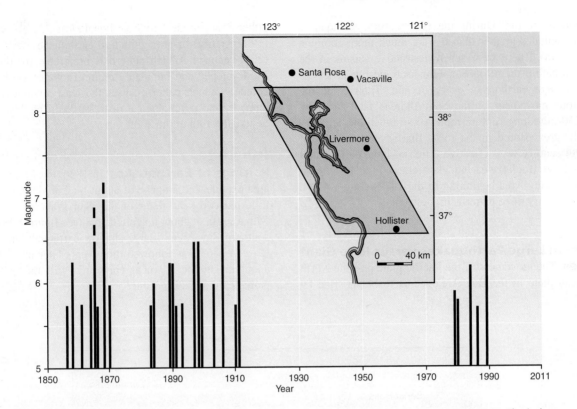

Figure 4.39 Distribution of earthquakes with magnitudes greater than 5.5 near San Francisco Bay, 1849–2011. The index map shows the area of earthquake epicenters.

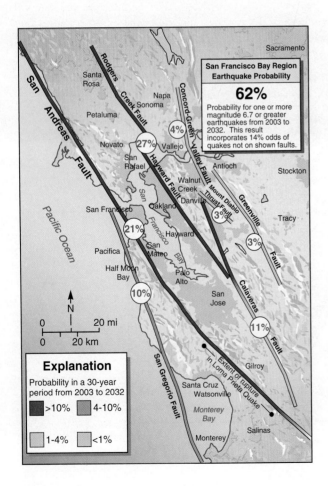

Figure 4.40 Probabilities of one or more magnitude 6.7 or larger earthquakes in the San Francisco Bay region, 2003–2032.

Source: Redrawn from Working Group on California Earthquake Probabilities, 2003.

Pattern 3

Northward Progression of Earthquakes The large earthquakes of 1865 and 1868 were preceded by five moderate earthquakes that moved northward up the Calaveras fault. Figure 4.38 shows moderate to large earthquakes that have moved from south to north up the Calaveras fault. Does this repeat pattern suggest an upcoming large seism on the Hayward fault? This region today is populated by more than 2 million people in 10 cities. The fault is covered by schools, hospitals, city halls, houses, and the University of California. If a seism like the 1868 earthquake occurs soon, the California Division of Mines and Geology estimates that up to 7,000 people might die. The number of deaths will depend in part on the time of day the earthquake occurs. When is the worst time for an earthquake? During the middle of the work and school day, when the maximum number of people are occupying the larger, older structures. When is the best time for an earthquake? During the night, when most people are home and asleep in their beds. In general, California houses handle earthquake shaking quite well because their wood frameworks are flexible and well tied together with nails, bolts, and braces.

Bay Area Earthquake Probabilities

For the combined San Francisco Bay region and its 6.8 million people, there is a 62% (+/−10%) chance that a 6.7 magnitude (Northridge-size) earthquake will occur on a fault crossing through the urban area before 2032 (figure 4.40).

The probability of the Hayward fault causing a magnitude 6.7 or greater earthquake before 2032 is estimated at 27%. The Hayward fault is expected to rupture for about 22 seconds with about 2 m (6 ft) of slip extending down about 13 km (8 mi). The next movement of the Hayward fault will cause tens of billions of dollars in property losses, and deaths may total in the thousands.

Summary

Most large earthquakes are caused by fault movements associated with tectonic plates. Plates have three types of moving edges: (1) divergent at spreading centers, (2) slide-past at transform faults, and (3) convergent at collision zones. The tensional (pull-apart) movements at spreading centers do not produce very large earthquakes. The dominantly horizontal (slide-past) movements at transform faults produce large earthquakes. The compressional movements at subduction zones and continent-continent collisions generate the largest tectonic earthquakes, and they affect the widest areas.

Subduction zones produce the largest number of great earthquakes. In 1923, a subduction movement of the Pacific plate destroyed nearly all of Tokyo and Yokohama; much of the devastation was caused by fires unleashed during building collapses. The largest earthquakes along western North America are due to subduction beneath the continent. The magnitude 9.2 Alaska earthquake in 1964 and magnitude 9.0 Japan earthquake in 2011 were due to subduction of the Pacific plate, and the magnitude 8.1 Mexico City event in 1985 was caused by subduction of the Cocos plate. The plates subducting beneath Oregon, Washington, and British Columbia generated a magnitude 9 earthquake on 26 January 1700 and will do so again in the future. In 2004, the Sumatra, Indonesia, subduction earthquake and tsunami killed more than 245,000 people.

Continent-continent collisions produce great earthquakes throughout Asia and Asia Minor. The 2005 Kashmir, Pakistan, earthquake killed 88,000; the 2008 Sichuan, China, earthquake killed 87,500; and the 2001 Gujarat, India, quake killed 20,000. The earthquakes did not kill directly; it was the collapse of human-built structures that was deadly. The deadliest earthquake in history occurred in 1556 in Shaanxi Province, China, when the loose, silty sediment into which cave homes had been dug collapsed and flowed, killing 830,000 people.

Deaths from earthquakes are mostly due to building failures. For example, for thousands of years, humans have built stone and mud-block houses along the Dead Sea fault zone (a major transform fault), and for thousands of years, these rigid houses have collapsed during earthquakes, causing many deaths. These geologic disasters have affected the teachings of Judaism, Christianity, and Islam.

The frequent earthquakes of western North America are mostly due to plate tectonics. The westerly moving North American plate has overrun the Pacific Ocean spreading center along most of California. To the south, ongoing spreading has torn Baja California from mainland Mexico, and Baja California, San Diego, Los Angeles, and Santa Cruz are now riding on the Pacific plate toward Alaska at 5.6 cm/yr. To the north, spreading still occurs offshore from northernmost California, Oregon, Washington, and southern British Columbia. Two separated spreading centers are connected by a long transform fault—the San Andreas fault. Its earthquakes include a magnitude 7.9 caused by a 225 mi long rupture in central California in 1857, a magnitude 7.8 due to a 250 mi long rupture passing through the San Francisco Bay region in 1906, and a magnitude 6.9 unleashed by a 25 mi long rupture near Santa Cruz in 1989.

Major losses of life and property damage are commonly due to problems with buildings. In San Francisco in 1906, unreinforced-masonry buildings collapsed, especially those built on artificial-fill foundations. Much damage was done after the earthquake by fires that raged unchecked because ground shaking had broken water pipes, rendering firefighters largely helpless. In Mexico City in 1985, 1- to 2-second-period shear waves caused shaking of 6- to 16-story buildings at the same 1- to 2-second frequency, and shaking was amplified in muddy, former lake-bottom sediments. The resonance of seismic waves and tall buildings, amplified by soft sediment foundations, caused numerous catastrophic failures 220 miles away from the epicenter.

Earthquake numbers and sizes have varied in the San Francisco Bay region. In the 19th century, magnitude 6.5 to 7 events occurred at an average of one per decade; the 20th century was dominated by the magnitude 7.8 event in 1906.

Southern California may have several large earthquakes in the 21st century. The southern segment of the San Andreas fault is the only one not to have a long rupture in historic time. In prehistory, it has ruptured every 250 years on average, but the last big movement was in 1690.

Terms to Remember

CE 86
cohesion 91
earthquake cluster 85
escape tectonics 94
fault 78
global positioning
   system (GPS) 84

liquefaction 101
rift 80
seismic-gap method 84
slurry 101
triple junction 80

Questions for Review

1. Draw a map of an idealized tectonic plate and explain the earthquake hazards along each type of plate edge.
2. Sketch a map of the Arabian plate and explain the origin of the Iranian, Holy Land, and Turkey earthquakes.
3. Explain why earthquakes at subduction zones are many times more powerful than spreading-center earthquakes.
4. Explain the seismic-gap method of forecasting earthquakes.
5. Which tectonic-plate edges fail most commonly in shear? In tension? In compression?
6. Why are fires in cities so commonly associated with major earthquakes?
7. Sketch a plate-tectonic map along western North America from Alaska through Mexico. Label the spreading centers, subduction zones, and transform faults. Label the maximum earthquakes expected along the coastal zones.
8. If present seafloor spreading trends continue, what will happen to Baja California, San Diego, Los Angeles, and Santa Cruz?
9. Which part of the United States sits in an opening ocean basin? Evaluate the earthquake threat there.
10. How long were the surface ruptures in the 1906 San Francisco and 1857 Fort Tejon earthquakes? What was the maximum offset of the ground surface during each quake?
11. Evaluate the earthquake hazards in locked versus creeping segments of the San Andreas fault. Are the biggest cities in locked or creeping segments?
12. Evaluate the seismic gap in the San Andreas fault south of San Francisco.
13. What factors combined to cause the resonance in Mexico City that was so deadly in the 1985 earthquake? How far was the city from the epicenter?
14. Sketch a Marina District (San Francisco) dwelling and explain why so many failed during the 1989 Loma Prieta earthquake.
15. What is usually the worst time of day for a big earthquake to strike a city in the western United States?
16. What are the four stages of formation of an ocean basin?
17. Can one large earthquake trigger others? What is the recent experience in Indonesia?
18. In the 2011 Japan earthquake, how large an area of plate moved? What was the maximum slip? What was the earthquake magnitude? When did the last earthquake of this size occur in the same area?
19. What is the largest earthquake measured (see Chile)?
20. Why do so many mega-killer earthquakes occur in the China, India, Pakistan region?
21. The 2010 Haiti and 1989 Loma Prieta (World Series) earthquakes had similar magnitudes. Why were 3,400 times more people killed in the Haiti earthquake?
22. Can we recognize that an earthquake is a foreshock before the mainshock occurs?
23. Sketch a sequence of cross-sections that shows how a continent is split, then separated to form an ocean basin.

Questions for Further Thought

1. How might people with no geologic knowledge, living in stone houses next to a major fault, explain a disastrous earthquake?
2. How might you use food to create a plate-tectonics model in your kitchen?
3. Which U.S. states are on the Pacific tectonic plate?
4. Which would be the better of two bad choices for an urban area: a magnitude 6.5 to 7 earthquake every 15 years or a magnitude 8 every century?
5. Why is the zone of active faults so much wider in southern than in northern California?
6. If a magnitude 9 earthquake occurred in the Cascadia subduction zone offshore from the Pacific Northwest, what might happen in Vancouver, Seattle, Portland, and other onshore sites?
7. On 19 September 1985, Mexico City was rocked by a magnitude 8.1 earthquake. Two days later, the city was shaken by a magnitude 7.5 earthquake. Would you consider this a mainshock and an aftershock or twin earthquakes?
8. Is East Africa likely to pull away from the rest of Africa to form a Somali plate?

CHAPTER 5

Earthquakes Throughout the United States and Canada

Eventually, everything east of the San Andreas fault will break off and fall into the Atlantic Ocean.

—MICHAEL GRANT, 1982, SAN DIEGO UNION

Highway 287 in Montana was destroyed by the Hebgen Lake earthquake on 17 August 1959.

R.B. Colton/U.S. Geological Survey.

LEARNING OUTCOMES

Earthquakes occur in many places not related to plate tectonics. After studying this chapter, you should:

- recognize the complexities of fault movements.
- realize the lack of connection between earthquakes and weather.
- know how the dates and magnitudes of prehistoric earthquakes can be determined.
- understand our inability to make short-term predictions of earthquakes.
- know the ways that humans trigger earthquakes.
- realize that earthquakes occur in every state.
- be familiar with the relationship between volcanism and earthquakes.

OUTLINE

At 1:51 p.m. on Tuesday, 23 August 2011, a 5.8M$_w$ earthquake near the town of Mineral in Virginia shook eastern North America and rattled the nerves of millions of people. Office workers in the Empire State Building ran down dozens of flights of stairs and poured out into the streets. Air traffic control towers were evacuated at busy eastern U.S. airports, fouling the travel plans of thousands of people. Cell phone service was overwhelmed. The earthquake was felt south to Atlanta; north to Montreal and New Brunswick Province, Canada; and west to Detroit and Chicago. Damages totaled about $300 million and were suffered as far as Washington, D.C., where the Washington Monument and the National Cathedral both cracked, and in Brooklyn, New York. And yet, there were no deaths or serious injuries. Easterners got a taste of what westerners frequently experience.

Because seismic waves move slower than Internet traffic, some Twitter users in New York City and Boston read about the earthquake before they felt it. The citizen-based earthquake intensity website Did You Feel It? received more than 100,000 reports within four hours.

How Faults Work

As our instrumentation and field equipment improve, we get better understanding of how faults work.

ELASTIC REBOUND

The popular explanation of how faults move has been the elastic-rebound theory developed after the 1906 San Francisco earthquake. Based on surveyor's measurements of ground along the San Andreas fault, it appears that Earth stresses cause deformation and movement on both sides of a fault (figure 5.1a, b). However, the rocks along the fault itself do not move in response to this stress because they are rough and irregular, resulting in strong interlocking bonds with **friction** that retards movement. But as the land-masses away from the fault continue to move, energy builds up and is stored as elastic strain in the rocks. When the applied stresses become overpowering, the rocks at the fault rupture, and both sides quickly move forward to catch up, and even pass, the rocks away from the fault (figure 5.1c). After a fault movement, all the elastic strain is removed from the area, and the buildup begins anew. The elastic-rebound theory is somewhat analogous to snapping a rubberband or twanging a guitar string; it even accounts for aftershocks. This idea has held sway for more than 100 years and is described in most textbooks. It still works as a first approximation to reality, but a better understanding has emerged in recent years.

NEWER VIEW

Movements along a fault may be better visualized as windows of opportunity. Fault movement begins at a hypocenter

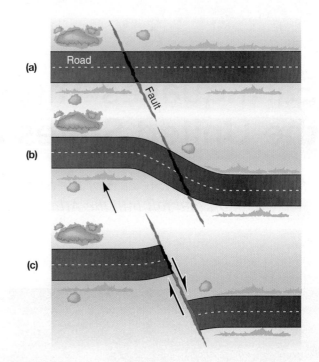

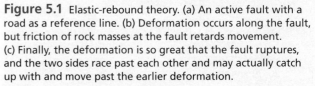

Figure 5.1 Elastic-rebound theory. (a) An active fault with a road as a reference line. (b) Deformation occurs along the fault, but friction of rock masses at the fault retards movement. (c) Finally, the deformation is so great that the fault ruptures, and the two sides race past each other and may actually catch up with and move past the earlier deformation.

and then propagates outward for a certain distance and length of time. How much of the stored energy is released during an earthquake depends on the number of seconds the fault moves. For example, if there were 12 m (40 ft) of unreleased movement along a section of fault and the rupture event, passing by from front to end, lasted long enough for only 6 m (20 ft) of movement, then only half of the energy would have been released. An analogous event might be opening a locked gate to a long line of people. If the gate is held open only long enough for half the people to enter and is then closed and locked, the other people will simply have to wait until the next time the gate opens. This is an important modification of elastic-rebound theory. The elastic-rebound theory has said that after a big earthquake, most of the elastic strain is removed from the rocks, and considerable time will be required for it to build again to a high enough level to create another big earthquake. We no longer think this is true.

Another way to visualize how faults move is to imagine rolling out a large carpet to cover an auditorium floor. Suppose that the carpet misses covering the floor to the far wall by a foot. You can't pull the rug the rest of the way to the wall; it won't move because the friction is simply too great. However, if you create a large ripple in the carpet and push the ripple across the auditorium floor, the carpet can be moved. Faults may act the same way. A small portion of a fault may slip, creating a ripple that concentrates elastic energy at its leading edge. The farther the ripple travels, the bigger the earthquake. The moving ripple may encounter different amounts of unreleased energy in different areas of the fault.

Landers, California, 1992

New insight on how faults work was provided by the Landers area earthquakes in 1992 and 1999. This earthquake sequence began on 22 April 1992 with the right-lateral movement of the magnitude 6.1 Joshua Tree earthquake (figure 5.2). Right-lateral movements along the fault trend resumed two months later at 4:58 a.m. on 28 June with the magnitude 7.3 Landers earthquake.

A third earthquake, triggered by the first two, broke loose a few hours later. At 8:04 a.m. on 28 June, the magnitude 6.3 Big Bear earthquake came from a left-lateral movement that ruptured northeast toward the center of the Landers ground rupture. The ruptures of the 28 June earthquakes form a triangle, with the San Andreas fault as the base (figure 5.2). These fault movements have acted to pull a triangle of crust away from the San Andreas fault, thus reducing the pressures that hold the fault together and keep it from slipping.

Activity continued along this trend on 16 October 1999 with the right-lateral movement of the Hector Mine earthquake in a magnitude 7.1 event (figure 5.2). Is this sequence of earthquakes finished? Probably not.

Examining the Landers earthquake records to see what happened during the 24 seconds that the faults moved 70 km (43 mi) can teach us a lot about *how faults move:*

1. Fault movements commonly are viewed as being restricted to one fault, with the rupture front stopping at large bends or steps in the fault, thus ending the earthquake. The Landers earthquake was different. It began right-lateral movement on the Johnson Valley fault and traveled northward about 20 km (12 mi) until reaching a right-step, pull-apart zone. The rupture front slowed, but it moved through the step and continued moving northward on successive faults for another 50 km (30 mi) until finally stopping within a straight segment of the Camp Rock fault (figure 5.3).

2. Rupture velocity on the Johnson Valley fault was 3.6 km/sec (8,000 mph), slowing almost to a stop in the right step and then continuing northward at varying speeds.

3. The amount of slip on the faults varied from centimeters to 6.3 m (21 ft) along the fault lengths and below the ground. Figure 5.4 shows the movements calculated by seismologists Dave Wald and Tom Heaton. Look at their cross-section and visualize the fault in movement as the rupture front snaked its way northward, up, down, and not always involving all the fault surface.

4. Notice how the amount of fault movement at the ground surface differs from that at depth (figure 5.4).

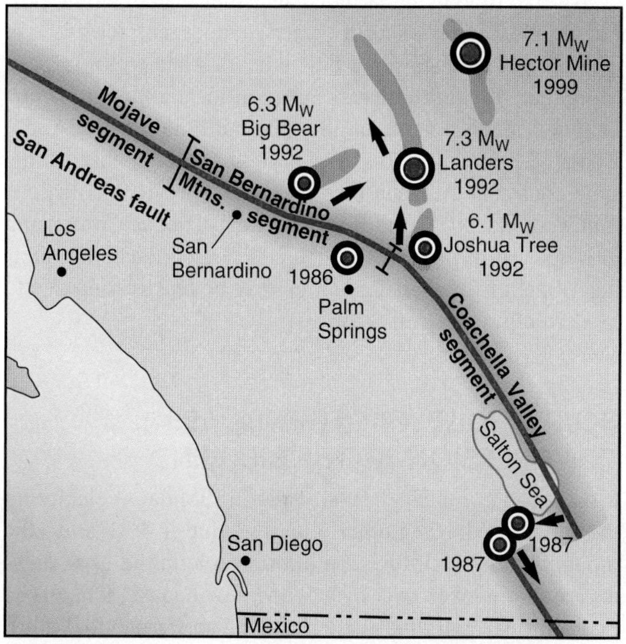

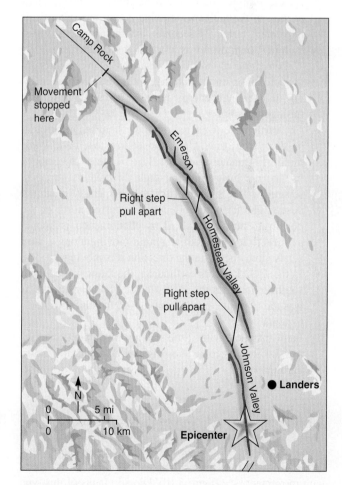

Figure 5.3 Northward-rupturing faults in the 1992 Landers earthquake. The rupture front slowed at right steps, and then moved onto adjacent faults before stopping in the middle of a straight segment.

Figure 5.2 Map of major earthquakes near the northern and southern ends of the Coachella Valley segment of the San Andreas fault. The triangular block of crust near the northern end has moved northward.

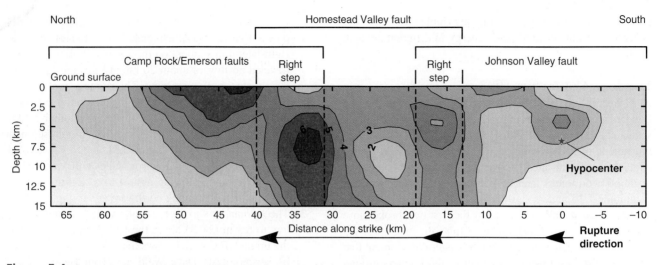

Figure 5.4 Slip on faults varied from centimeters to 6.3 m (21 ft) during movements of the 1992 Landers earthquake. The contour interval of slip areas is 1 m.

Source: Wald and Heaton in *Seismological Society of America Bulletin* 84:668–691, 1994.

5. Fault movement, and shaking, at the hypocenter was modest compared to what came later.

6. As the rupture front moved northward, only a small portion of the fault was slipping at any one time. Fault movement lasted 24 seconds, but the longest any one fault portion moved was less than 4 seconds.

7. Although the rupture front was slowed in a right step, the amount of slip behind the rupture front kept increasing until enough energy built up to cause movement through the step.

8. The earthquake triggered other earthquakes in Nevada, northern California, Utah, and Yellowstone Park, Wyoming. Fortunately for the Los Angeles megalopolis, the northward-moving fault directed its strongest seismic waves to the north into the sparsely inhabited desert. The triggered effects all occurred north of the northward-moving fault. This phenomenon is known as **directivity,** wherein a rupture moving along a fault sends more energy in the direction it is moving.

9. In the 1992 Landers earthquake, the faults moved from south to north; in the 1999 Hector Mine earthquake, it was the opposite, as the fault moved mostly from north to south.

10. Fault patches with little or no movement on figure 5.4 may become the origination points for future earthquakes.

Thrust-Fault Earthquakes

Some damaging earthquakes result when compressional forces push one rock mass up and over another in a reverse-fault movement (see figure 3.10). Dip-slip faults of this type are also known as **thrust faults,** especially when the fault surface is inclined at a shallow angle. Many of these thrust faults do not reach the ground surface; they are called **blind thrusts.**

VIRGINIA, 2011: ANCIENT FAULTS CAN REACTIVATE

The 23 August 2011 5.8 M_w event in Virginia occurred as a reverse-fault movement along a north-northeast striking fault about 10 km (6 mi) long. The hypocenter was only 6 km (3.7 mi) deep, which allowed seismic waves to reach the surface largely unweakened. The earthquake shook loose within the Central Virginia Seismic Zone, which has the same length and width, about 120 km (75 mi). Small earthquakes are frequent in this zone. Moderate-size events include a 4.5 M_w event on 9 December 2003 and a ~4.8 M in 1875.

We like to explain earthquakes using plate-tectonic processes. But eastern North America does not have active tectonic-plate edges now, but it has a plate-tectonic past. More than 480 million years ago, continent collision, which included much reverse faulting, began building the Appalachian Mountains. This was part of the assembly of the supercontinent Pangaea (see figure 2.24). More collisions followed, but by 220 million years ago Pangaea was being torn apart in a process that included much tensional faulting. Some of these ancient faults may be reactivating due to modern regional stresses.

NORTHRIDGE, CALIFORNIA, 1994: COMPRESSION AT THE BIG BEND

Monday, 17 January 1994, was a holiday celebrating the birth of Martin Luther King, Jr. But at 4:31 a.m., the thoughts of most of the 12 million people in the Los Angeles area were taken over by a 6.7 M_w earthquake. One of the many thrust faults that underlie the San Fernando Valley, the Pico blind thrust, ruptured at 19 km (11.8 mi) depth and moved 3.5 m (11.5 ft) northward as it pushed up the south-dipping fault surface (figure 5.5). Northridge and other cities setting on the upward-moving fault slab (hangingwall)

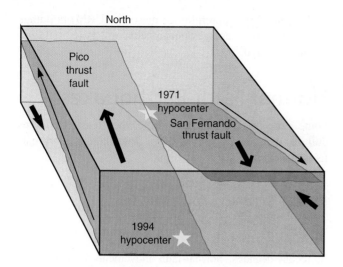

Figure 5.5 Block diagram of thrust-fault movements that created the 1994 Northridge and 1971 San Fernando earthquakes. In 1994, the Pico blind thrust fault moved 3.5 m (11.5 ft) *up to the north* from a 19 km (11.8 mi) deep hypocenter. The cities "riding piggyback" on the upward-moving thrust plate experienced intense ground shaking. In 1971, a block moved *up to the south* on the San Fernando thrust fault from a 15 km (9.3 mi) deep hypocenter.

Figure 5.6 The Kaiser-Permanente Hospital in Granada Hills collapsed during the 17 January 1994 earthquake. It was an older, non-ductile concrete-frame building.

NOAA/NGDC, M. Celebi, U.S. Geological Survey.

were subjected to some of the most intense ground shaking ever recorded. Ground acceleration was as high as 1.8 g (180% of gravity) horizontally and 1.2 g vertically. (At 1.0 g vertical acceleration, unattached objects on the ground are thrown up into the air.) This intense shaking caused the widespread failure of buildings (figure 5.6), parking garages (see figure 3.33), and bridges that killed 57 people, injured 9,000 more, and caused $40 billion in damages. The damages included the disabling of the world's busiest freeway system, creating months of problems for drivers (see figures 3.30 and 3.36). In terms of deaths, this earthquake can be viewed as a near miss due to its early morning occurrence.

Figure 5.7 Active faults in southern California are shown in yellow. The southern end of the San Andreas fault is on the east side of the Salton Sea (lower right). Follow the San Andreas fault up and to the left where it exits the photo in the upper left corner. The photo shows the San Andreas in its "Big Bend" that Southern California pushes against, creating mountains and thrust-fault earthquakes.

Jet Propulsion Laboratory/NASA.

Analysis of the failed buildings indicates that an estimated 3,000 people would have died if the seism had occurred during working hours.

The 1994 Northridge event was similar to the 1971 San Fernando earthquake, which had a magnitude of 6.6 and killed 67 people (see chapter 3). In 1971, the movement was up a north-dipping thrust fault that abuts the blind thrust that moved in 1994 (figure 5.5). In 1971, the energy was directed *toward* the city of Los Angeles, but in 1994, the energy was directed *away* from the city.

Southern California pushes northward against the "Big Bend" of the San Andreas fault, creating thrust faults that are mostly east-west oriented (figure 5.7). Satellite measurements of ground movement using the global positioning system (GPS) tell us that the Los Angeles region is experiencing a compressive shortening of 10 to 15 mm/yr. The measured deformation could generate an earthquake with a magnitude in the mid-6s every six years, plus a seism of magnitude 7 every 10 years.

The 1971 and 1994 earthquakes may be omens for a more earthquake-active 21st century. The death and destruction from numerous magnitude 6.5 to 7 earthquakes on thrust faults within the city of Los Angeles would exceed the problems caused by a magnitude 8 event on the San Andreas fault some 50 to 100 km away.

SEATTLE, WASHINGTON

The Seattle fault zone is oriented east-west and runs along the south side of Interstate 90 through the city of Seattle (figure 5.8). The fault zone is 4 to 6 km (2.5 to 3.7 mi) wide and has three or more south-dipping reverse faults. A major fault

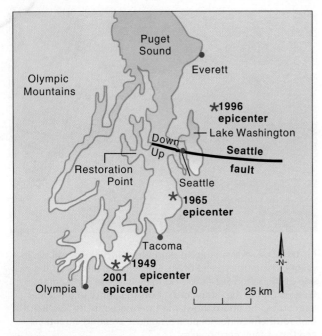

Figure 5.8 Map of the Puget Sound area. *Up* and *down* refer to movements on the Seattle fault.

movement occurred there about 1,100 years ago, as indicated by the following evidence: (1) The former shoreline at Restoration Point was uplifted about 7 m (23 ft) above the high-tide line in a single fault movement. This earthquake appears to have had a magnitude of around 7, about the size of the 1989 World Series event in the San Francisco Bay area. (2) Numerous large landslides occurred at this time, including some that carried trees in upright growth position to the bottom of Lake Washington. The age of these trees was determined by carbon-14 dating. (3) Several tsunami deposits have been recognized in the sediment layers of the area. Logs and trunks carried or buried by these large waves date to the same time period. (4) The same date appears in the ages of six major rock **avalanches** in the Olympic Mountains. The avalanches apparently were shaken into action by the earthquake. (5) Coarse sediment layers on the bottom of Lake Washington were formed by downslope movement and redeposition of sediment in deeper waters. These distinctive deposits appear to have been caused by the same earthquake.

Part of Seattle sits on a 10 km (6+ mi) deep basin filled with soft sediments that shake severely during an earthquake. Seattle residents were reminded of this earthquake hazard on 3 May 1996, when a magnitude 5.4 seism struck northeast of the city. Shaking in downtown Seattle's Kingdome was intense enough in the seventh inning of a baseball game between the Seattle Mariners and Cleveland Indians to cause postponement of the game. The owner of the Seattle Mariners then tried to use the earthquake as justification for breaking his lease with the Kingdome. (The team now plays in a new stadium.) When the next major earthquake (greater than magnitude 6.5) occurs on the Seattle fault, it may cause stunning levels of death and destruction. There are about 80

bridges and 1,000 unreinforced masonry (URM) buildings that could suffer damages plus a tsunami 2 m (6.5 ft) high could be created.

Normal-Fault Earthquakes

Some damaging earthquakes result when tensional forces pull one rock mass apart and down from another in normal-fault movements (see figure 3.9).

PUGET SOUND, WASHINGTON, 1949, 1965, 2001: SUBDUCTING PLATES CAN CRACK

In recent decades, normal-fault movements have brought seismic jolts to cities in the Puget Lowlands (figure 5.8). Three of these significant earthquakes were caused by down-to-the-east movements *within* the subducting Juan de Fuca plate.

At 11:55 a.m. on 13 April 1949, a jolt arose from a normal-fault movement 54 km (34 mi) below the Tacoma-Olympia area. The surface wave magnitude was $7.1M_s$, and eight people lost their lives. It could have been worse, since it happened during the day and badly damaged many schools, but luckily, it was the week of spring vacation, so the schools were largely vacant.

At 7:28 a.m. on 29 April 1965, the plate ruptured again—this time at 60 km (37 mi) depth below the Tacoma-Seattle area. The $6.5M_s$ seism killed seven people. In 2014 dollars, the destruction totaled $345 million in 1949 and $115 million in 1965.

At 10:54 a.m. on Wednesday, 28 February 2001, a normal-fault earthquake radiated out from a hypocenter 52 km (32 mi) below the Tacoma-Olympia area. The magnitude 6.8 event shook more than 3 million residents of the Puget Sound for 45 seconds. In Olympia, the earthquake cracked the dome of the State Capitol and made the legislators' offices unusable and the governor's home uninhabitable. In Seattle, 30 people were caught on top of the swaying Space Needle, bricks fell from the Starbucks headquarters building onto parked cars, and Bill Gates's talk at a hotel was interrupted as overhead lights crashed to the floor and frightened people knocked down others in their hurry to get outside. The earthquake killed no one, injured about 400 people, and caused about $2 billion in damages.

In each case, settling of soft sediments and artificial fill during the shaking caused major problems for structures built on them. There was substantial damage to older masonry buildings with inferior mortar and to buildings with inadequate ties between vertical and horizontal elements. Split-level homes suffered more than their share of damage as their different sections vibrated at different frequencies, helping tear them apart.

For all the damage the 2001 earthquake caused, the damage it did *not* do is even more significant. Following the

1965 earthquake, Washington improved its building codes and made many structural changes, such as tying homes to their foundations more securely, removing water-storage tanks from the tops of school buildings, and strengthening more than 300 highway bridges. These investments were more than repaid in damages prevented and lives saved during the 2001 earthquake.

Deep Earthquakes Beneath the Puget Sound

Primary emphasis on earthquake hazards in the Pacific Northwest has been focused on the subducting plates. The hypocenters in the 1949, 1965, and 2001 events were *within* the subducting plate at depth. Earthquakes 30 to 70 km (20 to 45 mi) deep occur beneath the Puget Sound about every 30 years. The subducting Juan de Fuca plate is only 10 to 15 million years old and is warm and buoyant. As the plate is pulled eastward, it reaches greater depths. The increasing temperature and pressure with depth cause the minerals making up the plate to become more dense and shrink. This builds up stresses that cause the plate to rupture, producing earthquakes with magnitudes as large as 7.5.

Neotectonics and Paleoseismology

Geologic history plays out on a longer timescale than human history. An active fault may have a large earthquake only once in several generations. How can the earthquake record be extended back further than the written historic record? Earthquake history can be read in sediments using the techniques of **neotectonics** (*neo* means "young") and **paleoseismology** (*paleo* means "ancient").

Faults slash through the land with compressive bends that cause land to uplift and with pull-apart bends that cause land to drop down (figures 5.9 and 5.10). The down-dropped or fault-dammed areas within the fault zone can become sites of ponds, receiving (1) sand washing in from heavy rains, (2) clays slowly settling from suspension in ponded water, and (3) vegetation that lives, dies, and is buried by clay and sand. These processes produce a delicate record of sediment layers that may be disturbed and offset by later fault movements. This is a record we can read.

Older, more deeply buried layers have existed longer (figure 5.11) and have been offset by more earthquake-generating

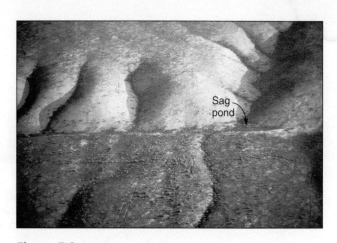

Figure 5.9 A close-up of the San Andreas fault at Wallace Creek in the Carrizo Plain. Notice the offset streams and the ponded depressions formed at the fault. Have the movements been right or left lateral?

Courtesy Pat Abbott

fault movements. The amount of fault offset is proportional to an earthquake's magnitude; the greater the offset of sediment layers, the bigger the earthquake. These principles suggest a method to determine the approximate sizes of prehistoric earthquakes. Simply dig a trench through the sediment infill of a fault-created pond and read the

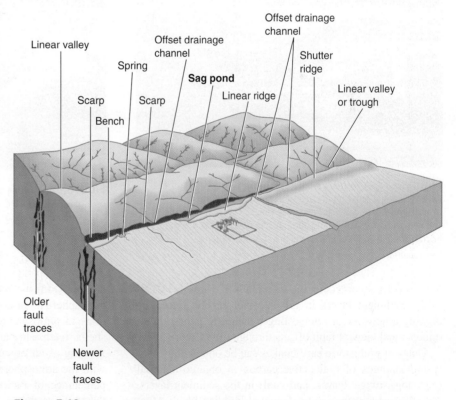

Figure 5.10 Schematic diagram of topography along the San Andreas fault in the Carrizo Plain. Notice the sag pond here and in figure 5.9. Sediments deposited in these depressions allow the prehistoric record of earthquakes to be read.

Source: Misc. Geol. Invest., *US Geological Survey.*

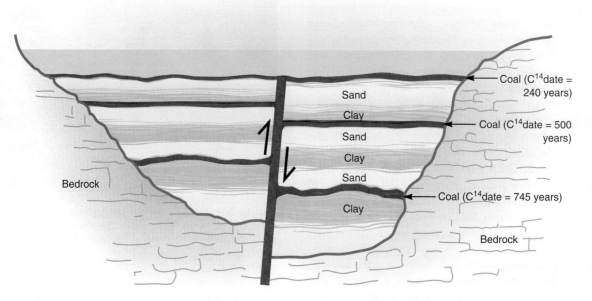

Figure 5.11 Schematic cross-section of trench wall cut through a fault-created pond. The fault offsets the once-continuous sediment layers. Notice that an upper layer of organic material formed 240 years ago is unbroken. At depth, a 500-year-old organic-rich layer has been offset. Deeper still, a 745-year-old organic-rich layer has been offset twice as much, indicating two major fault movements since it formed. What is the approximate recurrence interval between earthquakes at this site? When might the next big earthquake be expected here?

Figure 5.12 A trench wall across the San Andreas fault at Pallett Creek. Sandy layers are whitish, clay-rich layers are grayish, and organic-rich layers are black. The black layer in the center formed about 1500 CE. It has been offset 1.5 m (5 ft) horizontally and 30 cm (1 ft) vertically since 1500 CE.

Photo courtesy of Pat Abbott.

Figure 5.13 Maze of trenches dug to determine the offset of a gravel-filled stream channel by the Rose Canyon fault. The offset here is 10 m (33 ft), and the fault is active.

Photo courtesy of Pat Abbott.

fault offsets recorded in the sediments (figures 5.11 and 5.12). Sediment layers in trench walls can be traced by digging a network of intersecting trenches to gain a three-dimensional view of fault offsets through time (figure 5.13).

Dates of prehistoric earthquakes can be obtained by analyzing amounts of radioactive carbon in organic material (e.g., logs, twigs, leaves, and **coal**) in the sediment layers. All life uses carbon as a fundamental building block. Most carbon occurs in the isotope C^{12}, but a small percentage is radioactive carbon (C^{14}) produced in the atmosphere by bombardment of nitrogen atoms with subatomic particles

emitted from the Sun. Carbon is held in abundance in the atmosphere as carbon dioxide (CO_2). All plants and animals draw in atmospheric CO_2, and their wood, leaves, bones, shells, teeth, etc., are partly built with radioactive carbon. As long as an organism lives, it exchanges carbon dioxide with the atmosphere via **photosynthesis** or breathing. The percentage of radioactive carbon in a plant or animal is the same as that of the atmosphere during the organism's lifetime. However, when an organism dies, it ceases taking in radioactive carbon, and the radiocarbon in its dead tissues decays with a half-life of 5,730 years. The presence of organic material allows us to determine the time of death and hence the age of enclosing sediments. This places actual dates into faulted sedimentary layers. The determination of real dates allows us to estimate the recurrence intervals for earthquakes—that is, how many years pass between earthquakes at a given site.

The half-life of C^{14} is short, thus restricting its usage to the last 50,000 years or so. This short half-life is useful for determining events in human history.

Figure 5.11 is a schematic representation of a trench-wall exposure of faulted pond sediments, demonstrating how fault-rupture sizes and recurrence intervals may be determined. A real example of a faulted pile of ponded sediments is shown in figure 5.12. Here at Pallett Creek along the San Andreas fault, Caltech geologist Kerry Sieh has determined that fault movements with 6 m (20 ft) of horizontal offset recur about every 132 years. However, these 7+ magnitude earthquakes have occurred as close together as 44 years and as far apart as 330 years.

Earthquake Prediction

The public really wants to have earthquakes foretold in much the same style and accuracy as they receive with weather forecasts. Our ability to forecast earthquakes on longer timescales is fairly good, but on short timescales we have *no* ability at all.

LONG-TERM FORECASTS

Can we predict earthquakes on intermediate to long timescales using the paleoseismology approach? It seems to work well for some faults but not for others. Geologist Thomas K. Rockwell classifies fault-movement timing into three groups:

1. *Quasi-periodic movements.* These faults have major movements at roughly equal time intervals. This regular pattern can be defined using the trenching and radiocarbon dating of paleoseismology.
2. *Clustered movements.* Adjacent fault segments move during several decades, and then they cease movement for a century or millennium until the next cluster begins. A good example of clustered movements is occurring right now on the North Anatolian fault in Turkey (see figure 4.25).
3. *Random movements.* These faults are inherently unpredictable; they have no definable pattern for their major movements. The San Andreas fault seems to be in this category.

In December 1988, using paleoseismologic analysis, a group of geologists forecast earthquake sizes and probabilities for some major faults in California (figure 5.14). They placed a 30% probability on a magnitude 6.5 earthquake occurring on the Loma Prieta segment of the San Andreas fault within 30 years. Ten months later, the magnitude 6.9 World Series earthquake occurred there.

In 2003, the working groups stated that there is a 62% probability of at least one magnitude 6.7 earthquake striking

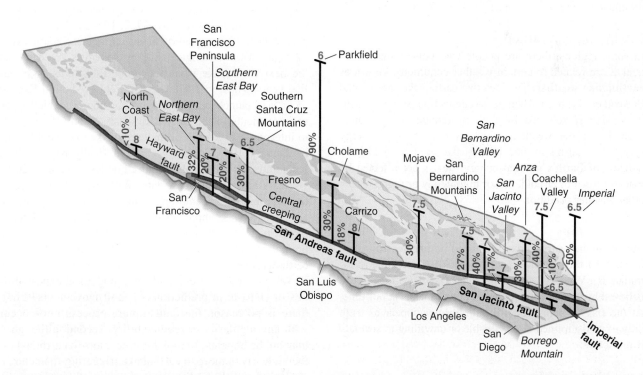

Figure 5.14 Working group analyses of expected earthquake magnitudes and their probabilities of occurring before the year 2032. Forecasts are based on historic records and trench-wall offsets of sediments dated by radiocarbon analyses and global positioning system measurements.

the San Francisco Bay region before 2032. Another forecast was an 85% probability of a magnitude 7 or higher earthquake in southern California before 2024. On Easter Sunday afternoon, 4 April 2010, a 7.2M$_w$ event occurred when a fault rupture began in Baja California, with its northward extent moving below ground into Southern California (see Laguna Salada fault in center of figure 4.8).

SHORT-TERM FORECASTS

Our knowledge of earthquakes is quite impressive. Plate tectonics tells us *why* and *where* they occur, mostly along plate edges. Neotectonic analysis allows us to know *how big* and *how often* earthquakes have occurred on any fault. However, many people are not satisfied; they want short-term prediction for earthquakes. Unfortunately, we are not even close to having that capability. We don't have a workable theory, and it seems quite possible that the detailed behavior of faults is too unpredictable to ever allow short-term prediction of earthquakes. Theories of earthquake prediction that seem logical have been developed, and they still receive coverage in textbooks, but all of them have been proved false. Science is a demanding thought process. Beautiful ideas may have no substance. Creative hypotheses may have no validity. The truth is elusive.

A public eager for short-term prediction of earthquakes includes many gullible people. In 1977, Charles Richter commented that "journalists and the general public rush to any suggestion of earthquake prediction like hogs toward a full trough . . . [Prediction] provides a happy hunting ground for amateurs, cranks, and outright publicity-seeking fakers."

Earthquake Weather

In some regions, there are people who believe that earthquakes are related to certain weather conditions, known as **earthquake weather.** The idea that earthquakes are related to weather is flawed. There is no connection between earthquake energy released by fault movements miles below ground and the weather, which is due to solar energy received at Earth's surface. Earthquakes are powered by the outflow of Earth's internal energy; this is not affected by whether it is hot or cold, dry or humid, day or night, or any other weather condition.

Nostradamus

Much ballyhoo surrounds the rhymed prophecies published in 1555 by the French doctor Michel de Notredame (Nostradamus). Vaguely worded statements by Nostradamus are believed by some people to predict earthquakes in our time. At the risk of being rude, the prophecies appear as truth only to undisciplined minds unable or unwilling to sort fact from fiction.

New Madrid, Missouri

An early 1990s prediction event occurred when a dying economist named Iben Browning filled his final days with personal excitement by predicting a major earthquake in the mid-United States similar to the earthquakes of 1811–1812. Scientists could readily see that his predictions were based on an old failed hypothesis, but an uncritical print and electronic media went on a binge of emotional coverage as a horde of television crews and reporters descended on New Madrid, Missouri, eagerly awaiting the earthquake that never came.

Psychic Predictions

Every so often we hear a psychic predict that a gigantic earthquake will cause California to break off and sink into the Pacific Ocean. Is this possible? No! This gigantic rupture-and-sink process is impossible; it is fantasy. Remember isostasy? Continents are made of less-dense rocks that float on top of denser mantle rocks. In fact, California did break off; it happened 5.5 million years ago as the Gulf of California began forming. California did not sink then and it won't sink in the future. The faulted slice of western California and Baja California will continue moving northwest toward a rendezvous with Alaska. If present trends continue, in a few tens of million years, the Californias will plow into Alaska and become part of its southern margin. Southern California will switch from surfing beaches to ski slopes.

Experiment at Parkfield, California

The U.S. Geological Survey forecast a magnitude 6 earthquake on the San Andreas fault in the Parkfield area based on the pattern of historical seismicity. Parkfield experienced magnitude 5.5 to 6 earthquakes six times in the historical period—in 1857, 1881, 1901, 1922, 1934, and 1966. Some people perceived a pattern of an earthquake about every 22 years. U.S. Geological Survey scientists forecast that the next earthquake would occur in 1988, plus or minus five years. Thus, in 1984, the Parkfield Prediction Experiment was launched by deploying an unprecedented array of instruments in the field with a large team of scientists to interpret every detail of the earthquake that would come by January 1993. *Breaking news:* It finally happened! A magnitude 6.0 earthquake occurred on 28 September 2004, 16 years after the forecast date. With more than 22 years of work and tens of millions of taxpayer dollars spent, the earthquake was unpredicted. The Parkfield Earthquake Experiment, the best-staffed and best-funded earthquake prediction experiment ever, was a total failure at short-term earthquake prediction.

What is our current understanding of the possibilities of short-term predictions of fault movements? First, there is no reason the fault-rupture process must occur with any regularity or predictability. Second, although it may not be hopeless to look for precursors to earthquakes, there clearly is more to earthquake triggering than can be explained simply by the steady loading of plate-tectonic stress onto faults that then rupture in evenly spaced, characteristic earthquakes. The bottom line for each person is this: short-term prediction of earthquakes is not forth-coming, so plan your life accordingly. Organize your home and office to withstand the biggest earthquake possible in your area, and then don't worry about when that day will come.

EARLY WARNING SYSTEM

An earthquake early warning system has been developed and is enjoying increasing success. Note: This system does *not* give predictions; it provides *warning* that a significant earthquake has begun and that seismic waves are already traveling toward you. The system is based on velocity differences in seismic waves.

1. P waves travel the fastest but they do little damage. S waves and surface waves travel slower but can cause immense damage.
2. Sensors record P wave arrivals and immediately transmit data to an alert center, where the earthquake location is determined and its magnitude is estimated and updated as P waves continue to flow in.
3. Messages from the alert center are immediately transferred to computers and mobile phones stating the expected intensity and arrival time of strong shaking at your location.

Even a few seconds of warning can prod people to drop, cover, and hold on; allow surgeons and dentists to stop delicate procedures; signal automated systems to place equipment into safe mode; and more.

The warning system will always have a blind zone around the epicentral area where there is little time difference between P & S wave arrivals. For example, the Silicon Valley region built on the San Francisco Peninsula along the San Andreas fault will receive little to no early warning when a major movement has begun. On the other hand, Portland, Seattle, and Vancouver could receive minutes of advance warning that an offshore magnitude 9 subduction movement was in progress.

Since 2006, the Japanese have used a system to automatically slow down and stop their 320 km/hr (200 mph) bullet trains. As of 2013, Japan was the only country with a nationwide earthquake early warning system. Installation of warning systems is slowly spreading into some regions, such as within the United States, Mexico, and China.

Human-Triggered Earthquakes

The Earth exists in a state of delicate balance that is readily adjusted to maintain equilibrium (see Isostasy in Chapter 2). Some of our human activities unintentionally upset the equilibrium and trigger earthquakes by adding or removing massive amounts of material at construction sites; by

forcing fluids underground under pressure; by extracting large volumes of groundwater; and by setting off underground explosions.

PUMPING FLUIDS UNDERGROUND

The cause-and-effect relationship between triggered earthquakes and pumping fluids underground under high pressure has long been understood in the petroleum industry. This fact was dramatically proven in the 1960s in the Denver area.

Disposal Wells Near Denver, Colorado

In early 1962, in secret, the Rocky Mountain Arsenal began pumping chemical warfare waste under pressure down a well into old rocks 3.7 km (2.3 mi) deep. Earthquakes began one month later and rose to more than 40 per month, causing alarm in Denver. Pumping stopped in September 1963 and earthquakes became minimal in number and magnitude. Pumping resumed in September 1964, again in secret, and so did earthquakes. In 1967, three of the earthquakes exceeded 5M. The suspected cause became public resulting in a hue and cry that forced pumping to stop. And the earthquakes stopped. The relationship is clear. Fluids pumped underground under pressure can be forced into ancient faults if they are present, adding stress and reducing friction and thereby causing the faults to begin moving again. The greater the amount of fluid pumped, the greater the number and magnitude of earthquakes.

Ashtabula Township, Ohio On 25 January 2001, Ashtabula Township in Ohio was rocked by a magnitude 4.5 earthquake. The shaking damaged 50 houses and businesses as ceiling tiles fell, plaster cracked, and gas lines ruptured, forcing people to evacuate. This earthquake was the biggest in a series that began on 13 July 1987 with a magnitude 3.8 event. Why did earthquakes begin and keep recurring in this industrial port city on the shores of Lake Erie? In 1986, a 1.8 km (1.1 mi) deep well was drilled to inject hazardous wastes underground. For seven years, beginning in 1986, millions of gallons of waste-carrying liquids were forced down this disposal well under pressure; earthquakes began in 1987. The pressurized fuilds pumped underground encounter faults at depth, causing movements big enough to do damage at the surface.

Hydraulic Fracturing

We humans have a huge thirst for fossil fuels to power our industries, transportation, and personal lives. This has led to increased use of **hydraulic fracturing,** commonly called **fracking,** wherein liquids are pumped down wells under high pressure in order to fracture and crack open rocks. The fractured rocks yield much greater volumes of natural gas and oil from deep underground. Hydraulic fracturing has significantly increased fossil-fuel energy production in

Side Note

the United States. For example, in 2005 the United States imported more than 12 million barrels of oil per day (1 barrel = 42 gallons). In 2014, thanks to fracking, U.S. imports were down to 6 million barrels per day. At $100 per barrel, this decrease in imports saved $220 billion in 2014; this money helped provide hundreds of thousands of good-paying jobs in the United States.

On the other side, there are environmental concerns such as triggering earthquakes, the use of enormous volumes of water pumped underground, the potential contamination of aquifers, and more. These worries have led some U.S. states and some countries to ban fracking.

Dallas–Fort Worth, Texas Hydraulic fracturing and new techniques of horizontal drilling are yielding enormous volumes of natural gas from rocks that previously were too "tight." For example, more than 200 wells drilled into the Barnett Shale (a tight mudstone) in the Dallas–Fort Worth area now yield huge volumes of natural gas. But more than 180 earthquakes up to 3.3M began in 2008 and continued into 2009. The earthquake source was traced to one wastewater disposal well drilled near an ancient fault. Abandonment of this well stopped most of the earthquakes. Natural gas production continues through the many other wells.

Oklahoma Earthquakes are increasing dramatically in recent years in U.S. midcontinent states such as Texas, Colorado, Arkansas, and Ohio but the biggest increase is in Oklahoma. Earthquakes of magnitude 3+ in Oklahoma averaged 1.6 per year between 1978 and 2008; rose to 39 per year between 2009 and 2012; jumped to 109 in 2013; and will exceed 500 in 2014, thus vaulting Oklahoma ahead of California as the earthquake leader in the 48 conterminous states. The Oklahoma earthquake swarms include a state record $5.7M_w$ event west of Prague toward Oklahoma City. It appears likely that the triggers for these earthquakes are the increased pore-water pressures in subsurface rocks receiving millions of barrels of water pumped into them under high pressure each month.

A significant concern exists about how large an earthquake could be triggered in the near future. Ancient lengthy faults below Oklahoma could be pressured into movements resulting in a magnitude 6 or 7 earthquake (see Central United States Earthquakes later in this chapter).

DAM EARTHQUAKES

The downwarping of the land beneath the filling Lake Mead triggered many small earthquakes beginning in 1935 (see figure 2.7). This is a common occurrence; build a dam and impound a reservoir of water, and then earthquakes follow. First, impounding a reservoir adds a huge weight on the surface, causing the earth to sink isostatically. Second, water seeping through the floor of the reservoir flows slowly underground throughout the region pushed by the large body of reservoir water above it. The underground water moves downward and outward as an advancing front of high pressure that may reach a fault and cause it to move. As an analogy, visualize what makes the water flow through the pipes in your house. In most cases, the water comes from a higher-elevation water tank or reservoir that pushes the water down through the pipes.

China, 2008

Monday, 12 May 2008, began peacefully, like so many other days near the Dragon's Gate Mountains in Sichuan, China. The 15 million people of the region were busy at work, their schools were full of children, and the giant pandas were at home in the Wolong Nature Reserve. But at 2:28 p.m., the earth ruptured along the base of the mountains and ripped northeastward along the Longmenshan fault for 250 km (155 mi) for about two minutes. When the shaking stopped, about 87,500 people were dead and 5 million were homeless. This massive earthquake was caused by the ongoing collision of India pushing into Asia. This rupture was a mountain-building thrust-fault event; similar movements over millions of years have built the Dragon's Gate Mountains.

Time of day is always a factor in earthquake deaths, and the timing of this seism was terrible. Many of the buildings were made of brittle concrete with little support steel, and at 2:28 p.m. on a Monday, the badly built schools and office buildings were full of people, resulting in a high death toll. The loss of so many children in the collapsed schools was especially tragic for families because of China's one-child policy.

Plate tectonics was the cause of the earthquake, but what was the trigger? A debate is in progress. A 156 m (512 ft) tall dam was built in 2005 to create the Zipingpu Reservoir. In 2008, only 2.5 years later, the reservoir held 900 million tons of water. The dam lies 500 m (1,600 ft) from the Longmenshan fault. The weight of the reservoir water, plus the pore pressure of the water seeping underground, beneath the reservoir, would have caused the land to warp downward. Was this the added stress that triggered the fault to move on 12 May 2008 rather than 100 or so years later?

BOMB BLASTS

Underground nuclear explosions in Nevada have triggered earthquakes. Some of the atomic-bomb blasts released energy equivalent to a magnitude 5 earthquake. The bomb explosions triggered significant increases in earthquakes in their region during the 32 hours after the blast.

Any time the level of stress or pressure is changed on rocks below the ground, earthquakes are possible. We humans can cause or trigger earthquakes.

Earthquake-Shaking Maps

Computers are being used to create maps of earthquake shaking in near-real time.

DID YOU FEEL IT?

Upon feeling an earthquake, a common response is to turn on the TV, radio, or computer to learn what just happened. Now you can help by sharing your shaking experience via your computer. Go through the USGS earthquake website to reach it, or simply google Did You Feel It? Click on Report Unknown Event, enter your ZIP code, and answer the questions about what you felt. In a matter of minutes, a Community Internet Intensity Map will show the intensities of shaking felt in affected ZIP-code areas. You can be an important part of creating these Mercalli intensity maps and can also learn about the earthquake via your own participation.

SHAKEMAPS

The intensities of seismic shaking are now recorded by instruments, and the data are fed into a computer that generates a ShakeMap (figure 5.15). The ShakeMap for the Northridge earthquake shows the effects of directivity, with most of the intense shaking occurring north of the north-moving fault. The ShakeMap also shows other areas of more intense shaking in land underlain by soft rocks.

The rock and sediment foundations beneath buildings may amplify seismic waves (figure 5.16). Seismic waves travel fast and with less amplitude in hard rocks. When seismic waves pass into soft rock or loose sediment, they slow down, but their amplitudes increase, and thus the shaking increases. Much of Los Angeles is built on soft rocks that amplify seismic shaking. In some areas, the soft rocks are 10 km (6 mi) thick, and seismic shaking may be amplified five times.

California Earthquake Scenario

An analysis of probabilities for earthquakes greater than magnitude 6.7 in California shows the event is most likely to occur next on the southern San Andreas fault. A model of this earthquake and its effects was constructed through 13 special studies and 6 expert panels. The scenario earthquake is a magnitude 7.8 event that first ruptures at 7.6 km (4.7 mi) depth next to the Salton Sea and then continues rupturing northward past Palm Springs and through San Bernardino for 300 km (185 mi) (figure 5.17). The modeled earthquake kills 1,800 people, injures another 50,000, and causes $213 billion in damages. The best way to reduce these numbers is to prepare in advance for an earthquake like this—and preparation begins with education and preparedness exercises.

An earthquake similar to the scenario event on the San Andreas fault broke loose in Alaska on 3 November 2002, mostly on the Denali fault; both are right-lateral faults. The

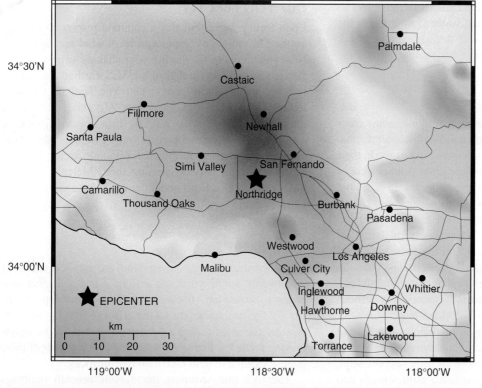

Figure 5.15 ShakeMap for the magnitude 6.7 North-ridge earthquake in 1994. The fault moved to the north, and the greatest shaking was north of the epicenter. The variations in intensity of shaking are due to distance and variations in rock foundations.

Perceived shaking	Not felt	Weak	Light	Moderate	Strong	Very strong	Severe	Violent	Extreme
Potential damage	None	None	None	Very light	Light	Moderate	Moderate/Heavy	Heavy	Very heavy
Peak ACCEL. (%g)	<.17	.17-1.4	1.4-3.9	3.9-9.2	9.2-18	18-34	34-65	65-124	>124
Peak VEL. (cm/s)	>0.1	0.1-1.1	1.1-3.4	3.4-8.1	8.1-16	16-31	31-60	60-116	>116
Instrumental intensity	I	II-III	IV	V	VI	VII	VIII	IX	X+

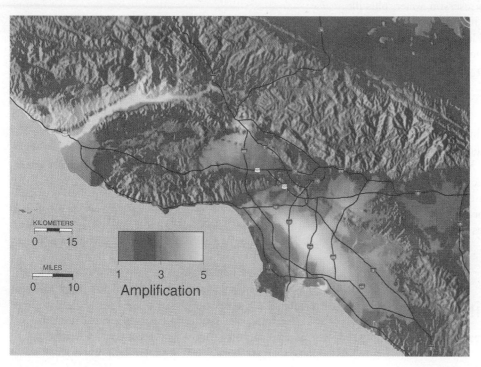

Figure 5.16 Amplification of ground motion during an earthquake in Los Angeles. Amplification is minimal in hard rocks (purple), significant in softer rocks (red), and greatest where the softer rocks are the thickest (yellow).

Ned Field/*US Geological Survey.*

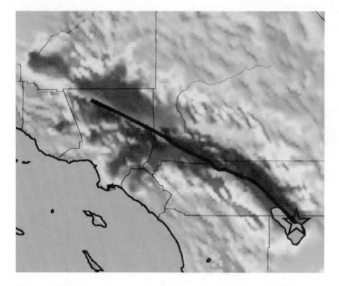

Figure 5.17 The big earthquake most likely to occur next in California is a magnitude 7.8 rupture on the southern San Andreas fault. Here it is assumed that the rupture will begin at the Salton Sea and move northward through Los Angeles. Colors indicate severity of ground shaking.

ShakeOut.org/scenario

GREAT SHAKEOUT EVENTS

A valuable way to prepare for earthquakes is to practice your response during a virtual earthquake. The first major virtual event was an effort by scientists and emergency managers to involve the public and schools in the "Great Southern California Shakeout" at 10 a.m. on 13 November 2008; it involved 5.4 million people. The concept has now spread to include great shakeout events in all of California, New Zealand, Nevada, Guam, British Columbia, Oregon, Idaho, and 10 states in the central United States. More countries have events planned.

One of the tips given to participants is that, upon feeling an earthquake, the immediate response that usually works the best is to:

Drop, Cover, and Hold on.

The most common hazard is having objects fall on you or be thrown at you. The best response is to "make like a turtle" and dive under a heavy table or desk to create a protective shell; then hang onto its legs to keep your shell in place until the shaking stops. The best way to remember this strategy is to practice it now so you can react instantly when everything starts shaking.

Alaskan earthquake began in the west with faults moving to the east and then southeast for 140 seconds, rupturing the ground for 340 km (210 mi) in a magnitude 7.9 seism with ground offsets up to 8.8 m (29 ft). Compare this earthquake to the San Andreas fault event of 9 January 1857 that moved southward for 130 seconds, rupturing the ground for 360 km (220 mi) in a magnitude 7.9 seism with ground offsets up to 9.5 m (31 ft). The 2002 Alaskan fault rupture had significant directivity; it was like a seismic shotgun aimed southeast, triggering earthquake swarms up to 3,660 km (2,270 mi) away in Washington, Wyoming, California, and Utah. In Lake Union in Seattle, the water sloshed back and forth and damaged houseboats. The good news about the 2002 Alaskan earthquake is that it happened in a remote area and had minimal effects on people. But this earthquake is like the fabled Big One of California, which will directly affect millions of people. For example, the San Andreas fault literally runs through the backyards of some of the 3 million people in the San Bernardino area.

ANNUALIZED EARTHQUAKE LOSSES

Although big earthquakes do not happen in the United States every year, we forecast potential future costs as annualized earthquake losses. Data on population, buildings, and shaking potential are analyzed in a software program called HAZUS. For the United States, $4.4 billion in annual earthquake losses are projected. Southern California accounts for almost half of the losses; to Los Angeles County alone are attributed more than $1 billion in losses each year.

Earthquakes in the United States and Canada

Awareness is growing that destructive and death-dealing earthquakes are a widespread problem, not just something that happens in California. Figure 5.18 is a map centered on the United States showing epicenters of significant earthquakes during a 92-year-long period. Compare the epicenter locations with the earthquake hazards map of the United States (figure 5.19). Figure 5.20 is a map of eight of the largest earthquakes in Canadian history. All these figures show that earthquakes cluster in certain areas.

In Alaska and California, earthquakes occur in such large numbers and large sizes that they tend to obscure the earthquake history of the rest of the United States. If Alaska and California are ignored, the list of 10 largest U.S. earthquakes shows that major seisms occur in numerous states—10 major earthquakes, 10 different states (table 5.1).

The history of earthquakes in Canada also shows variety (table 5.2). The list is dominated by events along the tectonically active west coast of British Columbia, yet the list of 11 largest earthquakes involves four provinces.

An expanded look at the earthquake history of the United States shows that all 50 states are hit by earthquakes, and many of the states have large earthquakes (table 5.3). At least 17 states have been rocked by magnitude 6 or greater earthquakes; some of the older seisms may have been as big, but scientific records are lacking.

The historic record shows that earthquakes are widespread, but when earthquake frequency is examined, a different picture emerges. The location of all U.S. earthquakes

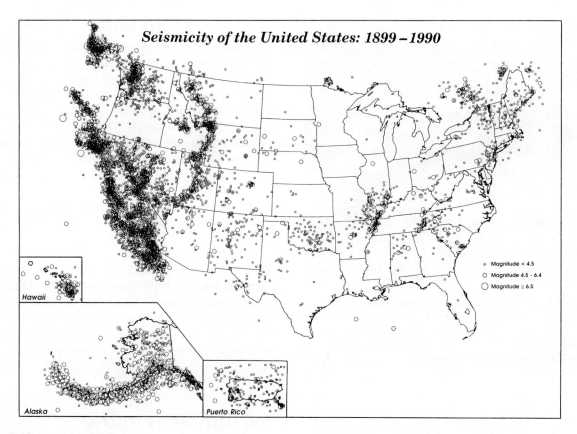

Figure 5.18 Epicenters of earthquakes in the United States, southern Canada, and northern Mexico, 1899–1990.

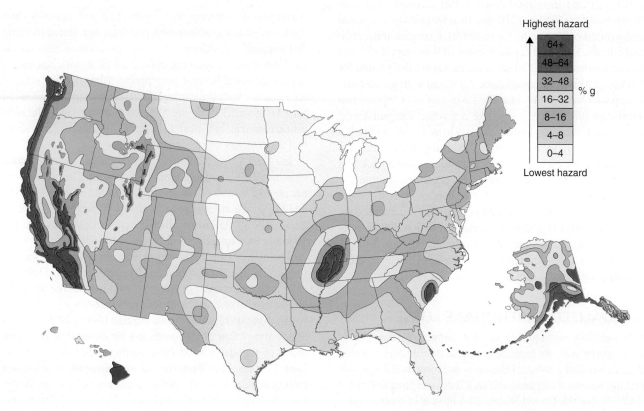

Figure 5.19 Earthquake hazards in the conterminous United States. Color show horizontal shaking, as a percentage of acceleration of gravity, that have a 2% probability of being exceeded in 50 years.

Source: *US Geological Survey,* 2008.

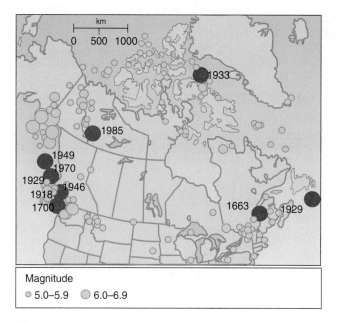

Figure 5.20 Eight of the largest earthquakes in Canadian history, 1660–2011, are shown by red circles. See table 5.2 for more data.

Source: EarthquakesCanada.

TABLE 5.1

Ten Largest Earthquakes in the United States (excluding Alaska and California)

Magnitude	Date	Location
9.0	26 Jan 1700	Washington, Oregon—Cascadia subduction
7.9	2 Apr 1868	Hawaii—Ka'u district
7.5	7 Feb 1812	Missouri—New Madrid
7.3	16 Dec 1811	Missouri, Arkansas—New Madrid
7.3	17 Aug 1959	Montana—Hebgen Lake
7.3	31 Aug 1886	South Carolina—Charleston
7.2	16 Dec 1954	Nevada—Dixie Valley
7.0	23 Jan 1812	Illinois—New Madrid zone
7.0	28 Oct 1983	Idaho—Borah Peak
6.8	28 Feb 2001	Washington—Nisqually

TABLE 5.2

Eleven Largest Earthquakes in Canada

Magnitude	Date	Location
9.0	26 Jan 1700	British Columbia—Cascadia subduction
8.1	22 Aug 1949	British Columbia—Queen Charlotte Island
7.7	27 Oct 2012	British Columbia—Queen Charlotte Island
7.4	24 Jun 1970	British Columbia—Queen Charlotte Island
7.3	20 Nov 1933	Northwest Territories—Baffin Bay
7.3	23 Jun 1946	British Columbia—Vancouver Island
7.2	18 Nov 1929	Newfoundland—Grand Banks
7.0	26 May 1929	British Columbia—Queen Charlotte Island
7.0	5 Feb 1663	Quebec—Charlevoix
6.9	23 Dec 1985	Northwest Territories—Nahanni
6.9	6 Dec 1918	British Columbia—Vancouver Island

Source: EarthquakesCanada (2006).

We will examine specific earthquakes and their causes in regional settings: western United States and Canada under the influence of plate tectonics and buoyancy forces; the stable (tectonically "inactive") central and eastern United States and Canada; and finally, the relationship between earthquakes and volcanism in Hawaii.

Western North America: Plate Boundary–Zone Earthquakes

Much of the earthquake hazard in western North America is due to the ongoing subduction of small plates, as well as the continuing effects of the overridden, but not forgotten, Farallon plate. When considering the size of the Pacific, North American, and Farallon plates, it is easy to appreciate why earthquakes affect the entirety of western North America. Consider that the Pacific plate is more than 13,000 km (8,000 mi) across and that it is grinding past the North American plate, which is more than 10,000 km (6,250 mi) wide. How broad a zone is affected by these passing giants? The affected zone must be large—as big as the entirety of western North America. The scale of these gigantic plates

of magnitude 3.5 and higher during a 30-year period shows a marked asymmetry (table 5.4). Alaska has 57% and California 23% of these earthquakes. Add in Hawaii and Nevada, and those four states received 91% of these earthquakes. Eight states had none (Connecticut, Delaware, Florida, Iowa, Maryland, North Dakota, Vermont, and Wisconsin).

A primary goal of the remainder of this chapter is to understand the large earthquakes in the United States and Canada that do not occur on the edge of a tectonic plate.

TABLE 5.3
Largest Earthquakes by State

State	Date	Magnitude or Intensity	State	Date	Magnitude or Intensity
Alabama	18 Oct 1916	5.1	Montana	17 Aug 1959	7.3
Alaska	27 Mar 1964	9.2	Nebraska	28 Mar 1964	5.1
Arizona	21 Jul 1959	5.6	Nevada	16 Dec 1954	7.2
Arkansas	16 Dec 1811	7.0	New Hampshire	24 Dec 1940	5.5
California	9 Jan 1857	7.9	New Jersey	30 Nov 1783	5.3
Colorado	8 Nov 1882	6.6	New Mexico	15 Nov 1906	VII
Connecticut	16 May 1791	VII	New York	5 Sep 1944	6
Delaware	9 Oct 1871	VII	North Carolina	21 Feb 1916	5.2
Florida	13 Jan 1879	VI	North Dakota	16 May 1909	5.5
Georgia	5 Mar 1914	4.5	Ohio	9 Mar 1937	5.4
Hawaii	2 Apr 1868	7.9	Oklahoma	6 Nov 2011	5.7
Idaho	28 Oct 1983	7.0	Oregon	5 Aug 1910	6.8
Illinois	23 Jan 1812	7.0	Pennsylvania	25 Sep 1998	5.2
Indiana	27 Sep 1909	5.1	Rhode Island	11 Mar 1976	3.5
Iowa	13 Apr 1905	V	South Carolina	31 Aug 1886	7.3
Kansas	24 Apr 1867	5.1	South Dakota	2 Jun 1911	4.5
Kentucky	27 Jul 1980	5.2	Tennessee	17 Aug 1865	5.0
Louisiana	19 Oct 1930	4.2	Texas	16 Aug 1931	5.8
Maine	21 Mar 1904	5.1	Utah	12 Mar 1934	6.6
Maryland	16 Jul 2010	3.4	Vermont	10 Apr 1962	4.2
Massachusetts	18 Nov 1755	6.3	Virginia	31 May 1897	5.9
Michigan	10 Aug 1947	4.6	Washington	26 Jan 1700	9
Minnesota	9 Jul 1975	4.6	West Virginia	20 Nov 1969	4.5
Mississippi	17 Dec 1931	4.6	Wisconsin	6 May 1947	V
Missouri	7 Feb 1812	7.5	Wyoming	17 Aug 1959	6.5

Source: US Geological Survey.

strongly suggests that their interactions are an underlying cause of earthquakes throughout the western United States, Canada, and Mexico.

WESTERN GREAT BASIN: EASTERN CALIFORNIA, WESTERN NEVADA
Owens Valley, California, 1872

The famous naturalist John Muir was in his cabin in Yosemite Valley when:

> At half past two o'clock of a moon-lit morning in March, I was awakened by a tremendous earthquake, and though I had never before enjoyed a storm of this sort, the strange thrilling motion could not be mistaken, and

> I ran out of my cabin, both glad and frightened, shouting, "A noble earthquake!" feeling sure I was going to learn something. The shocks were so violent and varied, and succeeded one another so closely, that I had to balance myself carefully in walking as if on the deck of a ship among waves, and it seemed impossible that the high cliffs of the Valley could escape being shattered. In particular, I feared that the sheer-fronted Sentinel Rock, towering above my cabin, would be shaken down, and I took shelter back of a large yellow pine, hoping that it might protect me from at least the smaller outbounding boulders. For a minute or two the shocks became more and more violent—flashing horizontal thrusts mixed with a few twists and battering, explosive, upheaving jolts—as if Nature were wrecking her Yosemite temple, and getting ready to build a still better one.

TABLE 5.4

Most Active Earthquake States (magnitudes 3.5 and above, 1974–2003)

State	Number of Earthquakes
1. Alaska	12,053
2. California	4,895
3. Hawaii	1,533
4. Nevada	778
5. Washington	424
6. Idaho	404
7. Wyoming	217
8. Montana	186
9. Utah	139
10. Oregon	73
Top 10 states total	20,702
Bottom 40 states total	378

Source: US Geological Survey.

Figure 5.21 View to the north in Owens Valley. Faults are subparallel and to the left of Highway 395; note that the town of Lone Pine (in center of photo) is down-dropped. There is a lake in the right-stepping pull-apart between two fault segments. The Alabama Hills are at left center, and the Sierra Nevada in upper left.

University of Washington Libraries, Special Collections, John Shelton Collection, Shelton 67-986

What happened on 26 March 1872? The fault zone on the western side of the Owens Valley broke loose along a length of 160 km (100 mi). This is the third longest fault rupture in California history after the 1906 San Francisco and 1857 Fort Tejon events (figure 5.21). Today, Highway 395 runs in a north-south direction, right along the faults. The 1872 faulted zone is up to 15 km (10 mi) wide, with vertical drops (normal faulting) of as much as 7 m (23 ft) and horizontal offsets (right lateral) up to 5 m (16 ft). The epicenter was near the town of Lone Pine, where 27 people, about 10% of the residents, were crushed to death in the collapse of their adobe (dried mud blocks) and stone houses. The seism is estimated to have had a magnitude of about 7.4. So, big earthquakes do happen far away from the coastal zone and the San Andreas fault.

The Western Great Basin Seismic Trend

This earthquake belt runs through eastern California and western Nevada and has a recognizable line of epicenters (see figure 5.18) and faults (figure 5.22). In historic time, Nevada has averaged one earthquake with a magnitude in the 6s per decade and one with a magnitude in the 7s every 27 years. Why so many earthquakes? In the last 30 million years, the region between the eastern Sierra Nevada in California and the Wasatch Mountain front in central Utah has expanded in an east-west direction, opening up by several hundred kilometers (figures 5.23 and 5.24). This extended area is known as the Great Basin, or the Basin and Range province. Nevada, in the heart of the extended province, has about doubled in width. As much as 20% of the relative motion between the Pacific and North American

plates may be accommodated in the Basin and Range province. Extensional, pull-apart tectonics stretch the area, leaving numerous north-south- oriented, back-tilted mountain ranges separated by down-dropped, sediment-filled basins (figure 5.23). The extension is accomplished with normal faulting, so vertical separation dominates over horizontal slippage.

Some major earthquakes of historic times have occurred in the western part of the Great Basin province (figure 5.22). (1) On 2 October 1915, a large earthquake occurred south of Winnemucca in Pleasant Valley, Nevada. This magnitude 7.7 event ruptured the surface for 59 km (37 mi). The slip was dominantly vertical (normal) with displacements up to 5.8 m (19 ft) (figure 5.25). Some fault strands had right-lateral components of offset up to 2 m (6.5 ft). (2) On 21 December 1932, a magnitude 7.2 event occurred near Cedar Mountain, Nevada, rupturing the ground for 61 km (38 mi).

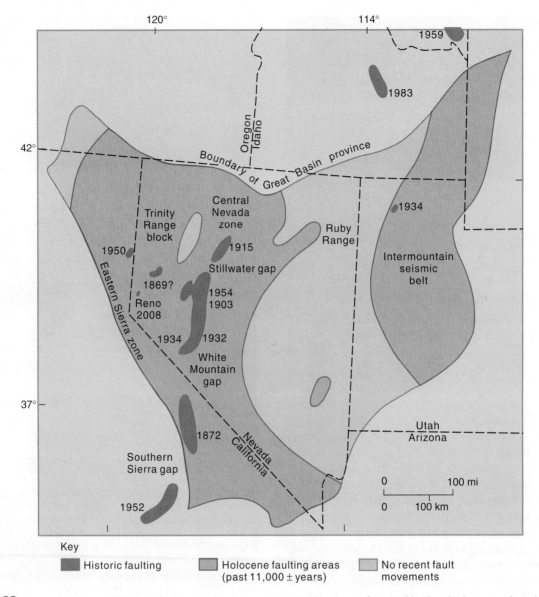

Figure 5.22 Generalized map of historic faulting in the western Great Basin. Areas of ground broken by large earthquakes are in dark orange; notice the seismic gaps in the trend. Areas with numerous smaller seisms are brown.

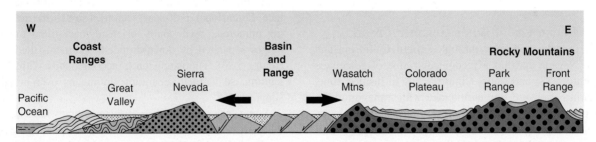

Figure 5.23 Schematic cross-section oriented west-east across the western United States. The Basin and Range province has stretched to double its initial width. This extension has created normal faults that generate earthquakes.

(3) The year 1954 was a big one for earthquakes in Nevada. Events included a magnitude 6.6 on 6 July and a 6.9 on 24 August near Fallon, as well as two shocks of 7.2 and 6.9 that rocked Dixie Valley on 16 December. Figure 5.22 shows several gaps in the trend of historic, long ruptures of faults. Residents in these seismic gaps may be in for some surprises.

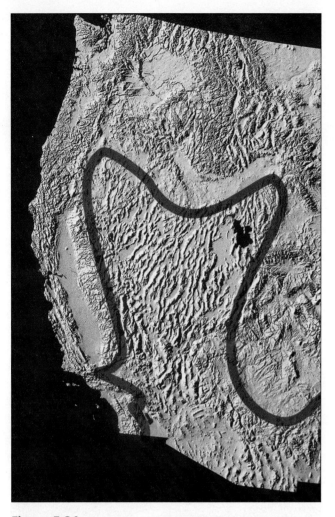

Figure 5.24 Computer-generated image of topography in the western United States. Notice in the center the north-south-oriented mountains separated by linear valleys. Basin and Range topography is outlined by the red line.

U.S. Geological Survey.

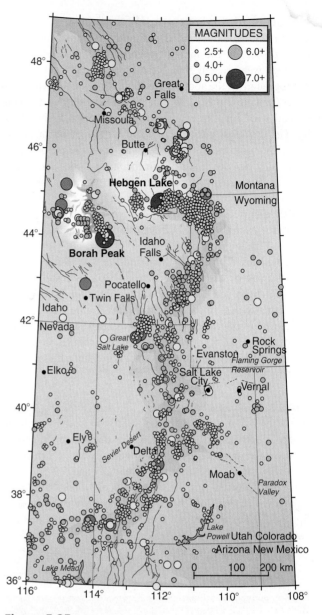

Figure 5.25 Earthquake epicenters in the Intermountain seismic belt, 1900–1985.

Reno, Nevada, 2008

Big earthquakes usually occur as a mainshock followed by numerous aftershocks. But sometimes quakes occur in a *swarm*, a cluster of earthquakes without a mainshock. Between 28 February and 3 June 2008, Reno experienced a swarm of 1,090 quakes of magnitude 2 and greater (figure 5.22). The peak of the swarm occurred in late April and early May when the numbers of earthquakes increased and magnitudes reached 4.2, 4.7, 4.2, and 3.8. The swarm occurred along a short fault and may well have been an interval of fault growth in response to Nevada being pulled apart by tectonic forces. The good news for Reno residents is that a short, poorly developed fault will not produce a large earthquake. The bad news is that the Reno area has other longer faults with a 65% chance of producing a magnitude 6 earthquake in the next 50 years.

THE INTERMOUNTAIN SEISMIC BELT: UTAH, IDAHO, WYOMING, MONTANA

The Intermountain seismic belt is a northerly trending zone at least 1,500 km (930 mi) long and about 100 to 200 km (60 to 125 mi) wide (figure 5.25). The belt extends in a curved pattern from southern Nevada and northern Arizona into northwestern Montana (see figure 5.22). In effect, the seismic belt is the eastern boundary of the extending Basin and Range province. The bounding faults on the eastern side of the Great Basin are mostly down-to-the-west, whereas the bounding faults on the western side (in eastern California and western Nevada) are mostly down-to-the-east. The earthquakes reaffirm that this part of the world is being stretched and pulled apart.

Hebgen Lake, Montana, 1959

The Rocky Mountains in the summertime are a beautiful place to be. On the moonlit evening of 17 August 1959, campers were settled into their spots at the Rock Creek Campground at the foot of the high walls of the Madison River Canyon. But at 11:37 p.m., the ground shook, and then an odd wind blew briefly down the canyon at high velocity. The wind was created by the push of an enormous rock slide. The south wall of the canyon dropped 43 million cubic yards of rock, which slid down the steep slope, across the Madison River, and moved about 150 m (500 ft) up the north wall (figure 5.26). It entombed 26 campers. The gigantic landslide buried the canyon to depths of 67 m (220 ft) and created a natural dam that began trapping a large body of water—Earthquake Lake.

What caused this life-ending landslide? At Hebgen Lake, directly west of Yellowstone National Park, two subparallel faults (Hebgen and Red Canyon) moved within 5 seconds of each other with $6.3 m_b$ and $7.5 M_s$ events (figure 5.25). These two normal faults had their southwestern sides drop 7 and 7.8 m (23 and 26 ft) down fault surfaces inclined 45° to 50° to the southwest. The fault movements created a huge seiche in Hebgen Lake.

Borah Peak, Idaho, 1983

Just after 7 a.m. on 28 October 1983, the Lost River fault broke free 16 km (10 mi) below the surface and ruptured northwestward 0.45 m (1.5 ft) horizontally and 2.7 m (9 ft) vertically for a $7.3 M_s$ event (figures 5.25 and 5.27). When the fault finished moving, Borah Peak, Idaho's highest point, was 0.3 m (1 ft) higher, and the floor of Thousand Springs Valley was several feet lower. The ground shaking caused Thousand Springs Valley to live up to its name as underground water, squeezed out by the subterranean pressures, spouted fountains 3 to 6 m (10 to 20 ft) high.

The Wasatch Fault

In historic times, large seisms have occurred in eastern California and western Nevada on the west and in Montana and Idaho on the east, but not on long sections of faults in Utah. Over 80% of Utah's population lives within sight of the **scarps,** or steep slopes, of the 370 km (230 mi) long Wasatch Front, the zone of normal faults separating the mountains from the Great Basin (figure 5.28). No large earthquakes have been reported along the Wasatch Front faults since the arrival of Brigham Young in 1847, but the sharply defined faults show obvious potential for earthquakes (figure 5.29). In an 1883 article in the *Salt Lake Tribune,* the famous geologist G. K. Gilbert warned the people of Utah of the earthquake threat and the danger for their towns. The fault segments shown in figure 5.28 are each capable of events

Figure 5.27 The 28 October 1983 Borah Peak earthquake was a 7.3 M_s event with 2.7 m (9 ft) of vertical offset. Notice that some left-lateral offset also occurred. Mount Borah (in background) was uplifted slightly by this event.

NOAA/NGDC, G. Reagor, U.S. Geological Survey.

Figure 5.26 Madison Canyon landslide and resulting lake, caused by the earthquake of 17 August 1959.

University of Washington Libraries, Special Collections, John Shelton Collection, Shelton 2940.

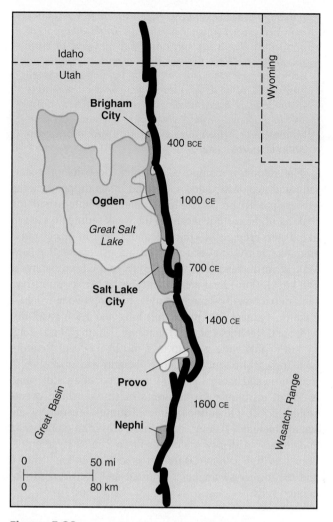

Figure 5.28 Map of faults along the Wasatch Front, Utah. The Wasatch fault has several segments. Dates of the most recent magnitude 6.5 or greater seisms are shown.

Figure 5.29 Aerial view eastward over Salt Lake City to the Wasatch fault running along the base of the mountains. The Wasatch fault zone is colored red.

University of Washington Libraries, Special Collections, John Shelton Collection, Shelton 3840.

like those of Hebgen Lake and Borah Peak. In the last 6,000 years, a magnitude 6.5 or stronger earthquake has occurred about once every 350 years on one of the Wasatch system faults. Parts of Salt Lake City, Provo, and Ogden lie on soft lake sediments that will shake violently during a large seism. The fault segment near Brigham City has not moved in the last 2,400 years and is a likely candidate for a major event.

RIO GRANDE RIFT: NEW MEXICO, COLORADO, WESTERNMOST TEXAS, MEXICO

The Rio Grande rift is one of the major continental rifts in the world. It is a series of interconnected, asymmetrical, fault-block valleys that extend for more than 1,000 km (620 mi) (figure 5.30). Here, it appears that the continental crust is being heated from below and is stretching. The crust responds by thinning and extending with accompanying normal faulting. In the last 26 million years, about 8 km (5 mi) of crustal extension has occurred near Albuquerque, New Mexico, a rate of about 0.3 mm/yr. The dominant motion on the faults is vertical, and the offset totals 9 km (5.5 mi). The rift basin is strikingly deep in places, yet most of the vertical relief created by fault offsets has been lessened by the copious quantities of volcanic materials and sediments that have poured into the rift over millions of years.

The topographic trough of the rift valley has attracted a major river (Rio Grande), which in turn has enticed human settlers in need of water. Today's settlements include

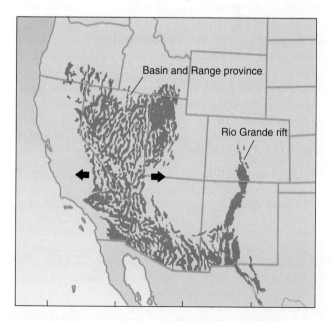

Figure 5.30 An east–west-oriented extension has pulled apart some of western North America to form the topography of Basin and Range province. The Rio Grande rift is a geologically youthful rift valley. The sediment-filled basins are shown in orange.

Albuquerque, Socorro, and Las Cruces in New Mexico, El Paso in Texas, and Ciudad Juarez in Mexico. Historic earthquakes have had only small to moderate magnitudes, but the continental lithosphere continues to extend, thus presenting a real hazard for large earthquakes.

Intraplate Earthquakes: "Stable" Central United States

The map of earthquake epicenters in the United States (see figure 5.18) shows that the western third of the country has an elevated level of seismic activity. But there are clusters of epicenters in the "stable" central and eastern United States, the intraplate regions away from the active plate edges. There are not as many epicenters, but some individual earthquakes are just as big. Seismic hazards are significant (see figure 5.19).

NEW MADRID, MISSOURI, 1811–1812

A succession of earthquakes rocked the sparsely settled central part of the Mississippi River Valley at the time of the War of 1812. Between 16 December 1811 and 15 March 1812, Jared Brooks, an amateur seismologist in Louisville, Kentucky, recorded 1,874 earthquakes. He classified eight of them as violent and another 10 as very severe. The four largest events occurred on 16 December 1811 (two), 23 January 1812, and 7 February 1812. The hypocenters were located below the thick pile of sediments where the Mississippi and Ohio rivers come together, at the upper end of the great Mississippi River **embayment** (figure 5.31). These major seisms are called the New Madrid earthquakes, taking their name from a Missouri town of 1,000 people. Although few people were killed, the destruction of ground and buildings at New Madrid tolled the end of its importance as "the Gateway to the West."

The following is excerpted from an eyewitness account of a New Madrid earthquake:

Accompanying the noise, the whole land was moved and waved like waves of the sea, violently enough to throw persons off their feet, the waves attaining a height of several feet, and at the highest point would burst, throwing up large volumes of sand, water, and in some cases a black bituminous shale, these being thrown to a considerable height, the extreme statements being forty feet, and to the top of the trees. With the explosions and bursting of the ground there were flashes, such as result from the explosion of gas, or from the passage of the electric fluid from one cloud to another, but no burning flames; there were also sulphuretted gases, which made the water unfit for use, and darkened the heavens, giving some the impression of its being steam, and so dense that no sunbeam could find its way through. With the

bursting of the waves, large fissures were formed, some of which closed again immediately, while others were of various widths, as much as thirty feet, and of various lengths. These fissures were generally parallel to each other, nearly north and south, but not all. In some cases instead of fissures extending for a considerable distance there were circular chasms, from five to thirty feet in diameter, around which were left sand and bituminous shale, which later would burn with a disagreeable sulphorous smell.

The region is composed of thick deposits of water-saturated, unconsolidated sands and muds dropped by the Mississippi River. These loose materials intensified the shaking of the earthquakes, and the weak sediments flowed like water, erupted as sand volcanoes, and in places quivered like Jell-O. Several long-lasting effects of the New Madrid earthquakes can still be seen in the topography. A 240 km (150 mi) long area alongside the Mississippi River sank into a broadly depressed area, forming two new lakes: Lake St. Francis, 60 km (37 mi) long and 1 km (0.62 mi) wide; and Reelfoot Lake in Tennessee, 30 km (19 mi) long, 11 km (7 mi) wide, and up to 7 m (23 ft) deep. Reelfoot Lake, now a bird sanctuary, hosts the gray trunks of cypress trees drowned more than 200 years ago; they still stand as silent testimony to the area's earth-wrenching events (figure 5.32). Other topographic features created by the seisms include (1) long, low cliffs across the countryside and streams with new waterfalls up to 2 m (7 ft) high; (2) domes as high as 6 m (20 ft) and as long as 24 km (15 mi); and (3) former swamplands uplifted and transformed into aerated soils.

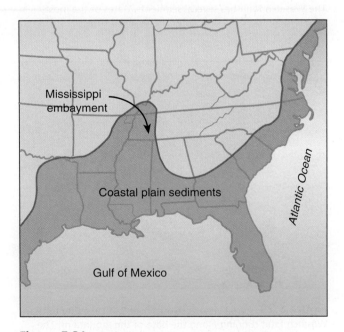

Figure 5.31 Map of coastal-plain sediments deposited by rivers eroding North America. The Mississippi River embayment contains a mass of soft sediments.

Figure 5.32 Trunks of cypress trees drowned in Reelfoot Lake after water was dammed by the New Madrid earthquakes. (These drowned trees are analogous to those drowned by the January 1700 Pacific Northwest magnitude 9 earthquake; see figure 4.17.)

© Guy E. Mitchell/National Geographic Creative/Corbis.

Felt Area

The New Madrid earthquakes have never been equaled in the history of the United States for the number of closely spaced, large seisms and for the size of the felt area (figure 5.33). The earthquakes were felt from Canada to the Gulf of Mexico and from the Rocky Mountains to the Atlantic seaboard, where clocks stopped, bells rang, and plaster cracked. These big earthquakes were not a freak occurrence.

The oral history of the local American Indians tells of earlier dramatic events.

Assessments of the earthquakes based on felt area yield magnitude estimates of 8 to 8.3. But are the sizes of the felt areas in figure 5.33 a good indicator of earthquake magnitude? Were the New Madrid seisms many times bigger than the 1906 San Francisco earthquake? Not necessarily. The size of the felt area is related to the types of rocks being vibrated. The New Madrid seisms shook the rigid basement rocks (more than 1 billion years old) of the continental interior. They rang like a bell, and the seismic energy was transmitted efficiently and far. The San Francisco earthquake took place in younger, tectonically fractured rocks that quickly damped out the seismic energy, thus confining the shaking to a smaller area.

Magnitudes

The first of four big earthquakes occurred on 16 December 1811 and seems to have occurred on the Cottonwood Grove fault as 13 ft (4 m) of slip along a 37 mi (60 km) rupture length (figure 5.34). This seism is likely to have triggered two ruptures on the Reelfoot blind-thrust fault at New Madrid, Missouri: one also on 16 December 1811 and the largest of the series on 7 February 1812. The large earthquake that occurred on 23 January 1812 has been the most difficult to locate. At present, it seems the earthquake epicenter was about 125 mi (200 km) to the north, around southern Illinois. If this interpretation holds up, the hazards associated with midcontinent earthquakes are more widely distributed than is commonly recognized.

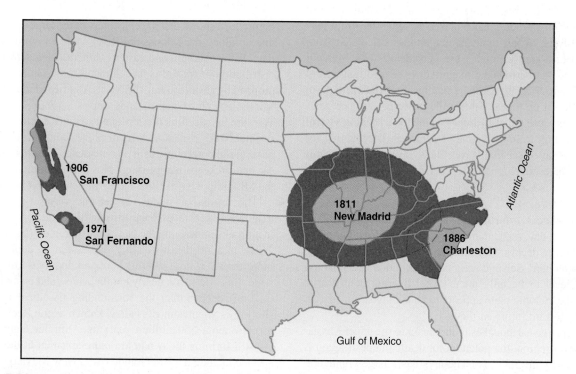

Figure 5.33 Felt areas of some large earthquakes in the United States. Orange areas are Mercalli intensities greater than VII; red areas are intensities of VI to VII.

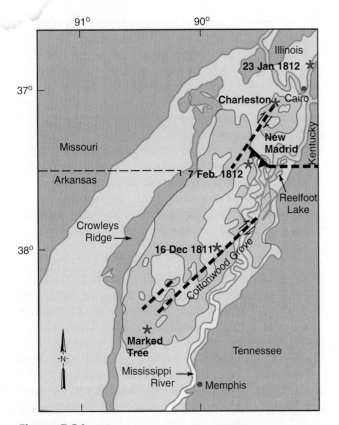

Figure 5.34 Map of the New Madrid region showing epicenters of large earthquakes and faults located using aftershocks. Marked Tree, Arkansas, was the site of a ~6M event in 1843; Charleston, Missouri, had a ~6M seism in 1895. Faults are indicated by dashed lines.

The defined fault-rupture lengths are too short to have generated magnitude 8 earthquakes as suggested from felt-area analysis. When moment magnitudes calculated from fault surface-area estimates are considered with Mercalli intensities, the earthquake magnitudes seem to range from about 7.3 to 7.7. However, remember that the earthquake epicenters sit on top of a thick pile of water-saturated sediment. This loose material amplifies the local shaking several times, leading to high Mercalli intensities. Discounting for amplification of seismic waves leads to earthquake magnitude estimates, in chronologic order, of 7.3, 7.0, 7.0, and 7.5.

The Future

Can aftershocks from the 1811–1812 earthquakes still be occurring today, more than 200 years later? Yes, but aftershock energy decays with time, and the expected decay rate is not seen here. Seismic activity in the region occurs at too high a level to be only aftershocks. Internal deformation within the tectonic plate appears to be adding strain energy, building toward future earthquakes.

When large earthquakes return again to the upper Mississippi River region, the potential for death and destruction is sobering (figure 5.35). (1) The area has a large population (e.g., Memphis, St. Louis, Nashville); (2) many buildings were not designed to withstand large seisms; (3) the wide

extent and great thickness of soft sediments will amplify seismic vibrations (remember the 1985 Mexico City and 1989 World Series events); and (4) a very large area will be subjected to strong shaking. The effects of a magnitude 7.5 earthquake could include deaths in the thousands and damages in the tens of billions of dollars.

How frequent are large earthquakes here? Paleoseismologic analyses of sediment and wood indicate major earthquake clusters occurred around 2350 BCE, 900 CE, 1450 CE, and 1812 CE. A magnitude 7 or greater earthquake could occur here about every 500 years. There is a 7–10% chance of one in the next 50 years. If there is good news, it is the low frequency of occurrence for these large earthquakes. The lessons of history here must be learned, and all new construction in the region should be built to withstand major earthquakes.

The earthquake threat in this region also includes lesser events. In 1843, it was an ~6M event at Marked Tree, Arkansas, and in 1895, it was an ~6M seism at Charleston, Missouri (figure 5.34). Paleoseismologic analyses in trenches cut across faults and folds has led the US Geological Survey to forecast a ~25% chance of a magnitude 6 to 7 earthquake here within the next 50 years. Although earthquakes in the central United States have low probability, they have high impact.

Since New Madrid sits in the continental interior, away from the active plate edges, why do big earthquakes occur here? The answer is unresolved, but a look at the geologic history of the region provides some understanding.

REELFOOT RIFT: MISSOURI, ARKANSAS, TENNESSEE, KENTUCKY, ILLINOIS

Figure 5.34 shows that the epicenters of the large earthquakes line up along the Mississippi River Valley. Figure 5.31 is a map of the southern and eastern United States depicting the distribution of coastal-plain sediments—the sands and muds dropped by rivers eroding the North American landmass. The Mississippi embayment stands out as a prominent feature. Why are these sediments deposited so much farther into continental North America? Why does the sediment distribution parallel the epicenters? Is it a coincidence that this same linear pattern keeps reappearing? No. Is it random chance that the Mississippi River flows along the course that it does? No.

The results of studies of seismic waves, gravity, and magnetism define a linear structural feature in the basement rocks underlying the New Madrid region (figure 5.36). There is a northeast-trending depression at depth that is more than 300 km (190 mi) long and about 70 km (43 mi) wide. It is linear, has nearly parallel sides, and is about 2 km (1.2 mi) deeper than the surrounding basement rocks. In short, it is an ancient rift valley, known as the Reelfoot rift, formed about 550 million years ago. Similar features that are still forming today and are more apparent at the Earth's surface include the Rio Grande rift in New Mexico (see figure 5.30) and the East African Rift Valley (see figure 4.6). The ancient Reelfoot rift was filled and covered by younger

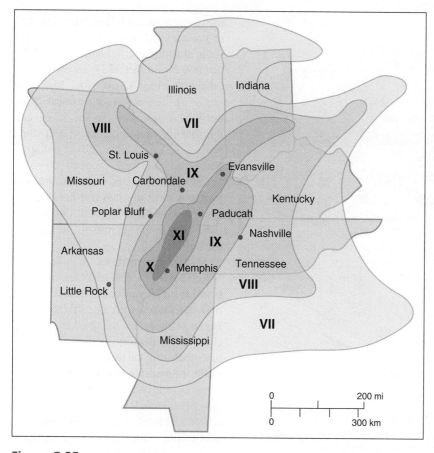

Figure 5.35 Map showing estimated Mercalli intensities expected from a recurrence of an 1811–1812 New Madrid earthquake. Intensity VIII and above indicates heavy structural damage.

Source: R. M. Hamilton and A. C. Johnston, "Tecumseh's prophecy: Preparing for the next New Madrid earthquake" in *US Geological Survey Circular 1066.*

sediments (figure 5.36). Today, the opening Atlantic Ocean basin pushes North America to the west-southwest, and some of the ancient faults of the Reelfoot rift are being reactivated to produce the region's earthquakes.

Isostatic Rebound as an Earthquake Trigger

Although stress from plate-tectonic movements elastically strains the rocks, what triggers the earthquakes? A new hypothesis by Purdue University geologist Eric Calais and colleagues suggests that isostatic rebound may be the trigger for movement of the ancient faults. As the North American ice sheet retreated northward from the United States and Canada between 16,000 to 10,000 years ago, enormous volumes of glacial meltwater poured through the Mississippi and Ohio river systems eroding huge volumes of sediment as they flowed. Examination of existing sediment layers shows that the upper Mississippi River eroded the region and carried away a 12 m (40 ft) thickness of sediments. With removal of this heavy load of sediments, the land rebounded upward, reducing the stresses that held the underlying faults

in place. Some faults that were close to failure have failed; other faults will follow.

ANCIENT RIFTS IN THE CENTRAL UNITED STATES

It is the fate of all continents to be ripped apart from below. Continents are rifted and then drifted and reassembled in different patterns. Sometimes the rifting process stops before separating a continent. The Reelfoot rift, now occupied by the Mississippi River, is a prominent **failed rift.** Other failed rifts, from different plate-tectonic histories, also exist beneath the surface in North America (figure 5.37).

Failed rifts remain as zones of weakness that may be reactivated by later plate-tectonic stresses to once again generate earthquakes. Because failed rifts are deeply buried, they are difficult to study. Yet they raise significant questions. What are the frequencies of their major earthquakes? In general, the recurrence intervals for major earthquakes appear to be from a few hundred to more than a thousand years. How great an earthquake might be produced at each rift? The New Madrid earthquake series offers a sobering benchmark. Approximately 83% of the large earthquakes recorded in the central United States are at or near the sites of ancient rifts.

The buried, ancient rifts in figure 5.37 correlate with active fault zones at the surface. There are several examples. (1) The St. Louis arm corresponds to the Ste. Genevieve fault zone. (2) The Rough Creek rift is expressed as the Rough Creek fault zone, appearing to continue eastward as the Kentucky River fault zone. Trenches dug across the Rough Creek fault have exposed the sedimentary records of reverse-fault movements with 1.1 m (more than 3.5 ft) of offset. (3) The southern Oklahoma rift corresponds with the frontal-fault system of the Wichita Mountains. Although this zone does not generate earthquakes at present, the land surface testifies to major earthquakes. The Meers fault is dramatic enough to make any Californian proud, but this fault strikes N 63°W across southwestern Oklahoma. Its fault scarp is 5 m (16 ft) high and 27 km (17 mi) long, and it has left-lateral offset up to 25 m (82 ft). At least two major fault ruptures have occurred there in geologically recent time. (4) The southern Indiana arm is overlain by the Wabash Valley fault zone, which appears to connect with the New Madrid zone. Prehistoric earthquakes read in the sedimentary record suggest seisms with m_b equal to 6.3 to more than 7. Damaging earthquakes occur in the area about once a decade. Examples include a magnitude 5.5 in November 1968, a magnitude 5.2 in June 1987, and a magnitude 5 in southwest Indiana on 18 June 2002.

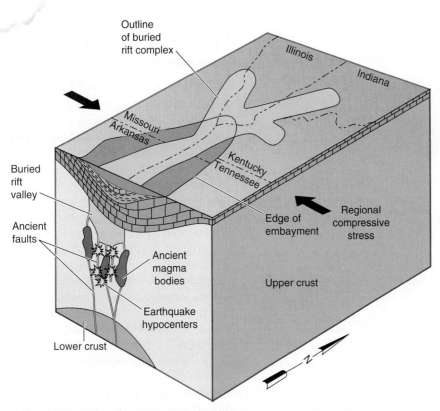

Illinois

Indiana

Outline
of buried
rift complex

Missouri

Arkansas

Kentucky
Tennessee

Edge of
embayment

Regional
compressive
stress

Buried
rift
valley

Ancient
faults

Ancient
magma
bodies

Upper crust

Earthquake
hypocenters

Lower crust

N

Figure 5.36 Schematic block diagram of the Reelfoot rift, the ancient failed rift valley beneath the upper Mississippi River embayment. Large earthquakes are likely caused by present tectonic stresses triggering failures on ancient faults.

Source: *US Geological Survey Professional Paper 1236L.*

Intraplate Earthquakes: Eastern North America

The large earthquakes of eastern North America share characteristics with those of central North America. Most occur at sites of ancient rift valleys. Most lack significant recent faults. The regions have low strain rates. Figure 5.38 shows some rift arms developed 220 to 180 million years ago as Pangaea was torn apart. Some rift arms succeeded, combining to create today's Atlantic Ocean basin. Other rift arms failed and left behind weakened zones within continents.

NEW ENGLAND

New England has a long record of significant earthquakes. On 11 June 1638, just 18 years after the Pilgrims landed in Plymouth, Massachusetts, a sizable earthquake rocked them. It rattled dishes, shook buildings, and in general frightened the Europeans, who were unfamiliar with earthquakes. Due to the limited number of settlements, it is difficult to pinpoint the location of the fault movement that generated this earthquake. However, a suggested epicentral site lies offshore from Cape Ann; estimates of Mercalli intensity range all the way up to IX, and a magnitude estimate based on felt area is 5.5. It has been suggested that this earthquake was

merely an aftershock from a magnitude 7 or greater earthquake that occurred offshore at a much earlier date.

On 9 November 1727, an earthquake rattled the East Coast from Maine to Delaware. The epicenter was near Newbury, Massachusetts (figure 5.39), and the shaking caused chimneys and stone walls to fall and cellar walls to collapse. Some uplands were dropped down to become wet lowlands, and some wet lowlands were uplifted and became dry enough to support grasses. **Quicksand** conditions were common during the earthquake.

Shortly before dawn on a frigid 18 November 1755, the entire eastern seaboard from Nova Scotia to South Carolina was shaken with an earthquake that began offshore from Cape Ann, Massachusetts. In Boston, so many chimneys reportedly toppled that some streets were made impassable by the debris. The seism is estimated to have had a magnitude of about 6.3, but the shaking was so severe that residents reported seeing the land rolling with waves like the surface of the sea. This earthquake occurred just 17 days after the epic earthquakes in Lisbon, Portugal, and it fired up the doom-and-gloom preachers who saw the seism as just punishment for the sins of New Englanders.

Many of these earthquakes may be related to the faults that bound former rift valleys (figure 5.38). The ancient faults may be reactivating and failing due to current stresses. Do all the rift-bounding faults have the potential for future seismic activity? The historic record is not long enough to properly answer this question. But if the answer is yes, then virtually the length of the Atlantic Coastal province could receive a significant shake sometime.

When the next magnitude 6 or greater earthquake strikes the eastern United States, the resultant destruction is likely to be proportionately greater than for a similar seism in the western part of the country. In the East, earthquake energy is transmitted more effectively in the older, more solid rocks, so damages may be experienced over a wider area. Also consider (1) the population density of the East, (2) the large number of older buildings not designed to withstand earthquake shaking, and (3) the concentration of industrial and power-generating facilities, including nuclear reactors.

ST. LAWRENCE RIVER VALLEY

The St. Lawrence is another river whose present path results from occupying an ancient tectonic structure. Some 600 to 500 million years ago, a major rift valley extended through

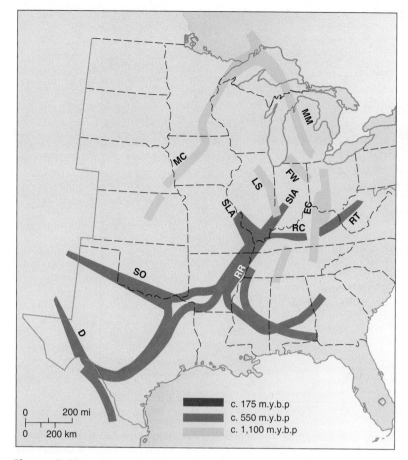

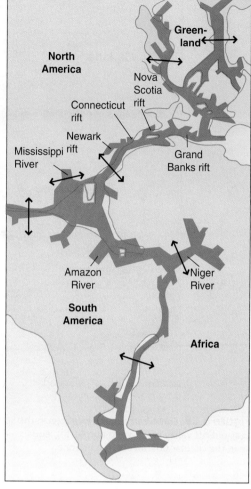

Figure 5.37 Map showing approximate locations of buried, ancient rifts in the central United States. Rifting occurred during three principal times—around 220 to 175 million years ago (red); 600 to 500 million years ago (purple); and 1,100 to 1,000 million years ago (green). Some older rifts were apparently rifted again under later plate-tectonic regimes. Rifts are: D, Delaware; EC, East Continent; FW, Fort Wayne; LS, La Salle; MC, Mid-Continent; MM, Mid-Michigan; RC, Rough Creek; RR, Reelfoot Rift; RT, Rome Trough; SIA, Southern Indiana Arm; SLA, St. Louis Arm; and SO, Southern Oklahoma.

Source: D. W. Gordon, *US Geological Survey Professional Paper 1364*.

Figure 5.38 Schematic map of rifts that tore at Pangaea about 220 million years ago. Successful rifts combined to open the Atlantic Ocean basin.

the region (figure 5.39). This now-buried rift coincides with most of the significant earthquakes in southeastern Canada. Seisms within the rift valley commonly reach magnitude 7, yet in unrifted continent nearby, the largest earthquakes are usually only in the magnitude 5 range.

The most active area along the St. Lawrence River Valley is an 80 km (50 mi) by 35 km (22 mi) zone near Charlevoix, northeast of Quebec City. Here, earthquakes of magnitudes 6 to 7 occurred in 1534, 1663, 1791, 1860, 1870, and 1925. Why the concentration of large seisms in this one relatively small area? Charlevoix was the site of a meteorite impact some 350 million years ago. The impact caused intensive fracturing of the area, including faults in curving patterns. These impact-caused fractures are perhaps being reactivated today under the stresses generated by the opening Atlantic Ocean basin.

CHARLESTON, SOUTH CAROLINA, 1886

Charleston sits alongside a beautiful bay, a charming city with distinguished buildings, wide boulevards, and inviting gardens. The presence of the port helped the city develop as a wealthy trading center. Yet Charleston has another side. In the mid-1800s, it was a hotbed of secessionist fervor; the first shots of the Civil War were fired over its harbor at Fort Sumter on 12 April 1861. After the war ended four years later, Charleston was

> a city of ruins, of desolation, of vacant houses, of widowed women, of rotting wharves, of deserted warehouses, of acres of pitiful and voiceless barrenness.

Yet by the mid-1880s, Charleston had been restored as a center of wealth, aesthetic buildings, and cultural achievement. Even the damages wrought by an 1885 hurricane were

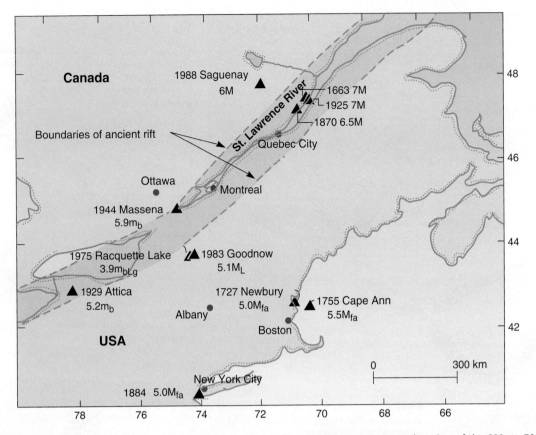

Figure 5.39 Some earthquake locations in the St. Lawrence River valley area. The approximate location of the 600- to 500-million-year-old rift valley is shown in purple. M$_{fa}$ equals magnitude estimated from the felt area. Large earthquakes northeast of Quebec City lie in the circular Charlevoix seismic zone.

Figure 5.40 Damage from the 1886 earthquake in Charleston, South Carolina.

J.K. Hiller/U.S. Geological Survey.

not enough to slow the city. Then came 31 August 1886, a typical sultry summer day. At 9:50 p.m., the quiet, breezeless evening was shattered by the largest earthquake to occur east of the Appalachian Mountains in historic time. Sixty seconds of shaking left 60 people dead, and once again, the remaining citizens had to put their city back together. About 90% of the buildings were damaged or destroyed (figure 5.40).

The earthquake had a magnitude estimated at 7.3. The event produced no surface faulting, so the fault movement may have occurred below 20 km (12 mi) depth. The large magnitude corresponds to a rupture length of about 30 km (19 mi) and a rupture width on the fault surface of about 19 km (12 mi). The seism was felt over a large area, and damages were widespread as well (figure 5.41). The large felt area is typical of the eastern United States and is largely attributable to the persistence of longer-period seismic waves (e.g., 1 second), which simply do not die down as quickly as they do in the western United States. The continental rocks at depth are geologically old and rigid,

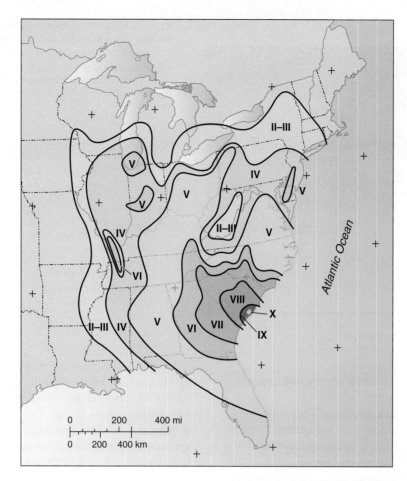

Figure 5.41 Mercalli intensity map for the 31 August 1886 earthquake near Charleston, South Carolina.

Source: G. A. Bollinger, *US Geological Survey Professional Paper 1028,* 1977.

causing the region to "ring like a bell" and transmit seismic waves far and wide.

How rare is an earthquake of this size for Charleston? Sediments exposed in trench walls, augmented by radiocarbon dates, tell of at least five other similar-sized earthquakes in the area in the past 3,000 to 3,600 years. Thus, large seisms may be expected about every 600 years.

Earthquakes and Volcanism in Hawaii

When we think about natural hazards in Hawaii, it is volcanism that comes to mind. But the movement of magma can cause earthquakes, including large ones (table 5.5). When rock liquefies, its volume expands, and neighboring brittle rock must fracture and move out of the way. The sudden breaks and slips of brittle rock are fault movements that produce earthquakes. When magma is on the move at shallow depths, it commonly generates a nearly continuous swarm of relatively small earthquakes referred to as **harmonic tremors.**

TABLE 5.5
Some Large Earthquakes in Hawaii

Date	Location	Intensity	Magnitude
2 Apr 1868	Southeast Hawaii	X	7.9
5 Oct 1929	Honualoa, Hawaii	VII	$6.5 \, M_s$
22 Jan 1938	North of Maui	VIII	$6.7 \, M_s$
25 Sep 1941	Mauna Loa, Hawaii	VII	$6.0 \, M_s$
22 Apr 1951	Kilauea, Hawaii	VII	6.5
21 Aug 1951	Kona, Hawaii	IX	6.9
30 Mar 1954	Kalapana, Hawaii	VII	6.5
26 Apr 1973	Southeast Hawaii	VIII	6.3
29 Nov 1975	Southeast Hawaii	VIII	$7.2 \, M_s$
16 Nov 1983	Mauna Loa, Hawaii	VII	$6.6 \, M_s$
25 Jun 1989	Kalapana, Hawaii	VIII	6.5
15 Oct 2006	Kalaoa, Hawaii	VIII	$6.7 \, M_w$

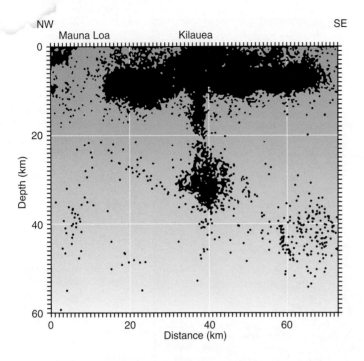

NW   SE
Mauna Loa   Kilauea

Figure 5.42 Cross-section showing hypocenters beneath Kilauea Volcano on the flank of the larger Mauna Loa Volcano, southeastern Hawaii, 1970–1983.

From F. W. Klein and R. Y. Koyanagi, "The Seismicity and Tectonics of Hawaii" in *Geological Society of America, Decade of North American Geology.* Vol. N. Copyright © Geological Society of America. Reprinted with permission.

Figure 5.42 shows that the earthquakes below Kilauea Volcano are dominantly near-surface events.

Magma movements also cause larger-scale topographic features and larger earthquakes with magnitudes in the 6s and 7s. The land surface is commonly uplifted due to the injection of magma below the ground surface. But the land surface is also commonly down-dropped due to withdrawal of magma. Figure 5.43 shows some down-dropped valleys on Kilauea; the valley walls are normal faults.

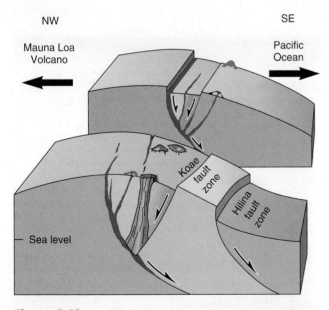

NW   SE

Mauna Loa   Pacific
Volcano   Ocean

Figure 5.43 Schematic block diagrams of the southeastern flank of Kilauea Volcano. Intruding magma (red) forces brittle rock to break and move, generating earthquakes. Gravity-aided sliding down normal faults causes more earthquakes as rock masses slide south-eastward into the ocean and cause rare mega-tsunami.

Kilauea is "supported" on the northwest by the gigantic Mauna Loa volcano and the mass of the Big Island of Hawaii. However, on its southeastern side, there is less support; Kilauea drops off into the Pacific Ocean. The effects of subsurface magma movement, both compressive during injection and extensional during removal, combine with gravitational pull to cause large movements on normal faults.

EARTHQUAKE IN 1975

On 29 November, one of the seaward-inclined normal faults moved suddenly in a $7.2M_s$ seism. It happened at 4:48 a.m., when a large mass slipped for 14 seconds with a movement of about 6 m (20 ft) seaward and 3.5 m (11.5 ft) downward. The movement of this mass into the sea caused tsunami up to 12 m (40 ft) high. Campers sleeping on the beach were rudely awakened by shaking ground; those who didn't immediately hustle to higher ground were subjected to crashing waves. Two people drowned. This fault movement had an effect on subsurface magma analogous to shaking a bottle of soda pop—gases escaping from magma unleashed an 18-hour eruption featuring magma fountains up to 50 m (165 ft) high.

EARTHQUAKES IN 2006

On 15 October, the Big Island of Hawaii was rocked by two large earthquakes just 7 minutes apart. The first seism was $6.7M_w$ at the hypocentral depth of 40 km (25 mi); the second was $6.0M_w$ at 20 km (12 mi) depth. The earthquakes seem to have resulted from the heavy load that the huge island of Hawaii places on the lithosphere. They occurred where the lithosphere is bent or flexed the most. The initial deeper earthquake resulted from tensional forces pulling rock apart. The following shallower earthquake occurred due to compressional forces pushing rock together.

Summary

Fault movement is complex; movement lasts only a few seconds at one spot, it speeds up and slows down, it slips different amounts in different sections, and it can trigger other activity in the direction of its movement.

The large left step in the San Andreas fault in the Los Angeles area causes compressive ruptures along east-west-oriented thrust faults, as in the 1971 San Fernando and 1994 Northridge events. More Northridge-type earthquakes are likely.

Prehistoric earthquakes may be interpreted using faulted pond sediments. The amount of offset of sediment layers is proportional to earthquake magnitude. Organic material in sediment layers can be dated by measuring the amount of radioactive carbon present. These techniques were used in 1988 to forecast a 30% probability of a magnitude 6.5 earthquake in the Loma Prieta area by the year 2018. In 1989, a magnitude 6.9 event occurred.

Southern California may have several large earthquakes in the 21st century. The southern segment of the San Andreas fault is the only one not to have a long rupture in historic time. In prehistory, it has ruptured every 250 years on average, but the last big movement was in 1690. The next big earthquake in California quite possibly will be a magnitude 7.8 event rupturing 300 km (185 mi) of the fault.

Humans have triggered earthquakes by pumping water underground under pressure; by building dams and impounding water, which seeps underground under pressure; and by underground explosions of atomic bombs.

Earthquakes occur throughout North America. Most are in the West along the edges of the active plates, but the central and eastern regions also have earthquakes—not as many, but some of them large.

In the Pacific Northwest, earthquakes occur 30 to 70 km (20 to 45 mi) deep within the subducting oceanic plate as minerals change form due to increased temperature and pressure. At the surface, strike-slip faults rupture the ground, as in Seattle.

The Basin and Range province between eastern California and central Utah is an actively extending area. For example, Nevada has about doubled in west-east width in the past 30-million years. Normal faults accommodate most of the extension, unleashing earthquakes up to magnitude 7.3.

In the central United States and eastern North America, ancient rift valleys remain from failed spreading centers. The ancient rifts today are zones of weakness whose faults can be reactivated due to long-distance effects of Atlantic plate spreading and Pacific plate collision. The Reelfoot rift, occupied today by the Mississippi River, had earthquakes in 1811 and 1812 with moment magnitudes (M_w) of 7.3, 7.0, 7.0, and 7.5. Other rift valleys are associated with earthquakes throughout North America.

Charlevoix, Quebec, has frequent earthquakes up to magnitude 7. The region was intensely fractured by the impact of an ancient asteroid, and the fractured rocks apparently move due to stresses within the moving plates.

The underground movement of magma in Hawaii generates earthquakes up to magnitude 7.9 by forcefully rupturing brittle rocks. Land is uplifted as magma is injected and dropped down when magma is removed; these land movements may be sudden, earthquake-generating events, and some create tsunami as well.

Terms to Remember

avalanche   112
blind thrust   110
coal   114
directivity   110
earthquake weather   116
embayment   130
failed rift   133
fracking   117
friction   108

harmonic tremors   137
hydraulic fracturing   117
neotectonics   113
paleoseismology   113
photosynthesis   114
quicksand   134
scarp   128
thrust fault   110

Questions for Review

1. Sketch a map, and explain the elastic-rebound theory of faulting. How has the theory been modified in recent years?
2. Draw a cross-section of a blind-thrust fault, such as the one that affected Northridge in 1994. Why was ground shaking so intense in this earthquake?
3. Draw a cross-section, and explain how faulted pond sediments can be used to tell the magnitudes and frequencies of ancient earthquakes.
4. During an earthquake, you probably will be safest if you _____, _____, and _____.
5. Explain three ways that humans have caused or triggered earthquakes.
6. Earthquakes are most abundant in which two U.S. states? In which Canadian province?
7. What types of evidence indicate major movements on surface faults in Washington?
8. As the depth to a hypocenter increases, how does that affect surface shaking?
9. What tectonic process has affected the Basin and Range province, from eastern California to Utah, during the past 30 million years? How much stretching has occurred across it? What is the orientation of the ranges and basins, and what does this tell us about the direction of stretching?

10. What type of fault movement best characterizes the Basin and Range province? What are the highest magnitude earthquakes generated there in the past century?

11. "Stable" central and eastern North America have earthquakes clustered in distinct areas. What is a likely control on these earthquake locations?

12. When and how did the Reelfoot rift form? Explain its history of earthquakes. What is the name of the strongest historical earthquake swarm that occurred here? What does the future hold?

13. How does intrusion of magma on the flanks of Kilauea volcano on Hawaii generate earthquakes?

14. What are the harmonic tremors experienced in Hawaii? What do they tell us?

15. Why are the western United States and Canada much more seismically active than the central and eastern regions? Does the smaller number of earthquakes in the central and eastern United States and Canada mean that we don't have to worry about a "big one" there?

16. Besides an actual downward subduction movement, how can a subducting plate generate earthquakes, as in Washington?

17. What is hydraulic fracturing? How can it trigger an earthquake?

18. What is meant by directivity of seismic waves? How significant is directivity?

19. Are seismic waves amplified more in hard rocks or in soft sediments? What happens to amplitude of seismic waves when they enter soft sediments? (See discussion under ShakeMaps.)

20. What event, unrelated to plate tectonics, weakened the crust along part of the St. Lawrence River Valley, making it easier to reactivate faults there?

Questions for Further Thought

1. So-called psychics are commonly quoted in the media predicting that California will break off along the San Andreas fault and sink beneath the sea. Is this possible? Why not?

2. Some people suggest that earthquakes usually occur at certain times of day. Does this make sense? Is there a pattern to the times of the earthquakes discussed in this text?

3. Assess the earthquake hazard in Salt Lake City.

4. What controls the course of the Rio Grande in New Mexico? Is there an earthquake threat also?

5. Compare the ability to withstand earthquake shaking of downtown buildings and bridges in West Coast cities to that of structures in the mid-continent or East Coast of the United States.

6. Humans can trigger earthquakes, but should we? Can we set off medium-size earthquakes in a controlled fashion that will prevent a large earthquake in an area? Make a list of pros and cons for earthquake control.

7. Could the Chinese earthquake of 2008 have been triggered by the recent filling of a nearby reservoir?

8. What are possible causes of the $5.8M_W$ earthquake in Virginia on 23 August 2011?

9. What are the pros and cons of using hydraulic fracturing to increase recovery of oil and natural gas?

CHAPTER 6

Volcanic Eruptions: Plate Tectonics and Magmas

"The simplest explanation that covers all the facts is the best one."

—OCCAM'S RAZOR, ATTRIBUTED TO *WILLIAM OF OCCAM*, C. 1295–1349

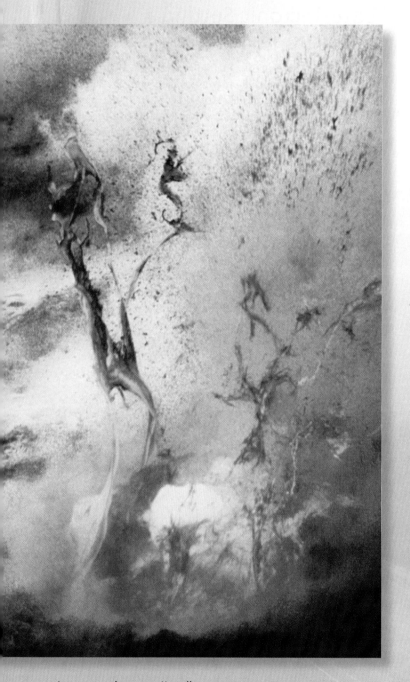

Lava meets the sea on Hawaii.
© StockTrek/Getty Images RF.

LEARNING OUTCOMES

Volcanic eruptions can be overwhelming events. Understanding of magma types and plate-tectonic settings explains a lot of volcanic behaviors. After studying this chapter, you should:

- know the plate-tectonic settings of volcanoes.
- understand the variations in magma characteristics that control peaceful versus catastrophic eruptions.
- comprehend the roles of magma gas content and pressure reductions in volcanic eruptions.
- be able to explain the three Vs of volcanism, and relate them to volcanic eruption styles, and to volcanic landforms.
- be familiar with the Volcanic Explosivity Index.
- be able to describe hot spots.

OUTLINE

- How We Understand Volcanic Eruptions
- Plate-Tectonic Setting of Volcanoes
- Chemical Composition of Magmas
- Viscosity, Temperature, and Water Content of Magmas
- How a Volcano Erupts
- The Three Vs of Volcanology: Viscosity, Volatiles, Volume

olcanoes deal out overwhelming doses of energy no human can survive. The dangers of volcanic eruption are obvious, but the quiet spells between active volcanism are seductive. Some people are lured to volcanoes like moths to a flame, even those who should know better. On 14 January 1993, volcanologists attending a workshop in Colombia, as part of the international decade of natural disaster reduction, hiked into the summit crater of Galeras Volcano to sample gases and measure gravity. They were looking for ways to predict imminent eruptions. The volcano had been quiet since July 1992, but during their visit, an unexpected, gas-powered secondary eruption killed six in the scientific party—four Colombians, a Russian, and an Englishman. Their deaths were not an unusual event (table 6.1). They serve as a small-scale example of the larger drama played out when a volcano suddenly buries an entire city. During long periods of volcanic quiescence, people tend to build cities near volcanoes. For example, 400,000 people live on the flanks of Galeras Volcano, defying the inevitability of a large, life-snuffing eruption.

An individual volcano may be active for millions of years, but its eruptive phases are commonly separated by centuries of inactivity, lulling some into a false sense of security. Around 410 BCE, Thucydides wrote, "History repeats itself." We know well that those who do not learn the lessons of history are doomed to repeat them. Every year, people inadvertently sacrifice their lives to volcanic eruptions.

How We Understand Volcanic Eruptions

Two primary building blocks of knowledge are paramount to understanding volcanic eruptions:

1. Plate tectonics has given us great insight into earthquakes; now it will help us understand volcanoes.

2. Magmas (liquid rocks) vary in their chemical composition, their ability to flow easily, their gas content, and their volume. These variations govern whether eruptions are peaceful or explosive.

In this chapter we first take a brief look at plate tectonics and volcanism. Then we examine magma variations and how they control eruptive style. Finally, we apply this knowledge globally to understand why volcanoes occur where they do, why only some volcanoes explode violently, and how volcanoes can be understood using the 3 Vs—viscosity, volatiles, and volume.

Plate-Tectonic Setting of Volcanoes

Convection of heat in the mantle drives plate tectonics. More than 90% of volcanism is associated with the edges of tectonic plates (figure 6.1). Most other volcanism occurs above hot spots (see figure 2.22). More than 80% of Earth's magma extruded through volcanism takes place at the oceanic spreading centers. Solid, but hot and ductile, mantle rock rises upward by convection into regions of lower pressure, where up to 30–40% of the rock can melt and flow easily as magma on the surface (figure 6.2). The worldwide rifting process releases enough magma to create 20 km³ (about 5 mi³) of new oceanic crust each year. Virtually all this volcanic activity takes place below sea level and is thus difficult to view.

Subduction zones cause the tall and beautiful volcanic mountains we see at the edges of the continents, but the volume of magma released at subduction zones is small compared to that of spreading centers. Subduction zones account for the eruption of 7–13% of all magma. The downgoing plate carries oceanic-plate rock covered with water-saturated sediments into much hotter zones (figure 6.2).

		Total	Dead
Year	Volcano	Deaths	Volcanologists
1951	Kelut, Indonesia	7	3
1952	Myojin-sho, Japan	31	9
1979	Karkar, New Guinea	2	2
1980	St. Helens, United States	62	2
1991	Unzen, Japan	44	3
1991	Lokon-Umpong, Indonesia	1	1
1993	Galeras, Colombia	9	6
1993	Guagua Pichincha, Ecuador	2	2
2000	Semeru, Indonesia	2	2

TABLE 6.1

Volcanologists Killed by Eruptions

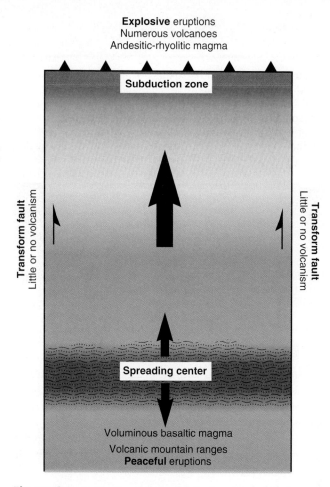

Figure 6.1 An idealized oceanic plate showing styles of volcanism.

The presence of water lowers the melting point of rock. Rising magma partially melts some of the continental crust it passes through. This adds new melt of different composition to rising plumes of magma. Each rising plume has its own unique chemical composition.

Transform faults and continent-continent collision zones have little or no associated volcanism. Thinking three-dimensionally, this is understandable. At a transform fault, the two plates simply slide past each other in a horizontal sense and at all times keep a quite effective "lid" on the hot asthenosphere some 100 km (60 mi) below. At continent-continent collisions, the continental rocks stack up into extra-thick masses that deeply bury the hot mantle rock, making it difficult for magma to rise to the surface.

From a volcanic disaster perspective, the differences are clear. Oceanic volcanoes are relatively peaceful, whereas subduction-zone volcanoes are explosive and dangerous. Ironically, humans tend to congregate at the seaward edges of the continents, where the most dangerous volcanoes operate.

People commonly speculate upon whether an individual volcano is active, dormant, or extinct. Because of the strong hope that a volcano is extinct and the nearby land is thus available for use, many dormant volcanoes are misclassified. But consider this: a subduction zone commonly lasts for tens of millions of years, and its province of volcanoes is active for the entire time. An individual volcano may be active for hundreds of thousands to several million years, despite "slumbers" of centuries between eruptions. As a general rule, if a volcano has a well-formed and aesthetic conical shape, it is active. A pretty shape is dangerous.

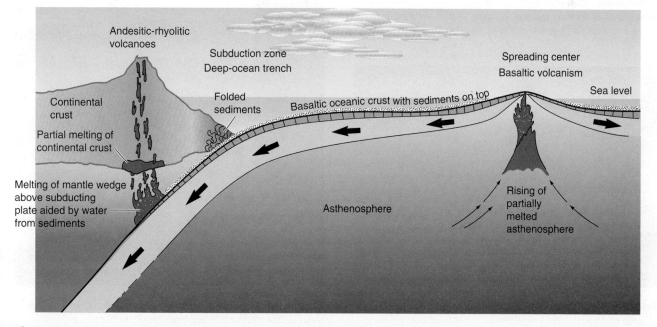

Figure 6.2 Idealized cross-section showing production of basaltic magma at spreading centers. Plates pull apart, and some asthenosphere liquefies and rises to fill the gap. Andesitic-rhyolitic magmas are created above subduction zones, where rising magma partially melts the continental crust on its way up, thus altering the melt by increasing SiO_2 content and viscosity.

A Classic Disaster

Eruption of Mount Vesuvius, 79 CE

The most famous of all volcanoes is probably Vesuvius in Italy, and the most famous of all its eruptions must be those of 79 CE. It was then that the cities of Pompeii and Herculaneum were buried and forgotten for more than 1,500 years. A warning of the natural hazards near Vesuvius arrived on 5 February 62 CE, when a major earthquake destroyed much of Pompeii and caused serious damage in Herculaneum and Neapolis (Naples). Additional earthquakes, although not as large as this first one, were a common occurrence for the next 17 years.

Pompeii had been a center of commerce for centuries. In 79 CE, the city had a population of about 20,000 people, 8,000 of whom were slaves. Robert Etienne described it as:

> An average city inhabited by average people, Pompeii would have achieved a comfortable mediocrity and passed peacefully into the silence of history, had the sudden catastrophe of the volcanic eruption not wiped it from the world of the living.

The 24th of August 79 CE was a warm summer day, but then Vesuvius began erupting and the day became even hotter. Vesuvius blew out 4 km³ (1 mi³) of volcanic material. About half of the old volcanic cone was destroyed. A modern example of a similar eruption occurred in 1991 at Mount Pinatubo in the Philippines (figure 6.3). In Pompeii, great clouds of hot gas and volcanic ash flowed across the city, killing the people who had not fled (figure 6.4). Today, the excavated city is a major attraction for tourists (figure 6.5).

In 79 CE, the fine volcanic ashes settling out from the great heights of the eruption cloud affected a large region. Pliny the Younger was at Misenum and wrote:

> And now came the ashes, but at first sparsely. I turned around. Behind us, an ominous thick smoke, spreading over the earth like a flood, followed us. "Let's go into the fields while we can still see the way," I told my mother—for I was afraid that we might be crushed by the mob on

the road in the midst of darkness. We had scarcely agreed when we were enveloped in night—not a moonless night or one dimmed by cloud, but the darkness of a sealed room without lights. To be heard were only the shrill cries of women, the wailing of children, the shouting of men. Some were calling to their parents, others to their children,

Figure 6.4 Body casts of a family trapped and killed by a pyroclastic flow of hot gas and volcanic ash from Vesuvius in late August 79 CE.
© Alan Morgan.

Figure 6.5 Tourists walk through the heart of Pompeii away from Vesuvius, the volcano that destroyed the city in 79 CE.
Photo by Pat Abbott.

Figure 6.3 The first big explosive blast from Mount Pinatubo occurred on 15 June 1991.
R.S. Culbreth/USAF/U.S. Geological Survey.

A Classic Disaster

others to their wives—knowing one another only by voice. Some wept for themselves, others for their relations. There were those who, in their very fear of death, invoked it. Many lifted up their hands to the gods, but a great number believed there were no gods, and that this was to be the world's last eternal night.

In the coastal city of Herculaneum, 300 skeletons were found in lifelike positions in boat chambers at the beach. The skeletons of these people killed by the eruption testify to the lethal energy they experienced. The people had not been battered or suffocated; they did not display any voluntary self-protection reactions or agony contortions. In other words, their vital organs stopped functioning in less than a second, in less time than they could consciously react. The types of bone fractures, tooth cracks, and bone coloration indicate the victims were covered by volcanic material at about 500°C (930°F). At this temperature, their soft tissues vaporized; their feet flexed in an instantaneous muscle contraction.

The timing of the major eruptions of Vesuvius offers an interesting lesson. Apparently Vesuvius did not have a major eruption from the 7th century BCE until 79 CE. People had at least 700 years to lose their fears and yield to the allure of the rich agricultural soils on Vesuvius. After 79 CE, large eruptions occurred more often: in the years 203, 472 (ash blown over much of Europe), 512, 685, 993, 1036 (first lava flows in historic time), 1049, and 1138–1139. Then nearly 500 years passed—plenty of time to forget the past and recolonize the mountain. But in 1631, Vesuvius poured out large volumes of lava that destroyed six towns; mudflows ruined another nine towns, and about 4,000 people perished. The two long periods of volcanic quiescence in the last 2,700 years seem like long times to short-living, land-hungry humans, but this is the time schedule of an active volcano. Humans' lack of appreciation for the time involved between eruptions leads them to falsely regard many active volcanoes as extinct.

Since 1631, the eruptions from Vesuvius have been smaller and not as dangerous. There were 18 eruption cycles between 1631 and 1944; each lasted from 2 to 37 years, with quiet intervals ranging from 0.5 to 6.8 years. Since 1944, Vesuvius has been quiet. Is this interval of calm setting the stage for another major eruption? We do not know for sure, but almost 3 million people live within reach of Vesuvius today, including about 1 million on the slopes of the volcano (figure 6.6).

Figure 6.6 Mount Vesuvius, near center of photo, is surrounded by the urbanized Naples region housing 3 million people.
NASA/GSFC/MITI/ERSDAC/JAROS, and U.S. Japan ASTER Science Team.

Why do spreading-center volcanoes have relatively peaceful eruptions? And why do subduction-zone volcanoes explode violently? The answers to these questions are found in knowing how different magmas behave.

Chemical Composition of Magmas

Although there are 92 naturally occurring elements, a mere eight make up more than 98% of the Earth's crust (table 6.2). The next four most abundant elements add another 1.2% to the crust, bringing the weight percent contributed by these 12 elements to 99.23%. The remaining 0.77% includes gold, silver, copper, carbon, sulfur, tin, and many other familiar elements.

Oxygen and silicon are so abundant that their percentages dwarf those of all other elements. Oxygen atoms carry negative charges (-2), while silicon atoms are positively charged ($+4$). As magma begins cooling, some silicon and oxygen atoms bond. Silicon and oxygen link up with four oxygen atoms ($4 \times -2 = -8$) surrounding a central silicon atom ($+4$) to form the silicon-oxygen tetrahedron (SiO_4) (figure 6.7). The SiO_4 tetrahedron presents a -4 charge on its exterior that attracts and ties up positively charged atoms. After negatively charged oxygen, the 11 elements of greatest abundance are all positively charged (table 6.2); they are attracted to, and bound up by, oxygen. This process is so common that elemental abundances in the crust are usually listed in combination with oxygen (as oxides). The weight percentages of elements are quite different for continental versus oceanic crust (table 6.3). Marked differences in silicon dioxide percentages produce magmas of variable eruptive behavior.

TABLE 6.2

Common Elements of the Earth's Crust (weight %)

Eight Most Common	
Oxygen (O^{-2})	45.20%
Silicon (Si^{+4})	27.20
Aluminum (Al^{+3})	8.00
Iron ($Fe^{+2,+3}$)	5.80
Calcium (Ca^{+2})	5.06
Magnesium (Mg^{+2})	2.77
Sodium (Na^{+1})	2.32
Potassium (K^{+1})	1.68
Total	98.03%
Next Four Most Common	
Titanium ($Ti^{+3,+4}$)	0.86%
Hydrogen (H^{+1})	0.14
Phosphorus (P^{+5})	0.10
Manganese ($Mn^{+2,+3,+4}$)	0.10
Total	99.23%

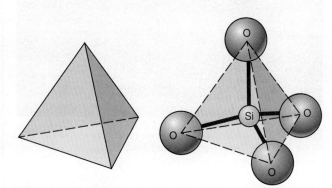

Figure 6.7 A silicon atom with +4 charge is linked to four oxygen atoms, each with a −2 charge.

TABLE 6.3

Crustal Elements in Weight-Percent Oxides

Continental Crust		Oceanic Crust	
SiO_2	60.2%	SiO_2	48.7%
Al_2O_3	15.2	Al_2O_3	16.5
Fe_2O_3	2.5	Fe_2O_3	2.3
FeO	3.8	FeO	6.2
CaO	5.5	CaO	12.3
MgO	3.1	MgO	6.8
Na_2O	3.0	Na_2O	2.6
K_2O	2.9	K_2O	0.4

Viscosity, Temperature, and Water Content of Magmas

Liquids flow freely; their volumes are fixed, but their shapes can change. Liquids vary in how they flow; some flow quickly, some flow slowly, and some barely flow at all. The fluidity of a liquid is measured by its **viscosity,** its internal resistance to flow; viscosity may be thought of as a measure of fluid friction. The lower the viscosity, the more fluid is the behavior. For example, tilt a glass of water and watch it flow quickly; water has low viscosity. Now tilt the same glass filled with honey and watch the slower flow; honey has higher viscosity. Low-viscosity magma flows somewhat like ice cream on a hot day. High-viscosity magma barely flows.

The viscosity of magma is changed by various means:

1. Higher temperature lowers viscosity; it causes atoms to spread farther apart and vibrate more vigorously. Thus atomic bonds break and deform more, resulting in increased fluidity. Consider the great effect of temperature on magma (table 6.5). At 600°C (1,100°F), magma viscosity is five orders of magnitude (100,000 times) more viscous than at 900°C (1,650°F).
2. Silicon and oxygen (SiO_2) increase the viscosity of magma because they form abundant silicon-oxygen tetrahedra (see figure 6.7) that link up in chains, sheets, and networks, creating more joins and bonds between atoms, which in turn make flow more difficult.
3. Increasing content of mineral crystals increases viscosity. Magma is a mixture of liquid and the minerals that have crystallized from it. The mineral content of magma varies from none to the majority of the mass.

Magma contains dissolved gases held as **volatiles;** their solubility increases as pressure increases and as temperature decreases. You can visualize the pressure–temperature relations using a bottle of soda pop. Carbon dioxide (CO_2) gas is dissolved in the soda pop and kept under pressure by the bottle cap. Pop the cap off the bottle, reducing pressure, and some volatiles escape. As the uncapped bottle warms, more volatiles are lost.

A good grasp of volcanic behavior can be gained by considering the properties of three types of magma (table 6.5). Notice that the highest temperatures and lowest SiO_2 contents are in basaltic magma, giving it the lowest viscosity and easiest fluid flow. The lowest temperatures and highest SiO_2 contents occur in rhyolitic magma, material so viscous that it commonly does not flow. Table 6.5 also states that about 80% of the magma reaching Earth's surface is basaltic, with only about 10% andesitic and 10% rhyolitic. Why the difference? Basaltic magma is produced in great abundance by partial melting of the mantle. The lower viscosity of basaltic magma helps it reach the surface, especially at spreading centers and other oceanic settings. Much basaltic magma is also produced at subduction

In Greater Depth

Minerals and Volcanic Rocks

The eight most common elements bond in different configurations to make up hundreds of different **minerals.** The process of mineral formation in a cooling magma is called **crystallization.** Just as with elements, a degree of simplicity occurs in a crystallizing magma because the overwhelming majority of Earth's crust is composed of just eight common rock-forming minerals. Laboratory experiments and microscopic examination of rock-forming minerals have shown the order in which these minerals crystallize from a cooling magma (figure 6.8).

Magmas at the surface with temperatures of around 800° to 1,300°C (1,470° to 2,370°F) have two separate lines of mineral growth:

1. Iron and magnesium link up with the silicon-oxygen tetrahedron as magma temperature decreases to sequentially form four distinct and discontinuous families of minerals—olivine, pyroxene, amphibole, and biotite mica.
2. Calcium combines with Al and SiO_4 to begin forming the plagioclase feldspar family, a continuous and gradational series of minerals. As temperature decreases, progressively more sodium (and less calcium) is locked within the plagioclase crystal structure. By the time magma has cooled down to the 800° to 1,000°C (1,470° to 1,830°F) range, it is largely depleted in Fe, Mg, and Ca. Now potassium crystallizes within muscovite mica and potassium-rich feldspar minerals, and excess Si and O combine without other elements to make the mineral quartz.

TABLE 6.4		
Igneous Rock Types		
Magma Type	**Plutonic Rock**	**Volcanic Rock**
$SiO_2 < 55\%$	Gabbro	**Basalt**
$SiO_2 = 55-65\%$	Diorite	**Andesite**
$SiO_2 > 65\%$	Granite	**Rhyolite**

Just as elements combine to make minerals, so minerals aggregate to make **rocks** (see page 4). Magmas have a broad range of compositions, resulting in many different types of igneous rocks that generations of geologists have classified into a dizzying array of rock names. Nonetheless, a working understanding can be gained by considering only three magma types and the three clans of igneous rocks that form from them. The rock types are based on their silicon and oxygen (SiO_2) percentages (table 6.4). If the magma cools and solidifies below the surface, it crystallizes as **plutonic rock,** named for Pluto, the Roman god of the underworld. If the magma reaches the surface, it forms **volcanic rock,** named for Vulcan, the Roman god of fire. The left side of figure 6.8 shows three main types of volcanic rock next to their respective mineral compositions. Table 6.4 tells more about the rocks, and figure 6.9 shows pictures of the rocks. Their magmas behave differently due to their varying temperatures, water contents, and viscosities.

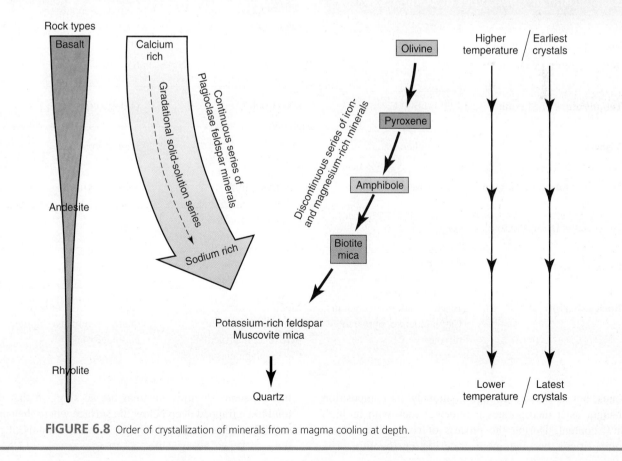

FIGURE 6.8 Order of crystallization of minerals from a magma cooling at depth.

In Greater Depth (Continued)

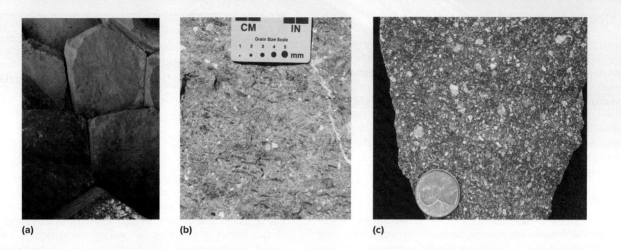

FIGURE 6.9 Volcanic rock types. (a) Columnar basalt at the Giant's Causeway in Northern Ireland, a UNESCO World Heritage site. (b) Andesite is medium-colored volcanic rock with plagioclase feldspar minerals and, in this sample, fragments of other volcanic rocks. Santiago Peak Volcanics, San Diego. (c) This sample of rhyolite contains quartz and feldspar minerals in great abundance, Poway, California.

Photos by Pat Abbott.

TABLE 6.5
Comparison of Three Types of Magma

	Basaltic	Andesitic	Rhyolitic
Volume at Earth's Surface	80%	10%	10%
SiO$_2$ Content	45–55%	55–65%	65–75%
	_____ Increasing SiO$_2$ ———————————————————→		
Temperature of Magma	1,000–1,300°C	800–1,000°C	600–900°C
	_____ Decreasing temperature ———————————————→		
Viscosity	Low (melted ice cream)		High (toothpaste)
	_____ Increasing viscosity ———————————————————→		
Water Dissolved in Magma	~0.1–1 weight %	~2–3 weight %	~4–6 weight %
	_____ Increasing water ———————————————————————→		
Gas Escape from Magma	Easy		Difficult
	_____ Increasing difficulty ——————————————————→		
Eruptive Style	Peaceful		Explosive
	_____ Increasing explosiveness ————————————————→		
Rock Description	Black to dark gray; contains Ca-plagioclase, pyroxene, olivine	Medium to dark gray; contains amphibole, pyroxene, intermediate Ca-Na-plagioclase	Light-colored; contains quartz, K-feldspar, biotite, Na-plagioclase

zones, but as it rises through continents, its composition changes as it incorporates continental rock with its high SiO$_2$ content. During the process of rising, the magma compositions become more andesitic or rhyolitic. The more viscous rhyolitic magmas are so sluggish that they tend to be trapped deep below the surface where they cool, solidify, and grow into the larger mineral crystals of plutonic rocks, such as granite.

Figure 6.10 Peaceful eruption of low-viscosity magma with easy separation of volatiles (gas) from magma, Surtsey, Iceland.

Courtesy Pat Abbott.

Figure 6.11 Eruption from Paricutin Volcano, Michoacan, Mexico.

Carl Fries/U.S. Geological Survey.

In magma, water is the most abundant dissolved gas. As magma rises toward the surface and pressure decreases, water dissolved in the hot magma becomes gas and forms steam bubbles. Basaltic magma is low in dissolved water content, helping make eruptions peaceful. Rhyolitic magma has dissolved water contents up to 6%; as it rises and steam bubbles form, they have difficulty escaping from the high-viscosity magma and have to burst their way out.

When the basaltic volcanoes of Hawaii and Iceland begin to erupt, it is a tourist event. Although such an eruption makes a thrilling show, it is a relatively peaceful happening. Why is it safe? Because it does not contain much dissolved water, and the dissolved gases escape from the low-viscosity magma with relative ease (figure 6.10). Compare this behavior to the eruption of a rhyolitic magma of lower temperature, greater dissolved water content, higher percentage of SiO_2, and very high viscosity. When rhyolitic magma oozes out onto the ground surface, the pressure within the magma is reduced and the dissolved gases expand in volume. But how do gases escape from their entrapment in sticky magma? By exploding (figure 6.11). Spectators at the eruption of rhyolitic magma frequently die. When it comes to volcanic hazards, the greatest problem is how much gas is in the magma and how easily the dissolved gases can escape from the magma. As Frank Perret stated: "Gas is the active agent, and magma is its vehicle."

PLATE-TECTONIC SETTING OF VOLCANOES REVISITED

Knowing about magma viscosity and volatile content allows us to revisit our earlier questions about plate tectonics and volcanism. Why does the vast majority of Earth's magma pour out at spreading centers and in relatively peaceful eruptions? And why does the magma above subduction zones commonly explode violently? Spreading centers operate in oceanic crust, and subducting plates commonly are pulled beneath continental crust. The chemistries of oceanic crust and continental crust are different (see table 6.3), their magmas are different (see table 6.5), and their volcanic behaviors differ.

Spreading centers are ideal locations for volcanism because (1) they sit above the high-temperature asthenosphere, (2) the asthenosphere rock has low percentages of SiO_2, and (3) the oceanic plates pull apart, causing hot asthenosphere rock to rise, experience lower pressure, and change to magma that continues to rise. This magma is high-temperature, low SiO_2, low-volatile content, low-viscosity basalt, allowing easy escape of gases (see figure 6.2). Spreading centers combine all the factors that promote the peaceful eruption of magma.

When a subducting oceanic plate reaches a depth of about 100 km (over 60 mi), magma is generated and rises toward the surface (see figure 6.2). The subducting plate stirs up the mantle, causing the hotter rock at depth to rise and then melt as pressure decreases. A significant reason magma forms here is that the subducting plate carries a cover of sediments, water, and hydrated minerals down with it. Water, even in slight amounts, promotes partial melting by lowering the temperature necessary for rock to melt. The partial melting process affects only those minerals with lower melting temperatures. As this partial melt rises upward, it in turn melts part of the overlying crust to produce magmas of highly variable compositions (figure 6.12). Magma compositions depend on the amount of crustal rock melted and incorporated into the rising magma. In general, in the subduction-zone setting, magma temperature decreases while SiO_2, water content, and viscosity increase. All these changes in magma add to its explosive potential.

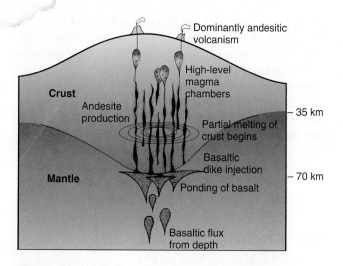

Figure 6.12 Schematic cross-section of magma rising from a subduction zone and being contaminated by crustal rocks en route.

How a Volcano Erupts

Earth's internal energy flows outward as heat (see chapter 2). The eruptions of volcanoes are rapid means for Earth to expel some of its internal heat.

A volcanic eruption begins with heat at depth. Superheated rock will rise to levels with lower pressure, and some solid rock may change phase to liquid magma, resulting in volume expansion and leading step-by-step to eruption.

Magma is generated by the melting of existing rock. Rock may melt by (1) lowering the pressure on it, (2) raising its temperature, or (3) increasing its water content. How do most rocks melt? The two most relevant melting agents are reductions in pressure (decompression) and increases in volatile content (mostly water).

Most magma is generated by decreasing the pressure on hot rock. For example, as the solid, but mobile, hot rock of the mantle rises upward, it experiences progressively less pressure and spontaneously melts, without the addition of more heat. Melting caused simply by a decrease in pressure is called **decompression melting.** The process of decompression melting is so important that it is worth restating: most of the rock that melts to form magma does so because the pressure on it decreases, not because more heat is added.

The largest nearby reservoir of superhot, ready-to-melt rock exists in the nearly molten asthenosphere. This rock, hot enough to flow without being liquid, is the main source of magma. As this superheated rock rises, the pressure on it decreases, allowing some rock to melt. The hot, rising rock-magma mixture also raises the temperature of rock it passes through, thus melting portions of the overlying rocks.

If pressure in the asthenosphere or lithosphere is decreased, some rock melts, with a resultant increase in volume that causes overlying rocks to fracture. The fractures allow more material to rise to lower pressure levels, causing more rock to liquefy. For example, at a depth of 32 km

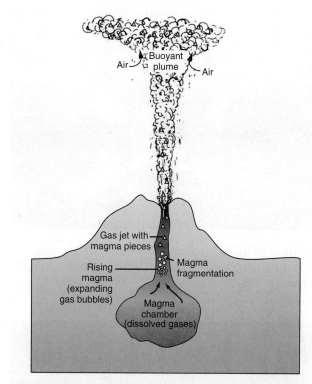

Figure 6.13 Anatomy of an eruption. As magma rises to levels of lower pressure, gas comes out of solution, forming bubbles that overwhelm the magma and create a gas jet leading to a buoyant plume.

(20 mi), basaltic rock melts at 1,430°C (2,600°F), but this same rock will melt at only 1,250°C (2,280°F) at the Earth's surface. Since upward-moving rock/magma reaches ever-lower pressures, rising rock can liquefy and magma can increase in fluidity, which in turn causes more superheated rock to become magma.

Magma at depth does not contain gas bubbles because the high pressure at depth keeps volatiles dissolved in solution. But as magma rises toward the surface, pressure continually decreases, and gases begin to come out of solution, forming bubbles that expand with decreasing pressure (figure 6.13). The added lift of the growing volume of gas bubbles helps propel magma upward through fractures or pipes toward an eruption. Gas bubbles continue increasing in number and volume as magma keeps rising upward to lower pressures. As gas-bubble volume increases, the gas can overwhelm magma, fragmenting the magma into pieces that are carried up and out by a powerful gas jet (figure 6.14). Upon escape from the volcano, the gas jet draws in air, which adds to buoyancy in the turbulent, rising plume (see figure 6.13).

ERUPTION STYLES AND THE ROLE OF WATER CONTENT

Whether a volcanic eruption is peaceful or explosive depends significantly on the concentration of water in the magma. For example, if all magmas contained low concentrations of

Figure 6.14 Remarkable view into the crater of Mount Pinatubo just as a major explosive eruption was beginning its upward blast, 1 August 1991.

T. J. Casadevall/U.S. Geological Survey.

Eruption Type	Composition
Icelandic type 	Low water content, low viscosity (basalt)
Hawaiian type 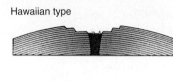	Low water content, low viscosity (basalt)
Strombolian type 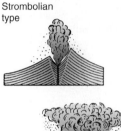	Moderate water content, low to moderate viscosity (basalt to andesite)
Vulcanian type 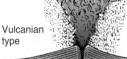	Moderate to high water content, moderate to high viscosity (basalt to rhyolite)
Plinian type 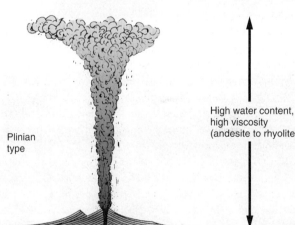	High water content, high viscosity (andesite to rhyolite)

Figure 6.15 Some types of volcanic eruptions.

water (such as 0.3 weight %; see table 6.5), there would be no highly explosive eruptions. Even a high-viscosity rhyolite magma with a low concentration of water only leads to slow flows or no flow as the magma oozes upward and builds a dome.

The most important requirement for explosive eruptions is high concentrations of volatiles (mainly water). Volatiles drive explosive eruptions. Given a high concentration of water, even a basalt magma can erupt violently, as occurred at Hawaii in 1790. Rhyolitic magma is often associated with explosive eruptions because of its high content of water (see table 6.5). Water concentration in magma plays a controlling role, and viscosity plays a secondary role, in determining the peaceful versus explosive style of eruption.

Different volcanic behaviors have been classified according to the eruptive style of individual volcanoes (figure 6.15). Nonexplosive eruptions are commonly subdivided into *Icelandic* and *Hawaiian* types. *Strombolian* types are somewhat explosive. Explosive eruptions can be described as *Vulcanian* or *Plinian* types. These classifications are just for general purposes; each volcano varies in its eruptive behavior over time.

SOME VOLCANIC MATERIALS

Magmas vary in their dissolved-gas (volatile) content and viscosity. Low-water-content, low-viscosity magma that reaches the surface typically moves as lava flows, with easy gas escape yielding nonexplosive eruptions. High-water-content, high-viscosity magma holds its volatiles, making gas escape difficult. Gas is forced to burst out of the magma, yielding explosive eruptions. Gas blasting into the atmosphere takes along chunks of magma and older rock known as **pyroclastic** debris (*pyro* = fire; *clastic* = fragments).

Nonexplosive Eruptions

Lava flows are especially typical of basaltic magma and exhibit a variety of textures (table 6.6). Highly liquid lava may cool with a smooth, ropy surface called **pahoehoe,** (pronounced pa-Hoy-Hoy) (figure 6.16). Slower-flowing, more viscous lava commonly has a rough, blocky texture called **aa** (pronounced ah-ah) (figure 6.17).

Explosive Eruptions

Gaseous explosions break rock and tear apart magma and older rock into pyroclastic debris with a wide range of sizes, from dust to huge blocks and bombs (table 6.6; figure 6.18).

TABLE 6.6
Volcanic Materials

Lava	Pahoehoe	Smooth, ropy surface
	Aa	Rough, blocky surface
	Pillow	Ellipsoidal masses formed in water
Pyroclastic Air-fall Fragments		
	Fine ash (dust)	Flour-size material
	Coarse ash	Sand size
	Cinders	Marble to baseball size
	Blocks	Big fragments, solid while airborne
	Bombs	Big fragments, liquid while airborne
	Volcanic tuff	Rock made of smaller fragments (e.g., deposit of a hot, gas-charged flow)
	Volcanic breccia	Rock made of coarse, angular fragments (e.g., deposit of a water-charged debris flow)
Glass	Obsidian	Nonporous glass
	Pumice	Porous glass (froth)

Figure 6.17 *Aa* flow from Puu Oo advances into Royal Gardens subdivision, Hawaii.

U.S. Geological Survey.

Magma reaching the surface can solidify so quickly that crystallization cannot take place because there is no time for atoms to arrange themselves into the ordered atomic structures of minerals. When magma cools this fast, it produces glass (table 6.6). Cooled nonporous volcanic glass is known as **obsidian** (figure 6.20a). When gas escapes quickly and violently from lava, it may produce a frothy glass full of holes left by former gas bubbles; this porous material, known as **pumice,** contains so many holes it can float on water (figure 6.20b). **Scoria** are rough crusts or chunks of basaltic rock full of holes made by expanding gases before solidification (figure 6.20c).

The Three Vs of Volcanology: Viscosity, Volatiles, Volume

We can understand volcanoes anywhere in the world using the three Vs of volcanology: viscosity, volatiles, volume. *Viscosity* may be low, medium, or high, and it controls whether magma flows away or piles up. *Volatile* abundance may be low, medium, or high, and volatiles may ooze out harmlessly or blast out explosively. *Volume* of magma may be small, large, or very large. Volume correlates fairly well with eruption intensity; the greater the volume, the more intense the eruption.

Consider the five eruptive styles in figure 6.15 in terms of viscosity and volatiles (table 6.7). The lower the volatile content and the viscosity, the more peaceful the eruption. As the volatile content increases, so can the explosiveness of the eruptions.

Applying what we have learned about magmas allows us to see linkages between eruptive behaviors and the landforms built by volcanic activity. By mixing and matching the values among the three Vs, you can define volcanic landforms (table 6.8) and forecast the eruptive styles that occur at each of them.

Figure 6.16 Small-scale *pahoehoe* near Halemaumau, Hawaii.

Photo by Pat Abbott.

Airborne pyroclasts have their coarsest grains fall from the atmosphere first, closest to the volcano, followed by progressively finer material at greater distances away (figure 6.19). An air-fall deposit can be recognized by the sorting of pyroclasts into layers of different sizes. Pyroclastic debris also can be blasted out over the ground surface as high-speed, gas-charged flows that dump material quickly, producing indistinct layering and little or no sorting of the various-size particles.

(a)

(a)

(b)

Figure 6.18 (a) This large blob of magma cooled while airborne and fell as a *volcanic bomb,* Irazu Volcano, Costa Rica. (b) *Pyroclastic bombs* kill people every year.

(a) Photo by Pat Abbott.

(b)

(c)

Figure 6.20 Volcanic glass and scoria. (a) Obsidian is a dense, dark glass. (b) Pumice, a porous, light glass, floats in water. (c) Scoria, a basaltic rock with large pores, sinks in water.

Photos (a) and (b) from NASA; (c) Photo by Pat Abbott.

Figure 6.19 *Volcanic ash* covers a house near Mount Pinatubo in the Philippines, June 1991.

R. P. Hoblitt/U.S. Geological Survey.

Side Note

How a Geyser Erupts

The eruption of water superheated by magma is called a **geyser** (figure 6.21). The name is from the Icelandic word *geysir,* meaning "to gush or rage." Geysers require subsurface water and abundant heat; these two requirements are common around the world, yet geyser areas are few. The third requirement is a subsurface plumbing system of fractures, **pore** spaces, and caves with *constrictions that confine* some water in cavities where it is superheated higher than the boiling point of water on the ground surface above (figure 6.22). When the temperature at depth is high enough to cause boiling, then steam bubbles rise upward toward lower pressure, expanding as they rise. The rising steam and bubbles carry some superhot water upward to lower levels of pressure, triggering continual conversion to steam. Bubbles and steam reaching the surface splash out some water, thus reducing the weight of water in the system, which triggers more superheated water to boil and gush upward into—the spectacular eruption.

Reduction of pressure on superheated water causes it to change from liquid to gas, triggering the geyser eruption. This process is analogous to the *reduction of pressure* on hot rock, causing it to change from solid to liquid, triggering a volcanic eruption.

Geysers around the world vary in many parameters such as depth of water circulation, temperature and volume of water, and frequency of eruption. Old Faithful Geyser in Yellowstone National Park is a famous example with its own unique parameters. Its spectacular high eruption lasts 2 to 5 minutes while ejecting about 23 m³ (>6,000 gallons) of water at about 110°C (230°F) (figure 6.21). Much of the erupted hot water sinks below ground during the 55 to 120 minutes it takes to refill the complex maze of fractures and cavities of the plumbing system. The next eruption cycle begins with up to 35 minutes of rising and falling small gushes of steam and hot water, which raise and lower the excitement of onlookers, before the next major eruption.

FIGURE 6.21 Old Faithful Geyser erupts skyward in Yellowstone National Park, Wyoming, 3 June 2014. The building is the Old Faithful Inn, a National Historical Landmark built in 1903–1904.

Photo by Deron Abbott.

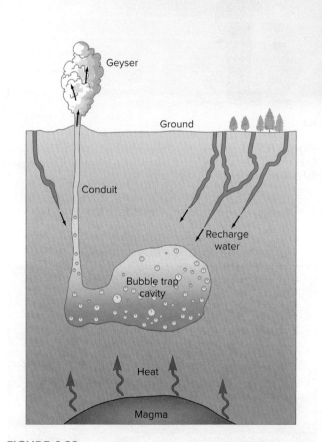

FIGURE 6.22 Oversimplified drawing of a geyser plumbing system, showing the three requirements: subsurface water; magma heat; system of constricted cavities to trap steam underground.

TABLE 6.7

Eruption Styles and Explosiveness

Eruption Style	Viscosity	Volatiles	Volcanic Explosivity Index
Icelandic	Low	Low	0–1 (very low)
Hawaiian	Low	Low	0–1 (very low)
Strombolian	Moderate	Moderate	1–3 (low)
Vulcanian	High	High	2–5 (high)
Plinian	High	High	3–8 (very high)

TABLE 6.8

Volcanism Control by the Three Vs (Viscosity, Volatiles, Volume)

Viscosity	+	Volatiles	+	Volume	=	Volcanic Landforms
Low		Low		Large		Shield volcanoes
Low		Low		Very large		Flood basalts
Low/medium		Medium/high		Small		Scoria cones
Medium/high		Medium/high		Large		Stratovolcanoes
High		Low		Small		Lava domes
High		High		Very large		Calderas

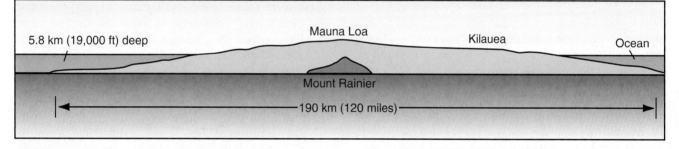

Figure 6.23 A shield volcano, such as Mauna Loa in Hawaii, has a great width compared to its height. A stratovolcano, such as Mount Rainier in Washington, has a great height compared to its width.

Data source: Tilling, R. I., et al., Eruptions of Hawaiian Volcanoes, *US Geological Survey,* 1987.

SHIELD VOLCANOES: LOW VISCOSITY, LOW VOLATILES, LARGE VOLUME

The rocks of **shield volcanoes** form mostly from the solidification of lava flows of basalt. These lava flows are low viscosity, contain less than one weight % volatiles, and are so fluid that they travel for great distances, somewhat analogous to pouring pancake batter on a griddle. Each basaltic flow cools to form a gently dipping, relatively thin volcanic rock layer. Many thousands of these lava flows must cool on top of each other over a long time to build a big volcano. A shield volcano, such as Mauna Loa in Hawaii, has a great width compared to its height, whereas a volcano built of high-viscosity magma, such as Mount Rainier in Washington, has a great height compared to its width (figure 6.23).

Hawaiian-type Eruptions

As with virtually all volcanic eruptions, Hawaiian-type eruptions are commonly preceded by a series of earthquakes as rock fractures and moves out of the way of swelling magma. When these fractures split the ground surface, they suddenly reduce pressure, allowing gas to escape from the top of the magma body. This can create a beautiful "curtain of fire" where escaping jets of gases form lines of lava fountains up to 300 m (1,000 ft) high. Also common in the Hawaiian eruption is formation of a low cone with high fountains of magma. After the initial venting of gas, great floods of basaltic lava spill out of the fissures and flow down the mountain slopes as red-hot rivers (figure 6.24). These eruptions may last from a few days to a year or more. Although few lives

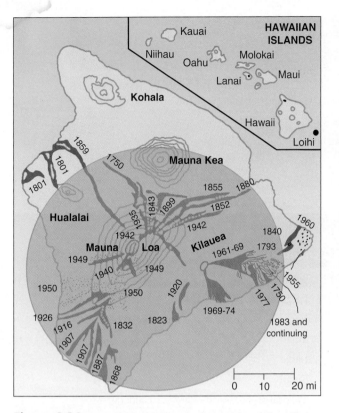

Figure 6.24 Map of Hawaii showing some historic lava flows. The circular color overlay shows boundaries of mantle plume rising from the hot spot at depth.

Figure 6.25 Lava flows caused the Wahalua Visitor's Center in Hawaii Volcanoes National Park to burn to the ground in 1989.

NOAA NGDC, Photograph by J.D. Griggs, Hawaiian Volcano Observatory, *US Geological Survey.*

are lost to Hawaiian volcanism, the ubiquitous lava flows engulf and incinerate buildings, bury highways, cause drops in property value of homes near the latest flow, and cause some homeowners to lose their peace of mind (figure 6.25).

Hawaiian volcanoes capable of eruption include Haleakala on the island of Maui, the five volcanoes that make up the island of Hawaii, and the growing but still subsea volcano of Loihi. Haleakala last erupted around 1790. Today, its 49 km^2 (19 mi^2) summit caldera is a major tourist attraction. In the last 200 years, eruptions have occurred only on the three southernmost volcanoes on the island of Hawaii and below sea level on Loihi. The island-to-be (Loihi) is located about 30 km (19 mi) off the southeastern shore of Hawaii. Loihi's peak is about 969 m (3,175 ft) below sea level, and the weight of the overlying ocean water suppresses the explosiveness of the eruptions for now, but the volcano is building upward impressively.

In general, the volcanism on Hawaii is relatively peaceful and acts as a magnet attracting tourists to witness nature's spectacle. But there are exceptions to this statement.

Killer Event of 1790

Although less than 0.5% of Hawaiian magma is blown out as pyroclastic material, rare killer events do occur. In 1790, traveling parties from King Keoua's army were caught and many of the people killed by a blast from Kilauea Volcano. The army was passing through the area but was stopped

by eruptions. After three days of waiting, it split into three parties of about 80 people each. As the parties marched southwest down the trail from Kilauea, disaster struck. An explosion column burst upward, with a **base surge** sweeping outward as a dense, basal cloud. Base surges can travel at hurricane speeds as masses of ground-hugging hot water and gases with or without magma fragments. The base surge in 1790 overtook King Keoua's middle party, killing them all. The victims huddled together, grasping each other to withstand the hurricane-force blast, but the hot gases seared their lungs and the intense heat scorched their skin. The base surge caught up with the lead party, but the surge had weakened, allowing most of those people to survive. The trailing party was alongside the blast and suffered no deaths or injuries. Although basaltic magma is not likely to explode, this case history shows how magma can heat groundwater to cause an eruption, including a base surge of superheated steam. The 1790 event is worth remembering for today's watchers of Hawaiian eruptions—at least seek the high ground during your viewing.

FLOOD BASALTS: LOW VISCOSITY, LOW VOLATILES, VERY LARGE VOLUME

Flood basalts are the largest volcanic events known on Earth. Two important characteristics are (1) the immense amounts of magma, gas, and heat they pour onto Earth's surface and (2) their geologically brief duration. Flood basalts erupt tremendous volumes of magma within a geologically short time—for example, less than 1 million years. Hot spots also bring up huge volumes of magma but do so during a long period of time—for example, 100 million years.

The volumes of magma erupted during a flood basalt event and the surface areas they bury with lava are so large

In Greater Depth

Volcanic Explosivity Index (VEI)

How often do big volcanic eruptions occur? On average, about once every three years, according to the volcanic explosivity index (VEI). Combining the historic record with the geologic information stored in the rock record, the major volcanic eruptions occurring between the years 1500 and 1980 were evaluated for their size. Factors evaluated include (1) volume of material erupted, (2) how high the eruption column reached, and (3) how long the major eruptive burst lasted (table 6.9). During the 481-year interval studied, 126 major eruptions occurred, with the number increasing in modern times. The increase in big eruptions in the 19th and 20th centuries is certainly due to better reporting of events, rather than an actual increase in major eruptions.

The VEI ranges from 0 to 8. The biggest event since 1500 CE was the VEI 7 eruption of Tambora in 1815 in Indonesia. This eruption caused a cooling of the world climate during the following year (see chapter 12). Four VEI 6 events occurred in the 481-year period, including the 1883 eruption of Krakatau, also in Indonesia (described in the section on calderas in this chapter). Volcanic events with high VEI values are those of Vulcanian and Plinian-type eruptions. A fifth VEI 6 event occurred in 1991 when Mount Pinatubo erupted.

TABLE 6.9
Volcanic Explosivity Index (VEI)

		VEI								
	0	**1**	**2**	**3**	**4**	**5**	**6**	**7**	**8**
Volume of ejecta (m^3)	$<10^4$	10^4–10^6	10^6–10^7	10^7–10^8	10^8–10^9	10^9–10^{10}	10^{10}–10^{11}	10^{11}–10^{12}	$>10^{12}$
Eruption column height (km)	<0.1	<0.1–1	1–5	3–15	10–25	>25			
Eruptive style	<----Hawaiian---->		<-----------Vulcanian----------->						
	<-------------Strombolian------------->			<---------------Plinian-------------->					
Duration of continuous blast (hours)	<----------------<1---------------->		<----------------1–6---------------->			<--------------12-------------->			
				<---------------6–12--------------->					

Source: After Newhall and Self (1982).

they are hard to visualize. For example, 252 million years ago (end of Permian time), up to 3 million km^3 (800,000 mi^3) of basalt flowed out and covered almost 4 million km^2 (1.5 million mi^2) of Siberia, Russia. Visualize basalt flows covering an area measuring about 1,200 mi by 1,200 mi with lava tens of meters thick. How does this area compare with the area of your state or province? Visualize your entire state or province buried beneath lava tens of meters thick.

Flood basalts occur on all continents and on all ocean floors, but none have occurred in historic times. Flood basalt eruptions obviously devastate a region, but can they have global effects? Yes, not from the lavas directly, but from the massive volumes of gases they bring up, such as carbon dioxide (CO_2) and sulfur dioxide (SO_2). During eruptions on the ocean floors, the overlying seawater absorbs huge volumes of gas, which can change seawater acidity and oxygen concentrations. Flood basalts that pour out onto the continents pump tremendous volumes of climate-changing gases into the atmosphere; CO_2 causes global warming, and SO_2 can cause global cooling. Either way, life on Earth suffers. Mass extinctions can occur.

The fossil record of life on Earth is clearest since shells, teeth, bones and other hard parts appeared during the last 541 million years (see Epilogue). During that time, five major mass extinctions of life occurred. Three of them coincide with flood basalt episodes.

- End Permian time, 252 million years ago, flood basalt buried much of Siberia
- End Triassic time, 201 million years ago, flood basalt occurred during opening of the north Atlantic Ocean basin
- End Cretaceous time, 66 million years ago, flood basalt in the Deccan region of India buried a 1.5 million km^2 (580,000 mi^2) area. During this time interval, life was also stressed by the impact of a 10 km (6+ mi) diameter asteroid on the Yucatan Peninsula of Mexico.

The famous asteroid impact is widely cited as causing the extinction of non-avian dinosaurs, but the flood basalt deserves part of the credit. Flood basalt occurred during three mass extinctions, whereas an asteroid impact occurred only once, along with a flood basalt.

SCORIA CONES: MEDIUM VISCOSITY, MEDIUM VOLATILES, SMALL VOLUME

Scoria cones are conical hills, typically of low height, formed of basaltic to andesitic pyroclastic debris piled up next to a volcanic vent. Scoria cones commonly are produced during a single eruptive interval lasting from a few hours to several years. The scoria, or **cinder cone,** has a summit **crater,** the basin on top of the cone that is usually less than 2 km (1.2 mi) in diameter. The summit crater may hold a lava lake during eruption. After the excess gas has been expelled from the magma body, the lava may drain and emerge from near the base of the cone. When that eruption ceases, scoria cones usually do not erupt again.

Strombolian-type Eruptions

Scoria cones are built mainly by Strombolian-type eruptions (see figure 6.15). The volcano Stromboli, offshore from southwestern Italy, has had almost daily eruptions for millennia. Its central lava lake is topped by a cooled crust. Even the tidal cycle disrupts the lava-lake crust, thus triggering eruptions. Gas pressure builds quickly beneath the crust, and eruptions occur as distinct and separate bursts up to a few times per hour. Each eruption tosses pyroclasts tens to hundreds of meters into the air. For many centuries, tourists have climbed Stromboli to thrill at the explosive blasts, but usually every year, a few of those tourists die when hit by large pyroclastic bombs. Strombolian eruptions are not strong enough to break the volcanic cone.

On 20 February 1943, a new volcano was born as eruptions blasted up through a farm field near the village of Paricutin in the state of Michoacan, Mexico (figure 6.26). The volcano erupted for nine years, building a distinctive scoria cone. Pyroclastic debris and lava flows buried about 260 km^2 (100 mi^2) of land and destroyed the towns of Paricutin and San Juan de Parangaricutiro.

STRATOVOLCANOES: HIGH VISCOSITY, HIGH VOLATILES, LARGE VOLUME

Stratovolcanoes, or **composite volcanoes,** commonly are steep-sided, symmetrical volcanic peaks built of alternating layers of pyroclastic debris successively capped by high-viscosity andesitic to rhyolitic lava flows that solidify to form protective caps. Stratovolcanoes may show marked variations in their magma compositions from eruption to eruption, and their eruptive styles include Vulcanian and Plinian. Some of Earth's most beautiful mountains are stratovolcanoes—for example, Mount Kilimanjaro in Tanzania, Mount Shasta in California, Mount Rainier in Washington, and Mount Fuji in Japan (figure 6.27).

Vulcanian-type Eruptions

All volcanoes take their name from Vulcan, the Roman god of fire and blacksmith for the gods. The prototypical volcano is one of the Aeolian Islands in the Tyrrhenian Sea north of

Figure 6.26 Paricutin Volcano erupting in 1943. This 400 m (1,300 ft) high scoria cone is in the state of Michoacan, Mexico.
K. Segerstrom, USGS/NOAA NGDC.

Sicily. The fire and smoke emitted from the top of the mountain reminded observers of the chimney of Vulcan's forge, so the mountain was named Vulcano. Vulcanian eruptions alternate between thick, highly viscous lavas and masses of pyroclastic material blown out of the volcano. Some Vulcanian eruptions are more violent blasts of high-viscosity magma loaded with trapped gases. The material blown out during eruptions covers wide areas. Vulcanian eruptions commonly are the early phase in the eruptions of other volcanoes as they "clear their throats" before emitting larger eruptions.

Plinian-type Eruptions

Plinian eruptions are named after the 17-year-old Pliny the Younger in honor of his detailed written observations of the 79 CE eruptions of Vesuvius that claimed the life of his well-known uncle Pliny the Elder. In Plinian eruptions, the volcano "throat is now clear," and incredible gas-powered vertical eruption columns carry pyroclastic debris, including lots of pumice, up to 50 km (30 mi) into the atmosphere (see figure 6.15). The Plinian eruption is a common final phase in a major eruptive sequence. About two to three Plinian eruptions occur each century.

Vesuvius, 79 CE

Vesuvius began as a submarine volcano in the Bay of Naples. It grew greatly in size, and its rocky debris filled in the waters that once separated it from mainland Italy (figure 6.28). What is the cause of the volcanism at Vesuvius and the neighboring volcanoes of Stromboli, Vulcano, Etna, and others? The subduction of Mediterranean seafloor beneath Europe to make room for the northward charge of Africa.

In 79 CE, when Mount Vesuvius began erupting, most residents fled from Pompeii. Those who stayed first experienced volcanic ash clouds dropping pumice. Pompeii lay downwind and was buried by pumice fragments accumulating up to 3 m (10 ft) deep (figure 6.28). Researchers estimate that about 60% of the people who remained in Pompeii survived the first flows of ash and pumice. About half of the survivors then fled, but many of them died when they were caught outside during later flows of ash and pumice. The people still inside houses remained alive, only to suffocate from breathing hot particles and gases seeping out of the volcanic debris. Death was not always quick. Some bodies were found inside houses on top of thick layers of pumice, giving evidence of hours of struggle by people fighting to stay alive. Their hands held cloths over their mouths as they tried to avoid asphyxiation from gases seeping out of the pumice.

Figure 6.27 Mount Fuji, a symmetrical stratovolcano rising 3,776 m (12,385 ft) above sea level, Honshu, Japan. The last major eruptions were in 1707–1708.

© Royalty-Free/Corbis.

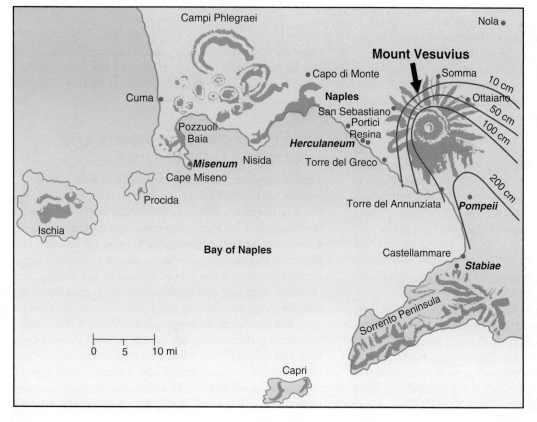

Figure 6.28 Map of the Bay of Naples area showing the location of Mount Vesuvius. Pumice fallout from the 79 CE eruption is contoured in centimeters. Pompeii and Stabiae were buried by pumice; Herculaneum by lahars.

Many other people were found near the sea. They escaped the falling pumice, but ground-hugging **pyroclastic flows,** full of hot gases, finished them off (for a modern example of a major Vulcanian-type eruption, see figure 6.3). About 4,000 people died. The more-distant town of Stabiae was also mostly destroyed. It was here that Pliny the Elder died; the weak heart of the overweight man failed at age 56 under the stress of the farthest-reaching gas-rich flow.

Testing of rocks formed during the 79 CE eruption, as well as roof tiles from Pompeii, indicates that the cloud of volcanic ash and pumice that smothered Pompeii erupted out of Vesuvius at about 850°C (1,550°F) and then cooled to less than 380°C (710°F) by the time it reached the city. Roof tiles in Pompeii were heated to maximum temperatures of 340°C (640°F), while some walls on the partially protected down-flow side of houses reached temperatures of around 180°C (350°F), presumably because cooler air mixed into the volcanic ash cloud.

Following the Vulcanian-type eruption, the volcano entered a second phase, the Plinian phase, where it blew immense volumes of pyroclasts up to 32 km (20 mi) high in the atmosphere. The height of the eruption column varied as the volcanic energy waxed and waned. During weakened intervals, the great vertical column of ashes would temporarily collapse, sending surges and pyroclastic flows down the volcano slopes. Pompeii was buried under an additional 6 to 7 ft of pyroclastic debris. These were the flows that finished off the surviving Pompeiians.

A Plinian eruption not only blows ashes to great heights but also volcanic gases. Water, as abundant steam, can be blown high into the atmosphere, cooling and condensing and then falling back down as rain—heavy rain. Some volcanic eruptions create their own "weather." Rain falling on thick piles of pyroclastic debris, sitting unstably on the steep slopes of Vesuvius, set off thick volcanic mud-flows (figure 6.29). Any gravity-pulled mass movements of muddy volcanic debris are known as **lahars,** an Indonesian word (for a modern example, see figure 6.30). Lahars buried the city of Herculaneum up to 20 m (65 ft) deep in pumice, ashes, and volcanic rock fragments jumbled together in a confused mass. However, this was during the second phase of the eruption, and most people had used the day or two before to clear out of the area, so the loss of life was not nearly as great as at Pompeii. Today, the town of Ercolano lies on top of the mudflows burying Herculaneum. The lessons of history have not been well learned here.

Recent seismic surveys have helped define the magma body underlying Vesuvius today. Seismic waves with lowered velocity define a 400 km² (150 mi²) horizontal, broad sheet of partially molten rock. This magma body lies at a depth of 8 km (5 mi) below the surface. Assuming a magma body thickness of 0.5 to 2 km (0.3 to 1.2 mi), the volume of magma is about 200 to 800 km³ (50 to 200 mi³). This magma

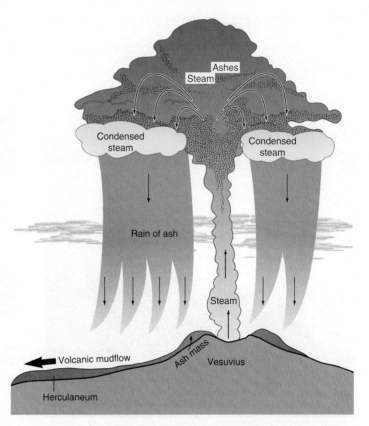

Figure 6.29 "Volcano weather" and the formation of lahars. Prolonged vertical eruption leads to accumulation of debris (ash mass) on steep slopes of the volcano. Steam blown upward into cold, high altitudes condenses and falls back as rain. The stage is set: steep slopes + loose volcanic debris + heavy rain = lahars. Some volcanic eruptions even generate their own lightning.

reservoir is fed from below, and it can supply magma to smaller layers closer to the surface, as it has in fueling past eruptions. The millions of people around the Bay of Naples live in real danger (see figure 6.6).

LAVA DOMES: HIGH VISCOSITY, LOW VOLATILES, SMALL VOLUME

Lava domes form when high-viscosity magma with a low content of volatiles cools quickly, producing a hardened dome or plug a few meters to several kilometers wide and a few meters to 1 km high. Lava domes can form in as little as a few hours, or they may continue to grow for decades. The formation of lava domes can be visualized as part of a larger eruptive process. When a large volume of hot rock/magma rises and undergoes decompression melting, dissolved volatiles are freed. Many of the freed gases rise and accumulate at or near the top of the magma mass. When a major eruption occurs, these gases power the initial Vulcanian-type blast and then the succeeding Plinian-type eruption, which lasts until the excess volatiles have escaped. What type of magma remains? Often it is a low-volatile, high-viscosity paste that oozes upward slowly and cools quickly, forming

Figure 6.30 Lahars from a Vulcanian-type eruption of Mount Pinatubo destroyed bridges to Angeles City, Philippines, 12 August 1991.

T.J. Casadevall/U.S. Geological Survey.

Figure 6.31 The Novarupta lava dome formed as hardened magma plugged the central magma pipe of the 1912 eruption of Katmai Volcano in southern Alaska. The dome is 244 m (800 ft) across and 61 m (200 ft) high.

Gene Iwatsubo/U.S. Geological Survey.

A Typical Eruption Sequence

A common pattern for a major eruptive episode is that gas-rich materials shoot out first as a Vulcanian blast, quickly followed by a longer-lasting, gas-driven Plinian eruption. When the gas is depleted, then gas-poor, high-viscosity magma slowly oozes out to build a lava dome over an extended period of time.

The volcanic sequence could be described as a Vulcanian precursor, a Plinian main event, and a lava dome conclusion.

CALDERAS: HIGH VISCOSITY, HIGH VOLATILES, VERY LARGE VOLUME

Caldera-forming eruptions are the largest of the violent, explosive volcanic behaviors. **Calderas** are large volcanic depressions formed by roof collapse into partially emptied magma reservoirs. Calderas differ from volcanic craters. Both are topographic depressions, but a crater is less than 2 km (1.2 mi) in diameter and forms by *outward explosion*. Calderas are larger; they range from 2 to 75 km (45 mi) in diameter and form by *inward collapse* (figure 6.32).

Calderas form in different settings: (1) Calderas that form at the *summits of shield volcanoes* include those at Mauna Loa and Kilauea on the island of Hawaii. A recent subsidence of the Kilauea caldera followed draining of magma to feed eruptions out of lower-elevation rift zones. (2) Calderas forming at the *summits of stratovolcanoes* include Crater Lake in Oregon, Krakatau in Indonesia, and Santorini in the Aegean Sea (figure 6.33a,b,c). Caldera collapses occurred following sustained Plinian eruptions of 55 km^3 (13 mi^3) of pyroclasts at Crater Lake, 18 km^3 (4.3 mi^3) at Krakatau,

a plug in the throat of the volcano. Figure 6.31 shows the lava dome emplaced in Mount Katmai in southern Alaska following its 1912 eruption, the biggest eruption of the 20th century.

Lava domes can provide spectacular sights. After the 1902 eruptions of Mont Pelée in the Caribbean killed more than 30,000 people (see chapter 7), a lava dome formed as a great spine that grew about 10 m/day (33 ft/day) and rose above the top of the volcano. The spine of hardened magma was forced upward by the pressure of magma below until it stood more than 300 m (more than 1,000 ft) higher than the mountaintop, like a giant cork rising out of a bottle.

Do lava domes present hazards? Yes, in the 1990s, they were responsible for 129 deaths: 19 from Soufriere Hills Volcano on Montserrat in 1997, 66 from Mount Merapi in Indonesia in 1994, and 44 from Mount Unzen in Japan in 1991. The hardened, brittle lava dome rock can fail in a gravity-pulled landslide from the mountain, or magma trapped below the brittle lava dome can break out in a violent eruption.

Side Note

British Airways Flight 9

On 24 June 1982, 247 people on British Airways flight 9 boarded a Boeing 747 for a night flight from Kuala Lumpur, Malaysia, to Perth, Australia. The night was moonless but clear, and the weather forecast was good. The crew took the big airplane up to its cruising altitude of 37,000 ft and then relaxed a bit. Weather radar showed that outside conditions were normal. But the pilot noticed puffs of "smoke" and an acrid or electrical odor. As he sat in the pilot's seat and peered through the front windscreens, the atmosphere seemed to be on fire as intense electricity danced about. Out the side windows, the engines were glowing as if they were lit inside. Then the flight engineer called out: "Engine failure number 4," followed shortly by:

"Engine failure number 2.

"Three's gone.

"They've all gone."

The pilot thought, "Four engines do not fail." The instrument panel was contradictory; some gauges read normal while others told of problems with a confusing lack of pattern.

Meanwhile, the plane was descending slowly. At 26,000 ft, the oxygen masks were released, but some didn't work; in this emergency, a steep descent was initiated to get down to atmospheric levels with more oxygen. When the plane reached 14,000 ft, the pilot said:

"Good evening ladies and gentlemen. This is your captain speaking. We have a small problem. All four engines have stopped. We are doing our darndest to get them going again. I trust you are not in too much distress." His words could not have brought much comfort to those passengers in window seats who had been watching the engines that seemed to be on fire.

What to do? Land on the ocean during a dark night? Too dangerous. Finally, at 12,000 ft, engine number 4 started, and 90 seconds later, the other three engines started. The pilot set the plane to climbing to avoid hitting the mountainous Indonesian topography, but at 15,000 ft, the bad atmospheric problems began again. Descent was once again initiated. Permission was granted for an emergency landing at Jakarta, but the approach was hazardous. The front and side windows were frosted and opaque, so the co-pilot had to look out a little side window and give instructions to the pilot landing the huge, fast-moving plane. At last, the plane landed smoothly, and the passengers cheered and clapped.

What happened that night to BA9? It flew during an eruption of Mount Galunggung and passed through its seething cloud of hot volcanic ash and larger pyroclastic debris. The volcanic ash clogged the engines, frosted the windscreen, and turned BA9 into a terror-filled flight. Airplanes must avoid volcanoes in eruption.

Figure 6.32 A crater (less than 2 km across) formed atop the volcano in the foreground during eruptions. The large caldera (more than 2 km across) at low elevation in the background formed when its volcano collapsed during a massive eruption, Kamchatka Peninsula, Russia.

C. Dan Miller/U.S. Geological Survey.

and 40 km^3 (10 mi^3) at Santorini. These eruptions opened void spaces that caused mountain peaks to collapse down into their magma chambers. (3) *Giant continental calderas* are huge negative landforms such as Lake Yellowstone in Wyoming or Long Valley in California. These broad and deep depressions formed following the rapid eruption of 2,000 km^3 (475 mi^3) of pyroclasts at Yellowstone and 600 km^3 (140 mi^3) at Long Valley. The huge volumes of magma pour out in short amounts of time as **ultra-Plinian** eruptions with extra-high ash columns and widespread sheets of outward-flowing ash and pumice.

The most recent example of an ultra-Plinian eruption occurred 74,000 years ago at Toba, on the island of Sumatra in Indonesia. The caldera at Toba is 30 km (20 mi) by 100 km (60 mi) long and has a central raised area inside it that is more than 1 km high. The raised area formed during the millennia following the giant eruption; this resurgent topography inside the caldera gives these features their name—**resurgent calderas.**

Crater Lake (Mount Mazama), Oregon

Crater Lake is one of the jewels in the U.S. national park system. Its intense blue waters are pure and lie cradled in a high-rimmed, nearly circular basin. Crater Lake is about 9.5 km (6 mi) across and as deep as 589 m (1,932 ft) (figure 6.34).

Several thousand years ago, the *stratovolcano* Mount Mazama stood about 3,660 m (12,000 ft) high as one of the Cascade Range volcanoes above the Cascadia subduction zone (figure 6.35a). More than 7,600 years ago, a major eruption began blowing sticky magma out of the mountain

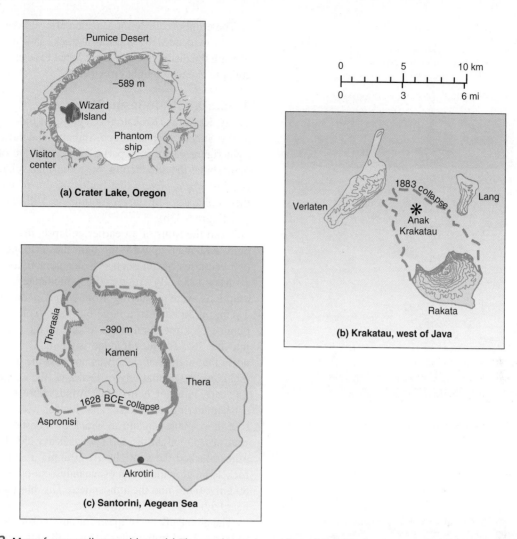

Figure 6.33 Map of some collapse calderas. (a) The nearly circular caldera of Crater Lake, Oregon, formed about 5677 BCE. (b) Island remnants of old volcano Krakatau; crudely ovoid shape of the 1883 collapse; and the new and growing volcano, Anak Krakatau. (c) Caldera in volcano Santorini that collapsed into the Aegean Sea about 1628 BCE.

Figure 6.34 Crater Lake, Oregon, fills the caldera of Mount Mazama, which collapsed in the year 5677 BCE. Wizard Island is visible.

© Robert Glusic/Getty Images RF.

as glassy, gas-bubble-filled pumice and ashes (figure 6.35b). The magma had too high a viscosity to flow as a liquid, so it erupted as pyroclastic flows and Plinian columns. As the erupted material grew in volume, its debris covered much of the U.S. Pacific Northwest and part of Canada with a thick, distinctive ash layer that is easily recognizable today. Mazama ash is found in the Greenland glacier within the ice layer formed during the snowfall season of the year 5677 BCE. About 40 km³ (10 mi³) of magma was ejected. Evacuation of this immense volume of magma left so tremendous a void below the surface that the weakened mountain peak collapsed and moved down in pistonlike fashion into the emptied magma chamber (figure 6.35c). The collapse produced a caldera about 10 km (6 mi) across that has collected the water for Crater Lake and hosted the growth of a 1,000-year-old successor volcanic cone called Wizard Island (figures 6.33a and 6.34).

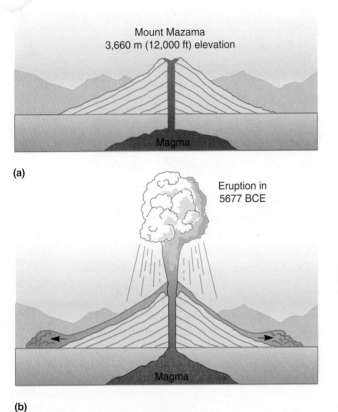

(a)

(b)

(c)

(d)

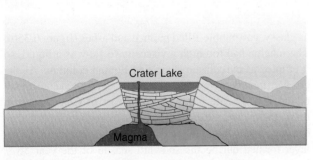

Figure 6.35 How Crater Lake formed. (a) The Mount Mazama volcano stood high. (b) A gaseous eruption in the year 5677 BCE emptied a huge volume of viscous magma. (c) The gigantic eruption left a void inside the weakened mountain, and the unsupported top fell down into the emptied magma chamber. (d) The waters of Crater Lake now fill the caldera, and a small new volcanic cone (Wizard Island) rises above lake level.

The eruption of Mount Mazama affected American Indians, as evidenced by moccasin tracks and artifacts found beneath the distinctive ash layer. What have caldera-forming collapse events wrought elsewhere?

Krakatau, Indonesia, 1883

Today, Krakatau (Krakatoa) is a group of Indonesian islands in the Sunda Strait between Sumatra and Java (see lower right figure 4.10). It is part of the grand arc of volcanoes built above the subducting Australia-India plate. Krakatau is a big *stratovolcano* that builds up out of the ocean and then collapses. Its larger outline is still distinguishable (see figure 6.33b).

From the ruins of an earlier collapse, magmatic activity built Krakatau upward through the 17th century. After two centuries of quiescence, volcanic activity resumed on 20 May 1883. By August 1883, moderate-size Vulcanian eruptions were occurring from about a dozen vents. At 2 p.m. on 26 August, a large blast shot volcanic ashes and pumice 28 km (17 mi) high as one of the cones collapsed into the sea, setting off huge tsunami. Eruptions were so noisy that night that sleep was not possible in western Java, including the capital city of Djakarta (then called Batavia). The early morning hours of 27 August were rocked by more ear-hammering eruptions, and further volcanic collapses sent more giant tsunami to wrack the coastal villages. Day was turned to nightlike darkness as heavy clouds of volcanic ashes blocked the sunlight. At 10 a.m., a stupendous blast rocketed a glowing cloud of incandescent pumice and ashes 80 km (50 mi) into the atmosphere. This blast was distinctly heard 5,000 km (3,000 mi) away.

The global effects of the eruptions in Oslo, Norway, seem to be recorded in the late August 1883 journal of the great painter Edvard Munch as "All at once the sky became blood red...and clouds like blood and tongues of fire hung above the blue-black fjord and the city...and I felt alone, trembling with anxiety...I felt a great unending scream piercing through nature." Munch's famous painting *The Scream* is thought to be his reaction to the skies made blood red in Europe by the Krakatau eruption.

The 10 a.m. volcano collapse sent tsunami higher than 35 m (115 ft) sweeping into bays along the low coastlines of Java and Sumatra. The volcanic eruptions caused tsunami that destroyed 295 towns and smashed or drowned an estimated 36,000 people. The eruption sequence blew out 18 km^3 (4.3 mi^3) of material (95% fresh magma and 5% pulverized older rock), creating a subterranean hole into which 23 km^2 (5.5 mi^2) of land collapsed. Where islands with elevations of 450 m (1,476 ft) had stood, there now was a hole in the seafloor 275 m (900 ft) deep.

The amount of magma erupted at Krakatau in 1883 was less than half that of the Mount Mazama eruption. But Krakatau collapsed into the sea, sending off tsunami.

In 1927, Krakatau began rebuilding a new volcanic cone called Anak Krakatau—"child of Krakatau"; it is still growing (see figure 6.33b). We will hear more from Krakatau.

Figure 6.36 Aesthetic view of the inside of the Santorini caldera. A classic white-washed Greek church and some homes have been built on the cliff edge.

© Adam Crowley/Getty Images RF.

Santorini, Greece

As the Mediterranean oceanic plate subducts beneath Europe, it causes numerous volcanoes. One of the biggest is the *stratovolcano* Santorini in the Aegean Sea. Today, Thera is the largest island in a circular group marking the sunken remains of Santorini (see figure 6.33c). Thera is one of the most popular tourist sites in the Greek islands (figure 6.36), but around 1628 BCE, Santorini underwent an explosive series of eruptions that buried the Bronze Age city of Akrotiri on Thera to depths of 70 m (230 ft) in four distinct phases. Where there had been a large island made of several volcanic cones, there now exists a huge caldera with depths of 390 m (1,280 ft) below sea level.

Caldera-forming eruptions are low-frequency, high-impact events. Santorini had a major eruption more than 21,000 years ago, then about 18,000 years passed without a huge eruption—until the ultra-Plinian eruption and volcano collapse during the Minoan civilization. A recent study of chemically zoned crystals of plagioclase that grew in the 855°C (1,570°F) rhyolitic magma reservoir determined how much time it took to recharge the magma reservoir before the overwhelming eruption of 1628 BCE. The results are surprising. After 18,000 years of waiting, it took less than 100 years to supply the 40–60 km^3 (10–15 mi^3) of magma that were

erupted. The huge supply of viscous magma forced relatively small-volume dikes or columns of magma to the surface, initiating the pressure drop that triggered the eruption.

It is sobering for us today to consider that a civilization-changing mega-eruption might take only a few decades of magma recharge before occurring.

Yellowstone National Park

A *giant continental caldera* exists in Yellowstone above a hot spot, a long-lived mantle plume that the North American continent is drifting across. The hot spot occupies a relatively fixed position above which the North American plate moves southwestward about 2 to 4 cm/yr (0.8 to 1.6 in/yr). Plate movement over the hot spot during the last 15 million years is recorded by a trail of surface volcanism cut across the Snake River plain in Idaho and on into Wyoming (figure 6.37). At present, Yellowstone National Park sits above the hot spot, and a large body of rhyolitic magma lies about 5 to 10 km (3 to 6 mi) beneath it.

In the past 2 million years, three catastrophic ultra-Plinian eruptions have occurred at Yellowstone at 2.06 million, 1.29 million, and 0.64 million years ago (figure 6.38). Such mega-eruptions do not come often, but in a few short weeks, they pour forth virtually unimaginable volumes of rhyolitic magma, mostly as pyroclastic flows. The oldest event erupted 2,500 km^3 (600 mi^3) of magma, the middle one emptied 280 km^3 (70 mi^3), and the youngest dumped out 1,000 km^3 (240 mi^3). (Compare these magma volumes to the 1980 eruption of Mount St. Helens, which totaled 1 km^3.) An eruption of 1,000 km^3 of rhyolitic pyroclastic flows would cover a surrounding area of 30,000 km^2 (11,500 mi^2) with a mass of pyroclastic debris ranging from a few to more than 100 m in thickness. The weight of volcanic material would cause a 500 km^2 (200 mi^2) area to sink isostatically.

The Yellowstone mega-eruption of 640,000 years ago created a giant caldera that is 75 km (47 mi) long and 45 km (28 mi) wide. Look again at figure 6.38 and consider the size of the giant caldera and the extent of its emitted pyroclastic flows: in a matter of days, all life in the area would have died and been deeply buried.

Eruptive Sequence of a Resurgent Caldera

Giant caldera-forming eruptions go through a characteristic sequence. They begin when a very large volume of rhyolitic magma rises to within a few kilometers below the surface, bowing the ground upward (figure 6.39a). The magma body accumulates a cap rich in volatiles and low-density components such as SiO_2.

A mega-eruption begins with a spectacular circular ring of fire as Plinian columns jet up from circular to ovoid fractures surrounding the magma body (figure 6.39b). The escaping magma erodes the fractures, thus increasing the size of the eruptive vents so that more and more magma escapes.

As greater volumes of gas "feel" the lessening pressure, the magma begins gushing out of the fractures in mind-boggling volumes (figure 6.39c). The outrushing magma is

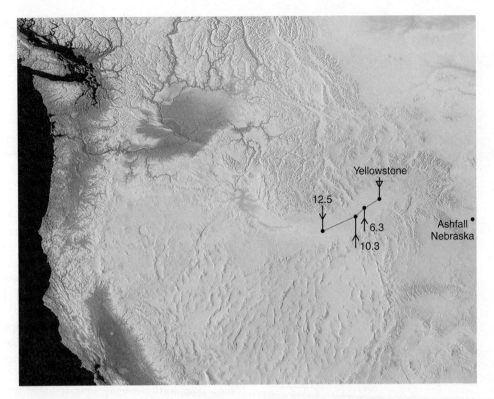

Figure 6.37 Track of the Yellowstone hot spot in western North America. See the 600 km (370 mi) long trail marking the path of the North American plate. Today, the hot spot underlies Yellowstone National Park. Numbers shown are millions of years since each site was over the hot spot.

NOAA.

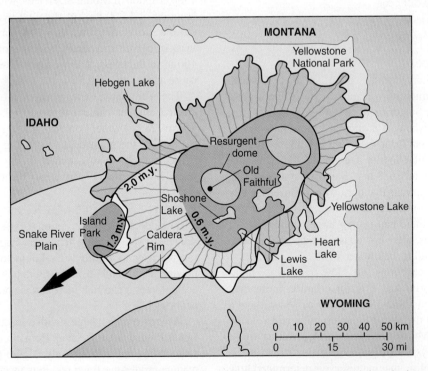

Figure 6.38 The Yellowstone hot-spot area. The North American plate is moving southwest, and thus the hot-spot magma plume erupts progressively farther northeast with time. Three giant calderas have erupted in the past 2 million years—at 2.06, 1.29, and 0.64 million years ago. The wiggly-lined area was covered by hot, killing pyroclastic flows during the eruption of 600,000 years ago. Notice the resurgent domes.

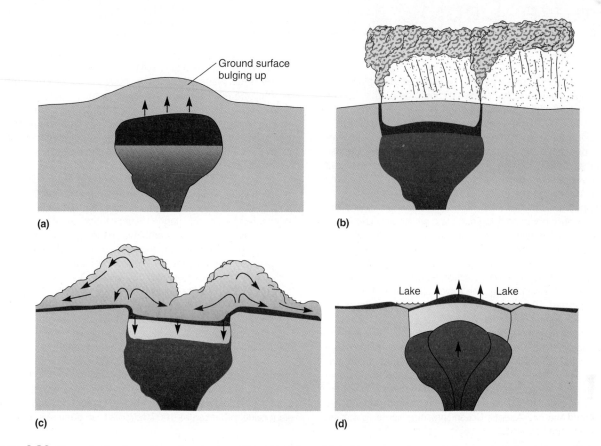

Figure 6.39 Stages in the formation of a giant continental caldera. (a) A rising mass of magma forms a low-density cap rich in SiO_2 and gases, causing the ground surface to bulge upward. (b) Plinian eruptions begin from circular fractures surrounding the bulge. (c) Magma pours out in pyroclastic flows of tremendous volume, causing the ground surface to sink into a giant caldera. (d) Removal of magma decreases the crustal pressure, allowing new magma to rise and cause the caldera floor to bulge up.

too voluminous to all go airborne, so most of it just pours away from the vents as pyroclastic flows, the fastest way to remove gas-laden, sticky magma. As the subsurface magma body shrinks, the land surface sinks as well, like a piston in a cylinder, creating a giant caldera (figure 6.39c).

The removal of 1,000 km^3 (240 mi^3) of magma creates a void, an isostatic imbalance, that is filled by a new mass of rising magma that bows up the caldera floor to create a **resurgent dome** (figures 6.39d and 6.40). Resurgent domes may be viewed as the reloading process whereby magma begins accumulating toward the critical volume that will trigger the next eruption.

The areas alongside resurgent domes are commonly occupied by lakes (see figure 6.38). Imagine driving the many miles from Yellowstone Lake to Old Faithful Geyser, all the time staying within the gigantic collapse caldera of the 640,000-year-old eruption. When will the next megaeruption occur?

Figure 6.40 View over the town of Mammoth Lakes to tree-covered hills in Long Valley. The hills compose a resurgent dome up to 500 m (1,600 ft) high above the caldera floor.

S.R. Brantley/U.S. Geological Survey.

In Greater Depth

Hot Spots

Hot spots are shallow masses of hot rocks/magmas or plumes of slowly rising mantle rock that create volcanism on Earth's surface. The temperature of the rising rock is hotter than the surrounding rock by about 300°C (570°F) in the plume center and only 100°C (212°F) along the outer margin of the plume head. But this temperature difference increases buoyancy enough to start the rise toward the surface, looking something like the blobs in a lava lamp. Most hot spots are visualized as rising plumes that operate for about 100 million years.

Hot spots do not move as much as tectonic plates and are used as reference points to help chart plate movements (see figure 2.20). They occur under the oceans and under the continents (e.g., Yellowstone), in the center of plates (e.g., Hawaii), and as part of spreading centers (e.g., Iceland). In the 1970s, a survey was made of hot spots that create elevated volcanic domes with diameters greater than 200 km (125 mi). The survey counted 122 hot spots active in the past 10 million years (figure 6.41): 53 under ocean basins and 69 under continents.

The largest number of hot spots lies beneath the African plate. The drifting of Africa has been slowed by its collision with Eurasia during the last 30 million years. The slowed African plate may be acting like a thermal blanket concentrating the mantle heat beneath it. With Africa effectively stopped from making large horizontal movements, the westward movement of South America has doubled, and the mid-Atlantic Ocean spreading center is moving westward also, leaving some hot spots behind, as at Tristan da Cunha and St. Helena (figure 6.41).

The explosiveness of volcanic eruptions above hot spots varies. They are relatively peaceful above oceanic hot spots, such as Hawaii, where low-volatile, low-viscosity, large-volume, basaltic magma flows easily, analogous to spreading-center volcanism, and builds shield volcanoes (see figure 6.23). The Hawaiian hot spot is about 80 km (50 mi) in diameter, as defined by earthquake hypocenters at 60 km (37 mi) depth (see figure 6.24).

A hot spot below a spreading center means that a much greater volume of basaltic magma can erupt. For example, at Iceland, the asthenosphere magma feeding the spreading process is augmented by deeper mantle magma to create an immense volume of basaltic rock. The combined magmas are basalt, and the eruptions are peaceful enough for the citizens of Iceland to live prosperously (see chapter 7). The mantle plume beneath Iceland is the most vigorous hot spot on Earth today. The rising plume has created crust beneath Iceland that is four to five times thicker than average.

Above continental hot spots, such as at Yellowstone National Park, the eruptions may be incredibly explosive because the rising magma breaks off and absorbs so much continental rock that it creates a volatile-rich, high-viscosity, very-large-volume magma. The mention of a big volcanic eruption may bring to mind a tall mountain emitting a powerful explosion, but the really big eruptions emit so much magma that they leave a hole bigger than a mountain, a giant caldera that can be 100 km (more than 60 mi) long.

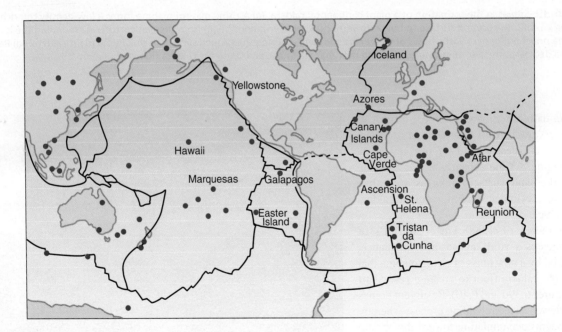

FIGURE 6.41 Hot spots active in the past 10 million years. Antarctica is not shown but lies above 11 hot spots, raising questions about the effects of melting massive volumes of ice.

A Classic Disaster

Santorini and the Lost Island of Atlantis

What were the effects of the eruption of Santorini on the Mediterranean world? So huge that they may be the basis of the Atlantis story.

Akrotiri was an important city on Thera (Santorini), a part of the advanced Minoan civilization based in Crete. In 1628 BCE, Akrotiri had three-story houses; paved streets with stone-lined sewers beneath them; advanced ceramic and jewelry work; regular trade with the Minoans' less-advanced neighbors in Cyprus, Syria, Egypt, and Greece; and colorful wall frescoes that depicted their wealthy and comfortable life. In short, these Minoans had a higher standard of living than many people in this part of the world today, more than 3,600 years later (figure 6.42).

The dramatic collapse of this piece of the Minoan civilization must have made an indelible impression on the people of that time. In fact, this may be the event passed down to us by Plato as the disappearance of the island empire of Atlantis, which after violent earthquakes and great floods "in a single day and night disappeared beneath the sea." Plato lived in Greece from 427 to 347 BCE. He told the tale in the dialogues of Critias, the historian, who recounted the visit of Solon to Egypt, where he learned the account of Atlantis from the Egyptian priests in their oral histories. About 1,200 years after the event, Plato wrote a reasonably good description of a caldera-forming collapse with attendant earthquakes, floods (steam surges or tsunami), and a landmass sinking below the sea in a day and a night.

The eruption and caldera-forming collapse into the sea at Santorini seem similar to the events at Krakatau 3,500 years later, except the Santorini event was bigger. The Santorini eruption is estimated to have blown out more than 40 km³ (10 mi³) of rhyolitic magma; Krakatau blew out 18 km³ (4.3 mi³). Krakatau sent out ocean waves 35 m (115 ft) high; Santorini must have done as much. The Aegean Sea region is one of the most island-rich areas on Earth. Tsunami in this region must have had a devastating effect on coastal towns and people, as well as leaving profound impressions on survivors, who passed these memories down to succeeding generations. The tales of Plato, the excavations by archaeologists, and the reconstructions by volcanologists all point to a remarkably consistent story.

Figure 6.42 View of a portion of Knossos, on the island of Crete, the center of the Minoan civilization. These buildings were hit hard by the Santorini eruption.

Photo by Pat Abbott.

Summary

Some of Earth's internal heat causes rock to rise via convection and then to melt near the surface and erupt as volcanoes. Spreading centers provide such ideal settings for volcanism that 80% of all extruded magma occurs there. Plates pull apart, and magma rises up the fractures with relatively peaceful eruptions. Subduction-zone eruptions involve magma contaminated by incorporated crustal rock, yielding high-viscosity, gas-rich magma that erupts explosively. Transform faults and continent-continent collision zones have little or no volcanism associated with them.

Hot rock at depth rises buoyantly. This hot rock may melt near the surface and become magma due to increased temperature, decreased pressure, and/or increased water content. Most magma is produced as pressure is lowered on rising hot rock via decompression melting or an increase in its water content. When magma nears the surface, gases come out of solution and help cause volcanic eruption. Whether magma erupts peacefully or explosively depends on magma types. Eruption styles and volcanic landforms can be understood via the three Vs of volcanology—viscosity, volatiles, volume.

Beneath the ocean basins, magmas are basaltic in composition with low contents of SiO_2, low weight % content of water, and high temperatures, producing low viscosity, easy escape of volatiles (gases), and peaceful eruptions. Beneath continents, rising basaltic magmas are contaminated by melting continental-crust rocks, thus altering magma compositions. The resultant andesitic-to-rhyolitic magmas have high contents of SiO_2, high weight % content of water, and relatively low temperatures, producing high viscosity, difficult escape for volatiles, and explosive eruptions.

When magma reaches the surface and gas escapes easily, lava flows result. Low-viscosity lava flows may build shield volcanoes much wider than they are tall, as are found in Hawaii, for example. If gas percentage is high and the gases are trapped in magma, explosions result, blasting pyroclastic debris into the air. A scoria cone may be built around a volcanic vent by the settling of pyroclastic debris (e.g., Paricutin). Tall symmetrical volcanic peaks are usually stratovolcanoes built of alternations of lava and pyroclastic material (e.g., Vesuvius).

The volcanic explosivity index (VEI) measures the size of volcanic eruptions on a scale of 0 to 8. Between the years 1500 and 2014, one VEI 7 eruption occurred (Tambora, 1815), along with five VEI 6 events (e.g., Krakatau, 1883; Pinatubo, 1991).

Calderas form when roofs collapse into partially emptied magma chambers. This can occur when a stratovolcano is too weak to stand and its peak collapses downward (e.g., Crater Lake, Oregon). If the peak falls into the ocean, major tsunami can result (e.g., Santorini, 1628 BCE; Krakatau, 1883). The biggest explosive eruptions occur on continents, where collapses may be bigger than mountains at resurgent calderas (e.g., Yellowstone).

Terms to Remember

aa   151
andesite   147
basalt   147
base surge   156
caldera   161
cinder cone   158
composite volcano   158
crater   158
crystallization   147
decompression melting   150
flood basalt   150
geyser   154
lahar   160
lava dome   150
mineral   147
obsidian   152
pahoehoe   151
Plinian eruption   158
plutonic rock   147
pore   154
pumice   152

pyroclastic   151
pyroclastic flow   160
resurgent caldera   162
resurgent dome   167
rhyolite   147
rock   147
scoria   152
scoria cone   158
shield volcano   155
stratovolcano   158
ultra-Plinian   162
viscosity   162
volatile   146
volcanic rock   147

Questions for Review

1. Mount Vesuvius is one of the world's most active volcanoes, yet it has quiet intervals lasting how long? Compare the times between major eruptions to a human life span.
2. Sketch a map of an idealized tectonic plate and evaluate the volcanic hazards along each type of plate edge.
3. What percentage of magma erupted each year comes out at spreading centers? At subduction zones? At hot spots?
4. What changes in temperature, pressure, and water content cause hot rock to melt? What are the two most relevant melting agents?
5. What common elements combine to form most igneous rocks?
6. What minerals combine to form most igneous rocks?
7. Contrast the differences between basaltic and rhyolitic magma in terms of SiO_2 percentage, weight % water content, temperature, viscosity, and mode of gas escape.
8. How does explosiveness vary between magmas that have a low versus a high weight % content of water?
9. If gas escapes easily from a magma, will the eruption be peaceful or explosive? If gas cannot escape easily, will the eruption be peaceful or explosive?
10. Volcanoes in the ocean tend to erupt peacefully, whereas volcanoes on continents tend to erupt explosively. What explains the differences?
11. Why do volcanoes above subduction zones erupt more explosively than volcanoes at spreading centers?
12. What determines whether volcanic activity will be a lava flow or a pyroclastic eruption?
13. Which magma will make a better lava flow—basalt or rhyolite?
14. Draw a cross-section illustrating a Plinian eruption.
15. Explain the factors controlling the volcanic explosivity index (VEI).
16. Play the three Vs game. Pick various low, medium, and high values for viscosity, volatiles, and volume, and then describe the resultant eruption styles and volcanic landforms.
17. Draw a cross-section showing the difference between a shield volcano and a stratovolcano.
18. Draw a cross-section and describe the collapse of an oceanic volcano, such as Krakatau. What usually is the biggest killer in this process?
19. Explain the eruptive behavior of a hot spot–fed volcano on a continent.
20. How might a volcanic eruption create its own weather?
21. Diagram and explain the sequence of events leading to a geyser eruption. Include temperature and pressure changes in your answer.

Questions for Further Thought

1. Why do people keep returning to a volcano, such as Vesuvius, and building new cities?
2. Evaluate this message: Plate boundaries are bad news.
3. Would you rather watch a volcano erupt in Hawaii or Washington? Why?
4. List the beneficial aspects of volcanoes and volcanism.
5. What evidence suggests that the eruption of the volcano Santorini led to the enduring tale of the lost civilization of Atlantis?

CHAPTER 7

Volcano Case Histories: Killer Events

Past civilizations are buried in the graveyards of their own mistakes.

—LORD RITCHIE-CALDER,
1970, "MORTGAGING THE OLD HOMESTEAD"

View over the harbor city of Catania, Sicily, toward Mount Etna. The city has been buried seven times by lavas from Etna. Each time, the city has been rebuilt on top of its predecessor to await its turn to be buried.

Photo by Pat Abbott.

LEARNING OUTCOMES

Active volcanoes are natural hazards, but understanding their processes helps us coexist with them. After studying this chapter, you should:

- recognize that humans can live successfully on oceanic hot-spot and spreading-center volcanoes.
- understand why subduction-zone volcanoes are so deadly.
- be able to explain the sequence of events in a catastrophic volcanic eruption as at Mount Saint Helens in 1980.
- be familiar with the leading causes of death by volcano.
- be able to explain pyroclastic flows and lahars.
- know the signs of impending volcanic eruption.

OUTLINE

- Volcanism at Spreading Centers
- Volcanism at Subduction Zones
- Volcanic Processes and Killer Events
- VEIs of Some Killer Eruptions
- Volcano Monitoring and Warning

ountaOntake is a sacred mountain; it is the second tallest mountain in Japan, after Mount Fuji, at 3,067 m (10,062 ft). Ontake is a popular site for hiking and Saturday, 27 September 2014, was a beautiful fall day with leaves changing color and welcoming weather that attracted visitors. Several hundred people were on the slopes at 11:52 a.m. when, with no warning, the volcano erupted rocks, ash, gas, and steam that killed 63 hikers. People high up the mountain died from inhaling hot gas and lung-choking ash, while others were hit by flying rocks. The iPhones of some deceased hikers contained photos of the oncoming ash flows that killed them.

Why was the eruption a surprise? Before the eruption there were no warning earthquakes because no magma was moving. This event was a hydrothermal explosion caused when underground water met magma and flashed to steam, powering the eruption. It is very difficult to predict when slowly flowing groundwater will come in contact with magma and explode.

In this chapter, we return to spreading-center volcanism at Iceland, examine subduction-caused explosive volcanism in the Cascade Range along the Pacific Coast of the United States and Canada, and then examine the historic record of volcano-related fatalities to understand the specific processes that kill people. Last, we look at failure and success in volcano monitoring and warning.

Volcanism at Spreading Centers

Most of the volcanism on Earth takes place along the oceanic ridge systems where seafloor spreading occurs. Solid, but hot and ductile mantle rock rises upward into regions of lower pressure, where up to 30–40% of the rock melts and flows as basaltic magma. The worldwide rifting process releases enough magma to create 20 km^3 (less than 5 mi^3) of new basaltic oceanic crust each year. Virtually all of this volcanic activity takes place below sea level and is thus difficult to view. We see and are impressed by the tall and beautiful volcanic mountains on the edges of the continents, but the volume of magma they release is small compared to that of spreading centers.

ICELAND

Iceland is a volcanic plateau built of basaltic lava erupted from a hot spot below the mid-Atlantic Ocean spreading center (see figure 4.4). The country is a little bit bigger than the state of Virginia; about 13% of its surface is covered by glaciers, and one-third consists of active volcanoes. During the nearly 1,000 years of human records, volcanic eruptions have occurred about every five years, on average. Most Icelandic eruptions do not cause deaths, but exceptions do occur (see famine of 1783 later in this chapter).

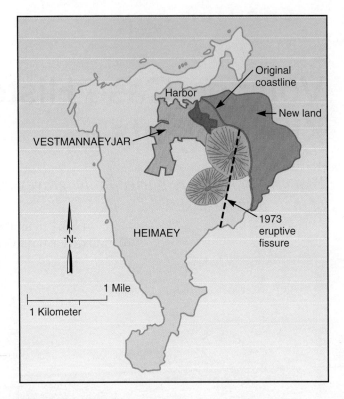

Figure 7.1 The island of Heimaey with the old coastline shown as an orange line. The dark gray area is new land formed by the 1973 lava flow. Note that the new harbor is bigger and better protected.

Data source: Williams, R. S., Jr., and Moore, J. G., "Man Against Volcano: The Eruption on Heimaey, Vestmannaeyjar, Iceland," *US Geological Survey,* 1983.

The most typical Icelandic eruptions are fissure eruptions, where lava pours out of long fractures up to 25 km (16 mi) long. To understand Icelandic eruptions, visualize the linear spreading center (see figure 4.5) that controls the rise of magma as it is fed upward through fractures. An Icelandic eruption can be beautiful to watch as an elongate "curtain of fire" shoots upward with varying intensity and height. Icelandic eruptions of low-viscosity, low-volatile lava flows can be so peaceful that their movement is almost waterlike.

Lava Flows of 1973

The recent story of Iceland shows that humans can make enough adjustments to live profitably and happily next to active basaltic volcanism. The 1973 eruptions on the small island of Heimaey south of the big island of Iceland illustrate the "peaceful" nature of these eruptions. The town of Vestmannaeyjar is built next to the premiere fishing port in Iceland. The safe harbor is itself a gift of volcanism; it was formed between ancient lava flows. On 23 January 1973, a fissure opened up only 1 km (3,300 ft) from the town of 5,300 people (figure 7.1). By early July, the eruption had emitted 230 million m^3 of lava (figure 7.2) and 26 million m^3 of pyroclastic material. The lava flows increased the size of the island by 20%. Gases vented during the eruptive sequence, other than water vapor, were dominantly CO_2

Figure 7.2 An aa lava flow stopped against and between two fish-factory buildings in Vestmannaeyjar, 23 July 1973.
© Óskar Elías Björnsson.

Figure 7.3 Seawater is being sprayed on the lava front to cool, harden, and stop it from closing off Vestmannaeyjar harbor, 4 May 1973.
US Geological Survey.

with lesser amounts of H_2, CO, and CH_4. The only fatality was a person asphyxiated inside a gas-filled building.

The early lava flows on Heimaey began filling in the harbor and destroying about 300 buildings; pyroclastic fallout buried another 70 buildings. But the volume of lava was not overwhelming, so the Icelanders took over. Pyroclastic material was bulldozed to create barriers that diverted and controlled the flow of later lavas and even controlled the flow paths of the dense volcanic gases. To save their harbor and economic livelihood, the Icelanders sprayed seawater on the lava flows, causing rapid cooling and hardening into wall-like features that forced the lava to flow off in another direction (figure 7.3). This action prevented the harbor from being filled and closed. Now, with its new shape and larger size, the harbor is better than before the 1973 eruption (see figure 7.1).

When the eruptions stopped, the people set up a pipe system that poured water into the 100 m (330 ft) thick mass of slowly cooling lava. Return pumps were installed to bring the water, which had been heated to 91°C (196°F), back to the surface and into town, where it was used to heat buildings. Basaltic eruptions do not have to be killers. Humans and volcanoes can coexist in harmony, with luck and with some exceptions.

Volcanism at Subduction Zones

Through newspapers and television, we learn of death-dealing volcanic eruptions at Galeras Volcano in Colombia, Mount Unzen in Japan, Mounts Pinatubo and Mayon in the Philippines, Mount St. Helens in Washington, and Soufriere Hills on Montserrat. They are all subduction-zone volcanoes. These stratovolcanoes have the biggest impact on humans. Many of the regions around subduction-zone volcanoes are heavily populated and feel the wrath of the eruptions. Also, because these volcanoes erupt directly into the atmosphere, they can affect climate worldwide (see chapter 12).

CASCADE RANGE, PACIFIC COAST OF UNITED STATES AND CANADA

Explosive eruptions are frequent at the numerous volcanoes in the Pacific Northwest region of the United States and in British Columbia (figure 7.4). The plate-tectonic process responsible for these volcanoes is identical to the cause of the region's great earthquakes—subduction. In fact, the frequent eruptions from the Cascade Range volcanoes provide clear evidence for active subduction. The melting of part of the mantle (asthenosphere) wedge above the subducting plate is aided by water released from sediments on top of the subducting plate. The rising basaltic magma partially melts overlying crustal rock as well, increasing its content of SiO_2 and water. Much of the magma changes its composition to andesite or rhyolite and increases its viscosity as it rises (figure 7.5). Some collects in great pods and cools underground, forming plutonic rocks, but some erupts explosively at the surface.

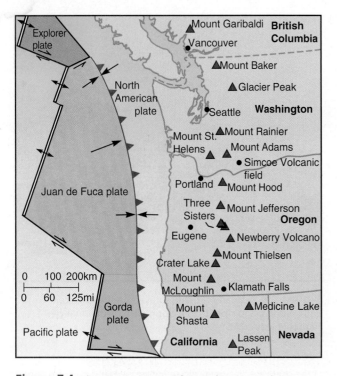

Figure 7.4 Plate-tectonic map of Cascade Range volcanoes. Volcanoes are subparallel to the subduction zone and spaced somewhat regularly.

How often do major eruptions occur? An example was documented in a 1975 study of Mount St. Helens by Dwight Crandell and colleagues. Their report stated that the latest large, volcanic mountain had formed in the last 2,500 years. Since then, Mount St. Helens has experienced major eruptions every century or two and has never been free from major volcanism for longer than 500 years. Their 1975 report stated, "Although dormant since 1857, St. Helens will erupt again, perhaps before the end of this century." The geologic analysis was prophetic (figure 7.6).

Figure 7.7 shows the age distribution and eruption history along the line of Cascade Range volcanoes. Basically, the volcanoes are all the same age; they sit above subducting plates and are active now. Volcanoes built above hot spots also line up—for example, Hawaii (see figure 2.22) and Yellowstone (see figure 6.37). In contrast to subduction-zone volcanoes, the ages along lines of hot-spot volcanoes range from young to old in orderly progressions.

How are prehistoric eruptions documented? The process is the same as that used to work out dates of prehistoric earthquakes. The slopes near a volcano reveal the remains of trees knocked down by volcanic blasts (figure 7.8). These trees may be buried by volcanic ash, incorporated in lahars, or otherwise preserved. Radiocarbon determinations of the dates when trees died also tell the dates of the volcanic eruptions that killed them (figure 7.9).

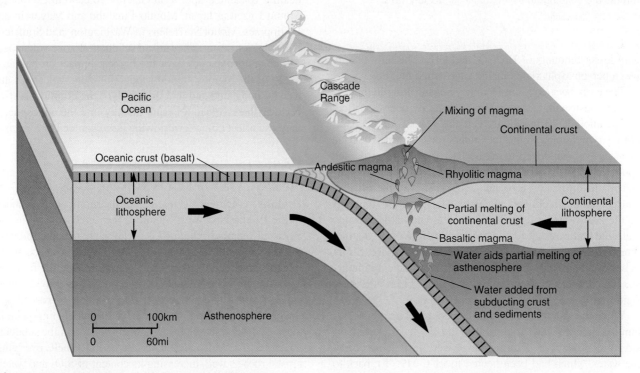

Figure 7.5 Subduction-zone volcano "factory." Basaltic magma forms in the upper asthenosphere where subducted water aids partial melting of oceanic crust and asthenosphere. Rising magma partially melts some continental crust, forming water-rich andesitic to rhyolitic magmas that erupt explosively.

Figure 7.6 Mount St. Helens, Washington. (a) *Before:* View to the northeast of the beautiful cone of Mount St. Helens on 25 August 1974. Mount Rainier is in the distance. (b) *After:* Same view on 24 August 1980, after the volcano had blown off its top 400 m (1,313 ft).

(a) University of Washington Libraries, Special Collections, John Shelton Collection, Shelton 6754; (b) University of Washington Libraries, Special Collections, John Shelton Collection, Shelton 67-706cr.

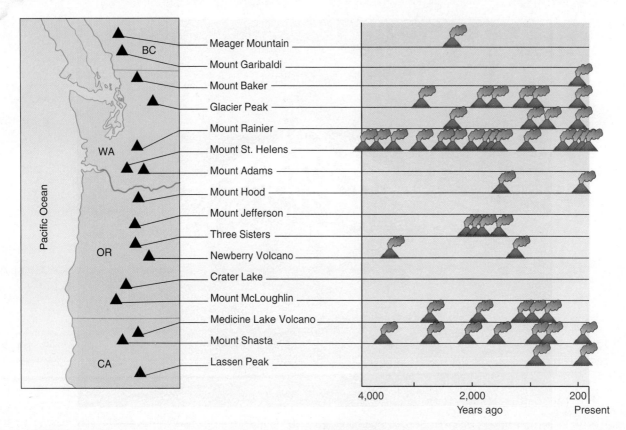

Figure 7.7 Eruption histories of Cascade Range volcanoes during the last 4,000 years.

Mount St. Helens, Washington, 1980

In late March 1980, Mount St. Helens awoke from a 123-year-long slumber. Dozens of magnitude 3 earthquakes occurred each day as magma pushed its way toward the surface. On 27 March, small explosions began as groundwater and magma came in contact. The spectacle of an erupting volcano was a tremendous lure for sightseers. People flocked to Mount St. Helens. The weekend traffic was so jammed that it reminded folks of rush hour in big cities. But this was an explosive giant just warming up its act, and all nearby life was in grave danger. The governor of Washington ordered blockades placed across the roads to Mount St. Helens to keep people away. Her action was unpopular. Then, at 8:32 a.m. on 18 May 1980, the volcano blew off the top 400 m (1,313 ft) of its cone during a spectacular blast that generated about 100 times the power of all U.S. electric-power plants combined. Most of the 62 people killed had found ways around the barricades to get a closer view of an erupting volcano. A look at the eruptive sequence provides a good example of how an explosive volcano does its thing (figure 7.10).

First, Mount St. Helens achieved its beautiful conical shape during the mid-1800s (figure 7.10a). In 1843, a SiO_2-rich lava dome grew at the volcano peak. In 1857, andesitic lava flows cooled high on the slopes. But these events also set up discontinuities, or weaknesses, within the volcanic cone.

Figure 7.8 What was once a mature forest is now a field of fallen trees pointing in the direction traveled by the volcanic blast from Mount St. Helens on 18 May 1980.
Photo by J. DeVine/U.S. Geological Survey.

Second, in 1980, rising magma began changing the shape of the volcano (figure 7.10b). Earthquake hypocenters were abundant at 1 to 3 km (0.6 to 2 mi) depth. The seisms were recording the injection and pooling of magma. With magma forcing its way upward, the northern side of the volcano began rising. The increasing volume of magma also caused the groundwater body to expand its volume. The effect on the volcano was dramatic.

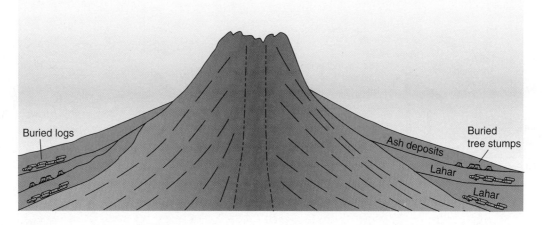

Figure 7.9 Schematic cross-section of a volcano and some of its eruptive deposits. Radiocarbon dates on buried wood tell when trees died—that is, when the volcano erupted.

By 12 April, a 2 km² (1.2 mi²) area on the north flank had risen upward and outward by 100 m (330 ft). This unstable situation grew worse as the "mega-blister" kept growing about 1.5 m (5 ft) per day.

Third, at 8:32 a.m. on 18 May 1980, the bulge failed. With magma injecting into the bulge from below and gravity pulling from outside, the huge mass of the bulge, with its weak strength, failed and pulled away as an avalanche. The shaking ground was recorded as a magnitude 5.1 earthquake. The avalanche material was 2.5 km³ of the north side of the mountain; it fell away at speeds up to 250 km/hr (150 mph) (figure 7.10c). The avalanche was a roiling mass of fragmented rock that once was the mountaintop and side, combined with ice blocks, snow, magma, soil, and broken trees; the internal temperature of the mass was about 100°C (212°F). Part of the avalanche slammed into Spirit Lake, causing waves 200 m (650 ft) high. Another part overrode a 360 m (1,180 ft) high ridge that lay 8 km (5 mi) to the north; then it turned and moved 23 km (14 mi) down the north fork of the Toutle River (figure 7.11). The resulting deposit was a chaotic mixture of broken rocks and loose debris that averaged 45 m (150 ft) in thickness and had a hummocky surface relief of 20 m (65 ft). Only a short time earlier, this material had been the top of the mountain. At the same time as the avalanche occurred, lahars were forming and flowing down the river valleys as rock particles mixed with water derived from melting snow and ice, from Spirit Lake, and from within the avalanche. These slurries continued to form and flow for many hours after the eruption began. Lahars moved long distances at speeds up to 40 km/hr (25 mph), carrying huge boulders and flowing with a consistency like wet concrete.

Fourth, as the landslide began to pull away, the dramatic drop in pressure on the gaseous magma and superheated groundwater caused a stupendous blast (figure 7.10d). The blast and surge roared outward at speeds up to 400 km/hr (250 mph). The blast overtook and passed the fast-moving avalanche, racing over four major ridges and scorching an area of 550 km² (210 mi²) with 0.18 km³ of volcanic rock fragments and swirling gases at about 300°C (572°F) (figure 7.11). The blast was a pyroclastic flow. It was denser than air, flowing along the ground as a dark cloud, with turbulent volcanic gases keeping solid rock fragments, magma bits, and splintered trees in suspension; it behaved as a very low-viscosity fluid.

Fifth, the big blast opened up the throat of the volcano, exposing an effervescing magma body. Rapidly escaping gases blew upward, carrying small pieces of magma to heights greater than 20 km (12 mi) during the Plinian phase, which lasted about nine hours (figure 7.10e). The boiling gases carried about 1 km³ (0.24 mi³) of volcanic ash up and away. About 0.25 km³ of ash was blown across the United States at different heights by various wind systems. Another 0.25 km³ formed pyroclastic flows by either spilling out of the volcano or falling down from the eruption cloud (figure 7.12). These pyroclastic flows had temperatures of 300° to 370°C (570° to 700°F) and moved at speeds up to 100 km/hr (more than 60 mph).

Today, the volcano is slowly repairing the damage done to its once-symmetrical cone as it builds an SiO₂-rich lava dome (figures 7.10f and 7.13). The magma building the lava dome has not erupted explosively, probably because it lost most of its volatiles during the big eruption on 18 May 1980.

Mount St. Helens looks very different these days (figure 7.14). Gone are the mountaintop, snowfields, forests, and lakes. The once tree-lined river valleys are clogged with volcanic debris (figure 7.15). But recovery is progressing well. Bacteria have eaten the sludge from dirty lakes, leaving pure water that has been stocked with trout. Plants have sprouted anew in devastated ground, and animals have returned to feed on them and on each other. Life is erasing the effects of the volcanic events.

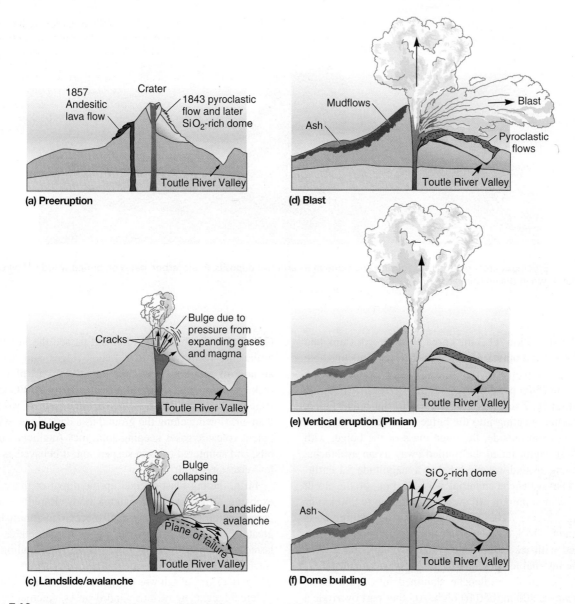

Figure 7.10 Eruptive sequence (VEI = 5) of Mount St. Helens in 1980. (a) The symmetrical volcanic cone was shaped in 1843 and 1857. (b) In late March, rising magma and expanding gases caused a growing bulge on the northern side. (c) At 8:32 a.m. on 18 May 1980, the bulge failed in a massive landslide/debris avalanche recorded as a magnitude 5.1 earthquake. (d) The landslide released pressure on the near-surface body of magma, causing an instantaneous blast of fragmented rock and magma. (e) The "throat" of the volcano was now clear, and the vertical eruption of gases and small blobs of magma shot up to heights of more than 20 km (12 mi) for nine hours. (f) Today, the mountain is slowly rebuilding a volcanic dome of low–water content, SiO_2-rich magma.

Were the explosive events at Mount St. Helens a rare occurrence? Are similar events likely at other Cascade Range volcanoes in our lifetimes?

Lassen Peak, California, 1914–1917

Lassen Peak is not the typical volcano; rather, it is an unusually large (about 1 mi^3) lava dome of SiO_2-rich volcanic rock analogous to that growing in Mount St. Helens today (see figure 7.13). Lava domes form when magma is too poor in volatiles and too viscous to flow away, so instead it oozes upward as a conduit-plugging mass (see figure 6.31).

Lassen Peak awakened in May 1914 with numerous eruptions, culminating on 18 July 1914 with a major episode that sent up an ash cloud more than 3,350 m (11,000 ft) high. Small-scale volcanic activity continued, but large events did not resume until May 1915. On 16–18 May 1915, the 300 m (1,000 ft) wide crater overfilled with water-deficient, sticky magma that stood higher than the rim. The magma was too viscous to flow over the lip, so instead, red-hot blocks broke off and rolled downslope. Meanwhile, to the east, the melting snow combined with rocky debris to set in motion a massive lahar that flowed outward 50 km (30 mi). On 19 May, on the north slope, the side of Lassen Peak split, and a pyroclastic flow blasted forth as a mixture of superhot gases, fragmental rock debris, trees, and water, devastating a triangular-shaped area 6.5 km (4 mi) long and

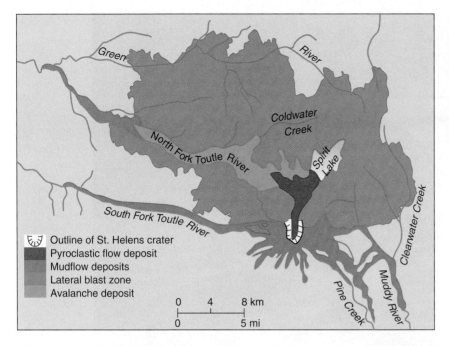

Figure 7.11 Map of materials dumped in the 18 May 1980 eruption of Mount St. Helens. Avalanche deposit was from the initial landslide. A lateral blast followed immediately. Then pyroclastic flows spilled out of the exposed magma body. Through it all, the superheated groundwater, plus melting snow and ice, fluidized sediments on the steep slopes as lahars (mudflows) that ran down the valleys.

Source: R. Tilling, "Eruptions of Mount St. Helens: Past, Present and Future," 1984, *US Geological Survey.*

Legend:
- Outline of St. Helens crater
- Pyroclastic flow deposit
- Mudflow deposits
- Lateral blast zone
- Avalanche deposit

Figure 7.12 High-temperature pyroclastic flow rolling down the side of Mount St. Helens, 7 August 1980.

Peter W. Lipman/U.S. Geological Survey.

Figure 7.13 Lava dome of high-viscosity, low-volatile magma growing in the central magma pipe of Mount St. Helens since its big eruption in 1980.

© PhotoLink/Photodisc/Getty Images RF.

Figure 7.14 View of the devastated northeastern side of Mount St. Helens, 20 August 1980. Mount Hood is in the background.

University of Washington Libraries, Special Collections, John Shelton Collection, Shelton 67-686.

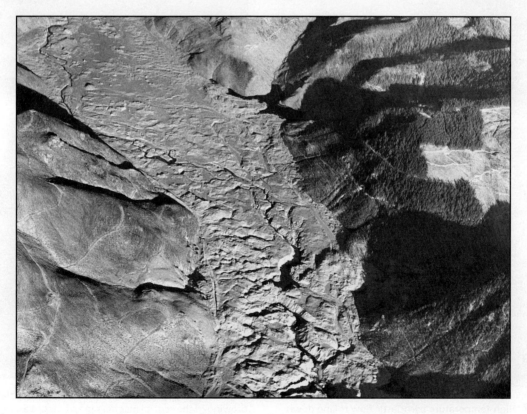

Figure 7.15 The Toutle River, choked with eruption debris.

University of Washington Libraries, Special Collections, John Shelton Collection, Shelton 67-704.

1.6 km (1 mi) wide (figure 7.16). Volcanic activity continued with more lahars and pyroclastic flows, and on 22 May, a broad mushroom cloud of ash was blasted 8 km (5 mi) high. Lassen remained relatively peaceful through 1916, but May and June 1917 brought renewed activity.

In three of four years, the month of May saw the start of extensive volcanic activity. Was this a coincidence? Maybe, but it is possible that as water from the melting snow sank and was heated underground, its volume expansion helped fracture Lassen Peak and reduce internal pressure enough to begin the eruptions. In the nonvolcanic year of 1916, Lassen Peak was too hot for snow to accumulate.

In the 20th century, two Cascade Range volcanoes underwent similar eruptions with sideward-directed blasts, pyroclastic flows, far-reaching volcanic mudflows (lahars), and great vertical eruptions of ash (Plinian phase). Luckily, each of these eruption sequences took place in sparsely inhabited areas. What are the prospects for similar eruptions near towns and cities?

Mount Shasta, California

Mount Shasta, at 4,318 m (14,162 ft) elevation, is the second tallest of the Cascade Range volcanoes (figure 7.17). The third highest is Shastina (3,759 m, or 12,330 ft), perched on its shoulder. The combined mountain mass is particularly impressive, standing over 3,000 m (10,000 ft) higher than its surroundings and visible from more than 160 km (100 mi) away. Mount Shasta is an active volcano, erupting 11 times in the last 3,400 years, including at least 3 times in the last 750 years. Its last eruption was probably in 1786.

The Mount Shasta area is a beautiful place to live, and the towns along the volcano base are growing. But how wise is this? The lower slopes of Mount Shasta are broad and smooth, allowing pyroclastic flows to spread widely as they move down the volcano flank (figure 7.18). Lahars are more prone to flow through valleys, and towns lie there (figure 7.19). The rock record gives further reason to pause and consider whether or not to build here. Figure 7.20 shows the distribution of a 300,000-year-old avalanche deposit that extends 43 km (27 mi) out from the volcano base. This catastrophic event deposited eight times more debris than Mount St. Helens did in 1980. This jumbled mass near Mount Shasta is the foundation for three towns and one large reservoir.

Would it be advisable to draw park boundaries around the hazardous Cascade Range volcanoes and not allow towns to be built there? Volcanologists Dwight Crandell, Donal Mullineaux, and Meyer Rubin point out that

The potential risk from future eruptions may be low in relation to the lifetime of a person or to the life expectancy of a specific building or other structure. But when dwelling places and other land uses are established, they tend to persist for centuries or even millennia.

Figure 7.16 View of the north side of Lassen Peak, devastated by the 19 May 1915 eruption.

University of Washington Libraries, Special Collections, John Shelton Collection, Shelton 2885.

Figure 7.17 View from the north to Mount Shasta and Shastina. Note the network of roads being used to develop towns on top of lava flows, lahars, and debris avalanche deposits. Note the debris avalanche deposit that flowed from photo center to base of photo.

University of Washington Libraries, Special Collections, John Shelton Collection, Shelton [no number].

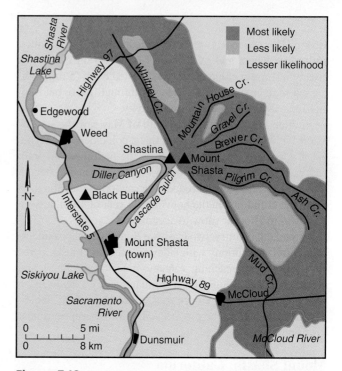

Figure 7.19 Map of the Mount Shasta–Shastina area showing the most likely paths for lahars. These volcanic mudflows tend to occupy the same river bottom flatlands where towns are built.

Source: D. R. Crandell and D. R. Nichols, "Volcanic Hazards at Mt. Shasta," 1989, *US Geological Survey.*

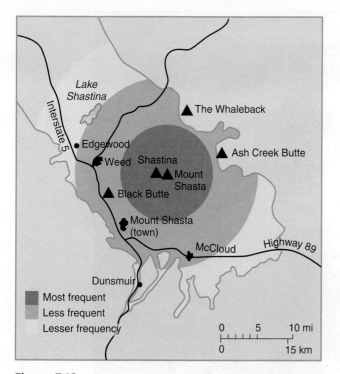

Figure 7.18 Map of the Mount Shasta–Shastina region, showing the areas most susceptible to lateral blasts and pyroclastic flows. Note the growing towns within the danger zones.

Source: D. R. Crandell and D. R. Nichols, "Volcanic Hazards at Mt. Shasta," 1989, *US Geological Survey.*

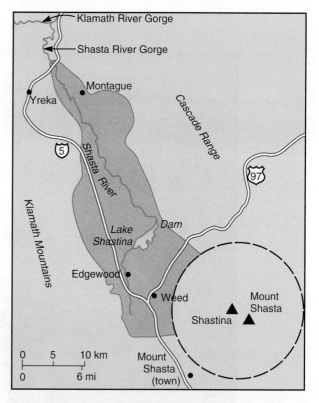

Figure 7.20 Map of a 300,000-year-old debris avalanche deposit at the base of Mount Shasta. The amount of material is eight times greater than the amount erupted at Mount St. Helens in 1980. It forms the foundation for three towns and one reservoir. See this avalanche deposit in the front center of figure 7.17.

In Greater Depth

Rapid Assembly and Rise of Magma

As 2 May 2008 began, a 5.2M earthquake shook southern Chile. Less than four hours later, Chaiten Volcano blew magma, ash, and steam 21 km (13 mi) up into the atmosphere in a Plinian eruption. The magma was rhyolite, the most viscous type. Unexpectedly, the toothpaste-like magma had risen 5 km (3 mi) in less than four hours, then erupted explosively. Apparently the earthquake opened **fissures** (cracks) to the magma reservoir that caused a pressure drop, triggering the eruption.

Volcanic events beginning and exploding so quickly bring into question how much warning time would there be before a super-eruption, a caldera-forming eruption. The volumes of magma extruded during super-eruptions are huge compared to recent events (figure 7.21). With thousands of years between caldera eruptions, it has seemed that long lengths of time would be necessary to accumulate a huge new body of magma; however, it now seems that it could occur in historical lengths of time rather than requiring geological time spans. Study of chemically zoned mineral crystals such as quartz at Yellowstone, Wyoming; plagioclase at Santorini, Greece; and olivine at Irazu in Costa Rica have distinguished pre-eruptive from eruptive processes. The mineral cores, their interiors, may be thousands of years old but their outermost layer may be less than 100 years old. This indicates that long-existing magma bodies may be rather quickly recharged and/or disturbed to connect and mobilize existing batches of magma that surpass the threshold conditions needed for a super-eruption.

The potential for a caldera eruption in our lifetimes seems intuitively unlikely, but it is not scientifically unreasonable. Two candidate sites for a super-eruption are Yellowstone National Park, USA, and Vesuvius (Campi Flegrei), Naples, Italy.

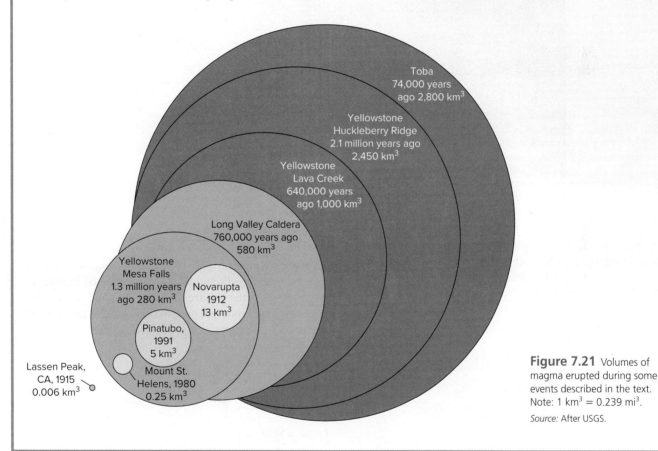

Figure 7.21 Volumes of magma erupted during some events described in the text. Note: 1 km³ = 0.239 mi³.
Source: After USGS.

Volcanic Processess and Killer Events

Volcanoes can kill in numerous ways (figure 7.22). They can burn you with a pyroclastic flow, slam and suffocate you with a lahar, batter and drown you with tsunami, poison you with gas, hit you with a pyroclastic bomb, fry you with a lava flow, or kill you with indirect events such as famine.

THE HISTORIC RECORD OF VOLCANO FATALITIES

Volcanoes operate all around the world. How many people do they kill? Which volcanic processes claim the most lives? The lack of written records for some time intervals and in some parts of the world makes these questions difficult to answer. Volcanologists Tom Simkin, Lee Siebert, and Russell Blong have studied the questions and given approximate

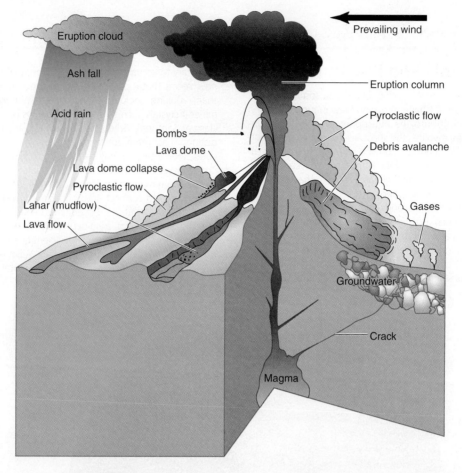

Figure 7.22 Volcanoes operate many life-threatening natural processes.

Source: US Geological Survey Fact Sheet 002–97 (1997).

answers. About 275,000 people have been killed by volcanic action during the past 500 years (figure 7.23). A dozen or so volcanic processes have done the killing (table 7.1). We will now individually examine some of the killer processes. As we do so, note that several of these each resulted in thousands of deaths (figure 7.23).

PYROCLASTIC ERUPTIONS

Explosive volcanic eruptions shatter magma into pieces by gas bubble growth and blast the fragmented magma as pyroclasts up to the surface and into the atmosphere. The pyroclasts may be brought back to the ground as **pyroclastic falls, pyroclastic flows,** and **pyroclastic surges.**

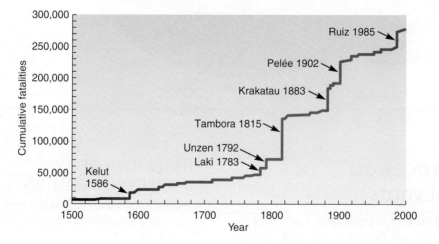

Figure 7.23 Cumulative fatalities from volcanoes during 500 years, 1500 to 2000.

Data Source: T. Simkin, L. Siebert, R. Blong, Science 291:255 (2001).

Pyroclastic Falls

During an explosive eruption, airborne pyroclasts fall down on the landscape with particles ranging in size from ash to bombs to huge blocks. The pyroclastic fall is similar to cannon bombardments during war, and it results in about 2% of volcanic deaths.

Pyroclastic Flows

Few experiences on Earth are as frightening as having a super-hot, turbulent cloud of ash, gas, and air come rolling toward you at high speed. History records numerous instances of pyroclastic flows killing thousands of people at each event.

A pyroclastic flow is an overwhelming mixture of hot hunks of magma, volcanic ash, volcanic gas, and mixed-in air that flows downslope at speeds greater than 10 m/sec (22 mph) and may exceed 100 m/sec (225 mph). Pyroclastic flows derive their energy from the volcanic eruption, gas expansion within the flowing mass, and the pull of gravity. Temperatures of 350°C (660°F) were measured inside the volcanic ash cloud at Mount Unzen, Japan, in 1992.

Pyroclastic flows are responsible for 29% of volcanic deaths; they are the deadliest volcanic process (table 7.1). Pyroclastic flows begin in a variety of ways (figure 7.24).

Dome Collapse A growing lava dome provides a unique combination of steady magma supply and the upward lift of unstable, overhanging topography. Big hunks of lava dome frequently break off and create pyroclastic flows (figure 7.24a). At Mount Unzen in Japan, between 1991 and 1994, more than 7,000 dome collapses were recorded.

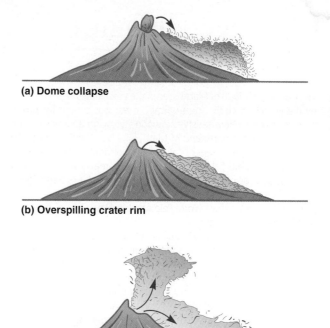

(a) Dome collapse

(b) Overspilling crater rim

(c) Direct blast

(d) Eruption column collapse

Figure 7.24 Ways of generating pyroclastic flows: (a) dome collapse as at Mount Unzen, 1991; (b) overspilling of crater rim as at Mont Pelée, 1902–1903; (c) direct blast as at Mount St. Helens, 1980, and Mount Pinatubo, 1991; (d) eruption column collapse as at Mount Mayon, 1984.

TABLE 7.1
Volcanic Causes of Deaths

	275,000 Deaths	530 Volcanic Events
Pyroclastic flow	29%	15%
Tsunami	21%	5%
Lahar	15%	17%
Indirect (famine)	23%	5%
Gas	1%	4%
Lava flow	<1%	4%
Pyroclastic fall (bombs)	2%	21%
Debris avalanche	2%	3%
Flood	1%	2%
Earthquake	<1%	2%
Lightning	<1%	1%
Unknown	7%	20%

Data source: Simkin, T., Siebert, L., and Blong, R., "Volcano fatalities" in *Science* 291:255, 2001.

In May 1991, the lava dome in Mount Unzen began a growth spurt that attracted international attention. As the unstable lava dome grew and towered 90 m (300 ft) above the crater rim, 15,000 residents were evacuated from villages and tea plantations around the mountain's base. As residents left, journalists and volcanologists arrived to record the numerous collapses of 200 to 300 ft high masses from the lava dome and watch the debris run downslope as glowing pyroclastic flows. At 4:09 p.m., on 3 June 1991, a much larger than usual mass fell off the lava dome and rolled downslope at about 60 mph, killing 44 observers, including the famed French volcano photographers, Maurice and Katya Krafft. All the deaths occurred in previously evacuated areas.

Overspilling Crater Rim A volcano crater may have its lake turned into a cauldron of boiling water, or the crater may

A Classic Disaster

Mont Pelée, Martinique, 1902

The Caribbean island of Martinique in the West Indies was colonized by the French in 1635. The tropical climate was superb for growing sugarcane to help satisfy the world's growing appetite for the sweetener. On the north end of Martinique is a 1,350 m (4,430 ft) high volcano. The French called the volcano Pelée, meaning "peeled" or "bald," to describe the bare area where volcanism had destroyed all plant life during the eruptions of 1792 and 1851. By coincidence, the pronunciation of the French word *Pelée* is the same as the Polynesian word *Pele* used in Hawaii to denote the goddess of volcanoes and fire.

In early spring of 1902, Vulcanian activity began. The crater atop Mont Pelée began filling with extremely viscous magma, displacing boiling lake waters through a V-shaped notch (figure 7.25). The extraordinarily sticky magma kept plugging the crater. At times, superhot pyroclastic flows would spill out of the crater; at other times, they would blast out. By late April, it was obvious to most people that this problem might get bigger. About 700 rural folks were migrating each day into St. Pierre, a city of picturesque, early 17th-century buildings that normally was home to 25,000 residents. Another 300 people a day were leaving St. Pierre, which lay only 10 km (6 mi) from Mont Pelée. At a little past noon on 5 May, a large pyroclastic flow sped down the Riviére Blanche,

destroying the sugar mill and 40 people. This further increased the anxiety level in St. Pierre. But there was an election coming up on 10 May, and the governor did not want everyone scattered from the island's largest city because that would likely change the election results. Governor Mouttet and his wife went to St. Pierre and used the militia to preserve order and halt the exodus of fleeing people. Bad decision. There was no election on 10 May anyway; all the voters, including the governor, died on 8 May (figure 7.26).

On the morning of 8 May 1902, a massive volume of gas-charged, ultrasticky magma had risen to the top of the crater. At about 7:50 a.m., witnesses heard sharp blasts that sounded like thousands of cannons being fired as trapped gas bubbles exploded and shattered magma into fine pieces. This spectacular pyroclastic flow moved as a red-hot avalanche of incandescent gases and glowing volcanic fragments (then called **nuée ardente,** which is French for "glowing cloud"). The mass moved as solid particles of magma suspended in gas. Its energy came from (1) the initial blast, (2) gravity, and (3) gas continuing to escape from the pieces of airborne magma, creating a "popcorn" effect. The momentum of the flow was aided and its friction reduced by internal turbulence and air mixed into the flow as it moved downward and outward. The temperature at the crater is estimated to have been about 1,200°C (2,200°F), and the glowing cloud was still hotter than 700°C (1,300°F) when it hit St. Pierre. The coarsest and heaviest part of the pyroclastic flow moved down the Riviére Blanche. The associated gas-ash clouds expanded in width and overwhelmed St. Pierre (see figure 7.25).

How was the town of St. Pierre destroyed? The pyroclastic flow moved with hurricane speeds of about 190 km/hr (115 mph),

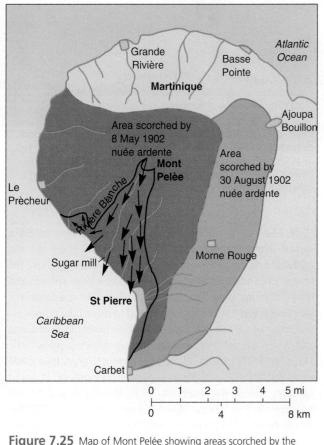

Figure 7.25 Map of Mont Pelée showing areas scorched by the largest pyroclastic flows of 1902.

Figure 7.26 The pyroclastic flow—charred remains of St. Pierre, May 1902. Mont Pelée is in the background.

Library of Congress, Prints and Photographs Division [LC-USZ62-76173].

but it was much denser than a hurricane because of its contained ash. The flow lifted roofs, knocked down most walls perpendicular to its path, twisted metal bars, and wrapped sheets of metal roofing around the scorched trunks of trees. Within the space of a couple of minutes, St. Pierre turned from a verdant tropical city to burned-out ruins covered by a foot of grey ash. Muddy ash also plastered any walls and tree trunks that were still standing.

What killed the people? Death was quick and came from one of three causes: (1) physical impact, (2) inhaling superhot gases, or (3) burns. The refugee-swollen population of St. Pierre was more than 30,000; only two people are known to have survived. One was Auguste Ciparis, a 25-year-old murderer locked in a stone-hut jail without windows and with only a small barred grating in his door. When hot gases entered his cell, he fell to the floor, suffering severe burns on his back and legs. Four days later, he was rescued; he then spent the rest of his life showing his scarred body at circus sideshows as "the prisoner of St. Pierre." The other survivor was a man inside the same house where his family members died.

Was it safe to be on a boat in the harbor? No. The fiery hot cloud did not stop when it hit the water. Of 18 boats in the harbor, only the British steamship *Roddam* survived, though it was badly burned and two-thirds of its crew were dead.

Pyroclastic flows continued rolling out of Mont Pelée. St. Pierre was overwhelmed again on 20 May, but it no longer mattered. On 30 August, a pyroclastic flow moved toward the southeast and scorched Morne Rouge and four other towns, killing another 2,000 people. Despite these tragic events, at present the area is fully settled once again.

---

fill with magma. If the crater overfills, hot water and magma can pour over the rim and flow downslope (figure 7.24b). This happened numerous times on Mont Peléee in Martinique in 1902.

Direct Blast In some eruptions, a pyroclastic flow may simply form as a direct blast from the volcano. In 1980, as a landslide moved down Mount St. Helens, the decrease in pressure on magma inside the volcano caused a tremendous direct blast (figure 7.24c). The direct blast traveled 150 m/sec (335 mph) and overwhelmed everything it encountered—lakes, trees, people.

Eruption Column Collapse At its greatest power, a volcano may send its eruption column of hot pyroclastic material, hot gas, and intermixed air high up into cooler air, providing time for heat to dissipate and for pyroclasts to cool and be spread far and wide. A dangerous phase of the eruption can occur in those moments when less energy is fed into the eruption column and the column begins to collapse, sending clouds of hot gases, ash, and pumice flowing as ground-hugging deadly pyroclastic flows (see figure 7.12).

Since 1616, more than 1,500 people have been killed during 40 recorded deadly eruptions of the subduction-caused stratovolcano Mount Mayon in the Philippines. In 1984, a series of Vulcanian eruptions sent magmatic debris 10 km (6 mi) into the atmosphere several times. Partial collapses of the eruption column sent pyroclastic flows rolling down the mountain slope at velocities ranging from 50 to 100 km/hr (30 to 60 mph) (figures 7.24d and 7.27).

Pyroclastic Flows over Water Can a pyroclastic flow travel across a body of water to kill you? Or does the water absorb heat from the hot, gas-rich cloud quickly enough to eliminate its ability to kill? A body of water does *not* eliminate the hazard. During the 1883 eruptions of Krakatau

Figure 7.27 Formation of pyroclastic flows as collapses from the vertical eruption column flow downhill, Mayon Volcano, Philippines, 1984.

Chris G. Newhall/U.S. Geological Survey.

in Indonesia, one remarkable blast on 27 August sent out a hot, gaseous pyroclastic flow that raced across the sea surface of the Sunda Straits for 40 km (25 mi) to reach the coastal province of Katimbang on Sumatra (see figure 8.20). It flowed onshore with enough heat to fatally burn more than 2,000 people.

Pyroclastic Surges

Pyroclastic surges occur when more steam and less pyroclastic material combine to produce a more-dilute, less-dense, high-velocity flow. Because of their low density, surges are not as easily controlled by topography. Some volcanic eruptions that involve magma and water interaction produce ground-hugging surges that may flow in all directions simultaneously as ring-shaped base surges. The deadliest pyroclastic surge in modern times occurred in Mexico on 4 April 1982.

(a)   **(b)**

Figure 7.28 El Chichón, Chiapas, Mexico. (a) *Before:* In September 1981, the lava dome–plugged volcano was not considered a big hazard. (b) *After:* During one week in 1982, the lava dome was destroyed, leaving a 1 km (0.6 mi) diameter crater.

(a) Courtesy René Canul D. (b) Robert I. Tilling/US Geological Survey

El Chichón Volcano sits in a remote part of Chiapas, the southernmost state in Mexico (figure 7.28a). The volcano had been dormant for at least 550 years and was not considered an imminent hazard. March 1982 was a month of numerous earthquakes leading up to 29 March, when an unexpected six-hour-long Plinian eruption blasted 1.4 km^3 of rock and magma into the atmosphere. The volcano had changed (figure 7.28b). The eruption was surprising and the pyroclastic debris settling from the atmosphere was uncomfortable, but the Plinian event was not enough to drive the rural farmers and villagers from their land. The next five days were calming for the residents, as only minor volcanic activity occurred. But suddenly, on 4 April, a pyroclastic surge flowed radially outward for 8 km (5 mi), overrunning nine villages and killing 2,000 people. Everyone within 8 km of the volcano, in any direction, was killed by the base surge. Following the surge, a Plinian column shot up 20 km (12 mi). On the same day, there were two more base surges and Plinian columns, but the last two base surges did not matter; everyone was already dead. In addition, the Plinian columns injected sulfur dioxide (SO_2) into the upper atmosphere, and the whole world felt the effect as global climate changed.

TSUNAMI

Volcanic tsunami can be created when some of the huge amounts of energy produced during volcanic eruptions are injected into large water bodies. Volcanic processes that generate tsunami include caldera collapse into the ocean; undersea eruptions; and travel of pyroclastic flows, lahars, and debris avalanches into the sea. Volcano-generated tsunami have been responsible for 21% of volcano-caused deaths.

Caldera Collapse

The collapse of Krakatau Volcano in 1883 killed more than 36,000 people. The volcanic eruptions directly killed less than 10% of the people; more than 90% of the fatalities were due to volcano-caused tsunami. The Krakatau tsunami are discussed in detail in chapter 8.

LAHARS

Lahars are volcanic mudflows and volcanic debris flows that are fluid when moving, but begin to solidify soon after stopping. These pyroclast-carrying flows can travel at speeds up to 65 km/hr (40 mph). The combination of water plus loose pyroclasts plus steep slopes plus the pull of gravity produces lahars. The word *lahar* comes from Indonesia and entered the scientific language after the deadly flows from the volcano Kelut in 1586 (see figure 7.23).

Lahars may occur as *primary* events during volcanic eruptions or as *secondary* events months or years after eruption. When steep slopes are covered with loose pyroclasts, it takes only the addition of water to create lahars. Water may be available as a crater lake during eruption, such as at Kelut Volcano, or it may come later from heavy rainfalls or melting glacial ice.

Lahars Due to Heavy Rainfall

The eruptions of Mount Pinatubo in the Philippines in June 1991 featured stupendous Vulcanian events (see figure 6.3). On 15 June 1991, Typhoon Yunya with its heavy rainfall passed over the erupting volcano, sending voluminous lahars downslope and through the cities below (see figure 6.30).

Lahars Due to Melting Glacial Ice

Does it take a huge eruption to kill a lot of people? No. Nevado del Ruiz in Colombia rises to an elevation of

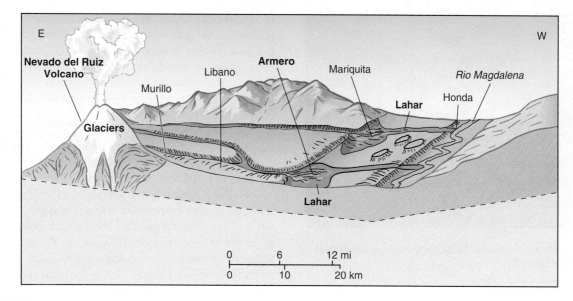

Figure 7.29 An eruption of Nevado del Ruiz in 1985 dropped hot pyroclastic debris onto glaciers, resulting in lahars.
Source: US Geological Survey.

5,400 m (more than 17,700 ft). A 19 km^2 (7 mi^2) area on top of the mountain is covered by an ice cap 10 to 30 m (30 to 100 ft) thick with an ice volume of about 337 million m^3. In November 1985, continuous harmonic tremors (earthquakes) foretold a coming eruption. On 13 November, at 9:37 p.m., a Plinian column rose several miles high. Hot pyroclastic debris began settling onto the ice cap, causing melting. By 10 p.m., condensing volcanic steam, ice melt, and pyroclastic debris combined to send lahars down the east slopes into Chinchina, destroying homes and killing 1,800 people.

But the worst was yet to come. Increasing eruption melted more ice, sending even larger lahars flowing down the canyons to the west and onto the floodplain of the Rio Magdalena (figure 7.29). At 11 p.m., the first wave of cool lahars reached the city of Armero and its 27,000 residents. These lahars had traveled 45 km (28 mi) from the mountaintop, dropping more than 5,000 m (16,400 ft) in elevation. In the steep-walled canyons, the lahars moved at rates up to 45 km/hr (28 mph), slowing as they flowed out onto the flatter land below.

A few minutes after 11 p.m., roaring noises announced the approach of successive waves of warm to hot lahars. Most of Armero, including 22,000 of its residents, ended up buried beneath lahars 8 m (26 ft) thick (figure 7.30). The 22,000 unlucky people were either crushed or suffocated by the muddy lahars.

But 5,000 people did escape. How? They were higher up the slopes. A memorable video showed a man's talking head, which appeared to be resting on top of the mudflows; the man was caught by lahars and buried to his chin as he tried to escape upslope. One step slower and he would have been completely buried and suffocated. But with a bit of digging, he was freed, shaken but unharmed.

The volcanic eruption at Nevado del Ruiz was actually rather minor. Had there not been an ice cap to melt, no

Figure 7.30 Most of the town of Armero, Colombia, and 22,000 of its residents lie beneath lahars up to 8 m (26 ft) thick.
Darrell G. Herd/U.S. Geological Survey

harm would have been done. The November 1985 lahars were a virtual rerun of the events that occurred in that area 140 years earlier, in February 1845. The same places were buried by the same types of lahars. In 1845, the death toll was about 1,000, but because Colombia's population has grown, the dead in 1985 numbered about 24,000.

Mount Rainier, Washington—On Alert

Should the Seattle-Tacoma metropolitan region be concerned about lahars? Yes. Nevado del Ruiz showed that a small eruption on a glacier-capped volcano can be big trouble. Mount Rainier is the tallest of the Cascade Range volcanoes, at 4,393 m (14,410 ft). It stands 2,150 to 2,450 m (7,000 to 8,000 ft) above its adjacent areas and is a beautiful

Figure 7.31 Mount Rainier looms on the skyline behind the Seattle-Tacoma region.

Lyn Topinka/U.S. Geological Survey.

sentinel readily seen from throughout the Seattle-Tacoma urban region (figure 7.31). Yet Mount Rainier is number one on the danger list of many U.S. volcanologists because of its (1) great height, (2) extensive glacial cap, (3) frequent earthquakes, and (4) active hot-water spring systems, which have weakened the mountain internally. Mount Rainier can be described as 33.6 mi^3 of structurally weak rock capped by 1 mi^3 of snow and ice; this volcanic mountain is inherently unstable. Mount Rainier is a national park and cannot

be densely developed, but it nonetheless presents distinct threats to heavily populated areas. The mountain itself may fail in a massive avalanche, and/or rapidly melted ice can cause floods or lahars. Mount Rainier supports the largest glacier system of any mountain in the lower 49 states. This ice can be melted by magma moving up inside the mountain, even without active volcanism.

The rock record shows numerous far-reaching lahars in the last several thousand years (figure 7.32). The Osceola mudflow moved about 5,600 years ago, flowing more than 120 km (75 mi) down the White River valley before spreading out onto the Puget Sound lowlands and into Puget Sound. It covers an area greater than 100 mi^2 to depths over 20 m (70 ft). The Osceola mudflow began as a water-saturated avalanche during summit eruptions of Mount Rainier. It transformed into a clay-rich lahar within 2 km (1.2 mi) of travel as it carried 3.8 km^3 (0.9 mi^3) of material at velocities up to 45 mph out across the Puget Sound lowlands. The affected area is now home to about 100,000 people. A repeat of an Osceola-size lahar could kill thousands of people. To visualize what could happen, see the 1985 lahar in Armero, Colombia (see figure 7.30); the Osceola event was 40 times larger than the Armero lahar.

The Electron mudflow is only 500 years old; it flowed down the Puyallup River valley for 48 km (30 mi) and also out onto the Puget Sound lowlands. Today, the region is a desirable place to live; the population is growing rapidly and building homes on top of these lahar deposits. Mount Rainier's next major eruption may bring staggering property damage and deaths.

Warnings are possible before lahars reach towns. Moving lahars may be detected in the upper reaches of valleys near Mount Rainier by acoustic flow monitors (AFMs). An AFM is a seismometer that records ground vibrations at different frequencies than those generated by earthquakes or most volcanic activity. An AFM concentrates on vibrations between 10 and 300 hertz (Hz), whereas seismometers recording earthquakes and volcanoes commonly focus on waves between 0.5 and 20 Hz. When data from AFMs cross critical values, they are transmitted by radio to emergency centers, and they can also trigger automatic warning devices.

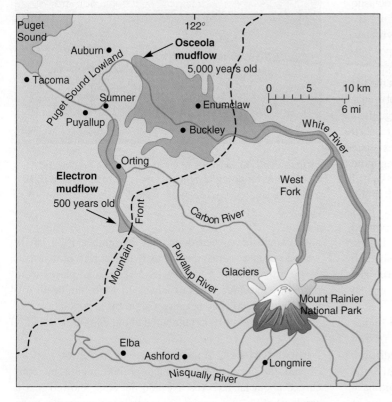

Figure 7.32 Map showing the area covered by two of the many lahars that have flowed from Mount Rainier.

Source: D. R. Crandell and D. R. Mullineaux, "Volcanic Hazards at Mt. Rainier, Washington," 1967, in *US Geological Survey Bulletin 1238.*

DEBRIS AVALANCHES

A tall stratovolcano is a beautiful sight and appears to be a mountain of strength. In reality, though, many centuries of forceful intrusions of magma into stratovolcanoes riddle them with fractures, creating planes of weakness. Hot water and gases rising through fractures chemically decompose the volcanic rock over time and weaken it. The fractures and rotten rock can lead to massive failures: **sector collapses** that flow downslope as debris avalanches (see Mount Shasta photo in figure 7.17 and map

Side Note

Death at Ashfall, Nebraska

Ten million years ago, the area around Ashfall, Nebraska, held water holes within a savanna setting, a warm, flat grassland similar to some classic wildlife areas in Africa today. Large herds of animals migrated to the water holes to drink: three-horned deer, giant camels, three-toed horses, oreodonts, four-tusked elephants, weasels, bear dogs, rhinoceroses, and many more species. Their daily routines changed for the worse one day when a huge volcanic eruption blasted forth 1,300 km (800 mi) away. The eruption came from the Yellowstone hot spot, but 10 million years ago, it sat beneath Idaho (see figure 6.37). Winds carried volcanic ash from Idaho and blanketed Nebraska with a layer of ash about 1 ft thick. After its initial deposition, local winds picked up and blew ash around in gray blizzards. Large amounts of reworked ash settled in the water holes.

What effects can cool, loose, ultrafine volcanic ash have on life? All the animals inhaled volcanic ash for days, weeks, and months, causing health problems. The high magnification of a scanning electron microscope reveals that volcanic ash is composed of sharp, jagged, angular pieces of glass and rock (figure 7.33), which are irritants inside living bodies. Breathing becomes difficult, and respiratory problems develop. Fossil bones of large animals at Ashfall show irregular growth, evidence that they were not getting enough oxygen to grow normal bones.

Fossil preservation at Ashfall is superb, with whole animal skeletons still joined together as they were in life. The layers of fossil-containing ash show the death sequence. In the lowest ash layer are the remains of the first to die—birds and turtles. In overlying ash layers are the fossils of musk deer and small carnivores. Some

Figure 7.34 Skeletons of rhinoceroses killed 10 million years ago by breathing fine volcanic ash for days, Ashfall, Nebraska.
Photo by Pat Abbott.

of the next animals to perish were the horses and camels. A herd of about 100 rhinoceroses kept returning to the water holes, kicking up and breathing volcanic ash clouds each time, until they too died (figure 7.34). Their fossils include a mother rhino who died before her suckling youngster lying next to her. As ash continued to blow about, it ultimately buried the water hole death sites. You can see the herds of animals, partially excavated and available for viewing, at Ashfall Fossil Beds Historical Park in Antelope County, Nebraska.

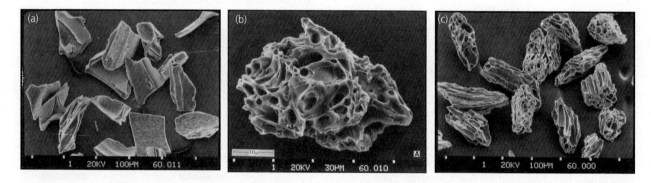

Figure 7.33 Volcanic ash in scanning-electron microscope photos. (a) Glass shards from walls around gas bubbles exploded during eruption of Yellowstone caldera 2.1 million years ago. Tic marks at bottom are 0.001 mm. (b) A single glass particle from 1980 Mount Saint Helens eruption. Voids and holes are from gas bubbles. Length of particle is 0.03 mm. (c) Tiny glass shards with many gas-bubble holes from Rockland Ash exploded 600,000 years ago from Brokeoff Volcano northwest of Lassen, California. Tic marks at bottom are 0.001 mm.
Photos by Robert Oscarson, Janet Slate, USGS, Glen Izett, USGS, Denver (Ret.), as marked, and A. M. Sarna-Wojcicki.

in figure 7.20). A debris avalanche deposit is composed of huge blocks of the volcano within a matrix of finer-grained material (see sector collapse of north side of Mount St. Helens in figure 7.10c and debris avalanche material choking the Toutle River in figure 7.15).

A volcano sector collapse may be triggered by the injection of fresh magma that inflates a volcano; by forceful expansion of water in contact with magma inside a volcano;

or by an earthquake. Debris avalanches are responsible for 2% of volcano-caused deaths.

INDIRECT—FAMINE

Volcanoes affect humans not only directly, but also indirectly. Volcanism can reduce agricultural output, weaken or kill livestock, and weaken humans, setting the stage for famine.

Laki, Iceland, Fissure Eruption of 1783

During the summer of 1783, the greatest lava eruption of historic times poured forth near Laki in Iceland. After a week of earthquakes, on the morning of 8 June 1783, a 25 km (16 mi) long fissure opened, and basaltic lavas gushed for 50 days at 5,000 m^3/sec. To better appreciate this volume of magma, consider that North America's mightiest river, the Mississippi, empties into the Gulf of Mexico at about three times this volume. When the eruption ended, an area of 565 km^2 (218 mi^2) was buried beneath 13 km^3 (3 mi^3) of basaltic lavas. The volume of ash and larger airborne fragments totaled another 0.3 km^3.

The 50 days of eruption were accompanied by the release of an enormous volume of gases that enshrouded Iceland and much of northern Europe in a "dry fog" or blue haze. This haze was rich in SO_2 (one of the visible components of today's urban smog) and an unusually large amount of fluorine. The gases slowed the growth of grasses and increased their fluorine content. An Icelandic farmer named Jon Steingrimmson wrote:

> The hairy sand-fall and sulfurous rain caused such unwholesomeness in the air and in the earth that the grass became yellow and pink and withered down to the roots. The animals that wandered around the fields got yellow-colored feet with open wounds, and yellow dots were seen on the skin of the newly shorn sheep, which had died.

The volcanic gases helped kill 75% of Iceland's horses and sheep and 50% of the cattle. The resulting famine weakened the Icelandic people, and about 20% of the population (10,000 people) died. In today's world of instant communication and rapid air transport, these deaths would have been avoided.

Tambora, Indonesia, 1815

The most violent and explosive eruption of the last 200 years was another Indonesian event; it came from Tambora Volcano on Sumbawa Island in April 1815. After three years of moderate activity, on 5 April, a Plinian eruption column shot up 33 km (20 mi) and carried out 12 km^3 (2.9 mi^3) of pumice in just two hours. On 10 April, an even more powerful Plinian eruption blasted up to 44 km (27 mi) high for three hours. The magma exited with so much force that it eroded and widened the vent in the volcano, thus cutting off the focused energy that drove the Plinian column. The eruption column stopped, and the widened vent lay open; with its insides exposed, the volcano now spilled its guts. On 11 April, about 50 km^3 (12 mi^3) of magma poured out of the caldera in overwhelming pyroclastic flows. The week-long eruption saw about 150 km^3 (36 mi^3) of magma burst forth. Tambora once stood 4,000 m (13,000 ft) high, but now its elevation was reduced to 2,650 m (8,700 ft) with a 6 km (3.7 mi) wide caldera that was over 1 km (0.6 mi) deep. The volcanic explosions were audible 2,600 km (1,600 mi) away, and volcanic ash fell 1,300 km (800 mi) from Tambora. On Sumbawa Island, pyroclastic flows killed at least 10,000 people. They also destroyed the feudal kingdoms of Sanggar and Tambora, leading to the erasure of the Tambora language, the easternmost Austro-Asiatic language.

The eruption of Tambora was responsible for an estimated 117,000 deaths: about 10% killed by the eruption and 90% dying slowly at the end of a chain reaction. Pyroclastic fallout devastated crops, which led to famine and weakened people, making them more susceptible to disease, and then the diseases killed them. But this was not just another Indonesian disaster. The Plinian eruptions of April 1815 so affected global climate that 1816 is known as "the year without a summer." The climatic effects of the eruption are discussed in chapter 12.

GAS

It is not just gas-powered magma that kills; gas can be deadly all by itself. Gases are a continuous product of volcanism but even nonerupting volcanoes can release significant volumes of gas.

Killer Lakes of Cameroon, Africa

Spreading centers commonly begin as three-armed rifts meeting at a triple junction (see figure 4.7). In northeast Africa, two rift arms have spread apart enough to create the Red Sea and the Gulf of Aden, while the third arm has failed, so far, to open the East African Rift Valley into another new ocean basin (see figure 4.5). **Failed rifts** that do not open up enough to become spreading centers are common (e.g., figure 7.35). If a rift fails to open a new ocean basin, must it stop all activity? No.

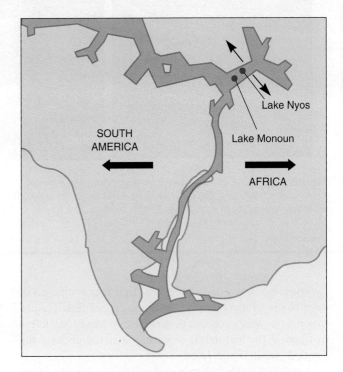

Figure 7.35 Schematic map of Africa and South America splitting apart 135 million years ago. Note the failed rift extending into Africa (upper right corner).

Cameroon sits near the equator in western Africa. It hosts a string of crater lakes running in a northeasterly trend. Prolific rainfalls fill the lakes and combine with the hot temperatures to cover the countryside with greenery. Lake Nyos is one of these crater lakes, filled with beautiful, deep-blue water. This topographically high crater is only several hundred years old. It was blasted into the country rock by explosions of volcanic gases and is 1,925 m (6,310 ft) across at its greatest width and as deep as 208 m (680 ft).

At about 9:30 p.m. on 21 August 1986, a loud noise rumbled through the Lake Nyos region as a gigantic volume of gas belched forth from the crater lake and swept down the adjacent valleys (figure 7.36). The dense, "smoky" rivers of gas were as much as 50 m (165 ft) thick and moving at rates up to 45 mph. The ground-hugging cloud swept outward for 25 km (16 mi). Residents of four villages overwhelmed by the gaseous cloud felt fatigue, light-headedness, warmth, and confusion before losing consciousness. After 6 to 36 hours, about half a dozen people awoke from their comas to find themselves in the midst of death: 1,700 asphyxiated people; 3,000 dead cattle; and not a bird or insect alive, nor any other animal. Yet the luxuriant plants of the region were unaffected.

This shocking event raised numerous questions. What was the death-dealing gas? What was the origin of the gas? How did the gas accumulate into such an immense volume? What triggered the gas avalanche? Is this event likely to happen again?

What was the death-dealing gas? After a lot of effort to identify some exotic lethal gas or toxic substance as the cause of the tragedy, the killer gas turned out to be simply carbon dioxide. This is the same gas we drink in sparkling spring water, soda pop, and champagne. Its toxicity at Nyos is explained by the principle set forth in 1529 by the German physician Theophrastus von Hohenheim (Paracelsus). The principle of Paracelsus states: *the dose alone determines the poison.* A gas does not have to be poisonous, just abundant. Life in the Nyos region was subjected to the same conditions we recreate inside the fire-extinguisher cylinders in our buildings. Fire extinguishers are loaded with carbon dioxide, which does not put out flames directly; because CO_2 is heavier than air, it deprives fire of oxygen, thus causing flames to die out. Animal life in the Nyos area was extinguished in the same fashion.

What was the origin of the gas? It had a volcanic origin, leaking upward from underlying basaltic magma. A 1,600 km (1,000 mi) long string of volcanoes, the Cameroon volcanic line, trends northeastward through several Atlantic Ocean islands and then on land through northeastern Nigeria and northwestern Cameroon. Interestingly, this is the location of the triple junction of spreading centers that ripped apart this section of Gondwanaland, helping give the distinct outlines to the Atlantic margins of South America and Africa (figure 7.35). The two successful spreading arms are still widening the South Atlantic Ocean. The failed rift is occupied by the line of volcanism that includes the crater that forms Lake Nyos; it is not a volcanic mountain but

a crater blasted through bedrock by largely gaseous explosions. The volcanic activity is not seafloor spreading per se; rather, it is a "wannabe" ocean basin that never made it but has not given up totally.

How did the gas accumulate into such an immense volume? Lakes by their nature are stratified bodies of water. Their water layers differ in density, one stacked on top of another. (This is a smaller-scale example of the density differentiation discussed for the whole Earth in chapter 2.) Carbon dioxide, given off by basaltic magma at depth, rises into the bottom waters of Lake Nyos, is dissolved into the heavier, lower water layer, and is held there under the pressure of the overlying water (figure 7.36). As the amount of CO_2 in the lake-bottom water increases, it becomes more unstable. When CO_2 bubbles form, they rise with increasing speed, setting off a positive feedback chain of events leading to more and more bubble formation and rise. Volcanologist Youxue Zhang calculated that the gas eruption was moving about 200 mph when it reached the lake surface. The event of 21 August 1986 released about 0.15 km³ of gas in about one hour. It was like a large-scale erupting champagne bottle, where removal of the cork causes a decrease in pressure, allowing CO_2 to escape in a gushing stream. About 66% of the dissolved gases escaped. After the event, the lake level was 1 m lower, and the water was brown from mud and dead vegetation stirred up from the bottom.

What triggered the gas avalanche? Many suggestions have been made, including volcanic eruption, landslide, earthquake, wind disturbance, or change in water temperature with resultant overturn of lake-water layers. It is interesting to note that a similar event occurred two years earlier at Lake Monoun on 15 August 1984. This was a smaller event, but it killed 37 people. Both events were in August, the time of minimum stability in Cameroon lake waters. Is this a coincidence, or is this a normal overturning of lake water during the rainy season?

Is this event likely to happen again? Definitely. The Lake Nyos gas escape left behind 33% of the CO_2, and more is constantly being fed through the lake bottom. In about

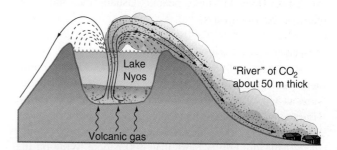

Figure 7.36 Schematic cross-section of Lake Nyos. Volcanic gas is absorbed by the deep-water layer. In 1986, when bottom water was disturbed, 0.15 km³ of CO_2 gas erupted out of the lake and poured down river valleys for an hour or more in a 50 m thick cloud. Virtually all animal life was killed; plants were unaffected. Solid lines show gas flow; dashed lines are water drops.

Diagram after Y. Zhang, 1996, *Nature*, 379, 57–59.

Figure 7.37 Future deadly gas eruptions are being stalled by venting the carbon dioxide gas in the lake-bottom water to the surface through a 200 m (650 ft) long pipe suspended down from a raft. The gas-rich water erupts 37 m (120 ft) up into the atmosphere and the gas simply blows away. There are no mechanical pumps involved here; the dissolved CO_2 powers the fountain.

Bill Evans/U.S. Geological Survey.

20 years, the lake water could again be oversaturated with CO_2. The same loss of life will occur again unless remedial actions are taken. Degassing pipes have been installed to allow high-pressure gas to shoot out of the lake as a fountain of gassy water (figure 7.37). This could prevent the CO_2 concentrations from building up to explosive levels.

As this situation has become better known, other similar lakes have been recognized. For example, the giant Lake Kivu that straddles the border between Rwanda and Congo holds more than 350 times as much gas as Lake Nyos.

LAVA FLOWS

Lava flows are common and impressive, but they are responsible for less than 1% of volcano-caused deaths. Why don't lava flows kill more people? Usually they move too slowly—but not always.

Nyiragongo, Zaire, 2002

As East Africa slowly rifts away from the African continent (see figure 4.5), magma rises to build stratovolcanoes such as Mount Nyiragongo in the East African Rift Valley. Nyiragongo has a long-lived lava lake in its summit crater. On 17 January 2002, lava flowed rapidly down the slopes of the volcano, killing more than 100 people living on the mountain. Upon reaching flatter ground, the lava flows slowed but moved relentlessly toward Lake Kivu. The city of Goma lay in the path of the oncoming lava: 500,000 residents plus uncounted thousands of civil war refugees from Rwanda lived there. Lava reached the lake, but it first flowed through the heart of Goma, destroying about 25% of the buildings and forcing the war refugees to flee again.

How were the lava flows able to catch and kill so many people? The lava had unusually low viscosity. In 1977, Nyiragongo lava flows had exceptionally low SiO_2 content, about 42%. (Compare this value to table 6.5.) The low-viscosity lava in 1977 flowed down the volcano slopes at about 60 km/hr (40 mph), killing an estimated 300 people.

An additional concern is the tremendous volume of carbon dioxide and methane gas held in the deep water of Lake Kivu. A large disruption of the bottom waters, as by an entering lava flow or eruption on the lake bottom, could cause a gas release affecting the 2 million people living along the shore of Lake Kivu.

VEIs of Some Killer Eruptions

Does the total energy involved in a volcanic eruption correlate well with number of deaths? Not necessarily. The volcanic explosivity index (VEI), in chapter 6, is a semi-quantitative approach to estimating the magnitude of explosive eruptions based on volume of material erupted and eruption-column height. Table 7.2 lists VEIs for some of the deadly events we have examined. Note that some of these events had low VEIs; they killed with a relatively small-volume pyroclastic flow, melted glacier ice, and gas escape without magma.

How frequent are eruptions at specific VEI magnitudes? Somewhere on Earth, a VEI 2 event occurs every few weeks, a VEI 3 happens several times a year, a VEI 4 erupts once or twice a year, a VEI 5 happens about once per decade, and a VEI 6 blasts forth once or twice a century. The bigger the eruptions, the less frequently they occur. As the human population continues its rapid growth, increasing numbers of people move into volcano hazard zones. The need for accurate prediction of eruptions is becoming ever more pressing.

TABLE 7.2

VEIs of Notable Volcanic Disasters (Volcanic Explosivity Index)

VEI	Volcano
8	Yellowstone, 600,000 years ago; Toba, 74,000 years ago
7	Tambora, 1815
6	Vesuvius, 79; Krakatau, 1883; Pinatubo, 1991
5	St. Helens, 1980
4	Pelée, 1902
3	Nevado del Ruiz, 1985
2	—
1	—
0	Lake Nyos, 1986

How much harm can a volcano do? The Indonesian volcano Toba may have driven the human race almost to extinction about 74,000 years ago. The eruption was a resurgent caldera event that ejected 2,800 km³ (670 mi³) of rock, magma, and ash in a mega-eruption with a VEI of 8 (table 7.2). The size of this eruption is equivalent to a combined 560 of the Mount Pinatubo eruptions in 1991 (see figure 7.21). The scar from the Toba eruption is the 100 km (60 mi) long Lake Toba lying near the equator at 2.5° North latitude on Sumatra. Low-latitude eruptions are the most dangerous because they spread debris around more of the world, as gases and fine ash choke the atmosphere and affect life globally.

DNA studies of humans alive today suggest we are descended from a small population of 1,000 to 10,000 people. The average rates of genetic mutation within our own DNA suggest that the time when human population was severely reduced was about the same time as the eruption of Toba. The regional devastation and global cooling of the climate caused by the Toba eruption may have made life so difficult that humans were almost forced into extinction.

Volcano Monitoring and Warning

Can we monitor the activity of a volcano and provide advance warning before a large eruption? Efforts to do so have met with both failure and success.

LONG VALLEY, CALIFORNIA, 1982

In the Long Valley–Mammoth Lakes area of California, abundant melting of crustal rock occurs, although no classic hot spot exists there. About 760,000 years ago, a colossal eruption blew out 580 km³ (140 mi³) of magma, generating pyroclastic flows that covered an east-central California area greater than 1,500 km² (580 mi²) with pyroclastic debris (called Bishop Tuff) up to hundreds of meters thick. Immediately after the magma blasted out, Earth's surface dropped nearly 2 km (more than 1 mi) into the void to form the Long Valley caldera (figures 7.38 and 7.39). One pyroclastic lobe flowed 65 km (40 mi) down the Owens Valley. Before the eruption, the magma body is estimated to have had a diameter of 19 km (12 mi), with its roof 5 km (3 mi) below the surface.

Huge eruptions are rare, but these giant continental calderas have fairly frequent small eruptions. There were eruptions in Long Valley 600 years ago and in Mono Lake just 150 to 250 years ago. Today, the main magma body is about 10 km (6 mi) in diameter and around 8 km (5 mi) deep (figure 7.40).

On 25–26 May 1980, one week after the catastrophic eruption of Mount St. Helens, Long Valley was shaken by

numerous earthquakes within 48 hours: four of magnitude 6, dozens of magnitude 4 to 5, and hundreds of smaller seisms. In the resort town of Mammoth Lakes, foundations and walls cracked, chimneys fell, and pantry and store shelves dumped their goods. Monitoring by the U.S. Geological Survey showed the resurgent dome had risen 25 cm (10 in) in late 1979–early 1980 (see dome in figure 6.40). The dome rose another several inches by early 1982, accompanied by swarms of earthquakes. Some magma that was 8 km (5 mi) deep in 1980 had risen to within 3 km (2 mi) of the surface by 1982.

Was a volcanic eruption imminent? What should be done? The affluent town of Mammoth Lakes draws most of its income from tourism; it has a year-round population of 5,500 but adds another 20,000 during winter ski season. Would issuing a formal warning of volcanic hazard do good or harm? On 27 May 1982, the U.S. Geological Survey issued a Notice of Potential Volcanic Hazard, the lowest level of alert. House prices fell 40% overnight and tourist visits dropped dramatically. Home and business owners erupted, but the volcano did not.

In the early 1990s, trees on Mammoth Mountain began dying as large amounts of carbon dioxide (CO_2) rose up from the underlying magma and killed them. At the same time, small earthquakes resumed and the ground surface

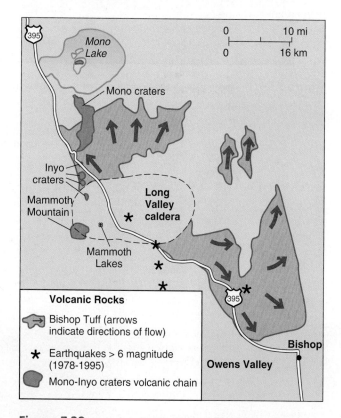

Figure 7.38 Map showing the Long Valley caldera formed by massive eruptions. The brown areas with red arrows are composed of Bishop Tuff, uneroded remains of pyroclastic debris from the last major eruption. The section of Highway 395 shown here lies just north of the section pictured in figure 5.20.

Figure 7.39 The large valley in the center and center-right is the caldera complex of Long Valley, California.

University of Washington Libraries, Special Collections, John Shelton Collection, Shelton 67-882.

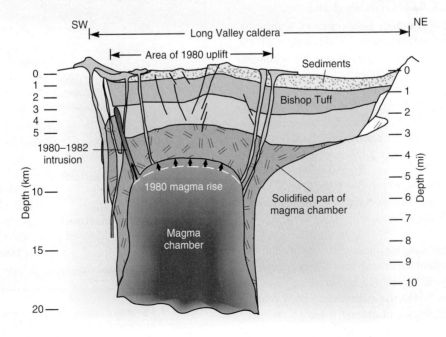

Figure 7.40 This cross-section oriented northeast–southwest through the Long Valley caldera shows the size and depth of the magma body. A tonguelike intrusion moved up the southern edge of the caldera in 1980–1982.

began rising. These phenomena can occur for decades or centuries; however, at large calderas, they do not always mean an eruption is imminent. Many residents remain angry about the "false alarm" of 1982, while many volcanologists and emergency planners are hesitant to issue another volcano warning. Residents are advised to follow the motto: prepare for the worst, but hope for the best.

To get a good view of the giant caldera that is Long Valley, look over your left shoulder as you ride up the chairlifts at the Mammoth Mountain ski resort. The big, dry valley below is the caldera (figure 7.39).

MOUNT PINATUBO, PHILIPPINES, 1991

A volcano-warning success story occurred in the Philippines in 1991 before the climactic eruption of Mount Pinatubo on 15 June. The volcanic eruption was the largest in the 20th century to occur near a heavily populated area. Nearly 1 million people, including 20,000 U.S. military personnel and their dependents, lived in the danger zone.

In March 1991, Mount Pinatubo awoke from a 500-year-long slumber as magma moved upward from a depth of 32 km (20 mi), causing thousands of small earthquakes, creating three small steam-blast craters, and emitting thousands of tons of sulfur dioxide–rich gas. U.S. and Philippine volcanologists and seismologists began an intense monitoring program to anticipate the size and date of a major eruption. On 7 June, magma reached the surface but had lost most of its gas (like a stale glass of soda pop), so the magma simply oozed out to form a lava dome (see figure 7.13). Then, on 12 June (Philippine Independence Day), millions of cubic meters of gas-charged magma reached the surface, causing large explosive eruptions. It was time to get out of the volcano's killing range! The message to speed up the evacuation spread quickly and loudly. Virtually every person, and every movable thing, left hurriedly. On 15 June, the cataclysmic eruption began (figure 7.41). It blew 5 km^3 (1 mi^3) of magma and rock up to 35 km (22 mi) into the atmosphere, forming an ash cloud that grew to more than 480 km

Figure 7.41 The 15 June 1991 Plinian-type eruption of Mount Pinatubo lasted 15 hours, sending pyroclastic flows downslope. VEI = 6.

Rick Hoblitt/U.S. Geological Survey.

(300 mi) across. The airborne ash blocked incoming sunlight and turned day into night. Pyroclastic flows of hot ash, pumice, and gas rolled down the volcano flanks (see figure 6.3) and filled valleys up to 200 m (660 ft) deep. Then, as luck would have it, a typhoon (hurricane) arrived and washed tremendous volumes of volcanic debris downslope as lahars (see figure 6.30).

How successful was the advance warning? Although almost 300 people died, it is estimated that up to 20,000 might have died without the forceful warnings. The scorecard for the monitoring program from March to June 1991 shows that a monitoring expense of about $1.5 million saved 20,000 lives and $500 million in evacuated property, including airplanes. What a dramatic and cost-effective success!

SIGNS OF IMPENDING ERUPTION

Several phenomena are being evaluated as signs of impending eruption. We need to determine if they are reliable enough to justify evacuating people out of a volcanic-hazard zone. Phenomena being studied include seismic waves, ground deformation, and gas emissions.

Seismic Waves

As magma rises up toward the surface, it causes rocks to snap and break, thus sending off short-period seismic waves (SP) with typical periods of 0.02 to 0.06 second. Magma on the move through an opened conduit generates longer-period seismic waves (LP) with periods of 0.2 to 2 seconds. In 1991, during the two weeks before Mount Pinatubo erupted, there were about 400 LP events a day coming from about 10 km (6 mi) deep. Apparently, the LP events were recording the arrival of new magma moving in and loading the volcano for eruption.

Further study shows that a volcano in action causes a variety of earthquakes that generate seismic waves with different periods. These seismic waves (acoustic emissions) produce a record similar to a symphony orchestra. Recognition of different seismic waves could develop into a way of forecasting eruptions. Volcanic processes include (1) creation and propagation of fractures in rock, (2) active injection and movement of magma, (3) degassing, and (4) changes in pore-fluid pressures. Seismic waves from volcanic activity include (1) high-frequency waves, (2) low-frequency waves, (3) very-low-frequency waves, (4) tremors of continuous low-frequency vibrations, and (5) hybrid mixtures. The goal is to relate each seismic-wave type to a specific volcanic activity. In effect, the work is like distinguishing the sound of the flute or the clarinet from the many sounds produced by a symphony orchestra.

Ground Deformation

The ground surface rises up and sinks down as magma moves up or withdraws. Ground deformation can be measured by tilt meters or strain meters placed in the ground and by electronic distance meters. In recent years, satellites have been using radar to measure movements of the ground over time. For example, between 1996 and 2000, a 15 km (9 mi) wide area on the flanks of the Three Sisters volcanoes in Oregon bulged upward 10 cm (4 in) as about 21 million m³ of magma rose to within 6 to 7 km (3.7 to 4.3 mi) below the ground surface (figure 7.42). Now, global positioning system (GPS) stations have been set in the area to add more data about ground deformation. The more the ground rises, the more likely it is that some magma will break through to the surface and erupt.

Gas Measurements

As magma rises toward the surface, the pressure on it drops and dissolved gases escape. For example, at Mammoth Mountain next to the Long Valley caldera in California (see figure 7.38), CO_2 is escaping from the magma. In the 1990s, more than 1,000 tons a day were oozing through the surface, killing trees and causing worry about an impending

eruption. Now, CO_2 releases have declined to about 300 tons a day, suggesting that an impending eruption is less likely. However, this interpretation could be misleading. In 1993, at Galeras Volcano in Colombia, a decrease in gas emissions was interpreted as meaning an eruption was less likely. But, in fact, it meant that the volcano had become plugged by its sticky magma, and gas pressure was building toward the eruption that killed seven volcanologists. So, either an increase or a decrease in gas emissions can be bad. More research must be done.

VOLCANO OBSERVATORIES

As volcanoes continue to burst out with damaging and killing eruptions, many countries are responding by establishing and staffing volcano observatories to provide warnings before big eruptions. In the 20th century, the United States experienced powerful eruptions in four states: Alaska, California, Hawaii, and Washington. There are at least 65 active or potentially active volcanoes in the United States. Watching these volcanoes for signs of activity has led the U.S. Geological Survey to establish a Volcano Hazards Program that includes five volcano observatories: Alaska (AVO), California (Cal VO), Cascades (CVO), Hawaiian (HVO), and Yellowstone (YVO). Each of these observatories maintains its own website to report current activity. Their observations are also presented using an alert system (table 7.3).

Figure 7.42 Uplift deformation of ground surface near Three Sisters volcanoes in central Oregon, 1 May 2001. Satellite radar interferometry image from InSAR (Interferometric Synthetic Aperture Radar) data.

U.S. Geological Survey.

TABLE 7.3	
Volcanic-Alert Levels, US Geological Survey	
Normal	**Typical background activity of a volcano in a noneruptive state**
Advisory	**Elevated unrest above known background activity**
Watch	**Heightened/escalating unrest with increased potential for eruptive activity or minor eruption underway**
Warning	**Highly hazardous eruption underway or imminent**

Summary

Spreading centers provide such ideal settings for volcanism that 80% of all extruded magma occurs there. Spreading centers sit on top of the asthenosphere, which yields basaltic magma that rises to fill fractures between diverging plates. Basaltic volcanoes may be successfully colonized by humans both at spreading centers (e.g., Iceland) and at oceanic hot spots (e.g., Hawaii).

Subduction-zone eruptions involve basaltic magma contaminated by crustal rock to yield water-rich, highly viscous magma containing trapped gases. Their explosive eruptions make the news (e.g., St. Helens, Unzen, and Pinatubo) and the history books (e.g., Santorini, Vesuvius, and Krakatau). Transform faults and continent-continent collisions have little or no volcanism associated with them.

The historic record tells of about 275,000 deaths by volcano in the last 500 years. The deadliest processes have been pyroclastic flows, tsunami, lahars, and indirect effects leading to famine. Gas-powered pyroclastic flows can move at speeds up to 150 mph with temperatures over 1,300°F and for distances over 30 mi; examples are Mont Pelée and El Chichón. Pyroclastic debris and water combine and flow downslope as lahars at speeds up to 30 mph and for distances up to 45 mi, killing thousands at Nevado del Ruiz and presenting a hazard to Seattle-Tacoma from Mount Rainier. Volcano-generated tsunami were mega-killers at Krakatau. Sectors of volcanic cones can collapse, producing giant debris avalanches that bury entire landscapes up to 30 mi away (e.g., Mount Shasta). Giant eruptions from continental calderas can erupt more than 1,000 times as much magma as a typical volcano (e.g., Long Valley).

A volcano may be active for millions of years, but centuries may pass between individual eruptions. The timescale of an active volcano must be considered by people living nearby.

It is possible to monitor a volcano and give advance warning of a major eruption. At Mount Pinatubo in the Philippines, early warning saved up to 20,000 lives before the 1991 eruption.

Terms to Remember

failed rift 192	pyroclastic flow 184
fissure 183	pyroclastic surge 184
nuée ardente 186	sector collapse 190
pyroclastic fall 185	

Questions for Review

1. How many years might one subduction zone operate? One volcano? How many years might pass between eruptions at an active volcano?
2. Explain why it is relatively safe to watch the eruption of a Hawaiian volcano but dangerous to watch a Cascade Range volcano.
3. What is sector collapse? What is a debris avalanche?
4. Draw a plate-tectonic map and explain the origin of the Cascade Range volcanoes.
5. Draw a series of cross-sections and explain the sequence of events in the Mount St. Helens eruption in 1980.
6. What volcanic processes have killed the most people in the last 500 years?
7. Name four ways of creating pyroclastic flows.
8. Why do pyroclastic flows travel so fast? How do they kill?
9. Is a Plinian eruption most dangerous when it is the strongest?
10. Can pyroclastic flows travel outward in all directions simultaneously? (See pyroclastic surges.)
11. How far can a pyroclastic flow travel over water and still be hot enough to kill people?
12. Draw a cross-section and explain how lahars form and move. How do they kill?
13. What four factors combine to produce lahars?
14. Explain the hazard that Mount Rainier presents to the Seattle-Tacoma region.
15. Draw a cross-section and explain the sequence of events at an African killer lake, such as Nyos.
16. How can an eruption with a low VEI (low magnitude explosivity) rating kill thousands of people?
17. How do the ages vary along a line of subduction-zone volcanoes compared to a line of hot-spot volcanoes?
18. What signs of impending eruption are produced by an active volcano?
19. What is column collapse? How does it generate pyroclastic flows?
20. What health hazards are associated with ash fall? What is the composition of most volcanic ash?

Questions for Further Thought

1. Is a Cascade Range volcano likely to have a major eruption during your lifetime?
2. Is it wise for towns near Mount Shasta to keep growing? What should be done about this situation?
3. Is it wise to build in river valleys below Mount Rainier, even tens of miles away?
4. Could a caldera-forming super-eruption occur during your lifetime? Where are possible sites for these events?

CHAPTER 8

Tsunami Versus Wind-Caused Waves

From out there came the sound. To the sea-beaten shore
We looked and saw a monstrous wave that soared
Into the sky, so lofty that my eyes
Were robbed of seeing the rugged cliffs.
It hid the isthmus and coastal rock.
Then seething up and bubbling all about
With foaming flood and breath from the deep sea,
Shoreward it came to where we stood.

AFTER EURIPIDES, 428 BCE, *HIPPOLYTUS*

An onrushing tsunami sets some people fleeing while others stay and stare at the building being destroyed in Koh Raya, Thailand, on 26 December 2004. This tsunami is the leading edge of a massive sheet of water, and it poured inland for several minutes, causing destruction and death. Standing and watching a tsunami was a life-ending decision for thousands of people. When tsunami approach, run fast and gain elevation—up a hill, upstairs in a strong building, or up a tree.

© John Russell/AFP/Getty Images.

LEARNING OUTCOMES

Tsunami and everyday waves have different origins and characteristics. After studying this chapter you should

- know how wind creates the everyday waves in water, how water moves within a wave, and how waves break on the shoreline.
- be able to explain four different origins for tsunami.
- understand the differences in wavelength and period between wind-caused waves and tsunami.
- comprehend how elastic strain builds up at subduction zones and finally is released creating killer tsunami.
- be able to explain, in order, the actions to take to survive a tsunami.
- be familiar with tsunami-warning systems.

OUTLINE

Mid-December 2004 found Tilly Smith, a 10-year-old English schoolgirl, watching video of a tsunami in her geography class in Oxshott, south of London. Two weeks later, the Smith family was on vacation on the island of Phuket in Thailand. On the morning of 26 December 2004, the family was walking along the beach near their hotel. Tilly remembers the morning:

"I saw this bubbling on the water, right on the edge, and foam sizzling just like in a frying pan. The water was coming in, but it wasn't going out again. It was coming in, and then in, and then in, towards the hotel."

She recognized the unusual phenomena as signs that powerful tsunami were on their way. Tilly told her mother, Penny:

"Mum, I know there's something wrong. I know it's going to happen—the tsunami."

Her father, Colin, recalls that *"Tilly went hysterical."*

Tilly, mum, dad, and little sister, Holly, returned to the hotel. While Colin Smith alerted the hotel staff to his daughter's observations, Tilly ran back to the beach to spread her tsunami warning to about 100 people. Tilly was convincing, and the beach was evacuated before the killer tsunami arrived. The beach near the Marriott hotel was a rarity in Phuket on that day—no one there was killed or seriously injured.

After meeting with Tilly, former U.S. President Bill Clinton said:

"Tilly's story is a simple reminder that education can make a difference between life and death. All children should be taught disaster reduction so they know what to do when natural hazards strike."

In 2004, nations throughout the Indian Ocean region were unprepared for tsunami, and 245,000 people were killed in 14 countries (figure 8.1).

Japanese Tsunami, 11 March 2011

Japan is the nation most prepared for tsunami. The Japanese have built high walls to stop tsunami along much of their coastline, constructed huge metal gates at the entrances to some harbors, and created warning systems to alert the people. Despite careful preparation, their defenses were overrun in many areas by powerful tsunami on 11 March 2011 (figure 8.2). The tsunami caused the world's most expensive natural disaster and killed 19,184 people.

The earthquake 70 km (43 mi) off Japan's northeast shore was a 9.0 M_w; it was felt by everyone in the region (see chapter 4). The Japanese know tsunami. After a huge earthquake they know what's coming. About nine minutes after the earthquake, warnings of a major tsunami were issued by the Japan Meteorological Agency at the highest end of the warning scale. Predicted tsunami heights were at least 3 m (10 ft) but for Miyagi Prefecture (district) predictions were up to 6 m (20 ft). Maximum readings of actual tsunami height began being recorded in nearby Iwate Prefecture 26 minutes after the earthquake, 17 minutes after the tsunami warnings. But the tsunami were bigger than predicted.

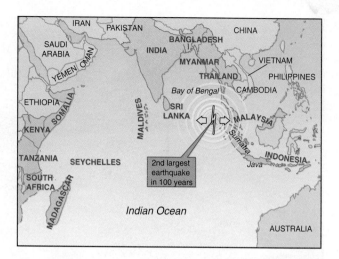

Figure 8.1 Third largest earthquake in more than 100 years. On 26 December 2004 a huge tearing of the seafloor off northern Sumatra sent tsunami throughout the Indian Ocean, killing about 245,000 people in 14 countries (names printed in red).

At least 101 designated tsunami evacuation sites, and their evacuees, were overrun. Tsunami overtopped coastal barriers and flowed inland as overpowering masses of seawater and debris that destroyed almost everything in their path.

Recapping the event At 2:46 p.m. on a Friday, a powerful earthquake broke loose 70 km (43 mi) offshore from northeastern Japan, with the most violent earth movements lasting 6 minutes. The seafloor thrust upward 5 to 8 m (16 to 26 ft), loading the seawater with energy that drove tsunami ashore less than 30 minutes later. At 69 minutes after the earthquake, as people around the world watched in horror, an NHK News helicopter broadcast live video of the tsunami as a massive sheet of water and debris rolling over the fields and through the city of Sendai, catching and overwhelming fleeing cars and their passengers. The unstoppable tsunami inflicted severe destruction and death along a 670 km (415 mi) length of coastline, pushed inland as far as 10 km (6 mi) in Sendai, ran up as high as 40.5 m (133 ft) above sea level in Miyako, inundated 560 km^2 (215 mi^2) of land, damaged or destroyed 330,000 buildings, ruined roads and railways, caused one dam to fail, triggered numerous fires, and caused nuclear-power plants to fail. The deaths were caused by tsunami drowning or blunt force (93%), earthquake-collapsed buildings (4%), burns (1%), and unknown (2%).

TSUNAMI TRAVEL THROUGH THE PACIFIC OCEAN

After the earthquake, a computer-generated map was issued predicting tsunami heights throughout the Pacific Ocean basin (figure 8.3). Actual tsunami heights and experiences include:

- Canada, Vancouver Island: 1 m (3.3 ft) high surges.
- Oregon, Curry County: 2.4 m (8 ft) high surges caused $7 million in harbor damages.

Figure 8.2 Tsunami destruction in Otsuchi, Iwate Prefecture, Japan, one week later, 18 March 2011.

NOAA/NGDC, Dylan McCord, U.S. Navy.

- California, Crescent City: 2.4 m (8 ft) high tsunami damaged docks and 35 boats, and killed one person.
- California, Santa Cruz: 2.4 m (8 ft) high surges did $17 million in damage to the harbor, sank 18 boats, and damaged others for another $4 million in losses.

- Hawaii: 3 m (10 ft) high waves did tens of millions of dollars in damage to boats, houses, and hotels, especially in Kona.
- Midway Atoll: 1.5 m (5 ft) high tsunami swept across Spit Island, killing more than 110,000 nesting sea birds.
- Chile: 18,000 km (11,000 mi) away, 3 m (10 ft) high tsunami damaged more than 200 houses.
- Antarctica: tsunami broke up a 125 km² (48 mi²) area of Sulzberger Ice Shelf into 80 m (260 ft) thick icebergs.

LAND SUBSIDENCE

Over the centuries, as the Pacific plate subducted downward beneath Japan, it caused the upper plate carrying Japan to warp upward. When the plates came unstuck, starting the earthquake, the seafloor was thrust upward at the same time as coastal land was moved downward along a 400 km (250 mi) length of coastline. Land subsidence was up to 1.2 m (4 ft) on the Oshika Peninsula in Miyagi Prefecture. The land subsidence is permanent. High tides and storm waves will now reach farther across the coastline and increase coastal flooding.

Subsidence of coastal land is common during great earthquakes caused by subduction, as occurred in Japan in 2011,

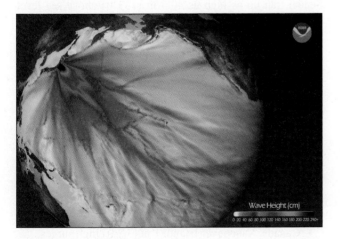

Figure 8.3 Tsunami wave height throughout the Pacific Ocean.

NOAA.

Side Note

in Indonesia in 2004, in Alaska in 1964, and in Chile in 1960. Land subsidence was a starting point for recognizing a ~9M earthquake in prehistoric time in the Pacific Northwest.

BRITISH COLUMBIA, WASHINGTON, OREGON, 26 JANUARY 1700

The ghost forests of gray, dead tree trunks still standing in tidal marshes today were described in the 1850s. In the 1980s, it was increasingly recognized that sea level had not just slowly risen and gradually poisoned these trees. The annual growth rings of the trees were thick and healthy to the end of their lives, which ended rather quickly. Sea level rising did not kill the trees, the culprit was the land subsiding rapidly below sea level during a great earthquake (see chapter 4 and figures 4.16 and 4.17).

As much as an 800 km (500 mi) length of coastal land—from Vancouver Island, Canada, through Washington, Oregon, and into northernmost California—dropped 1 to 2 m (6.5 ft) during an earthquake of about magnitude 9. The Japanese earthquake and tsunami of 2011 were not unique; similar events occurred 311 years earlier along the North American coastal zone. Excavations in the region expose coastal soils and land with long-dead, rooted plants still in place; they are buried by layers of sand containing marine fossils swept ashore and deposited by tsunami just like in Japan in 2011. Coastal land was dropped down into the tidal zone, plunging forests into salty seawater that killed them.

Subduction zones generate more killer tsunami than any other source. The subduction zone along the coastline reaching from British Columbia, Washington, Oregon, and northernmost California is active (figure 8.4). Subduction is still occurring, volcanoes are still erupting, and moderate-size earthquakes are shaking—as we build to the upcoming 9M earthquake and tsunami. Major cities such as Victoria, Vancouver, Seattle, Tacoma, and Portland are built along

water inlets; they will probably be most affected by the great earthquake and less by the tsunami. The small cities and towns along the Pacific Ocean coastline will be hammered by the tsunami, as seen in Japan in 2011.

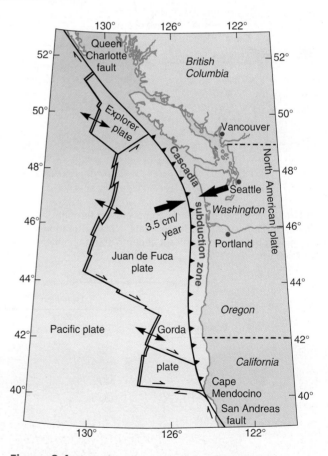

Figure 8.4 Map of small, young tectonic plates being subducted beneath the Pacific Northwest. On 26 January 1700, a subduction-caused earthquake here reached about magnitude 9 and sent powerful tsunami across these shorelines.

WAVES IN WATER

Tsunami can be the most overwhelming of all waves, but their origins and behaviors differ from those of the everyday waves we see at the seashore or lakeshore. The familiar waves are caused by wind blowing over the water surface. Our experience with these wind-caused waves misleads us in understanding tsunami. Let us first understand everyday, wind-caused waves and then contrast them with tsunami.

Wind-Caused Waves

Waves transfer energy away from some disturbance. Waves moving through a water mass cause water particles to rotate in place, similar to the passage of seismic waves (figure 8.5; see figure 3.18). You can feel the orbital motion within waves by standing chest-deep in the ocean. An incoming wave will pick you up and carry you shoreward and then drop you downward and back as it passes. At the water surface, the diameter of the water-particle orbit is the same as the wave height. The diameters of water orbits decrease rapidly as water deepens; wave orbital motion ceases at a depth of about one-half of the **wavelength.**

Waves vary from tiny, wind-blown ripples to monster rogue waves, but the rotational motion of water within a wave is quite similar. Most waves are created by the frictional drag of wind blowing across the water surface. A wave begins as a tiny ripple. Once formed, the side of a ripple increases the surface area of water, allowing the wind to push the ripple into a higher and higher wave. As a wave gets bigger, more wind energy is transferred to the wave. How tall a wave becomes depends on (1) the velocity of the wind, (2) the duration of time the wind blows, (3) the length of water surface (**fetch**) the wind blows across, and (4) the consistency of wind direction. Once waves are formed, their energy pulses can travel thousands of kilometers away from the winds that created them.

WHY A WIND-BLOWN WAVE BREAKS

Waves undergo changes when they move into shallow water—water with depths less than one-half their wavelength. Wave friction on the floor of the shallow ocean interferes with the orbital motions of water particles, so waves begin slowing (figure 8.6). Friction with the bottom flattens the circular motions of the water into elliptical and horizontal movements.

As waves slow down, their wavelengths decrease, thus concentrating water and energy into shorter lengths and causing the waves to grow higher. When the wave height-to-wavelength ratio (H:L) reaches about 1:7, the wave front has grown too steep, and it topples forward as a breaker (figures 8.7 and 8.8). Note that the 1:7 ratio is reached by

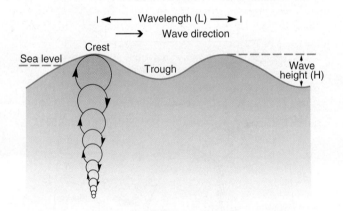

Figure 8.5 Waves are energy fronts passing through water, causing water particles to rotate in place. Rotational movement becomes insignificant at depth about one-half of the wavelength.

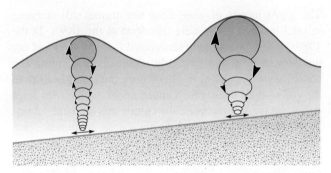

Figure 8.6 As a wave moves into shallower water, it rises higher. Circular rotating water touches bottom, flattening into a back-and-forth motion.

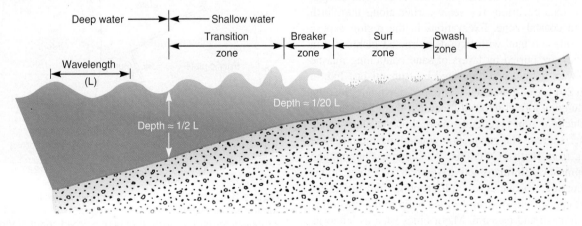

Figure 8.7 Schematic cross-section of deep-water waves entering shallow water. Wavelengths decrease and wave heights increase, causing the water to pitch forward as breakers.

Figure 8.8 A breaking wave. Circular rotating water is slowed at the base but rolls forward on top.

© Royalty-Free/Corbis.

changes in both wave height and wavelength; wave height is increasing at the same time that wavelength is decreasing.

The depth of water beneath a breaker is roughly 1.3 times the wave height as measured from the still-water level. At this depth, the velocity of water-particle motion in the wave crest is greater than the wave velocity, thus the faster-moving wave crest outraces its bottom and falls forward as a turbulent mass.

ROGUE WAVES

An ocean is such an extensive body of water that different storms are likely to be operating in different areas. Each storm creates its own wave sets. As waves from different storms collide, they interfere with each other and usually produce a sea **swell** that is the result of the constructive and destructive interference of multiple sets of ocean waves (figure 8.9a). However, every once in a while, the various waves become briefly synchronized, with their energies united to form a spectacular tall wave, the so-called **rogue wave** (figure 8.9b). The moving waves quickly disunite, and the short-lived rogue wave is but a memory. But if a ship is present at the wrong time, a disaster may occur.

During World War II, the *Queen Elizabeth* was operating as a troop transport passing Greenland when a rogue wave hit, causing numerous deaths and injuries. On 3 June 1984, the three-masted *Marques* was sailing 120 km (75 mi) north of Bermuda when two rogue waves quickly sent the ship under, drowning 19 of the 28 people on board. In 1987, the recreational fishing boat *Fish-n-Fool* sank beneath a sudden "wall of water" in the Pacific Ocean near a Baja California island.

On 10 April 2005 in New York, 2,300 eager passengers boarded the 295 m (965 ft) long *Norwegian Dawn* for a one-week vacation cruise to the Bahamas. On the return trip, the seas became rough. Then a thunderous disruption shocked people as a freak 22 m (70 ft) high wave slammed into the ship, breaking windows, sending furniture flying, flooding more than 60 cabins, and injuring four passengers. The wave even ripped out whirlpools on deck 10. Damage to the hull forced an emergency stop for inspection and repairs in Charleston, South Carolina. Spring vacation was interrupted by a rogue wave.

On occasion, rogue waves strike the shoreline and carry people away from the beach. On 4 July 1992, a rogue wave 5.5 m (18 ft) high rose out of a calm sea at Daytona Beach, Florida, crashed ashore, and smashed hundreds of cars parked on the beach, causing injuries to 75 of the fleeing people.

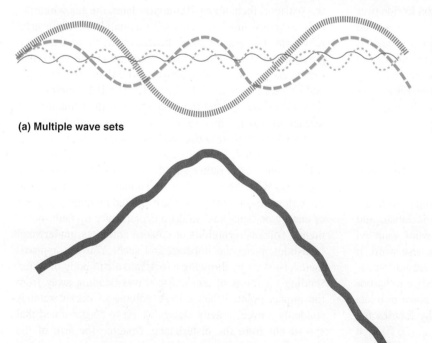

(a) Multiple wave sets

(b) Rogue wave

Figure 8.9 Waves on the sea surface. (a) At any time, there usually are several different storms, each producing its own waves of characteristic wavelength. Different wave sets usually interfere with each other. (b) On rare occasions, the various wave sets combine to produce an unexpected giant—a rogue wave. (a) and (b) are separate drawings; they do not directly correlate.

In Greater Depth

Deep-Water Wave Velocity, Length, Period, and Energy

Waves moving through water deeper than one-half their wavelength are essentially unaffected by friction with the bottom. The waves move as low, broad, evenly spaced, rounded swells with velocities related to wavelength by:

$$V_w = 1.25\sqrt{L}$$

where V_w equals wave velocity and L equals wavelength. A swell with a wavelength of 64 m would have a velocity of the square root of 64 (i.e., 8) times 1.25, or 10 m/sec (22.4 mph). The equation is telling us that wave velocity in deep water depends on the wave's length—as wavelength increases, so does velocity.

The period (T) is the amount of time it takes for two successive wave crests to pass a given point. Since the distance between successive wave crests is the wavelength, there must be a relationship between period (T) in seconds and wavelength (L) in meters. This relationship may be defined by:

$$V = \text{distance traveled/time} = L/T$$

which may be simplified to:

$$L = 1.56T^2$$

As a rule of thumb, the velocity of waves in miles per hour may be estimated as 3.5 times the wave period in seconds. For example, waves with a period of 10 seconds are moving about 35 mph.

Higher-velocity waves carry more energy, but how much more? Wave energies per unit length can be computed with the following relationship:

$$E_w = 0.125\rho\, gH^2L$$

where E_w equals wave energy, ρ (rho) equals density of water, g equals gravitational acceleration, H equals wave height in meters, and L equals wavelength. Some representative values computed from this equation are listed in table 8.1. Notice that doubling the wavelength doubles the wave energy, but that doubling the wave height quadruples the wave energy.

TABLE 8.1

Some Ocean-Wave Energies

Wave Period (T)	Wavelength (L)	Wave Height (H)	Energy in Joules
10 seconds	156 m	1 m	2.39×10^5
		2 m	$9.57 \times$ "
		3 m	$21.54 \times$ "
14.1 seconds	312 m	1 m	4.79×10^5
		2 m	$19.15 \times$ "
		3 m	$43.08 \times$ "

Rogue waves have been measured at 34 m (112 ft) in height. The problems they present also include the steepness of the wave front descending into the wave trough. A small, short boat is maneuverable and in good position to ride over the rogue wave, as long as it does not get hit sideways and rolled, or tossed from the front of one wave onto the back of the next wave. Large, long ships face either being uplifted at their midpoint, leaving both ends suspended in air, or having both ends uplifted with no support in their middle. Either case creates severe structural strains that break some ships apart.

Tsunami

The biggest, most feared waves of all pass mostly unnoticed across the open sea and then rear up and strike the shoreline with devastating blows. The country with the most detailed history of these killer waves is Japan, and the waves are known by the Japanese word *tsunami* (*tsu* = harbor; *nami* = waves). Being a Japanese word, it is the same in both singular and plural; there are no "tsunamis," just as there are no "sheeps." The reference to harbor waves emphasizes the greater heights waves reach in inlets and harbors because the narrowed topography focuses the waves into smaller spaces. For example, an 8 m (26 ft) high wave on the open coast may be forced to heights of 30 m (100 ft) as it crowds into a narrow harbor.

A deadly example hit Japan on 15 June 1896, a summer day when fishermen were out to sea and beaches were crowded with vacationers. An offshore earthquake swayed the seafloor; then, about 20 minutes later, the sea withdrew, only to return in 45 minutes with a sound like a powerful rainstorm. Tsunami hit all the beaches hard but reached their greatest heights of 29 m (95 ft) where they crowded into narrow inlets. The tsunami destroyed more than 10,000 homes and killed more than 27,000 people. The fishermen on the open ocean did not feel the earthquake or the tsunami; they learned of it when they sailed back into a bay littered with the wreckage of their houses and the bodies of their families.

In the United States, tsunami have been called "tidal waves," but this is rather silly because tsunami have nothing to do with the tides. Nor do tsunami have anything to do with winds or storms; they are created by huge injections of energy or "splashes" in deep ocean water by fault movements, volcanic eruptions or caldera collapses, underwater landslides, meteorite impacts, and such. You can approximate a tsunami by throwing a rock into a still body of water, sending off trains of concentric waves heading away from the impact point. When a large volume of ocean water is suddenly moved, gravity causes waves to be generated that spread out from the disturbance. Imagine the size of the waves formed when a really big rock drops in the water, such as the caldera collapse of the volcano Krakatau in 1883.

The biggest tsunami are caused by the rarest events, the impact of high-velocity asteroids and comets. Consider the amount of energy injected into the ocean when a 10 km (6 mi) diameter asteroid hits at 30,000 mph.

Tsunami are most commonly created during earthquakes, more specifically subsea fault movements with pronounced vertical offsets of the seafloor that disturb the deep ocean-water mass. Water is not compressible; it cannot easily absorb the fault-movement energy. Therefore, the water transmits the energy throughout the ocean in the waves we call tsunami.

The Indian plate subducts northward beneath the Burma plate (see figure 4.10). When a breaking point was reached on 26 December 2004, seafloor along a 1,200 km (740 mi) length snapped upward several meters, causing adjoining areas to move downward (figure 8.10). The uplifts and downwarps of the seafloor along a north–south trend set powerful long-wavelength tsunami in motion, with the greatest energy directed to the west and east. The tsunami water surface had shapes similar to those of the newly deformed seafloor topography that generated them. Notice that both troughs (downdrops) and crests (uplifts) formed.

Before a trough comes ashore, the sea retreats, but there is no retreating sea before a crest comes ashore.

It is the vertical-fault movements at subduction zones that most commonly cause tsunami (table 8.2). In the 20th century, 141 damaging tsunami combined to kill more than 70,000 people. Early in the 21st century, tsunami have killed more than 265,000 people.

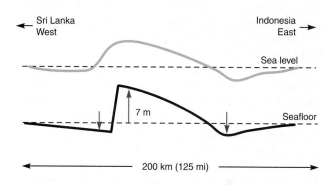
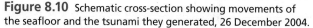

Figure 8.10 Schematic cross-section showing movements of the seafloor and the tsunami they generated, 26 December 2004.

TABLE 8.2
Notable Tsunami in Recent Times

Date		Cause	Height	Site	Deaths
1 November	1755	Earthquakes	10 m	Lisbon, Portugal	30,000
21 May	1792	Volcano avalanche	10 m	Japan (Unzen)	>14,000
11 April	1815	Volcano eruption	10 m	Indonesia (Tambora)	>10,000
8 August	1868	Earthquake	15 m	Chile	>25,000
27 August	1883	Volcano eruption	35 m	Indonesia (Krakatau)	36,000
15 June	1896	Earthquake	29 m	Japan	27,000
11 October	1918	Subsea landslide	6 m	Puerto Rico	116
2 March	1933	Earthquake	20 m	Japan	3,000
1 April	1946	Earthquake	15 m	Alaska	175
22 May	1960	Earthquake	10 m	Chile	>1,250
27 March	1964	Earthquake	6 m	Alaska	125
1 September	1992	Earthquake	10 m	Nicaragua	170
12 December	1992	Earthquake	26 m	Indonesia	>1,000
12 July	1993	Earthquake	31 m	Japan	239
2 June	1994	Earthquake	14 m	Indonesia	238
17 July	1998	Subsea landslide	15 m	Papua New Guinea	>2,200
26 December	2004	Earthquake	10 m	Indonesia, Sri Lanka, India	~245,000
17 July	2006	Earthquake	7 m	Indonesia	>600
29 September	2009	Earthquake	14 m	Samoa	190
27 February	2010	Earthquake	8 m	Chile	200
11 March	2011	Earthquake	13 m	Japan	19,184

Tsunami Versus Wind-Caused Waves

The typical ocean waves created by winds vary in size during the course of a year. Although the periods and wavelengths of wind-blown waves vary by storm and season, they are distinctly different from those of tsunami (table 8.3).

Wind-blown waves rise up as they near the beach, roll forward, run up the beach for several seconds, and then withdraw (figure 8.11a). Wind-blown waves not only come and go quickly, but the water run-up and retreat is confined to the beach (figure 8.11b).

Even huge wind-blown waves are different from tsunami. For example, at Waimea on the north shore of Oahu Island in Hawaii, the world-famous surfing waves may reach 15 m (50 ft) in height, but each wave is a solitary unit. These huge waves have short wavelengths and brief periods, meaning that each wave is an entity unto itself; there is no additional water mass behind the wave front. These waves are spectacular to view or ride, but what you see is what you get; the wave is the entire water mass.

Tsunami are different. Tsunami arrive as the leading edge of an elevated mass of water that rapidly runs up and *over* the beach and then floods inland for many minutes (figures 8.11c, d). Tsunami are dangerous because their tremendous momentum carries water and debris far inland. They may be no taller than the wind-blown waves we see at the beach every day, but they are much more powerful. Even a knee-high tsunami can kill you. The power of the fast-moving water can knock you down, then beat your body and head with debris, and then drown you.

The contrasts in velocities of wind waves versus tsunami are also great. A wind-blown wave moving through water deeper than half its wavelength (L) has its velocity (v) determined by

$$v = 1.25\sqrt{L}$$

(see In Greater Depth, page 206). From table 8.3, take a wave with 156 m wavelength and calculate its velocity as 15.6 m/sec = 35 mph.

Tsunami velocity requires a different calculation. Tsunami wavelengths are so long and so much greater than the depth of the deepest ocean that their velocity is calculated by:

$$v = \sqrt{gD}$$

where v equals wave velocity, g equals acceleration due to gravity (9.81 m/sec² or 32 ft/sec²), and D equals depth of ocean water. The Pacific Ocean has an average depth of 5,500 m (18,000 ft). Calculating the square root of g times D yields a tsunami velocity of 232 m/sec (518 mph).

The calculated velocities of tsunami are faster than are typically measured. The energy pulse that makes the wave also puts the water into a rotating motion to depths of about one-half the wavelength. Tsunami wavelengths can be as great as 780 km (485 mi) (table 8.3), meaning that ocean water would be disturbed to depths of 390 km (240 mi). But the ocean's average depth is only 3.7 km (2.3 mi), and the deepest trenches just exceed 11 km (6.9 mi). Therefore, the energy pulse of a tsunami moves the entire water column it passes through. Tsunami have such long wavelengths that they are always dragging across the ocean bottom, no matter how deep the water. The ocean basin has enough topography on its bottom to slow most tsunami down to the 420 to 480 mph range.

A tsunami of 1 m height in the deep ocean may be moving nearly 500 mph. As tsunami enter shallower water, the increasing friction with the seafloor and internal turbulence of the water slow their rush, but they still may be moving at freeway speeds. For example, when a tsunami is in water 50 m (165 ft) deep, using $v = \sqrt{gD}$ yields a velocity of 22 m/sec = 50 mph.

As the velocity of the tsunami front slows and the wavelength decreases, the water behind it begins to build up and increase in amplitude. If it reaches a height of 15 m (50 ft), it will not be like the Waimea, Hawaii, solitary wave with nothing behind it. The visible tsunami wave is only the leading edge of a tabular sheet of water that will flow on land for minutes (figure 8.11d).

TABLE 8.3
Representative Wave Periods and Lengths

		Periods		Lengths	
Wind-Blown	Short:	5 seconds		39 m	(130 ft)
Ocean	Medium:	10 seconds		156 m	(510 ft)
Waves	Long:	20 seconds		624 m	(2,050 ft)
Tsunami	Maximum: (deep ocean)	3,600 seconds (60 minutes)		780,000 m	(2,560,000 ft) (485 mi)
	Common: (nearer shore)	900 seconds (15 minutes)		20,000 m	(65,000 ft) (12 mi)

(a)

Figure 8.11 Ocean waves.
(a) A wind-blown wave rolls onto the beach in Natal, South Africa.
(b) Daily wind-blown waves break on the beach and do not flood higher areas. (c) Tsunami pour across the beach and flood inland for many minutes. Even small tsunami can knock you down, batter your body with debris, and kill you. (d) Tsunami overruns shoreline and a 3 m (10 ft) high tsunami barrier wall; then flows through Miyako, Iwate Prefecture. 11 March 2011.

Water rotates in circles

(b)

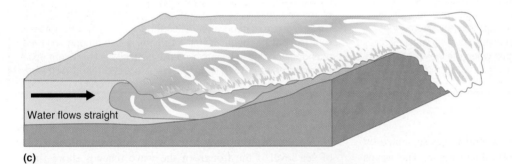

Water flows straight

(c)

(d)

A Classic Disaster

The Chile Tsunami of 1868

Several ships were moored in the harbor at Arica (then part of Bolivia), including the USS *Wateree*, a two-masted sidewheeler with a broad, flat bottom. About 4 p.m., the ship began vibrating and chains rattled as a huge earthquake shook down houses in Arica. Despite worries about tsunami, the desire to help people ashore caused the *Wateree* captain to drop extra anchors, close the hatches, lash the guns, rig lifelines, and then send a yawl with 13 men to the jetty to help. Here are some words written by Lieutenant L. G. Billings, who remained aboard the *Wateree*:

. . . survivors were coming down the beach and crowding on the little jetty, calling to the crews . . . to carry them to the apparent safety of the anchored vessels . . . all at once a hoarse murmuring noise made us look up; looking towards the land we saw, to our horror, that where a moment before there had been the jetty, all black with human beings, there was nothing: everything had been swallowed in a moment by the sudden rising of the sea, which the *Wateree*, floating upon it, had not noticed. At the same time we saw the yawl carried away by the irresistible wave towards the lofty, vertical cliff of the Morro, where it disappeared in the foam as the wave broke against the rock.

. . . there was another earthquake shock. Once more we saw the ground move in waves. This time the sea drew back from the land until we were stranded and the bottom of the sea was exposed, so that we saw what had never been seen before, fish struggling on the seabed and the monsters of the deep aground. The round-hulled ships rolled over on their sides, while our *Wateree* sat down upon her flat bottom; and when the sea came back, returning not as a wave, but rather like a huge tide, it made our unhappy companion [ships] turn turtle, whereas the *Wateree* rose unhurt on the churning water.

It had been dark for some time when the lookout hailed the deck and said that a breaking wave was coming. Staring into the night, we first made out a thin phosphorescent line which, like a strange kind of mirage, seemed to be rising higher and higher in the air: its crest, topped by the baleful light of that phosphorescent glitter, showed frightful masses of black water below. Heralded by the thunder of thousands of breakers all crashing together, the wave that we had dreaded for hours was at last upon us.

Of all the horrors, this seemed the worst. We were chained to the bed of the sea, powerless to escape. . . . We could only hold on to the rails and wait for the catastrophe. With a terrifying din, our ship was engulfed, buried under a half-liquid, half-solid mass of sand and water. We stayed under for a suffocating eternity; then, groaning in all her timbers, our solid old *Wateree* pushed her way to the surface, with her gasping crew still hanging on to the rails.

Our survival was certainly due to the construction of the ship . . . which allowed the water to pour off the deck almost as quickly as if she had been a raft. The ship had been carried along at a very great speed, but all at once she became motionless . . . we lowered a lantern over the side and discovered that we had run aground. . . . The sun rose upon such a spectacle of desolation as can rarely have been seen. We were high and dry, three miles from our anchorage and two miles inland. The wave had carried us at an unbelievable speed over the sand dunes which line the shore, across a valley, . . . leaving us at the foot of the coastal range of the Andes.

Note several items in this thrilling account: (1) After the first earthquake, the initial seawater reaction was a deadly onrushing mass of water that killed people on the jetty. (2) After the second earthquake, the initial seawater reaction was a huge withdrawal of the sea, leaving ships sitting on the seafloor. (3) The biggest tsunami occurred hours after the earthquakes. (4) The phosphorescent glow of the huge incoming tsunami is a commonly reported phenomenon due to bioluminescence of small sea life caught up in the monster wave. (5) A large U.S. warship was carried 2 mi inland.

Tsunami arrive as a series of several waves separated by periods usually in the 10- to 60-minute range. The waves are typically a meter high in the open ocean and 6 to 15 m (20 to 50 ft) high on reaching shallow water, except where topography, such as bays and harbors, focuses the energy to create much taller waves.

TSUNAMI AT THE SHORELINE

What does a tsunami look like when it comes onshore? Does it resemble the animation in the Hollywood movie *Deep Impact*, where a beautiful, symmetrical wave curves high above the buildings of New York City? No. A tsunami arriving at the shoreline does not look like a gigantic version of the breaking waves we see every day. A typical tsunami hits the coastline like a very rapidly rising tide or whitewater wave, but it does not stop on the beach; it keeps rushing inland (figure 8.11d).

A tsunami event may begin with a drawdown or retreat of sea level if the trough of the wave reaches shore first. The drawdown can cause strong currents that pull seawater, boats, and swimmers long distances out to sea. Or the tsunami wave crest or front may reach shore first. Then a strong surge of seawater, resembling a faster and stronger rising tide, rushes across the beach and pushes far inland.

Wavelength and Period Versus Height

People tend to attribute the destructive power of tsunami mostly to the great height of their waves, but the height of tsunami commonly is not as important as the momentum of their large masses separated by ultra-long wavelengths and periods (table 8.3). Visualize a flat or gently sloping coast hit by tsunami with a 60-minute period. The tsunami can rush inland, causing destruction for about 30 minutes before the water is pulled back to help form the next wave. A view of the aftermath of the 1960 Chilean tsunami in Hilo,

Figure 8.12 Tsunami damage to Hilo, Hawaii, following the magnitude 9.5 Chilean earthquake, 22 May 1960. Notice the Pacific Ocean in the background and how far inland the tsunami traveled, thanks to the long wavelengths and periods.

NOAA/NGDC, U.S. Navy.

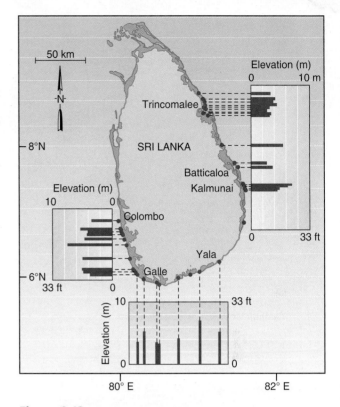

Figure 8.13 Tsunami from 4 to 7 m (13 to 23 ft) high were common all around Sri Lanka. The long wavelengths of the tsunami allowed them to encircle the island.

Source: *Science*, v. 308 (2005).

Hawaii, shows the effects of the long wavelengths and long periods between tsunami wave sets (figure 8.12). The powerful tsunami was able to charge upslope, through the city, for a long distance and a long time before receding to help form the next tsunami wave set.

The vastly different wavelengths and periods of wind-blown versus tsunami waves can be further appreciated by looking at islands. Huge wind-blown waves such as at Waimea, Hawaii, hammer the north shore of Oahu, the windward shore. The relatively short wavelengths and periods of the Waimea waves prevent them from affecting the other shores, the leeward or protected shores.

The long wavelengths and periods of tsunami allow them to bend around many islands and hit all shores with high waves. Tsunami wavelengths typically are longer than the dimensions of an island. During the 26 December 2004 Indian Ocean tsunami, the island nation of Sri Lanka was hit by 4 to 7 m (13 to 23 ft) high tsunami at sites on all shores (figure 8.13). The tsunami were directed at the east shore, where more than 14,000 people died, but more than 10,000 were killed on the south shore and more than 6,000 on the north shore. Nearly 100 people were killed in the capital city of Colombo on the "protected" west shore.

Earthquake-Caused Tsunami

Fault movements of the seafloor that generate large earthquakes may also cause powerful tsunami. In general, to create tsunami, the fault movements need to have a vertical component, either uplifting or downdropping the seafloor, and have an earthquake magnitude of at least 7.5M_W (table 8.4). Reverse (thrust) and normal fault movements of the seafloor can inject lots of energy into the overlying seawater.

In a subduction zone, the plates may become stuck together (figure 8.14a). Because the overriding plate is stuck to the subducting plate, its seaward (leading) edge is dragged downward while the area behind (landward) bulges upward (figure 8.14b). This movement goes on for centuries, building up elastic strain all the while. When the stuck area ruptures, causing an earthquake, the leading edge of the overriding plate breaks free, springs seaward and upward, causing a tsunami (figure 8.14c). At the same time, the landward bulge warps downward, lowering coastal land below sea level. Tsunami race through the ocean for hours, but the subsided land will remain down for generations (figure 8.14d).

TABLE 8.4
Fault Displacements of Seafloors

Earthquake Magnitude (M_W)	Fault Slip (m)	Rupture Duration (sec)	Vertical Movement of Seafloor (m)
7	0.6	23	0.2 (0.7 ft)
8	2.7	70	0.7 (2.3 ft)
9	9	200	2.3 (7.5 ft)
9.5	27	330	7 (23 ft)

Source: *Science*, v. 78 (1997).

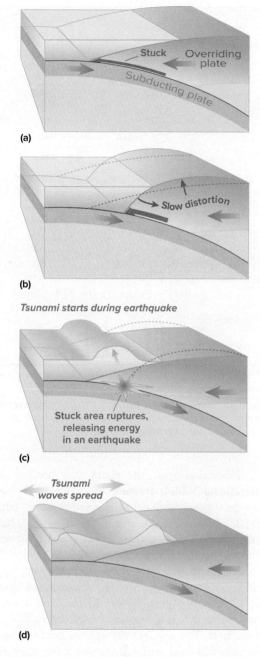

Figure 8.14 Subduction zones and tsunami. (a) Plates may stick together. (b) Seaward edge of the overriding plate is dragged downward. (c) When stuck area ruptures, the overriding plate springs upward starting a tsunami. (d) Tsunami race landward and seward.

Source: USGS Circular 1187.

INDIAN OCEAN 26 DECEMBER 2004

On 26 December 2004 (Boxing Day), a killer tsunami swept through the Indian Ocean and crossed Asian and African shorelines, causing death and destruction in 14 countries (see figure 8.1). The estimated death total was 245,000, but the true number is almost certainly higher and will never be known. Countries hit especially hard were Indonesia

Figure 8.15 The leading edge of a tsunami slams into a Thailand resort, 26 December 2004.
© Reuters/Corbis.

(about 198,000 dead), Sri Lanka (about 30,000 killed), India (about 11,000 dead), and Thailand (about 6,000 killed).

More than 3,000 of the deaths were European and North American tourists enjoying warm ocean water and sunny coastlines during the winter. Remember the tsunami threat when you are vacationing at a beach resort (figure 8.15).

What did Earth do to cause so many deaths? The seafloor west of northern Sumatra in Indonesia overcame frictional resistance, triggering faulting that ruptured northward for 1,200 km (740 mi) during almost 7 minutes; this created the third largest earthquake in the world in more than 100 years (see table 4.2). The rupture began 30 km (19 mi) below the seafloor and caused movements of up to 20 m (65 ft) that shifted the positions of some Indonesian islands and tilted other ones. The huge earthquake must have collapsed many nearby buildings that fell and killed many thousands of people, but the evidence of earthquake damage was largely erased by the powerful tsunami that swept across the land minutes later. The earthquake-causing earth rupture occurred due to compression caused by the subducting Indian-Australian plate (see figure 4.10) that raised the seafloor tens of feet, thus charging the seawater with energy that rapidly moved outward as tsunami racing throughout the Indian Ocean. The most powerful tsunami moved east to kill people in Indonesia and Thailand, and west to slaughter people in Sri Lanka and India. A second great earthquake occurred in Indonesia on 28 March 2005 as another subduction event created an 8.6M$_w$ seism just southeast of the 2004 rupture; it generated another tsunami (figure 8.16).

The last major tsunami event in the Indian Ocean occurred in 1883 when the volcano Krakatau collapsed into the sea, also offshore from Sumatra. The 1883 tsunami killed about 36,000 people. The world population of humans in 1883 was about 1.6 billion, but it had grown to 6.4 billion people in 2004. The dramatic growth of the human population during

Figure 8.16 Devastation caused by tsunami in Indonesia, 28 March 2005.

U.S. Air Force photo by Technical Sgt. Scott Reed.

those 121 years helped increase the death total to 245,000 people in the 2004 tsunami.

An overwhelming event such as this far-reaching tsunami results in many dramatic stories. In Sri Lanka, the train known as the Queen of the Sea left Colombo at 7:30 a.m. with upwards of 1,000 passengers. The train chugged slowly up the palm-fringed coast until about 9:30 a.m., when the tsunami struck. It knocked the railroad cars off the track and rolled them into a thick marsh, killing more than 800 passengers. The force of the tsunami was so great that wheels were torn off and tracks were twisted into odd shapes.

In India, it was Full Moon Day, and many Hindus were at the ocean's edge doing ritual bathing when they were pulled out to sea by the tsunami.

In Thailand, at the peak of the tourist season, the tsunami pushed snorkelers across sharp coral reefs only to pull them back out to sea, along with sunbathers. Meanwhile, farther offshore, scuba divers enjoyed the sights in deeper water unaware of the tsunami that raced past them.

Immediately following this great natural disaster, people and countries of the world mobilized to bring aid to the survivors. People needed shelter, food, clean water, and sanitary conditions. One of the big concerns was an outbreak of disease. With outside help, diseases such as cholera, typhoid, hepatitis A, and dysentery were prevented from causing another disaster.

ALASKA, 1 APRIL 1946: FIRST WAVE BIGGEST

As 1 April 1946 began in the Aleutian Islands, two large subduction movements occurred and shook the area severely. The five workers in the Scotch Gap lighthouse were shaken awake and wondered what lay ahead during the dark night. The lighthouse was built of steel-reinforced concrete, and its base sat 14 m (46 ft) above mean low-water level (figure 8.17a). About 20 minutes after the second earthquake, a tsunami approximately 30 m (100 ft) high swept the

(a)

(b)

Figure 8.17 The Scotch Gap lighthouse in the Aleutian Islands of Alaska (a) before and (b) after the tsunami unleashed by a magnitude 7.8 earthquake on 1 April 1946.

NOAA.

lighthouse away (figure 8.17b). This time, *the first wave was the biggest;* it killed all five men.

Tsunami are not just local events. The waves race across the entire Pacific Ocean. The April Fool's Day tsunami traveled about 485 mph in the deep ocean, slowing to about 35 mph as it neared shore in Hilo, Hawaii. Humans are no match for these massive waves. The long wavelengths allow the wave front to rush onland for long distances; there is no trough immediately behind, waiting to pull the water back to the ocean. This tsunami killed 159 people in Hilo, Hawaii. One of the ironies of this tragedy is that some of those killed were people who were warned. After being told a tsunami was coming, some folks laughed, said they knew it was April Fool's Day, and ignored the warning.

CHILE, 22 MAY 1960: THIRD WAVE BIGGEST

The most powerful earthquake ever measured occurred in Chile on 22 May 1960. Tsunami generated by this magnitude 9.5 subduction movement killed people throughout the Pacific Ocean basin. In Chile, the main seism broke loose at 3:11 p.m. on Sunday. Chileans are familiar with earthquakes, so many people headed for high ground in anticipation of tsunami. About 15 minutes after the seism, the sea rose like a rapidly rising tide, reaching 4.5 m (15 ft) above sea level. Then the sea retreated with speed and an incredible hissing and gurgling noise, dragging broken houses and boats out into the ocean. Some people took the "smooth wave" as a sign that these tsunami could be ridden out at sea, thus saving their boats. About 4:20 p.m., the second tsunami

arrived as an 8 m (26 ft) high wave traveling at 125 mph. The wave crushed boats and their terrified passengers, as well as wrecking coastal buildings. But *the third wave was the largest;* it rose 11 m (35 ft) high, but it traveled at only half the speed of the second wave. More than 1,000 Chileans died in these tsunami.

Since 1960, Hawaiians have been given warnings before tsunami arrive. The Pacific Tsunami Warning Center evaluates large earthquakes in the Pacific Ocean and then, using maps like the one in figure 8.18, provides people with hours of advance warning, including shrieking sirens. At 6:47 p.m. Hawaii time, the Chilean tsunami was predicted to arrive in Hilo, Hawaii, about midnight. The first wave arrived just after midnight and was 1.2 m (4 ft) high. At 12:46 a.m., the second wave came in about 2.7 m (9 ft) high. Many people thought the danger had passed and returned home. Then the sea level dropped 2 m (7 ft) below the low-tide level before the third wave, the largest wave, came in 6 m (20 ft) higher. The third wave killed 61 people in Hilo and seriously injured another 282.

The tsunami raced on to Japan, where it killed another 185 people, 22.5 hours after the earthquake. The energy in this set of tsunami was so great that it was recorded on Pacific Ocean tide gauges for a week as the energy pulses bounced back and forth across the entire ocean basin.

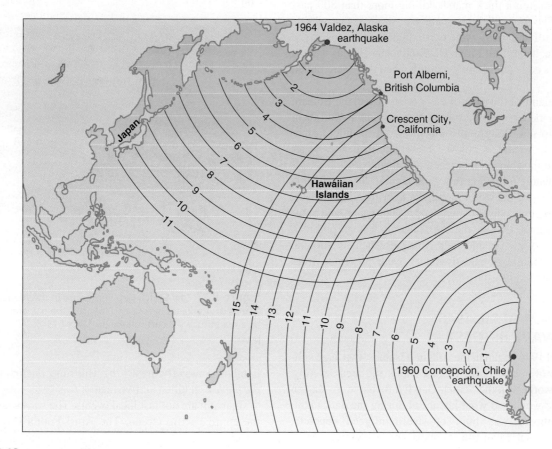

Figure 8.18 Tsunami travel times in hours.

Data source: Kious, W. J. and Tilling, R. I., The Dynamic Earth, p. 77, *US Geological Survey.*

ALASKA, 27 MARCH 1964: FIFTH WAVE BIGGEST

The Good Friday earthquake in Alaska was a magnitude 9.2 monster whose tsunami ravaged the sparsely populated Alaska coastline, killing 125 people. Imagine the energy in tsunami formed when these immense masses of land were raised up and dropped down (figure 8.19). Alaska sits on the upper plate and was shifted horizontally (seaward) up to 19.5 m (64 ft) and uplifted as much as 11.5 m (38 ft). Another landward block was down-dropped as much as 2.3 m (7.5 ft). Some of the seafloor offsets were even greater. More than 285,000 km² (110,000 mi²) of land and sea bottom were involved in these massive movements. More than 25,000 km³ (6,000 mi³) of seawater were jolted and moved. The sudden uplift of this huge volume of water resulted in tsunami racing through the entire Pacific Ocean.

Three hours after the earthquake (see figure 8.18), Port Alberni on Vancouver Island was hit by a 6.4 m (21 ft) high tsunami that destroyed 58 buildings and damaged 320 others. Thanks to advance warning, the Crescent City, California, waterfront area was evacuated, and residents waited upslope while tsunami arrived and did their damage. After watching four tsunami and seeing their sizes, many people could no longer stand the suspense of not knowing the condition of their properties. Some people went down to check them out—a big mistake. In this tsunami series, *the fifth wave was the biggest;* it was 6.3 m (21 ft) high, and it killed 12 of the curious people. All the fatalities of this event at Crescent City were caused by the fifth tsunami.

Tsunami arrive as a series of several waves over several hours. Which wave in the series will be the biggest? As the preceding events show, it is unpredictable. In 1946, 1960, and 1964, the biggest wave was the first, third, and fifth, respectively.

Volcano-Caused Tsunami

Volcanic action can create killer tsunami (see table 8.2). Volcanoes can put jolts of energy into a water body in several ways. They can explode, they can collapse, and they can send avalanches of debris into the water. It seems likely that tsunami were generated by all three mechanisms during the eruption of Krakatau in 1883.

KRAKATAU, INDONESIA, 26–27 AUGUST 1883

Krakatau sits in the sea between the major Indonesian islands of Sumatra and Java (figure 8.20). One of the most famous eruption sequences in history occurred here in 1883, including the killing events of 26–27 August. On Sunday afternoon, August 26, volcanic eruptions and explosions increased in frequency and strength. In the evening, some of the eruptions blasted large volumes of gas-charged rocky debris rapidly downslope, across the shoreline, and into the sea, putting energy into the water that radiated outward as tsunami that ravaged villages on distant shorelines. The highly irregular coastline in this region affects tsunami height and run-up in the various harbors, inlets, and peninsulas (figure 8.20).

On Monday morning, gigantic explosions occurred around 5:30, 6:45, and 8:20. Each explosion sent tsunami with their maximum energy focused in different directions,

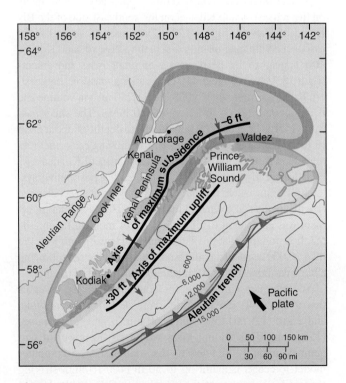

Figure 8.19 Map showing land uplifted and down-dropped during the 1964 Alaska earthquake.

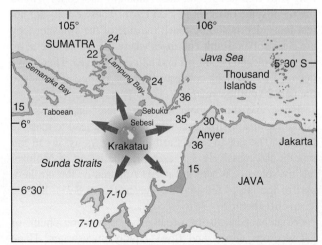

Figure 8.20 Many killer tsunami were sent off by the eruptive behaviors of Krakatau Volcano in Indonesia on 26–27 August 1883. Tsunami run-up heights are shown in meters. Tsunami-flooded areas are shaded in red.

wiping out different villages. These explosions may have been due to seawater coming in contact with the magma body and rapidly converting the thermal energy of the magma into the mechanical energy of tsunami.

The eruption sequence culminated about 10 a.m. with an overwhelming explosion that is commonly attributed to the volcano mountain collapsing into the void formed by its partly emptied magma chamber. The resulting blast was heard for thousands of miles. Tsunami pushed into harbors and ran up and over some coastal hills up to 40 m (135 ft) high. The tsunami during this 20-hour period destroyed 165 villages and killed more than 36,000 people. This tsunami death total was not exceeded until 26 December 2004, again in Indonesia.

Landslide-Caused Tsunami

Gravity pulls a variety of rock and sediment masses into and beneath the seas and lakes. Energy from these mass movements is transferred to the water, locally causing higher and larger run-ups of water than are caused by earthquake-generated tsunami.

VOLCANO COLLAPSES

Volcanic islands are huge and impressive features. Their bulk and beauty are inspiring and overwhelming, but they have weaknesses that lead to catastrophic failures.

Hawaii in the Pacific Ocean

The largest submarine mass movements were recognized first on the seafloor along the Hawaiian Islands volcanic chain. Slump and debris-avalanche deposits there cover more than five times the land area of the islands (figure 8.21). Some individual debris avalanches are more than 200 km (125 mi) long with volumes greater than 5,000 km³ (1,120 mi³), making them some of the largest on Earth. These events are not just loose debris sliding down the side of the volcano; they are catastrophic **flank collapses** where the whole side of an oceanic volcano breaks off and falls into the sea. There have been at least 70 flank collapses from the Hawaiian Islands in the past 20 million years.

Pause and think about this a moment: each Hawaiian Island has major structural weaknesses that lead to massive failures. The not-so-solid Earth here betrays us; it can fail rapidly and massively. For example, the island of Molokai has no volcano. Where did it go? Apparently, the northern part of the island fell into the ocean (figure 8.21), leaving steep cliffs behind.

What happens to the ocean when a gigantic chunk of island drops into it and flows rapidly underwater? Huge tsunami are created. For example, prehistoric giant waves washed coral, marine shells, and volcanic rocks inland, where they are found today as gravel layers on Lanai lying 365 m (1,120 ft) above sea level and on Molokai more than

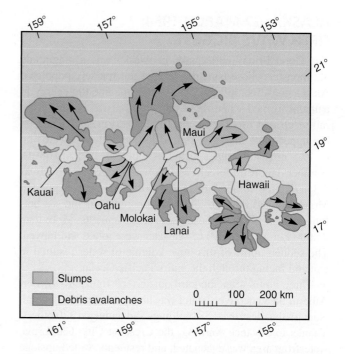

Figure 8.21 The Hawaiian Islands cover less surface area than the slumps and debris avalanches that have fallen from them.

2 km inland and more than 60 m (200 ft) above sea level. Tsunami of this size would not only ravage Hawaii but would cause death and destruction throughout the Pacific Ocean basin.

Where might an event like this happen next? Quite likely on the active volcano Kilauea on the southeast side of the Big Island of Hawaii (figure 8.22). The zones of normal faults associated with magma injection in the active volcano Kilauea also appear to be head scarps of giant mass movements (figure 8.23; see figure 5.44). One moving mass on southeastern Kilauea is more than 5,000 km³ (1,200 mi³) in volume and is sliding at rates up to 25 cm/yr (10 in/yr). The slide mass extends offshore about 60 km (37 mi) to depths of about 5 km (3 mi) (figure 8.23). The area on the moving mass includes the 80 km (50 mi) long coastal area southeast of Kilauea.

During early November 2000, the global positioning system (GPS) measured this large block moving at an accelerated rate that peaked at 6 cm (2.3 in) per day. The days-long movement was equivalent to an earthquake of moment magnitude 5.7. It is possible that this large hunk of Hawaii, with its more than 10,000 residents, could be plunged into the sea following a large earthquake or during a major movement of magma.

A computer simulation for southeast Hawaii assumes that a block 40 km (25 mi) long, 20 km (12 mi) wide, and 2 km (1.3 mi) thick slides 60 km (37 mi) at 70 m/sec (155 mph). This sliding mass would create powerful tsunami dominantly directed to the southeast. The computer model forecasts tsunami up to 30 m (100 ft) high striking susceptible segments of the coastline in Canada, the United States, and Mexico.

Canary Islands in the Atlantic Ocean

After scientists in Hawaii recognized that large volcanic islands experience major collapses that generate powerful tsunami, they began to search for other sites where flank collapses have occurred. A similar history was found in the Canary Islands in the Atlantic Ocean off the coast of northwest Africa at about 28° north latitude. At least three of the Canary Islands have had mega-collapses: Tenerife, La Palma, and Hierro. The last known major event happened on Hierro just 15,000 years ago. When the next collapse occurs, powerful tsunami could hit coastal cities along the east coasts of North and South America and along the west coasts of Europe and Africa. Although these events are rare, they are real and can be destructive.

A computer simulation was made by Steven Ward and Simon Day assuming a 500 km³ (120 mi³) flank collapse from Cumbre Vieja Volcano on La Palma in the Canary Islands. The model forecast that tsunami with heights of 10 to 20 m (30 to 65 ft) could travel across the Atlantic Ocean and strike the east coasts of the Americas (figure 8.24).

Flank collapses are not unique to Hawaii and the Canary Islands; they can occur around the world. As oceanic volcanoes grow, they tend to develop internal weaknesses and sides that are too steep, and thus they collapse in gigantic events. Worldwide, approximately one flank collapse occurs every 10,000 years.

EARTHQUAKE-TRIGGERED MASS MOVEMENTS

Earthquakes not only generate tsunami directly, but their energy can also trigger the movement of large masses of rock or sand whose kinetic energy causes tsunami.

Newfoundland, Canada, 18 November 1929

At 5:02 p.m. on 18 November 1929, an earthquake of magnitude $7.2M_w$ occurred offshore of eastern Canada (figure 8.25). The earthquake triggered the submarine movement of a sediment mass with an estimated volume of 200 km³ (50 mi³). The kinetic energy of the sediment mass moving downslope set off tsunami. Especially hard hit was the Burin Peninsula of Newfoundland, where about 40 villages were damaged and 28 people died. The tsunami began arriving about 2.5 hours after the earthquake; the waves came in three major pulses during a 30-minute interval. The long narrow bays caused 1 m high tsunami to build to 3 m (10 ft) in many inlets

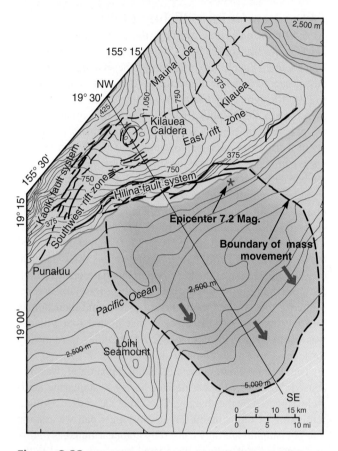

Figure 8.22 Map of the Kilauea area of Hawaii showing an outline of the mass sliding into the Pacific Ocean. The northwest–southeast line is shown in cross-sectional view in figure 8.23.

Data source: Lipman, P. W. et al., USGS Professional Paper 1276, "Ground deformation associated with the 1975 earthquake," US Geological Survey, 1985.

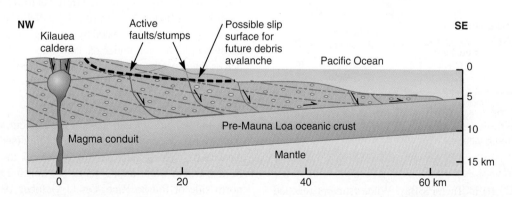
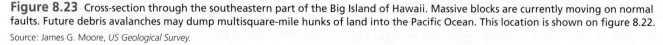

Figure 8.23 Cross-section through the southeastern part of the Big Island of Hawaii. Massive blocks are currently moving on normal faults. Future debris avalanches may dump multisquare-mile hunks of land into the Pacific Ocean. This location is shown on figure 8.22.

Source: James G. Moore, *US Geological Survey*.

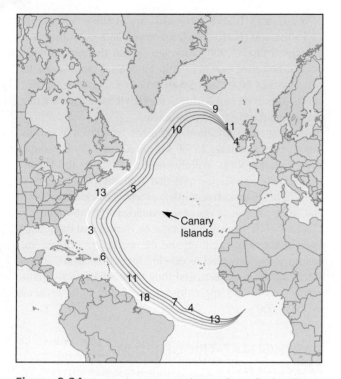

Figure 8.24 Tsunami wave set six hours after a flank collapse in the Canary Islands. Numbers denote the height in meters of the tsunami in various areas of the Atlantic Ocean. Tsunami striking the east coast of North America could locally be 13 m (40 ft) high.

Source: After Ward and Day.

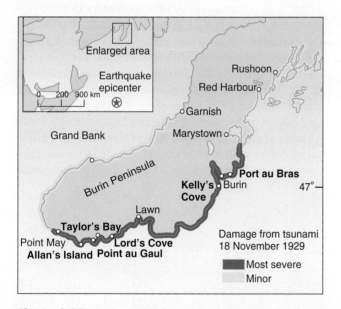

Figure 8.25 The irregular coastline of the Burin Peninsula in Newfoundland focuses the energy of tsunami. Deaths occurred in the bold-lettered villages.

and to 7 m (23 ft) in Taylor's Bay. When tsunami reached the heads of inlets and bays, they had so much energy and momentum that their run-up onto land reached 13 m (43 ft) elevation in some areas, causing significant damage. The

tsunami dealt a crippling blow to the local fishing industry and almost drove Newfoundland into bankruptcy.

The Newfoundland tsunami is not an isolated event. Beneath the Atlantic Ocean off the east coast of North America, new images of the seafloor show significant scars where big submarine landslides have occurred. Similar landslides in the future will generate tsunami.

Papua New Guinea, 17 July 1998

At 6:49 p.m. on Friday evening, the north shore of Papua New Guinea was rocked by a 7.1 magnitude earthquake occurring about 20 km (12 mi) offshore. As the shaking ended, witnesses saw the sea rise above the horizon and shoot spray 30 m (100 ft) high. They heard sounds like distant thunder, and then the sea slowly pulled back. About 4 to 5 minutes later, the people again heard a rumbling sound and saw a tsunami about 4 m (13 ft) high approaching. But if you can see the wave coming, it is too late to escape. Many people living on the barrier beach were washed into the lagoon (figure 8.26).

Several minutes later, a second wave approached, but this one was about 14 m (45 ft) tall. A tsunami does not have the shape of a typical wave; it is more like a pancake of water. This tsunami averaged about 10 m (33 ft) high and measured 4 to 5 km (2.5 to 3 mi) across. Visualize this thick pancake of water pouring over the heavily populated beach at 15 mph for more than a minute. This tsunami event was a three-wave sequence that washed thousands of people and their homes into the lagoon. A barrier beach that hosted four villages was swept clean. The estimated 2,200 fatalities were mostly those least able to swim—the children.

The Papua New Guinea tsunami apparently was not caused directly by the earthquake but by a submarine landslide triggered by the shaking. This event has caused a global rethinking of the tsunami threat. No longer can we assume that big tsunami are only caused by giant earthquakes in distant places. Tsunami can be created by smaller local faults that cause unstable sand and rock masses to slip and slide under water.

Numerous sites around the world could experience Papua New Guinea–style, landslide-generated tsunami. For example, Southern California is protected from Pacific Ocean basin tsunami by its offshore system of subparallel island ridges. But these ridges have been created by active faults, and their movements could trigger undersea landslides that send tsunami across the densely populated Southern California coastline.

Puerto Rico, 11 October 1918

Even the Atlantic Ocean has a few subduction zones along its Caribbean boundary and to the far south (see figure 2.13). The deepest part of the Atlantic Ocean is the Puerto Rico trench, where water depths reach 8.4 km (5.2 mi) along the north side of Puerto Rico. On 11 October 1918, an earthquake of magnitude $7.3M_s$ shook loose a 22 km (14 mi) wide submarine landslide. Tsunami up to 6 m (20 ft) high hit the northwestern Puerto Rico coastline, killing 116 people.

Figure 8.26 Tsunami swept villages from this sandbar into Sissano Lagoon, Papua New Guinea, drowning 2,200 people on 17 July 1998. Does this sandbar remind you of coastal New Jersey, North Carolina, Balboa Island in southern California, or other sites?

Courtesy of Hugh Davies, University of Papua New Guinea.

IN BAYS AND LAKES

The constricted topography of bays and lakes allows some landslides to create huge tsunami of local extent.

Lituya Bay, Alaska, 9 July 1958

The largest historic wave run-up known occurred on 9 July 1958, when a massive rockfall dropped into Lituya Bay

in Glacier Bay National Park, Alaska (figure 8.27). It was after 10 p.m. when the Fairweather fault moved in a 7.7M earthquake, causing about 90 million tons of rock and ice to drop more than 900 m (3,000 ft) into the water. Three boats were anchored in the bay. One was a 40 ft fishing boat operated by a father and son who later reported an earsplitting crash that caused them to look up the bay, only to see a huge wall of water about 30 m (100 ft) high roaring toward them faster than 100 mph. They only had enough time to turn the bow (front) of the boat toward the wave. The onrushing tsunami swept over 54 m (176 ft) high Cenotaph Island, and then it hit them. Their anchor chain snapped, and the boat soared near vertically upward like a high-speed elevator in a tall building. Reaching the crest of the wave, they dropped down the back side and survived. The second boat was carried across the sandbar beach into the ocean, and the crew survived. The third boat fired up its engine and tried to outrun the tsunami; this was a bad decision. The wave hit that boat on the stern (backside), flipped it, destroyed it, and killed the crew.

Looking at beautiful Lituya Bay after the traumatic evening showed that the rock-fall impact had sent a surge

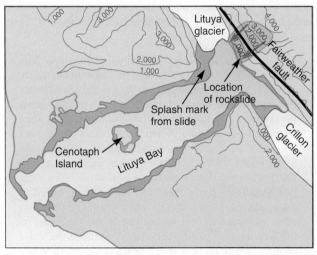

(a)

(b)

Figure 8.27 (a) A rockmass shaken loose by a Fairweather fault earthquake splashed into Lituya Bay, Alaska, sending a 30 m (100 ft) wave down the bay and into the ocean, 9 July 1958. (b) View up Lituya Bay. The light-colored areas around the bay were stripped of a mature forest by the wave.

D.J. Miller/U.S. Geological Survey.

of water up the opposite slope, stripping away mature trees up to 525 m (1,720 ft) above sea level. The tsunami destroyed and stripped away mature trees along both walls of the bay, 35 m (110 ft) above sea level, all the way to the open ocean.

Lake Tahoe, California and Nevada

Beautiful Lake Tahoe sits high in the Sierra Nevada in California and Nevada. The lake is 35 km (22 mi) long, 19 km (12 mi) wide, and over 500 m (1,600 ft) deep; it is the 10th deepest lake in the world. This broad and deep lake was created by subparallel normal faults dropping the land between them (figure 8.28). The faults are active. The underwater faults have a 4% probability of causing a magnitude 7 earthquake in the next 50 years. The likely fault movement could drop the lake bottom about 4 m (13 ft) and could generate 10 m (33 ft) high waves that rush over the populated shoreline.

Figure 8.28 also shows debris on the lake bottom from a huge landslide that would have generated even bigger tsunami. The tsunami hazard at Lake Tahoe is real, but its frequency is low.

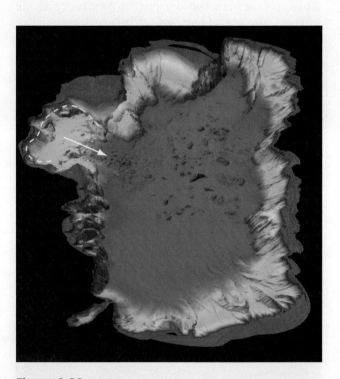

Figure 8.28 Lake Tahoe basin is created by down-dropping between active faults. Notice the amphitheater (bold dashed white lines) on the west side of the lake; it formed when a giant landslide pulled away and dumped debris onto the lake bottom in a tsunami-forming event.

Courtesy Graham Kent.

Seiches

Seiches are oscillating waves that slosh back and forth within an enclosed body of water, such as a sea, bay, lake, or swimming pool. The process was observed and named in Lake Geneva, Switzerland, in 1890. The word *seiche* (pronounced "SAY-sh") comes from a Swiss-French word that means to sway back and forth.

The energy to cause a seiche can come from a variety of sources. Winds blowing across a lake can cause the water body to oscillate at some natural period. Seiches are common in the Great Lakes of Canada and the United States, where they may be called sloshes. For example, Lake Erie with its elongate shape and relatively shallow water can experience seiches when strong winds blow. The oscillating water mass can form seiches up to 5 m (16 ft) high, alternating from one end of the lake to the other.

Earthquakes frequently cause seiches. For example, a large earthquake near Lake Tahoe may rock the land and water enough that the water body continues oscillating in seiches. This seiche is different from the tsunami caused by a large landslide moving into Lake Tahoe.

People living in seismically active areas commonly get to watch seiches during an earthquake when the water in swimming pools sloshes back and forth and overflows the sides.

HEBGEN LAKE, MONTANA, 17 AUGUST 1959

Shortly before midnight on August 17, two faults running beneath the northern end of Hebgen Lake moved in magnitude 6.3 and 7.5 earthquakes. These two normal faults had their southwestern sides drop 7 and 7.8 m (23 and 26 ft) down fault surfaces inclined 45° to 50° to the southwest, also dropping the northern end of Hebgen Lake (see figures 5.26 and 5.27). The foreman at the Hebgen Lake Dam was awakened by the earthquake and went outside for a look. In Foreman Hungerford's own words:

The dust was so intense you could hardly see. You could hardly breathe, or anything. It obscured the moon. We went to the river gagec. . . . Just as we got to it, we heard a roar and we saw this wall of water coming down the river. . . . We thought the dam had broken. . . . Then we went up to the dam. When we got there we couldn't see much, but I walked over to the edge of the dam and all we could see was blackness. There was no water. No water above the dam at all, and I couldn't imagine what had become of it. By that time the dust had started to clear, and the moon had come out a little. And then here came the water. It had all been up at the other end of the lake. . . . We rushed back when we heard the water coming. We could hear it before we could see it. When it came over the dam, it was a wall of water about three to four feet high completely across that dam, and it flowed like that for what seemed to me to be 20 minutes, but possibly it could have been 5 or 10. I have no idea of time. It flowed for a while, and then it started to subside. Then it all cleared away, and no water again. The lake was completely dry as far as we could see. All we could see down the dam was darkness again. It seemed like a period of maybe 10 to 15 minutes, and the water came back, and then it repeated the same thing over again.

Hungerford was eyewitness to a spectacular seiche event in which lake water sloshed back and forth for 11.5 hours. A seiche is analogous to what happens to your bathwater when you stand up quickly from a full tub—it sloshes back and forth. More than 130 km² (50 mi²) of land on the northern side of Hebgen Lake dropped down 3.3 m (11 ft). The warping of the lake floor set off a huge series of seiches.

Tsunami and You

If you are near the coast and feel an earthquake, think about the possibility of tsunami. A sharp jolt and shaking that lasts a few seconds suggest that the epicenter was nearby, but the earthquake was too small to create powerful tsunami. A relatively mild shaking that lasts for 25 or more seconds suggests that the epicenter was far away, but the energy released during that long shaking could create dangerous tsunami if the fault moved the seafloor. Remember that the most powerful earthquakes can create tsunami that kill many thousands of people who are too far away to feel the earthquake.

Other clues can alert us to the possibility of powerful tsunami. Before the first big wave of a tsunami, the sea may either withdraw significantly far from shore or suddenly rise. Sometimes the ocean water changes character or makes different sounds, or something else out of the ordinary may happen. Notice these changes in ocean behavior.

If you think tsunami might be coming—take action (table 8.5). If you wait until you see the tsunami, you have waited too long. No matter how strong a swimmer you are, it is not just the water that can hurt you, but also the debris it carries (figure 8.29).

SIMEULUE ISLAND, INDONESIA, 26 DECEMBER 2004

Simeulue Island is the inhabited land closest to the epicenter of the magnitude 9.1 earthquake in 2004. When the destructive seismic shaking stopped, the residents did not take time to check their houses or talk with their neighbors; they fled to the hills. Within 30 minutes of the earthquake, tsunami 10 m (33 ft) high ravaged their earthquake-damaged coastal villages and destroyed much of what remained. After the earthquake and tsunami, a count of the residents found that only 7 out of 75,000 inhabitants had died. Why were so few people killed? The islanders remembered the stories

TABLE 8.5
Surviving a Tsunami
Abandon your belongings.
Many lives are lost while trying to save possessions.
Head for high ground—and stay there.
If there is no high ground nearby, then
Climb to an upper floor or roof of a strong building.
If there is no sturdy building, then
Climb a tree.
If there are no climbable trees, then
Grab onto something that floats.
Look for something to use as a raft.

Figure 8.29 Tsunami breaking over Pier 1 in Hilo, Hawaii, on 1 April 1946. The man in the foreground was one of 159 killed. NOAA.

passed down as oral history from their ancestors: when the ground shakes, run to the hills before the giant waves arrive. Remembering the lessons of their history paid off big for the Simeulue islanders.

NICARAGUA, 1 SEPTEMBER 1992

Can a major earthquake occur nearby and offshore without you feeling it—and yet send killer tsunami? Yes, on a Tuesday evening at 6:16 p.m., an unusual earthquake occurred that was large (magnitude 7.6) but barely felt. About 45 minutes later, a 10 m (33 ft) high tsunami ravaged a 300 km (185 mi) long section of the Nicaraguan coastline. What happened?

Why did the ground shake very little, yet the ocean water still become agitated into large tsunami? The fault moved so slowly that the high-velocity, short-period seismic waves were relatively weak and shook the ground only slightly, about 1/100 of what is expected. However, the slow-moving fault released a lot of long-period energy into the water, thus creating powerful tsunami.

The earthquake was a subduction event in which a 100 km (62 mi) long segment of oceanic plate moved 1 m in 2 minutes. The slow-motion fault movement was especially efficient at pumping energy into water. The seawater absorbed the energy, sending tsunami onto the beach. The coastal residents were caught without warning; 13,000 homes were destroyed, and 170 people were killed, mostly sleeping children.

HUMANS CAN INCREASE THE HAZARD

Mapping of coastlines in Indonesia, India, and Sri Lanka following the killer tsunami of 2004 showed how human activities increased the damages and life loss in some areas. The coastal areas where forests had been removed suffered more extensive damage than neighboring areas with the natural vegetation intact. Trees and shrubs reduce the amplitude and energy of incoming waves. With the forest gone, houses, bridges, and other human-built structures were left to absorb the tsunami energy.

In Sri Lanka, many of the hardest-hit coastlines were ones where coral reefs had been removed. Coral reefs there are mined for souvenirs to sell, removed to open beaches for

tourists to use, and blown up to stun and catch the fish inside them. With coral barrier reefs removed, tsunami charge farther inland and with greater energy.

TSUNAMI WARNINGS

Sitting in the middle of the earthquake-prone Pacific Ocean basin, Hawaii is hit by numerous tsunami. The heights and run-ups of tsunami at the shoreline vary due to differences in local topography, both onshore and offshore. The broad-scale threats to homeowners and businesses are presented in a tsunami-hazard map for the Big Island of Hawaii (figure 8.30). As awareness of tsunami hazards increases, signs are being posted in coastal areas of many states to provide warning (figure 8.31).

Seismic Signals and Pressure Sensors

After the mega-killer Indian Ocean tsunami in 2004, the interest in and support for tsunami warning systems increased significantly. Now, rapid analyses of seismic waves for their tsunami-generating potential begin with the arrival of the first P waves. Quick initial determinations are made of epicenter location, depth to hypocenter, earthquake magnitude, and vertical versus horizontal components of fault movement. The National Oceanic and Atmospheric Administration (NOAA) has pre-computed scenarios for tsunami-generating earthquakes that could occur along the

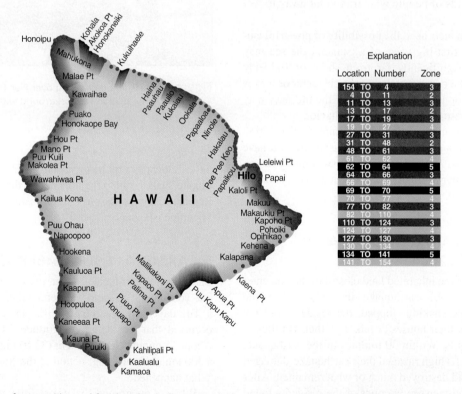

Figure 8.30 Map of tsunami hazard for the island of Hawaii. Black areas of coastline (zone 5) receive tsunami greater than 15 m (50 ft) high. Zone 4 tsunami (green) may be greater than 9 m (30 ft) high, zone 3 tsunami (purple) greater than 4.5 m (15 ft), and zone 2 tsunami (pink) greater than 1.5 m (5 ft) high.

Source: US Geological Survey Professional Paper 1240B.

Figure 8.31 Please follow this advice.
© John A. Karachewski.

subduction zones all around the Pacific Ocean. Early tsunami watches and warnings now can be issued within several minutes.

The NOAA earthquake/tsunami scenario closest to the epicenter becomes improved by measurements made by pressure sensors that are anchored to the seafloor. The pressure sensors are DART II (Deep-ocean Assessment and Reporting of Tsunami) stations (figure 8.32). As tsunami pass by, changes in water pressure are recorded by the bottom pressure sensors. An acoustic link sends the data up to the companion moored surface buoy. The data are then relayed through a satellite that transmits the data to a tsunami warning center. As tsunami race through the ocean, each DART II station records new measurements that allow tsunami forecasts to be updated.

NOAA has set in place 39 DART II stations close to subduction zones that can generate killer tsunami. Other countries combined have placed 19 stations on the sea floor. Data are handled and tsunami warnings issued for the Pacific Ocean by the Pacific Tsunami Warning Center in Hawaii and for the west coast of North America by the West Coast and Alaska Tsunami Warning Center in Alaska. The Japan Meteorological Agency also has a sophisticated tsunami warning system in place. Tsunami warning systems are being developed for the Indian Ocean, the Caribbean Sea, the Mediterranean Sea, and the northeast Atlantic Ocean.

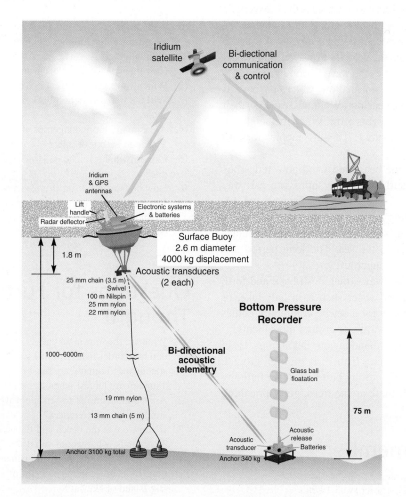

Figure 8.32 Diagram of a DART II tsunami warning system deployed by NOAA. A pressure sensor on the seafloor records water-pressure changes, and then sends the data up to the buoy, which relays the data through a satellite to a tsunami warning center.

Summary

Tsunami are the biggest waves of all. Earthquakes, volcanic eruptions, subsea landslides, and asteroid impacts disturb the deep ocean-water mass, sending off energetic waves. Earthquake-generated tsunami commonly travel almost 500 mph (800 km/hr) and may be spaced as far as 60 minutes apart. Tsunami slow down in shallow water but may still be moving at freeway speeds. Local topography, as in harbors and inlets, may focus tsunami energy, creating waves more than 30 m (100 ft) high that kill thousands of people.

In the U.S. media, tsunami are sometimes referred to as tidal waves, but this is silly. Tidal waves are powered by gravitational attraction of the Earth, Sun, and Moon, but tsunami are not.

Most waves are created by winds blowing across the water surface. The height of a wave depends on wind velocity, length of time the wind blows, length of water the wind travels across, and consistency of wind direction. On rare occasions, the various waves moving through the ocean will briefly synchronize and produce rogue waves up to 34 m (112 ft) high.

The distance between successive waves is the wavelength, and the time for two waves to pass a common point is the period. Ocean or lake waves cause orbital motion in water to depths of one-half wavelength. Velocity of waves in miles per hour is approximately 3.5 times the wave period in seconds.

When a wave moves into shallow water, its base is slowed by friction with the seafloor. When waves slow, their lengths decrease and their heights grow. When the height-to-wavelength ratio reaches about 1:7, a wave topples forward as a breaker.

The daily waves at the beach are wind-powered. They are solitary waves that curl over at the shore, run part way up the beach, and then pull back to the sea. Tsunami are different; they are the leading edge of a tabular mass of water that may keep running onshore for 5, 10, or even 30 minutes before being pulled back to the sea. The long period and momentum make tsunami dangerous.

Seiches are oscillating waves that slosh back and forth within an enclosed body of water such as a bay, lake, or swimming pool. The energy for a seiche commonly comes from strong winds, seismic waves, or ground movements.

If tsunami are coming: Abandon your belongings. Head for high ground and stay there. If there is no high ground, climb to an upper floor of a strong building. If no building is nearby, climb a tree. If there are no trees, grab something that floats.

Terms to Remember

fetch   204

flank collapse   216

rogue wave   205

seiche   220

swell   205

wavelength   204

Questions for Review

1. Can the terms *tsunami* and *tidal wave* be used to describe the same event?
2. What is the energy source of: (a) everyday waves at the beach? (b) tidal waves? (c) tsunami?
3. Draw a cross-section that defines wavelength, period, and water motion at depth.
4. As wavelength increases, what happens to wave velocity? Which change causes wind-blown ocean waves to be more powerful—doubling their wavelength or doubling their height?
5. Draw a cross-section and explain why an ocean wave breaks as it nears the shoreline.
6. How are rogue waves created?
7. What are four major causes of tsunami?
8. Which two types of fault movements of the seafloor can generate the most powerful tsunami—reverse (thrust), normal, or strike-slip?
9. How fast can tsunami travel? (a) in the deep ocean? (b) in shallow water? What are typical tsunami wavelengths and periods?
10. Is the killing power of tsunami due mostly to the wave heights or to the long wavelengths that allow them to travel far inland?
11. How many wave trains occur in a tsunami event?
12. Which wave is the biggest in a tsunami event?
13. Will the sea always pull back before a tsunami comes ashore?
14. Are tsunami more dangerous along a straight coastline or in a V-shaped harbor or inlet?
15. Explain what a seiche is. How could you create a small one?
16. How do tsunami differ from seiches?
17. In a tsunami set off by a subduction-zone earthquake, why might the seafloor rise up while coastal land warps down (see figure 8.14)?

Questions for Further Thought

1. What are the four main causes of tsunami (see text)? Make two rank-order lists (from #1 to #4) completing these statements: Tsunami are most frequently caused by _____. Tsunami are largest when caused by _____.
2. Are tsunami similar to seismic waves? What are major similarities or differences?
3. If you were at the beach and felt an earthquake lasting 30 seconds, what would you do?
4. How likely is it that powerful tsunami will be created by a huge mass movement southeast of Kilauea Volcano on the Big Island of Hawaii?

Disaster Simulation Game

Your challenge is to protect a Southeast Asian coastal village from tsunami disaster by constructing defenses and upgrading buildings. You are given a budget. Then you have real choices to make.

The village you must protect has a given population of people. You are provided with a map of the village and charged with protecting as many people, buildings, and livelihoods as possible. You must also build a hospital and a school; then construct two hotels to increase tourism. Are you ready for the challenge? Go to http://www.stopdisastersgame. org. Click on Play Game. On the next page, click on Play Game again. Select the Tsunami scenario. Then choose your preferred difficulty level: Easy (small map); Medium (medium-size map); or Large (large map).

Get to work! The tsunami is coming. Save as many people as you can.

CHAPTER 9

External Energy Fuels Weather and Climate

Great is the Sun and wide he goes
Through empty heaven with repose;
And in the blue and glowing days
More thick than rain he showers
*   his rays.*

—ROBERT LOUIS STEVENSON, 1885 *A CHILD'S GARDEN OF VERSES*

LEARNING OUTCOMES

Radiation received from the Sun supplies the energy that drives weather and climate. After studying this chapter you should

- know the difference between weather and climate.
- understand what happens to incoming solar radiation that reaches Earth.
- be able to explain the greenhouse effect.
- be able to explain the hydrologic cycle.
- appreciate the extraordinary properties of water.
- comprehend atmospheric pressure and winds.
- understand the general circulation of the atmosphere and ocean in both cross-section and map views.

OUTLINE

- External Sources of Energy
- Solar Radiation Received by Earth
- Outgoing Terrestrial Radiation
- The Hydrologic Cycle
- Water and Heat
- Energy Transfer in the Atmosphere
- Energy Transfer in the World Ocean
- Layering of the Lower Atmosphere
- Winds
- General Circulation of the Atmosphere
- General Circulation of the Oceans

Sunshine pours into Antelope Canyon, Arizona.
© Betty Wiley/Getty Images RF.

This chapter begins a major shift in energy sources as we move to those processes and disasters fueled by the Sun—weather, climate, flood, and fire.

Weather consists of the highly variable conditions in the atmosphere. Weather changes from hot to cold, wet to dry, clear to cloudy, calm to windy, and peaceful to violent. We experience the hour-to-hour, day-to-day, and season-to-season changes in the atmosphere as weather. Step outside and you experience weather.

When weather is viewed over the longer time spans of decades, centuries, even millions of years, it is called **climate.** Climate is the average pattern of weather in a region over a long period of time.

The news media bring us tales of lightning strikes, tornadoes, hurricanes, and floods; these processes all fall into the realm of weather. Climate involves phenomena such as ice ages, multiyear droughts, changes in gas contents in the atmosphere, and shifting circulation patterns in the ocean. Weather covers short-term processes, whereas climate refers to long-term conditions. Or, as Lazarus Long stated in Robert Heinlein's *Time Enough for Love,* "Climate is what you expect, weather is what you get."

The understanding of both weather and climate involves many of the same scientific principles, and this chapter begins with some of those principles.

TABLE 9.1
Energy Flow to and from Earth

	Energy Flow (×10^{12} joules per second)	Energy Flow (%)
Solar Radiation	*173,410*	*99.97*
Direct reflection	52,000	
Direct conversion to heat	81,000	
Evaporation	40,000	
Water transport in oceans and atmosphere	370	
Photosynthesis	40	
Heat Flow from Interior	*44.2*	*0.025*
General heat flow by conduction	43.9	
Volcanoes and hot springs	0.3	
Tidal Energy	*3*	*0.0017*

Source: Data from Hubbert (1971).

External Sources of Energy

Energy flowing from Earth's interior to the surface accomplishes impressive geologic work via plate tectonics, volcanism, and earthquakes. Yet the total amount of that energy is miniscule compared to the **solar radiation** received from the Sun. Only a minute percentage of the radiant energy of the Sun reaches Earth, but it is still about 4,000 times greater than the heat flow from Earth's interior (table 9.1).

Energy also is supplied externally via gravitational attractions between Earth, Moon, and Sun that add tidal energy to Earth. In addition, incoming meteorites, asteroids, and comets make high-energy impacts upon our planet.

THE SUN

Earth's climate is powered primarily by heat energy emitted from the Sun. The energy radiated from the Sun covers a broad spectrum of wavelengths, ranging from radio waves with wavelengths of tens of kilometers to X-rays and gamma rays measured in billionths of a meter; this is the electromagnetic spectrum (figure 9.1). Most of the solar radiation is concentrated in the part of the wavelength spectrum visible to humans (light) or nearly visible (infrared and ultraviolet). Visible light is about 43% of the solar radiation received

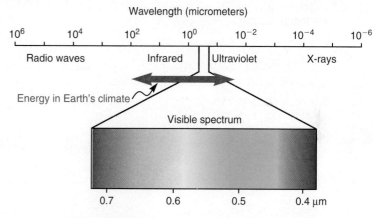

Figure 9.1 The electromagnetic spectrum. One micrometer (μm) = one-millionth of a meter = 0.001 millimeter.

on Earth; it ranges in wavelength from 0.0004 mm (violet) to 0.0007 mm (red). Of the almost-visible solar energy, 49% is received in near-infrared wavelengths we can detect as heat and 7% in the ultraviolet (UV) wavelengths that give us sunburn.

All objects radiate energy. The hotter the object, the more energy it radiates, and increasingly more of the energy is at shorter wavelengths. The Sun radiates hundreds of thousands times more energy than does Earth and mostly at shorter wavelengths. Solar radiation commonly is referred to as short wavelength, and radiation from Earth is called long wavelength.

Solar Radiation Received by Earth

Earth is a three-dimensional body, and it receives different amounts of solar energy at different latitudes. If we consider the planet as a whole, about 70% of the Sun's energy reaching Earth enters into Earth's climate system; this solar radiation is involved in accomplishing work (creating activity in and among Earth's systems; figure 9.2). About 30% is directly reflected back to space as short-wavelength radiation.

The Sun's energy heats Earth unequally. The equatorial area faces the Sun more directly than do the polar regions.

During the course of a year, the equatorial area receives about 2.4 times as much solar energy as the polar regions. The energy imbalance is lessened as Earth's spin helps set the heat-carrying oceans and atmosphere in motion. Gravity works to even out the unequal distribution of heat by pulling more forcefully on the colder, denser air and water masses. Circulation of the atmosphere and oceans is a major determinant of climate and weather all around the Earth.

Solar radiation is absorbed in massive amounts in the equatorial belt between about 38°N and 38°S latitudes (figure 9.3). Incoming sunlight strikes the surface at steep angles, allowing a high percentage of the energy to be absorbed,

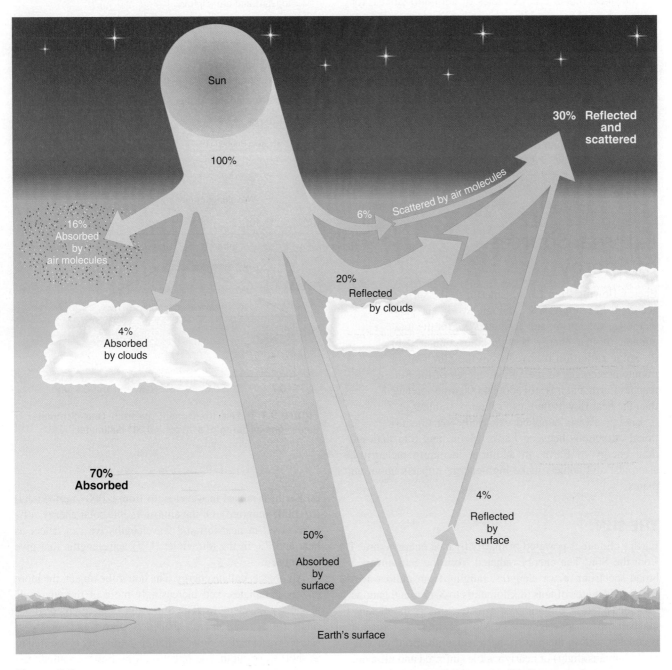

Figure 9.2 Solar radiation reaching Earth is 30% reflected and scattered and 70% absorbed into Earth's climate system.

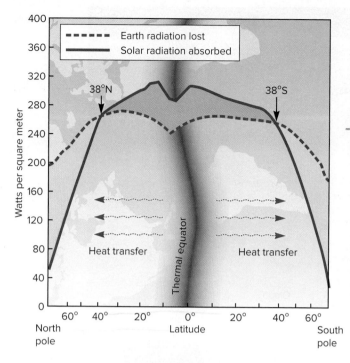

Figure 9.3 Energy radiated from Earth's surface and energy absorbed from solar radiation are plotted against latitude. Poleward from latitudes 38°N and 38°S, the energy loss deficit increases. Heat is transported poleward from the tropics via the ocean and the atmosphere, tending toward energy equilibrium.

Source: NOAA Meteorological Satellite Laboratory, Washington, DC.

especially in seawater. Polar latitudes receive far less of the Sun's energy because the incoming solar radiation arrives at a low angle, causing much to be reflected. Some of the excess heat of the low-latitude equatorial zone is transported to the high-latitude polar regions by moving seawater and air (figure 9.3). Despite the heat transfers, the high latitudes show a net cooling because the heat radiated back to space is greater than the amount received. The midlatitudes (such as the continental United States) are zones of energy transfer. Across the midlatitudes, cold air flows equatorward and hot winds move poleward, transferring much heat, especially carried in water vapor. The energy transported in moving air masses is often involved in severe storms.

Outgoing Terrestrial Radiation

Although Earth receives solar radiation every day, year after year, all this heat is not retained. An equivalent amount of heat is radiated back from Earth to space in the longer wavelengths of the infrared portion of the electromagnetic spectrum (figure 9.4). But all this energy is not returned to space as simply as it arrived. When the average surface temperature of Earth is calculated, the value is about 15°C (59°F).

This temperature seems reasonable from our day-to-day life experience. However, temperature measurements from satellites tell us that Earth is sending heat to space as if its average temperature were −16°C (3°F). Why is the average temperature of the surface of Earth 31°C (56°F) warmer than these measurements predict? Because of the **greenhouse effect.**

GREENHOUSE EFFECT

A glass-walled greenhouse (or your car with all the windows rolled up) admits incoming visible light—that is, solar radiation with short wavelengths (about 0.5 μm). The Sun's short-wavelength radiation warms objects inside the greenhouse (or car). Heat builds up inside the greenhouse and is given off as infrared radiation of longer wavelengths (about 10 μm). However, glass is rather opaque to long-wavelength radiation, and thus much of the outgoing radiant energy is trapped, producing a warm environment inside the greenhouse (or car).

Much of the solar radiation reaching Earth arrives in short wavelengths, where it is absorbed and thus raises the temperatures of land, water, and vegetation (see figure 9.2). Any body with a temperature above absolute zero (−273°C, or 0 K) radiates some heat toward cooler areas. When excess heat is radiated upward from Earth's surface, it is at long wavelengths that can be absorbed by gases in the atmosphere, such as water vapor (H_2O), carbon dioxide (CO_2), and methane (CH_4). Following this absorption, most of the energy is radiated back down to Earth's surface; this is the greenhouse effect, and it warms the climate of Earth (see figure 9.4). About 95% of long-wavelength radiated heat is held within Earth's climate, where it raises Earth's average surface temperature to about 15°C (59°F, or 288 K). Without the greenhouse effect, Earth's average surface temperature would be −16°C (3°F, or 257 K). In chapter 12 we examine the greenhouse effect on Earth twice: (1) the intense greenhouse in Earth's early history and (2) the human-increased greenhouse of the 20th and 21st centuries.

ALBEDO

Large amounts of incoming solar radiation are reflected by Earth rather than absorbed. Reflectivity is known as **albedo** and is usually measured as the percentage of solar radiation that is reflected (table 9.2). The albedo of Earth as a whole is 30%.

The polar zones receive less solar energy and are colder, thus helping snow and ice to form. But the presence of snow and ice increases albedo, making the cold polar climate even colder (figure 9.5).

The reverse is true during a warming cycle. Sea ice has high albedo, but as it melts, the exposed seawater absorbs solar energy. One of the biggest concerns about climate today is how rapidly Arctic Sea ice is disappearing and Arctic seawater is warming, thus adding significantly to global warming.

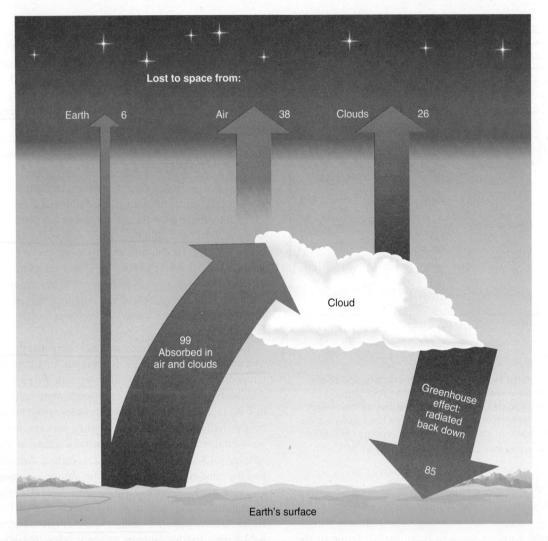

Lost to space from:

Earth 6   Air 38   Clouds 26

Cloud

99
Absorbed in
air and clouds

Greenhouse
effect:
radiated
back down

85

Earth's surface

Figure 9.4 Long-wavelength radiation to and from Earth's surface. The energy budget is not balanced here; see figure 9.11 for a balanced energy budget.

Source: Peixoto and Qort (1992).

TABLE 9.2	
Common Albedos of Earth Surfaces	
	Percent
Clouds —thick	50–85
—thin	5–50
Snow —fresh	70–90
Ice	60–70
Water —high Sun angle	5–10
—low Sun angle	10–50
Ground—bare	10–20
Sand	20–50
Grass	15–25
Forest	5–15
Cities	4–18

Source: Data from Smithsonosin Meteorological Tables (1966); Peixoto and Qort (1992).

The Hydrologic Cycle

About 24% of the solar radiation received by Earth is used to evaporate water and begin the **hydrologic cycle.** Evaporated water rises convectively, due to its lower density, up into the atmosphere, performing the critical initial work of the hydrologic cycle. The hydrologic cycle was in part recognized in the third century BCE (before the common era) in Ecclesiastes 1:7, which states: "Into the sea all the rivers go, and yet the sea is never filled, and still to their goal the rivers go." The Sun's radiant energy evaporates water, primarily from the oceans, which then drops on the land and flows back to the sea both above and below ground (figure 9.6). The same water, for the most part, has run through this same cycle, time and time again. Percy Bysshe Shelley described it in 1820 in "The Cloud":

I am the daughter of Earth and Water,
And the nursling of the sky;
I pass through the pores of the oceans and shores;
I change, but I cannot die.

The hydrologic cycle is a continuously operating, distilling-and-pumping system. The heat from the Sun evaporates water, while plants transpire (evaporate from living cells) water into the atmosphere. The atmospheric moisture condenses and precipitates as snow and rain. Some falls on the land and then is pulled back to the sea by gravity as glaciers or rivers, and via underground water flow. The system is more than 4 billion years old and will continue to operate as long as the Sun shines and water lies on the surface of Earth.

Water and Heat

Water has a remarkable ability to absorb heat (see In Greater Depth). It has the highest **heat capacity** of all solids and liquids (table 9.3), except liquid ammonia (NH_3). The amount of heat required to raise the temperature of water—its **specific heat**—is high. Sand and rock have smaller specific heats and heat capacities than water. Even though land heats to higher temperatures, it does so only to shallow depths because sand and rock are opaque to the Sun's rays; the resultant heat held is small compared to that absorbed by the same volume of water. Not only can water absorb more heat per unit volume, but solar radiation penetrates to depths of several hundred meters, where it is absorbed and carried away by moving water masses. The transmission of heat in flowing water, air, or deforming rock is the process of **convection.** The transmission of heat in a solid such as rock or metal occurs via **conduction** without the rock or metal moving. For example, stick one end of a metal rod into a fire

Figure 9.5 Albedo (reflectance) of solar radiation off snow is 70–90%.

Patrick Kelly/U.S. Coast Guard/U.S. Geological Survey.

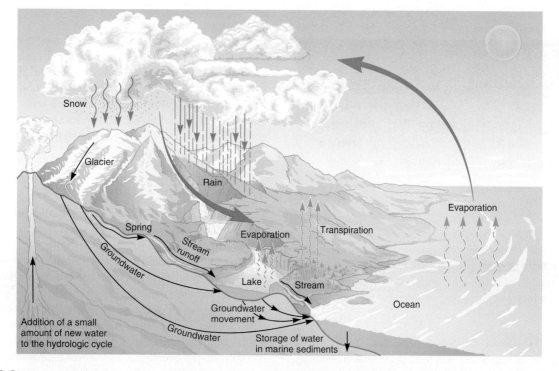

Figure 9.6 The hydrologic cycle. The Sun lifts water into the atmosphere by evaporation and transpiration. Atmospheric water condenses and falls under the pull of gravity. The water then flows as glaciers, streams, and groundwater, returning to the seas.

In Greater Depth

Water—The Most Peculiar Substance on Earth?

Humans tend to consider things that are common and abundant as ordinary and those that are uncommon and rare as extraordinary. The most common substance about the surface of Earth is water. It is so much a part of our daily lives that it seems ordinary. But such a bias will lead us astray. Water is a truly extraordinary chemical compound. Were it not such an odd substance, everything on Earth, from weather to life, would be radically different. Let's examine some of the characteristics of this most peculiar substance.

1. Water is the only substance on Earth that is present in vast quantities in solid, liquid, and gaseous states.
2. Water has the highest heat capacity of all solids and liquids except liquid ammonia. Because water stores so much heat, the circulation of water in the oceans transfers immense quantities of heat.
3. Water has the highest heat conduction of all liquids at normal Earth surface temperatures.
4. Water has the highest latent heat of vaporization of all substances. It takes about 600 calories to evaporate a gram of water. This latent heat is carried by water vapor into the atmosphere and is released when water vapor condenses to liquid rain. Much heat is transported about the atmosphere as air masses circulate.
5. Water has the second highest latent heat of fusion, exceeded only by ammonia. When ice melts, it absorbs about 80 calories per gram. When water freezes, it releases about 80 calories per gram.
6. Water is a bipolar molecule. The negative oxygen atoms and positive hydrogen atoms bond together, yielding a molecule

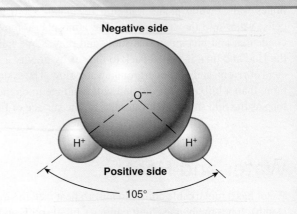

Figure 9.7 Water is a bipolar molecule exhibiting a negative and a positive side. This bipolarity greatly increases the activity of water.

with a negative and a positive side (figure 9.7). This + and − polarity allows water to readily bond with charged **ions.**
7. Water has the highest **dielectric constant** of all liquids. This property tends to keep ions apart and prevent their bonding, thus maintaining a solution. This is why water has been called the universal solvent.
8. Water has the highest **surface tension** of all liquids.
9. Water expands about 9% when it freezes. This is anomalous behavior. Usually, as a substance gets colder, it shrinks in volume and becomes denser. The maximum density for liquid water occurs at about 4°C (39°F). Imagine what lakes and oceans would be like if ice were heavier than liquid water and sank to the bottom.

TABLE 9.3
hermal Properties of Selected Materials

	Density ×	Specific Heat =	Heat Capacity
	(gm/cm³)	(cal/gm/°C)	(cal/cm³/°C)
Air	0.0013	0.24	0.00031
Quartz sand	1.65	0.19	0.31
Granite	2.7	0.19	0.51
Water	1.0	1.0	1.0

and heat moves through the rod, making it too hot to handle. Metal conducts heat readily but rock does not. In rock the daily fluctuation of heat energy supplied by the Sun reaches down only a meter or so.

CONVECTION

Convection is an important mechanism of heat transfer in many natural processes in the Earth, in the oceans, and in the atmosphere. Convective motion occurs as warmer, less-dense material rises and cooler, denser material sinks. Convection occurs in solids, liquids, and gases that flow. Examples are numerous. Hot rock inside Earth slowly flows upward toward the surface via convection. Hot spring water can flow upward through a cooler water body. Late afternoon air heated by the Sun and warm ground can rise convectively to form clouds. Air heated by fire will rise convectively in a smoky column.

WATER VAPOR AND HUMIDITY

Water vapor in the atmosphere ranges from near 0 to 4% by volume from place to place, but its importance in determining weather is great. The amount of water vapor in the air is measured as **humidity.** Relative humidity is the ratio of absolute humidity to saturation humidity. Saturation humidity is the maximum amount of water vapor that can exist in an air mass, and it increases with increasing temperature. If the temperature of an air mass is lowered without changing its absolute humidity, it will reach a relative humidity of 100% simply because at each lower temperature, a lower saturation humidity will apply. When relative humidity reaches 100%, excess water vapor condenses and forms

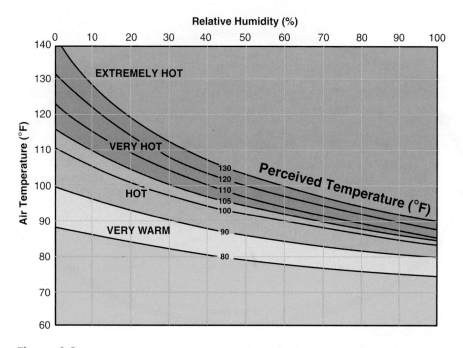

Figure 9.8 The heat index. As relative humidity increases, the temperature our body "feels" or perceives increases. Temperature values listed are for locations in the *shade*.

Source: National Weather Service.

liquid water; this temperature is the **dew point temperature** of the air mass.

Temperature and humidity combine to make us feel a perceived temperature. As humidity increases, we perceive that the temperature is rising. A heat index has been devised that combines temperature and relative humidity into a perceived temperature that more accurately describes what we feel on hot, humid days (figure 9.8).

LATENT HEAT

Water absorbs, stores, and releases tremendous amounts of energy when it changes phases between solid, liquid, and gas. Ice melts to water when supplied with about 80 calories of heat per gram (80 cal/g) of ice; this energy is stored as **latent heat** in the water (figure 9.9). Water evaporates to water vapor when it absorbs about 600 calories per gram of water; this energy is stored in water vapor as the **latent heat of vaporization.** Ice can change to water vapor directly without passing through a liquid state, but it requires both the 80 cal/g for melting and the 600 cal/g for evaporation. The process of changing directly from solid to gas is called **sublimation.** You probably have seen sublimation when dry ice (frozen CO_2) changes to vapor.

The latent heat carried in water vapor is released upon condensation to liquid water. The **latent heat of condensation** gives up the same number of calories as were absorbed during evaporation (figure 9.9). When water freezes to ice, it releases stored energy as the **latent heat of fusion;** it gives up the same number of calories as were absorbed during melting. Water vapor can change directly to ice in deposition; it gives up the latent heats of both melting and evaporation. You may have seen deposition of water vapor to ice in the buildup of "frost" in the freezer compartment of some refrigerators.

Water and water vapor are important for absorbing solar radiation, transporting heat about our planet, releasing that heat, and fueling convective processes. All of this heat transport helps prevent extreme ranges in temperature on Earth.

ADIABATIC PROCESSES

Adiabatic processes involve a change in the temperature of a mass without adding or subtracting heat. *Adiabatic cooling* occurs when an air mass rises and expands; *adiabatic warming* occurs when air descends and compresses.

A rising mass or parcel of air encounters progressively lower atmospheric pressures, causing the air to expand and cool at higher altitudes. The expanding mass of air (gas) has to do work to increase its volume and must extract energy from its own internal (thermal kinetic) energy, thus causing cooling. If the amount of heat in an air mass remains the same, then when the air mass expands, its temperature will drop adiabatically. The adiabatic cooling of rising air is a dominant cause of cloud formation.

When a parcel or mass of air sinks or descends into altitudes of greater pressure, the air is compressed and work is converted to heat, causing the air temperatures to rise adiabatically. The adiabatic warming of descending air is a common cause of hot, dry winds on Earth's surface.

LAPSE RATES

A rising air mass encounters lower pressure, expands, and cools. If the rising air mass is unsaturated with water, it will cool at a *dry adiabatic* **lapse rate** of about 10°C per km (~5.5°F per 1,000 ft) of rise.

As air cools, it has less ability to hold water vapor; thus, its relative humidity increases. When rising air attains 100% humidity, its dew point is reached, and excess water vapor condenses and forms clouds. The altitude where 100% humidity is reached is the **lifting condensation level.**

When water vapor condenses, it releases the latent heat it absorbed during evaporation. When latent heat is released into a rising air mass, it slows the rate of upward cooling to about 5°C per km (~2.7°F per 1,000 ft) of rise; this is known as the *moist adiabatic lapse rate.*

DIFFERENTIAL HEATING OF LAND AND WATER

Land heats up quickly, but the low heat capacity of rock causes the land to lose heat readily and cool down quickly. Water warms up more slowly, but its high heat capacity allows it to retain heat and cool down slowly.

As temperatures drop in the winter, the land cools quickly, but the ocean retains its warmth, causing warm, moist air to rise (figure 9.10a). Cool air over the land sinks toward the surface, forming a region of high-pressure air over the land. This cool, dry air flows out over the ocean.

As temperature rises in the summer, the land heats up quickly while the ocean warms up much more slowly. Hot, dry air forms over the land and rises, producing low-pressure air over the land (figure 9.10b). Cooler moist air above the ocean is drawn in to replace the warm air rising above the land. The moist air from above the oceans warms as it moves over the hot land, and then it rises and reaches colder levels of the atmosphere, where its water vapor condenses and falls as rain. This process creates the summer **monsoon** rains that are especially important in India and Bangladesh and to a much smaller degree in southwestern North America.

The atmosphere responds readily to the Sun's radiation; warm air rises and cool air sinks. But the vertical movement of air is small compared to its horizontal motion (i.e., wind). Air parcels want to flow horizontally from high pressure toward low pressure, seeking a balance or equilibrium condition.

Energy Transfer in the Atmosphere

Energy generated by nuclear fission within Earth flows to the surface constantly by conduction and, more importantly, by convection through hot rock, magma in volcanoes, and water in hot springs. Conduction and convection are commonly lumped together under the term **sensible heat.** Air and water receive even greater amounts of heat from solar radiation; the amounts received are measured as **insolation.** Some of this heat is transported as sensible heat in moving air and water.

The imbalances in heat received between the tropical and polar latitudes help cause ocean currents and winds that transfer heat from the tropics toward the poles. The ocean and the atmosphere act like gigantic heat engines that transfer energy around the world.

The evaporation of water to vapor (gas) requires heat. This heat energy is absorbed and stored in water vapor as latent (hidden) heat. When water vapor condenses, or changes back to a liquid such as rain or fog, it releases its stored latent

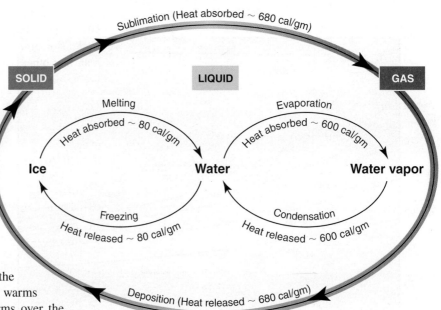

Figure 9.9 Water changing state from solid to liquid to gas absorbs heat. Water changing state from gas to liquid to solid releases heat.

heat. Although only a fraction of a percent of the water near Earth's surface is in the atmosphere, this water vapor is important because of the solar energy it holds, transports, and releases. On a broad scale, the latent heat of vaporization (evaporation) is an important factor in global climate, and on a local scale, it is the energy, the power—the "juice"—behind severe weather such as hurricanes and tornadoes.

After considering sensible heat and latent heat, we can now look at a global energy budget for the lower atmosphere and surface of Earth (figure 9.11). There is a balance between the amounts of energy entering and leaving the lower atmosphere as well as a balance at Earth's surface. On this grand world scale, the average annual temperature at and near the surface is relatively stable.

The atmosphere may be viewed as a heat engine that uses solar radiation to produce the mechanical energy of winds. The warm air masses of the equatorial region are less dense, rising buoyantly and flowing toward the poles, where they cool and sink. Cold, dense polar air masses flow away from the poles toward the equator. The rotation of Earth beneath its low-density fluid shell of atmosphere adds complexities to this simplistic model. The circulation of the atmosphere distributes heat around Earth.

Energy Transfer in the World Ocean

Where is the water on Earth? The oceans hold the greatest share: 97.2% of Earth's water covering 71% of Earth's

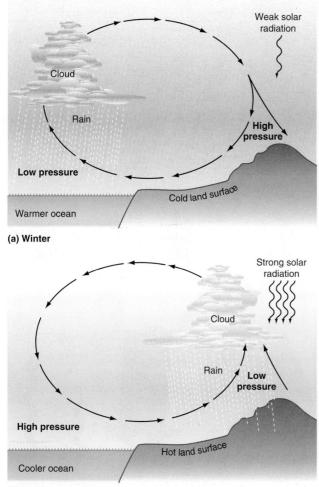

(a) Winter

(b) Summer

Figure 9.10 (a) In winter, rapid cooling of the land causes cool air that flows offshore to replace the air rising above the warmer ocean; this airflow produces rain on the ocean. (b) In summer, rapid heating of the land causes warm air that rises and is replaced by moist air drawn in from above the cooler ocean; this airflow brings rain to the land.

TABLE 9.4	
Where Is the Water?	
World oceans	97.2%
Glaciers	2.15
Groundwater	0.60
Lakes (fresh and saline)	0.017
Soil moisture	0.005
Atmosphere	0.001
Rivers	0.0001

surface. Of the remaining 2.8% of all water, 75% is locked up in glaciers (table 9.4). Satellite photos of Earth show the abundance of water in the oceans and moving as moisture in the atmosphere (figure 9.12). Water is uniquely qualified

to absorb and release solar energy. Water has an exceptionally high heat capacity, allowing storage of great amounts of heat (calories) in the ocean. Water gains and loses more heat than any other common substance on Earth. The unceasing motion of ocean water transfers its stored solar energy throughout the world. Winds blowing across the water surface cause circulation of surface water. Deep-ocean circulation is mostly driven by density differences caused by colder and/or saltier water masses sinking and flowing at depth in a global circuit. The ocean currents also are affected by Earth's rotation, which adds energy to water movements, and by continents, which block and divert the flow of warm ocean water, sending it up to colder latitudes. The solar energy stored in the oceans acts as a thermal regulator, strongly influencing global climate.

The global ocean is such an enormous body of water that the solar heat it stores influences weather around the world. On an average day, the amount of solar energy absorbed by the ocean and then radiated back as long-wavelength radiation would be enough heat to raise the temperature of the whole atmosphere almost 2°C (3°F). However, the atmosphere does not have the capacity to store this much additional heat. The total heat storage in the atmosphere is equivalent to that held in the upper 3 m (10 ft) of the ocean.

Layering of the Lower Atmosphere

Air is easily compressed. Gravity pulls on the atmosphere, causing each layer of air to press down ever more heavily on the air layer below it. Nearing the Earth's surface, air is increasingly compressed and becomes progressively denser. About 75% of the atmosphere lies within its bottom 10 km (6 mi).

Just like water and magma, air tends to flow from higher to lower pressure—from the higher pressures at the Earth's surface up through the progressively lower pressures high in the atmosphere. This tendency to flow upward from high to low pressure must overcome the opposing pull of gravity.

How can air near the surface overcome the pull of gravity and rise? The most common way is by adding heat to the air. The heated air rises by convection.

TEMPERATURE

The density of air decreases upward through the atmosphere, but temperature trends go through reversals, from cooling upward to warming upward. The atmosphere of Earth is separated into layers defined by temperature changes (figure 9.13). Earth's surface absorbs much solar radiation, causing average air temperatures to be highest near the ground. Rising from the surface, air temperature decreases with altitude up to a reversal called the **tropopause.** Below the tropopause is the lowest layer of the atmosphere, the

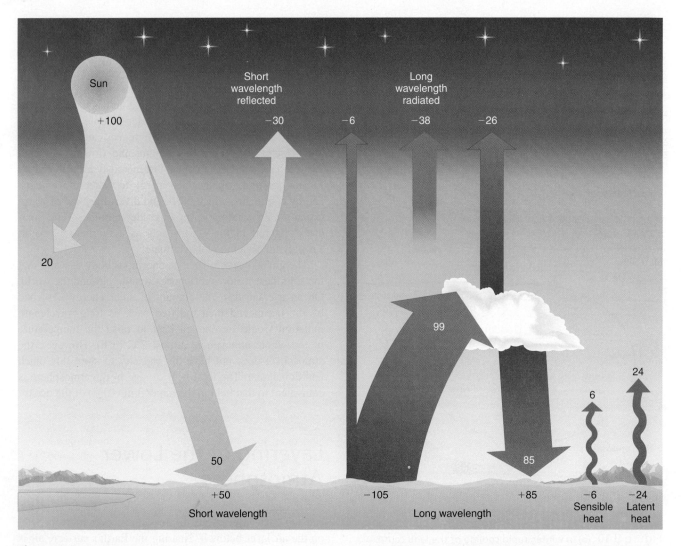

Figure 9.11 A global energy budget. The numbers balance for incoming and outgoing energy along the uppermost line in the atmosphere and along the ground surface.

Source: Peixoto and Qort (1992).

troposphere; most of the moisture and "weather" are found here. The troposphere ranges from about 8 km (5 mi) thick at the poles to 18 km (11 mi) at the equator. The troposphere is warmer at its base and colder above, thus creating a conditional instability as lower-level warm air rises and upper-level cold air sinks. The vertical contrasts in temperature set off a constant mixing of tropospheric air that is part of our changing weather pattern.

Above the tropopause, the cooling-upward trend reverses, and air begins to warm upward through the **stratosphere** (figure 9.13). The temperature inversion acts as a barrier or lid that mostly confines weather to the troposphere below. The stratosphere draws its name from its stratified condition where warmer air sits on top of cooler air; this is a stable configuration. The stratosphere is home to dry, ever-sunny conditions. The temperature trend reverses again at the stratopause as temperatures decrease upward through the lower **mesosphere.**

PRESSURE

Although barely perceptible to us in day-to-day life, air has weight. The atmosphere presses upon us at sea level with a pressure of 14.7 pounds per square inch (lbs/in²) or about 1,013 **millibars** or 101.3 **kilopascals.** Air pressure has been measured since 1643 using a mercury **barometer** (figure 9.14). Mercury is the only metal that is liquid at the daily temperatures we experience. The barometer is a tube with one end closed, filled with mercury, and turned upside down in a dish of mercury. The mercury in the tube rises and falls depending on the amount of air pressure on the dish. An air pressure of 14.7 lbs/in² is equivalent to a mercury level of 29.92 inches.

Winds

Air temperature differences create pressure differences, and pressure differences drive winds. Pressure within the

Figure 9.12 Satellite view showing the atmosphere and the ocean, two fluid masses in continual motion transporting energy about Earth.

© StockTrek/Getty Images RF.

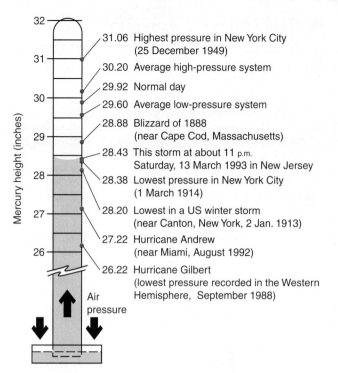

Figure 9.14 Air pressures recorded as mercury height (in inches) in a barometer. The barometer is a tube with its upper end closed; it is filled with mercury and turned upside down in a dish of mercury. The mercury in the tube rises and falls depending on the amount of air pressure on the mercury in the dish. Commonly, high pressure indicates clear weather, and low pressure means clouds and rain.

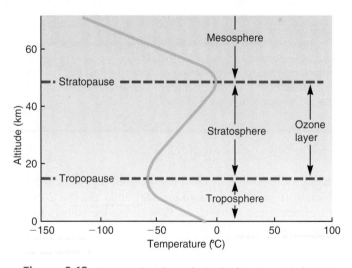

Figure 9.13 Cross-section through Earth's lower atmosphere. Ozone captures solar energy. The temperature inversion at the tropopause acts as a lid, holding moisture and "weather" in the troposphere.

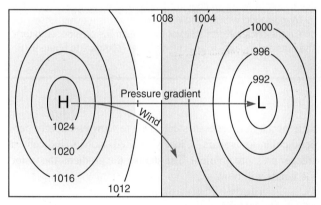

Figure 9.15 Map view showing that air flows along the pressure gradient from high to low pressure. The path of the wind is modified by the Coriolis effect. The high (H) and low (L) pressure air cells are measured in millibars.

atmosphere varies both horizontally and vertically. Near Earth's surface, air tends to flow from areas of higher pressure toward areas of lower pressure. It is the horizontal flow of air that we call *wind*. In regions where surface air pressure is low, the air tends to rise, or ascend; in areas where surface air pressure is high, the air tends to sink, or descend. Winds result from various forces.

PRESSURE GRADIENT FORCE

Air wants to flow along the pressure gradient from higher to lower pressure (figure 9.15). The **pressure gradient force** tries to equalize pressure differences. This force causes high pressure to push air toward low pressure. The high- and low-pressure air cells are defined by map lines called **isobars.** Each isobar line connects points of equal pressure. The

In Greater Depth

Coriolis Effect

Circulation of the atmosphere and oceans is inevitable because solar heat is received unevenly around the Earth. Earth rotates rapidly and sets cold and warm air and ocean masses into motions that are altered by topography. The velocity of rotation on Earth's surface varies by latitude from 465 m/sec (1,040 mph) at the equator to 0 mph at the poles (figure 9.16). Because there are different velocities at different latitudes, bodies moving across both latitudes and longitudes follow curved paths. This is the *Coriolis effect*, named for the French mathematician Gaspard Coriolis, who described it in 1835.

In the Northern Hemisphere, all moving masses sidle off to the right-hand side when viewed down the movement direction; in the Southern Hemisphere, moving bodies veer toward the left (figure 9.16). The magnitude of the Coriolis effect increases with the horizontal speed of the moving body and with latitude; it is zero at the equator. The Coriolis effect is important in determining the movement paths of ocean currents, large wind systems, and hurricanes; possibly important for a large thunderstorm; probably not important for individual tornadoes; but negligible for water draining down kitchen sinks or toilets.

The Coriolis effect causes winds to veer into arcuate (curved) paths. For example, look at the course hurricanes travel and see how prominently their paths curve (see chapter 11).

THE MERRY-GO-ROUND LIMITED ANALOGY

A good way to visualize the Coriolis effect is to go with friends to a local playground that has a merry-go-round, a large circular wheel that rotates horizontally around a pole. When the merry-go-round is spinning counterclockwise, visualize being above it and looking down onto it as an analogy for the Northern Hemisphere. As the merry-go-round whirls rapidly, which person is moving faster, the one in the center (North Pole) or the one on the outside (equator)? The outside (equator) rotates much faster (compare to figure 9.16).

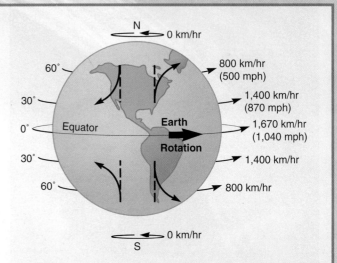

Figure 9.16 The Coriolis effect describes how air and ocean masses tend to follow curving paths because of the rotating Earth. Looking down the direction of movement (dashed lines), paths veer toward the right (solid line) in the Northern Hemisphere and toward the left (solid line) in the Southern Hemisphere.

If the person riding in the center tosses a ball directly at the rider on the outer rim, will the ball reach the targeted person? Probably not; the person will have rotated away. If you are standing on the ground and watching only the flight of the ball, it obviously moves in a straight line. But if you change your frame of reference and plot the path of the ball on top of the moving merry-go-round, the path curves to the right.

Now spin the merry-go-round in the opposite direction, in a clockwise pattern analogous to the Southern Hemisphere viewed from above the South Pole. All other factors are the same, but the ball tossed from the center to the outside now appears to curve to the left (compare to figure 9.16).

---

differences in air pressure cause air to flow essentially perpendicular to isobars. Closely spaced isobars indicate a steeper pressure gradient. The steeper the gradient, the faster the winds will blow.

Coriolis Effect

If the pressure gradient force were the only force acting on the air, then air would simply flow from high to low pressure—but there are other factors. Earth's rotation creates an apparent force known as the **Coriolis effect;** it deflects the paths of large, moving bodies to the right in the Northern Hemisphere. Ignoring other factors, the combination of pressure gradient force and Coriolis effect would cause air to flow parallel to isobars with low pressure on the left (see figure 9.15).

Friction

The flow of surface air is influenced by frictional drag with the ground, hills, trees, and such. Surface friction is a force acting in opposition to the pressure gradient force. When friction is added to pressure gradient force and the Coriolis effect, it causes winds to flow across the isobars (see figure 9.15).

ROTATING AIR BODIES

High-pressure air zones enclosed by roughly circular isobars (see figure 9.15) are called **anticyclones.** The air within high-pressure zones sinks downward, descending to the ground, where it flows outward (figure 9.17). In the Northern Hemisphere, the winds blow clockwise as the pressure gradient force causes air to flow outward and the Coriolis effect deflects the airflow to the right. The air flowing outward at the ground surface is replaced by sinking air from above. The descending air warms and usually creates dry conditions on the ground. In the Southern Hemisphere, the airflow is downward, but the winds blow counterclockwise.

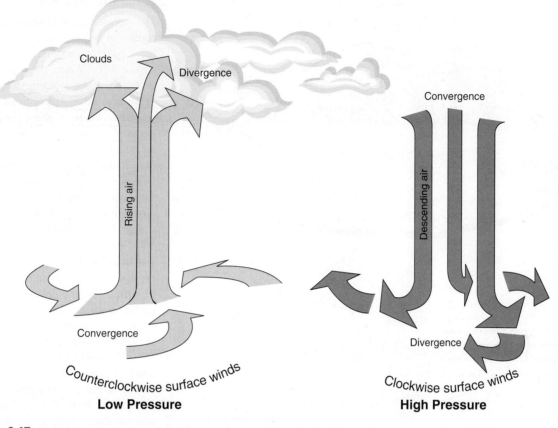

Figure 9.17 Air rises at a low-pressure zone in the Northern Hemisphere; it is fed by counterclockwise surface winds. Descending air at a high-pressure zone flows over the ground surface as clockwise winds.

Low-pressure air zones enclosed by somewhat circular isobars are called **cyclones.** In the Northern Hemisphere, counterclockwise-blowing wind at ground level flows into the center of the low-pressure zone, collides, and rises (figures 9.17 and 9.18). The surface inflow of winds toward the low-pressure core feeds a large updraft of rising air that cools, forming clouds and possibly producing rainy weather, along with an upper-level outflow of counterclockwise-moving air. In the Southern Hemisphere, the Coriolis effect causes the air to flow clockwise.

General Circulation of the Atmosphere

Earth's atmosphere circulates to distribute solar heat received in different amounts at different latitudes (see figure 9.3). The general circulation of Earth's atmosphere transports heat from the low latitudes around the equator to the high latitudes of the poles (figure 9.19). In simplest form, this redistribution of energy could be accomplished by one large convection cell flowing between the equator and pole in each hemisphere. Heated equatorial air would rise and flow toward the poles at upper levels, becoming progressively cooler until reaching the

Figure 9.18 Space shuttle photo in the Northern Hemisphere of a low-pressure air mass rotating counterclockwise.

© Royalty-Free/Corbis.

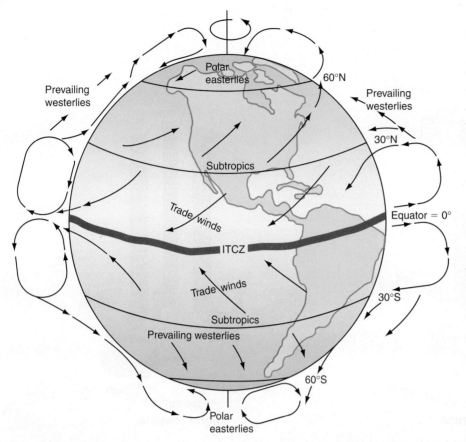

Figure 9.19 General circulation of the atmosphere. Warm air rises at the equator at the intertropical convergence zone (ITCZ) and sinks in the subtropics. Cold air at the poles sinks and flows toward the equator. The middle latitudes are transfer zones where warm air moves toward the poles and cold air flows toward the equator.

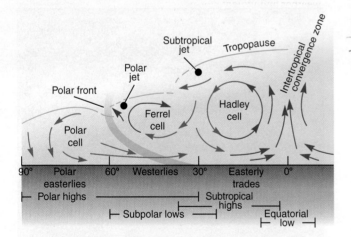

Figure 9.20 A cross-section showing average air circulation between the equator and a pole.

Source: Adapted from J. Eagerman, Severe and Unusual Weather, 1983; Van Nostrand Reinhold, New York.

poles, where it would sink and flow as cold air over the surface, and then become progressively warmer on its return to the equator. But the rapid rotation of the Earth complicates the process; it reduces the size of the convection cells and increases their number to three in each hemisphere (figure 9.20).

LOW LATITUDES

The intense sunshine received in the equatorial belt powers huge air circulation patterns known as **Hadley cells** (figure 9.20). Warm, moist equatorial air rises in giant columns to high altitudes, where it cools and drops its condensed moisture as abundant rain on the tropics. The rising limb of a Hadley cell carries an enormous amount of heat from low to higher altitudes.

After rising, the upper-elevation air is cooler, and after dropping rain on the tropics, it is drier. The air then spreads both north and south. Around 30°N and S latitudes, the now-denser air sinks at the Subtropical High Pressure Zone, warming adiabatically as it descends and returns to the surface as a warm, dry air mass. Some of the descending air flows poleward as westerly winds (westerlies), and some flows equatorward as the trade winds (figure 9.19).

The warm air descending in the Subtropical High has low moisture content; thus, precipitation is scarce. The warm, dry winds of the subtropical belts pick up moisture as they flow across the Earth's surface; between 30° and 20°N and S latitudes, they are responsible for many of the world's great deserts, such as the Sahara and Kalahari of Africa, the Sonora of North America, and the Great Australian.

Hadley cell circulations are completed where the trade winds from the Northern and Southern Hemispheres meet in the tropics at the **intertropical convergence zone** (**ITCZ**) (see figure 9.19). Water vapor picked up by the trade winds as they flowed over land and sea is carried upward at the ITCZ in the rising limbs of Hadley cells, where the water vapor condenses and contributes to the most intense rainfalls on Earth.

Solar radiation is the energy source that powers Hadley cell circulation. Because the amount of solar energy received in the Northern and Southern Hemispheres varies with the seasons, the location of the ITCZ moves also; it shifts northward during the Northern Hemisphere summer (June to September) and southward during the Southern Hemisphere summer (December to March).

HIGH LATITUDES

Cold, dense air in the polar region flows along the surface toward lower latitudes in both hemispheres (figure 9.19). At around 60°N or S latitude, the polar air collides with warmer air from the westerlies. The warmer air is lifted and rises via convection to help drive a thermal loop, the **Polar cell,** somewhat analogous to a Hadley cell (figure 9.20). The slightly warmer, rising air helps create *subpolar low-pressure zones*. This air rising around 60° latitude moves poleward at high altitudes in the upper troposphere. When the upper-level air reaches a polar region, the air is cold and dry and descends in *polar high-pressure zones*. The sinking air flows along the ground surface with a westward deflection due to the Coriolis effect, thus producing the winds of the *polar easterlies* in both hemispheres (see figure 9.19).

MIDDLE LATITUDES

The Hadley cell and the Polar cell are both thermally driven loops that operate pretty much as described. The middle latitudes have neither a strong heat source nor a strong cold sink to drive an air-circulation loop. The middle latitudes feature a secondary air circulation, an indirect circulation called the **Ferrel cell,** which exists between the Hadley and Polar cells and is driven by those adjoining cells (see figure 9.20).

In a Ferrel cell, surface air flows poleward from the subtropical highs toward the subpolar lows. These airflows in both hemispheres are deflected by the Coriolis effect to produce the wind belts called the *westerlies* (see figure 9.19). Overall movement of surface air is from about 30° to 60° latitude; however, the upper-level airflow is not well defined. When the westerly winds reach the 60° latitude regions, they converge with the polar easterly winds at the *polar front*. At the polar front there is no great vertical wall of rising air analogous to that at the ITCZ. Instead, some of the cold, dense polar air collides with and flows under the warmer air of the midlatitudes, helping form a complex pattern of regional air masses. The midlatitudes are a much more turbulent zone where competing tropical and polar air masses transfer their energies back and forth as the seasons vary, commonly creating severe weather conditions (see figure 9.20). The never-ending passage of high- and low-pressure air masses in the midlatitudes is largely unknown above the 60th and below the 30th latitudes.

Air Masses

Air masses are large bodies of air that have little horizontal variation in moisture content or temperature. The air masses that move across North America come mainly from several large source areas (figure 9.21). The polar air masses are cool to cold, whereas the tropical air bodies are warm to hot. Air masses that gather over land are dry, whereas those that form over water are moist. The dominant direction of air-body movement is from west to east under the influence of Earth's rotation. For the most part, air masses that build over the northern Pacific Ocean have a much greater chance of affecting the United States than those that form over the North Atlantic Ocean.

Fronts

The boundaries between different air masses are called **fronts.** (The term *front* came out of World War I, describing the battlefronts where armies clashed.) Many of the clouds and much precipitation in the midlatitudes are associated

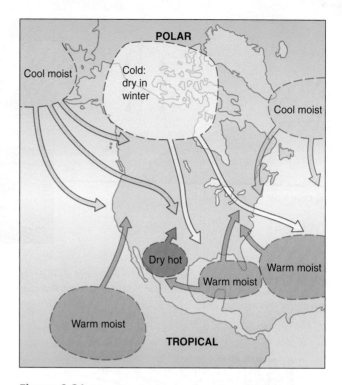

Figure 9.21 Map showing areas where large air masses acquire their temperature and moisture characteristics before moving about North America.

Source: Adapted from J. Eagerman, *Severe and Unusual Weather,* 1983; Van Nostrand Reinhold, New York.

with fronts. A front is a sloping surface separating air masses that differ in temperature and moisture content. Fronts often can be dry and cloud free, but sometimes they trigger severe weather and violent storms. The largest frontal system in the Northern Hemisphere separates cold polar air from warm tropical air along the polar front (see figure 9.20).

The advance of a cold front acts like a wedge, lifting warm air up to higher altitudes. When rising warm air is moist and unstable, it often forms tall clouds that may grow into thunderstorms (figure 9.22).

You can create and observe your own weather front at home. Open the door to the freezer compartment of your refrigerator and watch the cold air mass flow out. The cold air body moves into the warmer, moister air of the room, causing clouds.

A warm front leads the advance of a warm air mass in a flatter wedge. The lighter-weight warm air flows up and over a cooler air mass along a gentle slope. The warm air cools as it rises along the broad and gentle front, commonly producing widespread clouds and steady rain or drizzle (figure 9.23).

Jet Streams

Jet streams, or jets, are high-energy, elongated flows common in Earth's atmosphere and oceans. In the atmosphere, jet streams are relatively narrow bands containing high-velocity winds that flow from west to east at high altitudes. There are two main jet streams in both the Northern and

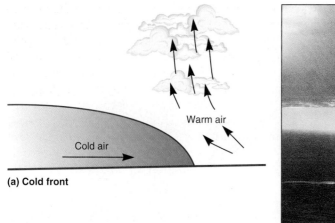

(a) Cold front

(b)

Figure 9.22 Schematic cross-section of a front and air mass. (a) A cold front wedges under warm air, forcing it upward. (b) A cold front moving south (to the right) near Sioux Falls, South Dakota, causes thunderstorm clouds and heavy rain showers.

University of Washington Libraries, Special Collections, John Shelton Collection, Shelton [no number].

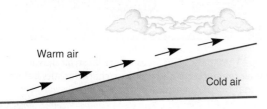

(a) Warm front

Figure 9.23 Schematic cross-section of a front and air mass. (a) A warm front runs up and on top of a cold air mass. (b) A warm front moving north (to the right) runs up a shallow, sloping cold air mass, producing clouds over the north shore of Lake Erie, Ontario, Canada.

University of Washington Libraries, Special Collections, John Shelton Collection, Shelton 1285.

(b)

Southern Hemispheres, a polar-front jet stream and a subtropical jet stream. They occur high in the atmosphere, near the tropopause, near the boundaries of the Hadley and Ferrel, and the Ferrel and Polar cells where there are significant differences in temperature (see figure 9.20). The jet streams occur at the sharp pressure drops created by the steep thermal gradients at the boundaries of the cells. These poleward flows are directed to the east by the spin of Earth and the Coriolis effect. Jet streams follow meandering paths that change readily.

The subtropical jet stream runs north of the tropics, at about 30° latitude at 10 to 16 km (6 to 10 mi) above the ground. The tropical atmosphere absorbs heat; the air expands and rises to begin circulating in a Hadley cell (see figure 9.20). Traveling up through a column of warm air, temperature decreases more slowly than moving upward through a column of cold air. The result aloft is that higher-pressure tropical air flows toward the lower-pressure polar air in the upper atmosphere, creating strong poleward flows of air. But these poleward airflows occur on a rapidly

spinning Earth, which turns them into belts of high-speed jet-stream winds from the west.

The most powerful and variable jet stream is the polar jet, which flows from west to east in meandering paths between latitudes 30° to 60°N (figures 9.24 and 9.25). Polar jet streams flow from west to east over the midlatitudes at elevations of about 7 to 12 km (4 to 7.5 mi). A polar jet is a belt of winds about 1,000 km wide (over 600 mi) and a few kilometers thick, flowing as fast as 400 km/hr (250 mph) in its central "core." A polar-front jet stream's path is ever-changing, like that of a meandering river. Meanders in the flow can bend so much that, locally, jet-stream flow directions may be to the north, south, or west. As the flow path twists and turns, it may cut off and abandon some flow loops, temporarily achieving a straighter west-to-east flow.

A polar-front jet stream also changes position with the seasons. In the Northern Hemisphere, it flows over Canada during the summer, when the warm air volume is greatest; during the winter, it migrates equatorward over the United States, closer to 30°N latitude, as the volume of Northern

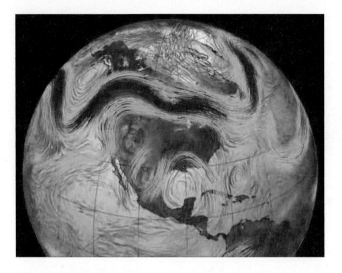

Figure 9.24 Polar-front jet stream at high latitudes during the winter season, 24 January 2012. Fastest winds are colored red.

NASA/Goddard Space Flight Center.

Hemisphere cold air increases to its maximum. During each hemisphere's winter, the atmospheric temperature contrasts between pole and equator are greatest, and each polar jet races its fastest. When polar jets reach speeds around 200 km/hr (120 mph), they have significant effects in moving heat and air masses, as well as in provoking storms.

The polar-front jet stream results from temperature differences, but its existence in turn influences the movement and behavior of warm and cold air masses. The polar jet flows from west to east, under the influence of Earth's rotation, in both the Northern and Southern Hemispheres.

Troughs and Ridges

The meanders in the polar-front jet stream can help create rotating air bodies. In the Northern Hemisphere, the bends in the west-to-east flow of the polar jet create areas of diverging air east of *troughs* of lower pressure, and regions of converging air east of *ridges* of higher pressure (figure 9.26). A trough in the jet stream refers to a bend that is concave northward, whereas a ridge is a bend that is convex northward.

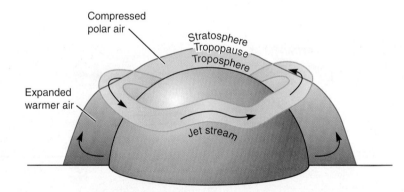

Figure 9.25 The polar-front jet stream flows at high altitudes in a meandering path from west to east. It exists where the expanded volume of warm air slopes down toward the compressed volume of colder air.

Adapted from J. Eagleman, *Severe and Unusual Weather,* 1983; Van Nostrand Reinhold, New York.

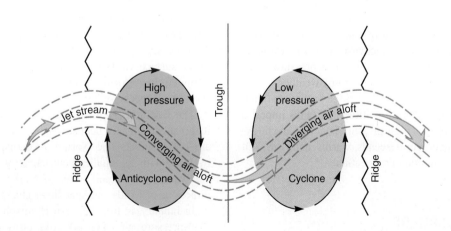

Figure 9.26 Influence of the polar-front jet stream aloft in creating counterclockwise cyclonic flow around a low-pressure zone and clockwise anticyclonic winds around a high-pressure core at the surface.

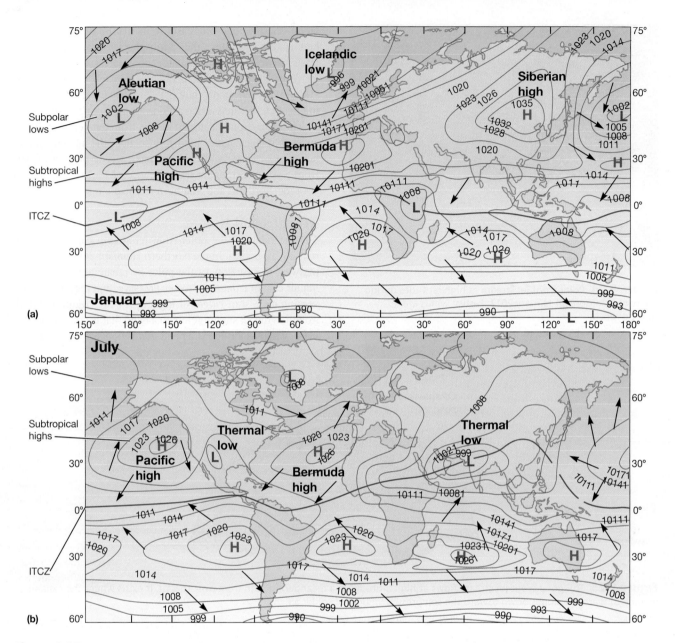

Figure 9.27 Average surface pressure (in millibars) and global wind patterns for January (top) and July (bottom).

The lower-pressure zone at a trough forms the core of a cyclonic circulation, a counterclockwise flow. In a cyclone, the winds at the surface flow into the low-pressure zone, then rise and cool, forming clouds and rainy weather.

The above process is reversed at a ridge in the jet stream (see figure 9.26). Here, the high-altitude jet helps impart a clockwise rotation about a high-pressure zone. At an anticyclone, upper atmosphere air descends, warms, and flows out over Earth's surface, creating dry weather.

OBSERVED CIRCULATION OF THE ATMOSPHERE

Air pressure and wind patterns on Earth show some consistency but also have significant variations by hemisphere and season (figure 9.27). Both hemispheres have subtropical high-pressure zones around 30° latitude, where air descends to the surface via Hadley cells (see figures 9.19 and 9.20). Subpolar lows form around 60° latitude, where polar air (Polar cell) meets midlatitude air (Ferrel cell) and rises.

The Southern Hemisphere is dominated by water with its great capacity for heat storage, so seasonal changes there are not as great. The Northern Hemisphere is dominated by land, with its smaller heat capacity, leading to significant variations in wind patterns. As the seasons change, the directions of winds and heat flows change.

In January, in the Northern Hemisphere winter, a strong high-pressure air mass of cold air known as the *Siberian high* influences Eurasian weather; an analogous, but smaller, high-pressure system exists over wintry North

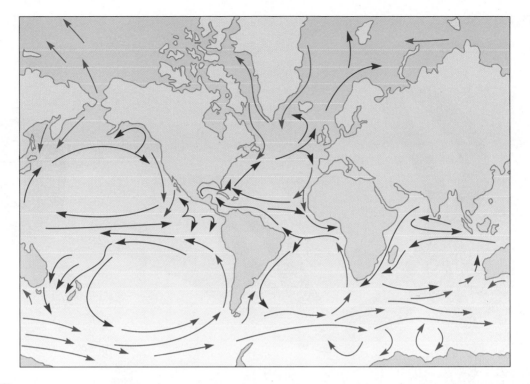

Figure 9.28 Circulation of the wind-blown surface waters of the oceans. Notice how the equatorial waters are deflected both northward and southward by the continents, thus sending warmer waters (red arrows) toward the poles. Also note that the only latitude not blocked by continents (60°S) has latitudinal flow; this is the Southern Ocean, and it encircles Antarctica. Blue arrows indicate cold water.

America (figure 9.27a). Cold air flows off the continents as increasing volumes of air rise in subpolar low-pressure zones known as the *Aleutian* and *Icelandic lows.*

In July, in the Northern Hemisphere summer, the Eurasian and North American continents have warmed, and hot air rises in thermal lows (figure 9.27b). This aids the strength of the high-pressure systems over the oceans known as the *Pacific* and *Bermuda highs.* The onshore flow of moist air brings monsoonal rains and hurricanes onto the land, especially in South Asia.

The seasonal shifts of the Pacific and Bermuda highs and the Aleutian and Icelandic lows are major determinants of the paths of jet streams and hurricanes.

General Circulation of the Oceans

The surface and near-surface waters of the oceans absorb and store huge quantities of solar energy, especially in the low latitudes. Some of this heat penetrates downward into deeper water in low latitudes when denser, salty, warm water undergoes turbulent mixing by tides and winds. Circulation of surface- and deep-ocean waters distributes heat throughout the oceans and affects climate around the world.

SURFACE CIRCULATION

The surface circulation of water through the ocean basins is mostly driven by winds (figure 9.28). Blowing winds drag on the sea surface and push against swells to move water. When the top layer of water moves, it drags on the underlying water layer, causing it to move, and so forth; this process moves water down to a depth of about 100 m (325 ft). The flow directions of surface water are modified by the Coriolis effect and by deflection off continents.

Surface circulation carries heat from the warm low-latitude waters toward the poles. For example, look at the North Atlantic Ocean in figure 9.28. Warm surface water is blown westward from Africa into the Caribbean Sea and Gulf of Mexico, where its westward path is blocked by land, forcing the seawater to escape northward along the eastern side of North America and over to Europe. The heat in this seawater adds significant warmth to the climate of northern Europe.

DEEP-OCEAN CIRCULATION

The oceans are layered bodies of water with less-dense water layers floating on top of progressively denser water layers. In general, the density of water is increased ~~~~~~~ ~~~~~~-ing its temperature or (2) increasing its conte~~~~~~~~ salts. The deep-ocean waters flow in an ov~~~~~~ lation called **thermohaline flow** (figure ~~~~~~

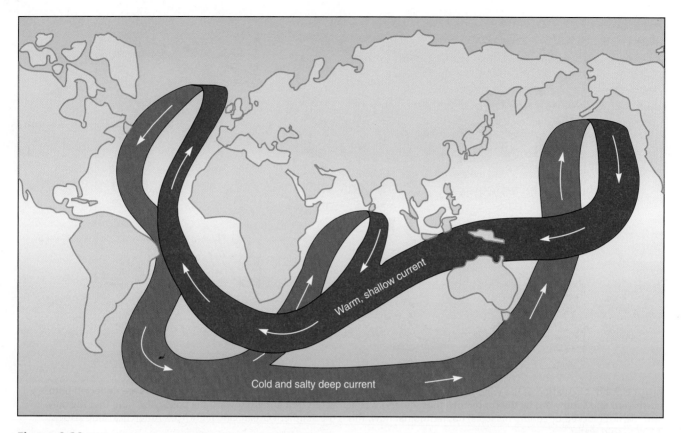

Figure 9.29 This schematic drawing of the ocean circulation system shows warm, shallow, wind-blown water moving into the North Atlantic, thus keeping Europe 5° to 10°C (9° to 18°F) warmer. Cooling in the Arctic increases ocean-water density, causing it to sink and move as thermohaline flow at depth southward out of the Atlantic Ocean. This ocean flow system is the equivalent of 100 Amazon Rivers.

thermohaline is formed from *thermo* for heat and *haline* for halite, the name for rock salt.

Seawater density is increased (1) at high latitudes, where water temperature is lowered, (2) in the Arctic Ocean and around Antarctica, where freezing seawater forms pure ice but excludes salts into the remaining seawater, and (3) in warm climates, where evaporation of water excludes salts into the remaining seawater. Most of the deepest and densest ocean water today forms in the high-latitude North Atlantic Ocean and in the Southern Ocean.

Summary

The amount of solar energy received by Earth varies over time and with geographic position. The cold water and air of the polar areas and the warm water and air of the equatorial belt are in motion aided by Earth's spin and the pull of gravity. Short-term changes in atmospheric conditions are known as weather; longer-term changes are referred to as climate.

Solar radiation is received in abundance between 38°N and 38°S latitudes. Much of this heat is transferred as latent heat in water vapor by winds that rise about the equator and then move aloft toward the poles. Cold polar air is dense and flows equatorward. The midlatitudes are the transfer zone between the equatorial and polar air masses; they have the most severe weather. From the equator to the poles, air circulates in three major cells: Hadley, Ferrel, and Polar.

Solar radiation is received by Earth in short wavelengths. As the surface warms, it radiates heat in longer wavelengths; much of this heat is trapped by atmospheric gases such as water vapor, carbon dioxide, and methane in the greenhouse effect.

Nearly one-fourth of the Sun's energy that reaches Earth is used to evaporate water to begin the hydrologic cycle. Under the pull of gravity, snow and rain fall back to the land and then run downslope as glaciers, streams, and groundwater until the water is returned to the ocean to complete the cycle. While in motion, ice, water, and wind act as agents of erosion that wear down the land and dump the debris into the ocean basins.

Water has a remarkable ability to absorb and release heat. Energy is absorbed in water vapor during evaporation (latent heat of vaporization) and released during condensation. Energy is absorbed in liquid water during melting (latent heat of fusion) and released during freezing.

Air masses flow from high pressure to low pressure along a pressure gradient, creating winds. Air rises at a low-pressure zone, cools, and forms clouds that may produce rain. Air descends within a high-pressure zone and flows out over the ground; this usually creates dry conditions.

The paths of large, moving air and water masses are deflected by the force of Earth's rotation. In the Northern Hemisphere, moving masses are pushed to the right of their movement direction; this is the Coriolis effect. In the Southern Hemisphere, moving objects are pushed to their left.

Where tropical air in the heat-expanded troposphere meets the cold, compressed polar air, a west-to-east high-level airflow exists—the fast-moving, polar-front jet stream. The position of the polar jet migrates across North America with the seasons. In the summer, it is over Canada; in the winter, it is over the central and southern United States, sometimes dipping to near the Gulf of Mexico. The polar jet stream plays a large role in moving heat and air masses and is involved in many severe weather situations.

Air masses vary in their temperature and water-vapor content. Different air masses do not readily mix; they are separated along boundaries called fronts. Much severe weather occurs along fronts.

Surface waters of the ocean absorb immense amounts of solar radiation. Surface waters are blown by winds and move in huge patterns. Ocean water made denser by cold or dissolved salts sinks and flows at depths in a global circuit.

Terms to Remember

adiabatic process   233
albedo   229
anticyclone   238
barometer   236
BCE   230
climate   227
conduction   231
convection   231
Coriolis effect   238
cyclone   239
dew point temperature   233
dielectric constant   232
Ferrel cell   241
front   241
greenhouse effect   229
Hadley cell   240
heat capacity   231
humidity   232
hydrologic cycle   230
intertropical convergence
   zone (ITCZ)   240

insolation   234
ion   232
isobar   237
jet stream   241
kilopascal   236
lapse rate   233
latent heat   233
latent heat of
   condensation   233
latent heat of fusion   233
latent heat of
   vaporization   233
lifting condensation
   level   233
mesosphere   236
millibar   236
monsoon   234
Polar cell   240
pressure gradient force   237
sensible heat   234
solar radiation   227

specific heat   231
stratosphere   236
sublimation   233
surface tension   232

thermohaline flow   245
tropopause   235
troposphere   236
weather   227

Questions for Review

1. What are the differences between climate and weather?
2. How does the amount of energy flowing from the interior of Earth compare to the energy received from the Sun?
3. Explain the greenhouse effect in some detail.
4. Name some important greenhouse gases.
5. What is the albedo (reflectance) of snow compared to that of bare ground?
6. Explain how the hydrologic cycle operates. What are the roles of the Sun and gravity?
7. What properties make water so peculiar?
8. What is latent heat? How is it important in moving energy through the atmosphere?
9. Is latent heat absorbed or released during: (a) melting, (b) freezing, (c) evaporation, (d) condensation?
10. Explain the adiabatic cooling that occurs in a rising air mass.
11. Is the lapse rate in a rising mass of air greater in dry or moist air?
12. Draw a cross-section through the atmosphere that defines troposphere, tropopause, and stratosphere.
13. Use the terms high pressure, low pressure, and pressure gradient force to explain how winds form.
14. Why is the Coriolis effect always to the right in the Northern Hemisphere and to the left in the Southern Hemisphere?
15. For the Northern Hemisphere, sketch a cross-sectional view of the Hadley, Ferrel, and Polar cells, using arrows to show air movements.
16. Why is there so much rainfall at the intertropical convergence zone (ITCZ)?
17. Draw a north–south vertical cross-section, and explain the cause of the polar-front jet stream. Why does its position vary across the United States and Canada during a year?
18. What are the relationships between high- and low-pressure zones and between cyclones and anticyclones?
19. Draw a map and explain the relationship between troughs and ridges in the polar-front jet stream and air circulation as cyclones and anticyclones.
20. What causes the surface circulation of water throughout the oceans?
21. What causes the global circulation of water deep within the ocean?

Questions for Further Thought

1. Why are air temperatures near the seashore cooler during the day and warmer during the night than in inland areas?
2. In the current discussions about global warming, does weather change sometimes get confused with climate change?

CHAPTER 10

External Energy

Tornadoes, Lightning, Heat, and Cold

It rained and it rained and it rained. Piglet told himself that never in all his life, and he was goodness knows how old—three, was it, or four?—never had he seen so much rain. Days and days and days.

—A. A. MILNE, 1926, WINNIE-THE-POOH

LEARNING OUTCOMES

Severe weather brings many phenomena that we must understand in order to protect ourselves. After studying this chapter, you should:

- recognize how cold can lead to hypothermia.
- be able to explain how an air-mass thunderstorm initiates, matures, and dissipates.
- understand how a supercell thunderstorm can generate tornadoes, rain, hail, and lightning all at the same time.
- be able to describe how tornadoes form.
- understand the weather conditions that produce tornado outbreaks.
- be able to explain what to do when a tornado warning is issued.
- understand how lightning forms, and how to protect yourself from being struck.
- recognize how prolonged heat can lead to hyperthermia.

OUTLINE

- Severe Weather
- Winter Storms
- How Thunderstorms Work
- Air-Mass Thunderstorms
- Severe Thunderstorms
- Tornadoes
- How Tornadoes Form
- Tornado Outbreaks
- Tornado Safety
- Lightning
- Heat

The terror of a tornado.
© Mike Agliolo/Science Source.

On 22 May 2011, the residents of Joplin, Missouri, were issued an official **tornado warning** by the National Weather Service at 5:17 p.m. Then, 24 minutes later, a monster tornado entered the densely populated part of the city with rotating winds speeding more than 320 km/hr (200 mph). The tornado spun across the ground for 35.6 km (22 mi) during 38 minutes of fury, destroying most things in its path (figure 10.1). On that day Joplin was hit by the 7th deadliest tornado in U.S. history; it killed 160 people directly and caused another death due to lightning during cleanup the next day.

Adopting a positive approach after the disaster, the citizens of Joplin established a committee to explore how to improve the livability of their city via its rebuilding. They want to create a better living environment by making changes such as widening sidewalks, turning some tornado-cleared land into hiking trails, and more.

Severe Weather

Weather kills. People drown in floods, are struck down by random bolts of lightning, are battered and drowned in hurricanes, are chased and tossed by tornadoes, and pass away during heat or cold waves. Severe weather causes about 75% of the yearly deaths and damages from natural disasters. The destruction wrought by storms and associated phenomena kills hundreds of people in the United States each year (table 10.1), more than are killed by earthquakes, volcanoes, and mass movements combined.

Severe weather is expensive. During the 32 years from 1980 to 2011, the United States suffered 134 weather-related disasters in which each event caused more than $1 billion in damages (figure 10.2). The losses from these 134 events

TABLE 10.1
Deaths Due to Severe Weather in the United States, 1977–2013

Event	Average Yearly Deaths
Heat	161
Winter storm/cold	102
Flood	96
Tornado	72
Hurricane	63
Lightning	55
Wind	53
	602 fatalities per average year

Source: National Weather Service (2014) Hazstats.

Figure 10.1 The impressive brick buildings of Joplin High School were a second home to 2,223 students until the tornado passed through. The damaged Joplin High School sign had the "OP" made into HOPE by adding duct tape.

John Daves/U.S. Army Photo.

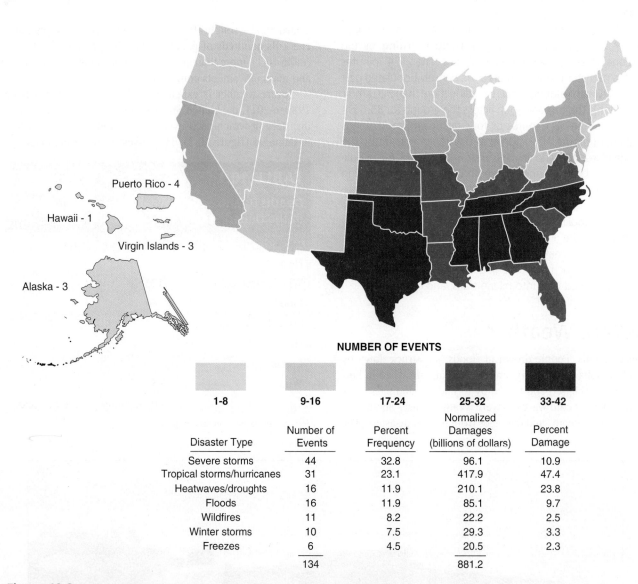

NUMBER OF EVENTS

	1-8	9-16	17-24	25-32	33-42

Disaster Type	Number of Events	Percent Frequency	Normalized Damages (billions of dollars)	Percent Damage
Severe storms	44	32.8	96.1	10.9
Tropical storms/hurricanes	31	23.1	417.9	47.4
Heatwaves/droughts	16	11.9	210.1	23.8
Floods	16	11.9	85.1	9.7
Wildfires	11	8.2	22.2	2.5
Winter storms	10	7.5	29.3	3.3
Freezes	6	4.5	20.5	2.3
	134		881.2	

Figure 10.2 Billion-dollar climate and weather disasters in the United States and its territories, 1980–2011.

Source: NOAA National Climatic Data Center.

exceeded US$880 billion. In 2012 and 2013, another 20 of the billion-dollar loss events occurred, bringing the 34-year total to 154 events, with losses exceeding US$1 trillion. The loss figures would be higher if disasters costing less than $1 billion per event were included.

The severe-weather categories discussed in this chapter are winter storms/cold, tornadoes, **lightning,** and heat (table 10.1). Hurricanes and floods, also listed in the table, are each discussed in separate chapters.

Winter Storms

Winter storms bring wind, snow, and ice with their life-draining cold. Winter storm times are tough times; they test your ability to live. Fatalities occur (1) from heart attacks while shoveling snow and pushing stuck cars, (2) when automobiles slide and collide, (3) when people slip on ice and fall, or (4) when folks get lost and freeze.

COLD

Our bodies maintain an internal temperature around 37°C (98.6°F), allowing our brains and internal organs to function efficiently. But when exposure to cold lowers our body temperature below 35°C (95°F), **hypothermia** may set in. If body temperature stays low, your heart, respiratory, and other systems cannot function properly. This can lead to failure of your heart or respiratory system—and death.

Warning signs of hypothermia include uncontrollable shivering, slurred speech, memory loss, incoherence, and apparent exhaustion. The first stage of treatment is to warm the body core (not the arms and legs) to protect the heart and respiratory system.

In past centuries, hypothermia mainly affected homeless people or others accidentally caught outside in the cold. Nowadays, with the popularity of cold-weather recreation, most of the people whose bodies cool into hypothermia are hikers, skiers/snowboarders, snowmobilers, mountain climbers, and the like. Many additional hypothermia deaths happen to people inside their disabled or stuck automobiles.

Wind chill commonly plays a role in hypothermia. Wind can strip heat from exposed skin and thus drive down body temperature. The National Weather Service has developed a Wind Chill Chart that states the apparent temperature felt on skin due to the effect of wind (figure 10.3). The wind speeds on the chart are measured at 1.5 m (5 ft) above the ground, at face level.

PRECIPITATION

When it is cold, clouds may drop precipitation as snowflakes or ice particles (figure 10.4). Falling snow and ice commonly pass downward through air warm enough to melt the snow/ice and turn it into rain. If the falling rain then enters a below-freezing air layer near the ground, the rain may refreeze into tiny (1.6 mm or 1/16 in) ice particles called **sleet.** If falling rain is not in subfreezing air long enough to freeze to ice, it may be super-cooled and become **freezing rain** that freezes upon impact on colder, subfreezing surfaces. If you are outdoors in freezing rain, it will not freeze on your exposed, relatively warm arm. Finally, what starts falling as snow may remain in cold air and build up on the ground as snow.

NOR'EASTERS

Much of the severe winter weather in the midlatitudes of the Northern Hemisphere occurs via cyclones: air masses rotating counterclockwise about a low-pressure core. In the northeastern United States and southeastern Canada, these cyclones can create potent storms known as **nor'easters.** When an Atlantic low-pressure system moves up the eastern U.S.–Canada coastal region, its eastern or seaward side is supplied with warm and moist tropical air brought into the cyclone by southeast winds. On its western or landward side, the surface winds blow in from the northeast, yielding the name

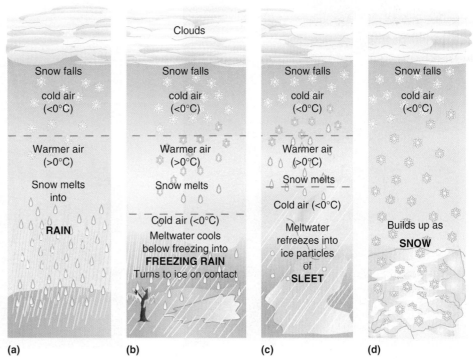

Figure 10.3 Wind chill chart. Combine the thermometer temperature (at top) with the wind speed (on left side) to find the temperature felt on your skin.

Source: Chart from National Weather Service.

Figure 10.4 Snowflakes fall and then:
(a) melt into RAIN;
(b) melt and then cool below freezing, becoming FREEZING RAIN;
(c) melt and refreeze as SLEET; or
(d) stay as SNOW.

nor'easters. In winter, when these cyclones collide with cold Arctic air, they can cause death and destruction via heavy snowfall, hurricane-force winds, and blizzards.

The "Storm of the Century," 12–15 March 1993
Shortly before spring began in 1993, an immense cyclone moved in and covered eastern North America from Canada

In Greater Depth

Doppler Radar

In 1842, the Austrian physicist Christian Doppler explained an effect now named after him. The *Doppler effect* describes the changes in observed frequencies of sound, electromagnetic, or other waves caused by the relative motion between the source and the observer. We experience this effect when an emergency vehicle (source) is moving near us (observer) with its siren blaring. The frequency of

Figure 10.5 Doppler effect. Waves emitted by a source (red dot) moving to the left. Wave frequency is higher on left (ahead of moving source) than on the right (behind).

sound we receive (compared to emitted frequency) increases as the emergency vehicle approaches us, is identical as it passes by, and decreases as it pulls away from us (figure 10.5). In similar fashion, when electromagnetic waves reflect off a moving object, the light shifts to shorter wavelengths as the object moves toward us, and to longer wavelengths as it moves away from us (see figure 9.1).

Doppler radar allows us to measure velocity between two objects. Doppler radars were first developed for the military, but have since been adapted for use in police radar guns to catch speeders, for measuring velocities in sports, and for describing weather systems. Weather-radar systems emit pulses of electromagnetic energy. Some of that energy is reflected back from raindrops and ice particles. Radar returns are commonly converted to colors ranging from blue (light precipitation) to magenta (extremely heavy precipitation). As precipitation increases, so does reflectivity (figure 10.6a). The radar can also measure the *radial velocity* of targets moving toward (blue-green) or away from (red) the radar site (figure 10.6b). These data not only indicate intensity of rainfall, but can also show wind speeds and regions of strong rotation that could spawn tornadoes. The advance warnings provided by Doppler radar have saved many lives.

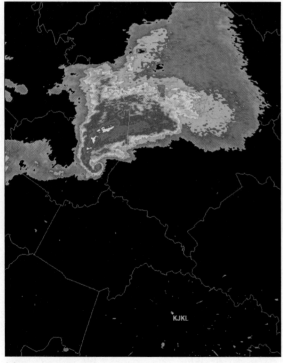

(a)

(b)

Figure 10.6 Doppler radar images of the weather system that spawned the deadly West Liberty, Kentucky, tornado, 2 March 2012. (a) The intensity of the precipitation is shown by colors increasing from blue (light precipitation) through green to yellow to orange and then red and magenta (extremely heavy precipitation) in the central area. Note the counterclockwise rotation. (b) The radial velocity in left center of the image shows different parts of the storm moving toward (blue-green) and away (red) from the KJKL radar site, indicating the intense rotation of the storm.

NOAA Radar image from Jackson, KY (KJKL) of a supercell thunderstorm using GR2 Analyst Software, from Michael Landin and Ross Lazear at the University at Albany, SUNY.

to Florida and on over Cuba (figure 10.7). Between 12 and 15 March, normal life was put on hold for most people living near the path of the cyclone. Winds gusted to more than 160 km/hr (100 mph), driving snow and sleet into the eastern seaboard and sending shallow seas surging into coastal communities. When the cyclone passed, 270 people had died from Cuba to

the Maritime Provinces of Canada, 48 sailors had been lost, and damages exceeded US$6 billion. The hardest-hit area was Florida, where more than 50 tornadoes spun out of the storm and killed more people than Hurricane Andrew did in 1992.

Why did such a huge winter storm hit such an unusually large area so late in the season? How did it combine some

Figure 10.7 This "storm of the century" covering eastern North America killed 270 people, with another 48 sailors lost at sea, and caused more than $6 billion in damages.

NOAA.

The scene was set with an unusually strong cyclone in the Gulf of Mexico; it caused big trouble with its warm, moist air and line of violent thunderstorms rotating around it. A trough in the jet stream interacted with the surface low and created enhanced rotation that drew in a fast-moving mass of frigid arctic air from the north as well as a rainy and snowy east-moving air mass from the Pacific Ocean. The interactions among the three air masses began in Florida and the powerful cyclone rode up the eastern seaboard with the jet stream, savaging everything in its path (figure 10.7).

This late-winter cyclone was more intense than some hurricanes and was even bigger than the legendary blizzard that struck the northeastern United States on the same days of March in 1888. The 1888 blizzard brought wind gusts up to 135 km/hr (85 mph) and snowdrifts 6 m (20 ft) deep. In New York City, the heavy snow immobilized the city, leaving about 400 people dead.

BLIZZARDS

One of winter's most unpleasant events is the blizzard. In an average year, as many people are killed by blizzards as die during tornadoes. Blizzards occur when strong, cold winds blow at least 56 km/hr (35 mph) and falling or blowing snow reduces visibility to less than 400 m (0.25 mi) for at least three hours. The winter storm may travel slowly, although its winds blow rapidly (figure 10.8). A blizzard can occur without snowfall; on freezing days, strong winds can pick up and blow snow dropped by an earlier storm. Fatalities occur when people become lost in "whiteout conditions."

Northeastern United States, 6–8 January 1996

A powerful Canadian blizzard blew into the northeastern United States, dumping record snowfalls in Ohio, Pennsylvania, West Virginia, and New Jersey. Wind speeds commonly exceeding 80 km/hr (50 mph) blew blinding attacks of snow. The blizzard killed 154 people; 80 of those fatalities occurred in Pennsylvania. In this case, the bitter conditions of the blizzard were immediately followed by warm weather with moderate to heavy rains that melted the record snows, unleashing destructive floods. In a few days, the region was pushed from one extreme to another, resulting in 187 deaths and US$3 billion in damages.

Figure 10.8 A blizzard blows through Negaunee, Michigan.

NOAA/NWS.

of the worst aspects of both a **blizzard** and a hurricane? The weather map for 12 March 1993 shows a large trough in the jet stream and three air masses migrating toward it. The collision of two of these air masses would have caused a significant storm, but the conflict involving all three created a "storm of the century."

ICE STORMS

Ice storms bring large volumes of freezing rain—that is, rain super-cooled below 0°C (32°F) that freezes on impact on sub-freezing surfaces. Coatings of ice a few inches thick add so much weight to tree limbs, power lines, and even house roofs that they may snap and collapse (figure 10.9).

Figure 10.9 Freezing rain built up on powerlines in Elora, Ontario, Canada in the early 1900s.

NOAA/Department of Commerce.

Ice Storm 98, 5–9 January 1998

Ice Storm 98 was the most expensive natural disaster in the history of Canada. During 80 hours of freezing rain, at least 25 people died of hypothermia and US$7 billion in damages occurred in Ontario, Quebec, New Brunswick, and Newfoundland as the power system collapsed beneath the weight of the clinging ice. Ice Storm 98 also severely affected Maine, New Hampshire, Vermont, and upstate New York. Millions of trees were split and downed by the ice as well. The destruction included 130 major power transmission towers, 30,000 wooden utility poles, and 120,000 km (74,000 mi) of power lines and telephone cables. In the dead of winter, hundreds of thousands of people were without power for up to four weeks. The damage to the Canadian life-support systems was so great

that repair was impossible; they had to be rebuilt. Due to our dependence on fragile power lines, the disruption of lives was greater in 1998 than if the same storm had occurred in 1898.

LAKE-EFFECT SNOW

Lake-effect snow can produce huge volumes of snow. World-famous sites for lake-effect snow are the southern and eastern shores of the Great Lakes in the United States and Canada. These snows fall after cold, dry air moves long distances over the warmer waters of the lakes and then crosses the downwind shoreline (figure 10.10). Lake-water temperatures may be only 1°C (34°F), but an Arctic air mass blowing across the water may be −18°C (0°F). The warmer surface waters of the lakes add heat and water vapor to the colder air mass. Note on figure 10.10b that the cold air moves across the lake for a long distance (**fetch**) before becoming saturated with the water vapor that condenses to form clouds. When the clouds reach the downwind shore, they encounter colder earth with rough land surfaces that increase friction, causing the air to slow, rise, and drop huge volumes of snow.

Lake-effect snow is common within 40 km (25 mi) of shorelines. For example, the southern shores of Lake Erie commonly receive thick blankets of lake-effect snow in cities ranging from eastern Cleveland, Ohio, to Erie, Pennsylvania,

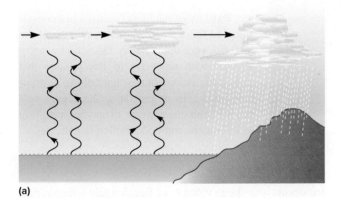

(a)

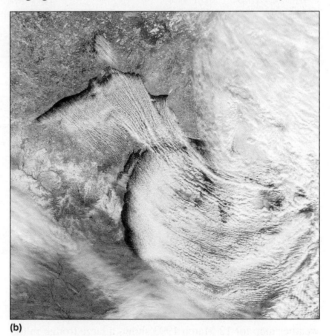

(b)

Figure 10.10 Lake-effect snow. (a) Cold dry air blows over warmer water. Rising water vapor forms clouds that move over cold, rough land, causing lake-effect snow to fall. (b) Cold, dry air from the northwest (upper left) blows over the warmer waters of Lake Superior and Lake Michigan, picks up water vapor, and forms clouds that drop lake-effect snow on the downwind sides of the lakes, 5 December 2000.

(b) NASA images courtesy the MODIS Rapid Response Team at NASA GSFC.

(a) Early: growth of cumulus clouds

(b) Late in the mature stage

Figure 10.11 Stages in thunderstorm growth.
(a) © PhotoLink/Getty Images RF (b) © Corbis RF.

to Buffalo, New York. These snowfalls are most likely to occur in late fall or early winter when lake-water temperatures are warmer. When lakes freeze over, lake-effect snows are no longer possible.

The "lake"-effect phenomenon occurs around the world, including on the downwind shores of some bays and seas. Another U.S. example occurs over the Great Salt Lake in Utah, where the lake effect helps create the deep, dry, powdery snows of the Wasatch Range.

How Thunderstorms Work

Thunderstorms are tall, buoyant clouds of rising moist air (figure 10.11a). They generate lightning and thunder, commonly with rain, gusty winds, and sometimes hail.

Air temperature normally decreases upward from the ground surface through the troposphere. The degree of atmospheric instability increases as the temperature differences increase between warm bottom air and overlying cool air. Warm, low-altitude air is less dense and it tends to rise upward by convection. Once convective upwelling begins, the warm air mass continues to rise as long as it is less dense than the surrounding air.

The warm air may rise high enough to pass through the lifting condensation level, allowing condensation of water vapor to begin and **cumulus** clouds to build. Once condensation is occurring, the rising cloud also is fueled by a large and important energy source—release of the latent

heat of condensation. This latent heat provides fuel to help form thunderstorms, tornadoes, hurricanes, and other severe weather.

The continuing rise of warm air builds clouds up to 20 km (12 mi) high. These are **cumulonimbus** clouds (figure 10.12b).

LIFTING OF AIR

Thunderstorms can develop from air lifted high by different mechanisms. In *convectional lifting,* air heated near the ground rises vertically due to buoyancy and forms local clouds and air-mass thunderstorms (see figures 10.11 and 10.12).

In *frontal lifting,* when air masses collide at frontal boundaries, the warmer, less dense air mass rises to form clouds, including lines of severe thunderstorms (see figure 9.22).

In *orographic lifting,* an air mass flowing up a steep land surface or mountain expands as it rises and cools adiabatically, thus raising its relative humidity; this can create clouds, including thunderstorms.

Air-Mass Thunderstorms

Air-mass thunderstorms are the most common; they occur locally, are the least destructive, and last for short lengths of time. They form most frequently in tropical locations where sunshine and moisture are abundant all year. In the midlatitudes, they form most commonly during late afternoons in the summer when surface heating is significant.

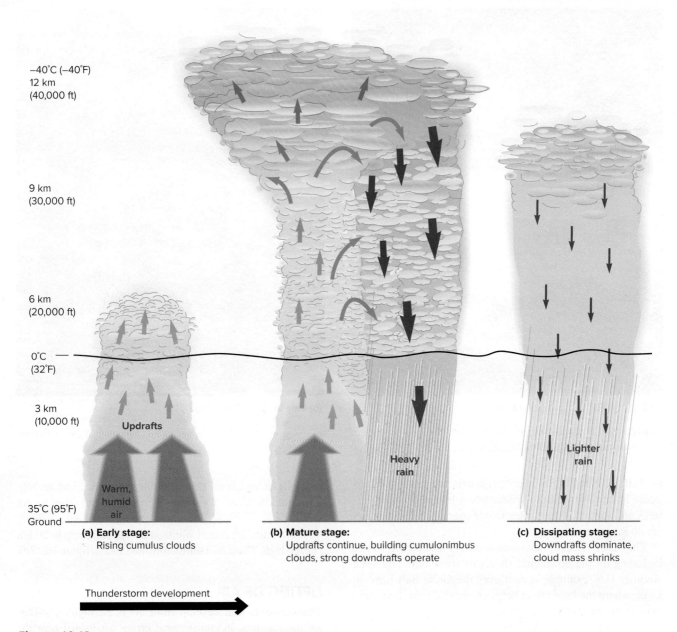

−40°C (−40°F)
12 km
(40,000 ft)

9 km
(30,000 ft)

6 km
(20,000 ft)

0°C
(32°F)

3 km
(10,000 ft)

Updrafts

Warm, humid air

35°C (95°F)
Ground

Heavy rain

Lighter rain

(a) Early stage:
Rising cumulus clouds

(b) Mature stage:
Updrafts continue, building cumulonimbus clouds, strong downdrafts operate

(c) Dissipating stage:
Downdrafts dominate, cloud mass shrinks

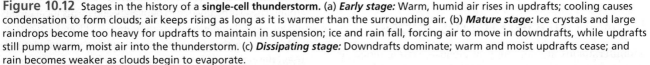

Thunderstorm development

Figure 10.12 Stages in the history of a **single-cell thunderstorm.** (a) *Early stage:* Warm, humid air rises in updrafts; cooling causes condensation to form clouds; air keeps rising as long as it is warmer than the surrounding air. (b) *Mature stage:* Ice crystals and large raindrops become too heavy for updrafts to maintain in suspension; ice and rain fall, forcing air to move in downdrafts, while updrafts still pump warm, moist air into the thunderstorm. (c) *Dissipating stage:* Downdrafts dominate; warm and moist updrafts cease; and rain becomes weaker as clouds begin to evaporate.

It is a common experience to feel the daily warming of Earth's surface and watch warm, moist air rise by convection and build thunderstorms in a three-stage cycle (figure 10.12).

Most individual thunderstorms form on sunny days in the late afternoon or early evening, when temperatures of the ground surface and lower troposphere are the highest. A thundercloud begins with an initial updraft of warm, moist air, perhaps aided by wind pushing up a hill slope, by surface-wind convergence, or by frontal lifting. The early stage of thunderstorm development requires a continuous supply of warm, moist air rising by convection.

The parcel of moist air cools as it rises and expands causing the moisture to condense into cumulus clouds (figures 10.11a and 10.12a). The condensation of water vapor into water drops releases latent heat, adding warmth to the air. The area below the thundercloud is a low-pressure zone.

Warm air rises until it can rise no farther, forming a cumulonimbus cloud. Water drops may grow or freeze. When the amount of ice crystals and water drops becomes too heavy for the updrafts to support, some upper-level precipitation begins. Falling rain causes downward drag,

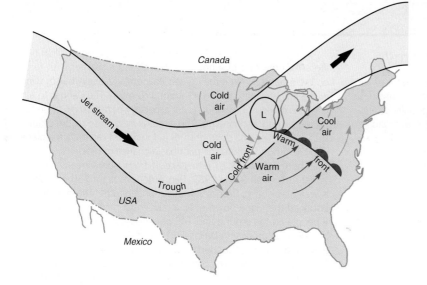

Figure 10.13 A trough in the jet stream helps cause a large-scale frontal cyclone formed of horizontally rotating winds around a low-pressure core. The cold front sweeping down from the north wedges beneath warm air, lifting it to form a line of thunderstorms. The warm front coming from the southwest flows up a gentle slope on top of the cooler air mass to the east to form widespread clouds and rain.

developing downdrafts and pulling in cooler, dryer air surrounding the tall cloud mass. An air-mass thunderstorm is basically an "up and down" storm where warm, moist air rises up, then cools and the condensed moisture falls back down as rain. In the mature stage of a larger, more complex thunderstorm, the cloud-mass top commonly spreads out as an icy cap or anvil (figures 10.11b and 10.12b). Updrafts and downdrafts operate side by side as warm, moist air is rising high at the same time that cool, dry air is descending rapidly. This is the most violent stage of the thunderstorm, with gusty winds and rain.

The dissipating or self-extinguishing stage is reached as precipitation falls into the core of the updraft. The cool air created by evaporation of the falling precipitation, and the cool air dragged down by the precipitation itself, reach the surface and expand as outflow cutting off the supply of warm, moist air at the surface (figure 10.12c). Without new moisture, the tall thundercloud mass evaporates in the surrounding dry air.

Severe Thunderstorms

Severe thunderstorms occur when any of these three conditions are met: (1) winds exceed 50 knots (93 km/hr or 58 mph); (2) hail is 25 mm (1 in) in diameter or larger; or (3) a tornado is reported. Severe thunderstorms commonly form where weather-system fronts collide, sending up numerous large thunderclouds along a 100 to 1,000 km (60 to 600 mi)

long zone. The large areas covered by severe thunderstorms allow updrafts and downdrafts of air to occur simultaneously and to reinforce each other.

Surface cyclones in the Northern Hemisphere commonly form at locations east of the upper-level trough in the jet stream. Often extending from the center of these cyclones are a cold front heading south and a warm front heading east (figure 10.13). A cold front may wedge under a warm, moist air mass, sending it upward to form thick clouds and, if the air is unstable enough, a line of thunderstorms.

Cyclonic airflow also characterizes individual thunderstorms within a large frontal cyclone, but the radius of an individual thunderstorm is much smaller (e.g., 10 km versus 1,000 km). Although the thunderstorm radius is only about 1% that of a frontal cyclone, its momentum is greater, and its wind speeds are higher. Within a cyclonic thunderstorm, an even smaller-radius rotation, a **tornado,** may spring forth. The radius of a tornado funnel cloud may be only a few percent that of a thunderstorm, but again, the smaller radius of the spinning tornado brings its mass closer to the axis of rotation, thus causing even higher wind speeds.

SUPERCELLS

A **supercell thunderstorm** is a particularly violent type of severe thunderstorm that forms from a huge updraft of air. A supercell can cover an area with a 20–50 km (12–30 mi) diameter (figure 10.14) and may move in large-scale rotation lasting 2–4 hours. The rotation within the supercell is known as the **mesocyclone.** This large-scale rotation of the updraft creates a spiraling column of air, or **vortex,** rotating around a vertical axis. Complexities within the huge thunderstorm, such as updrafts, downdrafts, and wind shear, can tilt the supercell, allowing different processes to develop side-by-side (figure 10.15b). Rain and **hail** commonly fall from the leading part of the supercell, while tornadoes may spin out of the trailing portion. Many of the most powerful tornadoes develop from the mesocyclones within supercell thunderstorms.

THUNDERSTORMS IN NORTH AMERICA

The distribution of thunderstorms in the United States shows a nonuniform pattern (figure 10.16). Florida has the most thunderstorms; west-central Florida averages more than 100 days of thunderstorms per year. The low-lying Florida peninsula is a place for air masses above the warm waters of the Atlantic Ocean to meet those from the Gulf of Mexico. The converging warm, moist air masses are commonly forced upward, triggering the ubiquitous thunderstorms.

In Greater Depth

Downbursts: An Airplane's Enemy

In the mature stage of a thunderstorm, violent downdrafts of air may occur (see figure 10.12). These sudden, strong downrushes of wind are known as **downbursts;** they are especially strong downdrafts of air that can hit the ground and spread outward in all directions with straight-line winds up to 240 km/hr (150 mph). Downbursts may be referred to as *microbursts* if they affect an area with less than a 4 km (2.5 mi) diameter; downbursts that affect larger areas are called *macrobursts*.

Downbursts may be dry or wet. A thunderstorm with abundant rainfall concentrated in shafts may drop a wet macroburst because of its heavy load of water and hail. Or some air parcels surrounding a thundercloud may become entrapped within the cloud where the water droplets and ice crystals in the trapped air readily evaporate, cooling the cloud further, increasing its density, and thus triggering a downburst. When a wet macroburst hits the ground, it does so with enough violence to leave a splattered pattern of damage.

Airplane safety also is challenged during downbursts by a radical change in horizontal wind direction known as *horizontal wind shear.* As a plane moves into a downburst, the winds shift from head winds needed to maintain lift of the airplane, to tail winds as the plane exits the downward-rushing wind area.

A crude analogy can be made to dropping a water-filled balloon from the roof of a house and watching it splatter with force on the ground. What if the "water balloon" is a heavy ball of wind with rainwater descending at about 170 mph? The danger to airplanes is obvious. Downbursts are especially hazardous to airplanes during takeoff and landing. The airplane is so close to the ground that the unexpected downdraft of a downburst can push the plane into the ground before the pilot has a chance to react.

Shortly after a 4:10 p.m. takeoff from New Orleans International Airport on 9 July 1982, Pan American flight 759 was hit by a wet macroburst from a mature thunderstorm. The macroburst bounced the plane down onto a residential neighborhood, killing 8 people on the ground and 145 persons on board the aircraft.

While approaching Dallas–Fort Worth International Airport for a landing during a thunderstorm at 5:06 p.m. on 2 August 1985, Delta flight 191 was just coming out of a wall of extremely heavy rain and lightning when a rounded bulge was seen dropping down from the cloud. This was a wet macroburst that shoved the plane onto the ground, killing 133 of 163 passengers. The 30 survivors were seated in the plane's tail.

Many other downburst incidents have been reported. Many airports in downburst-prone regions now use Doppler radar to detect downburst conditions in time to warn pilots. Training is also given to pilots on how to avoid downbursts.

Figure 10.14 Warm, moist, buoyant air rises high, forming a huge supercell thunderstorm over Africa with a prominent cap or anvil.

Earth Sciences and Image Analysis Laboratory, NASA-JSC.

masses moving down from the north, which underride the warm air masses, helping lift the warm air upward.

Thunderstorms also develop in the mountainous west-central United States, where warm, moist air rises upward, sometimes with the help of the locally mountainous topography.

Thunderstorms wreak havoc with their (1) heavy rains, (2) flash floods, (3) hail, (4) high-speed winds, which can be either straight-line blasts or rotating tornadoes, and (5) lightning-caused deaths and fires.

HEAVY RAINS AND FLASH FLOODS

Thunderstorms are major suppliers of water to many areas of the United States. The abundance of water they deliver makes them positive events, yet there is a price to be paid. Thunderstorms are high-intensity, localized, short-term events that can also bring hail, lightning, and flash floods. A prime example of thunderstorm activity is found in central Texas.

Frequent thunderstorms affect a broad region in the central and southern United States (figure 10.16). Most of these events occur in spring and early summer, when warm, moist air from the Gulf of Mexico advances northward over the United States. The Gulf air masses meet strong, cold air

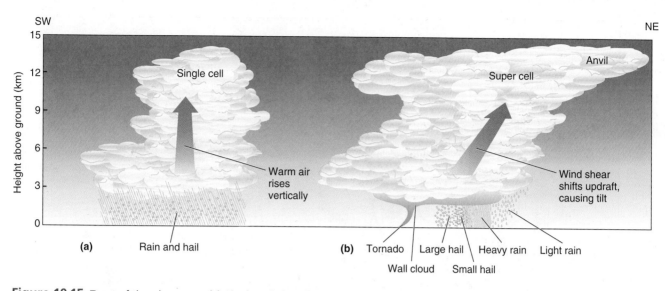

Figure 10.15 Types of thunderstorms. (a) *Single-cell thunderstorm:* An **air-mass thunderstorm** in which warm, moist air rises vertically, cools via adiabatic expansion, and then produces rain. (b) *Supercell thunderstorm:* A thunderstorm accompanied by wind shear, tilting, rotation, and possibly tornadoes. A tilted thunderstorm has rain and hail on the leading side with tornadoes on the trailing side.

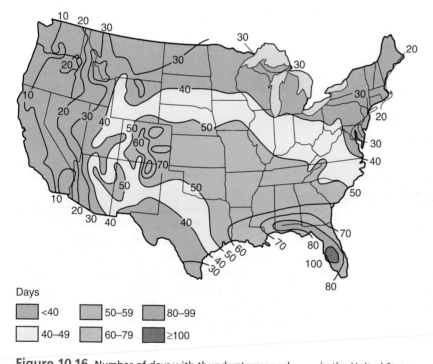

Figure 10.16 Number of days with thunderstorms each year in the United States.

Adapted from J. Eagleman, *Severe and Unusual Weather,* 1983; Van Nostrand Reinhold, New York.

Central Texas

Some of the most intense precipitation in the world occurs in short lengths of time over small areas of central Texas (figure 10.17). Many of the U.S. and world record rainfalls in the 1- to 24-hour range were established here. For example, a May 1935 thunderstorm dumped 56 cm (22 in) of rain on the town of D'Hanis in just 2.75 hours. In September 1921,

the region around Thrall was drenched with 81 cm (32 in) of rain in only 12 hours; flash flooding from this deluge drowned 215 people.

Why do such heavy rainfalls occur here? First, there is a ready source of precipitation in the warm, moist winds that flow in from the Gulf of Mexico. Second, the hot, dry air to the west often meets the warm, humid air to the east, forming a thunderstorm-triggering boundary called the *dry line.* Third, the airflow over the gently sloping coastal plain may be interrupted and turned upward along the escarpment 30 to 150 m (100 to 500 ft) high formed by the Balcones fault zone. The updraft of warm, moist air creates an ideal situation for the formation of thunderstorms. The Balcones Escarpment runs for 545 km (340 mi) from Del Rio on the Mexican border through San Antonio and Austin and north to Waco. It stands roughly at right angles to the prevailing winds that blow in from the Gulf of Mexico. The escarpment was formed by down-to-the-coast normal fault movements, primarily between 19 and 16 million years ago. Today, the height of the eroded ancient fault scarp is still high enough to influence the weather.

The Balcones Escarpment is also the dividing line between two grand physiographic divisions of North America, the Great Plains to the west and the Coastal Plain to the east. The geologic fault is also a "cultural fault" separating the cotton economy of the Old South on the fertile soils of

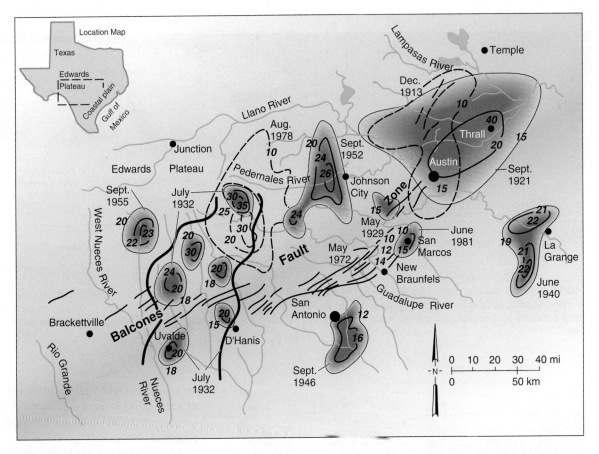

Figure 10.17 Map showing localized areas of high rainfall for some Balcones Escarpment thunderstorms in Texas. Rainfall depths are contoured in inches.

Data summarized from Breeding and Montgomery, 1954; Colwick and others, 1973; U.S. Army Corps of Engineers, 1964; U.S. Soil Conservation Service, 1954, 1958; and Williams and Lowry, 1929.

the Coastal Plain from the cattle economy of the Old West on the deeply eroded and dissected limestone of the plateaus. The faults themselves act as conduits for spring water. The springs enticed the original settlers into starting the towns that have grown into the large cities of today.

The growing urban areas along the Balcones Escarpment are randomly visited by torrential downpours from thunderstorms. For example, on 11 May 1972, a thermally unstable air mass lifted upward to form a localized thunderstorm near New Braunfels, Texas. The storm dumped 30 cm (12 in) of rain in only one hour; up to 41 cm (16 in) fell in only four hours (figure 10.17). The Guadalupe River rose from 0.9 to 9.5 m (3 to 31 ft) deep in just two hours. Flash flooding claimed 17 lives in New Braunfels. The large Canyon Reservoir with its flood-control potential sits just 17 km (11 mi) upstream from New Braunfels, yet the localized thunderstorm dumped its heavy rains on the small, unprotected area below Canyon Dam, wreaking havoc.

HAIL

Hailstones are layered ice balls that drop from some thunderstorms (figure 10.18). In North America, hail kills humans only occasionally, but it takes a tremendous toll on agriculture, as well as roofs and automobiles. Some estimates of U.S. crop damages due to hail range as high as 2% of total crop value.

Thunderstorms that drop large, damaging hailstones are irregularly distributed in the United States (figure 10.19). Important requirements for hail are (1) large thunderstorms with buoyant, warm, moist air rising and (2) mid- and upper-level freezing cold air, creating maximum temperature contrasts and resulting in (3) the strong updrafts needed to keep hailstones suspended aloft while adding coatings of ice onto ever-growing cores. While suspended in clouds, hailstones rise and fall as lifting updrafts of air vary in strength from powerful to weak. Comparing thunderstorm abundance in figure 10.16 with the abundance of large hailstorms in figure 10.19 reveals a marked difference. Thunderstorms are common in Florida, but the cold air aloft necessary for hail formation is uncommon. Destructive, large hail abounds in the colder midcontinent.

A cross-section through a hailstone (figure 10.18a) reveals accretionary layers of ice like the layers of an onion. The layering indicates that the hailstone traveled through parts of the thunderstorm cloud with greater and lesser amounts of supercooled liquid-water content. On 23 July 2010, at

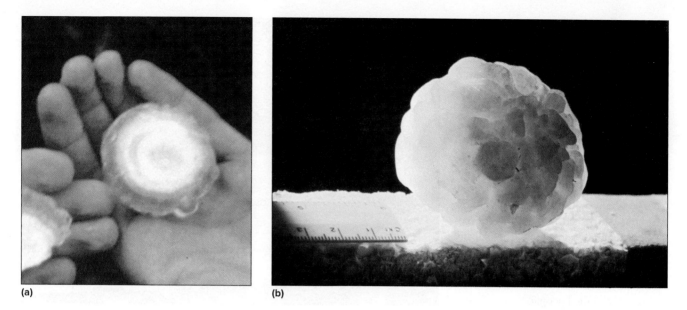

Figure 10.18 (a) A broken hailstone shows the layers of its growth, near Roosevelt, Oklahoma, 9 April 1978. (b) An aggregate hailstone showing its tumultuous growth history. Hailstone is 6 cm (2.4 in) diameter.

(a) OAR/ERL/National Severe Storms Laboratory (NSSL), (b) OAR/ERL/National Severe Storms Laboratory (NSSL).

Vivian, South Dakota, a hailstone set new U.S. records for largest diameter (20 cm = 8 in), largest circumference (47.3 cm = 18.62 in) and heaviest (0.88 kg = 1.93 lbs). Updrafts must be very powerful to keep hailstones this large and heavy suspended in the air.

Hail is most common in the late spring and summer. The timing of hailstorms is related to the position of the polar-front jet stream. In April, Texas and Oklahoma receive much hail, but by June, most of the hailstorms have migrated northward into Montana, Wyoming, and South Dakota. In July, much of the hailstorm activity has moved into Canada. For example, in July of 1996 and 1998, four hailstorms hit Calgary, Alberta, causing total damages in excess of US$450 million.

What can be done to protect against hail damage? An American Society of Civil Engineers study calls for development of economical materials for roofs and exterior walls that are resistant to impacts by hailstones.

DERECHOS

Winds can arrive with tremendous power; they may be rotating or straight-line. The Spanish word **derecho,** meaning "straight ahead," is applied to widespread, powerful, straight-line winds that last at least three hours. Derechos, or straight-line winds, can be as damaging as a small tornado; they kill people, extensively damage mobile homes and lightly constructed buildings, and cause tragic incidents with airplanes, trucks and cars.

Thunderstorms advancing on a line can combine their individual energies to form a line of ferocious winds with

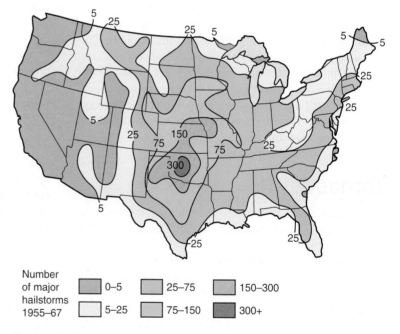

| Number of major hailstorms 1955–67 | 0–5 | 25–75 | 150–300 |
| | 5–25 | 75–150 | 300+ |

Figure 10.19 Major hailstorm frequency, 1955–1967. Contours indicate the number of storms producing hail with 3/4 inch or larger diameter.

Data from Pautz, 1969.

hurricane-force gusts. Derechos commonly extend along a line at least 400 km (250 mi) long, with wind gusts of at least 26 m/sec (58 mph); maximum recorded wind speeds are 67.5 m/sec (151 mph). In the United States, derechos mostly occur in the middle and eastern states. Three states (New York, Michigan, Ohio) suffered 37% of derecho deaths between 1986 and 2003. Most of the deaths occur in vehicles or while boating. Derechos are primarily warm-season events; about 60% of the annual total occur in May, June, and July.

In the United States, the National Weather Service has been tracking wind deaths since 1995. From 1995 through 2013, there were 1,015 wind deaths, averaging 53 deaths per year (see table 10.1).

Ontario to New York Derecho, 15 July 1995

A 160 km (100 mi) wide mass of thunderstorms traveling 130 km/hr (80 mph) in Ontario, Canada, moved south into New York at about 5:30 a.m. During the next two hours, it rushed through the Adirondack Mountains and across the Albany area, before losing strength in New York City. Wind gusts measured at 170 km/hr (106 mph) blew down millions of trees, five of which killed people.

Because severe thunderstorms feed off solar energy, they usually occur during the hottest part of the day, in the late afternoon or early evening. Why did this derecho occur in the early morning hours? In mid-July 1995, a heat wave was in progress. Late-evening temperatures on 14 July were still in the mid-80s, as was the humidity. The hot, humid air just above the ground supplied energy to the mass of thunderstorms traveling through the night. As the cloud mass raced across New York, it drew the hot, humid air upward until it condensed, releasing its contained energy. Derecho winds flowed down from the fast-moving cloud mass, causing extensive damage.

Tornadoes

The rotating winds of tornadoes capture our imagination; they instill fear in the strongest people. Tornadoes derive their name from the Spanish verb *tornar,* which means "to turn." A tornado is a rapidly rotating column of air, descending from a large thunderstorm. Tornadoes have the highest wind speeds of any weather phenomena. The strongest tornadoes are more intense than the biggest hurricanes, but they affect smaller areas. The Great Plains region of the central United States hosts about 70% of the tornadoes that occur on Earth. The most violent U.S. tornadoes mostly move from southwest to northeast at speeds up to 100 km/hr (62 mph), with rotating wind speeds sometimes in excess of 500 km/hr (310 mph).

Only slightly more than 1% of U.S. tornadoes have wind speeds in excess of 320 km/hr (200 mph), but they are responsible for more than 70% of deaths. The core of the whirling vortex is usually less than 1 km wide and acts like a giant vacuum cleaner, sucking up air and objects (figure 10.20).

Funnel clouds initially form hundreds of meters up in the atmosphere, and many never touch the ground. Tornadoes

Figure 10.20 A tornado rips up the region of Dimmitt, Texas, 2 June 1995.
NOAA Photo Library, NOAA Central Library; OAR/ERL/National Severe Storms Laboratory (NSSL).

may touch ground only briefly, or they may stay in contact for many kilometers, moving along an irregular path but making abrupt changes in direction.

The rotating wind speeds of a tornado are highest a hundred meters or so above the ground. This is most likely due to the winds at ground level being slowed by the drag resistance of earth, trees, buildings, cars, and such.

TORNADOES IN 2011

In the United States, the year 2011 was the second deadliest tornado year on record, with 553 people killed. The 1,706 confirmed U.S. tornadoes included six EF5 monsters, the maximum category, where rotating wind speeds exceed 200 mph. EF5s cut paths of destruction through Alabama (two), Mississippi (two), Oklahoma and Missouri.

In 2011, the rest of the world had a total of 24 tornado deaths. The United States is the tornado capital of the world.

How Tornadoes Form

What is necessary to turn an ordinary thunderstorm into a tornado-spinning monster? Since about 70% of Earth's tornadoes occur in the central United States, let us consider their topography and weather conditions on different scales.

REGIONAL SCALE

The central United States and Canada have low-lying topography, with no mountain ranges to disrupt the flow of air masses. Ideal conditions for thunderstorms to develop, including the possibility of tornadoes, occur when these air masses meet simultaneously: (1) a low-altitude, northward flow of marine tropical air from the Gulf of Mexico that has temperatures at the ground in excess of 24°C (75°F) and high humidity with dew-point temperatures in excess of 18°C (64°F); (2) a cold, dry air mass moving down from Canada or out from the Rocky Mountains at speeds in excess of 80 km/hr (50 mph); and (3) high-altitude jet-stream winds racing east at speeds in excess of 240 km/hr (150 mph). These three air masses, all moving different directions, can create wind-shear conditions that impart spin to a thundercloud (figure 10.21).

The warm, moist Gulf air rises vertically, releasing its latent heat and forming a strong updraft that is sheared and spun by the fast-moving polar air and by the jet stream. Any corkscrew motion may be enhanced by vertical movements of air: warm air rising on the forward flank or leading side, with cool air descending on the rear flank or trailing side. The rotation of the winds is achieved without requiring the Coriolis effect.

SUPERCELL THUNDERSTORM SCALE

Most large thunderstorms do not spin off tornadoes. In a common *single-cell thunderstorm* (see figure 10.15a), warm air rises by convection to build a nearly vertical cloud mass. In the colder air at high altitudes, water vapor condenses, and rain falls back down through the thunderstorm, creating a cool downdraft that blocks the upward flow of energy-carrying warm air. A single-cell thunderstorm does not grow large enough, or energetic enough, to create violent tornadoes.

Sometimes a huge updraft of air builds a massive thunderstorm that is tilted by wind shear and develops into a *supercell thunderstorm* (see figure 10.15b). The tilt of the supercell allows warm air to continue rising into central portions of the storm while most of the rain and hail fall in the front flank. Most tornadoes are produced within supercell thunderstorms, but only about 30% of supercells produce tornadoes.

The supercell has rotating updraft through the depth of the thunderstorm. It develops in a wide zone within the supercell. The cloud mass rotating around the updraft sets up conditions that could lead to tornadoes.

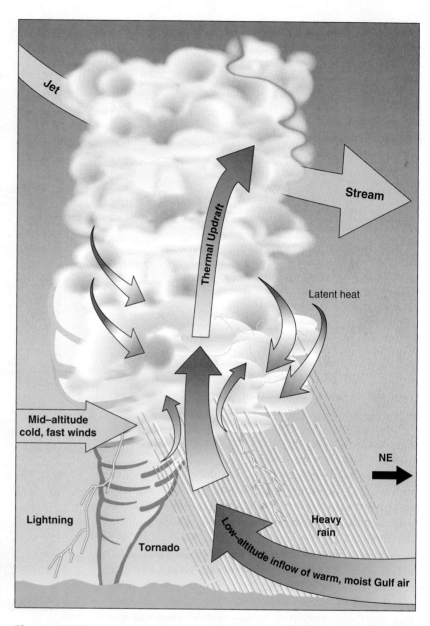

Figure 10.21 Components of a tornado. A warm, moist mass of Gulf of Mexico air collides with a fast-moving mass of polar air, causing updrafts that build up into the polar-front jet stream. Three fast-moving air masses, all going different directions, impart shears, causing rotation. A tornado receives additional energy via lightning and latent heat released by rainfall.

Adapted from J. Eagleman, *Severe and Unusual Weather,* 1983; Van Nostrand Reinhold, New York.

Wall Clouds

A wall cloud may form as a markedly lower cloud beneath the main mass of the mesocyclone (figure 10.22); wall clouds are sometimes called pedestal clouds. Wall clouds form in a supercell where updrafts are the strongest, causing water vapor to condense at lower altitudes. Powerful tornadoes, which develop in the mesocyclone, will emerge from the wall cloud. The diameter of a rotating mesocyclone is larger than the diameter of its rotating wall cloud, and an even smaller-diameter vortex can develop with

Figure 10.22 Wall cloud (center foreground) formed below the main cloud mass of a supercell thunderstorm. Heavy rain and lightning are in the background, 19 June 1980 near Miami, Texas.

National Severe Storms Laboratory/NOAA.

greater wind speeds. As a rotating core of air pulls into a tighter spiral, its wind speed increases dramatically as its angular momentum is conserved. This principle is analogous to ice skaters spinning with arms outstretched who can spin even faster by pulling their arms toward their bodies. In other words, the smaller the diameter of a rotating mass, the faster it spins.

If a vortex is stretched, not only does its wind speed increase, but the vortex may be stretched in length far enough to touch the ground—and then it can wreak havoc as a tornado. Many tornadoes form as a result of vortex stretching (figure 10.23).

Another hypothesis holds that some tornadoes are created within supercells where increasing rainfall creates a volume of descending air in the rear flank. This downdraft in the rear flank of the storm accelerates as it nears the ground and may drag the rotating air of the mesocyclone down with it to form a tornado.

Hook Echo

For more than 50 years, radar imagery has been able to define a **hook echo** within a supercell thunderstorm. A hook echo is a swath of high radar reflectivity in the shape of a hook on a radar image. The hook forms from the weak echo region in the updraft (figure 10.24). A radar image defines the concentrated, rotating band of heavy rain, hail, or airborne debris that exists within the hook echo. When the National Weather Service sees a hook echo, they may issue a tornado warning, an alert that a tornado is occurring or is imminent. A tornado warning is accompanied by loud sirens, and by alerts from media and phone apps.

Newer generations of radar can not only see a hook echo, but can also recognize a mesocyclone. Doppler radar detects different wind velocities occurring in different areas of the storm, including areas of rotation.

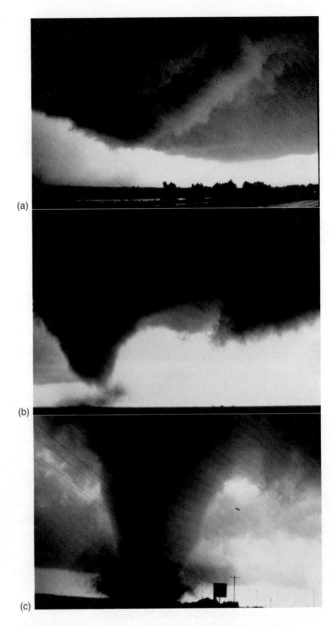

(a)

(b)

(c)

Figure 10.23 A tornado is born. (a) In upper left, a dark-gray spinning cloud begins. (b) The gray cloud lengthens into a vortex that stirs dust as it nears the ground. (c) The vortex reaches the ground as the violent Dimmit, Texas, tornado, 2 June 1995.

(a) NOAA (b) NOAA (c) NOAA Photo Library, NOAA Central Library; OAR/ERL/National Severe Storms Laboratory (NSSL).

In those instances when a mobile team of tornado chasers is able to get a Doppler radar unit close enough to a tornado, the radar data tell some interesting tales. At least some tornadoes have downward-moving air in the center surrounded by a cone- or cylinder-shaped funnel that rapidly spirals upward. This indicates that, as the tornado is sucking up huge volumes of air plus debris from the ground, some of the air supplied to this mega-vacuum cleaner comes from a central downdraft.

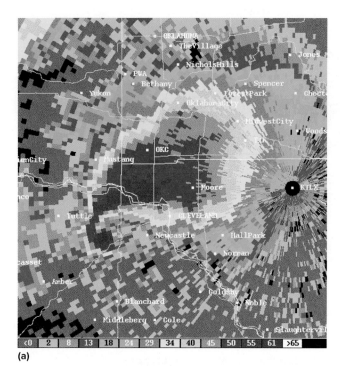

(a)

(b)

Figure 10.24 Hook echo. (a) Radar image of a hook echo associated with a supercell thunderstorm. The bright red color marks a heavy rainfall band rotating around the supercell, just south of Oklahoma City, 3 May 1999. (b) One of the tornadoes that assaulted Oklahoma on 3 May 1999.

(a) NSW/NOAA (b) OAR/ERL/National Severe Storms Laboratory (NSSL).

VORTEX SCALE

Another hypothesis that explains how some tornadoes form involves a three-stage sequence of events (figure 10.25). The collision of air masses low in the atmosphere can create a nearly invisible, horizontal tube of air, a vortex tube rotating parallel to the ground (figure 10.25a). The warmer air rising as an updraft supplies the vertical wind shear that tilts the rotating tube of air (figure 10.25b). As the vortex is uplifted, it is stretched, thus decreasing its diameter and increasing its velocity to mesocyclone status (figure 10.25c).

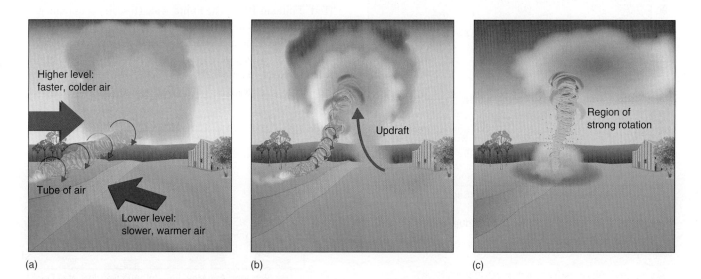

(a)　　　　　　　　　(b)　　　　　　　　　(c)

Figure 10.25 An hypothesis of mesocyclone formation: (a) Colliding masses of air create a horizontal, spinning tube of air, a vortex, in the lower atmosphere. (b) Rising air within the updraft tilts the rotating tube of air and stretches it upward. (c) As a vortex is uplifted and stretched, its spin velocity increases and it may develop into a mesocyclone.

Source: NOAA.

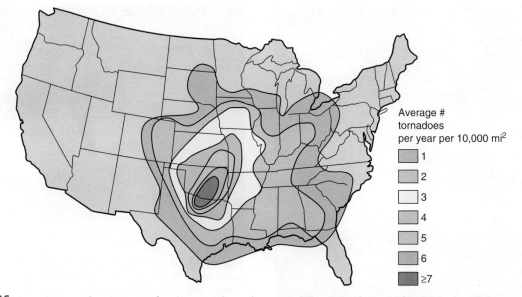

Average #
tornadoes
per year per 10,000 mi^2

- 1
- 2
- 3
- 4
- 5
- 6
- ≥7

Figure 10.26 Average annual occurrence of major tornadoes. The general direction of travel of individual tornadoes is toward the northeast. The elongate heart of the contoured area is Tornado Alley.

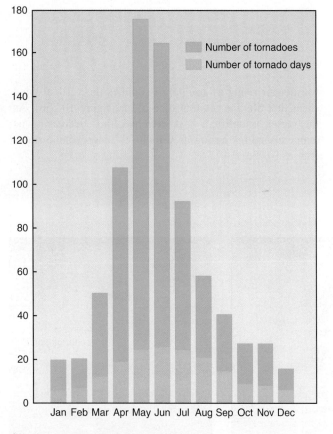

Number of tornadoes
Number of tornado days

Figure 10.27 Average number of tornadoes each month in the United States, 1950–1999.

Based on data from NOAA.

After a mesocyclone and tornado form from a supercell thunderstorm and then dissipate, a new mesocyclone may arise near the center of the storm. Then, one or more tornadoes may develop from the newer mesocyclone.

TORNADOES IN THE UNITED STATES AND CANADA

The interior of the United States is the tornado capital of the world. Over these gently sloping lands, known as *Tornado Alley,* air masses collide, spinning out tornadoes that usually head northeast under the influence of the jet stream flowing northeast out of a trough (figure 10.26). Tornadoes can occur at any time of the year, but are most common in late spring and early summer (figure 10.27). Tornado deaths in the United States are most common in the spring (figure 10.28). The furious energy of these twisters generates wind speeds that stretch our imagination.

A tornado wind damage scale was developed by Tetsuya "Ted" Fujita in 1971. The Fujita scale (F scale) uses damages to structures and trees to estimate the wind speed of tornadoes just above the ground. The F scale ranges from F0 to F5 and has been widely used (table 10.2, left). However, over the years, it has become necessary to better define the types and severity of damages done by varying tornado wind speeds. For

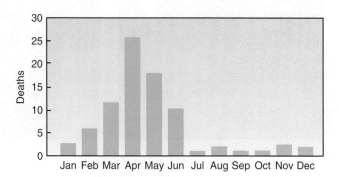

Figure 10.28 Average number of tornado deaths each month in the United States, 1950–1999.

Based on data from NOAA.

TABLE 10.2
Fujita Wind Damage Scales

Fujita Scale	mph	(m/sec)	Enhanced Fujita Scale	mph	(m/sec)
F0 Little damage	Under 72	(32)	**EF0** Siding and shingle damage	65–85	(29–38)
F1 Roof damage	73–112	(33–50)	**EF1** Uprooted trees and overturned single-wides	86–110	(38–49)
F2 Roof gone	113–157	(51–70)	**EF2** Permanent houses off foundations	111–135	(50–60)
F3 Wall collapse	158–206	(71–92)	**EF3** Severe damage. Houses mostly destroyed.	136–165	(61–74)
F4 House blown down	207–260	(93–116)	**EF4** Devastating. Large sections of school buildings destroyed.	166–200	(74–89)
F5 House blown away	261–318	(117–142)	**EF5** Incredible damage. Deformation of mid- and high-rise buildings.	Over 200	(89+)

example, simply stating that house walls were blown down ignores the varying construction styles and ages of houses. Therefore, an extensive study defined more criteria for tornado wind damage and resulted in an Enhanced Fujita scale (EF scale) that was accepted in 2007 (table 10.2, right). Now 28 types of structures and trees have been identified, each evaluated based on 9 levels of damage. The new EF scale is comparable to the original F scale, thus maintaining continuity of historic records. The incredible damages caused by an EF5 tornado are shown in figure 10.29. Greensburg, Kansas, and its population of 1,574 people suffered 11 deaths and 95% of buildings destroyed, with the remaining 5% severely damaged.

The number of fatalities goes up dramatically as F-scale winds increase (table 10.3). The United States suffered 1,706 tornadoes in 2011; 59 were killer tornadoes and, of these, the six EF5 tornadoes claimed 51% of the victims.

Tornadoes cause destruction in several ways: (1) High wind speeds blow away buildings and trees. (2) The furious winds throw debris that acts like bullets or shrapnel,

Figure 10.29 Greensburg, Kansas, 12 days after being struck by an EF5 tornado on 4 May 2007. Five years later, Greensburg was largely rebuilt as an environmentally friendly city with more eco-architecture awards per capita than any other U.S. city.
Greg Henshall/FEMA.

breaking windows and killing people. (3) When the fast winds lift and blow away a roof, the exposed walls are easily knocked over and the furniture removed.

Tornado deaths in the United States are dealt out unequally (figure 10.30). A belt of states running from the Gulf of Mexico into Canada experiences most of the carnage, while

TABLE 10.3

U.S. Tornado Fatalities versus Fujita Scale, 2011

Fujita Scale	Number of Killer Tornadoes	Fatalities
EF0	1	1
EF1	3	4
EF2	16	29
EF3	21	75
EF4	12	160
EF5	6	284
Total	59	553

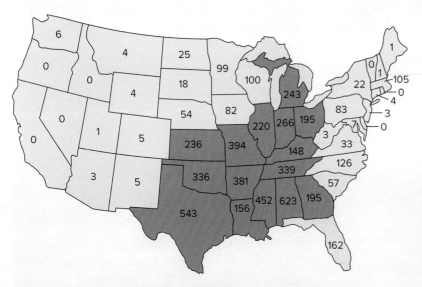

Figure 10.30 Distribution of U.S. tornado deaths, 1950–2013.

the coastal states are largely spared the agony. Note that 9 states suffered no tornado fatalities between 1950 and 2013, including Alaska and Hawaii. The deadly dozen states are listed in table 10.4 (left column). But if the number of tornado deaths by state is divided into the population of each state, it yields a per capita ranking of deaths that rearranges the deadly dozen states (table 10.4, right column). On a per capita basis, Texas drops from second deadliest to out of the top 12.

The litany of tornado deaths skips no years. But there is some encouragement to be found in the declining numbers of tornado deaths during the last eight decades (table 10.5). Even though the number of humans in the United States keeps increasing, we are getting earlier warnings before tornadoes, allowing us time to take safety actions. Tornadoes are a threat that must be faced. Either we play Russian roulette and hope they don't hit buildings, or else buildings must be designed and built to withstand tornado forces. High winds are a difficult test for any structure. A tornado passing over a building can destroy it and suck up some of the debris with the updrafting air, which may be rising at more than 45 m/sec (100 mph). We must either design with nature, or run the risk of dying by nature.

Who dies during tornadoes? Tornadoes most commonly kill (1) old people, (2) mobile-home residents, (3) occupants of exterior rooms with windows, and (4) people unaware of broadcast tornado warnings or sirens. Residents of frame houses can run into interior rooms to gain some protection. But where do mobile-home dwellers run to hide? There are no interior rooms, and who wants to run

TABLE 10.4

Deadly Dozen States for Tornado Deaths, 1950–2013

Total Deaths	Deaths per Capita
1. Alabama	1. Mississippi
2. Texas	2. Alabama
3. Mississippi	3. Arkansas
4. Missouri	4. Oklahoma
5. Arkansas	5. Kansas
6. Tennessee	6. Missouri
7. Oklahoma	7. Tennessee
8. Indiana	8. Indiana
9. Michigan	9. Kentucky
10. Kansas	10. Louisiana
11. Illinois	11. Nebraska
12. Georgia	12. Iowa

TABLE 10.5

U.S. Tornado Deaths by Decade, 1920–2009

Decade	Recorded Deaths
1920–1929	3,169
1930–1939	1,944
1940–1949	1,788
1950–1959	1,409
1960–1969	935
1970–1979	987
1980–1989	522
1990–1999	580
2000–2009	557

Tornado deaths average for 90 years, 1920–2009 = 133 deaths per year

Data source: NOAA.

A Classic Disaster

The Tri-State Tornado of 1925

A little before 1 p.m. on 18 March 1925, the residents of Annapolis, Missouri, heard a sound "like a thousand freight trains" heading their way. It was an unusually broad tornado, moving at about 100 km/hr (60 mph) along a northeasterly path. After traveling down Main Street and leveling the town, the tornado spun on toward Murphysboro, Illinois, and its 11,000 residents. The extrawide tornado destroyed everything in a nearly 1-mile-wide path. Some of the wrecked buildings caught on fire, and just as in an earthquake, the event destroyed the water-supply system. Firefighters watched helplessly as flames devoured people trapped in wrecked buildings. In Murphysboro, 210 people were killed either by the tornado force or the ensuing fires.

Along its path, the tornado never seemed to lose energy. Residents of DeSoto, Illinois, watched the advancing tornado with its suspended load of house and auto parts and uprooted trees, seemingly defying the law of gravity. The tornado spun toward the town school, lifting off its roof and knocking down the walls onto 125 students and teachers, 88 of whom died. The dead students were laid out on the school lawn, but many parents were unable to claim their bodies—they too were dead.

A railroad engineer faced the tornado coming straight down the tracks toward him. He responded by accelerating as fast as possible and trying to blast through the tornado. He succeeded, but the roofs were stripped off his train cars like lids off sardine cans.

The "Tri-State Tornado" had the longest track known and is the longest lived. It traveled on a N 21°E path for 353 km (219 mi) across parts of Missouri, Illinois, and Indiana, devastating 23 cities and towns in a broad swath up to 1 mile wide. This single tornado claimed 689 lives.

TABLE 10.6
U.S. Tornado Fatalities by Location, 1985–2008

Mobile home	46%
Permanent home	26%
Vehicle	10%
School/church	5%
Outdoors	5%
Business	4%
Unknown	4%

Data source: NOAA.

outside into a severe thunderstorm with heavy rain, lightning, hail, and flying debris? Almost half of the Americans killed by tornadoes die inside their disintegrating mobile homes (table 10.6). Many towns now require mobile-home communities to provide tornado protection with a common basement shelter or an above-ground reinforced concrete building.

Are you safer in a car or a mobile home? Many mobile homes can be tipped over by 35 m/sec (80 mph) winds, whereas many modern cars (excluding many SUVs) have low centers of gravity and streamlined shapes that require wind speeds of about 55 m/sec (120 mph) to tip over. Also, most cars are built to provide some protection during a rollover or collision. Many mobile-home dwellers facing a tornado threat would do better to run outside and sit in their cars.

It is an interesting commentary on human nature that so many people in the central United States are not worried about the dangers posed by tornadoes yet wonder how Californians can live with the threat of earthquakes. Meanwhile, most Californians are quite comfortable with earthquakes, but many do not want the stress of living with potential tornadoes. Many people seem to be most comfortable with the events common in the area where they grew up.

Tornado Outbreaks

Tornadoes are often associated with a cold front meeting a high-energy warm air mass. A lengthy collision zone allows the birth and growth of numerous supercell thunderstorms, which collectively can produce several to many dozens of tornadoes in a **tornado outbreak.** Three of the five biggest tornado outbreaks on record hit during six weeks of April and May 2011.

Outbreak of 14–16 April 2011 One of the largest recorded U.S. tornado outbreaks swept eastward across 16 states as 178 tornadoes caused death and destruction on all three days. Worst hit were North Carolina, with 24 deaths, and Alabama, with 7 killed. A total of 43 people died; 38 from tornadoes and 5 by straight-line winds associated with the storm system. This was the largest number of fatalities in a U.S. tornado outbreak since the February 2008 Super Tuesday outbreak. However, within the following two weeks, an even larger tornado outbreak hammered the eastern United States and Canada.

Outbreak of 25–28 April 2011 The largest outbreak of tornadoes in U.S. history occurred between 25–28 April 2011, as a record-setting 358 tornadoes were confirmed over 21 states from Texas to New York to southern Ontario, Canada. The tornadoes inflicted damages in excess of

$11 billion and killed 348 people in six states: Alabama, Tennessee, Mississippi, Georgia, Arkansas, and Virginia. Hardest hit was Alabama with 239 fatalities, including those caused by the EF4 tornado that cut through Tuscaloosa. After the tornado, so many thousands of good-hearted volunteers drove into Tuscaloosa to help with the recovery that the traffic tie-ups were reminiscent of fall football weekends at the University of Alabama.

On 27 April, a U.S. record for single-day tornadoes was set as 199 confirmed tornadoes touched down. The 27 April onslaught included four EF5 tornadoes; two in Alabama (Hackleburg and Rainsville) and two in Mississippi (Philadelphia and Smithville). The April 2011 total of 770 tornadoes in a month set a record for most tornadoes in any single month in U.S. history, easily breaking the old record of 542 in May 2003.

Outbreak of 21–27 May 2011 The 21–27 May outbreak brought 242 confirmed tornadoes spinning across states from Minnesota to Texas. Damages totaled $7 billion, making it the second costliest tornado outbreak in U.S. history, behind only the late April outbreak less than four weeks earlier. The tornadoes combined to kill 180 people in five states: Missouri, Oklahoma, Arkansas, Kansas, and Minnesota. Most of the deaths resulted from the two EF5 tornadoes: 160 dead in Joplin, Missouri, on 22 May; and 9 dead in the Piedmont, Oklahoma, area on 24 May.

The Joplin EF5 tornado was a multivortex tornado with multiple smaller-diameter, intense centers of rotation within the larger supercell thunderstorm (figure 10.31). About one-third of Joplin suffered heavy damage as houses were swept off foundations, cars were crushed like soda cans, and the hospital was hit hard in part of the wide path of destruction cut across Joplin (figure 10.32).

Super Outbreak of 3–4 April 1974

Prior to 2011, the record for most overwhelming tornado outbreak was set during a 16-hour period on 3 April 1974, when 148 tornadoes touched ground in 13 states and Ontario, Canada, killing 339 people (figure 10.33). The outbreak included seven tornadoes of F5 force: two in Ohio, three in Alabama, and one each in Indiana and Kentucky, as well as 24 F4 tornadoes. The mighty seven F5s were a decade's worth, all in a few hours; each touched ground for more than

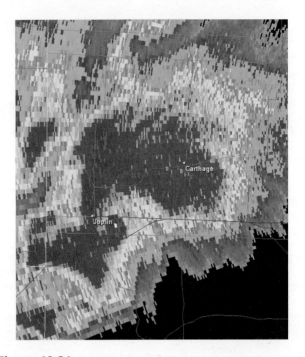

Figure 10.31 Radar image of base reflectivity of the supercell thunderstorm that produced the Joplin tornado of 2011. Note how the colors define the high-velocity, EF5 Joplin tornado.

Courtesy image from NWS radar data, Gibson Ridge Radar software.

50 km (30 mi), and two stayed down for more than 160 km (100 mi). Most of the tornadoes touched down during the warm hours between 4 and 9 p.m.; these are typical hours for tornado touchdown.

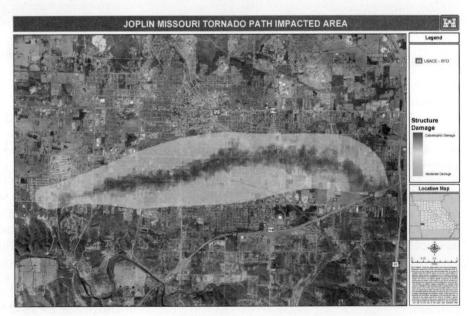

Figure 10.32 Tornado-damage map of Joplin, Missouri. Red area suffered catastrophic damage; green area had moderate damage.

Courtesy U.S. Army Corps of Engineers, Kansas City.

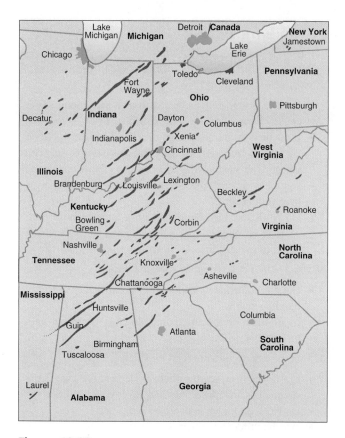

Figure 10.33 Paths etched across the ground by 148 tornadoes on 3–4 April 1974. The northeasterly trend is typical; it is the direction the jet stream travels coming out of a trough.

Based on a map prepared by T. Fujita at the University of Chicago.

The seven F5 tornadoes are still a record for one outbreak, but as an annual record it was almost tied by the six EF5 tornadoes in 2011.

TORNADOES AND CITIES

A commonly heard tale is that tornadoes never strike big cities. It is a myth. The United States averages about 1,200 tornadoes per year. Most of these tornadoes occur in the central states where there are few big cities to strike. Cities cover very little ground in a large region, thus they are small targets. Yet in the three years from spring 1997 to spring 2000, approximately 3,600 tornadoes occurred, and 10 of them struck large metropolitan areas (table 10.7).

The recent tornado strikes on cities include assaults on downtowns and their skyscraper buildings. The Nashville tornado sent pedestrians scurrying into high-rise buildings for protection. In Salt Lake City, a tornado ripped across the Delta Center arena and through the Salt Palace Convention Center in the early afternoon (figure 10.34). Downtown

TABLE 10.7
Cities Struck by Tornadoes, May 1997–March 2000

Date	City	Fujita Scale Intensity	Deaths
12 May 97	Miami, Florida	F1	—
8 Apr 98	Birmingham, Alabama	F5	33
16 Apr 98	Nashville, Tennessee	F3	1
21 Jan 99	Little Rock, Arkansas	F3	3
9 Apr 99	Cincinnati, Ohio	F4	7
3 May 99	Oklahoma City, Oklahoma	F5	42
3 May 99	Wichita, Kansas	F4	6
11 Aug 99	Salt Lake City, Utah	F2	1
8 Mar 00	Milwaukee, Wisconsin	F2	—
28 Mar 00	Fort Worth, Texas	F2	4

Fort Worth was battered at 6:30 p.m. by wind and hail that knocked the windows out of tall buildings and destroyed the offices and businesses inside. The Oklahoma City tornado was noteworthy for its record-high wind speeds of 140 m/sec (318 mph) measured 20 m (65 ft) above ground using Doppler radar (see figure 10.24). If there is a capital city for tornadoes in the United States, it might be Oklahoma City. Sitting in the heart of Tornado Alley, Oklahoma City has felt the wrath of at least 103 tornadoes between 1893 and 1999.

Figure 10.34 Tornado moves through Salt Lake City on 11 August 1999. The orange fireball is a power substation exploding.

Photo given to NOAA for public use by KTVX News 4 Utah.

HOW A TORNADO DESTROYS A HOUSE

A wood-frame house built on top of a concrete slab can be destroyed by a tornado in just a few seconds. A typical destruction sequence is:

1. Flying debris shatters windows.
2. Wind blowing under the eaves and through the broken windows into the house increases the interior air pressure, helping lift the roof (similar to blowing up a balloon).
3. Wind blowing over the roof creates uplift similar to how air flows over an airplane wing, providing the lift it needs to fly.
4. With the roof gone, the walls lose significant support. Exterior side walls typically are sucked out first, then the front wall blows in, and the back wall blows out.
5. With outside walls gone, the wind progressively blows down walls inside the house.

In sequence, a tornado breaks windows, lifts the roof, and knocks down walls progressively inward. Where is the safest place in the house? A small room in the middle of the house is commonly the last to fail. With a little luck, the fast-moving tornado will move on before knocking down all the interior walls.

Tornado Safety

If you see or hear a tornado warning, evacuate or seek shelter—immediately (table 10.8). The value of quickly evacuating or sheltering was dramatically shown by a survey of Golden Spur Mobile Home Park residents following the Andover, Kansas, F5 tornado of 26 April 1991. A tornado warning was issued 45 minutes in advance and the tornado was visible for 14 minutes before it hit. The 333 residents made the following choices:

- 146 evacuated the area (no deaths, no injuries)
- 149 entered the community shelter (no deaths, no injuries)
- 38 stayed in their homes (13 died, 17 hospitalized, 8 injured)

SAFE ROOMS

The traditional protection against dying in a tornado has been to go underground, into a cellar. But more and more houses are being built without cellars. What can a homeowner do? The Oklahoma City tornado showed the value of a new type of shelter, the safe room. In the interiors of houses, closets or small rooms are being built as safe rooms with 12-inch-thick concrete walls, a steel door, and a concrete roof. These safe rooms are reminiscent of bank vaults. Heeding the Oklahoma City tornado warning, some residents went into their safe rooms and emerged later to find their homes blown away. Although their houses had disappeared, they suffered no deaths or even injuries.

TABLE 10.8
What to Do During a Tornado

If you are in a building

- Go to a designated shelter, such as a safe room, storm cellar, or basement.
- If no shelters are nearby, go to a small interior room on the bottom floor. Put as many walls as possible between you and the outside. Stay away from windows and doors.
- Get under a heavy piece of furniture and put your arms over your head.
- Do NOT stay in a building with a large-expanse roof, such as an auditorium, cafeteria, mall, manufacturing plant, or gymnasium.

If you are in a car, truck, or mobile home

- Get out immediately and go to a storm shelter or sturdy building nearby.
- In an urban or congested area, do NOT try to outrun a tornado.

If you are caught outside with no shelter

- Lie down in a nearby ditch or depression away from streambeds and powerlines and cover your head.
- Do NOT get under an overpass or bridge.
- Watch out for flying debris.

Lightning

Thunderstorms generate lightning (figure 10.35); **thunder** is caused by lightning. Where does lightning occur in the world? Lightning is concentrated over land; it is relatively uncommon over the oceans (figure 10.36). Land warms up faster and gets hotter than the oceans; more warm, unstable air at the land surface means more convective upwelling to form thunderstorms and thus more lightning. Lightning is not evenly distributed over the continents. Topography, rainfall, and air temperatures are all controlling factors (figure 10.36).

Lightning is the leading natural cause of forest fires and is one of the major causes of weather-related deaths in the United States (see table 10.1). Hurricanes, tornadoes, and floods are dramatic events that take many human lives in brief, dramatic episodes, whereas a lightning bolt kills only one person in 91% of deadly events.

Where does lightning occur in the United States? Its distribution is the same as for thunderstorms (see figure 10.16). More thunderstorms develop in central Florida than anywhere else in the United States, and most lightning deaths happen in Florida (figure 10.37). Lightning claims the majority of its victims in late spring and summer, when thunderstorms occur after moist air warmed by the Sun-heated ground builds tall, unstable clouds. Lightning deaths

Figure 10.35 Lightning strikes near the U.S. Capitol in Washington, D.C., 14 May 2005.

Photo by U.S. Air Force Technical Sgt. Cherie A. Thurlby.

Figure 10.36 Global lightning. Annual averages for flashes per km² for 1995–2002. Deep red = 70 flashes/ km²/year, red = 20, green-yellow = 5, light blue = 1, purple = <1.

NASA.

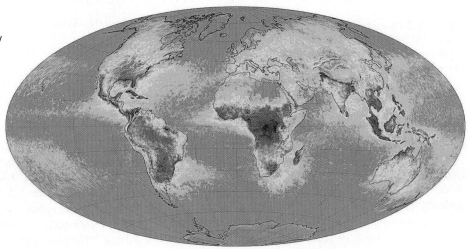

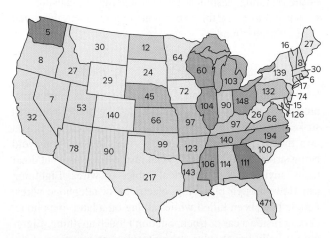

Figure 10.37 Lightning deaths in the United States, 1959–2013. In the 55 years of data, only two states escaped lightning deaths—Alaska and Hawaii.

plotted by month show that thunderstorms are most active in summertime (figure 10.38). Where and whom does lightning kill? Mostly men who are outdoors, especially outside workers and golfers (table 10.9).

HOW LIGHTNING WORKS

You can make your own "lightning." Drag your feet across a carpet and become a negatively charged "thundercloud." Now touch a metal door handle and feel the "lightning" as the negative charges bolt from you to the positive charges on the metal.

The lightning of thunderstorms involves a similar flow of electric current as areas having excess positive charges seek a balance with places having excess negative charges. During the buildup of tall clouds, charged particles separate,

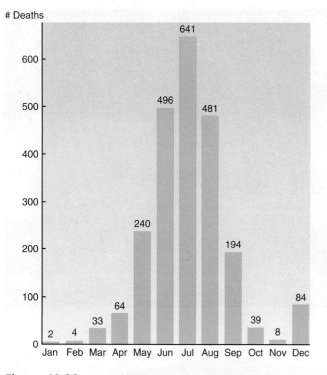

Deaths

Jan	2	
Feb	4	
Mar	33	
Apr	64	
May	240	
Jun	496	
Jul	641	
Aug	481	
Sep	194	
Oct	39	
Nov	8	
Dec	84	

Figure 10.38 Lightning deaths in the United States by month, 1959–1980. Note: The 84 deaths in December include 81 killed in Maryland on 8 December 1963, when lightning caused the crash of a Boeing 707 jetliner.

TABLE 10.9
U.S. Lightning Deaths, 1959–1994

Locations	Frequency (percent)
Open fields, ball fields	27
Under trees	14
On boats and in water-related activities	8
Golf courses	5
On tractors and heavy road equipment	3
Telephone-related	3
Unspecified	40
Total	100
Genders	
Male	84
Female	16
Time of Day (most common)	
2 to 6 p.m.	

Source: NOAA (1997).

creating an abundance of positive charges up high and an excess of negative charges down low (figure 10.39).

The charge imbalance apparently comes about as the freezing and shattering of super-cooled raindrops in the cloud initiate charge separations that are distributed by updrafts and downdrafts within the thundercloud. The charge separations occur during the cloud buildup of the early stage (see figure 10.12a), and then lightning bolts forth during the mature stage (see figures 10.11b and 10.12b).

A thundercloud interacts electrically with the ground. The abundance of negative charges in the basal part of the cloud induces a buildup of positive charges on the ground surface, because the opposite charges attract each other. Lightning can move from cloud to earth, earth to cloud, cloud to cloud, or within a cloud (figure 10.40).

Lightning moves at speeds more than 6,000 miles per second and typically includes several strokes, all occurring within 0.5 to 2 seconds. Thanks to high-speed photography, it is now possible to explain the basic sequence within a lightning flash: (1) Static electricity builds up within the lower thundercloud and induces opposite charges on the ground. (2) Discharge begins within the cloud and initiates a dimly visible, negatively charged stream of electrons propagating downward (figure 10.41a). (3) The conductive stream moves earthward in 50 m (165 ft) jumps as a stepped leader (figure 10.41b). (4) As the stepped leader nears the ground, the electric field at the surface increases greatly, attracting streamers

of positive sparks upward and connecting with the stepped leader about 50 m (165 ft) above ground (figure 10.41c). (5) The connection closes the electrical circuit and initiates the return stroke, sending positive charges up to the cloud with a brilliant flash (figure 10.41d). (6) More lightning strokes occur as charges flow between the cloud and earth.

Several different strokes all occur within the 1- or 2-second event we call a lightning bolt. If you have seen a lightning bolt appear to flicker, you have witnessed the several different up-and-down strokes that constitute a given "bolt." The electrical discharge of lightning can briefly create temperatures as high as 55,000°F. The high temperatures of the lightning flash heat the surrounding air, causing it to expand explosively; this expansion of air produces the sound waves we call thunder.

DON'T GET STRUCK

Mark Twain said, "Thunder is good, thunder is impressive; but it is lightning that does the work." Lightning bolts can strike 15 km (10 mi) away from the thundercloud. If you hear thunder, you are in an area of risk. How can you avoid being struck by lightning? (1) Get inside the house, but don't touch anything with electrical cables or wires. Lightning can flow through plumbing, electrical, or telephone wires. People have been killed while talking on a land-line phone. (2) Get inside a car or truck, but don't touch anything. Lightning usually travels along the outside metal surface of the vehicle and then jumps to the ground through the air, a wet tire surface, or the tire itself, causing it to blow out. (3) If

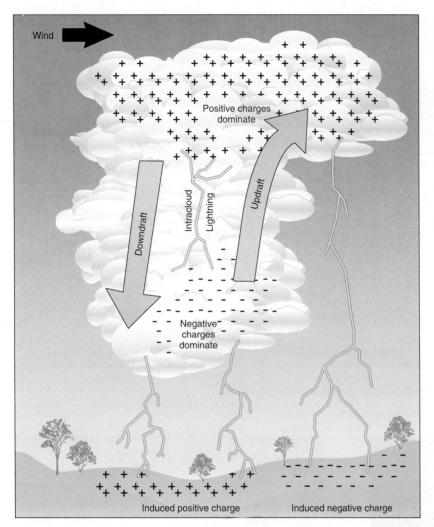

Figure 10.39 Schematic view of charge separation within a thundercloud and the induced charges on the earth beneath. When electric potential is great enough, narrow channels of air become ionized, allowing them to conduct the electricity that we see as lightning.

Figure 10.40 Triple lightning modes. Left: cloud-to-ground followed by extra-bright return stroke from ground. Lower right: ground toward cloud seeking to connect. Middle right: cloud-to-cloud.

NOAA Photo Library, NOAA Central Library; OAR/ERL/National Severe Storms Laboratory (NSSL).

you are caught outside, move to a low place. Avoid tall structures, including trees (figure 10.42). Don't lie down; lightning can flow along the surface of the earth. Assume the lightning position: crouch down on the balls of your feet with your hands over your ears. (4) Personal electronic devices are causing lightning burns and injuries. When lightning strikes nearby and flashes outward, it can pass over your skin and contact metal in iPods, jewelry, and such, causing burns. For example, a Canadian jogger suffered burns, ruptured eardrums, and a broken jaw when lightning traveled through his iPod wires. A good guideline to follow is: *When thunder roars, go indoors.*

Progress is being made in reducing the lightning threat. More structures are fitted with lightning rods and conductors to convey the lightning energy to the earth. Also, increasing numbers of people know what actions to take to avoid lightning. The death toll from lightning in the United States has been dropping dramatically (table 10.10). The U.S. population more than doubled between 1940 and 2014 (from 130 million to 322 million people), yet lightning deaths now are only 15% of the 1940s numbers.

Heat

The days of high temperatures during a heat wave may be one of the least appreciated weather disasters. The heat wave is an invisible, silent killer, and it can kill in large numbers. Data for the past 30 years in the United States show that heat is the biggest killer of all severe weather (see table 10.1). Hurricanes, tornadoes, and floods grab the headlines, but heat waves kill more people.

Our normal body temperature remains around 37°C (98.6°F). But when exposure to heat raises our body temperature and the body's temperature-control system is overwhelmed, **hyperthermia** (high heat) may set in. Body temperatures above 40°C (104°F) can be life threatening.

The human body works to control temperature by perspiring. If humidity is low, perspiration evaporates easily and carries heat away from the body (see figure 9.8). But when humidity is high, evaporation slows, decreasing the cooling effect. Sometimes perspiration is not enough. Prolonged exposure to excessive heat and/or humidity during a heat wave can lead to heat exhaustion or heat stroke. Warning signs include: (1) high body temperature, (2) mental confusion, staggering, strange actions, (3) fainting, and (4) dry skin with rapid or slow pulse.

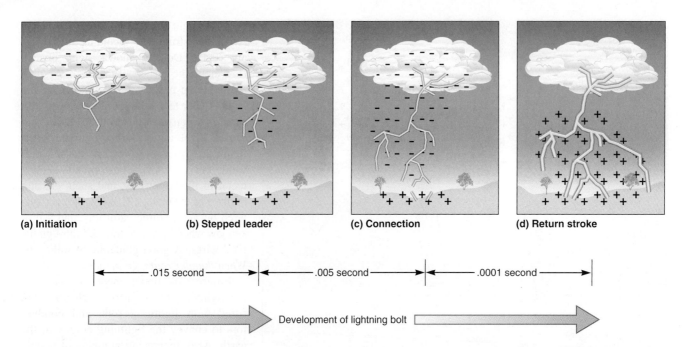

Figure 10.41 Steps in creating a lightning bolt. (a) *Initiation:* Charge separation in a cloud builds up static electricity. (b) *Stepped leader:* Negative charges move in a dimly visible stream downward in intermittent steps. (c) *Connection:* When the leader nears ground, a positive discharge leaps up, completing the attachment. (d) *Return stroke:* The connected path flashes bright as charges exchange between the cloud and the ground in several events, all totaling about 0.5 second.

Figure 10.42 Lightning strikes and electrifies a 20 m (65 ft) tall sycamore tree. Do not seek shelter beneath a tree during a thunderstorm! Lightning has even been known to volatilize tree sap, causing a tree to explode.

© Johnny Autery.

TABLE 10.10
U.S. Lightning Deaths by Decade

Decade	Deaths
1940–49	3,293
1950–59	1,841
1960–69	1,332
1970–79	978
1980–89	726
1990–99	570
2000–09	412

Source: NOAA data sets.

HEAT WAVE IN CHICAGO, JULY 1995

Dry, hot weather in the central United States and Canada is commonly associated with high-pressure atmospheric conditions (figure 10.43). In July 1995, a strong, upper-level ridge of high pressure sat on top of a slow-moving, hot, humid air mass on the surface. During the three-day period from 13 to 15 July, heat records were broken at numerous locations in the central and northern Great Plains. What made this heat wave especially difficult was its combination of both high maximum and high minimum temperatures (figure 10.44). The surface air mass did not cool much at night because its high humidity (high water content) held so much heat.

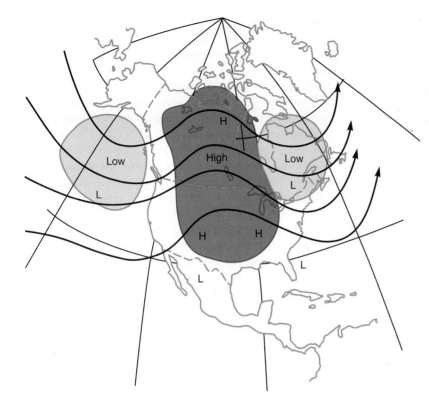

Figure 10.43 Dry conditions in the central United States and Canada are commonly caused by a long-lasting, high-pressure ridge in the upper troposphere. The ridge in the upper-level wind-flow path over the central parts of Canada and the United States causes anticyclones (clockwise rotations) where warm air aloft descends, warming further and lowering its humidity, and then sucking up moisture from the land below. The midcontinent high also blocks the northward flow of moist Gulf of Mexico air.

The health effects of the heat wave became apparent as bodies were recovered from overheated dwellings. Notice in figure 10.44 that most deaths do not occur at the peak of the heat wave, but rather happen for days afterward as bodies weaken and fail. From 13 to 27 July, 465 people in Chicago died of heat-related causes. By comparison, there were no heat-related fatalities from 4 to 10 July.

The people most affected were those without access to air conditioning and especially older people; the greater their ages, the higher their death percentages. An interesting statistic is that more than half the deceased lived on the top floor of their buildings, where heat buildup was greatest.

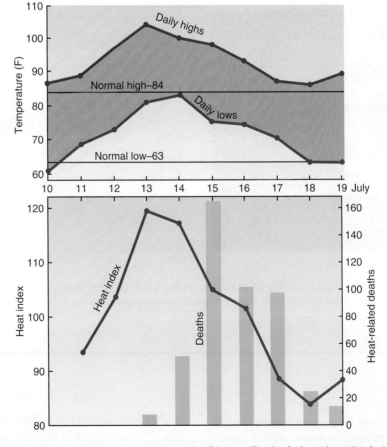

Figure 10.44 Top: Daily maximum and minimum temperatures in Chicago, Illinois, during 10 to 19 July 1995. Bottom: Numbers of heat-related deaths during that same time period. The heat index involves both temperature and humidity.

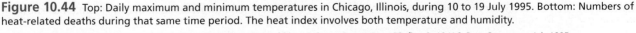

Data source: NOAA/U.S. Department of Commerce, "Heat Wave Natural Disaster Survey Report," pg. 23, figs. 9, 10;U.S. Dept Commerce, July 1995.

CITY WEATHER

Cities create their own weather. In the 1800s, it was recognized that Berlin, London, and Paris were warmer than the surrounding countryside. A study of Atlanta, Georgia, shows it to be an *urban heat island* with nighttime temperatures up to 10°C (18°F) warmer than the adjoining areas (figure 10.45). Atlanta has developed by cutting down 380,000 acres of forest and replacing it with buildings and roads. The urban concrete, asphalt, and stone absorb heat during the day and radiate heat at night. The warm air rising above a city can create its own low-pressure cell, a convecting plume of heat that can rise, cool, condense, and form thunderstorms.

City areas are urban heat islands where buildings and streets absorb solar heat all day and then release stored heat at night, thus robbing residents of the relief that cool nights bring. Night-time temperatures in the city may be much warmer than in the nearby countryside. City dwellers' body temperatures stay higher, and they are under greater stress.

EUROPEAN HEAT WAVES, 2003 AND 2010

The summer of 2003 brought record-breaking heat waves to Europe, and a delayed recognition of how many people it killed—more than 35,000. Countries especially hard hit with heat-wave deaths were France (14,802), Germany (7,000), Spain (4,230), and Italy (4,175). The 2003 European summer heat wave was unique. Statistical analysis of temperature records of the last 150 years shows that the 2003 temperatures were unprecedented.

In western Russia, the summer of 2010 brought early warmth in June—but when July arrived, the heat really began. Daytime high temperatures around 40°C (104°F) were common and numerous daily high-temperature records were set. July was the hottest known since good record keeping began in 1880. Persistent atmospheric high pressure stayed over western Russia, analogous to conditions in the United States in 1995 (see figure 10.43). The descending warm air also dried out the vegetation, setting the stage for fires to break out, and they broke out in abundance. More than six weeks of heat stress and elevated air pollution levels caused deadly conditions. Munich Re estimates that 56,000 people died.

As climate continues to warm through the 21st century, more and more heat waves are expected to occur at the same time as world population grows rapidly, with huge numbers of people migrating into cities. We need to learn more about how to deal with sustained heat waves with temperatures commonly up to 10°C (18°F) above average; they are expected to happen around the world with greater frequency.

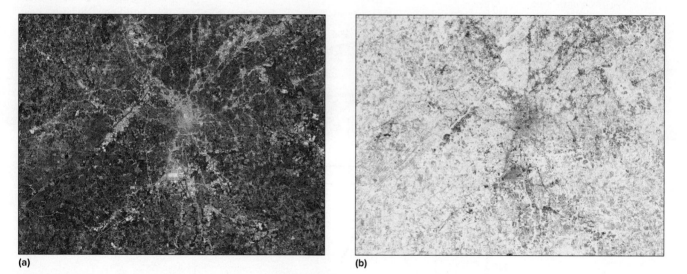

(a)   (b)

Figure 10.45 Urban heat island, Atlanta, Georgia, 28 September 2000. (a) In this photo, green indicates trees, and gray represents developed areas. (b) On this land surface temperature map, the hottest temperature is shown in red.
NASA.

Summary

Severe weather kills an average of 557 people in the United States each year. When exposure to winter storms lowers body temperature below 35°C (95°F), hypothermia may set in and death may follow. Falling snow may: (1) melt to form rain, (2) melt and then refreeze as sleet, (3) melt and then be supercooled into freezing rain, or (4) remain as snow. In winter, moist low-pressure systems moving up the Atlantic Coast may meet cold arctic air, producing powerful storms called nor'easters. Cold air blowing over warmer water, such as the Great Lakes, picks up water vapor and dumps it in huge volumes on downwind shores as lake-effect snow.

Air-mass thunderstorms commonly form in three stages on hot days in late afternoons: (1) In the early stage, warm humid air rises in updrafts, then cools into clouds. (2) In the mature stage, ice and rain fall in downdrafts. (3) In the dissipating stage, downdrafts dominate and cut off the updraft supply of moisture.

Severe thunderstorms occur when winds are faster than 93 km/hr (58 mph) or hail 25 mm (1 in) falls or a tornado is reported; they commonly form where warm and cold weather fronts collide. A supercell is a violent thunderstorm that forms from a huge updraft of air; it must develop large-scale rotation. Supercells are tilted, thus allowing rain and hail to fall from their leading parts while tornadoes may spin out of their trailing flanks. Layered ice balls called hail may fall from thunderstorms, especially in the central United States.

Tornadoes are rotating columns of air with wind speeds that can exceed 300 mph. Tornado strengths are classified by the Enhanced Fujita scale (EF). EF5 tornadoes have wind speeds exceeding 200 mph and are the deadliest. About 70% of Earth's tornadoes occur over the flatlands of the central United States where air masses can collide without interruption. Tornado-causing collisions often involve: (1) warm moist air moving north from the Gulf of Mexico, (2) fast-moving cold dry air moving in from the west, and (3) the jet stream moving east.

Wall clouds may form below a mesocyclone where the updraft is strongest; tornadoes commonly develop there. If the mesocyclone rotation is stretched within the wall cloud, its narrowed diameter of rotation may increase its velocity into a tornado. Tornadoes also form where the low-level collision of air masses creates a horizontal vortex of spinning air; the vortex tube can be lifted and stretched by an updraft, forming a tornado. The smaller the radius of a rotating air mass, the faster the wind speeds.

Lightning is a flow of electric current connecting positive charges with negative charges. Thunder is the sound made by rapidly expanding air flowing away from a lightning bolt. In the past 30 years in the United States, more people were killed by lightning than tornadoes, and Florida is the deadliest state for lightning.

Heat waves have been the deadliest weather phenomenon in the United States in recent decades. When body temperature exceeds 40°C (104°F), hyperthermia may set in and threaten life.

Terms to Remember

air-mass thunderstorm 255
blizzard 253
cumulonimbus 255
cumulus 255
derecho 261
downburst 258
fetch 254
freezing rain 251
hail 257
hook echo 264
hyperthermia 275
hypothermia 250
lake-effect snow 254
lightning 250
mesocyclone 257
nor'easter 251
severe thunderstorm 257
single-cell thunderstorm 256
sleet 251
supercell thunderstorm 257
thunder 272
thunderstorm 255
tornado 257
tornado outbreak 269
tornado warning 249
vortex 257
wall cloud 263

Questions for Review

1. On average, what kills more people in the United States each year: heat or tornadoes? Winter storms or hurricanes?
2. What weather events create billion-dollar disasters?
3. At what human body temperature can hypothermia set in?
4. Draw a cross-section through the lowermost atmosphere and explain how freezing rain forms.
5. Draw and label a cross-section that explains how lake-effect snow forms.
6. Sketch a series of vertical cross-sections showing the stages of development of a late-afternoon thundercloud. Label the processes occurring in the cloud during each stage.
7. Draw a cross-section through a supercell thunderstorm showing where rain, hail, lightning, and tornadoes are likely to occur.
8. What is a downburst?
9. How does hail form? Where does most of the large hail fall in the United States? How large can a hailstone grow?
10. What are derechos?
11. Draw a map and explain how a Gulf of Mexico warm air mass, a Canadian cold air mass, and the polar-front jet stream can meet and generate tornadoes.
12. How might a horizontal, spinning tube of air form and then be transformed into a tornado?
13. Describe the Enhanced Fujita Scale used to classify tornadoes.
14. What land and air conditions make the central United States the tornado capital of the world?
15. In what direction do most U.S. tornadoes travel? What controls this?
16. Why do higher wind speeds develop in a tornado than in a hurricane?
17. Why is lightning concentrated over the land and relatively uncommon over the oceans?
18. Which of the 48 conterminous states has the most hailstorms? Which state has the most lightning fatalities? Why are there different locations for these thunderstorm phenomena?
19. How do electrical-charge imbalances lead to lightning?
20. What is the relationship between thunder and lightning?
21. Does lightning move from cloud to ground, from ground to cloud, from cloud to cloud, or all?

Questions for Further Thought

1. Is it more dangerous to live in earthquake country or tornado country?
2. Where is the safest place to be during a tornado? Where is the most dangerous?
3. Do tornadoes have major impacts on large cities?
4. Sketch where and how a safe room could least expensively be added to your house in tornado country.
5. If you were on a site with a lot of flattened trees, how could you tell whether the damage was done by a tornado or a downburst?
6. You are hiking in the countryside when a lightning bolt flashes, followed quickly by loud thunder. What should you do?

CHAPTER 11

Hurricanes

Miracles appear to be so, according to our ignorance of nature, and not according to the essence of nature.

—MICHAEL DE MONTAIGNE, 1580, *ESSAYS I*

LEARNING OUTCOMES

Hurricanes grow over warm ocean waters and slam into coastal regions, killing many people and causing grievous destruction. After studying this chapter, you should:

- know the requirements for hurricane development.
- recognize the global distribution of hurricanes, typhoons, and cyclones.
- understand how Cape Verde-type and Caribbean-type hurricanes form.
- be familiar with the deadliest and most destructive processes wrought by hurricanes.
- be able to explain what to do when threatened by a hurricane.

OUTLINE

Debris from houses destroyed by Hurricane Katrina in Gulfport, Mississippi, in August 2005.

John Fleck/FEMA.

H urricane Sandy began in the western Caribbean Sea on 22 October 2012, taking the lives of 119 people while moving northward through the islands toward the mainland (figure 11.1). As Sandy traveled, its strength weakened at times but grew back to hurricane strength traveling over warm seawater along eastern North America. During the days that Sandy moved northward, hurricane warnings for the eastern United States were frequent and clear. Surveys of the effectiveness of the warnings were done by the Wharton School of Business at the University of Pennsylvania; they found that

- Some 90% of people made some preparations, mostly filling gas tanks in their cars and buying extra water.
- Only 50% of people who owned storm shutters put them up.
- Only 20% of people living in areas under evacuation orders planned to go.
- Most people thought the damages would come from wind, not flooding.
- More than 75% of respondents felt safe staying in their own homes.
- The people most worried were those who had never experienced a hurricane.

Hurricane Sandy covered a broad region with a wind diameter of 1,850 km (1,150 mi). Landfall occurred on the morning of 29 October near Atlantic City, New Jersey, with wind speeds up to 150 km/hr (90 mph) (figure 11.2). The biggest problem was not the winds but the flooding caused by the 4 m (13 ft) high storm surge, which rode on top of a high tide, helping raise an already high sea level even higher. Seawater flowed across the coastline and inland; cutting off power; flooding streets, tunnels, and subway lines; and covering 17% of New York City. Sandy ended the lives of 162 people in the United States and 2 in Canada; it is the 2nd costliest hurricane in U.S. history, causing more than $60 billion in damages.

As Hurricane Sandy moved northward off the eastern coast of North America, it could not take the usual path curving to the right out into the greater Atlantic Ocean; that path was blocked by a high-pressure atmospheric ridge over Greenland, which forced Sandy westward into the United States, where it combined with a stalled arctic front, increasing the fury of the total storm event (see later section on hurricane transition to post-tropical cyclone).

In 2011, Hurricane Irene also followed a path close to the North American coastline affecting every U.S. state and Canadian province along the eastern seaboard. Irene crossed the coastline three times; on the Outer Banks of North Carolina on Saturday, 27 August and in southern New Jersey and in Brooklyn, New York City, both on Sunday 28 August. There were 50 direct and 7 indirect deaths across 11 states and the province of Quebec. Deaths came mostly from wind blowing down trees and inland flooding caused by heavy rains. U.S. damages totaled $15.6 billion making Irene the 8th most expensive hurricane to hit the United States. Due to the extensive damages, the name

Figure 11.1 Track of Hurricane Sandy from Origin in Caribbean Sea on 22 October to landfall on New Jersey on 29 October. Symbols are circle = tropical cyclone; triangle = post-tropical cyclone; bright blue = tropical storm; light yellow = category 1 hurricane; yellow = category 2 hurricane.

NASA Goddard Space Flight Center Image by Reto Stöckli. Enhancements by Robert Simmon. Data and technical support: MODIS Land Group. Topography: USGS EROS Data Center.

Irene was retired by the World Meteorological Organization; the name will be replaced by Irma in 2017.

- In 2010, no hurricanes crossed the shorelines of the United States or Canada.
- In 2009, no hurricanes crossed the shorelines of the United States or Canada.

In 2008, there were six hurricane/tropical storm landfalls in the United States and one in Nova Scotia, Canada. The deadliest and most damaging was Hurricane Ike, the

Figure 11.2 Hurricane Sandy ravaged the Casino Pier and roller coaster in Seaside Heights, New Jersey, 29 October 2012.

U.S. Air Force photo by Master Sgt. Mark C. Olsen.

Hurricanes

Being caught by a hurricane while at sea was described in 1785 by Philip Freneau in *The Hurricane:*

> While death and darkness both surround,
> And tempests rage with lawless power,
> Of friendships voice I hear no sound,
> No comfort in this dreadful hour—
> What friendships can in tempests be,
> What comforts on this raging sea?
>
> The barque, accustomed to obey,
> No more the trembling pilots guide:
> Alone she gropes her trackless way,
> While mountains burst on either side—
> Thus, skill and science both must fall;
> And ruin is the lot of all.

A **hurricane** is a natural hazard that is given a human name. Andrew, Camille, Hugo, Ike, Iniki, Katrina, Mitch, Rita, Sandy and their kin share family traits, but each has its own personality. Each hurricane "lives" for enough days that we get to know its individual characteristics. Hurricanes (or typhoons) were described by Joseph Conrad in 1903 in his well-known book *Typhoon:*

> This is the disintegrating power of a great wind: it isolates one from one's own kind. An earthquake, a landslip, an avalanche, overtake a man incidentally, as it were—without passion. A furious gale attacks him like a personal enemy, tries to grasp his limbs, fastens upon his mind, seeks to rout his very spirit out of him.

Hurricanes are huge **tropical cyclones.** They are heat engines that convert the heat energy of the tropical ocean into winds and waves. These powerful storms can generate winds more than 240 km/hr (150 mph) (figure 11.3). Hurricanes can push massive volumes of seawater onshore as **surges** that temporarily raise sea level more than 6 m (20 ft). Their heavy rains can cause dangerous floods, killing people well away from the coastline, and tornadoes may spin out from their clouds.

Tropical Storm Allison poured rain onto Houston, Texas, on 7 June 2001, and then left the city only to return unexpectedly on 9–10 June. On its return, Allison dumped 36 cm (14 in) of rain onto the already saturated ground. Flooding drowned 2,500 animals involved in medical research at the Texas Medical Center and claimed the lives of 24 people in the region. One poignant story occurred in the Bank of America building after employees were warned that their cars in the underground parking garage should be moved before they were flooded. A 42-year-old law clerk took an elevator down to retrieve her car from the fourth underground level. As she reached the third level, flood water shut off power to the elevator. The slowly rising water drowned the unfortunate woman trapped in her elevator jail. Remember: Do *not* use elevators during natural disasters.

4th most expensive hurricane in U.S. history. Ike began as a storm crossing the Atlantic Ocean, roaring through the Caribbean killing 83 people, then moving over the Gulf of Mexico heading for Texas. Almost 1 million people along the Texas coast were told to evacuate ahead of Hurricane Ike's landfall. Forecasters said Ike could come ashore with 210 km/hr (130 mph) winds pushing a wall of water 6 m (20 ft) high, with waves up to 15 m (50 ft) high, and more than 25 cm (10 in) of rain. Most residents evacuated. But some folks waited; they had been through hurricane warnings before when nothing serious happened. George Helmond, a 72-year-old native of Galveston, planned to ride out this hurricane in his house as he had done for other storms. But at 2:30 a.m. on the 13th, during the full fury of the hurricane, George panicked, abandoned his home, and drove off in his pickup truck. His body was found the next day inside his battered truck; his abandoned home suffered little damage. Gail Ettinger, a 58-year-old Exxon chemist, decided to weather the storm in her house. As Ike pushed seawater higher and higher, she phoned a friend saying that she had made a big mistake by not evacuating. Her body was found 10 days later in a pile of debris 10 miles from her home. Warnings to evacuate before hurricane landfalls should be taken seriously.

Figure 11.3 Counterclockwise-rotating hurricane heading to Florida. Note the well-developed eye in its center.
© StockTrek/Getty Images RF.

Figure 11.4 Hurricane Katrina grew in strength traveling across unusually warm waters in the Gulf of Mexico. Water temperatures in yellow 28°C (82°F) and in red 32°C (90°F and above).
NASA/Goddard Space Flight Center Scientific Visualization Studio.

How a Hurricane Forms

A hurricane is a storm of the tropics. Heat builds up in the tropics during long, hot summers, and hurricanes are one means of exporting excess tropical heat to the midlatitudes. Before a hurricane can develop, several requirements should be met: (1) Seawater should be at least 27°C (80°F) in the upper 60 m (200 ft) of the ocean; (2) air must be warm, humid, and unstable enough to sustain convection; (3) the storm must be far enough (~500 km = 300 mi) from the equator for a Coriolis effect to be strong enough to spin the system; and (4) upper-level winds should be weak and preferably blowing in the same direction the developing storm is moving.

The 27°C (80°F) temperature is an ideal threshold for hurricane development. As the sea-surface temperature increases, the amount of water vapor that air can hold increases exponentially. When 27°C is exceeded, the amount of latent heat lifted from the tropical ocean easily becomes large enough to fuel a hurricane. Hurricane Katrina picked up tremendous amounts of heat from the warm Gulf of Mexico water (figure 11.4).

The development of a hurricane begins with a low-pressure zone that draws in a poorly organized cluster of thunderstorms with weak surface winds; this is a **tropical disturbance.** As surface winds strengthen and flow more efficiently around and into the center of the growing storm, it becomes a **tropical depression** and receives an identifying number. The Coriolis effect is the mechanism that spins the storm in a counterclockwise (cyclonic) rotation around a central core in the Northern Hemisphere. The converging surface winds meet at the central core, which acts like a chimney, sending warm, moist air flowing rapidly upward toward the stratosphere (figure 11.5). The rising moist air cools and reaches its dew point temperature where water vapor condenses, thus releasing prodigious quantities of latent heat. The released heat warms the surrounding air, causing stronger updrafts, which in turn increase the rate of upward flow of warm, moist air from below.

If the converging winds continue to spiral up the core wall at ever-increasing speeds, then the cyclonic system grows in strength. When sustained surface-wind speeds exceed 63 km/hr (39 mph), it has become a **tropical storm** (surface winds from 39 to 74 mph).

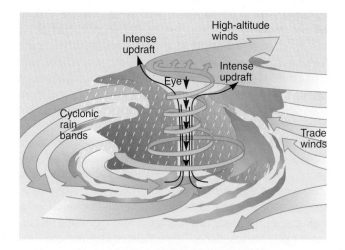

Figure 11.5 Schematic drawing through a hurricane. Low-altitude trade winds feed moisture and heat to the eye. Updrafts rise rapidly up the core (eye) wall and are helped away by high-altitude winds.
Source: U.S. Department of Commerce, 1971.

As increasing amounts of wind blow faster into the center of a tropical storm, it becomes difficult for all winds to reach the center. The result is a spiraling upward cylindrical wind mass in the center of the storm. When surface-wind speeds exceed 119 km/hr (74 mph), none of the wind reaches the center of the storm, resulting in the calmer clear area known as the **eye** (figures 11.3 and 11.6). Because the distinctive eye forms at wind speeds of about 119 km/hr (74 mph), this wind speed defines the threshold where a tropical storm has grown strong enough to be called a hurricane.

The strength of a hurricane depends on the speed with which surface winds can flow into the central core, race up its sides, and easily flow out and away in the upper atmosphere. As the central core or column becomes a more efficient "chimney," the hurricane grows stronger.

Consider the tropical weather system in figure 11.6. Where in the world was it?

How a Hurricane Works

A hurricane is not a single, uniform convective storm. Instead, it is composed of numerous thunderstorms in bands that move in rotating spirals around the hurricane center (see figure 11.3). The general wind flow pattern is convergent toward the center at low levels, rising in convective clouds, and divergent at upper levels (figure 11.7a).

Clouds and rainfall are not evenly distributed within a hurricane. They are most intense in **rain bands,** the spiraling

Figure 11.6 Is this a hurricane in the Northern Hemisphere (counterclockwise rotation) or a cyclone in the Southern Hemisphere (clockwise rotation)? For answer, see Questions for Review at the end of the chapter.

NASA image courtesy Jeff Schmaltz, MODIS Land Rapid Response Team at NASA GSFC.

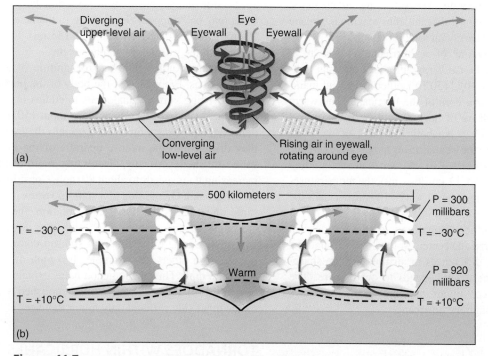

Figure 11.7 **Structure of a hurricane.** (a) Low-level air converges toward the storm center, rises into clouds and rotates up the eyewall, and diverges aloft. (b) Lines of temperature (dashed) and pressure (solid black) define the low-pressure warm core. The steepest pressure gradient at low levels of the eyewall causes the greatest wind speeds.

Figure 11.8 Rain bands of Hurricane Frances are emphasized by removing other clouds. Note the tall and powerful thunderstorms within the bands. Colors indicate rainfall per hour; green is 0.5 in and blue is 0.25 in.

Goddard Space Flight Center/NASA.

Figure 11.9 Inside the eyewall of Hurricane Katrina, 29 August 2005.

Dewie Floyd/NOAA.

bands of clouds around the eye (see figures 11.3, 11.4, and 11.6). The dense clouds in the curving rain bands contain thunderstorms with heavy rainfall and strong wind gusts (figure 11.8). Tornadoes associated with hurricanes often form within the rain bands. Between the spiraling rain bands, the weather is relatively calmer and rainfall is less intense.

Beyond the hurricane's mass of clouds is a region of subsiding air. Divergent air aloft descends, warms adiabatically, and makes clouds evaporate. The subsiding air creates a clear region surrounding a hurricane, thus making hurricane boundaries stand out sharply on satellite photos.

More of the structure of a hurricane is shown in figure 11.7b. Follow the *isotherms,* the lines of equal temperature (T). The T = 10°C (50°F) isotherm rises up markedly to define the *warm core* of the hurricane. The warm core is due to the release of huge quantities of latent heat from rising winds and to adiabatic warming of the sinking air in the eye. At the sea surface, temperatures in the eye may be only 0–2°C (0–3°F) warmer than surrounding air, but at altitudes of 10–12 km (6–7 mi), the temperatures within the warm core may be 11°C (20°F) warmer than the surrounding environment.

Follow the 920-millibar isobar and see the big drop in air pressure at the hurricane center. At low levels around the hurricane center, the differences in air pressure are sharpest, causing the highest wind speeds to occur there. The pronounced differences in air pressure and temperature decrease upward through the hurricane.

EYEWALL AND EYE

As inward-flowing air gets closer to the hurricane center, the rotational wind speed increases. At a 5 to 15 km (3 to 9 mi) distance from the center, the wind speed can no longer

increase, and the converging winds rise up the **eyewall** (figure 11.9). The eyewall is an eye-encircling ring of tall thunderstorms that typically have the highest wind speeds and heaviest rainfall in a hurricane. Air in the eyewall spirals upward and outward. To replace this outflow, cool air aloft sinks into the center of the core up high. As this air descends, it warms adiabatically and absorbs moisture, leaving the core clear and relatively cloud free, thus forming the eye. Inside an eye, blue sky or stars may be seen, and seabirds may be "trapped." Eyes range in diameter from about 8 km (5 mi) to more than 200 km (125 mi). Eyes and eyewalls form the warm core of a hurricane.

Hurricane Andrew allowed recognition of a new twist in hurricanes. Embedded in the body of the hurricane were small twisting vortices analogous to eddies in a river or to tornadoes. The twisting eddies were about 150 m (500 ft) in diameter, and many were sucked up into the eyewall. When they drifted into the intense updrafts of the eyewall, they were stretched vertically, decreasing their diameters and increasing their speeds to about 80 mph. Consider the effects on the ground. On the side of a spin-up vortex moving in the same direction as the hurricane's rotation, the two speeds are additive—for example, 130 mph plus 80 mph equals 210 mph. On the other side of a spin-up vortex, the winds oppose each other—for example, 130 mph minus 80 mph equals 50 mph. This phenomenon helps explain why houses on one side of a Florida street were demolished while houses on the other side suffered only minor damage.

TORNADOES WITHIN HURRICANES

Tornadoes form within some hurricanes—most commonly in the right-front quadrant of the hurricane in outer rain bands some 80–500 km (50–300 mi) from the hurricane center.

Tornadoes are most likely to form in hurricanes that are: (1) large, intense, and strongly curving; (2) moving forward at 12–30 km/hr (8–18 mph); (3) interacting with old, weakened fronts; and (4) over land. Tornadoes are more likely when hurricanes interact with land. The surface winds slow down due to friction with the land, while the winds aloft keep up their momentum, creating the necessary vertical wind shear for rotation.

In 2004, Hurricane Ivan caused a multiday outbreak of 127 tornadoes, killing 7 people in 9 eastern states from Florida to Pennsylvania. Also in 2004, Hurricane Frances spun out 106 tornadoes. In contrast, some hurricanes produce no tornadoes.

ENERGY FLOW IN A HURRICANE

Follow the flow of energy through a hurricane (figure 11.10). First, warm, moist surface winds converging toward the eyewall pick up ever more heat from the ocean, and wind velocity keeps increasing. Upon reaching the eyewall, the air rises rapidly, cools, and releases latent heat of condensation that adds to buoyancy and upward velocity. At the top of the hurricane, airflow diverges and loses energy through long-wavelength radiation to space.

Returning to the sea surface in figure 11.10, winds increase their speed inward as the pressure gradient increases, and as ever more heat and water vapor are picked up by the turbulent wind from the ever rougher seas.

These energy increases in converging air are reduced by greater frictional losses of energy to the rougher sea surface. Also, air flowing toward the very low pressure in the storm center undergoes adiabatic expansion and cooling.

How do these increases and decreases in energy balance out? The inward drop in air pressure causes converging air to expand and cool slightly below the sea-surface temperature. This temperature difference generates a flow of sensible and latent heat from the sea into the air in order to reach equilibrium. These ongoing processes allow the air to pick up and store increasing amounts of energy from the sea. This energy is carried up into the eyewall and released to power the hurricane.

HURRICANE ENERGY RELEASE

A hurricane acts as a heat engine, transferring heat from the warm, moist air above tropical seas into the core of the hurricane. As air rises into the hurricane, latent heat is released in staggering quantities. The average hurricane generates energy at a rate 200 times greater than our worldwide capacity to generate electricity. The kinetic energy of winds in a typical hurricane is about half our global electrical

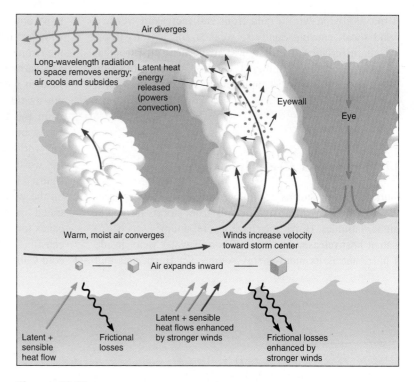

Figure 11.10 Energy flow within a hurricane.

capacity. Summing up, the energy released in a hurricane by forming clouds and rain is 400 times greater than the energy of its winds.

HURRICANE TRANSITION TO POST-TROPICAL CYCLONE

Hurricanes may transform into **post-tropical cyclones,** often between 30° and 40° latitude, if there is sufficient collision with upper-level troughs in the westerly wind belt. The collision forces a hurricane to tilt back into a colder air mass, disrupting convection up the eyewall and thus cutting off its primary source of energy—the latent heat of condensation supplied by thunderstorms near the hurricane center. When a hurricane ceases to have warm, moist air flowing up its *warm core,* it can turn into a *cold core* and the hurricane body can merge with the colliding weather front. The combined weather system may increase in size as the hurricane transforms.

In 2012, Sandy provided a good example of a hurricane transforming into a post-tropical cyclone before slamming ashore in New Jersey. As Sandy moved northward over the warm seawater of the Gulf Stream, it was supplied enough latent heat to maintain hurricane status (see circles along travel path in figure 11.1). However, when Sandy turned westward toward land, it passed over cold seawater, transforming its warm, energy-feeding core into the cold core of a post-tropical cyclone (see triangles along travel path in figure 11.1). The westward turn of Sandy also brought it into collision with a cold front and increasing vertical wind shear, causing Sandy to weaken and lose its tropical characteristics. Landfall was made as a post-tropical cyclone. The huge combined weather system of Sandy plus the cold front was unofficially called "Superstorm Sandy" by the media.

Hurricane Origins

Hurricanes are storms from low latitudes—that is, from the tropics. They differ significantly from storms formed in higher latitudes. Hurricanes have several unique aspects: (1) Latent heat released by condensation of water vapor inside a hurricane is its main energy source. (2) Hurricanes that move onto land weaken rapidly. (3) Fronts are not associated with hurricanes. (4) The weaker the high-altitude winds, the stronger a hurricane can become. (5) Hurricane centers are warmer than their surroundings. (6) Hurricane winds weaken with height. (7) Air in the center of the eye sinks downward.

In the United States we know these storms as hurricanes, but they go by different names in different parts of the world. In the Indian Ocean, they are **cyclones,** and in the western Pacific Ocean, they are **typhoons.** Notice in figure 11.11 and figure 11.12 that tropical cyclones form on the west sides of oceans where warm water is concentrated, with two informative exceptions. Why do hurricanes form off the Pacific Coast of Mexico? A bend in the coastline isolates a pool of warm coastal water from the cold California current. Why do hurricanes rarely form off

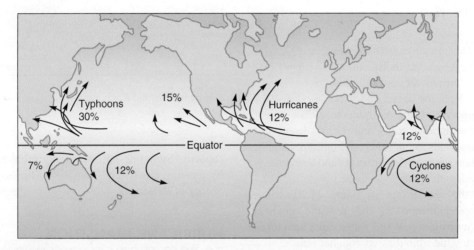

Figure 11.11 Map of common areas where hurricanes form, typical paths they travel, and annual percentage of Earth's large cyclones occurring in each region. Note that they are called cyclones in the Indian and South Pacific Oceans and typhoons in the west Pacific Ocean.

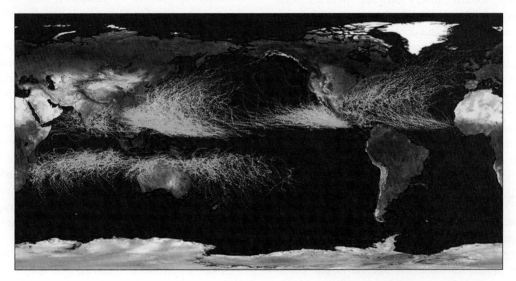

Figure 11.12 Tracks of all tropical cyclones in the world, 1985–2005. Blues indicate tropical depression and tropical storm strengths, yellows are hurricane categories 1 and 2, and oranges to reds are categories 3–5.

Tracks from nilfanion on NASA image.

TABLE 11.1

Saffir-Simpson Hurricane Damage Potential Scale

	Barometric Pressure		Wind Speed		Storm Surge		
	(millibars)	(inches)	(km/hr)	(mph)	(meters)	(feet)	Damages
Category 1	≥980	Over 28.92	119–154	74–95	1.2–1.5	4–5	Minimal
Category 2	965–979	28.50–28.91	155–178	96–110	1.8–2.4	6–8	Moderate
Category 3	945–964	27.91–28.49	179–210	111–130	2.7–3.7	9–12	Extensive
Category 4	920–944	27.17–27.90	211–250	131–155	4–5.5	13–18	Extreme
Category 5	<920	Less than 27.17	>250	Over 155	>5.5	Over 18	Catastrophic

Brazil? The Atlantic Ocean is too narrow and there is not enough warm water.

All hurricanes, cyclones, and typhoons are rotating, low-pressure weather systems with warm cores that generally form over warm seawater between 5° and 20° latitude and then travel off to deliver their heat to higher latitudes. Hurricanes do not form along the equator because there the Coriolis effect is zero. The Coriolis effect is so weak within 5°N or S of the equator that there is not enough rotation to build hurricanes. Even an already formed hurricane could not cross the equator because without the Coriolis effect it would lose its rotation (see figure 11.12).

Each year, about 84 tropical cyclones (hurricanes, typhoons, cyclones) form on Earth. U.S. residents think the North Atlantic Ocean–Caribbean Sea–Gulf of Mexico area is where the action is, but on a global scale, it accounts for only about 10 of the roughly 84 events. The typhoons of the northwest Pacific Ocean hit Japan, China, and the Philippines about three times as often, and the storms can be stronger.

The strength of tropical cyclones and the damages they inflict are assessed by the Saffir-Simpson scale (table 11.1). In category 1, winds damage trees and unanchored mobile homes. Category 2 winds blow some trees down and do major damage to mobile homes and some roofs. In category 3, winds blow down large trees and strip foliage, destroy mobile homes, and cause structural damage to small buildings. In category 4, all signs are blown down; damages are heavy to windows, doors, and roofs; flooding extends miles inland; and coastal buildings suffer major damage. In category 5, damages are severe to windows, doors, and roofs; small buildings are overturned and blown away; and damages are major to all buildings less than 5 m (15 ft) above sea level and within 500 m (1,640 ft) of the shoreline.

Hurricanes of category 3 to 5 strength hitting the United States cause 86% of the damages but constitute only 24% of the landfalling hurricanes. Because of their damaging power, they are called **major hurricanes.**

North Atlantic Ocean Hurricanes

Hurricanes in the North Atlantic Ocean are large, mobile, and long-lasting (table 11.2). Each year, from 4 to 28 tropical storms and hurricanes occur in the North Atlantic–Caribbean Sea–Gulf of Mexico region (table 11.3). The coastline of the United States is frequently crossed by landfalling tropical cyclones (table 11.4).

The arrival of rotating tropical-weather systems is an annual event in the United States. In the 156 years from 1851 to 2006, the U.S. Gulf and Atlantic coastlines were hit by 279 tropical cyclones of hurricane strength, averaging 1.8 per year. Of these 279 hurricanes, only 3 were category 5 in strength and 18 were category 4. These figures exclude the additional large number of destructive hurricanes that hit the Caribbean islands and Central America.

Hurricanes form when sea-surface temperatures are warmest; in the North Atlantic Ocean, this occurs in late summer (figure 11.13). The warmest weather occurs earlier in the summer, but sea-surface temperatures are highest at the end of summer because the ocean water with its high heat capacity keeps absorbing solar energy all summer long. Looking at the arrival times of the deadliest hurricanes in the United States reveals an abundance in September.

TABLE 11.2

General Characteristics of North Atlantic Hurricanes

Storm diameter	200 to 1,300 km (125 to 800 mi)
Eye diameter	16 to 70 km (10 to 44 mi)
Surface wind speed	≥119 km/hr (≥74 mph)
Direction of motion	Westward, then north, then east
Life span	1 to 30 days

TABLE 11.3

North Atlantic Tropical Storms and Hurricanes, 1851–2014

Category	Frequency (year)	
	Maximum	Minimum
Tropical storms and hurricanes	28 (2005)	4 (1983)
Hurricanes	15 (2005)	2 (2013)
Major hurricanes (winds >110 mph)	8 (1950)	0 many times (e.g., 2013)
U.S. landfalling tropical storms and hurricanes	8 (1916)	1 many times (e.g., 1997)
U.S. landfalling hurricanes	6 (1985, 2004, 2005)	0 many times (e.g., 2013)
U.S. landfalling major hurricanes	4 (2005)	0 many times (e.g., 2014)

Data from www.nbc.noaa.gov.

TABLE 11.4

Hurricane Landfalls on U.S. Coastline, 1851–2014

State	Hurricane Hits
Florida	114
Texas	64
Louisiana	58
North Carolina	53
South Carolina	30
Alabama	27
Georgia	23
Mississippi	19
New York	13
Massachusetts	11
Connecticut	11
Virginia	10

Data Source: NOAA.

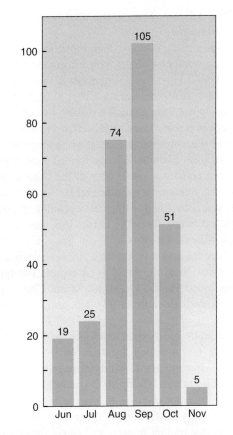

Figure 11.13 Hurricanes striking the U.S. mainland, 1851–2006.
Data source: NOAA.

CAPE VERDE-TYPE HURRICANES

Tropical cyclones may form and begin growing toward hurricane status in the waters near the Cape Verde Islands off northwest Africa. Hurricanes may develop from either a preexisting convective storm or, more commonly, from an **easterly wave** (figure 11.14). Easterly waves are disturbances or mega-ripples that develop within the trade winds; they are elongate, north–south oriented, and extend 2,000 to 3,000 km (1,200 to 1,900 mi). On the east side of the wave axis, the low-level winds converge and rise, forming clouds with rain. On the west side of the axis, the upper-level winds diverge and sink, resulting in clear skies. The instabilities in the atmosphere created by the easterly waves lead to tropical disturbances that sometimes grow energetic enough to be a hurricane.

Upon reaching the warm water of the subtropical Atlantic Ocean, some of these tropical disturbances strengthen rapidly and may even reach tropical storm status near the Cape Verde Islands. These Cape Verde–type tropical cyclones are blown westward across the Atlantic Ocean by the trade winds between 5° and 20°N latitude, picking up heat from warm ocean water. The great distances that these tropical cyclones

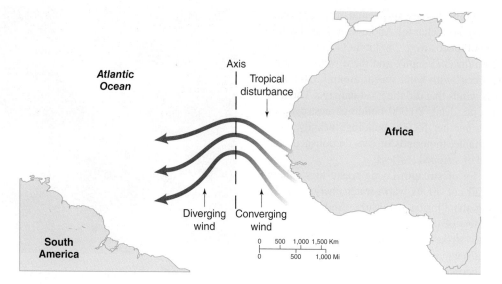

Figure 11.14 An easterly wave within the trade winds. To the east (right) of an axis, the winds converge and rise, and clouds with rain occur. To the west (left) of an axis, the winds diverge and sink, and clear skies occur.

travel over water warmer than 27°C (80°F) are a major factor in growing to hurricane strength. And it is late summer when the low-latitude North Atlantic Ocean is at its warmest; summer is when westward-traveling storms have the opportunity to get well-organized as heat engines using abundant fuel (latent heat) derived from evaporation of sea-surface water to grow into Cape Verde–type hurricanes. Approaching the Western Hemisphere, they commonly move north on clockwise-curving paths due to the Coriolis effect. As a hurricane moves poleward, the Coriolis effect strengthens.

Hurricane Andrew, August 1992

Andrew was born in Africa. On 13 August, it had developed into thunderstorms over West Africa. It then moved out over the Atlantic Ocean as rainy, low-pressure wind waves that converged at low angles to form a rotating air mass (figure 11.15). By 17 August, the central circulation

had intensified into a tropical storm, but high-level winds disrupted the upward growth of the rotating core of clouds. Weak and disorganized, Andrew drifted west across the Atlantic Ocean. By Friday, 21 August, Andrew had moved to 1,600 km (1,000 mi) off Florida when the upper-level winds died down, allowing growth in cloud height and wind strength. At the same time, a high-pressure zone built to the north, forcing Andrew to travel westward over energy-yielding warmer waters (figure 11.15). On Saturday, its wind speeds blew above 119 km/hr (74 mph); Andrew had grown to hurricane strength. On Sunday, 23 August, the revitalized Andrew moved through the northern Bahamas with wind speeds of 240 km/hr (150 mph), killing four people. But the worst was yet to come.

After 3 a.m. on Monday, 24 August, Andrew crossed southern Florida with a vengeance. In Andrew's 40 km (25 mi) wide path of destruction lay the residences of more than

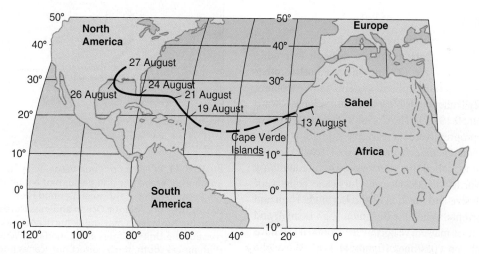

Figure 11.15 The odyssey of Hurricane Andrew, August 1992.

350,000 people. Mobile homes and abundant poorly constructed houses were no match for Andrew's winds. By this time, sustained wind speeds were 250 km/hr (155 mph) with gusts up to 282 km/hr (175 mph), and there were tornado-like "spin-up vortices" with wind speeds around 320 km/hr (200 mph). It was winds that did the most damage. The hurricane left 33 people dead, 80,000 buildings destroyed or severely damaged, another 55,000 buildings heavily damaged but still habitable, thousands of cars demolished, and most trees downed or stripped leafless.

Andrew lost a lot of strength laying waste to southern Florida; the landfall cut off its warm-water energy supply. But after crossing the Florida peninsula, Andrew moved onto the warm waters of the Gulf of Mexico and recouped enough energy to strike the Louisiana coastline with 190 km/hr (120 mph) winds early on Wednesday, 26 August, killing another 15 people. Andrew spent most of its remaining energy as heavy rains dropped on Mississippi on 27 August, and then it dissipated.

Andrew was the 3rd most destructive hurricane in U.S. history, with $46 billion in damages; it was the 3rd strongest hurricane since 1900. The only stronger hurricanes were one on Labor Day 1935, which crossed the Florida Keys, killing 408 people; and Camille, which slammed into Mississippi in mid-August 1969 with a 7.3 m (24 ft) high sea surge, leaving 256 dead.

Many hurricane-related deaths are due to poorly constructed buildings, not the natural event itself. Older mobile homes are dangerous places to be during a hurricane. Many houses in southern Florida have roofs covered with shingles that are only stapled down, building frameworks that are weakly supported, and windows that are not covered by storm shutters. Once a window breaks, hurricane winds enter a house, tear up the inside, and lift off its roof. A study of Andrew's damages concluded that up to 40% of the losses could have been avoided if buildings had been constructed to meet the wind-resistance standards of the South Florida Building Code. Poor construction and lax enforcement of building codes caused much unnecessary suffering.

Hurricane Paths

The paths followed by Atlantic hurricanes have been plotted (see figure 11.12). Hurricane paths are difficult to predict in detail for several reasons: (1) They must adjust to other high- and low-pressure atmospheric systems they encounter (figure 11.16). (2) Trade winds blow tropical cyclones toward the west (see figure 9.19). (3) The Coriolis effect adds a curve to the right that progressively increases in strength with distance from the equator (see figure 9.16). (4) An extensive high-pressure zone called the Bermuda High commonly sits above the North Atlantic Ocean. Hurricane paths vary depending on the size and position of the Bermuda High and other steering currents. When the Bermuda High is small and to the east, hurricanes may curve northward around it and have little or no effects on coastlines (figure 11.17a). When the Bermuda High is strong and extensive, it may guide hurricanes

Figure 11.16 A cold front moving east holds Hurricane Erin over the Atlantic Ocean, preventing landfall on the United States or Canada.
NASA.

along the east coast of the United States (figure 11.17b), causing widespread death and destruction as resulted from the New England hurricane of 1938, Diane in 1955, Donna in 1960, and Agnes in 1972. Sometimes the Bermuda High drifts southwestward toward Florida and helps direct hurricanes into the Caribbean Sea and Gulf of Mexico (figure 11.17c). When a hurricane travels far enough to the north, the westerly winds push it to the northeast (see figure 9.19).

The position and strength of the Bermuda High may help explain the historic pattern of hurricane landfalls.

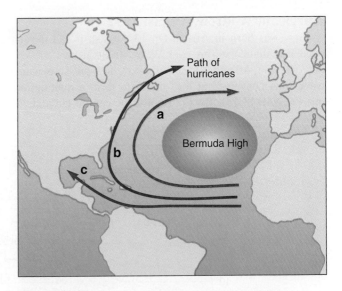

Figure 11.17 Paths of Cape Verde–type hurricanes are influenced by the size and position of a high-pressure zone, the Bermuda High. (a) A small Bermuda High allows hurricanes to stay over the Atlantic Ocean and miss North America. (b) A large Bermuda High may guide hurricanes along the eastern coast of the United States and Canada. (c) When the Bermuda High moves south, it may direct hurricanes into the Caribbean Sea and Gulf of Mexico.

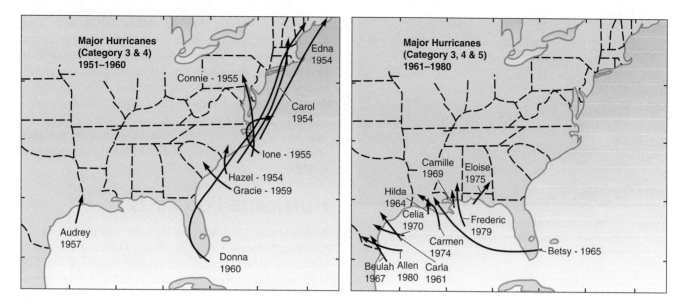

Figure 11.18 Category 3 and greater hurricanes cross the U.S. coastline at varied places. (a) In the 1950s, hurricanes mostly hit the East Coast. (b) In the 1960s and 1970s, the Gulf Coast was most affected.

Figures after Pielke and Pielke (1997).

For example, in the 1950s, the East Coast of the United States was hammered by major hurricanes, but in the 1960s and 1970s, it was the Gulf Coast that was hit most frequently (figure 11.18).

CARIBBEAN SEA AND GULF OF MEXICO–TYPE HURRICANES

Another way tropical cyclones can originate and then grow to hurricane status occurs at the Intertropical Convergence Zone (ITCZ) (see figure 9.19). The ITCZ is a major site of convergence of warm, moist, low-level winds that rise high, forming a nearly continuous ring of thunderstorms surrounding the Earth. Thunderstorms are a daily event along the ITCZ, and occasionally a cluster of thunderstorms breaks away and organizes into a more unified system. Add some rotation to these convective columns and a tropical cyclone is born.

Hurricanes commonly form above the very warm waters of the Caribbean Sea. Water warm enough to spawn hurricanes is carried by the Loop Current from the Caribbean Sea, northward between Mexico and Cuba, into the Gulf of Mexico, adding to the hurricane potential there.

The convergence zone (ITCZ) occurs where the trade winds meet near the equator (see figures 9.19 and 9.27b). The southwestward-blowing trade winds of the Northern Hemisphere (northeast trades of figure 9.19) converge or collide near the equator, with the Southern Hemisphere trade winds flowing northwestward (southeast trades). The location of the ITCZ moves with the seasons. The average position of the ITCZ is about 5°N latitude. In January, it mostly lies south of the equator, but by July, all the ITCZ is north of the equator. Where the airflows converge, low-pressure areas and tropical depressions form, with thunderstorms, a large core or center, rising moist air, and rotation

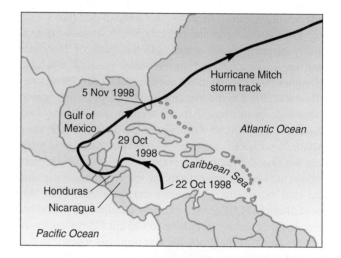

Figure 11.19 Hurricane Mitch began in the Caribbean Sea on 22 October 1998. Mitch stalled offshore from 27 to 29 October, dumping enormous volumes of rain on Honduras and Nicaragua. On 3 November, Mitch entered the Gulf of Mexico, picked up strength from the warm water, and headed northeast across the Atlantic, traveling north of England on 9 November.

combining to create a tropical cyclone that can strengthen to a hurricane, such as Mitch in 1998.

Hurricane Mitch, October 1998

In the early morning hours of 22 October 1998, Tropical Depression 13 formed at the ITCZ over the Caribbean Sea north of the Panama-Colombia border (figure 11.19). The warm Caribbean water supplied so much energy that within 18 hours, it was Tropical Storm Mitch, and in another 36 hours, it was Hurricane Mitch. On 26 October, Mitch had grown to be one of the strongest category 5 hurricanes on record, with sustained winds of 290 km/hr (180 mph) and gusts greater

Figure 11.20 Hurricane Mitch stalled offshore and poured huge volumes of rain onto Honduras and Nicaragua, October 1998.
Hal Pierce, Laboratory for Atmospheres, NASA GSFC.

than 320 km/hr (200 mph). Wind speeds remained at more than 250 km/hr (155 mph) for 33 consecutive hours, which is the second longest in the North Atlantic region.

Mitch was heading toward Cuba but then turned sharply left toward Central America. As landfall for the 37 km (23 mi) diameter eye and the category 5 winds were anxiously awaited, Mitch stalled off the coast of Honduras late on 27 October and stayed there, pouring rain onto Honduras and Nicaragua all day long on 28 and 29 October (figure 11.20). On the evening of 29 October, wind speeds slowed down to tropical storm strength, and Mitch slowly came onshore on 30 October. In the case of Hurricane Mitch, wind speeds and landfall were not the biggest problem—it was the rainfall. In effect, Mitch acted like a giant siphon sucking up water from the sea and dumping it onto the land, especially in Honduras and Nicaragua. Three-day rainfall totals of 64 cm (25 in) were common, and in some mountainous areas, rainfalls up to 190 cm (75 in) were estimated. Think of the problems that occurred when 2 to 6 ft of rainfall had to run off the land.

In Honduras, about 6,500 people were killed, 20% of the population became homeless, about 60% of roads and bridges were made unusable, and 70% of the crops were destroyed. The president of Honduras, Carlos Flores Facusse, stated that Mitch had wiped out 50 years of progress.

In Nicaragua, about 3,800 people were killed. The worst incident occurred 360 km (225 mi) inland from the Caribbean Sea when the crater lake atop Casitas Volcano filled with rainwater and the crater wall failed, sending lahars (mudflows) 23 km (14 mi) downslope to the Pacific Ocean. Four villages were overwhelmed, and about 2,000 people were buried beneath mud 2 to 6 m (6 to 20 ft) thick.

While many of the people of Central America were still fighting for their lives, Mitch moved out onto the warm water of the southern Gulf of Mexico (figure 11.19), grew in strength to a tropical storm, crossed southern Florida, and carried its heat and energy across the Atlantic Ocean north of the British Isles on 9 November.

During a 15-day rampage, Mitch killed more than 11,000 people, making it the second deadliest hurricane in the Americas, behind only the Great Hurricane of October 1780, which killed an estimated 22,000 people on several Caribbean islands. As survivors worked to restore the Nicaraguan economy in the Casitas Volcano area in late 1998 and 1999, they had to cope psychologically with "the return of their dead." New rains eroding the mud deposits kept exposing the bodies of family members and neighbors.

Hurricane Forecasts

Atlantic basin hurricanes are forecast and tracked by the National Hurricane Center in Miami, Florida. We see their constantly updated reports and satellite images on our TVs and computers as they follow the growth and development of tropical storms and hurricanes and forecast their paths. The advance warnings of hurricane landfalls provided by the National Hurricane Center permit timely evacuations that have saved thousands of lives. For example, in September 1900, a hurricane hit Galveston, Texas; residents had little advance warning, and about 8,000 people died. In September 2008, Hurricane Ike hit Galveston, but due to ample warnings most of the residents evacuated before Ike's landfall; the death toll was 112 of the people who ignored the warnings and stayed. Same severity of hurricane, same place: with minimal warning, 8,000 died; with maximum warning, 112 died.

Can we forecast how many Cape Verde– and Caribbean/ Gulf of Mexico–type hurricanes are likely to form each year? Progress is being made. Both William Gray of Colorado State University and the National Hurricane Center have had some success, based on numerous variables: (1) When the western Sahel region of Africa (see figure 11.15) is wet, its greater number of thunderstorms provides more nuclei for hurricanes (figure 11.21). (2) The warmer the sea-surface temperatures, the more energy is available to help tropical depressions grow into hurricanes. (3) Low atmospheric pressure in the Caribbean region aids the formation of tropical cyclones. (4) If La Niña conditions are present in the Pacific Ocean, west-blowing trade winds help move hurricanes over warm water. But if El Niño exists, its east-blowing, high-level winds tend to disrupt and break apart tropical cyclones. Remember that a tropical low-pressure zone can grow into a hurricane only if rapidly rising moist air releases huge quantities of latent heat into the growing storm. If upper-level winds are cutting into the tall storm clouds, they disrupt the vertically rising air and make it difficult for hurricanes to form. Most Northern Hemisphere hurricanes travel westward, whereas El Niño brings eastward-blowing winds that disrupt storms and help make a quiet season with few hurricanes. However, if La Niña conditions are present, then westward-blowing trade winds aid the growth of hurricanes. During a La Niña in the Pacific Ocean, hurricanes may be more common and more intense in the North Atlantic Ocean, Gulf of Mexico, and Caribbean Sea.

Wherever they begin, many tropical cyclones reach hurricane strength above the warm waters of the westernmost Atlantic Ocean, Caribbean Sea, and Gulf of Mexico. The annual probabilities of a named storm occurring over the waters of this region are sobering (figure 11.22). Many of these hurricanes will attack the United States. The annual probabilities of hurricane landfalls on the U.S. coastline are shown in figure 11.23.

Hurricanes are

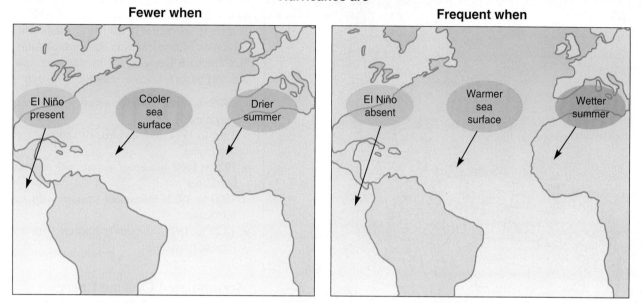

Figure 11.21 The frequency of North Atlantic region hurricanes is affected by climatic conditions.

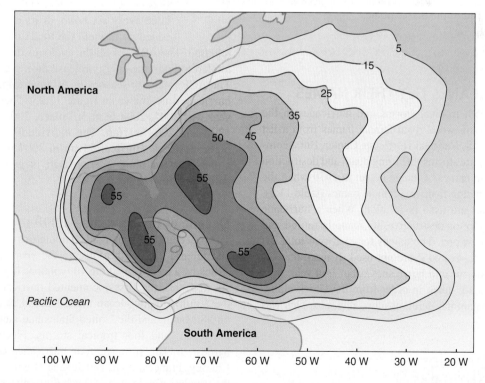

Figure 11.22 Annual probability of a named storm being within each area during the June to November season for tropical storms and hurricanes, 1944–1999.

Map after Chris Lansea of NOAA.

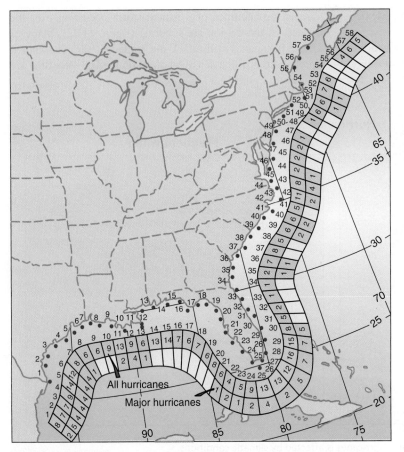

Figure 11.23 Percent probability that a hurricane (74 mph or faster) or a major hurricane (125 mph or faster) will occur in a given year along 80 km (50 mi) long segments of the U.S. coastline.

After Simpson and Lawrence, 1971.

HOW HURRICANES GET THEIR NAMES

Beginning in 1953, tropical storms and hurricanes in the North Atlantic basin were given female names from a list prepared by the U.S. National Hurricane Center. But a complaint arose. Why are storms that bring death and destruction only given female names? Change began in 1979 when the lists started alternating male and female names (table 11.5). Six lists of names are used in rotation. When a hurricane is especially deadly or destructive, its name is retired and replaced with a new one. The names for each year total 21; the letters Q, U, X, Y, and Z are not used. If more than 21 named tropical storms and hurricanes occur in a year, then additional names are taken in order from the Greek alphabet: alpha, beta, gamma, delta, epsilon, zeta, etc. This was necessary in 2005.

HURRICANE TRENDS IN THE ATLANTIC BASIN

Attempts to understand hurricane frequency and intensity lead into other data and thought paths.

Atlantic Multidecadal Oscillation

An **Atlantic Multidecadal Oscillation (AMO)** has been recognized in recent years based mainly on sea-surface temperatures in the North Atlantic Ocean (figure 11.24). A reasonably distinct pattern appears to exist, showing alternating times of colder versus warmer water. Scientists debate the accuracy of the variations in amplitude. They question the relevance of AMO sea-surface temperatures in the North Atlantic to the formation of hurricanes in the tropical Atlantic. But compare figure 11.24 to this overview of Atlantic basin hurricane seasons:

- 1995 to now: common seasons with intense hurricanes
- 1961 to 1994: few seasons with above-average hurricanes
- 1926 to 1960: numerous seasons with destructive hurricanes
- 1900 to 1925: infrequent seasons with many hurricanes
- 1870 to 1899: numerous seasons with many intense hurricanes

Accumulated Cyclone Energy

The **Accumulated Cyclone Energy (ACE)** index is a compilation by NOAA of the energy expended in each tropical cyclone and in each season. A tropical cyclone's energy is calculated every six hours of its existence and then added up to yield the total amount of energy it released. The ACEs of all the cyclones during a season are totaled to learn the energy release during an entire season. The ACE totals for seasons of Atlantic tropical storms and hurricanes are shown in figure 11.25. The most energetic year was 2005, 2004 is in 3rd place, 2003 is 9th, 2010 is 12th, and 2008 is 16th. The individual cyclone with the highest ACE is Hurricane Ivan in 2004 with a 70.4. Ivan alone released more energy than is generated in some entire seasons.

Decade of the Naughts—2000 to 2009

The past decade was marked by numerous powerful cyclones and intense cyclone seasons. In the year 2004, Japan was struck by a record-breaking 10 typhoons (the old record was six in 1993), the first documented hurricane (Catarina) in the South Atlantic Ocean made landfall in southern Brazil on 28 March, and the United States had landfalls from five hurricanes and four tropical storms. Florida was hit the hardest. In late summer, four hurricanes crossed Florida: Charley, Frances, Ivan, and Jeanne (figure 11.26). Perhaps the cruelest was Jeanne. After killing more than 3,000 people in Haiti, Jeanne took a clockwise turn out into the open Atlantic Ocean and away from the United States, only to

TABLE 11.5

Hurricane Name Lists for the North Atlantic Basin Seasons, 2015–2020

2015	2016	2017	2018	2019	2020
Ana	Alex	Arlene	Alberto	Andrea	Arthur
Bill	Bonnie	Bret	Beryl	Barry	Bertha
Claudette	Colin	Cindy	Chris	Chantal	Cristobal
Danny	Danielle	Don	Debby	Dorian	Dolly
Erika	Earl	Emily	Ernesto	Erin	Edouard
Fred	Fiona	Franklin	Florence	Fernand	Fay
Grace	Gaston	Gert	Gordon	Gabrielle	Gonzalo
Henri	Hermine	Harvey	Helene	Humberto	Hanna
Ida	Ian	Irma	Isaac	Imelda	Isaias
Joaquin	Julia	Jose	Joyce	Jerry	Josephine
Kate	Karl	Katia	Kirk	Karen	Kyle
Larry	Lisa	Lee	Leslie	Lorenzo	Laura
Mindy	Matthew	Maria	Michael	Melissa	Marco
Nicholas	Nicole	Nate	Nadine	Nestor	Nana
Odette	Otto	Ophelia	Oscar	Olga	Omar
Peter	Paula	Philippe	Patty	Pablo	Paulette
Rose	Richard	Rina	Rafael	Rebekah	Rene
Sam	Shary	Sean	Sara	Sebastien	Sally
Teresa	Tobias	Tammy	Tony	Tanya	Teddy
Victor	Virginie	Vince	Valerie	Van	Vicky
Wanda	Walter	Whitney	William	Wendy	Wilfred

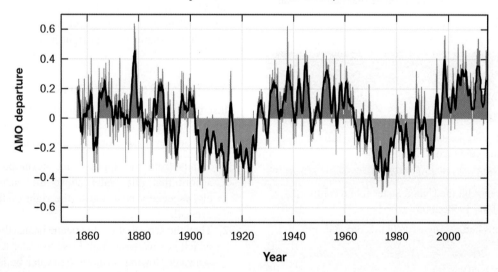

Monthly values for the AMO index, 1856–2013

Figure 11.24 Atlantic Multidecadal Oscillation (AMO) time series with a 12-month moving average, 1856–2013.
Source: NOAA.

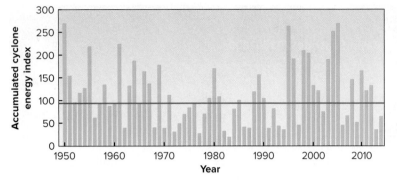

Figure 11.25 Annual energy release by Atlantic hurricanes and tropical storms compared to the 1950 to 2000 median (red line). The 1970–1994 interval is below average, and 1995 and ongoing is above average. Source: NOAA.

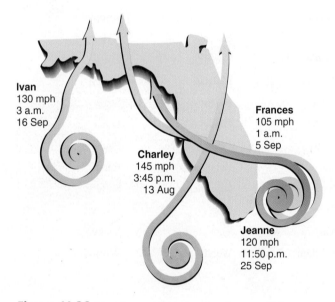

Ivan
130 mph
3 a.m.
16 Sep

Frances
105 mph
1 a.m.
5 Sep

Charley
145 mph
3:45 p.m.
13 Aug

Jeanne
120 mph
11:50 p.m.
25 Sep

Figure 11.26 Florida was hit by four hurricanes within a 44-day-long interval in 2004. The last time that happened was in Texas in 1886.

TABLE 11.6

Records Made During the 2005 Atlantic Hurricane Season

Most Numerous

28 named storms (old record: 21 in 1933)

15 hurricanes (old record: 12 in 1969)

4 major hurricanes hit the United States (old record: 3 in 1909, 1933, 1954, 2004)

7 tropical storms before August 1 (old record: 5 in 1997)

Costliest

U.S. hurricane: Katrina

Mexican hurricane: Wilma

keep turning in a circular path and then hammer Florida following the same track that Frances had cut just 20 days earlier. In Florida, the hurricanes combined to kill more than 110 people, damage more than 20% of all the houses, and inflict more than $55 billion in losses. Crops were destroyed, tourists were scared away, and the economy suffered. But Florida quickly rebounded; it has a dynamic economy. Natural disasters have their longest effects on stagnant economies.

In 2005, many storm records fell (table 11.6). The ACE for the year was 250% of average, damages exceeded $115 billion, and almost 1,900 people lost their lives in the United States.

Hurricane Damages

Since 1900, the catalog of landfalling hurricane strengths and their effects on the United States is fairly complete. The deadliest hurricanes (table 11.7) occurred mostly in the earlier part of this century (table 11.8). Deaths by hurricanes in the United States have dropped dramatically due to the advance warnings that are now broadcast widely before a hurricane makes landfall. Most people evacuate the low-lying areas and save themselves. Since naming of hurricanes began in 1953, only three named hurricanes have been really big killers in the United States.

Although hurricane deaths are down, the damages they cause are up (table 11.8). An ongoing trend is for Americans to move to the coastline and to build larger and more expensive homes filled with costlier possessions. It is interesting to reevaluate the damages caused by past hurricanes after converting their losses to 2010 dollars (table 11.9). It is even more instructive to consider present populations and wealth at the sites of past hurricane landfalls and then estimate the damages that the same hurricane would cause today. Normalizing hurricane damages in this way shows that the future will bring more frequent hurricane disasters to the United States (table 11.8, far right-hand column).

STORM-SURGE HAZARDS

Most tropical cyclone deaths in the world are associated with sea surges occurring when a cyclone nears and moves on land. Sea level rises for several reasons:

1. Winds from the approaching storm push sea swells ashore that pile water above the normal tidal levels, especially near and to the right of the landfalling eyewall.

2. Then the arrival of the hurricane brings the storm surge, which is a relatively rapid rise in water level caused by seawater flowing onshore. Seawater builds up beneath the eye because it is a low-pressure zone. Winds racing into the eye and into the right forward quadrant push

TABLE 11.7

The 13 Deadliest Hurricanes in the United States, 1900–2014

When	Where	Number of Deaths
8 September 1900	Galveston, Texas	8,000
Mid-September 1928	South Florida—Lake Okeechobee	2,500
29 August 2005	Hurricane Katrina—Louisiana/Mississippi	1,836
Mid-September 1919	Florida Keys/Corpus Christi, Texas	600
29 September 1915	New Orleans, Louisiana	600
27 June 1957	Hurricane Audrey—Morgan City, Louisiana	416
2 September 1935	Florida Keys	408
14–15 September 1944	East Coast—Virginia to Massachusetts	390
Mid-September 1926	Miami, Florida/Alabama	372
21 September 1909	Grand Isle, Louisiana	350
17 August 1915	Galveston, Texas	275
21 September 1938	New England, especially Rhode Island	256
17–18 August 1969	Hurricane Camille—Mississippi	256

TABLE 11.8

Years of the 17 Deadliest and Costliest U.S. Hurricanes, 1900–2005

Years	Deadliest	Costliest (in 2006 US$)	Normalized Costs* (in 2006 US$)
1900–15	5	0	2
1916–30	3	0	2
1931–45	2	1	2
1946–60	3	2	4
1961–75	3	3	3
1976–90	0	3	1
1991–2005	1	8	3

*Normalized costs adjusted for increases in population and wealth.

TABLE 11.9

The 22 Costliest Hurricanes Striking the United States, 1915–2014 (in 2010 US$)

When	Where	Damage in Billions	Hurricane Category
2005	LA/MS—Katrina	106	3
2012	NJ/NY—Sandy	60	1
1992	FL—Andrew	46	5
2008	TX—Ike	28	2
2005	FL—Wilma	21	3
2004	FL/AL—Ivan	20	3
2004	FL—Charley	16	4
2011	NC/NJ/NY—Irene	15	1
1989	SC—Hugo	14	4
1972	FL/NE US—Agnes	12	1
2005	TX—Rita	12	3
1965	FL/LA—Betsy	11	3
2001	TX—Allison	11	TS
2004	FL—Frances	10	2
1999	E US—Floyd	9	2
1969	MS—Camille	9	5
2004	FL—Jeanne	8	3
1995	FL/AL—Opal	8	3
1955	NE US—Diane	7	1
1979	AL/MS—Frederic	7	3
1938	New England	6	3
1996	NC—Fran	6	3

Data from NOAA.

seawater that pours onto land as the eye reaches the shoreline. About 95% of the seawater that surges ashore is wind driven; 5% is sea-level rise under the low atmospheric pressure.

3. Not all coastal residents are hit by the same wind velocities; wind speeds vary along the coastline (figure 11.27). If you are on the "right-hand side" of the tropical cyclone, you experience the speed of the storm body *plus* the wind speeds. If you are on the "left-hand side," you feel the wind speed *minus* the storm motion. Also, on the left-hand side, the winds come off the land, while on the right-hand side, the

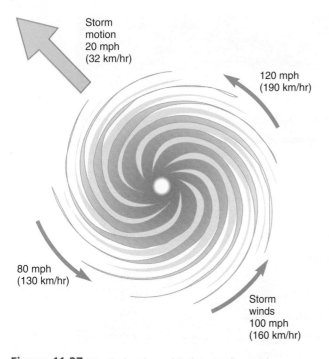

Storm motion 20 mph (32 km/hr)

120 mph (190 km/hr)

80 mph (130 km/hr)

Storm winds 100 mph (160 km/hr)

Figure 11.27 Tropical cyclones hit the coastline with different wind speeds. Storm motion and wind velocity may combine or subtract.

winds come off the ocean, pushing much more seawater onto the land. The highest sea levels occur along the coastline on the right side of the eye landfall (figure 11.28). In the Northern Hemisphere, maximum storm surges occur about 15 to 30 km (10 to 20 mi) to the right of the path of the eye.

4. On top of the already elevated sea level come the large waves blown by the hurricane winds.

5. And don't forget about the astronomical tides that raise and lower sea level each day. When is the worst time for a storm surge to come onshore? During the already high sea level of an astronomical high tide.

Why do impacting waves kill so many people? A cubic yard of water weighs 1,685 lb (760 kg), almost a ton, and water is almost incompressible. Being hit by a wall of water is not much different from being hit by a solid mass.

Hurricane Camille is one of three category 5 hurricanes to hit the United States since 1900. Camille brought winds gusting over 320 km/hr (200 mph) that hit Mississippi in 1969 with a sea surge of 7.3 m (24 ft); this caused many of the 256 fatalities. Fatal sites included the three-story brick buildings of the seaside Richelieu apartment complex where, instead of evacuating, 32 party-hearty people held a "hurricane party" to celebrate the event—it was the last party for 30 of them (figure 11.29).

Hurricane Ike crossed the Texas shoreline on 13 September 2008 with a ferocious storm surge responsible for many of the 112 deaths. The town of Gilchrist had 99% of its houses—more than 1,000 homes—swept off their foundations (figure 11.30).

HEAVY RAINS AND INLAND FLOODING

A hurricane holds massive amounts of water vapor in its cloud mass. After moving onto land, no more water vapor is fed into the hurricane and its energy decreases. But there remains aloft in a dying hurricane a massive volume of atmospheric water, and when it falls it can cause severe flooding. Notice in table 11.9 that five of the 22 costliest cyclones were only category 1 or tropical storm events. Much of their damages were inflicted by heavy rainfall and flooding away from the coast. In the United States between 1970 and 1999, 59% of the hurricane deaths were due to inland flooding and

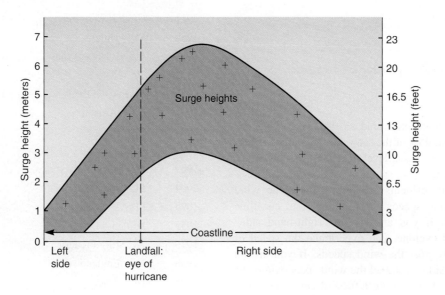

Figure 11.28 Elevation of storm surges along a coastline. Notice that the highest water levels occur to the right of the hurricane eye where the winds are the strongest.

(a)

(b)

Figure 11.29 Storm surge of Hurricane Camille. (a) *Before:* Richelieu Apartments in Pass Christian, Mississippi, where a hurricane party was held. (b) *After:* Remains of the building where 30 of 32 partygoers died.
NOAA.

Figure 11.30 One lone house on stilts remains standing on the Gilchrist shore after Hurricane Ike surged inland, 13 September 2008.
Jocelyn Augostino/FEMA.

TABLE 11.10
U.S. Hurricane Deaths, 1970–1999

Inland flooding	59%
Wind	12
Storm surge	12
Offshore	11
Tornado	4
Other	2
	100%

Source: Tropical Cyclones & Inland Flooding (2001); NOAA.

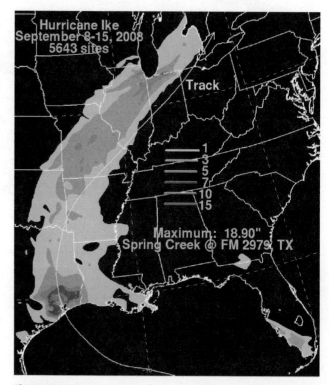

Figure 11.31 Hurricane Ike dropped many inches of rain and caused flooding in states from Texas to Michigan, 8–15 September 2008.
NOAA.

only 12% were due to storm surge (table 11.10). The rainfall map from Hurricane Ike shows how much water can be brought inland (figure 11.31). And don't forget the 11,000 people killed in 1998 by inland flooding from Hurricane Mitch (see figures 11.19 and 11.20).

A Classic Disaster

The Galveston Hurricane of 1900

On 8 September 1900, the deadliest natural disaster in U.S. history struck Galveston, Texas. Galveston is built on a low-lying island, a sandy barrier beach (figure 11.32). Behind the sandbar island lies Galveston Bay, where trading ships made Galveston the wealthiest city in Texas early in the 20th century. In 1900, the 38,000 residents were warned of a possible hurricane, and many thousands evacuated the island but many thousands remained.

The category 4 hurricane arrived in late afternoon. A high tide and the hurricane surge combined to flood the highest point on the island to a depth of 0.3 m (1 ft). Moving on top of this elevated sea level were storm waves blown by 210 km/hr (130 mph) winds. No place was safe. Wooden buildings were destroyed quickly. Even many of the big brick buildings fell to the high winds and ferocious waves that used the debris of other broken buildings and ships as battering rams to beat down their sturdy walls. Many people crammed into the Bolivar lighthouse to find refuge, sitting on the laps of strangers on the curving metal staircase. They were jammed so close together that no one could move; they had no water to drink and no facilities for relief. Surrounding them was an air of fear permeated with the stench of human waste excreted where people sat or stood for the many hours that the 9 m (30 ft) high waves kept them confined. When finally they could open the massive door, they viewed a scene of smashed buildings and boats, and thousands of bodies.

The morning of 9 September brought pleasant weather, but half of the buildings had been destroyed and most of the rest were heavily damaged. There was no water, no food, no electricity, and no medical supplies; all bridges to the mainland were down, and all boats wrecked. The survivors were marooned. But mainland Texans came as quickly as possible to help in the overwhelming task of cleaning up. The 8,000 decaying human bodies presented a serious problem—the spread of disease. With much unhappiness, thousands of bodies were barged out to sea and dumped to avoid an epidemic. However, the tides and waves carried the floating bodies back to shore. The survivors had to pile up wood from wrecked buildings and build funeral pyres to consume the bodies.

Afterwards, the city constructed a 5 km (3 mi) long seawall that was 5.2 m (17 ft) high and 4.9 m (16 ft) wide at its base (see figure 11.32). Sand was brought in to elevate the island. Then the city rebuilt and began again. Nonetheless, another hurricane arrived on 17 August 1915, claiming another 275 lives.

Figure 11.32 An early version of the Galveston seawall. Waves reflecting off the seawall have carried away beach sand.
W. T. Lee/U.S. Geological Survey.

Hurricanes and the Gulf of Mexico Coastline

HURRICANE KATRINA, AUGUST 2005

On 24 August 2005, a tropical air mass over the Bahamas grew powerful enough to be given a name—Katrina. Two hours before reaching Florida on 25 August, Katrina had grown to be a hurricane. Katrina lost strength crossing Florida, but after reaching the warm waters of the Gulf of Mexico, the storm grew rapidly, with wind speeds reaching 280 km/hr (175 mph) and its size nearly doubling. By 26 August, computer models identified New Orleans as a possible target with a 17% chance of a direct hit. Many residents were nervous; they knew that most of the city of New Orleans lay below sea level and that Katrina would bring huge volumes of water. As 27 August dawned, residents saw that Katrina had moved ever closer. More warnings were issued. People began shuttering their homes, grabbing prized possessions, and fleeing. On the morning of 28 August, the situation looked even worse, causing a mandatory evacuation to be ordered for 1.2 million residents. All lanes of all roads out of the area were filled with evacuating vehicles.

Katrina came ashore 55 km (35 mi) east of New Orleans on 29 August in Mississippi with wind speeds of 195 km/hr (120 mph) (figure 11.33) and hurricane-force winds reaching across Louisiana, Mississippi, and Alabama (figure 11.34). Katrina pushed in enough water to breach levees (figure 11.35) and to overflow canals, flooding low-lying areas that make up 80% of New Orleans. At least 100,000 people did not evacuate from New Orleans; more than 1,500 of them died there. Katrina is the costliest natural disaster in U.S. history and the deadliest since the 1928 Okeechobee hurricane (see table 11.7).

Were the Katrina-Caused Deaths and Destruction a Surprise?

No one should have been surprised by Hurricane Katrina. New Orleans has a long history of flooding, destruction, and death from hurricanes. A hurricane like Katrina has been anticipated for decades. The major newspaper in the region, the *New Orleans Times Picayune*, ran an award-winning series of detailed articles from 23 to 27 June 2002 called "Washing Away." The newspaper predicted: *"It's only a matter of time before South Louisiana takes a direct hit from a major hurricane. Billions have been spent to protect us, but we grow more vulnerable every day."* The special reports included:

- "In Harm's Way: Levees, our best protection from flooding, may turn against us"
- "The Big One: a major hurricane could decimate the region, but flooding from even a moderate storm could kill thousands. It's just a matter of time"

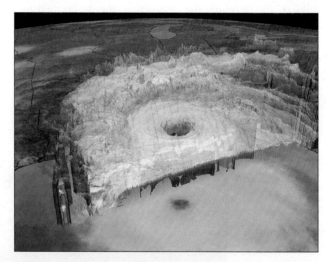

Figure 11.34 Cutaway view into Hurricane Katrina showing variable rainfalls. Rainfall amounts are shown in inches per hour: blue = 0.25, green = 0.5, yellow = 1, red = 2.

NASA/JAXA.

Figure 11.33 Hurricane Katrina turned the Interstate 90 bridge connecting Pass Christian to Bay St. Louis, Mississippi, into a stack of dominoes.

John Fleck/FEMA.

Figure 11.35 Water rushes through a failed levee to flood New Orleans on 30 August 2005.

Jocelyn Augustino/FEMA.

- "Evacuation: it's the best chance for survival, but it's a bumpy road, and 100,000 will be left to face the fury"

Less than a year before Katrina, in September 2004, New Orleans lay in the path of Hurricane Ivan with winds of 240 km/hr (150 mph). New Orleans was ordered to evacuate. Upwards of 1 million people fled, mostly in bumper-to-bumper traffic on highways turned into one-way evacuation routes for 2.5 days. Many older and sick residents unable to evacuate were moved to high levels inside the 72,000-seat Superdome, home of the NFL New Orleans Saints. But luck favored New Orleans that time as Ivan veered to the right and came ashore 160 km (100 mi) east of New Orleans. Luck also favored the thousands of people who did not evacuate. Thanks to Hurricane Ivan, New Orleans had a practice evacuation to learn from one year prior to Katrina.

It is a sad thing to say, but there were no surprises before, during, or after Hurricane Katrina passed alongside New Orleans. The storm, the evacuation difficulties, the levee failures, the flooding, the destruction, the deaths—all had been accurately predicted for many years. But the years of advance warnings were to no avail; the drama played out as if scripted.

New Orleans: Can This Setting Be Protected?

Water has always been part of life in New Orleans. The original French settlement was built on Mississippi River delta swampland surrounded by huge bodies of water—the Mississippi River, the Gulf of Mexico, Lake Pontchartrain, and more (figure 11.36). Three centuries of floods brought by the river and hurricanes have led to spending billions of dollars to build levees to try to keep water out. The land forming the foundation of New Orleans is a loose mixture of mud, sand, and water deposited by Mississippi River floods onto its river delta. It is the nature of deltas to compact and subside below sea level. Subsidence has lowered part of the city 6 m (20 ft) below sea level, and the subsidence continues. The levees built to keep Mississippi River water out of New Orleans also prevent deposition of additional mud and sand that would build new land. As levees are built higher and the city sinks lower, a bowl has been created (figure 11.37), with lake and river water levels higher than city land levels.

What does the future hold for New Orleans? A sinking city. An increasingly elevated Mississippi River. High lake levels. Much destroyed marshland between the Gulf of Mexico and the city. Globally rising sea level. And more hurricanes.

What should be done about New Orleans? Suggestions are numerous. Reduce the size of the city. Rebuild only on ground above sea level. Make New Orleans into an island. Restore marshlands between the Gulf of Mexico and the city. Raise ground level in the city. Construct huge levees and floodgates. Build bigger water-pumping systems.

Figure 11.36 Mississippi River delta and the low-lying city of New Orleans (top center). Hurricanes flood much of the city up to 6 m (20 ft) deep.

NASA.

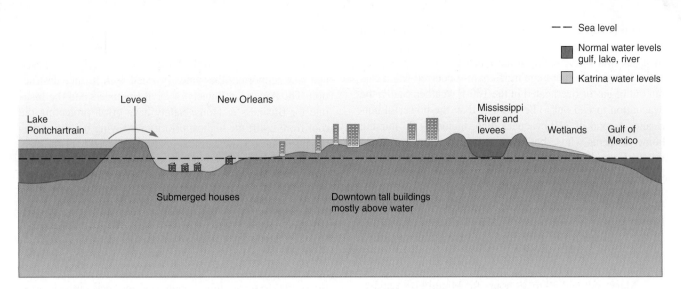

Figure 11.37 New Orleans sits in a bowl between the levees of the Mississippi River and Lake Pontchartrain. Much of the city is below sea level. The land continues to sink. The river bed is building higher, so levees are built higher. The normal lake level is 1.2 m (4 ft) above sea level.

Hurricanes and the Atlantic Coastline

HURRICANE HUGO, SEPTEMBER 1989

Charleston, South Carolina, has experienced many disastrous events. On 12 April 1861, the Confederate army began bombarding Charleston's Fort Sumter and started the Civil War, which ultimately left Charleston in ruins. On 31 August 1886, a major earthquake severely damaged the city (see chapter 5). Then, at about midnight on Thursday, 21 September 1989, Hugo came to town as the 10th strongest hurricane in the 20th-century United States.

Hugo brought a 5.2 m (17 ft) high surge to Fort Sumter and 220 km/hr (135 mph) winds; it was a category 4 hurricane (figure 11.38). The move onto land cut Hugo off from its warm-water energy supply. With decreasing energy, the storm adopted a curving path; a prominent northward hook led past the west side of Charlotte, North Carolina, at tropical-storm intensity at 6 a.m. Friday. Hugo then moved up the Appalachian Mountains and lost its status as a tropical weather system by 6 p.m. Friday near the junction of West Virginia, Ohio, and Pennsylvania. The remains of Hugo were rains that crossed New York into Canada.

Although Hugo was a powerful hurricane, only 11 lives were claimed; however, property damages exceeded $14 billion. Again, here is the century-long trend—decreasing deaths, increasing damages.

The Evacuation Dilemma

Thanks to advance warning, the number of people killed by hurricanes in the United States has dropped dramatically since the beginning of the 20th century (see table 11.8). When

Figure 11.38 Beach houses destroyed by storm surge of Hurricane Hugo, September 1989.

S.J. Williams/U.S. Geological Survey.

satellite photos show a hurricane is coming to town, people evacuate inland to safety before it makes landfall. However, this way of saving lives is threatened as a new migration within America has brought about 50 million people to live in counties on the Atlantic Ocean or Gulf of Mexico coastline.

An evacuation dilemma is here. Population grows far faster than new roads and bridges are built. Moving large numbers of people out of harm's way is a time-consuming operation complicated by the limits of weather forecasting, by human psychology, and by transportation systems that do not allow free-flowing movements even with fewer vehicles in perfect weather. A September 1999 warning in South Carolina that Hurricane Floyd was approaching resulted in a massive gridlock that tied up the highway system and stopped people from evacuating. A potential catastrophe was avoided when Floyd hit North Carolina instead.

But people were not as lucky in 2005 after being warned to evacuate coastal Texas before Hurricane Rita arrived. So many people were fleeing in so many slowly moving cars on so many roads that huge traffic jams occurred when cars ran out of gas or overheated in the 100°F weather. Sadly, the evacuation turned out to be deadlier than the powerful hurricane. A total of 110 evacuees died, including 23 nursing home residents whose bus burst into flames while stuck in traffic. If you live above the storm-surge level, then you may be safest in an interior room of a well-built home.

Consider the difficulty of the evacuation process. It is estimated that evacuating some cities could take 72 hours. Do we know three days in advance where a hurricane will strike? No, the detailed path of a hurricane is still unpredictable. Predicted evacuation times for other areas include 50 to 60 hours for Ft. Myers, Florida; 40 to 49 hours for Ocean City, Maryland; and 30 to 39 hours for Miami/Fort Lauderdale and Cape May County, New Jersey.

If people cannot run far and high enough to escape an oncoming hurricane, what should they do? They must seek out local shelters built strong enough to withstand the hurricane attack and high enough to avoid flooding by the storm surge. As a general rule:

- Run from the water.
- Hide from the wind.

Reduction of Hurricane Damages

In 1992, the then costliest hurricane in U.S. history occurred when Hurricane Andrew tore through South Florida with sustained winds of 250 km/hr (155 mph), ripping apart 130,000 homes and causing more than $46 billion in damages. But these dollar losses were exceeded by Hurricanes Katrina in 2005 and Sandy in 2012. Even more, the next five most expensive U.S. hurricanes all hit after Andrew with Ike (2008), Wilma (2005), Ivan (2004), Charley (2004), and Irene (2011). Is this just a fact of life that a powerful hurricane hitting growing populated areas will cause ever-increasing amounts of damage? Not necessarily. We need better planning and design. There are also numerous shortcomings common in our human-built coastal structures.

Manufactured Homes
In 1994, two years after Hurricane Andrew destroyed countless manufactured homes, commonly known as mobile homes, tougher building codes were enacted. The manufactured-home industry filed a lawsuit in an unsuccessful attempt to block the stiffened rules. When Florida was hit by four hurricanes in 2004, countless old mobile homes were destroyed, but the new ones built to meet the tougher standards fared well. The Florida Manufactured Housing Association admitted that it had been wrong to oppose the new rules.

Roofs
When hurricane winds destroy a building, they usually lift the roof off first. Then, with the building walls standing exposed and less supported, the winds proceed with further destruction. The American Society of Civil Engineers and the Institute for Business & Home Safety have given protecting the roof from uplift the highest priority for saving buildings during hurricanes. How can this goal be achieved? (1) Builders should eliminate or strengthen eaves that project out from roofs, thus making it harder for winds to lift off roofs. (2) Strap roofs to walls. Inside the attic where the roof meets the walls, add numerous hurricane straps to help hold the roof to the walls. Straps are heavy belts of material similar to those used on suitcases or backpacks. Each strap is wrapped around a roof rafter and a heavy stud in the wall, then fastened by numerous nails. (3) Ban the common practice of using rapid-fire staples to secure thin asphalt roofing sheets onto plywood roofs. Hurricane winds easily strip off these lightweight, flexible materials and thus gain entry to the house.

Wind-Borne Debris
Hurricane winds pick up and hurl debris that breaks through windows and pierces entry and garage doors (figure 11.39).

Figure 11.39 Hurricane Andrew flung a 1-by-4 timber like an arrow through the trunk of this royal palm tree, 24 August 1992. This could happen to your house.

NOAA/Department of Commerce.

In Greater Depth

How to Build a Home Near the Coastline

Houses built on or near the beach must be able to withstand two powerful forces: (1) high wind speeds and (2) overwhelming storm-surge water flow. Wind can tear a building apart piece-by-piece, and storm-surge water can knock down a house by brute force and erode away the sandy land beneath the home. The elements of design to withstand these forces are shown in figure 11.40.

- A house must rest on vertical pilings 3 to 10 m (10 to 30 ft) deep.
- It must be elevated 2 to 8 m (6.5 to 26 ft) above land level so storm water can flow underneath.

- Pilings and house must be bound together by horizontal steel beams held by steel bolts.
- Roof, roof rafters, and walls must be securely tied together.
- Buildings should not have eaves or overhangs so wind lift is reduced.
- Exterior boards should be large, heavy, and securely fastened to reduce wind ripping.

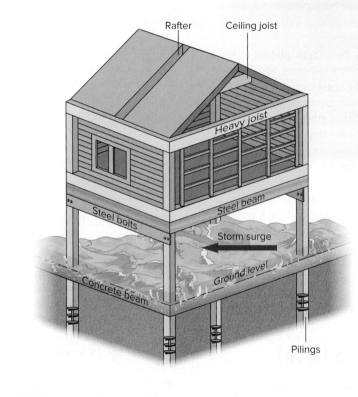

Figure 11.40 House-design elements that can lessen damages from hurricane winds and storm surges.

Once an opening exists in a building, the winds can come inside and intensify their destruction. To prevent penetration, all exterior glass windows and sliding-glass doors should be made of shatterproof glass or be protected by shutters. If the building does not have built-in shutters, plywood should be precut to fit over all windows and sliding-glass doors.

Protecting your house against wind-borne debris also means eliminating potential debris. Pick up and store all outside items that could be used by the wind to attack your house, such as patio furniture, trash cans, barbecues, awnings, potted plants, and toys. Discuss picking up items with your neighbors; their belongings can be thrown against your house also. Shrubs and trees in your yard can be pulled out of the ground and flung against your house.

LAND-USE PLANNING

Beaches are aesthetic vacation sites. So many people enjoy the beach that a building boom is under way along the

Atlantic Ocean and Gulf of Mexico shorelines. Many thousands of expensive new homes are crowding the shoreline, waiting for the inevitable hurricane attack of winds, storm surges, waves, and heavy rains. A recent study for the Federal Emergency Management Agency (FEMA) estimates that during the next 60 years, 25% of houses within 150 m (500 ft) of the shoreline will fall into the water unless mitigating actions are taken, such as adding sand to beaches, adding riprap, or building hard walls.

Decisions made about land use before development takes place can prevent a lot of damage. Think of the destruction that could be avoided if cities and counties designated low-lying coastal land for use as parks, farm fields, golf courses, nature preserves, or other uses where flooding will cause fewer problems. At the same time, the higher and more protected land could be zoned for house and business construction.

After the Fact

After Hurricane Sandy struck New York and New Jersey especially hard, the U.S. Congress allocated $60 billion to aid the recovery effort. The U.S. Army Corps of Engineers is using about half the funds for infrastructure repair, construction of barriers and levees, and strengthening of beaches.

But there is also the realization that it (another Sandy) will happen again and could be worse next time. Much of the recovery funds are being used to relocate entire communities inland. This allows wetlands to be restored and to provide their benefits, including protection from hurricanes. For example, the community of Oakwood Beach on Staten Island in New York City suffered overwhelming damages. Government-funded *buyouts* have been offered to 300 homeowners, giving each the pre-storm value of their home plus a 10% bonus for community consensus to relocate plus a 5% bonus to stay within New York City. The average buyout is about $375,000, costing a total of about $160 million.

Global Rise in Sea Level

Another problem facing shoreline residents is the global rise in sea level. It rose about 1 ft in the 20th century and is likely to rise 2 ft (or more) in the 21st century. One or two feet may not sound too threatening, but consider its effect on low-lying lands such as the Gulf and Atlantic coasts. A 1 ft rise in sea level may cause a beach to move inland 1,000 ft in some areas (figure 11.41).

Hurricanes and the Pacific Coastline

Each year, about 15% of Earth's tropical cyclones of hurricane strength form in the eastern Pacific Ocean, mostly offshore from southern Mexico/Guatemala/El Salvador (see figures 11.11 and 11.12). There are about 25% more hurricanes per year in the eastern Pacific Ocean than in the North Atlantic/Caribbean Sea/Gulf of Mexico. Why don't Pacific Ocean hurricanes strike the West Coast of the United States as often as Atlantic Ocean hurricanes do the East Coast? First, the trade winds blow most of the hurricanes westward out into the Pacific Ocean. Second, there is a marked difference in seawater temperatures. Along the eastern United States, the northward-flowing Gulf Stream current brings warm water from the Gulf of Mexico up along the East Coast, while the West Coast is bathed by cold water of the California Current coming down from Alaska (see figure 9.28). The cold-water current acts as a hurricane defense line. Even in the summer, water temperatures off California are usually only 18° to 20°C (64° to 69°F). As a hurricane moves over cooler water, the evaporation rate and energy supply slow down.

HURRICANE INIKI, SEPTEMBER 1992

The Hawaiian Islands lie between 19° and 22°N latitude, on the northern edge of the warm, hurricane-generating waters of the Pacific Ocean. Tropical storms form to the south of Hawaii and then follow the right-hooking, clockwise paths that the Coriolis effect produces in the Northern Hemisphere (see figures 11.11 and 11.12). Some of the north-hooking storms grow into hurricanes.

On Friday, 11 September 1992, the fast-moving category 4 Hurricane Iniki roared across the western side of Kauai, "the Garden Isle." Iniki means "sharp and piercing," and that well describes the damage done by its 210 km/hr (130 mph) sustained wind speeds, with wind gusts in excess of 260 km/hr (160 mph). No buildings escaped damage, more than 6,000 utility poles snapped like matchsticks, and bark was

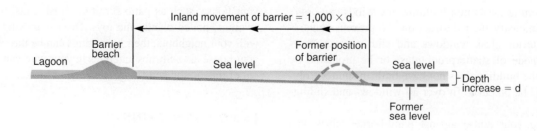
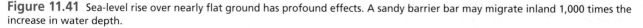

Figure 11.41 Sea-level rise over nearly flat ground has profound effects. A sandy barrier bar may migrate inland 1,000 times the increase in water depth.

even stripped from many trees. Kauai has a population of 52,000, and 80% of its economy is derived from the year-round population of 20,000 tourists. Although only two people were killed, the island economy suffered more than $2 billion in damages. Iniki was reminiscent of Hurricane Iwa, which ran across Kauai on 23 November 1982. Iwa killed one person and left $234 million in damages.

Cyclones and Bangladesh

In the 20th century, seven of the nine deadliest weather events in the world were cyclones striking Bangladesh (figure 11.42). The country sits mostly on sediments eroded from the Himalaya Mountains and dumped into the Bay of Bengal as the delta of the Ganges and Brahmaputra Rivers (figure 11.43).

The nation has densely populated a comparatively small area (table 11.11). Because Bangladesh has a rapidly growing population and scarce land and food, it is little wonder that many millions of people are driven to the rich delta soils that yield three rice crops per year. The delta country is low-lying, most of it a foot or less above sea level; more than 35% of Bangladesh is at less than 6 m (20 ft) elevation.

Bangladesh is a nation of water. More than 20% of the entire country is submerged beneath river floods in an average year; in 1988, 67% of the country was covered by river flood-waters. Then come the cyclones that bring surges of seawater 6 m (20 ft) high; they can flood 35% of the nation. It is little surprise that the national flower is the water lily.

Bangladesh has a 575 km (360 mi) long coastline shaped like a funnel that catches the cyclones roaring up and over the warm waters of the Bay of Bengal (figure 11.42). About five cyclones per year enter the Bay of Bengal both before (April–May) and after (October–November) the southwest monsoon season.

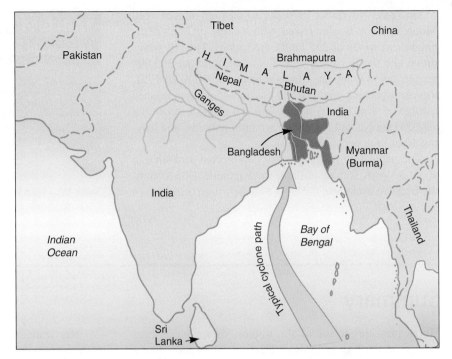

Figure 11.42 Bangladesh sits largely on the low-lying delta of the Ganges and Brahmaputra Rivers built into the Indian Ocean. Cyclones commonly move up the Bay of Bengal into Bangladesh.

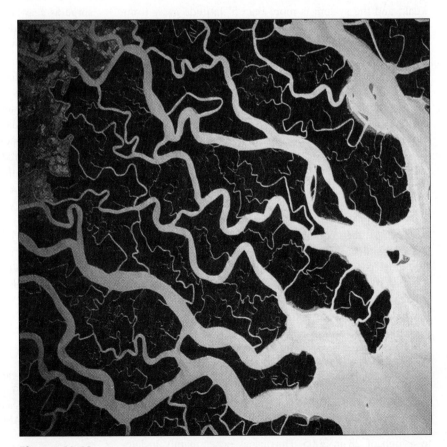

Figure 11.43 Much of Bangladesh is low-lying river-delta land where cyclone storm surges kill many thousands of people. This photo from space shows multiple rivers winding through heavily populated mud flats.

NASA/JSC.

On 12–13 November 1970, during the high tides of a full Moon, cyclone Bhola arrived with a surge of 7 m (23 ft) height and winds of 255 km/hr (155 mph). The tall waves drove into the low-lying delta land, killing about 300,000 people and as many large farm animals. On 30 April 1991, a cyclone packing winds of 235 km/hr (145 mph) unleashed a surge of 6 m (20 ft) height into Bangladesh, drowning 140,000 people and 500,000 large farm animals, and leaving 10 million people homeless.

The population of Bangladesh is projected to double in about 30 years, forcing millions more into the delta to await the frequent cyclones.

TABLE 11.11
Comparative Statistics of Bangladesh

	Area (km²)	Population (2010)	Median Household Income (2010)
Bangladesh	143,998	162,221,000	$690
Wisconsin	145,436	5,711,767	$50,522
Iowa	145,753	3,062,309	$49,177

Source: U.S. Census Bureau; United Nations.

Summary

Hurricanes are storms of the tropics; they convert heat energy into winds and waves. Hurricanes are most likely to develop where seawater is at least 27°C (80°F) in the upper 60 m (200 ft) of the ocean; air is warm, humid, and unstable enough for convection; upper-level winds are weak; and the system is more than 5° latitude away from the equator.

A hurricane is not a single convective storm; it is composed of numerous thunderstorms in spiral bands that rotate around a core. Low-level winds converge near the center, spiral up the eyewall, and flow outward as upper-level winds. The fastest winds and greatest precipitation occur in the eyewall. Air descends in the eye, warms, and clears the air.

Hurricanes in the Northern Hemisphere rotate counterclockwise. The Coriolis effect bends their travel paths to the right.

Hurricanes commonly form in the North Atlantic basin from easterly waves that occur as disturbances within the trade winds; they contain converging airflow that can create tropical disturbances. They can grow to be Cape Verde–type hurricanes. As trade winds blow storms westward from Africa across the ocean, they can increase in energy to tropical depression status and then to tropical storms with names. If wind speeds exceed 120 km/hr (74 mph), they are hurricanes, with clear eyes in their centers.

Hurricanes also form at the Intertropical Convergence Zone, including in the Caribbean Sea. Here the trade winds collide, yielding huge convective storms that can acquire rotation and grow to hurricane status.

The numbers of hurricanes and tropical storms in the North Atlantic basin have ranged from a low of 4 in 1983 to a high of 28 in 2005. Hurricanes occur mainly in late summer when the ocean and the air are warmest. Rainfall is concentrated in spiraling bands of clouds (rain bands), especially near the eye.

Hurricane strengths are described by the Saffir-Simpson scale; category 3 begins with wind speeds greater than 179 km/hr (111 mph). Major hurricanes are category 3 through 5; they constitute only 24% of landfalls in the United States, but they cause 86% of damages. Damages come from storm surge as seawater flows ashore driven by winds especially on the right side of the eyewall; inland flooding as a hurricane moves over land, cutting it off from its energy source and thus causing heavy rainfall; and strong winds in which tornadoes can form.

Hurricanes are called typhoons in the northwestern Pacific Ocean and cyclones in the Indian Ocean. Typhoons are more abundant and commonly stronger than hurricanes and cyclones because western North Pacific seawater tends to be warmest there most of the year.

The strength of each hurricane and the energy released in an entire season are measured by the Accumulated Cyclone Energy (ACE) index. The highest-energy season was 2005.

Globally rising sea level will create more problems for millions of people living on coastlines. Hurricanes will arrive closer to their front doors. Evacuations are becoming more difficult as more people try to use the same roads in a short time and gridlock occurs. The evacuation from Hurricane Rita in 2005 killed more people than the hurricane.

Terms to Remember

Accumulated Cyclone Energy (ACE)   296
Atlantic Multidecadal Oscillation (AMO)   296
cyclone   288
easterly wave   290
eye   285
eyewall   286
hurricane   283

major hurricane   289
post-tropical cyclone   287
rain bands   286
surge   283
tropical cyclone   283
tropical depression   284
tropical disturbance   284
tropical storm   285
typhoon   288

Questions for Review

Answer to question in figure 11.6: Southern Hemisphere (see the clockwise rotation). The photo is of Cyclone Adeline-Juliet, a category 4 cyclone in the southern Indian Ocean, April 2005.

1. Rank the following in order of increasing strength: hurricane, tropical depression, tropical storm, tropical disturbance.
2. Draw a cross-section through a hurricane and explain how it operates. Label the internal flow of winds. Be sure to explain the eye.
3. Explain how latent heat helps hurricanes grow.
4. What is the difference between a hurricane, a typhoon, and a cyclone? Where in the world does each occur?
5. What are the most common months for hurricanes to strike Canada and the United States? Why?
6. Explain the sequence of events that turns an African storm into a hurricane hitting the United States.
7. What factors control the path of a Cape Verde–type hurricane from Africa to North America?
8. Do hurricanes in the Northern Hemisphere rotate clockwise or counterclockwise?
9. Draw a cross-section showing how a hurricane produces a sea surge that floods the land. How high can surges be?
10. Why do hurricanes strike the East Coast of Canada and the United States but not the West Coast?
11. Compare a tornado to a hurricane. Which has the most total energy? Which has the highest wind speeds?
12. When hurricanes come ashore, which area receives the highest wind speed and storm surge—the right side or left side of the eye?
13. Which part of the world is hit by the most tropical cyclones?
14. What is the warm core of a hurricane?
15. Draw a map and explain how easterly waves become tropical cyclones.
16. What is the role of the Intertropical Convergence Zone (ITCZ) in initiating Caribbean hurricanes?
17. What role does the Coriolis effect play in hurricane formation?
18. Can a hurricane develop at 0° latitude on the equator?
19. At what wind speed does a tropical storm become a hurricane?
20. Why is the eye of a hurricane so clear?
21. Where are tornadoes most likely to form in a hurricane?
22. What hurricane-defensive features should be incorporated into houses built near the Atlantic and Gulf of Mexico coastlines?

Questions for Further Thought

1. What can be done to strengthen buildings enough to withstand hurricanes?
2. Should buildings be allowed next to the coastline on Atlantic and Gulf of Mexico barrier-island beaches? Should federal relief funds (taxpayer monies) go to homeowners who suffer losses from hurricane surge and waves?
3. If global ocean temperature continues to rise, what effect will it have on hurricane strength?
4. Many hurricanes form north of the equator in the North Atlantic, Caribbean Sea, and Gulf of Mexico. Why don't many hurricanes form south of the equator in the South Atlantic Ocean?
5. Compare how many hours in advance we know where a hurricane will strike with how many hours it takes to evacuate cities along the Atlantic and Gulf coasts.

Disaster Simulation Game

Your challenge is to protect a Caribbean town that relies on fishing and tourism from the deadly and devastating effects of a hurricane. You can construct or upgrade buildings and provide anchors or defenses for boats. You are given a budget. Then you have real choices to make.

The city and port you must protect have a specified population of people. You are provided with a map of the town and port and charged with protecting as many people, buildings, boats, and livelihoods as possible. Are you ready for the challenge? Go to http://www.stopdisastersgame.org. Click on Play Game. On the next page, click on Play Game again. Select the Hurricane scenario. Then choose your preferred level of difficulty: Easy (small map); Medium (medium-size map); or Large (large map).

Good luck! Choose your best options and save as many people and as much property as you can afford.

CHAPTER 12

Climate Change

All truth passes through three stages. First, it is ridiculed. Second, it is violently opposed. Third, it is accepted as being self-evident.
—Arthur Schopenhauer, 1788–1860

Iceberg near Pleneau Island, Antarctica.
Photo by Pat Abbott.

LEARNING OUTCOMES

Climate change has been occurring throughout Earth's history. Today there is the added concern that human beings are accelerating climate change. After studying this chapter, you should:

- know the conditions needed for Earth to experience an Ice Age.
- be able to explain why continental ice sheets advance and retreat during an Ice Age.
- know the conditions needed for a greenhouse to build and warm Earth into a Torrid Age.
- understand the multiyear climate changes of El Niño and La Niña and their global effects.
- be familiar with how volcanism can change the climate.
- recognize the striking increase in global warming during the past century.
- be familiar with the greenhouse gases in the atmosphere and know how they are being increased.
- understand the concepts of tipping points and lag times for 21st-century climate change.

OUTLINE

- Early Earth Climate—An Intense Greenhouse
- Climate History of Earth: Timescale in Millions of Years
- Glacial Advance and Retreat: Timescale in Thousands of Years
- Climate Variations: Timescale in Hundreds of Years
- Shorter-Term Climate Changes: Timescale in Multiple Years
- The Past Thousand Years
- The 20th Century
- Solar Energy Variation
- Greenhouse Gases and Aerosols
- The 21st Century
- Mitigation Options

External Energy

Climate is a realm of change. Climate change occurs at nearly every historic and geologic timescale examined. Many processes affect climate, each with its own operating principles and timetable. Climate changes as:

- Ocean basins open and continents drift
- Earth's orbit around the Sun changes
- Volcanism pumps ash and gas up into the stratosphere
- The Sun burns hotter or colder
- Global sea level rises or falls
- Humans burn huge volumes of wood, oil, gas, and coal

This chapter examines climate change on several timescales: billions of years, millions of years, thousands of years, hundreds of years, tens of years, and a few years. We will begin with Earth's climate on the long timescales and finish with a close look at the 20th and 21st centuries.

Early Earth Climate— An Intense Greenhouse

Billions of years ago, Earth's climate was dramatically different from today. The climatic regime of the early Earth, the third planet from the Sun, can be appreciated by looking at the atmospheric compositions of the inner planets (table 12.1).

Venus is the second planet from the Sun and thus receives intense solar radiation. Much of that solar energy is trapped by its dense, carbon dioxide–rich atmosphere, which helps create surface temperatures of about 477°C (890°F). Life as we know it on Venus is difficult to imagine when temperatures are so high that surface rocks glow red like those in a campfire ring.

Mars is the fourth planet from the Sun, and its greater distance causes it to receive much less solar energy. However, the thin Martian atmosphere is also relatively rich in carbon dioxide (CO_2), helping hold the heat it does receive and maximizing its surface temperature, although its temperatures of about −53°C (−63°F) are still cold. The atmospheres of Venus and Mars may be little changed over more than 4 billion years, yet Earth's atmosphere has undergone a radical change from being CO_2-rich to CO_2-poor.

Why has Earth's atmosphere changed? The changes have been caused in large part by life processes. Plants remove CO_2 from the atmosphere via photosynthesis and respire O_2 as a by-product that has built up O_2 in the atmosphere. But the total amount of CO_2 locked up in plants, dead or alive, is small compared to the amounts originally in the atmosphere.

Where has all the CO_2 in the early Earth's atmosphere gone? It is stored in physical form in several ways, but about 80% of that CO_2 is now chemically tied up in limestone (table 12.2). Limestone is rock composed of $CaCO_3$. Most limestone is made from the hard parts of oceanic life, such as shells and reefs. (The process by which carbon [C] is precipitated from seawater is described by the equations in the In Greater Depth feature, "How to Create a Cave," in chapter 15).

Returning to the earliest Earth (table 12.1), when no life was present and the atmosphere was full of CO_2, the surface temperature of our planet would have been about 290°C (550°F). Why was Earth so hot? This global warming was due in part to the greenhouse effect. Because of its abundance, CO_2 (an important greenhouse gas), kept Earth's climate hot. Over geologic time, as CO_2 dissolved in water and $CaCO_3$ sediments formed, the amount of atmospheric CO_2 declined. Early photosynthesizing life on Earth pulled CO_2 out of the atmosphere, causing a decline. Innumerable species of animals made skeletal material out of $CaCO_3$, reducing the amount of atmospheric CO_2 even further. Biologic use of CO_2 has lessened the greenhouse effect to yield the present congenial temperatures on Earth.

Today, carbon dioxide makes up only 0.04% of the Earth's atmosphere, but it helps create the greenhouse effect that keeps Earth's average temperature at 16°C (61°F); this

TABLE 12.1
Atmospheres of the Inner Planets

	Venus	Early Earth	Mars	Earth Today
CO_2	96.5%	98%	95.3%	0.04%
N_2	3.4%	1.9%	2.7%	78%
O_2	Trace	Trace	0.13%	21%
Ar	0.01%	0.1%	1.6%	0.93%
Temperature °C	477	290	−53	16
Pressure (bars)	92	60	0.006	1

TABLE 12.2
Carbon on the Earth (in gigatons = 10^9 metric tons)

Atmosphere	720
Oceans	
Total organic	1,000
Seawater layers	
Surface	670
Deep	36,730
Continents	
Living biologic mass	~800
Dead biologic mass	1,200
Fossil fuels (oil, coal, gas)	4,130
Organic matter in mudstone	15,000,000
Limestone	>60,000,000

Source: P. Falkowski and others, The Global Carbon Cycle. *Science,* 290 (2000): 291–96.

is 34°C higher than it would be if CO_2 and other greenhouse gases were absent. If CO_2 were not present in the atmosphere, the average temperature at Earth's surface would be about −18°C (0°F), and much of life would be different from what we know.

The Earth has always been influenced by a greenhouse effect, and life has always been in dynamic equilibrium with it. However, humans are now changing the CO_2 concentration in the atmosphere by burning tremendous volumes of plants, both living (trees and shrubs) and dead (the fossil fuels of coal, oil, and natural gas). Combining the C in plants with O_2 via fire returns large amounts of CO_2 to the atmosphere (figure 12.1). About 6 gigatons (1 gigaton equals 10^9 metric tons) are returned to the atmosphere each year by burning fossil fuels; about 5,000 gigatons remain to be burned. The human contribution is small compared to the natural fluxes between the atmosphere and ocean, and between the atmosphere and continents, each of which exchange in excess of 100 gigatons annually. Although human changes in CO_2 and other gases are relatively small, they can be enough to trigger climate shifts that cause major problems.

Climate History of Earth: Timescale in Millions of Years

Many sedimentary rocks contain information about the climate at the time they formed. Warm climates are indicated by (1) fossil reefs and most limestones; (2) aluminum ores, which form only in tropical soils; and (3) beds of salts that crystallize when water bodies evaporate under high-temperature, arid climates.

Cold climates may be marked by the powerful erosion of glaciers that sculpt the landscape (figure 12.2), leaving polished and grooved surfaces beneath them (figure 12.3) and dumping massive piles of debris. The distribution of fossil organisms tells much about ancient climates. For example, when fossil shells of organisms that live only in polar seas also are found in abundance in rocks formed in midlatitudes, it suggests that the world climate must have been colder at that time.

The rocks and fossils tell of extreme variations and changes in world temperature and precipitation throughout

Figure 12.2 Aiyalik glacier flows in southern Alaska, as the present Ice Age continues.
© Dr. Parvinder Sethi.

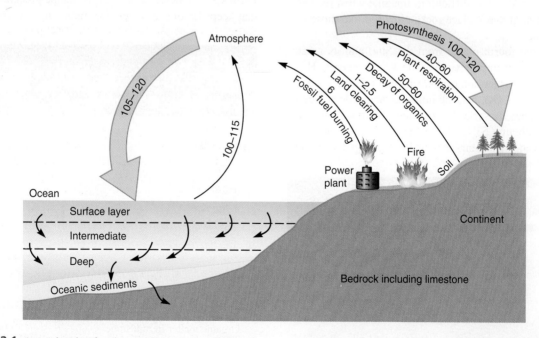

Figure 12.1 Annual cycle of carbon exchange measured in gigatons. Plants take in CO_2 from the atmosphere during photosynthesis. CO_2 is returned during plant respiration, during decay after death, and by burning of forests and fossil fuels. Near equilibrium exists between CO_2 in ocean surface water and the atmosphere. The ocean "pumps" some CO_2 into deep-water storage. Organisms remove dissolved CO_2 to build shells, which end up in sediments.

In Greater Depth

Equilibrium Between Tectonics, Rock Weathering, and Climate

Are plate tectonics and climate related? At first glance this may seem unlikely, but an intriguing hypothesis holds that major climatic changes of the last several hundred million years were caused in good part by variations in rates of CO_2 input to the atmosphere and oceans by plate-tectonic processes. Here is how it works: The amount of CO_2 in the atmosphere is a significant control on climate. A major source of CO_2 to the atmosphere is volcanism. The greater the volume of magma poured onto the land and ocean floors, the greater the volume of CO_2 released to the atmosphere. During much of Earth's history, the most abundant volcanism occurred at spreading centers; today that figure is about 80% of erupted magma. But rates of seafloor spreading are not constant; they vary markedly. When spreading is more rapid than now, huge volumes of CO_2 are released. Increased spreading also means the rate of subduction must increase, with accompanying greater releases of magma and CO_2 through subduction-zone volcanoes. So, fast rates of seafloor spreading increase the CO_2 in the atmosphere, which increases the greenhouse effect and warms the climate.

Increased rates of plate-tectonic activity build mountains as plates collide and volcanoes grow. But the increasing volumes of CO_2 entering the atmosphere increase the negative feedback that begins removing CO_2 via mountain destruction due to increased chemical weathering of rocks. The uplifting of mountains exposes fresh, unweathered rock that is heavily fractured by the uplift process. Mountains receive large volumes of rainfall and undergo disproportionately high rates of rock decomposition by water charged with CO_2—that is, carbonic acid ($H_2O + CO_2 = H_2CO_3$). Rock weathering can be illustrated using the common mineral feldspar (albite) in contact with carbonic acid, which yields a weathered solid plus sodium, bicarbonate, and silica held in water. A simplified chemical equation illustrating this rock-weathering process is:

$$2NaAlSi_3O_8 + H_2O + \mathbf{CO_2} \rightarrow AlSi_2O_5(OH) + 2Na^+$$
$$+ 2HCO_3^- + 4SiO_2$$

Notice that CO_2 was consumed in the weathering process, thus lessening CO_2 in the atmosphere and reducing the greenhouse effect. Multiply this reaction by millions of years, and major climate changes occur.

A geologic example of this process involves the tectonic collision of India pushing into Asia and uplifting the Himalaya Mountains, beginning about 40 million years ago (see figure 4.18). Before the collision, the atmospheric concentration of CO_2 was about 1,400 parts per million, but by 24 million years ago, CO_2 was down to about 200 parts per million. Uplifting the huge masses of fractured, fresh rock to form the Himalayas increased the rate of rock weathering, thus drawing down CO_2 levels in the atmosphere. Plate tectonics keeps supplying CO_2 (source) to the atmosphere, and weathering of uplifted rocks keeps removing CO_2 (sink), thus tending toward a quasi-equilibrium that helps control climate warming and cooling over millions of years.

---

geologic time. Not only do warm and cold intervals come and go, but they do not necessarily correlate with wet and dry periods, nor is there a pattern to the arrivals and departures of various climates.

Earth's climate depends on the balance between incoming and outgoing heat. At any given time, the atmosphere-ocean-continent system may be gaining or losing in its overall heat budget. Global heat supply has a profound effect on water, which exists on Earth's surface at the transition between its three phases of solid (ice), liquid, and vapor (gas). Water has such a tremendous capacity to either absorb or release heat that it acts as a powerful control on global climate.

The surface of the Earth is divided into temperature zones of frigid, temperate, and torrid by latitude (figure 12.4). Climate seems to swing like an irregular pendulum, from ages where cold temperatures of the frigid zone dominate Earth to other times when warmth covers most of the world. During a frigid period, an Ice Age, the colder climates of the high latitudes expand in area while the area of warmer climate in the low latitudes shrinks but does not disappear. Conversely, in an era of warmth, a Torrid Age, the globe is marked by expansion of the subtropical climatic zones, while the cold-climate belts shrink back toward the poles. Let's look at examples of extreme climates.

LATE PALEOZOIC ICE AGE

One of the major Ice Ages in Earth history began around 360 million years ago and lasted until 260 million years ago. For a glacial interval to last so long, broad-scale and long-lasting conditions are required. The major factors appear to be changes in the shapes, sizes, and orientations of the continents and oceans.

1. An initial, absolute requirement for an Ice Age is having one or more large continental masses near the poles. A polar landmass is necessary to collect the snowfall that allows the buildup of immense, 3 km (2 mi) thick ice sheets that bury continents. Massive glaciers cannot be built on top of ocean water. In Late Paleozoic time, the continents were largely united as the single landmass Pangaea. The southern portion of the Pangaea supercontinent is Gondwanaland; it moved across the south polar region and was progressively covered by major ice sheets. South America–Africa probably first supported the great ice sheet, then Antarctica, and finally Australia (figure 12.5).

2. Another important consideration is ocean-water circulation. No matter how much the Sun's brightness varies, equatorial waters receive more solar energy than polar waters. Without continents present to block ocean-water flow, the warm equatorial waters simply circulate latitudinally (east–west) due to the spin of the Earth.

Figure 12.3 Glacial grooves (striations) carved in rocks by a former glacier. This rock is exposed west of New Haven, Connecticut.
University of Washington Libraries, Special Collections, John Shelton Collection, Shelton KC10418.

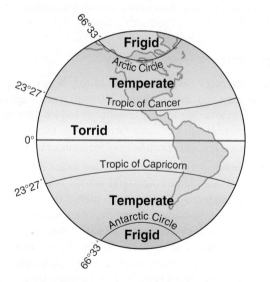

Figure 12.4 Five great divisions of Earth's surface today with respect to temperature and latitude.

What does warm water have to do with building massive ice sheets? In the hydrologic cycle, water must first evaporate from the ocean before clouds can form and move over cold landmasses to drop snow. Cold water is extremely difficult to evaporate; warmer water is a great help in promoting evaporation. The geologic record shows that Ice Ages are favored when oceanic circulation is more longitudinal (north–south) than latitudinal (east–west). When the continents are aligned in a north–south direction, they block latitudinal circulation of ocean water, thus sending warmer equatorial waters toward the poles, where evaporation can form the clouds that yield the snowfall that builds up on polar landmasses as glaciers. The continents had a north–south alignment during the Late Paleozoic time, as they do today (see figure 9.28) in the current Ice Age.

Why did the Late Paleozoic Ice Age end? Possibly because Gondwanaland began to break up and disperse. As the continents moved apart, ocean circulation patterns around the world changed. Warm waters stayed near the equator, and cold waters encircled the poles, thus drastically reducing the moisture supplied to polar landmasses. Additionally, when continents move away from the poles, no platform exists for the accumulation of snow and the building of glaciers.

LATE PALEOCENE TORRID AGE

The world was warming during Paleocene time (66 to 55 million years ago). There was more heat in the Paleocene oceans and atmosphere than at any time since. The equatorial zones had tropical temperatures and rainfalls higher than, but similar to, those they enjoy today; however, more poleward latitudes were markedly warmer. Sea-surface temperatures in the Southern Ocean around Antarctica were

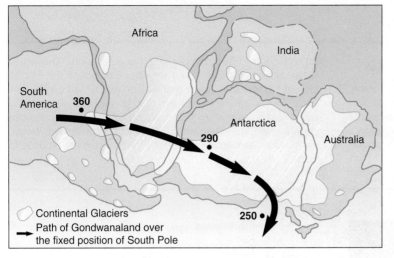

Figure 12.5 Late Paleozoic ice masses on Gondwanaland. Not all of these continental glaciers existed at the same time. Arrows approximate the path of drifting Gondwanaland over the fixed position of the South Pole. Numbers mark positions of the South Pole in millions of years ago.

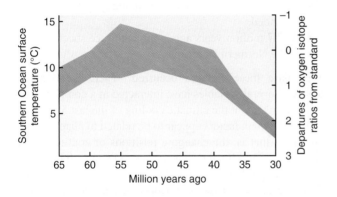

Figure 12.6 Surface temperature of the Southern Ocean over time, based on oxygen isotope measurements.

10° to 15°C (18° to 27°F) warmer than today based on measurements of oxygen isotopes (figure 12.6).

What was the world like during the Late Paleocene Torrid Age? There was less difference in temperature between tropical and polar waters; an absence of cold, dense, sinking water at the poles; and less difference in temperature between surface and deep-ocean waters. This means that the pull of gravity was less effective and ocean circulation probably was more sluggish.

Temperature differences in the atmosphere also decreased, resulting in more peaceful weather worldwide. There was an absence of strong seasons, weather was more constant, and rainfall was more evenly distributed throughout the year. Most of the world was wetter and warmer.

The conterminous United States was covered by either tropical or subtropical climates. Along the coastal zones, subtropical conditions existed above the Arctic Circle, as evidenced by fossil crocodiles and palm trees found there. Continental ice sheets apparently did not exist anywhere in

the world. Evergreen (coniferous) and warm deciduous forests covered much of the land. Hot deserts and arctic tundra covered smaller percentages of the ground. The world climate lacked extremes.

How did Earth's climate become so dominated by warmth? Several factors apparently combined to turn up the heat: (1) The equatorial zones were largely covered by oceans, allowing more absorption of solar radiation. (2) As oceans warmed, areas covered by snow and ice decreased, thus exposing more land. Snow and ice reflect the Sun's rays; land absorbs heat. (3) Enormous outpourings of lavas from the opening North Atlantic Ocean are likely to have released large volumes of gases to the atmosphere, which may have increased global warming via the greenhouse effect. (4) The oceans changed their style of density differentiation. At present, cold Antarctic and Arctic waters are the densest of all waters; they sink and flow along the ocean deeps. By Paleocene time, the polar water became so much warmer that the heaviest waters may have been tropical waters that had become saltier due to evaporation.

Warm, oxygen-deficient, salty waters apparently sank, flowing through the ocean deeps and warming up the oceans from surface to bottom. Warm, salty water masses moving along the ocean bottoms likely affected deep-ocean life. Organisms used to living in cold, oxygen-rich bottom waters experienced the shock of their environment becoming warm and oxygen-poor. At about 55 million years ago, the massive change in deep-sea water temperature reached a peak, causing up to 50% of unicellular deep-sea animal species to become extinct—a natural disaster.

What was responsible for the final increase in warmth? The warming of ocean bottom waters about 8°C (14°F) probably caused melting of icy **methane hydrates** on the seafloor, thus releasing **methane** gas to the atmosphere. What are methane hydrates? Bacteria living on the deep-ocean floor release methane (CH_4) as part of their life process, but the overlying water is so cold and the pressure from the weight of the overlying water is so great that the methane is locked up inside linked, near-freezing water molecules to form an icelike deposit (figure 12.7). Methane hydrate holds more energy than all of Earth's oil, coal, and natural gas combined. It becomes unstable if temperature rises a few degrees above freezing or if pressure is less than that of 500 m (1,640 ft) of overlying ocean. Today, about 15 trillion tons exist on the seafloor; if it were to melt, the world would see a sharp greenhouse increase in temperature. Recent analyses of carbon isotopes in Late Paleocene sedimentary rocks suggest that a major release of methane occurred about 55 million years ago during a 10,000-year-long interval. This is a very short time for the atmosphere to receive such a large volume of a powerful greenhouse gas (methane has a 21-times-stronger capacity to trap heat than carbon dioxide). During about 250,000 years of excess methane in the

In Greater Depth

Oxygen Isotopes And Temperature

Ancient temperatures can be determined from the ratio of stable isotopes of oxygen in the calcium carbonate ($CaCO_3$) shells (fossils) of single-celled sea life. An atom of oxygen may have either 16, 17, or 18 protons and neutrons in its nucleus. Water evaporated from oceans removes more of the lighter common oxygen (^{16}O) and less of the heavier ^{18}O. This ^{16}O-enriched water is locked up on land as ice and snow, leaving the ocean with ^{18}O-enriched water. Shells constructed from seawater incorporate the $^{18}O/^{16}O$ ratio of the seawater during their lifetime within their $CaCO_3$ shell walls. Thus, measurement of the $^{18}O/^{16}O$ ratio in shells acts as a paleothermometer, which is used to estimate the temperatures of ancient seas. Heavier seawater (^{18}O-enriched) corresponds to glacial buildups and lighter seawater (^{18}O-depleted) corresponds to warmer climates.

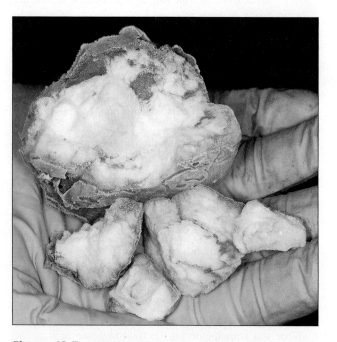

Figure 12.7 Methane hydrate is an icelike deposit of methane trapped within near-freezing water on and under the ocean floor. This sample is from offshore Oregon.

International Ocean Discovery Program.

atmosphere, the Earth experienced its warmest climate of the last 66 million years. Over time, the methane oxidized to CO_2, which was withdrawn and used by life, and then finally global temperatures began to decrease.

Earth changed from a Late Paleozoic icehouse to a Late Paleocene hothouse. But climatic change is the way of the world.

LATE CENOZOIC ICE AGE

Beginning from the temperature peak at 55 million years ago, Earth began the long-term cooling trend that has carried us into our current Ice Age (see figure 12.6). The sequence of events included:

- After 55 million years ago, intervals of high temperature still occurred, but the torrid climate had begun a long-term cooling trend.

- At 40 million years ago, Antarctica was surrounded by cold water.
- At 34 million years ago, glaciers were widespread in Antarctica.
- At 14 million years ago, a continental ice sheet existed on Antarctica, and mountain glaciers were in the Northern Hemisphere.
- At 5 million years ago, the Antarctic ice sheet had expanded.
- At 2.7 million years ago, continental ice sheets existed in the Northern Hemisphere.

Why have these changes occurred? There is no single answer. Several variables have interacted in a complex fashion to bring about the climatic cooling of the last 55 million years. The main factors appear to be related to plate-tectonic changes—that is, the changing positions of continents and oceans. (1) The climatic change is associated with the ongoing breakup of Pangaea into separate continents (see figures 2.24 and 2.25). (2) As continents drifted, seaways opened and closed, thus altering the circulation patterns within the oceans and the distribution of heat about the globe. (3) Continental masses have moved into polar latitudes, with Antarctica centering on and rotating about the South Pole, while North America and Eurasia have moved to encircle the North Pole region. (4) As snow and ice began accumulating on polar landmasses, they reflected more sunlight (increased albedo), and thus heat, back to space. (5) Circulation of the ocean water around the equator was restricted at about 23 million years ago with the closure of the eastern Mediterranean Sea and ended at 3 million years ago when volcanism completed building Central America as a continuous north–south barrier that blocked east–west ocean-water flow. (6) The area of shallow oceans has been reduced, so less water surface is available to absorb sunlight. (7) The uplifts of the Tibetan Plateau/Himalaya Mountains in Asia and the Colorado Plateau in the western United States have deflected west-to-east atmospheric circulation in the midlatitudes with resultant airflows to the north and return flows to the south.

The Last 3 Million Years

The ice sheet on Antarctica is older and more stable than ice in the Arctic. The cold ocean water circulating around

Antarctica (see figure 9.28) helps isolate the continent from major changes. The ice sheets on North America and Eurasia have a greater effect on global climate change because they expand and shrink in a more dynamic fashion. Their initial growth as continental glaciers occurred between 3.0 and 2.7 million years ago and coincided with Central America forming a continuous link between North and South America, it blocked westward-flowing ocean water and began diverting the warm water of the Caribbean Sea and Gulf of Mexico and forcing it to flow northward along the western Atlantic Ocean. The warm water delivered to Canada, Greenland, and Europe caused greater evaporation and formation of water vapor, which resulted in greater snowfall that accumulated to build glaciers.

Once continental ice sheets existed in the Northern Hemisphere, they underwent complex cycles of glacial advance and retreat. The cycles appear to have been present during earlier Ice Ages and are strongly linked to regular variations in the Earth's orbit and rotation. These cycles continue today.

Glacial Advance and Retreat: Timescale in Thousands of Years

As an Ice Age begins, ocean surface water evaporates, and some precipitates on the continents as snow. Snow accumulates, and burial pressure converts it into ice (figure 12.8). Continental glaciers reach thicknesses of about 3 km (2 mi), deeply burying the land. The immense volumes of ice deform internally under their own weight and slowly flow out over the countryside like mega-bulldozers, scarring and reshaping the land (figure 12.9). The record of glacially deposited sediments tells of numerous glacial advances and retreats. Starting in the 1970s, our knowledge of the advance-retreat history has been leaping ahead, thanks to cores of sediments taken from the ocean floor and cores of ice removed from the Greenland and Antarctic continental glaciers. Each core holds the cumulative record of the annual deposits of sediment or snow (ice) that may be read like the pages of a history book using techniques such as the ratios of oxygen isotopes.

The emerging story for the past 1 million years is of worldwide glacial advances that last about 100,000 years followed by retreats that take place more rapidly—withdrawing over periods from decades to a few thousand years (figure 12.10). What causes the cycles of slow buildup and advance of glaciers followed by rapid shrinkage and retreat? The answer lies in the cyclic peculiarities of Earth's spin and its orbit around the Sun, each of which affects the amount of solar energy received by Earth. Verification of the importance of orbit and rotation cycles came in the 1980s, when computer analyses of data from sediment and ice cores were shown

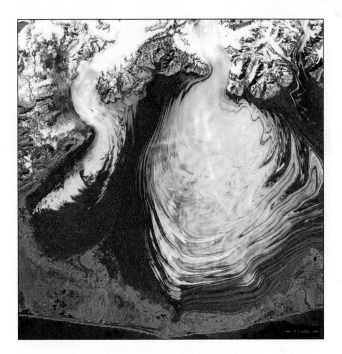

Figure 12.8 Part of the Malaspina Glacier west of Yakutat Bay in Alaska. A river of solid water flowing over the land.

Earth Resources Observation and Science (EROS) Center/USGS.

Figure 12.9 The southern limit of the last glacial advance is shown by the dumped glacial debris (irregular, hilly land on left) on the Waterville Plateau, Washington.

University of Washington Libraries, Special Collections, John Shelton Collection, Shelton 5568.

to match the theoretical astronomical framework erected by the Serbian astronomer Milutin Milankovitch in the 1920s and 1930s. Milankovitch defined astronomical changes in Earth's orbit, tilt, and wobble and how they affect the amount of solar radiation received by Earth.

An important factor in continental glacier formation is the amount of solar radiation received at high latitudes on

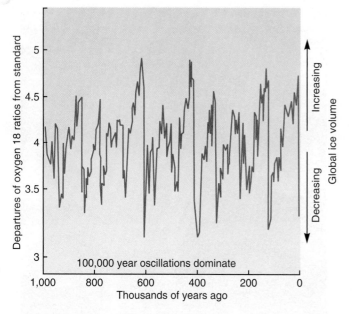

Figure 12.10 World ice-mass volumes of the past 1 million years based on oxygen isotope measurements.

Data from J. W. C. White, 2004.

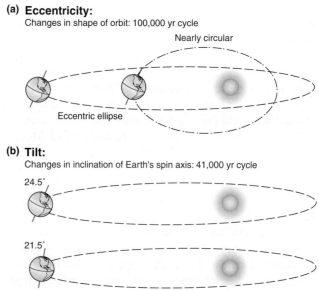

(a) Eccentricity:
Changes in shape of orbit: 100,000 yr cycle

Nearly circular

Eccentric ellipse

(b) Tilt:
Changes in inclination of Earth's spin axis: 41,000 yr cycle

24.5°

21.5°

(c) Wobble:
Precession of the equinoxes: 19,000 and 23,000 yr double cycle
Changes in direction of spin axis (same tilt)

Figure 12.11 Astronomical peculiarities that affect the amount of solar energy received by the Earth. The ellipticity of orbit is exaggerated.

Adapted from *Planet Earth* by V. A. Schmit, 1998; Kendall-Hunt Publishing Company.

Earth each summer. During a warm summer, all the snowfall from the previous winter can melt. But if the winter snowfall of one year persists until the next winter snowfall begins, then glaciers can start to form. And when glaciers grow for thousands of years, continents can become buried by ice.

The **Milankovitch theory** explains that glaciers advance and retreat due to variations in solar radiation received at high latitudes during summer and that these variations are due to the following factors:

1. Eccentricity of Earth's orbit around the Sun. The more elliptical the orbit, the less solar radiation is received and the less snow melts. The shape of the orbit varies every 100,000 years, from nearly circular to a slightly eccentric ellipse (figure 12.11a). The eccentricity time cycle is similar to the broad-scale, primary length of time for each glacial advance and retreat pairing, suggesting that eccentricity sets the fundamental frequency of the cycles.

2. Tilt of the Earth's axis. The spin axis of the Earth tilts away from the orbital plane in a 41,000-year-long cycle, where tilt varies from 21.5° to 24.5° (figure 12.11b). Greater tilt angles cause more increased seasonal extremes, including more snow melt. At present, the tilt is about 23.5°.

3. Precession of the equinoxes where the direction of the tilt changes even though the angle stays the same. The effect is a wobble roughly analogous to what you see in the spin of a toy top. The wobble has a double cycle with periodicities of 23,000 and 19,000 years (figure 12.11c). Currently, the wobble places Earth closest to the Sun during the Northern Hemisphere winter,

giving it milder winters and summers than the Southern Hemisphere.

The changes over time of the eccentricity, tilt, and wobble cycles have been calculated for the past and into the future (figure 12.12). At present, the tilt contributes to cooling while the eccentricity and wobble (precession) work to warm the climate.

The sediment- and ice-core records seem to show that glacial advances and retreats are synchronous in both the Northern and Southern Hemispheres. How are ice masses around the opposing poles affected simultaneously by astronomical tilts and wobbles? Probably by heat transfer within the world ocean and atmosphere. Any increased heat received in one hemisphere is shared with the other.

THE LAST GLACIAL MAXIMUM

Growing continental glaciers reached their last maximum positions between 33,000 and 26,500 years ago due to decreases in (1) summer insolation in the Northern Hemisphere, (2) tropical Pacific sea-surface temperatures, and

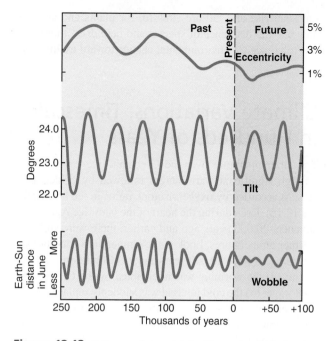

Figure 12.12 Patterns of eccentricity, tilt, and wobble for the past, present, and future.

Data from A. Berger.

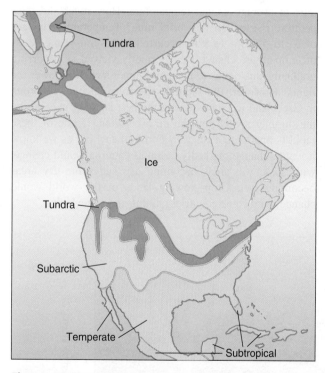

Figure 12.13 The maximum extent of glacial ice during the present Ice Age.

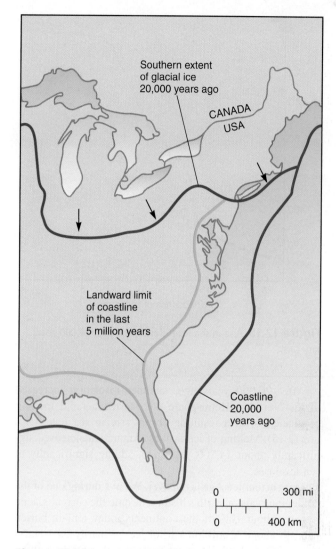

Figure 12.14 Some past positions of coastlines.

Source: S. J. Williams, et al. "Coasts in Crisis," *U.S. Geological Survey Circular 1075,* 1990.

(3) atmospheric CO_2. The glaciers were still near their maximum positions 20,000 years ago.

What was Earth like around 20,000 years ago when glacial ice masses were at peak extent? The continental ice sheets contained about 70 million km³ (17 million mi³) of ice, and the ice masses had spread out to cover about 27%

of today's land, including virtually all of Canada and part of the northeastern United States (figure 12.13). Each ice sheet had its own cell of atmospheric high pressure that displaced midlatitude storm systems to the south. The displacement of storm systems increased midlatitude rain, turned the desert basins of the southwestern United States into a series of lakes, and produced much heavier rainfall over the Mediterranean region. The high pressure over Asia kept away its monsoonal rains, thus increasing the aridity of the Indian subcontinent.

The amount of seawater required to build the glaciers caused sea level to drop 130 m (425 ft) below the level of today (figure 12.14). But extensive ice masses carry some of the seeds of their own destruction: shrinking ocean surface area plus colder ocean water mean less water evaporation with less snowfall. Cutting down on evaporation reduces the supply of snow necessary to maintain glaciers, and thus they shrink.

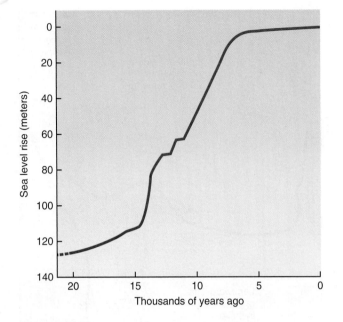

Figure 12.15 Rise in sea level during the past 20,000 years.
After NASA.

After 20,000 years ago, summer insolation increased in the Northern Hemisphere and started a rapid melting of glaciers with a resulting abrupt rise in sea level (figure 12.15). Melting of ice in West Antarctica increased significantly about 14,700 years ago, adding significantly to sea-level rise.

Our current Ice Age is not over. We live during one of the colder intervals in Earth's history, despite the current glacial retreat. About 10% of the continents today remain buried beneath about 32 million km³ of ice, primarily on Antarctica and Greenland. If this ice were to melt, with or without human help, sea level would rise about 65 m (210 ft), and many of the world's major cities would be submerged. The changes associated with marked climatic fluctuations and sea level rise and fall are stressful for plants and animals; they test the resilience of species and help drive those with low tolerance for environmental change toward **extinction.**

Climate Variations: Timescale in Hundreds of Years

The air temperature over the Greenland ice cap proceeds through significant warm stages followed by colder intervals, as recorded by oxygen-isotope ratios in ice layers (figure 12.16). Even during the heart of the latest Ice Age glacial advance, 20,000 years ago and earlier, there were spikes of warmer temperatures. Look at the temperature conditions in figure 12.16 since 20,000 years ago: (1) Conditions began to warm; (2) the warming was interrupted by the Older Dryas cold stage; (3) the cold interval was suddenly replaced by the elevated temperatures of the Bølling period; (4) the higher temperatures deteriorated through the Allerød interval; (5) temperatures plunged back into the depths of the Ice Age during the Younger Dryas stage from 12,900 to 11,700 years ago; (6) last came the current interglacial period.

Look again at figure 12.16 and note the sharp rises and falls in average annual temperature. How much did temperature rise or fall in a brief time? Temperature changes of 3° to 5°C (5° to 9°F) occurred in several years. Rates of temperature change were formerly viewed as occurring gradually, analogous to using a dimmer switch to lower the lights. However, the rapid temperature changes recorded in Greenland ice show us that the best analogy may not be the dimmer switch but the on-off switch. If one of these rapid temperature changes occurred today, life as we know it would change markedly. Rainfall patterns would change, with some wet areas becoming dry and some dry areas becoming wet. Crop-growing lands of the world would change, with some countries gaining and some losing.

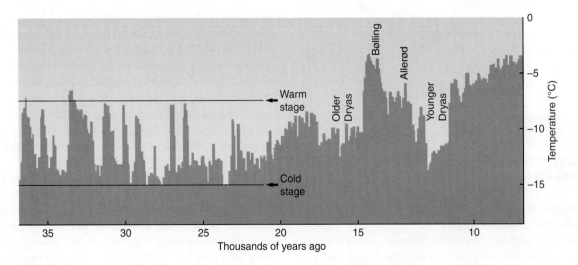

Figure 12.16 Air temperature over the Greenland continental glacier as recorded by oxygen-isotope compositions of ice at the glacier summit.

Why the sudden jumps or drops in temperature? One suggested cause relies on changes in the North Atlantic Ocean. As the massive ice sheets on the continents were melting, enormous lakes of pure, cold water formed, held back by ice dams. The shape of the land surface (see last section of chapter 13) and the sediment record tell of enormous floods produced by the failure of the ice dams. Floods of cold, glacial meltwater flowed through the Columbia River and other rivers out on top of the northern Pacific Ocean, creating a surface layer of cold, nonsaline water. Research conducted in 2010 documented that mega-floods flowed around 13,000 years ago through northwest Canada in the Mackenzie River into the Arctic Ocean. These huge pulses of cold freshwater in the Arctic would ultimately have flowed into the North Atlantic Ocean. Megafloods also flowed down the Saint Lawrence River into the North Atlantic. These cold surface-water layers could alter the ocean-circulation pattern shown in figure 9.29 by stopping Arctic seawater from sinking and by blocking the northward inflow of Gulf Stream warm water.

As long as North Atlantic cold, salty seawater sinks and flows away as deep-ocean currents, it is replaced by warm surface water flowing up from the Gulf of Mexico and the Caribbean Sea. In northern latitudes, the winds pick up tremendous amounts of heat from warmer surface water and warm the adjacent lands of Greenland and Europe. But when Arctic surface water is fresh and cold, its low density prevents sinking, and its cold temperature results in colder air temperatures. This is apparently what happened during the Younger Dryas (figure 12.16). Glacial meltwater floods from 12,900 to 11,700 years ago put cold freshwater on top of the North Atlantic Ocean, shutting down the circulation system of figure 9.29. It took centuries for solar energy to return the ocean surface to its warmer, saltier condition and the present circulation system.

Remember that sea level was 130 m (425 ft) lower at the peak of continental glaciation and that the removed water was stored on land as glacial ice. When the glaciers retreated, sea level rose by the inflow of cold freshwater freed by the melting of glacial ice. The return of this massive volume of meltwater affected the oceanic distribution of heat; this is a climate-modifying process. The last major melting of the ice sheets is recorded by the sea-level rise curve (see figure 12.15).

At about 7,000 years ago, average global temperatures were warmer, and rainfall totals had risen. At this time, known as the "climatic optimum," even North Africa had enough rainfall to support civilizations. Since then, there has been a 7,000-year-long lowering of global average temperature totaling about 2°C. However, the cooling trend has had several smaller cycles of glacial expansion and contraction superimposed on it (figure 12.17). It seems that climatic cycles can be found at any timescale we choose to use.

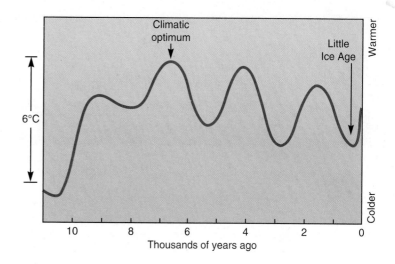

Figure 12.17 Generalized trends in global temperature for the past 11,000 years.

Data from J. Imbrie and K. P. Imbrie, *Ice Ages,* 1979; Harvard University Press, Cambridge.

Shorter-Term Climate Changes: Timescale in Multiple Years

Several processes change climate on timescales of 1 to 20 to 30 years. Let's look at examples.

EL NIÑO

The high heat capacity of water gives it the ability to absorb and store tremendous volumes of heat upon warming and to release copious quantities of heat upon cooling. The heat supply in the world ocean has a major effect on the atmosphere and on world climate. An example of ocean—atmosphere coupling is the phenomenon commonly marked in South America by the arrival of warm ocean water to Peru and Ecuador near Christmastime. This phenomenon is known as **El Niño** (Spanish for "the child").

Typical conditions in the central Pacific Ocean find high atmospheric pressure over the eastern Pacific, resulting in trade winds that blow toward the equator from the north and south. The trade winds push Pacific Ocean surface waters to the west within the equatorial zone, where they absorb solar energy (figure 12.18a). The winds push so hard that sea level is not level; it is about 1.5 ft higher on the western side of the ocean. The warm water piled up on the west side forms a pool of heated water that evaporates readily, helping produce heavy rainfalls for the tropical jungles of Indonesia and Southeast Asia and providing the environment for the Great Barrier Reef of Australia. Meanwhile, on the eastern side of the Pacific Ocean, the warm surface water blown west is replaced by cold waters rising from depth (figure 12.18b) and from the polar regions. The colder waters along the coast evaporate less readily, and thus deserts are common

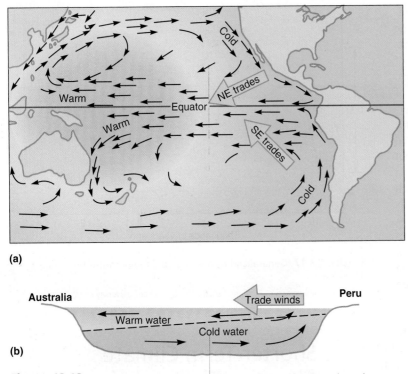

(a)

(b)

Figure 12.18 Pacific Ocean circulation. (a) Map. The northeast and southeast trade winds combine to push warming surface water westward across the ocean in the equatorial belt. After circling near the poles, the return water flowing along the North and South American coasts is cold. (b) Schematic cross-section. The trade winds stack up warm surface water on the western side of the Pacific Ocean.

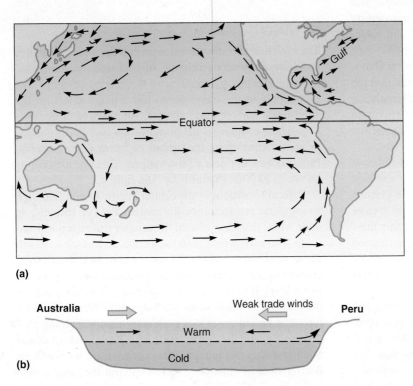

(a)

(b)

Figure 12.19 Pacific Ocean circulation during El Niño. (a) Map. Equatorial winds blow toward the center of the ocean from both sides. (b) Schematic cross-section. Weakened trade winds plus winds from the west cause warm water to accumulate along the equatorial Americas.

along the coasts of Ecuador, Peru, Baja California, and California because of the shortage of cloud-producing water vapor.

Every two to seven years, the typical ocean-atmosphere pattern breaks down for about 6 to 18 months. The trade winds weaken, the atmospheric low pressure over Indonesia moves out over the central Pacific Ocean, and winds blow into the Pacific basin from the west (figure 12.19). The warm surface waters then flow "downhill" toward South and Central America. Some surface currents are reversed as some winds from the west blow surface water to the east. The reversal places a huge mass of warm water against the Americas (figure 12.20), where it evaporates more readily and produces more clouds. The warm, moist air flowing eastward off the ocean into the Americas commonly leads to heavier rains than the coastal deserts can handle.

The El Niño of 1982–83 was especially strong. On the eastern side of the Pacific Ocean, the cold-water fisheries off Peru and Ecuador collapsed as ocean water warmed as much as 8°C (14°F) above average. The warm water led to greater than average evaporation, which fed torrential rainfalls. The rains caused overwhelming floods and mass movements from steep hillsides that killed 600 people in Peru and Ecuador and severely punished the economies of those nations. The warm coastal waters also promoted heavy rainstorms in the western United States. For example, California suffered $300 million in damages, 10,000 people were evacuated due to flooding and landsliding, and 12 people were killed. Meanwhile, out in the ocean, the tropical rain belt shifted to the central Pacific Ocean, helping form hurricanes that hit Hawaii and Tahiti. On the western side of the Pacific Ocean, Australia and Indonesia were covered by high-pressure air and below-average rainfalls. Australia suffered its worst drought of the century, and out-of-control bushfires whipped by high winds killed 75 people and many domestic and wild animals and caused $2.5 billion in damages (see chapter 14).

The high-level atmospheric winds are also affected by reversal of flow direction during an El Niño. The change brings negative and positive results for the United States. For example, in the 1997–98 El Niño, winds flowing eastward caused heavy rains and floods in California and brought higher rainfall with

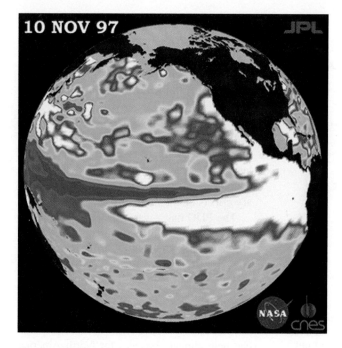

TABLE 12.3

Estimated Impacts on the United States of the 1997–98 El Niño

Losses	
Human lives lost: 189	
Economic losses: US$4.5 billion	
Benefits	
Human lives *NOT* lost: 850	
Economic gains: US$19.5 billion	

Source: American Meteorological Society Bulletin.

Figure 12.20 Pacific Ocean water temperatures during a massive El Niño, 10 November 1997. Warmer water is white or red; colder water is purple or blue.

NASA/JPL.

some tornadoes to the southeastern United States; on the other side, El Niño winds helped break apart Atlantic and Caribbean storms, resulting in fewer hurricanes. Meanwhile, the midwestern and northern states had a warmer winter with reduced damages to crops, increased economic productivity, and a below-average number of deaths. The 1997–98 El Niño brought more economic gains than losses as well as fewer fatalities to the United States as a whole (table 12.3).

The El Niño effects in the Pacific Ocean basin are impressive, but the ocean-atmosphere system is linked on a larger scale. In the South Pacific Ocean, the shifting of weather patterns is known as the Southern Oscillation; it occurs when the usual low-pressure atmosphere is replaced by high-pressure air, as measured at Darwin on the north coast of Australia (figure 12.21). For example, in late 1996, strong weather systems in the Indian Ocean migrated into the western Pacific Ocean before the strong El Niño of 1997–98. The combined system is called the El Niño/Southern Oscillation (ENSO).

It is increasingly evident that unusually high rains in one area and drought in another are not isolated events; rather, they are both parts of a globally connected weather system. A change that occurs in one area can trigger other changes around the world, somewhat like a falling line of dominoes—knock over one and it starts the process whereby they all fall down. Every several years, the tropical atmosphere goes through changes that link up around the world. The tropical Indian Ocean warms, and its weather pattern may blow into the western Pacific Ocean, setting off an El

Niño. After the El Niño wind shifts cross the Pacific Ocean and South America, the tropical Atlantic Ocean may begin to warm. This global circuit takes about four years to move around the Earth.

LA NIÑA

The surface waters of the ocean are a mosaic of warmer, intermediate, and cooler water masses that exert strong controls on regional weather. The Pacific Ocean surface-water masses of different temperature are measured by satellite. When warmer-than-normal surface waters extend along the west coast of the Americas during an El Niño, they usually bring high rainfalls and accompanying floods. But El Niño has a sister called **La Niña** ("the girl"), and she has a different personality. La Niña occurs when cooler water moves into the equatorial Pacific Ocean (figure 12.22).

During a La Niña, trade winds are stronger and other wind systems change their paths, bringing different weather patterns across North America. A typical La Niña winter brings cold air with high rainfall to the northwestern United States and western Canada but causes below-average rainfall in most of North America. The winter of 1999–2000 was typical, with heavy rainfalls in the northwest and below-average rainfalls elsewhere, accompanied by numerous wildfires in the west.

There are hazards associated with the La Niña cooling in the Pacific Ocean. La Niña allows the growth of hurricanes in the Atlantic Ocean, spelling trouble for the eastern United States and Gulf of Mexico coastal areas. La Niña leads to decreased rainfall in the American Southwest, helping dry out the El Niño–fed vegetation and leading to wildfires.

El Niño and La Niña are the extreme conditions in which warm or cold water masses strongly influence the distribution of rainfall. Extreme ocean conditions make weather prediction easier. However, there are also times when the tropical Pacific Ocean is neither excessively warm nor markedly cool but neutral. The weaker signal from the ocean makes weather more difficult to predict. NASA oceanographer William Patzert has suggested the term **La Nada** for this neutral condition.

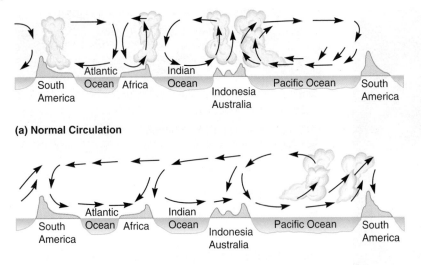

(a) Normal Circulation

(b) El Niño Condition

Figure 12.21 Cross-sections of Southern Hemisphere atmospheric circulation. (a) With "normal circulation," moist air rises over the landmasses, condenses, and falls as rain on eastern South America, eastern Africa, and Indonesia-Australia. (b) With El Niño conditions, eastern Africa and Indonesia-Australia do not get their customary rain, while western South America receives heavy rainfall.

PACIFIC DECADAL OSCILLATION

In recent years, another weather-influencing cycle, the **Pacific Decadal Oscillation (PDO)**, has been recognized by studying sea-surface temperatures in the north Pacific Ocean. The PDO cycles last 20 to 30 years (figure 12.23) and occur as midlatitude conditions of the Pacific Ocean that have secondary effects on the tropics. Compare this to El Niño, which lasts 6 to 18 months as low-latitude (tropical) conditions of the Pacific Ocean with secondary effects on the midlatitudes.

The PDO has a warm phase accompanied by increased storminess and rainfall. Warm phases occurred from 1925 to 1946 and from 1977 to 1998. Cool phases with decreased numbers of storms were in effect from 1890 to 1924 and from 1947 to 1976.

The cause of the PDO is not known. But even without a theoretical understanding, PDO data can help climate forecasts for North America because PDOs tend to persist for multiple years.

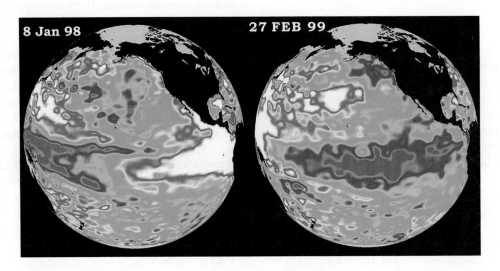

Figure 12.22 Pacific Ocean water temperatures during an El Niño in January 1998 (left) and during a La Niña in February 1999 (right).

NASA/JPL.

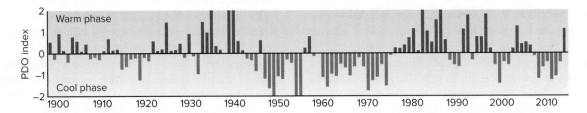

Figure 12.23 The monthly Pacific Decadal Oscillation index from 1900 to 2014. Warm phases have more storms and rains, whereas cool phases bring the opposite.

VOLCANISM AND CLIMATE

Benjamin Franklin recognized in 1784 that volcanism can affect the weather. He suggested that the haze and cold weather in Europe during 1783–84 were due to the massive outpourings of lava and gas at Laki, Iceland (see chapter 7). How else can volcanism affect the climate? Large, explosive Plinian eruptions can blast fine ash and gas high enough to be above the normal zone of weather. Free from the cleansing effects of rainfall in the stratosphere, the volcanic products can float about for years and interfere with incoming sunlight.

The finest volcanic ash (0.001 mm) can stay suspended for years. Most gases blown into the stratosphere disappear into space, but sulfur dioxide (SO_2) picks up oxygen and water to form an aerosol of sulfuric acid (H_2SO_4) that may stay aloft for years (table 12.4). The combined ash and sulfuric acid produce **haze,** reducing the amount of sunshine that reaches the troposphere and the ground surface; thus, climatic cooling results.

El Chichón, 1982

Located in the state of Chiapas in southern Mexico is the relatively small volcano called El Chichón (which translates loosely as "bump"). Four big Plinian eruptions from El Chichón on 29 March through 4 April 1982 blew out about 0.6 km³ of material, leaving a 1 km diameter crater and killing 2,000 people (see figure 7.28). Although the eruptions were not as big as the Mount St. Helens event in 1980, more than 100 times the volume of SO_2 gases was pumped into the stratosphere, along with volcanic ash. The cloud of stratospheric gases took 23 days to circle the globe (figure 12.24). The SO_2 gas combined with O_2 and water vapor, converting to sulfuric acid (H_2SO_4) aerosol. Sunsets were spectacular for months, beginning with a purple glow high over the horizon, changing gradually to surreal yellows and oranges as the Sun set, and ending with a red afterglow when the sky normally would have been dark. World temperature from this event was lowered 0.2°C.

Mount Pinatubo, 1991

After a slumber of 635 years, Mount Pinatubo awoke to disrupt life on the Philippine island of Luzon in the spring of 1991. The 1,745 m (5,724 ft) high summit was blasted to bits and replaced by a 2 km (1.25 mi) wide caldera as up to 5 km³ (1.2 mi³) of dense magma blew out in the form of pyroclastic debris. Despite ample advance warnings, more than 300 people died in pyroclastic flows. Adding to the tremendous destruction of property and life loss was a major storm that poured torrential rains on the loose pyroclastic debris, setting in motion numerous large-volume lahars.

Of climatic importance were the 30 million tons of SO_2 gas blasted into the stratosphere (figure 12.25)—triple the volume of SO_2 released by El Chichón. The H_2SO_4 aerosols reflected 2–4% of incoming short-wavelength solar radiation back to space, causing a 20–30% decline in solar radiation directly reaching the ground. Mean global temperatures at the ground surface dropped 0.5°C. The greatest cooling occurred in the midlatitudes of the Northern Hemisphere, including the United States, where temperatures declined by 1°C. The volcanically induced cooling from SO_2 in the stratosphere more than offset the greenhouse warming that was expected in 1991–92 due to the CO_2 added to the troposphere by humans burning wood, oil, coal, and natural gas.

TABLE 12.4
Volcanic Eruptions into Stratosphere

Volcano	Date	Estimated Loading of SO₄ in Stratosphere (10,000,000 metric tons)
Tambora	1815	>100
Krakatau	1883	~50
Pinatubo	1991	30
El Chichón	1982	12
Stratospheric background	1979	<1

Source: M. Mills in *Volcanoes* (2000).

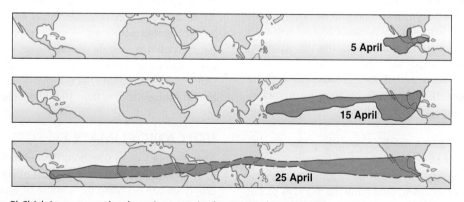

Figure 12.24 The El Chichón gaseous cloud moving west in the stratosphere in 1982.

Adapted from Robock, Alan, and Michael Matson, "Circumglobal transport of the El Chichón volcanic dust cloud" in Science 221 (1983);195–97.

Figure 12.25 Vertical eruptions from Mount Pinatubo in June 1991 injected 30 million tons of sulfur dioxide (SO_2) into the stratosphere.

Karin Jackson/U.S. Air Force.

Tambora, 1815

In the early 1800s, Mount Tambora, on the island of Sumbawa in Indonesia, stood 4,000 m (13,000 ft) tall. After the explosive eruptions of 10–11 April 1815, Tambora was only 2,650 m (8,700 ft) high and had a caldera 7 km (4+ mi) wide and 1.1 km (0.7 mi) deep. About 150 km^3 (36 mi^3) of rock and magma were blasted out during the eruption, producing 175 km^3 of ash and other pyroclastic debris. The eruption has been called the greatest in historic times, killing about 10,000 people outright by pyroclastic flows and another 117,000 indirectly through famine and disease.

The volcanic ash and aerosols, especially sulfur dioxide (SO_2), blown into the stratosphere blocked enough sunshine to make 1816 known as "the year without a summer." Agricultural production was down throughout the world as global temperatures lowered another 0.3°C during an already cold series of years. Lord Byron spent a cold, dark summer in 1816 on the shores of Lake Geneva and described it in his poem "Darkness."

> The bright Sun was extinguish'd, and the stars
>   Did wander darkling in the eternal space
> Rayless and pathless, and the icy earth
>   Swung blind and blackening in the moonless air;
> Morn came and went—and came,
>   and brought no day . . .

On the other side of the world in the northeastern United States, snow or frost occurred in every month of the year. An estimated 8% of the residents of Vermont were forced to leave the state due to agricultural failures. In India, the Tambora-cooled climate induced famine and weakened the population, possibly triggering the cholera epidemic that soon followed. The disease then slowly migrated around the world, killing people who lived under the harshest, least sanitary conditions.

A Cold Decade: 1810–1819

In late 2009, analyses of ice cores from Greenland and Antarctica identified unique sulfur isotopes and ash in the year 1809 ice layers; they came from the same volcanic eruption. To spread volcanic debris over the whole Earth, the eruption must have been huge and at a near-tropical latitude. Which volcano erupted? We don't know yet, but the search is on. Newspapers, personal letters, and such from late 1809 should provide clues for locating the volcano.

The 1809 eruption from the mystery volcano would have made the years 1810 and 1811 significantly colder than average. Then the eruption of Tambora in 1815 made the years 1815–1817 much colder. The decade from 1810 to 1819 must have been a cold one, with reduced agricultural production and increased human suffering.

Toba, Indonesia, About 74,000 Years Ago

Tambora erupted an impressive 150 km^3 of material, but if we go back 74,000 years, the eruption of Toba on Sumatra expelled about 2,800 km^3 (670 mi^3) of material. The Toba event is the youngest known resurgent caldera eruption. Scientists estimate that the Toba ash and H_2SO_4 aerosols formed a dense cloud in the stratosphere lasting for up to six years. Global cooling may have been 3° to 5°C (5° to 9°F) for several years. A *volcanic winter* of this magnitude may have triggered additional climate responses that prolonged the cold climate and increased the severe drought, ecological disasters, and famine. Some researchers even speculate that the Toba eruption effects drove down the global population of humans to less than 10,000 people.

VOLCANIC CLIMATE EFFECTS

The eruptions of El Chichón, Pinatubo, Tambora, and Toba give an idea of how volcanism can affect climate. Gas-rich eruptions can decrease the amount of incoming solar radiation and thus cause agricultural production to decline, which in turn can lead to famine, disease, and death.

In Greater Depth

The Mayan Civilization and Climate Change

Weather brings us local events that can cause death and destruction, whereas climate change brings regional or global events that can lead to the fall of civilizations. For example, the Mayan civilization of Mesoamerica made great accomplishments in agriculture, irrigation, social organization, mathematics, and astronomy during a thousand-year period. However, a centuries-long pattern of decreased rainfall caused problems that intensified during multi-year **droughts,** especially occurring between the years 800 to 1100 CE. The droughts set off a chain of events that led the Mayans to permanently abandon many urban areas in the southern and central lowlands (figure 12.26), to stop construction of monuments, and to experience the breakdown of social and political order leading to wars. Population in some cities dropped by 90%, and society became increasingly decentralized, forcing many people to return to a life of rural subsistence. The advanced Mayan civilization declined significantly during a string of events triggered by long-term climate change.

Climate change is not just something that affected ancient civilizations. Earth's climate today is responding to significant changes. During the 21st century, modern civilizations may be seriously affected by climate change.

FIGURE 12.26 The Mayan ruins at Uxmal in Yucatan, Mexico, are a favorite destination for tourists.
Photo by Pat Abbott.

What if eruptions from different volcanoes kept occurring at closely spaced intervals for many years? The Greenland ice cores record many years with high acid content during the Little Ice Age. Is the acidic ice a record of abundant SO_2-rich volcanic eruptions that helped cause the Little Ice Age? More work must be done to answer this question. Conversely, the record of global warming from 1912 to 1952 has been partly attributed to the absence of major SO_2-rich volcanic eruptions.

What are the main variables that come into play when volcanism affects climate? Principal factors include:

1. Size and rate of eruptions
2. Heights of eruption columns

3. Types of gases and the atmospheric level where they are placed. (For example, sulfur dioxide in the stratosphere reflects sunlight and cools the climate below, whereas carbon dioxide in the atmosphere creates a greenhouse effect.)
4. Latitude. Low-latitude eruptions spread atmospheric debris across more of the world and have greater global effects than high-latitude eruptions.

What is a worst-case scenario for volcanic effects on climate? Probably the hypothesis that blames flood-basalt volcanism for the great die-off of species, including non-avian dinosaurs, that occurred 66 million years ago. Some scientists suggest that the extinctions resulted from the climatic effects of the voluminous flood basalts erupted to form the Deccan Plateau in India. This hypothesis is based on the fact that about 2.6 million km^3 (625,000 mi^3) of basaltic lavas poured forth in as short a time as 700,000 years. What might the climatic effects have been? Proponents of this hypothesis estimate that atmospheric CO_2 increased significantly, with consequent global warming that raised the average world temperature as much as 10°C (18°F). The elevated surface temperatures, more acidic ocean waters (carbonic, sulfuric, and nitric acids), and possibly a depleted ozone layer all combined to deal punishing blows to life on land and in the uppermost layer of the ocean.

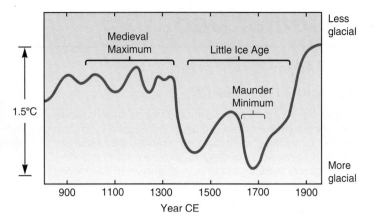

Figure 12.27 Climate of the past 1,000 years, based on European winters.

Data from J. Imbrie and K. P. Imbrie, *Ice Ages,* 1979; Harvard University Press, Cambridge.

The Past Thousand Years

During the past thousand years, the combined effects of Earth's orbital patterns of eccentricity, tilt, and wobble caused a cooling trend, but climate records show numerous variations, testifying to other processes also at work. The climate variations seen in figure 12.27 are actively being studied to learn more about (1) the extent of the temperature fluctuations, (2) whether they were regional or occurred simultaneously around the world, and (3) the causes of the changes. Information is being sought by analyzing glacial ice layers for oxygen isotopes, air bubbles, and volcanic debris; and annual growth rings of corals and trees. The past thousand years find us in the realm of human history with ever-improving observational records. Historic records studied to learn about past climates include (1) tax records of grain and grape crops; (2) advances and retreats of mountain glaciers; (3) paintings of winter scenes showing frozen lakes, rivers, and ports; and (4) numbers of weeks per year of sea ice around Iceland.

Looking at figure 12.27 reveals a warm period from about 1000 to 1300 CE referred to as the **Medieval Maximum.** During this time, northern Europeans emigrated to Iceland, where the almost ice-free coast helped fishers build thriving industries. The coastal plains of Greenland also were settled by Europeans who farmed the land. In England, wine grapes were grown and harvested. But this did not last.

The **Little Ice Age** affected Europe from about 1350 to the mid-1800s CE. It was originally defined by Francois Matthes in 1939 as an "epoch of renewed but moderate glaciation." Late in the Little Ice Age, part of northeastern Canada had accumulated permanent snowfields and the beginning of an ice sheet. Cold winters in Europe led to shorter growing seasons, with reduced crop yields leading to local famine. Mountain glaciers advanced throughout Europe. The fishing industry in Iceland was slowed by many weeks of sea ice each year.

Possible causes of the Little Ice Age include increased volcanic eruptions, changed ocean circulation, and reduced solar energy. One hypothesis now being tested suggests that low-latitude volcanic eruptions in the late 1200s put **aerosols** in the atmosphere that reduced the amount of solar energy reaching the surface; this caused climate cooling that allowed glaciers and sea ice to grow. The sea ice cooled North Atlantic

seawater, thus reducing northward transport of heat in seawater flowing to North America and Europe; this cooled the climate. Direct effects of volcanism last only years, but the effects on ocean circulation of heat last decades.

Climatic conditions during the Little Ice Age were far from constant, as smaller-scale warmings and coolings occurred (figure 12.27). One colder interval between 1645 and 1715 CE is known as the **Maunder Minimum.** During this time, minimal sunspot activity was noted by astronomers, and the Sun may have been 0.25% weaker.

Look at figure 12.27 again, at arm's length, and notice the thousand-year pattern—a warm interval for the first few hundred years, followed by a cooling until around 1700 when temperatures began rising. At the start of the 20th century, temperatures began exceeding the 1,000-year high; this warming trend continues today.

Statistical analyses of the Northern Hemisphere temperatures of the past 1,000 years, conducted by Michael Mann, Raymond Bradley, and Malcolm Hughes, have produced curves with lower-amplitude temperature variations (figure 12.28). Notice the range of pre-1900 temperatures in figure 12.27; they rise and fall 1.2°C. In figure 12.28, the pre-1900 temperature range is only 0.4°C. The difference between 1.2°C and 0.4°C is significant because it is a natural range of temperatures; it sets a baseline for assessing the human-caused temperature changes of the 20th and 21st centuries. At the beginning of the 20th century, both curves show sharp increases in temperatures that rise above the 1,000-year high. Much of this marked increase in temperature is climate change due to human activities.

Several processes were involved in the climate changes of the past thousand years: (1) changes in Earth's orbital pattern, causing cooling; (2) volcanic eruptions; (3) changes in solar-energy output; (4) changes in sea-surface temperatures in the tropics; (5) changes in aerosol concentrations; and (6) increases in greenhouse gases in the atmosphere. There probably were also interactions between the ocean, the atmosphere, and the ice sheets that are yet to be understood.

The 20th Century

When viewed in terms of the past 1,100 years, the 20th-century warming trend was unprecedented in both amount and rate (figure 12.29). What processes of the 20th century caused this dramatic increase in global temperature? Changes in Earth's orbit and plate-tectonic processes are both too slow to have affected climate significantly in one century. Volcanic eruptions and La Niñas cool global climate, and El Niños warm it, but only for a year or two each time; they do not dominate the century. The processes that can change significantly and last for a long time are changes in solar-energy output and increased contents of greenhouse gases and aerosols in the atmosphere. A global assessment of satellite data finds that changes in solar-energy output are responsible for less than 10% of recent warming. Thus, the major causes of global warming are changes in the greenhouse gas and aerosol contents of the atmosphere.

How much did global temperature rise in the 20th century? Average global surface temperature rose 0.6°C (+/−0.2°C), or 1.1°F. Could a person feel the climate warming of the 20th century? No, because the climate warming is small compared to the day-to-day temperature fluctuations of weather.

Did it warm continuously through the 20th century? No, most of the warming occurred during two time intervals: 1910 to 1944 and since 1977. It appears that the early warming was largely due to a hotter Sun and a lack of global volcanism. The present warming is likely due to increases in greenhouse gases in the atmosphere.

How much of the 20th century warming was due to natural processes and how much was due to human activities? Natural processes appear to have caused a net increase in temperature of about 0.2°C. Changes in Earth's orbital patterns caused a cooling of −0.02°C that was offset by a hotter Sun, netting a warming of +0.2°C. Human activities probably were responsible for the remaining 0.4°C increase in global temperature.

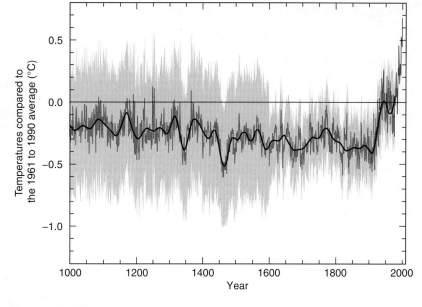

Figure 12.28 Trends in average annual temperatures in the Northern Hemisphere during the last 1,000 years. The whole pattern has been called a "hockey stick."

Source: IPCC *Assessment Report 3* (2001).

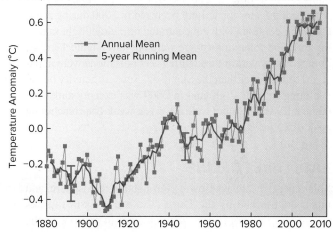

Figure 12.29 Global land-ocean temperature index from 1880 to 2014 compared to the 1951–1980 average. Note that the warmest year was 2014 and that 14 of the first 15 years of the 21st century are the warmest on record.

Source: NASA/Goddard Institute for Space Studies.

Solar Energy Variation

Solar radiation is not a constant. One type of variation is marked by the **sunspot** cycle. Dark areas called sunspots occur on the surface of the Sun; they are regions of intense magnetic activity. The abundance of sunspots varies in an 11-year cycle (figure 12.30) wherein solar radiation varies about 0.15%. Earth's average temperature warms and cools less than 0.1°C during a cycle. The cycle is too brief for Earth to come to temperature equilibrium

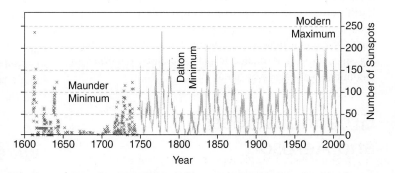

Figure 12.30 Number of sunspots since 1600. Notice the 11-year cycle.

Side Note

Stradivari Violins

The most famous violins probably are those made by the Italian Antonio Stradivari (1644–1737). He learned his craft as a pupil of Nicolas Amati (1596–1684). In 1684, Stradivari made changes in his violins such as increasing the size of the instrument and using a secret varnish. The reasons for the superior tones of these old violins are still debated. Now it is suggested that Stradivari, Amati, and their contemporaries benefited from the Maunder Minimum that occurred from 1645 to 1715, beginning one year after the birth of Stradivari. During this 70-year-long interval of reduced sunspots and lesser output of solar energy, Earth was in a cold spell, with longer winters, cooler summers, and slow, even tree growth. The unique climatic conditions produced dense wood with narrow tree rings. This wood may be the reason for the superior acoustical properties of violins made during the late 1600s and early 1700s.

---

with the changes in solar radiation. If the cycles were longer, the temperature could vary about 0.3°C.

Sunspots were nearly absent from the Sun between 1645 and 1715 during the Maunder Minimum. During those decades, the Sun was about 0.25% weaker, causing Earth's temperature to fall 0.2° to 0.5°C (see figure 12.27).

A peak of sunspot activity occurred in 2000 during Solar Cycle 23. Regions with concentrated ultraviolet light radiating from the solar atmosphere were obvious (figure 12.31a). Temperatures within these ultraviolet pulsations were about 1,000 times hotter than the Sun itself. Sunspot activity began declining in early 2002 and in 2009 was exceptionally low (figure 12.31b). Solar Cycle 24 was a weak one, reaching a peak of 82 in 2014.

RADIATIVE FORCING

Solar energy is always flowing through the atmosphere, and much is absorbed by the Earth. The warm Earth radiates energy out into cold space. Energy flow in and out of the atmosphere is measured in watts per square meter (W/m^2) at the boundary between the troposphere (lower atmosphere) and the stratosphere. **Radiative forcing** is a measure of changes in these energy flows. Since the year 1750, many of the changes in the energy balance reflect human-caused changes in greenhouse gas and aerosol concentrations, albedo, cloud cover, and more. A positive forcing warms Earth's climate, and a negative forcing cools it. The biggest change since 1750 has been positive forcing due to increasing CO_2 content (table 12.5). The total of this radiative forcing since 1750 is estimated to have increased the effect of incoming solar energy by about 1%.

Greenhouse Gases and Aerosols

The major greenhouse gases are water vapor, carbon dioxide (CO_2), methane (CH_4), nitrous oxide (N_2O), ozone (O_3),

TABLE 12.5
Radiative Forcing from Human Activities

Component	Watts/m^2
Carbon dioxide	+1.79
Methane	+0.54
Chlorofluorocarbons	+0.34
Nitrous oxide	+0.18
Ozone	
In troposphere	+0.35
In stratosphere	−0.05
Surface albedo	
Black carbon on snow	+0.1
Land use	−0.2
Aerosols	
Direct effect	−0.5
Cloud albedo	−0.7
Net human effect	+1.8

Source: IPCC *Assessment Report 4* (2007); NOAA (2011).

and human-made chlorofluorocarbons (CFCs). These gases absorb energy radiated from the Earth and "trap" it in the lower atmosphere (see figures 9.4 and 9.11). Concentrations of gases have been rising following the Industrial Revolution and the rapid growth of the human population. The warming caused by increasing concentrations of gases such as CO_2 and CH_4 is well understood (table 12.6). Less well known are the size of the feedbacks from increasing water vapor and aerosols.

WATER VAPOR

Water vapor is Earth's most abundant greenhouse gas. We feel its presence in our daily lives; for example, think

In Greater Depth

When Did Humans Begin Adding to Greenhouse Warming?

During recent decades, the greenhouse warming caused by humans has become well known. For example, we release large quantities of CO_2 to the atmosphere by burning oil, natural gas, coal, and wood. But when did human activities begin producing the greenhouse gases that warmed the climate? According to William Ruddiman of the University of Virginia, humans began warming the climate about 8,000 years ago by cutting and burning forests to clear land for agriculture; the widespread forest destruction added CO_2 to the atmosphere. Then about 5,000 years ago, rice began to be extensively grown with techniques that formed artificial wetlands, which gave off methane to the atmosphere.

Ruddiman calculates that these land uses caused the climate to warm about 0.8°C. If this estimate is correct, then the amount of human warming of climate has about tripled: from about 0.8°C by our ancestors over thousands of years to 0.4°C by us in tens of years.

If our prehistoric ancestors did warm the climate over thousands of years, they probably prevented some Little Ice Ages and helped keep global climate warmer and more stable. Even if we owe them thanks for a warmer climate, remember that their changes took place slowly over thousands of years. Today we are increasing the amount of greenhouse gases in the atmosphere at rates many times faster than our ancestors did, and we may inadvertently push the climate across some threshold, causing an abrupt climate change that presents major problems.

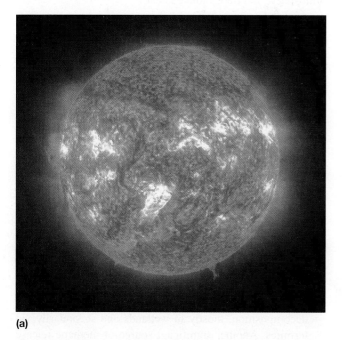
(a)

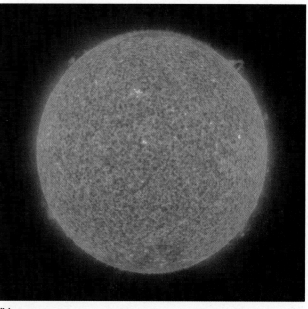
(b)

Figure 12.31 The Sun. (a) Ultraviolet light radiating from the Sun's atmosphere during a peak of sunspot activity, 19 July 2000. (b) During a sunspot minimum, ultraviolet activity was subdued, 18 March 2009.
NASA.

about how little the nighttime temperature drops in humid air versus how much it drops in low-humidity air such as in the desert. Water vapor is a vast, natural control on temperature and climate. As the atmosphere warms, more water vapor can be absorbed into the warmer air in a positive feedback cycle that leads to ever greater warmth. The warmer the air, the greater the percentage of water vapor it can hold, and this includes the warmer air resulting from the CO_2 and CH_4 put into the atmosphere by human activities.

Water is a *condensing* greenhouse gas (GHG); it responds quickly to changes in air temperature and pressure by condensing and precipitating, or evaporating. Although water vapor and clouds are highly variable, they provide about 75% of Earth's greenhouse effect. The *non-condensing* greenhouse gases account for the remaining 25% of the greenhouse effect; they do not condense and precipitate from the atmosphere. Instead, the non-condensing GHGs are well mixed in the atmosphere and individual molecules may remain up in the atmosphere for centuries; they provide a stable temperature structure that sustains the water vapor and clouds (table 12.6). Without the radiative forcing provided by the non-condensing GHGs, especially CO_2, Earth's global climate could cool into an icy state.

TABLE 12.6

Noncondensing Greenhouse Gases

Gas	Relative Percentage Responsible for Greenhouse Warming	Ability to Trap Heat (compared to $CO_2 = 1$)
Carbon dioxide (CO_2)	60	1
Methane (CH_4)	16	21
Nitrous oxide (N_2O)	5	310
Ozone (O_3)	8	2,000
Chlorofluorocarbons (CFCs)	11	~12,000

CARBON DIOXIDE (CO_2)

About 60% of greenhouse warming caused by humans comes from releasing CO_2 into the atmosphere (table 12.6). How does carbon cycle through Earth's surface environments? The element carbon is a major building block of life on Earth. CO_2 is removed from the atmosphere by plants during photosynthesis to build their tissue. Upon death, much of the organic tissue is oxidized, and CO_2 both returns to the atmosphere and dissolves in water.

In the year 1800, the CO_2 concentration in the atmosphere was about 280 parts per million (ppm). Looking for a place to monitor atmospheric CO_2 without much influence from plants or industry, Charles D. Keeling installed instruments high up on Mauna Loa in Hawaii in 1958 when CO_2 was 315 ppm. By the end of 2014 it had increased to 400 ppm (figure 12.32). The increase since 1800 is greater than 40%, and 70% of that increase has occurred since 1958.

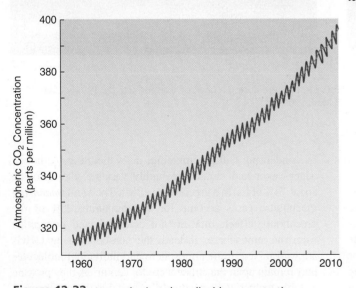

Figure 12.32 Atmospheric carbon-dioxide concentrations measured at Mauna Loa, Hawaii, 1958–2011. Compare these CO_2 increases with the temperature increases in figure 12.29.

Source: C. D. Keeling, Scripps Institution of Oceanography, and NOAA.

CO_2 concentration at 400 ppm is the highest value in the past 800,000 years, according to analyses of air bubbles in Antarctic ice.

How do we decompose plants to release CO_2? In two main ways: burning wood and burning fossil fuels. About 15% of the increased atmospheric CO_2 is due to humans burning wood to clear land for agriculture, heat homes, and make charcoal for furnaces. Fossil fuels are coal, oil, and natural gas. They form during transformation of dead plant material in swamps, river deltas, and other organic-rich environments after burial beneath sediments. More than 80% of the energy that powers our global societies is generated by burning fossil fuels, and these energy-producing processes have added about 80% of the excess CO_2 now in the atmosphere.

How is CO_2 removed from the atmosphere? The scorecard for the latter part of the 20th century shows about 20% is removed by plants during photosynthesis, about 30% dissolves in ocean water, and the remaining 50% stays in the atmosphere and traps radiated heat.

METHANE (CH_4)

About 16% of modern greenhouse warming has come from increasing methane in the atmosphere. Notice in table 12.6 that the **global warming potential (GWP),** or heat-trapping ability, of methane is 21 times greater than that of carbon dioxide.

Air trapped in ice tells us that methane concentration in the year 1750 was about 700 parts per billion (ppb), but it has risen to more than 1,815 ppb in 2014. The increase in atmospheric methane was slow in the 19th century and rapid in the 20th century.

How is methane released to the atmosphere? It is released during decomposition of vegetation in oxygen-poor environments such as swamps, rice paddies, and cattle digestive systems. Bacteria remove carbon (C) from dead vegetation, and if oxygen is absent, the carbon combines with hydrogen (H) to make methane (CH_4).

About 40% of methane release occurs by natural decomposition, mostly in wetlands and secondarily via termites. Another significant source of methane release is through "mud volcanoes," cone-shaped piles of mud and rock built by methane rising from depth through the ground surface and into the atmosphere. More than 900 mud volcanoes have been located in 26 countries. Azerbaijan has the most and the largest mud volcanoes, with one standing 700 m (2,300 ft) tall. About 60% of methane is given off by human activities, listed in order of importance: burning fossil fuels, growing rice, and maintaining livestock, with lesser amounts emitted from landfills, burning wood, and rotting animal waste and human sewage.

Remember that the hottest climate in the past 66 million years occurred when deep-ocean water warmed enough to melt icy methane hydrates on the seafloor, thus releasing a huge volume of methane gas into the atmosphere. If the warming of the deep oceans occurring today continues for enough decades to melt methane hydrates, a warm climate could become a torrid climate.

NITROUS OXIDE (N₂O)

Nitrous oxide is another contributor to the greenhouse effect (see table 12.6). About 70% of N_2O is produced naturally, mostly by bacteria removing nitrogen from organic matter, especially within soils. Human activities cause the remaining 30% of nitrous oxide release via our agricultural practices, including use of chemical fertilizers. The second important way humans release N_2O is by combustion burning of fuels in car and truck engines. N_2O is increasing in the atmosphere regularly and rapidly.

OZONE (O₃)

Ozone is a greenhouse gas in the troposphere, but in the stratosphere it provides negative feedback (table 12.5). It is a gaseous molecule composed of three atoms of oxygen rather than the usual two-atom molecule (O_2).

Ozone is also a principal component of the smog that chokes urban atmospheres. Our automobiles and industries emit gases, some of which react with sunlight to produce the ozone that makes our eyes water and our lungs ache. The ozone story is well described by the saying that pollutants are merely resources that are in the wrong place. Ozone in the stratosphere shields us from killing UV rays, but ozone in the air we breathe weakens us and shortens our lives.

CHLOROFLUOROCARBONS (CFCs)

Chlorofluorocarbons do not occur naturally. They are examples of gases produced solely by humans. CFCs are used as coolants in refrigerators and air conditioners, as foam insulation in buildings, and as solvents, among other purposes. Chlorofluorocarbons are not only greenhouse gases (table 12.6), but they also aid in the destruction of the ozone in the stratosphere, which helps shield life from damaging ultraviolet (UV) rays. CFCs may remain in the atmosphere for a century, causing so many problems that international treaties have been signed restricting their use.

20TH-CENTURY GREENHOUSE GAS INCREASES

Why did we humans release such great volumes of greenhouse gases in the 20th century? The gases were a by-product of many well-intentioned activities, such as providing energy for industries, homes, and personal automobiles; growing rice; and raising livestock for human consumption. It took a long time to recognize how much the climate could be changed by these activities that raised the standard of human existence.

Another significant factor in the increase of greenhouse gases was the 20th-century growth of the human population; it doubled twice, from 1.5 billion in 1900 to 3 billion in 1960, and then to 6 billion in 1999. Even conservative estimates for 21st-century population growth forecast another doubling to 12 billion people. Most people desire the affluent lifestyle of the industrialized world; this means the billions of people alive today who don't have that lifestyle, plus the billions of people yet to be born. All will be seeking a higher standard of living. Greenhouse-gas–caused global warming will be a growing political issue throughout your lifetime.

In the 19th century a famous quotation attributed to Mark Twain stated, "Everybody talks about the weather but nobody does anything about it." But in the 21st century we humans are doing something about the weather; we are warming the planet.

AEROSOLS

Aerosols are microscopic or very small particles of solids and liquids suspended in air. Our understanding of how aerosols affect Earth's climate is much less certain than our knowledge of the effects of greenhouse gases. The effects of aerosols are complex, and sometimes they seem contradictory. For example, an urban atmosphere of brownish air can block some incoming sunlight, causing cooling, but particles of black soot in that same air absorb heat, causing warming. The scattering of light by aerosols is visible as haze; an example is the spraying of whitish salts into the air by breaking ocean waves. Aerosols mostly reflect sunlight, thus cooling the Earth's atmosphere, as measured by the *direct effect* in table 12.5.

Aerosols also affect the volume of clouds, their distribution, and their albedo as measured by *cloud albedo* in table 12.5. Clouds are a problem for computer modelers trying to forecast 21st-century climate. Greenhouse gases will continue to warm the atmosphere. Will the warmer air cause clouds to thicken and spread, thus shading Earth's surface with a cooling effect? Or will the warmer air hold more water vapor, with thinner and smaller clouds causing a warming effect? To date, studies of cloud effects suggest that warming will dominate.

The 21st Century

Global warming has become a topic of great concern to people in all walks of life all around the world. The magnitude of concern is shown by the number of scientists working together to evaluate the risk of climate change caused by human activity. The largest coordinated effort is the Intergovernmental Panel on Climate Change (IPCC), which delivered its *Assessment Report 4* in 2007. The effort involved more than 450 lead authors, 800 contributing authors, and 2,500 expert reviewers representing more than 130 countries. Their consensus marks a milestone in scientific effort and was judged to be so significant that they shared the 2007

TABLE 12.7

Climate Consensus for 21st Century

- Warming of climate is unequivocal.
- Most of the increase in global temperatures since 1950 is very likely (>90%) due to human greenhouse-gas emissions.
- Atmospheric concentrations of CO_2, CH_4, and N_2O have increased markedly as a result of human activities since 1750 and now far exceed preindustrial values of the past 650,000 years.
- The probability that global warming occurs by natural changes alone is <5%.
- World temperatures could rise between 1.1° and 6.4°C (2° and 11.5°F) during the 21st century.
- There is >90% probability of more frequent warm spells, heat waves, and heavy rainfall.
- There is >66% probability of increased droughts, tropical cyclones, and extreme high tides.
- Sea level will probably rise 18 to 59 cm (7 to 23 in).
- Past and future CO_2 emissions will continue to contribute to warming and sea-level rise for more than 1,000 years.

Source: IPCC *Assessment Report 4* (2007).

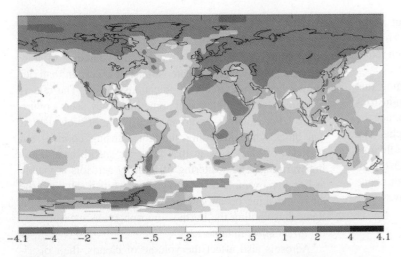

Figure 12.33 Map showing 10-year average (2000–2009) temperature anomaly relative to the 1951–1980 mean. The temperature scale at the bottom is in degrees centigrade.

NASA/GISS.

Nobel Peace Prize. Some of their main, unanimous conclusions are listed in table 12.7.

The warming trend of the 20th century continues in the 21st century (see figure 12.29). The decade from January 2000 through December 2009 was the warmest on record, according to NASA's Goddard Institute for Space Studies (GISS) (figure 12.33).

For the past three decades, the GISS surface temperature record shows an increase of about 0.2°C (0.36°F) per decade. Since 1880, a distinct warming trend has occurred, albeit with leveling-off trends. In total, average global surface temperatures have increased about 0.9°C (1.6°F) since 1880.

GLOBAL CLIMATE MODELS

What changes will occur in the global climate in the 21st century? This is a complex question because the answers involve predicting many variables through time, such as greenhouse gas and aerosol contents of the atmosphere, temperatures around the Earth, ocean warmth and circulation patterns, wind strengths and positions, cloud cover, and more. Added complications arise because a change in one variable may cause other variables to change in unknown ways. Questions are addressed by constructing **global climate models (GCMs)** involving complex computer simulations. The IPCC *Assessment Report 4* analysis set up different economic and political scenarios for computers to consider in modeling climate (table 12.8). Which model will best fit the 21st century? Will economic development be global and make the people of the world more homogeneous? Or will economic development occur in regions with different policies and goals, keeping more heterogeneity among people? Will growth and development focus on economics or environment? In table 12.8, compare the ranges of different results for 21st-century temperature changes among the different groups of models.

IPCC *Assessment Report 5* (AR5) was released in three parts from late 2013 to 2014. Documentation of ongoing changes shows the atmosphere and ocean warming, snow and ice diminishing, and sea level rising. The changes are stressing human communities, agriculture, and natural ecosystems; these effects are highly likely to increase throughout the 21st century. The latest report especially emphasizes the major climate risks for humans, the greatest reasons for concern.

Temperature

All the computer-model scenarios of 21st-century climate are provisional and subject to change. As real measurements of climate continue to be made, and the amounts of data and time covered increase, the computer models will increase in accuracy. Nonetheless, some major results are evident. All computer models forecast increasing volumes of greenhouse gases and increasing Earth-surface

TABLE 12.8
Model Scenarios for Changes in Average Global Surface Temperatures by Years 2090–2099

	More Economic Focus	More Environmental Focus
Globalization (homogeneous world)	Scenario A1 — Rapid economic growth 1.4° to 6.4°C temperature rise	Scenario B1 — Global environmental sustainability 1.1° to 2.9°C temperature rise
Regionalization (heterogeneous world)	Scenario A2 — Regionally oriented economic development 2.0° to 5.4°C temperature rise	Scenario B2 — Local environmental sustainability 1.4° to 3.8°C temperature rise

Source: IPCC *Assessment Report 4* (2007).

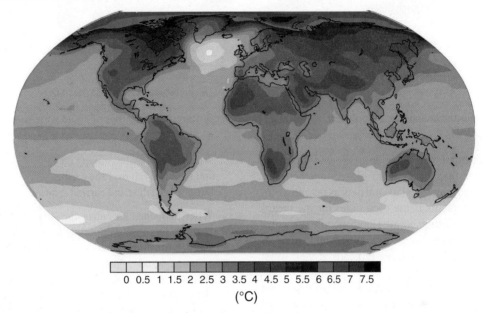

0 0.5 1 1.5 2 2.5 3 3.5 4 4.5 5 5.5 6 6.5 7 7.5
(°C)

Figure 12.34 Estimated surface temperature increases by year 2090–2099, relative to 1980–1999 average.

Source: An IPCC *Assessment Report 4* A1 scenario (2007).

temperatures. Figure 12.34 shows a geographic pattern of surface warming expected by the end of the 21st century in an IPCC rapid growth of the global economy (A1) model. Regional changes include the greatest warming occurring over northern lands in high latitudes; this means there will be less area covered by snow and ice; decreased sea ice; increased thawing of frozen ground; disappearance of more mountain glaciers and late summer water runoff; and increased evaporation from farmland. It is very likely that heat waves and droughts will increase in the middle and low latitudes. Wildfire seasons will be longer and fires more frequent. Changes in global precipitation amounts and locations are another concern.

Precipitation

It is very likely that precipitation will increase in high latitudes and decrease in most subtropical lands (figure 12.35). As precipitation and temperature patterns change around the world, there will be regions of winners and regions of losers. And a winner in one decade may become a loser in following decades as temperatures continue to rise and precipitation decreases, increases, or shifts location. In regions where temperatures rise and precipitation falls, drought and famine can become problems.

DROUGHT AND FAMINE

The 21st century is very likely to experience more frequent and longer duration heat waves. When extended heat waves are coupled with decreased rainfall, drought can set in. Drought does not equal desert. Drought describes times of abnormal dryness in a region when the usual rains do not appear and all life must adjust to the unexpected shortage of water. The lack, or reduction, of moisture can cause agricultural collapse or shortfalls, bringing famine, disease-causing deaths, and mass migrations to wetter areas. Famine is the slowest-moving of all disasters. Earthquakes, volcanic eruptions, tornadoes, and the like all hit suddenly and with great force and then quickly are gone. But famine is slow. First, the expected rains do not arrive, and then vegetation begins to wither, food supplies shrink, and finally famine sets in.

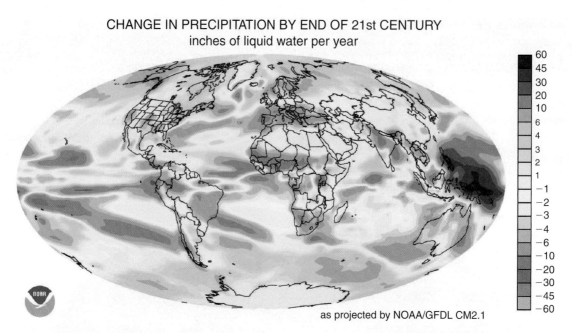

CHANGE IN PRECIPITATION BY END OF 21st CENTURY
inches of liquid water per year

60
45
30
20
10
6
4
3
2
1
−1
−2
−3
−4
−6
−10
−20
−30
−45
−60

as projected by NOAA/GFDL CM2.1

Figure 12.35 Projected changes in precipitation by the end of the 21st century. Scale is inches of liquid water per year.
Source: NOAA/GFDL CM2.1.

Unlike other natural disasters, drought tends to drive people apart rather than bring them closer together. The shortages of food and water lead to conflicts as people, communities, and governments battle each other for the means to survive. After an earthquake or flood, people are commonly at their best as they aid their neighbors and strangers in need; during a drought, people are typically at their worst as they fight for survival.

In the *early stage* of a famine, food is still available, but there is not enough. Healthy people can lose up to 10% of their body weight and still remain mentally alert and physically vigorous. In the *advanced stage,* body weight decreases by around 20%, and the body reacts to preserve life itself. Body cells lower their activity levels, reducing the energy needed to keep vital functions going. People sink into apathy. In the *near-death stage,* when 30% or more of body weight has been lost, people become indifferent to their surroundings and to the sufferings of others, and death approaches.

U.S. Dust Bowl, 1930s

Might the worst conditions of the recent past serve as a prologue to the future? One of the greatest weather disasters in U.S. history occurred during the 1930s, when several years of drought turned grain-growing areas in the center of the nation into the "Dust Bowl." Failed crops and malnutrition caused abandonment of thousands of farms and the broad-scale migration of displaced people, mostly to California and other western states. This human drama was captured in many articles and books, including *The Grapes of Wrath* by John Steinbeck:

Now the wind grew strong and hard,
it worked at the rain crust
in the corn fields.
Little by little the sky

was darkened by the mixing dust,
and the wind felt over the earth,
loosened the dust and carried it away.

What happened to cause the drought? Recurrent large-scale meanders in the upper-air flow created ridges of high pressure with clockwise flows resulting in descending air (see figure 10.43). The upper-level high-pressure air was already dry, but as it sank, it became warmer, thus reaching the ground hot, dry, and thirsty. As the winds blew across the ground surface, they sucked up moisture, killing plants and exposing bare soil to erosion. Wind-blown clouds of dust built into towering masses of turbulent air and dust called rollers (figure 12.36). When they rolled across an area, the Sun was darkened, and dust invaded every possible opening on a human body and came through every crack in a home.

Figure 12.36 A dust storm (a haboob) rolls into Stratford, Texas, in 1935.

NOAA George E. Marsh Album.

Dust even blew as visible masses across East Coast cities and blanketed ships at sea.

The drought began in 1930, a particularly bad time. Only months before, in October 1929, the U.S. stock market had crashed, and the nation's economy began sinking into the Great Depression. By 1931, farmers were becoming desperate. For example, a group of 500 armed farmers went to the town of England in Arkansas to seek food from the Red Cross. They were denied aid, so they went to the town's stores and took the food they needed. The event drew worldwide attention. Here were farmers from the U.S. heartland who had helped feed the world during the early 20th century and through the ravages of World War I; now they could not even feed themselves.

The dust storms became even worse in 1934 and 1936. Some of the blame for the Dust Bowl was heaped onto the farmers for plowing deeply through drought-tolerant native grasses and exposing bare soil to the winds. The plowed lands were sowed with seeds of plants that could not handle drought and thus died, exposing more soil. The farming practices were not the best, but they did not cause the drought; they just accentuated its effects. Evidence showing that droughts are common is found in the archaeological record. For example, droughts in the past probably led to the downfall of the Anasazi civilization of the southwestern United States and the migration of its people to areas with more dependable water supplies.

The Dust Bowl drought affected more than just local agriculture. Combined with the stock market crash, it caused fundamental changes in the economic, social, and political systems of the United States.

Haboobs

Overwhelming dust storms (see figure 12.36) are becoming more common. Their name is **haboob,** which means "strong winds" in Arabic. Haboobs occur in arid regions around the world, including the southwestern United States. For example, Phoenix, Arizona, was overrun by several haboobs in summer 2011, causing reduced visibility, dangerous driving conditions, downed power lines, delayed airplane flights, and some structural damage. Haboobs range up to 100 km (60 mi) wide and several kilometers high, move 35–100 km/hr (20–60 mph), may arrive with little warning, and last up to three hours. Haboobs form when thunderstorm downbursts or collapse cause air to rush down to the ground and flow outward, pushing a gust front and carrying dust, silt, and sand (figure 12.37).

ICE MELTING

Glacial ice holds 2.15% of the water on Earth. If all glacial ice melted, global sea level would rise about 67 m (220 ft) and flood vast areas of land. Nothing this drastic will happen in the foreseeable future, but there is concern, about certain ice masses. Mountain glaciers, summer snow cover, and Arctic sea ice are declining in volume. This trend is very likely to continue.

Arctic Sea Ice

Throughout recorded history the Arctic Ocean has been mostly covered with sea ice. European explorers sailing to the Arctic Ocean looking for a direct passage to Asia were blocked by sea ice from the 15th through the 20th centuries, even during the warmest summer months. But the 500+ year barrier is lifed in 21st-century summers as enough sea ice melts to open long-sought passages (figure 12.38). Now dozens of the world's largest ships travel across the Arctic Ocean each summer using either the Northwest Passage along Canada or the Northeast Passage along Russia. Today, you personally can book a cabin on a luxury cruise ship and

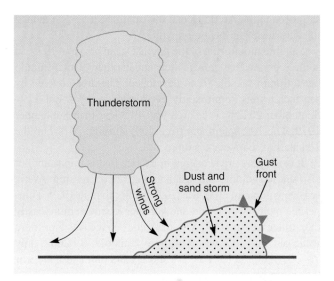

Figure 12.37 Haboobs form when downbursts from thunderstorms flow as strong winds over dry ground, carrying huge clouds of dust and sand.

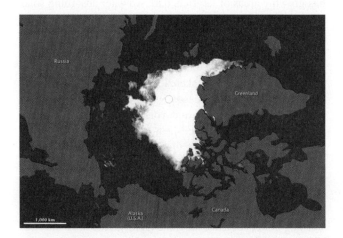

Figure 12.38 Sea ice on the Arctic Ocean, 13 September 2012. The circle in the center is the North Pole; Greenland sits to the right of the ice. Canada and the northern United States lie at bottom of image.

Jesse Allen/NASA Earth Observatory.

In Greater Depth

Tipping Points

Change is commonly viewed as a gradual process, but sometimes it is not. The concept of a **tipping point** recognizes that there are points at which small changes suddenly produce large effects. Although changes through the years may be slow and gradual, the response to these changes, at some point, may be abrupt and disproportionate. One of the concerns with rising temperatures due to increasing greenhouse gases in the atmosphere or the increasing rate of glacial melting is that their history of change may not predict the future. Thresholds may be crossed that lead to dramatic changes in short lengths of time.

sail through the Arctic Ocean this summer, a voyage denied to centuries of sailors before you.

Satellite monitoring of the Arctic Ocean began in 1979. In some years, sea ice grew in area, and in other years it shrank, but an unmistakable trend developed. The area covered by sea ice shrinks significantly every decade, by about 12%. September 2012 sea-ice area was 49% less than September 1979. The 10 Septembers from 2005 through 2014 had the 10 smallest sea-ice areas on record.

It is not only area covered by sea ice that is important but also sea-ice volume and age, and both are decreasing. Arctic sea-ice volume in 2012 was 76% smaller than in 1979. Thinner ice melts away faster, and younger ice melts more easily than older ice. In 1987, 57% of Arctic sea ice was 5 or more years old, and 14% was 9 or more years old. By 2007, only 7% of Arctic sea ice was 5 or more years old, and none was 9 or more years old.

In the recent past, Arctic sea ice covered an area about 1.5 times larger than the United States. This huge sheet of ice with high albedo reflected sunlight back to space, acting as a huge air conditioner for planet Earth. Now the Arctic Ocean has growing areas of ice-free, dark seawater absorbing solar energy and warming rapidly. Further sea-ice loss and warming are accelerated by positive feedbacks in the system such as thinner and weaker sea ice, which is more easily broken into smaller pieces by storms; better access for the warmer water brought by the Mackenzie River ($325 \text{ km}^3 = 78 \text{ mi}^3$) each year; and an increased water-vapor feedback cycle wherein the greater area and volume of warmer seawater release moisture and stored heat to the atmosphere. Has a *tipping point* been crossed? Only time will tell.

Arctic sea ice can be called the canary in the global-warming coal mine. Caged canaries have long been carried into mines because they die quickly when exposed to gases such as carbon monoxide (CO) and methane (CH_4) thus warning miners to leave the mine rapidly. And now, the Arctic sea-ice cover is dying quickly as the polar climate warms due to buildup of gases such as CO_2 and CH_4 in the atmosphere. Some computer models suggest that Arctic summer sea ice could disappear as early as 2030. Global warming is occurring before our eyes.

Permafrost

Permafrost is perennially frozen ground occurring where mean annual temperature is at or below 0°C (32°F). In the Northern Hemisphere, permafrost underlies about 24% of exposed land from depths of several meters to 150 m (500 ft) or more, primarily in Siberia, northern Canada, Alaska, and Greenland. It is estimated that permafrost contains about 0.022% of the water on Earth. During summer, the ice may melt at and just below the surface, creating an active soil zone where plant growth occurs, such as stunted shrubs, mosses, and lichens. Winter ice entombs the plant material. Over thousands of years, a tremendous reservoir of organic remains has been preserved within permafrost ice.

Today, near-surface air temperatures in the high latitudes are warming about twice as fast as the global average (see figure 12.33). The increasing warmth is being referred to as *Arctic amplification* of global warming. Warmer summers are thawing more permafrost, thus exposing greater amounts of carbon-rich organic material. As soils defrost, microscopic organisms decompose the ancient organics and release CO_2 and methane (CH_4) to the atmosphere. Some surface areas are drying out, allowing wildfires to release huge volumes of CO_2 to the atmosphere after hundreds of years of storage in ice. A positive feedback system is operating in which warming air leads to warmer seawater, and warming land thaws more permafrost, releasing more greenhouse gases to the atmosphere which warms the air; and the cycle continues.

SEA-LEVEL RISE

How much will global sea level rise in the 21st century? The IPCC *Assessment Report 5* model that continues present trends through the 21st century yields a sea-level rise up to 0.82 m (2.7 ft). The sea-level rise mostly comes from (1) melting of glaciers, sending water to the seas, and (2) thermal expansion of seawater as warming causes an increase in water volume. Isostatic adjustments also occur where (1) seawater floods the edges of continents causing land to sink and the shoreline to move farther inland, and where (2) glaciers melt and remove their weight, causing the land to rise.

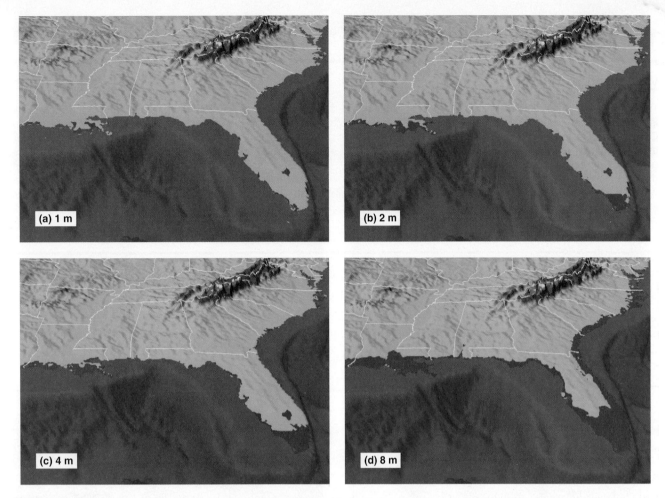

Figure 12.39 The southeastern United States is shown, with red indicating areas below sea level at (a) 1 m sea-level rise, (b) 2 m rise, (c) 4 m rise, and (d) 8 m rise, which would occur if the Greenland ice cap melted.

Geophysical Fluid Dynamics Laboratory/NOAA.

A 0.82 m (2.7 ft) sea-level rise will affect tens of millions of people around the world. But it could be worse. The IPCC *Assessment Report 4* also warned of potentially large sea-level rises in a sobering analysis: "There is medium confidence that at least partial deglaciation of the Greenland ice sheet, and possibly the West Antarctic ice sheet, would occur over a period of time ranging from centuries to millennia for a global average temperature increase to 1–4°C (relative to 1990–2000), causing a contribution to sea-level rise of 4–6 m or more." A sea-level rise of 4–6 m (13–20 ft) would cause major problems worldwide for coastal cities. It seems likely that a 1°C temperature rise will be reached about mid-21st century, and a 3°C increase may be here by the end of the century. If these temperatures are reached, the sea-level rise will not be immediate because of *lag times* for full response. Nonetheless, it will be too late to reverse; the world will be committed to flooding huge areas of low-lying land upon which hundreds of millions of people now live (figure 12.39).

Greenland

Greenland is buried beneath 3 million km^3 of ice. If all that ice melted, global sea level would rise 7 m (23 ft). The Northern Hemisphere continental glaciers have retreated from Canada, Scandinavia, and Siberia, but the Greenland continental glacier remains; it is a relic left over from the last major glacial advance.

How stable is the Greenland ice sheet? This is difficult to assess, and there are several concerns: (1) The rate of melting has increased (figure 12.40). (2) The major glaciers entering the ocean are moving faster. (3) There are more than a thousand large meltwater lakes on top of the ice sheet, and some of them are pouring huge volumes of water down through holes in the ice where it reaches the bedrock below and then flows seaward. Glaciers sitting on top of running water move faster than glaciers sitting on bedrock. (4) Common falls of large ice masses into the ocean reduce support for the toes of glaciers and raise the possibility of large-scale catastrophic collapses. So far, the computer scenarios have only modeled ice melting, not ice masses collapsing.

Greenland is losing ice faster than expected. Is this just a normal short-term cycle? Or have we passed a tipping point (see In Greater Depth) and entered a new regime? We don't know. The computer models of Greenland ice melting are

In Greater Depth

Lag Times

The human race is provoking Earth's climate system by pouring immense volumes of greenhouse gases into the atmosphere. Changes in climate are occurring slowly, but the full effects of gases emitted today will not be felt for decades or centuries. The atmosphere is warming, but slowly because much of the greenhouse-trapped heat is being absorbed by the oceans with their tremendous capacity for energy storage. IPCC AR4 estimates that the greenhouse gases we have already emitted will continue to warm the oceans throughout the 21st century, by about 0.6°C (1.1°F). The climate system contains significant lag times.

Another concern about rising temperatures in the atmosphere and ocean is melting of the massive ice sheets on Greenland and Antarctica. How much warmth is enough to cause the melting? How will we know when we reach the critical temperature? Consider these questions by trying an experiment. Take an ice cube out of your freezer and set it on the kitchen counter. The room temperature is warm enough to melt the ice cube, but how much time passes before significant melting begins? Once the melting begins, what is going to stop it?

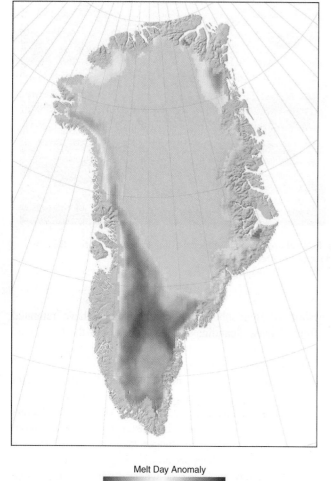

(a)

Melt Day Anomaly

-30   -15   0   +15   +30

(b)

Figure 12.40 Ice melt. (a) Greenland in 2007 experienced up to 30 more days of melting than the average number of melt days in 1988–2006. If the tipping point for melting the Greenland ice sheet was ever passed, how would we know it? How could we stop the melting? What would be the lag time for all the ice to melt? (b) Ice cubes taken out of your freezer have passed their tipping point for melting. But what is the lag time for them to melt completely?

(a) NASA/Earth Observatory, (b) © Comstock Images/PictureQuest RF.

conservative. There remains the risk of rare but extreme events. If Greenland ice loss accelerates, then global sea level will rise faster than anticipated.

Antarctica

Antarctica is the fifth largest continent; it supports a continental ice sheet holding 29 million km³ of ice. If all that ice melted, global sea level would rise about 60 m (200 ft). The ice sheet on the continental interior seems relatively stable, but ice losses are occurring along the coastlines where glaciers flow faster and ice shelves exist. Ice shelves are wide bodies of ice, hundreds of meters thick, flowing off the continent and now partially floating on seawater. An example is the West Antarctic ice sheet, a Texas-size mass of ice containing enough water to raise global sea level 5 m (16 ft). West Antarctica warmed 2.4°C (4.3°F) between 1958 and 2010, making it one of the fastest-warming areas on Earth. Here again is the worry about tipping points. As this ice sheet retreats, what if it crosses a threshold leading to runaway retreat? We don't know how to recognize if and when a runaway retreat may occur, but the world will pay a big price if one happens.

OCEAN CHANGES

Much of the increasing global warmth is being held in seawater. Oceans store nine times as much of the Sun's heat as do the atmosphere and land combined. This heat influences weather, but many other changes are occurring within the oceans.

Acidification

The increasing volume of CO_2 in the atmosphere is more than just a climate-changing greenhouse gas; it is a reservoir in contact with the oceans that cover 71% of Earth's surface. CO_2 readily dissolves in water, creating an acid:

$$CO_2 + H_2O \text{ yields } H_2CO_3 \text{ (carbonic acid)}$$

Ocean acidification has been called the evil twin of global warming.

Regular, direct measurements of CO_2 in the atmosphere document its greater than 25% rise since 1958 (see figure 12.32). The acidity of the oceans has been measured consistently since the late 1980s. Measurements are referred to the pH scale, which defines the acidity or alkalinity of a liquid by measuring its hydrogen ion activity. The pH scale ranges from 0 to 14, with pH 7 being neutral. Smaller numbers are acidic; larger numbers are basic (more alkaline). The pH scale is logarithmic, with each whole number being an increase or a decrease of 10 times. In the past 30 years, ocean pH has changed from 8.11 to 8.06, which is a rise in acidity of 12%. Projections of rising CO_2 emissions indicate that by 2100 ocean acidity could increase 170%. As acidity increases, so does the dissolving of coral reefs, shells of clams and oysters, planktonic larvae, and more.

Fisheries

Food harvested from the sea by humans provides about 15% of the protein eaten by 60% of the human population. The IPCC *Assessment Report 5* projects that climate change by the mid-21st century and beyond will cause global redistribution of marine species and reductions in biodiversity that will challenge the sustained productivity of fisheries (figure 12.41). Algae are shifting poleward 10 km (6 mi) per decade and plankton are shifting 400 km (250 mi) per decade. As the base of the food pyramid shifts, other marine organisms will change with them. The fish yields in temperate latitudes could be 30–70% higher than in 2005 but tropical yields could fall 40–60%. These potential yields do not consider the effects of overfishing and ocean acidification. Although the computer model projections may not be accurate in detail, they do foretell changes— major changes with unknown effects. Climate change may undermine food security.

Circulation

Major climatic shifts will occur if the present deep-ocean circulation system is altered (see figure 9.29). At present,

CHANGE IN MAXIMUM CATCH POTENTIAL (2051-2060 COMPARED TO 2001-2010, SRES A1B)

| <50% | -21–50% | -6–20% | -1–5% | No data | 0–4% | 5–19% | 20–49% | 50–100% | >100% |

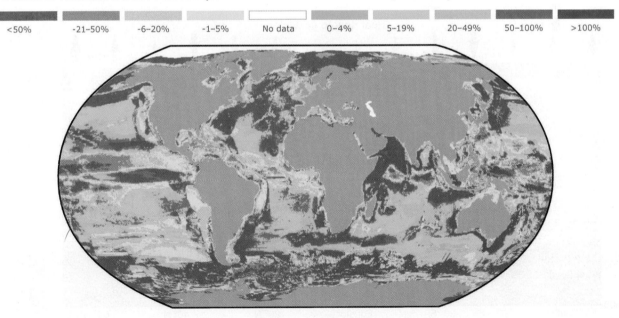

Figure 12.41 Changes in catch potential of ~1,000 exploited marine fish and invertebrate species in 2051–2060 using a computer model projecting moderately high warming. This analysis does *not* account for overfishing or ocean acidification.

Figure 6.14 From Pörtner, H.-O., et al. 2014: Ocean systems. In: Climate Change 2014: Impacts, Adaptation, and Vulnerability. Contribution of Working Group II to the Fifth Assessment Report of the Intergovernmental Panel on Climate Change

in the North Atlantic Ocean, the surface water carries warm tropical water to the north, where it releases heat to the atmosphere. As it travels north, the seawater cools and becomes denser, sinks, and then flows south at intermediate depths almost to Antarctica (figure 12.42). This deep-water circulation is driven by density contrasts in seawater. As the global ocean warms, the density contrasts become less, raising the possibility that the deep-water circulation system could slow or stop. Changing the heat-carrying ocean currents would cause significant climate change, affecting life in the sea and on land.

SIGNS OF CHANGE

Have you seen them? Have you seen the signs of climate change? The IPCC report evaluated 75 studies that reported more than 29,000 observational data series of significant changes in physical and biological systems; more than 89% of them are consistent with the change expected from warming (table 12.9).

Mitigation Options

There is a widely perceived need to slow global warming. How can this be done? Various schemes to control climate have been proposed; they fall into two main categories: (1) controlling CO_2 content in the atmosphere and (2) managing solar radiation received by Earth. A third category proposes some schemes that could achieve rapid results; they are (3) fast-action strategies.

CONTROLLING CO_2 CONTENT OF ATMOSPHERE

The CO_2 concentration in the atmosphere can be managed by reducing emissions and/or by removal. Following are some proposed methods:

- **Changes in energy-usage technologies.** There is a tremendous need for carbon-free energy technologies. But significant gains could be achieved just with changes, such as reducing CO_2 emissions from our gasoline-powered cars and our fossil-fuel power plants. Changing established infrastructure is not quick or easy. The first requirement is for people and governments to feel a sense of urgency. There are considerable lag times before political systems act. After the policymakers act, it takes many years to develop technologies and then build and implement new types of cars, power plants, and more.

- **Cap-and-trade.** A cap-and-trade plan to limit emissions of CO_2 has been set in action as the European Union emissions-trading system. It is an economy-wide, market-based system that places CO_2 emission allowances on companies. The companies deal with the situation by buying or selling emissions credits. If a company emits *less* CO_2 than allowed, it can sell allowances on the open market and earn money. If a company emits *more* than its CO_2 emission cap, it can buy credits. Cap-and-trade plans proposed for the United States are not popular.

- **Air scrubbing.** Machines and methods could be developed to remove (scrub) CO_2 from the atmosphere. This is technologically possible, but very expensive.

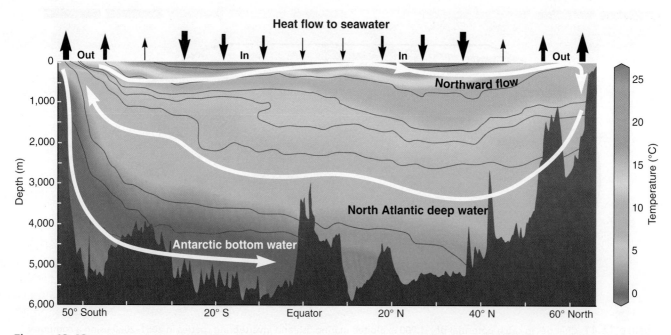

Figure 12.42 Deep-water flow in the Atlantic Ocean. Warm equatorial surface water flows north to the Arctic and releases heat to the atmosphere. As water density increases northward, it sinks and flows at depth to Antarctica.

Adapted from G. Hegerl and N. Bindoff, *Science* (2005).

TABLE 12.9

Some Observed Effects of Global Warming in Recent Decades

Freeze-free periods lengthening in mid- and high latitudes

Asymmetrical warming in many regions. Daily low temperatures are increasing at twice the rate of daily high temperatures

Longer growing seasons

Mosquitoes carrying diseases are migrating farther north and south

Earlier arrival of spring climate in Europe and North America

1. Earlier breeding of birds
2. Earlier arrival of migrant birds
3. Earlier appearance of butterflies
4. Earlier spawning of amphibians
5. Earlier flowering of plants

Population shifts in latitude and altitude

1. Greenland: becoming green again as plants sprout and grow
2. Europe and New Zealand: treeline climbing to higher altitudes
3. Alaska: expansion of shrub-covered area
4. North Atlantic Ocean and offshore California: increasing abundance of warm-water species
5. Europe and North America: 39 butterfly species extended their ranges northward up to 200 km (125 mi)
6. Costa Rica: lowland birds extended their ranges to higher elevations
7. Britain: 12 bird species extended their ranges northward an average of 19 km (12 mi)
8. Canada: red foxes extended their range northward while arctic fox range retreated

- **Ocean fertilization.** The ocean could be fertilized to stimulate massive blooms of photosynthetic algae, which would draw down atmospheric CO_2 as part of their life processes. This is easy to accomplish, but what other effects would result from this massive tampering with life in the oceans?

- **Rock weathering.** Mountain building, with its uplift and fracturing of rocks, triggers accelerated weathering of the rocks. This chemical decomposition process works by drawing down atmospheric CO_2 to make carbonic acid, which does the rock weathering. We could fracture and pulverize large masses of igneous rock to increase rock weathering and draw down atmospheric CO_2. The process works, but it is slow—geologically slow.

MANAGING INCOMING SOLAR RADIATION

Slowing down global warming by reducing solar radiation into Earth's climate is achievable within months. But beware—there is uncertainty about unintended effects caused by changes we initiate:

- **Sulfates in the stratosphere.** We could imitate volcanoes, such as El Chichón in 1982 and Mount Pinatubo in 1991. Their Plinian eruptions shot sulfur dioxide (SO_2) into the stratosphere and caused global cooling for many months. We could send millions of tons of sulfate into the stratosphere by balloon, plane, or cannon. Placing these tiny reflective particles in the stratosphere would cool the planet quickly. But what are the risks?

- **Cloud brightening.** Sea salts could be sprayed into clouds to brighten them and increase their albedo. The effects would be real, but local.

FAST-ACTION STRATEGIES

Attention on global warming has focused on CO_2, which makes up about 75% of greenhouse-gas emissions. Slowing CO_2 emissions requires basing our global energy systems on new technologies instead of fossil fuels. These changes are both politically and technologically difficult; they will take a long time to accomplish. But the world can gain some time by reducing emissions of four pollutants that are potent warmers of climate but have short residence times in the atmosphere:

- **Black carbon.** Black carbon is a component of soot; it absorbs solar radiation. Black carbon is produced by incomplete combustion of diesel fuels and biofuels such as wood. Reductions can be achieved by (1) placing particulate filters on diesel engines and (2) using clean-burning cookstoves for burning wood or dung.

- **Ozone in the troposphere.** Ozone is a component of smog; it absorbs solar radiation. Ozone is formed when sunlight interacts with hydrocarbons and nitrogen oxides emitted by cars and fossil-fuel plants and as gasoline vapors. Improving combustion processes is necessary.

- **Methane.** Methane is a potent greenhouse gas (see table 12.6). Reduction of human-caused emissions from landfills, farming, and coal mining is necessary.
- **Hydrofluorocarbons (HFCs).** Hydrofluorocarbons are widely used refrigerants. Leakage of HFCs to the atmosphere is significant, and they are greenhouse gases about 1,400 times more potent than CO_2.

All these climate-modification plans bring with them the danger of unintended consequences. It is possible that even the best intentions and the best engineering plans might create bigger problems than they solve. Unpredictable responses could include changes in ocean currents, drying up the tropics, acidifying the oceans, and disturbing plant and animal ecology.

Summary

Change occurs at nearly every historic and geologic time-scale, from hours or days to millions of years. Many processes affect climate and weather, each with its own operating principles and timetable.

Early in Earth's history, the atmosphere was rich in carbon dioxide, and surface temperatures were about 290°C. Now, more than 99% of CO_2 is locked up in limestone, and CO_2 constitutes only 0.04% of the atmosphere. The presence of atmospheric gases, such as CO_2, water vapor, and methane, creates a greenhouse effect whereby incoming, short-wavelength solar radiation passes through the atmosphere, but heat radiated by the Earth is in longer wavelengths that are unable to pass back through the atmosphere. Thus, Earth's surface temperature rises with the increasing abundance of greenhouse gases. At present, humans are burning wood, coal, natural gas, and oil, returning large amounts of CO_2 to the atmosphere. This leads to global warming. Global surface temperature rose 0.6°C (1°F) in the 20th century to reach the highest temperature in more than a thousand years.

Our knowledge about warm climates of the past is deduced from evidence in fossil reefs, tropical soils, evaporite mineral bodies, and widespread fossils of tropical and subtropical organisms. Our understanding of ancient cold climates is based on features such as glacially deposited debris, ice-polished and grooved rock surfaces, and wide distribution of the fossils of cold-water organisms.

Torrid Ages are times when tropical and subtropical conditions cover much of the Earth. They commonly involve buildup of greenhouse gases and extensive equatorial seas that absorb solar energy.

Ice Ages have occurred several times in Earth's history and last for millions of years. They require (1) large continents at the poles to support 3 km thick glaciers and (2) continents aligned to deflect warm ocean water toward the poles, where it can evaporate and then fall as snow on land to build glaciers.

Advances and retreats of glaciers during an Ice Age occur on a timescale of thousands of years. They are largely due to the pattern of incoming solar radiation. Glaciers advance when winter snowfall is greater and summer snowmelt is less. These changes are set up by changes in the orbit and rotation of Earth, affecting the amount of solar energy received. Earth's orbit is eccentric, moving closer to and then farther away from the Sun in a 100,000-year cycle. The angle of Earth's spin axis varies in its tilt toward the Sun on a 41,000-year cycle. The tilt direction changes over a double cycle with periodicities of 23,000 and 19,000 years. The effects of variations in orbit, tilt, and wobble may reinforce or cancel each other. At present, orbit and tilt contribute to cooling, while wobble aids warming.

Climate change occurs on hundred-year timescales also. For example, sea level can rise and fall by 200 m (650 ft) as glaciers wax and wane. When glaciers retreat rapidly, massive volumes of cold, fresh meltwater cover the ocean surface, causing big changes in ocean circulation, weather, and life.

Climate also changes on the scale of years. In the El Niño condition, warm water in the Pacific Ocean shifts positions. For example, the trade winds usually push warm water to the west in the central Pacific Ocean, producing tropical conditions and the Great Barrier Reef on the western side of the ocean but leaving coastal deserts on the eastern side. When the system shifts during an El Niño, cooler water off Australia results, yielding less rainfall and leading to massive bushfires. At the same time, warm water off the west coast of the Americas may bring heavy rains.

Volcanism has major effects on climate. For example, when ash and SO_2 are blasted through the troposphere into the stratosphere, they block some incoming solar radiation, leading to cooling. The Mount Pinatubo eruption in 1991 produced so much SO_2 that its cooling effect more than offset the greenhouse warming caused by humans for two years.

In the past thousand years, Earth's average surface temperature has fluctuated about 1.5°C. At its warmest, this created the Medieval Maximum, and at its coldest, it set off the Little Ice Age. Entering the 20th century, the average temperature was higher than at any time in the past 1,000 years. Then, during the 20th century, humans increased global warming via the greenhouse effect as they poured carbon dioxide (CO_2), methane (CH_4), nitrous oxide (N_2O), ozone (O_3), and chlorofluorocarbons into the atmosphere. The 21st century began with the warmest decade on record and will most likely bring much more warming. Living conditions on Earth will change as glaciers retreat, sea level rises, and rainfall patterns change. There is worry about reaching tipping points at which unexpected dramatic changes occur.

The sunspot cycle of energy variation lasts about 11 years, resulting in $\sim 0.15\%$ less solar radiation, thus causing Earth's surface temperature to drop $\sim 0.1°$ C. Longer sunspot cycles in the 1600s caused Earth's temperature to drop $0.2°$ to $0.5°C$.

The Intergovernmental Panel on Climate Change (IPCC) issued its *Assessment Report 4* in 2007 and *Assessment Report 5* in 2013 and 2014. They were written by more than 800 scientists and agreed upon by more than 130 countries. The consensus was that global warming is real and mostly due to humans pouring greenhouse gases into the atmosphere. In the 21st century, global surface temperatures are expected to rise between $1.1°$ and $6.4°C$. Sea level may rise up to 82 cm (2.7 ft), and there will be more frequent heavy rainfalls, extreme high tides, droughts, and heat waves.

Terms to Remember

aerosols 330	Little Ice Age 330
drought 329	Maunder Minimum 330
El Niño 323	Medieval Maximum 330
extinction 322	methane 317
global climate model	methane hydrate 317
(GCM) 336	Milankovitch theory 320
global warming	ozone 335
potential (GWP) 334	Pacific Decadal Oscillation
haboob 339	(PDO) 326
haze 327	radiative forcing 332
La Nada 325	sunspot 331
La Niña 325	tipping point 340

Questions for Review

1. The earliest Earth had an atmosphere loaded with CO_2 in an intense greenhouse climate. Explain where that atmospheric CO_2 has gone.
2. What information about ancient climates is suggested by (1) fossil reefs? (2) aluminum ore? (3) bodies of sea salt? (4) the area covered by a fossil species? (5) polished and grooved surfaces in rocks?
3. Climate is related to the amount of solar radiation received on Earth. How is incoming solar radiation affected by (1) continental ice sheets? (2) volcanic ash in the stratosphere? (3) elevated levels of atmospheric CO_2? (4) SO_2 blown into the stratosphere by volcanic eruption?
4. How can oxygen-isotope ratios be used as an ancient thermometer?
5. What causes glacial advances and retreats during an Ice Age?
6. How much can sea level drop during an Ice Age? What effect does this have on ocean-water temperature and evaporation rates?

7. When massive continental ice sheets melt, what happens to (1) sea level? (2) deep-ocean-water circulation? (3) salinity of sea-surface water? (4) organisms living near the sea surface?
8. Describe what happens to produce the El Niño phenomenon. What changes does it bring to Australia? To California?
9. How long does a sunspot cycle typically last? What percentage increase or decrease in solar radiation does Earth receive?
10. Name six greenhouse gases. How do they cause Earth's average surface temperature to rise?
11. How much did global surface temperature rise in the 20th century? How much is temperature projected to rise in the 21st century?
12. Instrumental records of Earth's surface temperature have been reliable since about 1880. When did the 10 warmest years occur?
13. Explain radiative forcing.
14. Give examples of aerosols that warm Earth's climate. Give examples that cool Earth's climate.
15. What are the stages of a famine? How do people react during famine compared to during a tornado, flood, or hurricane?
16. How much might sea level rise during the 21st century?
17. How does water circulate from top to bottom and north to south in the Atlantic Ocean?
18. What is meant by the term *tipping point?*
19. What is meant by the term *lag time?*
20. How is the acidification of the oceans caused by our adding massive volumes of CO_2 into the atomosphere each year?
21. Sea level is rising because of _____, _____, and _____.

Questions for Further Thought

1. Considering the climatic history of the Earth, are you living at a typical time?
2. What is the relationship between continental drift and the existence of an Ice Age?
3. If major volcanic eruptions occurred nearly every year for a century, what might happen to global climate?
4. What would be the global effect of sea-level rises of 1 m, 5 m, and 10 m? What would happen to New Orleans, Miami, and other coastal cities?
5. Humans are causing global warming by burning wood, coal, natural gas, and oil and thus returning CO_2 to the atmosphere. What global changes may result?
6. What changes in the natural environment are likely to happen in the next 40 years? How might your life change because of them?
7. What is the difference between a drought and a desert?
8. Using the concept of lag time, once a warm enough temperature has been reached to melt a glacier, how much time will it take to actually melt away?
9. Begin an annual list of climate-related events where you live—for example, dates of first flowering; arrivals of migratory birds; dates of butterfly emergence; number of above-average hot days.
10. How will we know when a climate tipping point has been passed?

CHAPTER 13

Floods

In the world there is nothing more submissive and meek than water. Yet for attacking that which is hard and strong nothing can surpass it.

—LEO TZU, 6TH CENTURY BCE

LEARNING OUTCOMES

All around the world, when large volumes of rain fall, floods will flow. After studying this chapter, you should:

- understand that streams are equilibrium systems.
- be able to explain positive versus negative-feedback.
- appreciate how and why streams construct floodplains.
- be familiar with flood-frequency curves.
- know the differences between flash floods and regional floods.
- be familiar with levees and the control of flood flow.
- understand the effects of urbanization on flood peak heights and duration.

OUTLINE

- How Rivers and Streams Work
- Flood Frequency
- Flood Styles
- Societal Responses to Flood Hazards
- Urbanization and Floods
- The Biggest Floods

Iowa Army National Guard personnel search for stranded people in Cedar Rapids, Iowa, 12 June 2008.

DoD photo by Staff Sgt. Oscar Sanchez, U.S. Air Force.

Saturday, 30 August 2003, was a rainy day in Kansas. As a dark night unfolded, some drivers on Interstate Highway 35 near Emporia never imagined their impending fates. Floodwater swept seven vehicles off the modern interstate highway, and seven people died. It was a **flash flood,** and they are deadly (table 13.1). An eyewitness, the Reverend Steve Gordon, said: "It happened really fast; there was nothing that could be done. It was a sick feeling just watching them go under." The floodwater was so powerful that the concrete barriers between highway lanes, each weighing 5 to 6 tons, were washed off the road and rolled 45 to 55 m (50 to 60 yd) down the creek.

Within small drainage basins, brief, localized downpours can cause fast-moving but short-lasting flash floods, as occurred near Emporia in 2003 and on the Guadalupe River of central Texas in 1972 (see chapter 10). Within large drainage basins, such as the Mississippi River, maximum floods result from widespread rains that last for many weeks, producing high waters that also last for many weeks. These are **regional floods.** Wide areas flooded for weeks cause tremendous damage (table 13.1).

La Niña conditions in the Pacific Ocean commonly result in wetter weather in midwestern North America in spring and early summer. The year 2008 was a La Niña year. Powerful cold-weather fronts moving slowly to the east and warm, moist air flowing north from the Gulf of Mexico collided over the Midwest again and again, week after week, and flood records were broken. In Iowa, 83 of its 99 counties were declared disaster areas. In Cedar Rapids, Iowa, the Cedar River rose to 31.12 ft, more than 11 ft higher than its previous record in 1929. High-water levels forced the evacuation of a hospital, a prison, and thousands of homes (figure 13.1). Property damages during regional floods are high (table 13.1), and high-water levels take weeks to subside to nonflood levels.

Rainfalls vary in intensity and duration, and so does the volume of rainwater that runs across the land. When rains are heavy, floods can result. No matter where you live—be it the tropics, the plains, or the desert—floods occur. Within a human lifetime, everyone will have a flood pass near them. Many people have stories to tell of their flood experiences, and the tales of the truly largest floods are passed on to succeeding generations. To live successfully with floods, we need to understand how rivers and streams operate.

How Rivers and Streams Work

A river is simply a large stream. Streams reveal much about their behavior when examined over their total length. A cross-sectional plot of a stream's bottom elevation versus the stream's distance from its source yields a fairly consistent and revealing relationship (figure 13.2). When we exaggerate the vertical scale on the graph to emphasize the relationship, the profile of the stream bottom appears relatively smooth and concave upward, with a steeper bottom or slope (higher **gradient**) near the stream source and a flatter bottom (lower gradient) near the stream mouth. Figure 13.2 shows the Arkansas River, but by changing the scales of elevation and length, this longitudinal cross-section could serve for virtually any stream in the world.

The lessening of gradient in a stream's lower reaches is partly due to the limitations of **base level,** the level below which a stream cannot erode. For many streams, base level is the ocean, but base level for the Arkansas River is where it joins the Mississippi River. For a small stream, base level may be a lake or pond into which the stream drains.

Streams have similar longitudinal cross-sections whether they run through the tropics or the deserts, whether they

Figure 13.1 Homes washed off their foundations are floating on the Cedar River jammed against an 1898 Union Pacific railroad bridge in Cedar Rapids, Iowa, 25 June 2008.

Greg Henshall/FEMA.

Side Note

A Different Kind of Killer Flood

The 15th of January 1919 was unusually hot in Boston. As the temperature climbed higher, the pressure of 27 million pounds of expanding molasses was too much for its heated tank to hold at the Purity Distilling Company. Steel bolts popped with a sound like gunfire, and the steel panels of the tank burst apart, releasing a flood of crude molasses; 2.3 million gallons flowed forth as a 4.5 m (15 ft) high brown wave. The flood knocked down supports for the elevated train, pushed houses off foundations, and smothered employees at the Public Works Department, as well as teamsters and their horses unloading goods at the Boston & Worcester and Eastern Massachusetts railroads (figure 13.3). As the molasses flowed, it cooled and congealed, holding people so tightly in its sticky grasp that rescue workers spent hours freeing them. The great molasses flood of 1919 killed 21 people and injured 150. Flood threats are not always obvious.

Figure 13.3 Damage in Boston from the molasses flood of 15 January 1919.

Boston Public Library, Print Department [77103].

TABLE 13.1		
U.S. Flood Deaths and Damages by Decade, 1960–2009		
Decade	**Deaths**	**Damages (in billions of 2011$)**
1960–1969	1,297	32.249
1970–1979	1,819	72.287
1980–1989	1,097	49.125
1990–1999	992	82.026
2000–2009*	1,378	102.466
Totals 50 Years	6,752 Deaths	$338.153 Billion
Average Deaths per Year = 135		
Average Damages per Year = $6.76 billion		

* Includes Katrina adjustment.
Data from National Weather Service, NOAA.

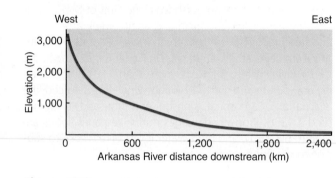

Figure 13.2 Longitudinal cross-section of the Arkansas River bottom from its origin in the mountains of Colorado to its mouth at the Mississippi River. Note the extreme exaggeration of the vertical scale necessary to emphasize the concave-upward profile of the bottom.

Source: Henry Gannett, *U.S. Geological Society.*

are long or short, and whether they run through hard or soft rocks. The worldwide similarity of bottom profiles of streams implies that some equilibrium processes must be at work.

THE EQUILIBRIUM STREAM

Numerous factors interact to make streams seek **equilibrium,** a state of balance in which a change causes compensating actions. To grasp the fundamentals of how streams work,

a few key variables must be understood: (1) **discharge,** the rate of water flow expressed as volume per unit of time; (2) available sediment, the amount of sediment waiting to be moved; (3) gradient, the slope of the stream bottom; and (4) channel pattern, the **sinuosity** of the stream path. Streams occupy less than 1% of the land surface but convey the rainfall runoff (discharge) from all the land and carry away loose sediment (**load**).

The U.S. Geological Survey maintains more than 7,000 stream-gauging stations that measure streamflow. Some of these stream gauges have been operating for more than a century. Each stream-gauging station measures water depths, channel width, and water velocity, allowing calculation of the discharge or flow volume. The greater the discharge, the greater the load of sediment carried. Both discharge and available sediment are independent variables—that is, the stream has no control over how much water it will receive or how much sediment is present. Nonetheless, a stream's task is to move the sediment present with the water provided. How can a stream accomplish this task? Excesses in discharge or load are managed by changing dependent variables, such as gradient and channel pattern.

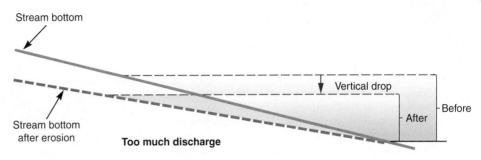

Figure 13.4 Schematic cross-section of a stream with too much discharge. The excess water erodes the bottom, flattening the gradient and thus slowing water flow.

Case 1—Too Much Discharge

If a stream has too much water, it will flow more rapidly and energetically. The move away from equilibrium triggers negative-feedback responses which work to correct the imbalance: (1) Some of the excess energy is used in eroding the stream bottom (figure 13.4). (2) The sediment picked up by erosion adds to the load carried by the stream, thus consuming more of the excess energy. (3) Notice in figure 13.4 that the slope of the stream bottom is lowered by erosion, reducing the vertical drop downstream and causing slower and less energetic water flow.

The stream also responds by increasing the sinuosity of its channel pattern through **meandering** (figure 13.5). A meandering stream cuts into its banks, thus using some of its excess energy to erode and transport sediment. Notice how the meandering pattern lengthens the flow path, lowering the stream's gradient and thereby slowing water flow (figure 13.6).

A close look at the meandering process tells us much of value (figure 13.7). Water does not flow at even depth and

Figure 13.5 The meandering San Juan River has eroded its channel over 300 m (1,000 ft) deep in Goosenecks State Park, Utah, 9 May 2004.

© DigitalGlobe/ScapeWare3d/Getty Images.

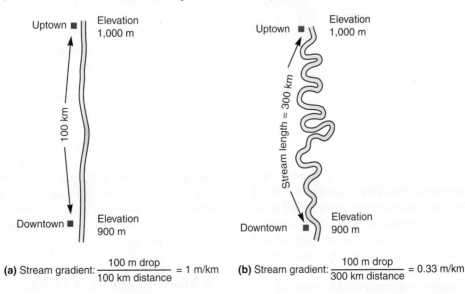

(a) Stream gradient: $\dfrac{100 \text{ m drop}}{100 \text{ km distance}} = 1$ m/km   **(b)** Stream gradient: $\dfrac{100 \text{ m drop}}{300 \text{ km distance}} = 0.33$ m/km

Figure 13.6 (a) A straight stream is the shortest path between two points. (b) A meandering channel lengthens the stream, thus reducing gradient and slowing water flow.

In Greater Depth

Stream Velocity Profile

In three dimensions, water flows downstream in an elongated spiral, causing the line of maximum velocity to move from side to side, from one stream bank to the other. Water velocity is slowed by friction with the stream bottom and sides and with the atmosphere. The variation in water velocity across a meandering stream is shown in a cross-section (figure 13.8).

The width and depth of the water define an area (blue) in figure 13.8; this area is one of the variables used to calculate discharge. The other variable is the average velocity of the stream water. The equation is:

$$Q = AV$$

where Q = discharge, the water flow (m^3/sec); A = cross-sectional area (m^2); and V = velocity (m/sec).

The safest place to cross a meandering stream is in the straight lengths between the stream bends. Here the stream cross-sectional profile is nearly symmetrical, with water at relatively constant depth and velocity all the way across the stream.

Figure 13.8 Stream velocity profile. Friction with stream bottom, sides, and the atmosphere slow the water flow. Water flow is fastest where deepest (Max V); erosion occurs on the outside bend. Water flow is slowest where shallow (Min V); deposition of sediment occurs on the inside bend. The curving arrow is a 2D representation of the actual 3D water flow which, over distance, is a spiral.

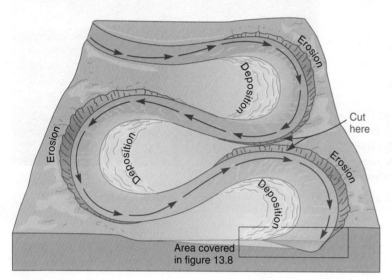

Figure 13.7 In a meandering stream, the outside bank is eroded and steep, while the inside bank grows as sediment is deposited. In the lower meander, notice how the erosional banks have almost met; in many cases, humans have cut through adjacent banks to straighten a river.

Drawing suggested by Gabi Laske.

power across a stream. Instead, a deeper, more powerful volume of water flows from side-to-side, eroding the river bank on one side and then on the other. This lengthens the path of the stream from left to right, decreases the gradient, and slows the water flow. Notice that deposition of sediment occurs on the inside bend of each meander where water is shallower and less powerful. The meandering process can proceed so far that two erosional banks can merge and create a shortcut that straightens the river locally (see center right of figure 13.7).

Human intervention in the meandering process has been common. Ships spending time and energy sailing long routes around meanders have been "helped" by humans excavating through meanders to shorten sailing paths. This human intervention has had unintended yet profound effects on rivers such as the Mississippi.

Case 2—Too Much Load

If a stream is choked with sediment and has insufficient water to carry it away, this also triggers negative feedback. The excess sediment builds up on the stream bottom, increasing the gradient and causing streamwater to flow faster and thus have more load-carrying capacity (figure 13.9). The channel pattern responds by straightening to shorten the flow distance and increase the gradient. The straighter stream still contains excess sediment, causing the water to pick its way through as a **braided stream** (figure 13.10).

Another "too much load" situation for a stream is the presence of a lake. For example, if a landslide dams a stream, it adds excess load that the stream will attempt to carry away. The stream will gradually fill in the lake basin with its load of sediment until flow can reach the dam (figure 13.11). When the stream is able to flow rapidly over the steep-gradient face of the dam, it does so with heightened erosive power, allowing it to carry away the landslide dam as well as the sediment fill in the lake. In a geologic sense, lakes are temporary features that streams are striving to eliminate.

GRADED-STREAM THEORY

All streams operate in a state of delicate equilibrium maintained by constantly changing the gradient of the stream

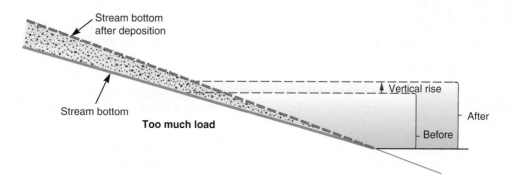

Figure 13.9 Schematic cross-section of a stream with too much load. Excess sediment is dropped on the stream bottom, increasing the gradient and thus speeding water flow.

Figure 13.10 A braided channel pattern occurs as water flows through excess sediment within a fairly straight valley, Chichagof Island, Alaska.

Alaska ShoreZone Program NOAA/NMFS/AKFSC; Courtesy of Mandy Lindeberg, NOAA/NMFS/AKFSC.

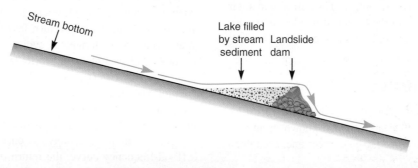

Figure 13.11 Schematic cross-section of a landslide-dammed valley. Over time, a stream fills the lake with sediment and then flows across the infilled lake. Flow down the dam's steep face is at high velocity, causing erosion and transport of dam and lake sediment.

bottom, thus sustaining a **graded stream.** Every change in the system triggers compensating changes that work toward equilibrium. A typical stream has too much load and too little discharge in its upstream portions, thus it maintains a braided channel pattern there. In its downstream segments, the typical stream has too much discharge, finer sediments in its load, and less friction, thus it runs in a meandering pattern there. Streams also change their equilibrium states from one season to another and in response to global changes in sea level and to tectonic events. That most streams have the same longitudinal cross-section is a testimony to the effectiveness of their negative-feedback mechanisms.

Side Note

Feedback Mechanisms

Many systems display either negative or positive **feedback. Negative-feedback** occurs in self-regulating systems and works to maintain a system in equilibrium. In the case of a stream, when too much water pours into the channel, it triggers negative-feedback responses whereby increased erosion lowers the gradient to slow the water flow and maintain equilibrium.

Positive feedback is also known as the "vicious cycle." It occurs where one change leads to more of the same, and the whole system changes dramatically in one direction. In a human

sense, positive feedback can have positive results if, for example, the system is your money in the bank; the interest you earn gathers more interest, and the system (your money) grows rapidly. But positive feedback can have negative results if, for example, you have credit-card debt and the resultant interest charges add to your debt, causing you to pay interest on earlier interest charges. Positive feedback can make you rich or poor depending on whether you choose to invest your money or incur debt.

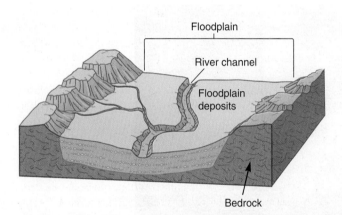

Figure 13.12 Floodplains are stream floors during floods.

THE FLOODPLAIN

Floodplains are the floors of streams during floods (figure 13.12). Streams build floodplains by erosion and deposition, and streams reserve the right to reoccupy their floodplains whenever they see fit. Humans who decide to build on a floodplain are gamblers. They may win their gamble for many years, but the stream still rules the floodplain, and every so often it comes back to collect all bets.

Flood Frequency

What has happened in the past provides the best forecast for future events. But even the largest flood known in an area is likely to be exceeded someday by a larger one, even in an area with a long history.

FLORENCE, ITALY, 1333 AND 1966

Florence, Italy, has been described as Europe's greatest cultural storehouse, the site of amazing artistic and scientific accomplishments in the 13th through 16th centuries. The people of Florence knew their largest flood; it occurred on 4 November 1333, when the River Arno inundated the city to 4.22 m (14 ft) depth. The flood is memorialized by several

artistic creations. However, 633 years later, to the day, the River Arno flowed through the city at 6.2 m (more than 20 ft) depth, causing heart-wrenching destruction and damage to ancient and priceless paintings, sculptures, tapestries, books, maps, musical and scientific instruments, and more.

The 1966 events started on 3 November in the River Arno headwaters, when enough rain fell to equal one-third of the average annual rainfall. Upstream dams could not hold all the water. The river raged through villages, killing people, ripping apart buildings, and bursting open drums of oil. Some of the treasures of Florence were smashed, some were buried beneath the 500,000 tons of mud deposited by the flood, and many more were coated with a smelly, oily slime. Will an even-larger flood arrive in Florence's future?

Even though Florence has good records of its flood history, it still is hit with nasty surprises. The United States and Canada, unlike Europe, do not have lengthy historical records to guide development.

FLOOD-FREQUENCY CURVES

Everyone living near a stream needs to understand the frequency with which floods occur. Small floods happen every year or so. Large floods return less often—every score of years, century, or longer. A typical analysis of flood frequency involves a plot of historic data on flood sizes versus recurrence interval (figure 13.13). Flood-discharge volumes are plotted on the vertical axis, and the **recurrence intervals** are plotted on the horizontal axis in years between floods of each size. The longer the historical record of floods in an area, the more accurately the curve can be drawn. With a flood-frequency curve, the return times of floods can be estimated. You can try this in figure 13.13: move upward from 100 years, intercept a curve, and then read to the left to obtain the expected flood size. The U.S. Federal Emergency Management Agency (FEMA) uses the 100-year flood in establishing regulatory requirements.

Individual flood-frequency curves must be constructed for each stream because each stream has its own characteristic floods. A flood-frequency curve should serve as the basis for designing all structures built on a floodplain

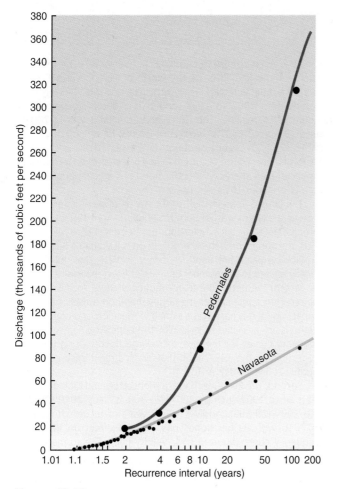

Figure 13.13 Flood-frequency curves for the Pedernales River at Johnson City and the Navasota River at Easterly, Texas.

and determining where buildings should be located for the highest probability of safety. Planners can decide what size flood (how many years of protection) to accommodate when determining how the land is to be used.

When designing roads, bridges, and buildings, it is seductive to consider only the smaller floods and save large amounts of money on initial construction costs. However, these initial savings can be eaten away by higher maintenance and repair costs. In the long run, it is commonly cheaper to build in anticipation of large floods; this can save money in the future and also eliminate much of the human suffering that occurs when homes and other buildings are flooded.

A planner needs to know the likelihood of a given size flood occurring during the expected usage time of a structure. Flood frequency can also be expressed as the statistical probability of stream discharges of a given size arriving in any year or number of years (table 13.2). The bigger the flood, the longer the return period and the smaller the probability of experiencing it in any one year. Statistically, the 100-year flood has a 1% chance of occurring any year. What is the probability that a 100-year flood will occur once in 100 years? The obvious answer (once) is wrong; from table 13.2, the probability is only 63%. No flood has a 100% chance of occurring.

We must distinguish between yearly versus cumulative probability. In cumulative probability, the longer the wait for a 100-year flood, the more likely its occurrence becomes. Nevertheless, the yearly probability of a flood is the same for any year regardless of when the last flood occurred. The confusion that commonly arises when hearing of a "100-year flood" has led some people to stop using the term and to replace it with "1%-chance flood." A 1%-chance flood is a flood event that has a 1% chance of occurring or being exceeded in any given year.

TABLE 13.2
Cumulative Probabilities of Floods

Percentage Chance of This Size Flood Occurring					
Return Period (years)	Any 1 Year	10 Years	25 Years	50 Years	100 Years
2	50				
5	20				
10	10	65	94	99.9	
20	5	40	71	90.5	
50	2	18	40	63	86
100	1	9.6	22	39	63
200	0.5	5	12	22	39
500	0.2	2	5	9.5	18
1,000	0.1	1	2.5	4.8	9.5
2,000	0.05	0.5	1.2	2.3	5
5,000	0.02	0.2	0.5	1	2
10,000	0.01	0.1	0.25	0.5	1

Source: B. M. Reich, *Water Resources Bulletin,* 9 (1973), 187–88.

Constructing Flood-Frequency Curves

Floods are random events. It is not possible to predict just when a flood will occur or what its discharge will be. But we need to know how often a given-size flood may occur in order to intelligently develop land. The process used is statistical. If enough runoff records exist for a river, then probable flood frequencies can be estimated.

Here is a relatively simple method for constructing a flood-frequency curve: Begin by determining the peak discharge for each rainfall year. Ignoring the chronologic order of the rainfall years, rank each annual flood in order, from biggest discharge (= 1), second biggest (= 2), etc., on down to the smallest. To plot a curve such as figure 13.13 for a river of interest, first calculate recurrence intervals for each year's maximum flood using the formula

$$\text{recurrence interval} = \frac{(N + 1)}{M}$$

where N = the number of years of flood records, and M = numerical rank of each year's maximum flood discharge. After calculating a recurrence interval for each year, locate that value (in years) on the horizontal axis (which usually is a logarithmic axis as in figure 13.13); then move upward until reaching the appropriate discharge value. Stop and plot a point marking the intersection of the recurrence interval and discharge values. After plotting a discharge versus recurrence interval point for each year, draw a best-fit line through your plotted points.

As examples of calculating flood frequency, let's use the flood record of the Red River of the North at Fargo, North Dakota, over two time intervals. First, let's examine the 115 years of records from 1882 to 1997, and plot a flood-frequency curve (figure 13.14, red line). The 12th highest discharge was 11,200 ft³/sec on 3 April 1994. The recurrence interval for a flood of this size is 116/12 = 9.67 years, or about once every 10 years. The probability of a given flood occurring in any one year is the reciprocal of the recurrence interval. The 1994 flood has an annual probability of 1/10 or about 10%.

The largest flood in the first 115 years of records for the Red River of the North was 28,000 ft³/sec on 18 April 1997. Using the formula, the recurrence interval for this flood is (115 + 1)/1 = 116. Thus, the annual probability of this flood is less than 1%.

How valuable are flood-frequency curves? Their reliability is directly related to the number of years of flood records; the longer the record, the better the flood-frequency curve. In this method, the recurrence interval for the largest flood on a river is the most suspect point; it is based on a sample population of one. Statistical methods are available to help plot the upper segments of flood-frequency curves for the rare, extra-large floods. Using a statistical technique, the 1997 flood plots with a recurrence interval of roughly 200 years (figure 13.14).

Second, the Red River has experienced several high discharge years since 1997 such as 2001 (20,300 ft³/sec), 2010 (20,000) and 2009 (29,500). Adding the 13 years of records from 1998 to 2010 with its big floods has a major effect on the flood-frequency curve (figure 13.14, yellow line). The former record

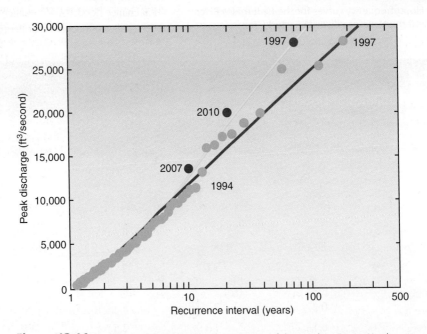

Figure 13.14 Flood-frequency curves for the Red River of the North at Fargo, North Dakota, for years 1882–2010.

Data from National Water Storage and Retrieval System.

In Greater Depth

flood in 1997 is now the second largest in 128 years of records. Its recurrence interval recalculates as (128 + 1) / 2 = 65 years. The new, steepened flood-frequency curve suggests that the record 2009 flood is about a 90-year event.

Flood conditions in the Red River of the North are changing (figure 13.15). The deep blue colors show the times of highest daily flow; they commonly begin in March and may last into the summer. Go to the year 1900 at the bottom of figure 13.15 and follow the blue color upward to 2014. Note the increase in frequency of large floods starting in the early 1990s and continuing through 2014. Flood-frequency curve shapes and forecasts are changing rapidly here.

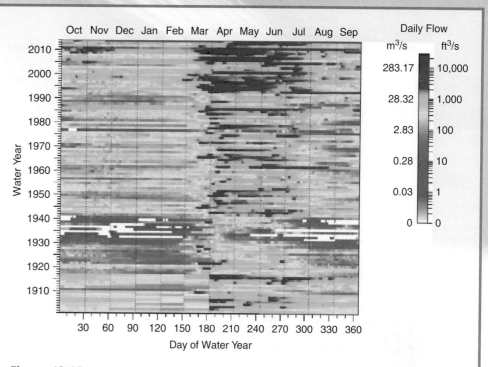

Figure 13.15 Daily water flow of the Red River of the North at Fargo, North Dakota, 1900–2014. From NOAA/NWS.

By similar analysis, even though a "150-year flood" may occur one year, it is still possible for another of the same size to come again in the following year or even in the same year. For example, in 1971, the Patuxent River between Baltimore and Washington, DC, had a flood that was 1.6 times bigger than its calculated 100-year flood. The next year, in 1972, the Patuxent River conveyed a flood that was 1.04 times as big as its 100-year flood. A 100-year flood, or any other size flood, is a statistically average event that occurs by chance, not at regular intervals. As the adage states: "Nature has neither a memory nor a conscience."

Flood Styles

Killer floods are unleashed by several phenomena: (1) A local thundercloud can form and unleash a flash flood in just a few hours. (2) Abundant rainfall lasting for days can cause regional floods that last for weeks. (3) The storm surges of tropical cyclones flood the coasts. (4) The breakup of winter ice on rivers can pile up and temporarily block the water flow, and then fail in an ice-jam flood. (5) Hot weather can cause rapid melting of snow. (6) Short-lived natural dams made by landslides, log jams, or lahars fail and unleash floods. (7) Human-built dams and levees fail, causing voluminous floods.

One useful way to distinguish flood styles is by contrasting flash floods with regional floods. A typical flash flood results when rain falls intensely for hours in a small area, causing the runoff of a fast-moving, powerful flood of water that lasts only a short time. Regional floods occur when lots of rain falls over a large area for days or weeks, causing river flood levels to rise slowly and then to fall slowly. Flash floods cover small areas and are sometimes called *upstream floods*, whereas regional floods inundate large areas and are called *downstream floods*. Flash floods can be deadly, whereas regional floods cause widespread economic losses. In 2011, U.S. flash floods killed 69 people and caused damages of $1.44 billion; regional floods killed 44 people and caused $6.77 billion damages.

FLASH FLOODS

Large convective thunderstorms can build up in a matter of hours and quickly set loose the terrifying walls of water known as flash floods. Steep topography helps thunderstorms build and then provides the rugged valleys that channelize the killer floods.

Most flood-related deaths in the United States are caused by flash floods. About 50% of these deaths are vehicle-related. Not enough people appreciate what a shallow-water flood can do to a car (figures 13.16 and 13.17). Flowing water about 0.3 m (1 ft) deep exerts about 225 kg (500 lb) of lateral force. If 0.6 m (2 ft) deep water reaches the bottom of a car, there will be a buoyant uplift of about 680 kg (1,500 lb),

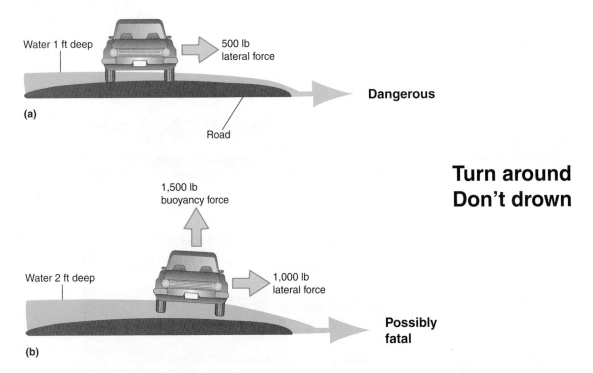

Water 1 ft deep

500 lb
lateral force

Dangerous

(a)

Road

1,500 lb
buoyancy force

Water 2 ft deep

1,000 lb
lateral force

**Possibly
fatal**

(b)

**Turn around
Don't drown**

Figure 13.16 Do not drive through a flood. (a) Floodwater 1 ft deep exerts 500 lb of lateral force. (b) Floodwater 2 ft deep both buoyantly lifts and laterally pushes a car.

Modified from *U.S. Geological Survey Fact Sheet 024–00.*

Figure 13.17 A flash flood traps two cars.

NOAA.

which helps the 450 kg (1,000 lb) lateral force push or roll the car off the road. Many automobile drivers and riders die in floodwater that is only 2 ft deep.

Antelope Canyon, Arizona, 1997

The plateau country of the southwestern United States is world renowned for the canyons cut into it—for example, Grand Canyon, Glen Canyon, and numerous tributary canyons leading to the Colorado River. Some of these tributary canyons are wonder-inspiring narrow clefts. Imagine walking down a slot canyon so narrow that you

Figure 13.18 A slot canyon has steep walls that allow no escape for hikers caught in a flash flood, Antelope Canyon, Arizona.

© Image Ideas/PictureQuest RF.

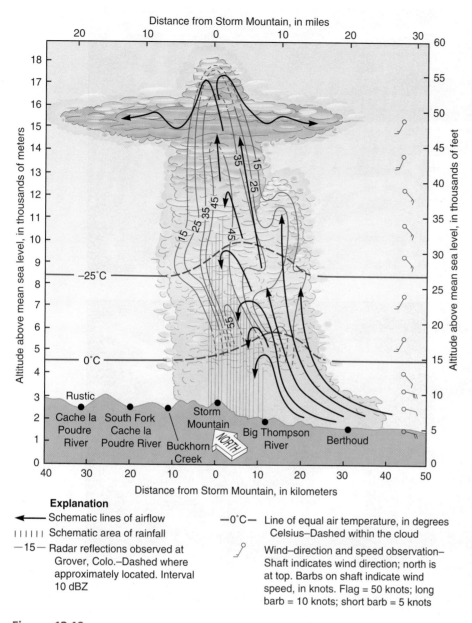

Figure 13.19 Model of the thunderstorm that fed the Big Thompson Canyon flood. Drawing depicts conditions at 6:45 p.m., 31 July 1976.

Source: *U.S. Geological Survey Professional Paper 1115.*

Explanation

◄— Schematic lines of airflow

||||| Schematic area of rainfall

—15— Radar reflections observed at Grover, Colo.–Dashed where approximately located. Interval 10 dBZ

—0°C— Line of equal air temperature, in degrees Celsius–Dashed within the cloud

⌐ Wind–direction and speed observation– Shaft indicates wind direction; north is at top. Barbs on shaft indicate wind speed, in knots. Flag = 50 knots; long barb = 10 knots; short barb = 5 knots

Big Thompson Canyon, Colorado, 1976

Saturday, 31 July 1976 was the eve of the 100th birthday of Colorado statehood and the start of a three-day centennial weekend. The weather was pleasant, and about 3,500 people were staying in or passing through the recreational area of Big Thompson Canyon about 80 km (50 mi) northwest of Denver, near Estes Park, in the Front Range of the Rocky Mountains. About 6 p.m., strong low-level winds from the east pushed a moist air mass upslope into the mountains. The unstable air kept rising, powered by heat released from condensing water vapor. Late-afternoon cloud building is a common phenomenon and usually concludes with upper-level winds pushing the thunder-cloud eastward over the plains. On this day, the mid and upper-level winds were weak, the thunder-storm remained stationary, and a "cloudburst" ensued (figure 13.19). The slightly tilted updraft structure allowed rain to fall profusely. From 7:30 to 8:40 p.m., rainfall was as heavy as 19 cm (7.5 in); in four hours, it equaled a typical year's total.

Rain runoff from the steep, rocky slopes fed a flash flood that roared down the canyon, with an initial wall of water reaching 6 m (20 ft) high in the narrows at the eastern end of Big Thompson Can-yon. The flood crest moved 25 km/hour (15 mph) through the entire canyon, which did not allow much time to spread warnings. More than 400 automobiles were on Highway 34 within the canyon. Drivers were presented with a quick choice—aban-don their cars and run upslope or stay inside and try to out-race the flood. Those who abandoned their cars spent an uncomfortable night on the rain-swept canyon walls; those who stayed with their cars died. The new road signs now advise: "Climb to safety! in case of a flash flood."

At 6 p.m., the Big Thompson River flowed with 137 ft³/sec of water; at 9 p.m., the flow was 31,200 ft³/sec. Flow volume was 3.8 times greater than the estimated 100-year flood. The flash flood killed 139 people, and 6 more were never found; 418 houses were destroyed, along with 52 businesses and more than 400 cars, and damages totaled $36 million (figures 13.20 and 13.21).

can simultaneously touch both walls with your hands even though the walls rise near-vertical for 30 m (100 ft) above your head (figure 13.18). The experience of walking down these slot canyons is so magical that tourists from around the world come to hike between the exquisitely beautiful, sculpted walls with their pink and orange colors; the setting is like a natural cathedral. Now imagine a localized, late afternoon thundercloud 18 km (11 mi) away—too far away for you to see or hear it. What can happen in the slot canyon? On 12 August 1997, in Antelope Canyon in Arizona, a flash flood roared down the canyon as a 3.3 m (11 ft) high wall of water that picked up 12 hikers and tumbled them along with it while helpless viewers watched from the canyon rim. The natural cathedral turned into a death trap; only one hiker survived.

(a) Before

(b) After

Figure 13.20 The community of Waltonia before (a) and after (b) the 1976 flood in Big Thompson Canyon, Colorado.
U.S. Geological Survey.

Figure 13.21 A house dropped onto a bridge during the Big Thompson Canyon flood in Colorado, 1976.
W.R. Hansen/U.S. Geological Survey.

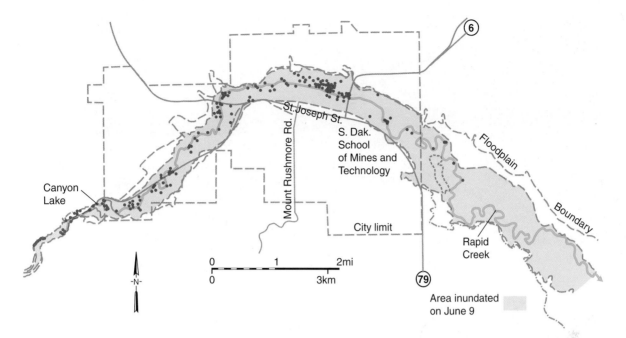

Figure 13.22 Map of Rapid City, South Dakota, showing the area inundated by the flood of 9 June 1972. Red dots mark locations of bodies found after the flood.

Rapid Creek, Black Hills, South Dakota, 1972

The Black Hills are one of the most beautiful sites in the United States; they host nearly 3 million tourists a year. At the foot of the Black Hills sits Rapid City, first settled south of Rapid Creek in 1876. The early inhabitants were wary of the Rapid Creek floodplain, and the wisdom of their caution was borne out by a large flood in 1907. As the city's population grew, the peaceful, meandering stream became a magnet that induced development on the floodplain. In 1952, Pactola Dam was built 16 km (10 mi) upstream to provide flood protection and a reserve water supply for Rapid City. The dam eliminated most small floods, giving some people a false sense of security. By 1972, the floodplain was host to numerous houses, mobile-home parks, shopping centers, car lots, and other urban structures serving residents in a city approaching 50,000 population.

On Friday, 9 June 1972, southeast winds bringing moist air from the Gulf of Mexico met a cold front coming from the northwest. Under conditions similar to those at Big Thompson Canyon, the moist air turned upward to build 16 km (10 mi) high thunderclouds that remained stationary due to weak upper-level winds. Shortly after 6 p.m., heavy rain began to fall; up to 38 cm (15 in) fell in less than six hours, but most of the rain fell downstream from Pactola Dam. Rain runoff filled Canyon Lake, built on the western edge of Rapid City. The spillway at Canyon Lake Dam became plugged by automobiles and house debris, causing the lake level to rise an additional 3.6 m (12 ft). Then the dam failed at about the same time as the natural flood crest arrived, unleashing a torrent of water on Rapid City (figure 13.22). The river reoccupied its floodplain with vengeance, leaving destruction totaling $664 million in 2002 dollars. Floods in the region killed 238 people, mostly in Rapid City, and destroyed 1,335 homes and 5,000 automobiles.

This time the lesson was learned by many. Canyon Lake Dam and many bridges were redesigned and rebuilt to prevent debris accumulations. The portion of Rapid Creek floodplain inundated in 1907 and 1972 was declared a floodway, and rebuilding was not permitted. Even most buildings that survived the flood were moved out of the floodway. In their place lies an 8 km (5 mi) long greenway featuring a golf course, picnic areas, bike and jogging paths, recreation areas, ponds and ice-skating rinks, low-maintenance grasslands, and an area reseeded with native vegetation: in short, the floodway is now being used for activities that will not be harmed by the occasional flood. A Rapid City slogan is: "No one should sleep on the floodway." As long as this policy is maintained, the next big flood will not cause deaths, destruction, and heartache.

REGIONAL FLOODS

Regional floods are different from flash floods. High waters may inundate an extensive region for weeks, causing few deaths but extensive damages and severe tests of human endurance. Regional or inundation floods occur in large river valleys with low topography when prolonged, heavy rains result from widespread cyclonic systems.

In the United States, about 2.5% of the land is floodplain and home to about 6.5% of the population. Floodplains contain much valuable property that is periodically flooded.

Red River of the North

The Red River of the North is unusual because it flows northward (figure 13.23), draining parts of South Dakota, North Dakota, and Minnesota before flowing into the Canadian province of Manitoba. Floods are likely here each spring. As described by John McCormick of *Newsweek*,

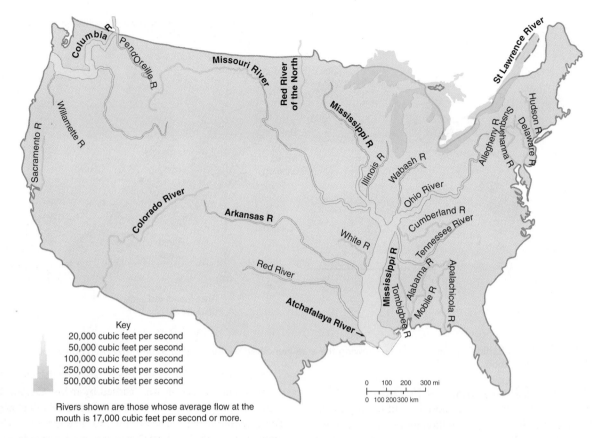

Key
20,000 cubic feet per second
50,000 cubic feet per second
100,000 cubic feet per second
250,000 cubic feet per second
500,000 cubic feet per second

Rivers shown are those whose average flow at the
mouth is 17,000 cubic feet per second or more.

0   100   200   300 mi
0 100 200300 km

Figure 13.23 Large rivers in the conterminous United States.

Source: K.T. Sei and W.B. Langben, "Large Rivers of the United States," in U.S. Geological Survey Circular 685, 1974.

"On the northern plains, nature is less an enemy than a sparring partner, trading rounds in a grudge bout that never ends." In 1997, nature won the round as flood levels set records throughout the region. A Presidential Disaster Proclamation was declared for all 53 counties in North Dakota.

Why are floods so common along the Red River of the North? Several factors combine to create this situation: (1) The Red River valley is geologically young—only about 9,000 years old—and the river has not carved a deep valley. (2) The gradient or slope of the riverbed is very low, averaging only an 8 cm/km (5 in/mi) drop in elevation. The gradient is only 2.4 cm/km (1.5 in/mi) just south of the Canadian border. The nearly flat river bottom causes slow-flowing water that tends to pool into a broad and shallow lake during high-water flow. (3) River flow increases as winter snow melts. The meltwater runs northward into still-frozen lengths of the river, where ice jams up and impedes water flow, causing floods.

In 1997, several variables combined to unleash record floods: (1) Fall 1996 rainfall was about four times greater than average. (2) Winter 1996 freezing temperatures began earlier than normal, thus freezing the water saturated in the soil. (3) Winter 1996–97 snowfalls were 3 to 3.5 times greater than average, bringing and storing more moisture in the region. (4) Spring 1997 began cold, including a blizzard on 4–6 April that brought 25 to 30 cm (10 to 12 in)

of snow and record low temperatures, which delayed melting and draining. (5) Finally, flooding began as a rapid rise in air temperature melted snow and soil ice, sending water flowing.

On 8 April 1997, the average temperature was −13° C (9°F), but on 18 April, the average was 14°C (58°F). Snow melted, soil ice melted, and the ground everywhere was covered by overland waterflow that overwhelmed the water-transporting ability of the flat-bottomed Red River. As the floodwaters slowly flowed northward, they progressively inundated farmland and towns; one incongruous sight was of flooded downtown Grand Forks with its buildings on fire. In North Dakota, the flood caused damages exceeding $1 billion; forced the evacuation of 50,000 people; destroyed potato, sugar beet, and grain crops; prevented planting of the next crop; and drowned farm animals, including 123,000 cattle. In Manitoba, the flood caused damages exceeding $815 million and forced the evacuation of 25,000 people.

How common is a flood this size? The Red River flood-frequency curve (see figure 13.14) indicates that the 1997 flood has a recurrence interval of almost 200 years. But this is an estimate; there are not enough controls on where to plot the largest event on a flood-frequency curve. The 1997 flood discharge set a record, but records are made to be broken. The 1997 record fell in 2009, and the flood-frequency curve was recalculated.

Mississippi River System

The greatest inundation floods in the United States occur within the Mississippi River basin, the third largest river basin in the world; it drains all or part of 31 states and two Canadian provinces (figure 13.23). Of the 28 biggest rivers in the United States, 11 are part of the Mississippi River system. In the lower reaches of the river, the average water flow is 18,250 m³/sec (645,000 ft³/sec); this may be increased fourfold during an inundation flood.

The Mississippi River carries immense volumes of sediment to the sea, where it is deposited within the Birdfoot delta lobe building out into the Gulf of Mexico (figure 13.24). As the delta lobe builds seaward, the river's length increases, and sediment is deposited within the river channel. When a river deposits sediment on its channel floor, its channel-bottom elevation grows higher. Over time, the channel bottom may build to an elevation higher than its adjoining floodplain. When major floods occur and water overtopping banks and levees flows to lower elevations outside the channel, the river may adopt a new lower-elevation course and abandon its old channel; this is the process of **avulsion**. Look at the delta lobes in figure 13.24; each represents an avulsion event, a changing of the course of the Mississippi River.

Today, you can stand on the Mississippi River floodplain north of New Orleans and look *up* at the river and see ships passing *above* you. This is an unstable position for a river. It leads to river-channel abandonment and establishment of a new path to the sea. The Mississippi River has long "wanted" to abandon its course (avulse) and flow down the shorter and steeper path to the ocean offered by the Atchafalaya River (figure 13.24). The U.S. Army Corps of Engineers works hard to make the Mississippi River stay put and continue flowing through the important cities of Baton Rouge (the capital of Louisiana) and New Orleans (a major U.S. port). It takes a huge commitment of time, money, and effort to control a river's course.

This switching course from the Mississippi to the Atchafalaya has not been entirely a natural one; humans have inadvertently aided the process by cutting through meanders (see figure 13.7). A particularly lengthy meander used to exist where the Mississippi and Atchafalaya rivers met. This meander so annoyed ship captains in the early 1800s that it was cut through to shorten their trips. The human alteration caused more floodwater to travel down the Atchafalaya, and the process of avulsion was accelerated. Congress directed the Army Corps of Engineers to build flood-control structures that allow 30% of the water to flow down the Atchafalaya but to keep 70% in the Mississippi River (figure 13.25). Nature prefers the Atchafalaya; humans prefer the Mississippi. This is an ongoing battle that never can be finished.

Some Historic Mississippi River Floods New Orleans was founded in 1717 in the lower Mississippi River basin. It experienced its first large flood in the same year. The response to the flood in 1717 was the same as today—**levees** were built higher to keep the river water inside its channel. The word *levee* is derived from the French verb

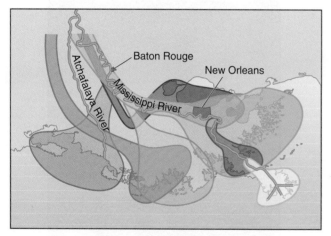

Salé-Cypremort until 4500 Years Before Present (YBP)
Cocodrie	4600-3700YBP
Teche	3900-2800YBP
St. Bernard	2800-1700YBP
Lafourche	1900-800YBP
Plaquemines	1100-500YBP
Birdfoot	550YBP

Figure 13.24 The Mississippi River has changed its course seven times in the past 4,600 years. Its various delta lobes have been named and are distinguished here by color. The river's present course is held in position only by active engineering projects run by humans.

Figure 13.25 Old River Control Structure Complex viewed to the east-southeast; the State of Mississippi is at top of image; Louisiana is the lower majority. The Mississippi River flows downstream from upper left curving to the right in the distance. The Atchafalaya River meets the Mississippi River at three points and flows downstream to the bottom right. There is a control structure (dam) on each of the three forks of the uppermost Atchafalaya River. The structures are designed to keep 70% of the water in the Mississippi River and 30% in the Atchafalaya River.

U.S. Army Corps of Engineers.

lever, which means to raise. Humans commonly increase the height of natural levees built by stream overbank flow, or we construct new levees where none exist.

In 1879, the Mississippi River Commission placed a major emphasis on building levees, yet the 1882 flood broke through the levees in 284 places. By 1926, there were 2,900 km (1,800 mi) of levees averaging 6 m (20 ft) in height. But the 1927 floods breached the levees in 225 places, inundating 50,000 km² (19,300 mi²) and drowning 183 people.

After 1927, a *project design flood* was developed as a hypothetical "maximum probable" flood of the Mississippi River. Flood peak discharges were calculated along the river at about 25% greater than the 1927 flood. To control and manage floods: (1) levees were raised higher and built in new places, (2) the river bottom was dredged to increase the volume of water that could flow in the river, (3) dams were constructed on tributary rivers to capture and control flood-water, (4) river meanders were straightened to shorten the river by 270 km (170 mi), and (5) diversions were established to funnel part of the major river flow into **floodways** to utilize smaller rivers and to flood lowlands. Despite all these efforts, in the spring of 1973, the Mississippi River system flooded along 1,930 km (1,200 mi) of rivers in 10 states, again inundating 50,000 km², reaching record flood heights at numerous sites, and staying above flood stage for 97 days at Chester, Illinois, 77 days at St. Louis, and 63 days at Memphis.

The Great Midwestern Flood of 1993 The summer of 1993 saw the biggest flood in 140 years of gauged measurements for the upper Mississippi River basin (table 13.3). Notice how late in the year the 1993 peak flood occurred at St. Louis (table 13.3).

Rising water goes through **flood stages** where water level is high enough to overtop the river banks/levees. In the *action stage,* water begins overtopping the banks. In the *minor flood stage,* roads, parks, and yards may be covered by water. In the *moderate flood stage,* building inundation occurs; roads are closed and evacuations may be necessary. In the *major flood stage,* buildings may be completely submerged; lives are threatened and large-scale evacuations may be necessary.

A wet winter and spring passed into an even wetter summer of 1993 for Iowa and parts of eight other upper Midwest states, causing record flood levels on the lower Missouri and upper Mississippi Rivers (figures 13.26 and 13.27). High-water levels began in April and continued through August. Some towns had more than 160 consecutive days of flooding. At the end of August, the upper Mississippi River basin had endured record high floods, yet this floodwater mass did not significantly affect the lower Mississippi River basin because the input from the Ohio River was low (see figure 13.23).

In 1993, floodwaters occupied more than 20 million acres. The entire state of Iowa was declared a federal disaster area, as

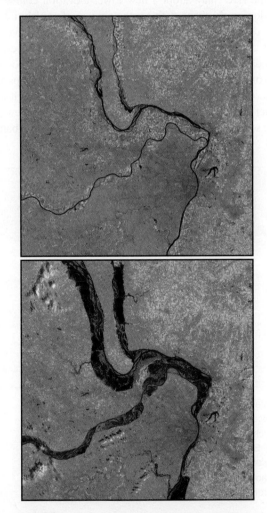

TABLE 13.3

Top Ten Mississippi River Floods at St. Louis, 1861–2010

Date	Discharges (cubic feet/second)
1993 August	1,030,000
1903 June	1,019,000
1892 May	926,500
1927 April	889,300
1883 June	862,800
1909 July	860,600
1973 April	852,000
1908 June	850,000
1944 April	844,000
1943 May	840,000

Source: Illinois State Water Survey Miscellaneous Publication 151 (1994), Champaign.

Figure 13.26 Images of states of Illinois (right side) and Missouri (left side). The Illinois River (upper right), Missouri River (lower left), and Mississippi River (middle) are shown. Top photo shows typical river levels. Bottom photo shows 1993 flood level.

Jesse Allen/NASA Earth Observatory.

Figure 13.27 The Missouri River flood of 1993 surrounds the state capitol in Jefferson City, Missouri.

Missouri Highways and Transportation Department.

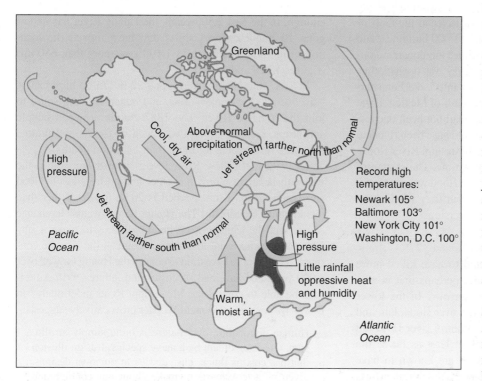

Greenland

Above-normal precipitation

Cool, dry air

Jet stream farther north than normal

High pressure

Jet stream farther south than normal

Pacific Ocean

Record high temperatures:
Newark 105°
Baltimore 103°
New York City 101°
Washington, D.C. 100°

High pressure

Little rainfall oppressive heat and humidity

Warm, moist air

Atlantic Ocean

Figure 13.28 A typical weather map for the United States in July 1993.

were sections of eight other states—North Dakota, South Dakota, Minnesota, Wisconsin, Illinois, Missouri, Nebraska, and Kansas. Flooding killed 48 people, completely submerged 75 towns, destroyed or damaged 50,000 homes, closed 12 commercial airports, and shut down four interstate highways (I-29, I-35, I-70, and I-635). Damages totaled about $12 billion, making this flood one of the most expensive disasters in U.S. history.

Weather Conditions for 1993 Floods Why were the 1993 floods so big? The weather pattern was remarkably similar for the big floods of 1927, 1973, and 1993. In each case, the preceding autumn and winter were wet, thus leaving the ground saturated as the new year began. In each of the years, the polar jet stream bent southward, forming huge troughs of low pressure that attracted cyclonic systems into the Mississippi River basin to drop their moisture. In 1993, a cold-air mass over Greenland and a high-pressure ridge over the northeastern United States resulted in large-scale bends in the polar jet stream (figure 13.28). The high-pressure ridge over the eastern United States produced record-breaking high temperatures. At the same time, the low-pressure trough over the Mississippi River basin brought moist air from the Gulf of Mexico into contact with cool, polar air masses, resulting in long-term heavy rainfall. The persistence of the polar jet stream pattern caused storm after storm to dump its water in the nation's heartland, producing high flood levels that went on week after week. Was this a new and unusual weather pattern? No, it is a reasonably common occurrence.

2011 Flood Management
April and May 2011 storm systems, including three with major tornado outbreaks (see Chapter 10), brought voluminous rains to the Mississippi River system, resulting

in record or near-record river levels in Illinois, Missouri, Kentucky, Tennessee, Arkansas, Mississippi, and Louisiana. The people in charge of the Mississippi River flood-control system faced agonizing decisions. They needed to protect the southern region, including Baton Rouge, New Orleans, and the numerous oil refineries and chemical plants along the lower Mississippi River; this meant giving orders in regions to the north to divert massive flows of Mississippi River water into floodways and out onto historic floodplains. These actions were designated in the 1930s to serve as relief valves whenever major river flood stages became too high, but now these floodways have farms and towns upon them. The goal of minimizing overall flood damage means deliberately causing harm and economic damage to many people so that even greater numbers of people may avoid harm and economic damage. Cost/benefit ratios must be evaluated.

The first suggested action was to save Cairo, Illinois, by blasting a gap in the Birds Rock levee, thus sending floodwater out through the New Madrid floodway to submerge 530 km² (130,000 acres) of Missouri farmland and to evacuate about 300 residents. The Missourians challenged the action in the U.S. Supreme Court, but lost. A 3 km (2 mi) gap was blasted through the levee on 3 May, sending out a wall of water up to 15 feet high that washed away 2011 crop prospects and damaged about 100 Missouri homes, but Cairo, Illinois, was spared.

High water continued flowing downstream in a *slow-motion disaster*. Even though the highest water levels were a week or two away, people along the lower Mississippi River began evacuating as their anxiety rose along with the water level. On 14 May, 10-ton steel floodgates were slowly raised at the Morganza Spillway, sending Mississippi River water westward into the Atchafalaya Basin floodway. Slowly rising water flowing through the Atchafalaya River to the Gulf of Mexico affected about 25,000 people, 11,000 buildings, and 600 oil and gas wells. At the same time, the concrete control structure at the Bonnet Carre Spillway was opened, thus diverting some Mississippi River water into Lake Pontchartrain, which then moved south into the Gulf of Mexico. If the Morganza and Bonne Carre Spillways had not been opened, both Baton Rouge and New Orleans would have been flooded.

This was the first time that three Mississippi River flood-control systems were open at the same time. The system as a whole performed as designed. A huge flood was managed. Overall losses were about $4 billion, but they would have been much higher without the water diversions.

Flooding in China

The Huang (Yellow) River is reputed to have killed more people than any other natural feature; perhaps that is why it became known as the "River of Sorrow." In the lower 800 km (500 mi) of its course, it flows over floodplain and coastal-plain sediments. Attempts to control river flow and protect people and property go back at least as far as the channel dredging of 2356 BCE. Levees are known to have been constructed since at least 602 BCE. Have these efforts controlled the river and protected the people? Not well

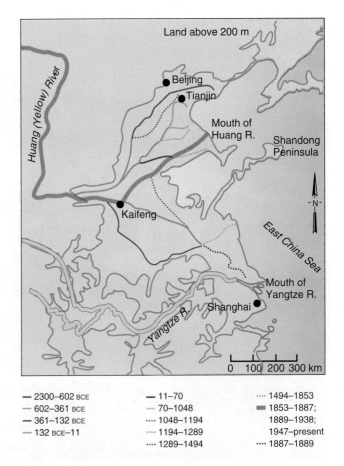

— 2300–602 BCE	— 11–70	···· 1494–1853
— 602–361 BCE	— 70–1048	▬ 1853–1887;
— 361–132 BCE	···· 1048–1194	1889–1938;
— 132 BCE–11	···· 1194–1289	1947–present
	···· 1289–1494	···· 1887–1889

Figure 13.29 The present course of the Huang River is shown in dark blue. Its previous paths are represented by other colors.
Data from Czaya, 1981.

enough. In the past 2,500 years, the Huang River has undergone 10 major channel shifts that have moved the location of its mouth as much as 1,100 km (more than 680 mi) (figures 13.29 and 13.30).

In 1887, the Huang overtopped 22 m (75 ft) high banks, "discovered" lower elevations, and began flowing south to join the Yangtze River. The 1887 floods drowned people and crops, creating a one-two punch of floods and famine that were responsible for more than 1 million deaths.

In 1938, the Yellow River levees were dynamited in the war with Japan, resulting in another million lives lost to flood and famine. Today, the riverbed is 20 m (65 ft) higher than the adjoining floodplain! The Huang River "wants" to change course, but the Chinese keep building levees to make it stay where it does not want to be. How long can the river be confined? The Chinese have tried to control the Huang for well over 4,000 years and have had successes and failures. Will Americans be able to control the Mississippi River indefinitely? In *Under Western Skies*, Donald Worster provocatively suggests:

Human domination over nature is quite simply an illusion, a passing dream by a naive species. It is an illusion that has cost us much, ensnared us in our own designs, given us a few boasts to make about our courage and genius, but all the same it is an illusion.

Figure 13.30 Delta of the Yellow River (Huang), China.

NASA image created by Jesse Allen, using Landsat data provided by the United States Geological Survey.

Societal Responses to Flood Hazards

People like to be near rivers. Rivers provide food and drink, business and transportation, arable land and irrigation, power, and an aesthetic environment. But being near rivers also means being subjected to floods. Human responses to flood hazards have been in two main categories: structural and nonstructural.

Structural responses include constructing dams to trap floodwater; building levees along rivers to contain floodwater inside a taller and larger channel; engineering projects designed to increase the water-carrying ability of a river channel via straightening, widening, deepening, and removing debris; and short-term actions such as sandbagging.

Nonstructural responses include more accurate flood forecasting through the use of satellites and high-tech equipment, zoning and land-use policies, insurance programs, evacuation planning, and education.

DAMS, RESERVOIRS, AND NATURAL STORAGE AREAS

Flood sizes may be lessened by storing some floodwater in reservoirs behind human-built dams or by diverting water from the river into natural storage areas with low topography. For example, after the 1937 flood, the building of storage reservoirs began in the Ohio River basin. There are now

88 reservoirs that can hold enough water to have a significant impact on flood levels.

All dams have life spans limited by the durability of their construction materials and style, and by the rate at which stream-delivered sediment fills in their reservoirs. Despite all the massive dams and extensive reservoirs that have been built, major floods still occur downstream due to overtopping (e.g., Rapid City, South Dakota) and to heavy rains that fall below the dam (e.g., Guadalupe River, Texas, and Big Thompson Canyon, Colorado). In addition, major killing floods have been unleashed by failed dams. In the United States alone, hundreds of dams have failed (table 13.4).

The problems with dam integrity have not all been solved. In 1981, the U.S. Army Corps of Engineers studied 8,639 dams and judged that 2,884 of them were unsafe.

Johnstown, Pennsylvania, Flood, 1889

The deadliest dam-failure flood in U.S. history occurred on 31 May 1889. At that point in the Industrial Revolution, Pennsylvania was prospering as a steel-producing state, and Johnstown was a steel-company town of 30,000 people. Unfortunately, it was located on the floodplain at the fork of the Little Conemaugh and Stony Creek Rivers. Beginning in 1838, the state built the South Fork Dam 23 km (14 mi) up the Little Conemaugh River from Johnstown. The dam site was later abandoned, reused, and abandoned again; then, in 1881, it was purchased privately. The old dam was modified, cottages were built, and the South Fork Fishing and Hunting Club opened for its wealthy owners. Everyone knew the dam

TABLE 13.4
Some U.S. Dam Failures

Date		Place	Deaths
16 May	1874	Connecticut River near Williamsburg, Massachusetts	143
31 May	1889	Little Conemaugh River, Johnstown, Pennsylvania	Over 2,200
27 Jan	1916	Otay River, San Diego	22
12 Mar	1928	St. Francis dam, north of Los Angeles	420
14 Dec	1963	Baldwin Hills, Los Angeles	5
26 Feb	1972	Buffalo Creek, West Virginia	118
9 Jun	1972	Rapid Creek, South Dakota	238
5 Jun	1976	Teton River, Idaho	14

was low quality; its frequent leaks were usually patched with mud and straw. The dam impounded a 3 km (2 mi) long lake that was up to 1.6 km (1 mi) wide and 18 m (60 ft) deep at the dam. Then, truly bad weather struck. On 30–31 May, up to 25 cm (10 in) of rain fell in 24 hours. Johnstown was flooded up to 3 m (10 ft) deep. But a bad situation became worse. At 4:07 p.m. on 31 May, the dam failed. A flood crest up to 11 m (36 ft) high swept into town carrying debris, including the stock of a barbed-wire factory. The worst of the flood was over in 10 minutes, but more than 2,200 people were dead, which amounted to about 7% of the residents and included the elimination of 99 entire families.

On 19 July 1977, another big rainstorm flooded Johnstown up to 2.4 m (8 ft) deep. No dam failed this time; nonetheless, 80 people were killed by floods in the region.

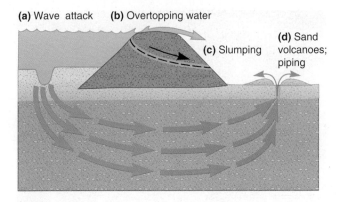

Figure 13.31 Levees are attacked by several processes: (a) wave attack; (b) erosion by overtopping; (c) slumping of the levee mass; and (d) subsurface erosion (piping).

LEVEES

Levees have opponents and proponents. Opponents suggest that the costs of building more levees and dams may be higher than the value of the buildings they are protecting. They recommend lowering or removing levees along some farmland to allow river floods to spread out and dissipate their energies over a wide expanse of land, thus also lowering water levels in levee-protected major cities and towns. In contrast, proponents of levees say that the increased carrying capacity of a river allows more years of high-water flow to occur without flooding. They state that flood damages would have been many billions of dollars more without the levees. The management of the 2011 floodwater in the Mississippi River is cited as an example of use of levees to minimize economic losses in a major flood.

Figure 13.32 Breach in Kaskaskia Island levee, Illinois, August 1993.
U.S. Army Corps of Engineers.

Levees are commonly built of soil and sediment and thus are not strong; levees can fail. As levees become water saturated, the river finds weak spots, compromising the levees by wave attack, erosion by overtopping, failing by slumping, and undermining by piping (figure 13.31). The floodwater that escapes confinement by levees spreads out, inundating farms and buildings (figure 13.32). In 1993, 1,083 out of 1,576 levees in the upper Mississippi River system were overtopped or damaged.

The debate about the value of levees will go on for generations. How many miles of levees? How high? Where should they be located? Should taxpayers provide the funds to replace flood losses for people who build in the floodplain? The distribution of some federal funds has been restricted to people willing to either move to higher elevations or raise the floor level of their homes. With or without federal funds many homeowners are elevating their own houses and/or building their own levees (figure 13.33). Some towns have voted to move in their entirety to sites above the floodplain.

A study by Robert Criss and Everett Shock shows that peak floodwater heights in the past 150 years have increased 2 to 4 m (6 to 13 ft) for the same water volume in upper Mississippi River sections with levees and engineered channels. Meanwhile, there has been no increase in flood heights per water volume on the mostly nonengineered upper Missouri River. Where human engineering has deepened and narrowed rivers, flood crests are commonly higher than in nonengineered sections. For example, the water volumes of the Mississippi River floods in St. Louis in 1903 and 1993 were almost identical, but in 1903, the river crested at 11.6 m (38 ft), while in 1993, it reached 15.1 m (50 ft).

Dams have long been designed with emergency spillways to divert floodwater that exceeds the design of the dam, thus helping prevent dam failure and catastrophe. Some levees

(a)

(b)

Figure 13.33 Homeowners taking charge. (a) The Grover House in Freeport, New York, filled with flood water 5 feet deep. Here, workers are in the process of elevating the house 12 feet above ground as hydraulic lifts raise the house onto cribbing and helical piles, 20 May 2013. (b) An arcuate mass of sandbags has been placed around a wood framework to keep the Missouri River away from this house in Jefferson, South Dakota, 12 June 2011.

(a) Kenneth Wilsey/FEMA, (b) Jeannie Mooney/FEMA.

are now being built with a similar philosophy of planned failure points. Certain levee sections are built lower, allowing extreme floods to flow out over open land or minimally developed land. This helps prevent levee failure at sections protecting cities or highly developed land.

SANDBAGGING

When a big flood is on the way, a common response is to quickly build temporary levees using hastily filled bags of sand and mud. During the 1993 Mississippi River flood, an estimated 26.5 million sandbags were filled and set in place. Some of the sandbag levees did lessen the damages, while others did not. But even where sandbag levees failed,

Figure 13.34 Volunteers build a levee of sandbags trying to keep floodwater from the Cache River out of Biscoe, Arkansas, 26 March 2008.

Jocelyn Augostino/FEMA.

a real therapeutic value was observed in the people working together for the common good (figure 13.34).

FORECASTING

Thanks to technologic advances, our growing knowledge of weather and floods allows better forecasts of the time and height of regional floodwaters. These forecasts have significantly reduced the loss of life. But it is interesting to note that the twin trends of better forecasting and engineering are offset by ever-greater dollar losses during big floods. We know more, yet suffer greater damages.

ZONING AND LAND USE

A standard approach to lessening flood losses is to ban building on the portion of the floodplain that will be covered by the 100-year flood. This policy was adopted by the National Flood Insurance Program in the early 1970s, was issued as Executive Order 11988 in 1977, and has been used by the Federal Emergency Management Agency (FEMA) since 1982. Notice in table 13.2 that the 100-year flood has a 1% chance of occurring in any year, a 9.6% chance of happening each 10 years, and a 22% chance every 25 years. Although adoption of the 100-year flood standard does not prevent all structures from being flooded, it does discourage some construction at frequently flooded sites.

INSURANCE

Flood insurance has been available to farmers and towns-people through the National Flood Insurance Program since the 1950s, but it has not been popular. For example, when the Red River of the North flooded Grand Forks, North Dakota, in 1997, only 946 out of more than 10,000 house-holds were covered by flood insurance. Four months before the flood, FEMA had spent $300,000 on a media campaign warning citizens of the ominous snow-melting conditions in 1997. The FEMA ad campaign motivated only 73 Grand Forks homeowners to buy flood insurance.

Why is flood insurance purchased by such small percentages of people? It is an expense, and some homeowners may be hoping that politicians will provide federal dollars to help disaster victims. For example, the U.S. Congress passed a bill providing $6.3 billion to aid people hit by the 1993 flood in the upper Midwest.

PRESIDENTIAL DISASTER DECLARATIONS

Under the Disaster Relief and Emergency Assistance Act, federal disaster relief is provided to states and communities if they receive a Presidential Disaster Declaration (PDD). Declarations are made at a president's discretion, but a criterion for issuing a PDD is "a finding that the disaster is of such severity and magnitude that effective response is beyond the capabilities of the state and the affected local governments." The frequency of disastrous floods, storm surges, and rains is shown by these declarations; floods provoked 64% of the PDDs in 50 years of record (table 13.5).

TABLE 13.5
Presidential Disaster Declarations, 1961–2010

Decade	Major Disaster Declaration	Flood-Related Disasters*
1961–1970	191	142
1971–1980	337	255
1981–1990	252	166
1991–2000	467	251
2001–2010	597	363
Total	1,844	1,177 (63.8%)

*Includes floods, storm surges, rain.
Source: FEMA.

Urbanization and Floods

As more people move near rivers and streams, they encounter unexpected problems. Some of the problems are due to human activities that increase both flood heights and frequencies. Various methods have been developed to analyze the effects of urbanization on floods.

HYDROGRAPHS

A **hydrograph** plots the volume of water or stream depth versus time; it records the passage of water volumes flowing downstream (figure 13.35). There is a time lag for rainwater to flow over the ground surface and reach a stream channel, but stream surface height usually rises quickly once surface runoff reaches a channel; that is, the rising limb of the hydrograph is steep. When a flood crest passes downstream, stream level does not fall as rapidly as it rose. That is because the stream was fed water by underground flow of rain that soaked into the ground and moved slowly to the stream; that is, the falling limb of the hydrograph has a gentle slope.

The flood hydrograph in figure 13.35 is typical of rural, unurbanized areas. But what happens in an urban setting? Humans cover much of the ground with houses and other buildings, pave the ground for streets and parking lots, and build storm-sewer systems to take rainwater runoff directly to streams. Covering the ground with an impervious seal prevents rainwater from soaking into the ground and causes rainwater to flow rapidly across the surface, thus reaching the stream ever more quickly.

Figure 13.36 shows flood hydrographs from similar-size rainstorms in Brays Bayou in Houston, Texas, both before and after urbanization in the drainage basin. The rainstorm of October 1949 mostly soaked below the surface and flowed slowly underground to feed the stream running through the bayou. Following urbanization, the rainstorm of June 1960

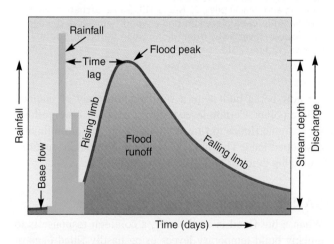

Figure 13.35 A hydrograph charts flood runoff (read on right-hand *y* axis). Commonly, stream flow rapidly increases from surface runoff, as shown by a steep rising limb reaching a peak flow. From the peak, discharge decreases slowly as infiltrated rain flows underground and feeds the stream.

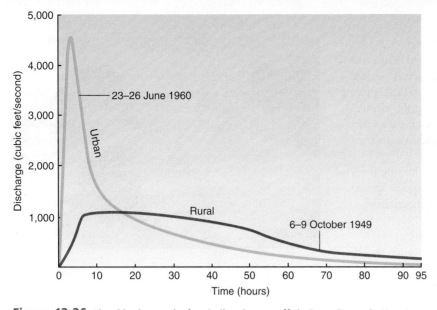

Figure 13.36 Flood hydrographs for similar-size runoffs in Brays Bayou in Houston, Texas, before and after urbanization. In the rural setting, much of the rainfall soaked into the ground and moved slowly to the bayou. Now, after urbanization, the roofs and pavement covering the land prevent infiltration, and most rain quickly runs over the surface to the bayou, creating much higher floods.

Source: Keith Young.

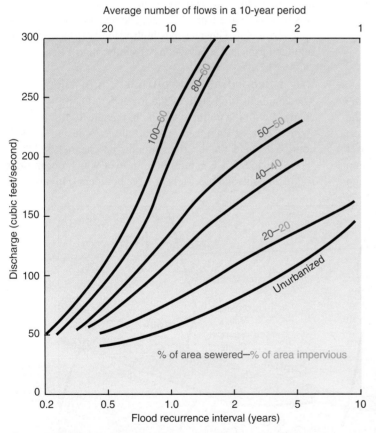
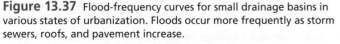

Figure 13.37 Flood-frequency curves for small drainage basins in various states of urbanization. Floods occur more frequently as storm sewers, roofs, and pavement increase.

Source: L.B. Leopold, "Hydrology for Urban Land Planning," in *U.S. Geological Survey Circular 554*, 1968.

produced a flood hydrograph with a very different shape. This is a proverbial good news–bad news situation for city dwellers. The good news is that the urban flood lasted only 20% as long; the bad news is that it was four times higher. The roofing and paving that accompany growing urbanization cause many areas to receive more frequent floods.

FLOOD FREQUENCIES

Another way of looking at flood runoff within urban areas is to see how urbanization affects the frequency of floods. Roofs and pavement increase the surface runoff of rainwater, thus causing higher stream levels in shorter times; that is, runoffs become flash floods. Figure 13.37 shows the effects of building storm sewers (percent of area sewered) and of roofing and paving (percent impervious). For example, notice on the vertical axis that a discharge of 100 ft³/sec occurred about once every four years in the rural (unurbanized) setting but now happens about three times per year after urbanization.

Even well-planned communities can experience unanticipated floods. As population grows and development moves up the river basin, the increase in the amount of ground covered by buildings, roads, and parking lots means that floods will occur more frequently. Remember that wherever urbanization is increasing, the floods of today will be smaller than the floods of tomorrow.

Strategies to Reduce Runoff

There are actions that can be taken within urbanized areas to lessen urban runoff and floods. Roofs of buildings can be built to support soil and plants that capture rainwater, which is held for plant transpiration and later evaporation. Roads and sidewalks can be built using permeable materials, thus allowing much rainwater to soak into the ground. Storm-sewer pipes that carry away rain runoff can be diverted into human-dug ponds and basins in city parks and open spaces, where the water can soak underground.

CHANNELIZATION

Humans try to control floodwaters by making channels (1) clear of debris, (2) deeper, (3) wider, and (4) straighter. A typical channelization project involves clearing the channel of trees, debris, and large boulders to reduce channel roughness and then increasing the channel capacity by digging it deeper, wider, and straighter. All these activities make it easier for water to flow through the channel. There are more than 60,000 km (37,000 mi) of channelized streams in the United States.

Think about channelization in terms of graded-stream theory. Most steps in the process push the system further into the "too much discharge" case. Straightening the channel increases the gradient of the stream bottom, making the water flow faster. Clearing the channel of obstructions reduces friction, also making water flow faster. These actions trigger a stream's negative-feedback mechanism, making it try to slow down by eroding its bottom and banks to pick up load and try to return to equilibrium by decreasing its gradient.

The Concrete Approach: Los Angeles

In response to flood damage, Los Angeles not only cleared, straightened, and deepened its river channels, but it also lined them with concrete to further reduce friction and speed up flow. As long as flood volumes stay smaller than channel capacity, there are no urban floods. But feel sorry for the person who falls into the waters charging through these channels; television news has shown people drowning in the racing water despite the frantic efforts of would-be rescuers.

It should also be noted that concrete-lined channels obliterate the habitat of all riverine plants and animals (figure 13.38). The "soul" of a community is less well served by a concrete ditch than a tree-lined stream.

The Binational Approach: Tijuana and San Diego

Rivers commonly serve as boundaries between nations. But the boundary between westernmost Mexico and the United States is drawn as a straight line, cutting through the 6,039 km^2 (2,325 mi^2) drainage basin of the Tijuana River: 80% in Baja California, 20% in California. The river's lower reaches pass through urban Tijuana, Mexico (population 2 million). Water crosses the border, enters the San Diego

Figure 13.38 The concrete-lined channel of Forester Creek, El Cajon, California.
Photo by Pat Abbott.

region (population 3 million), and runs its last few miles to the Pacific Ocean.

The two countries agreed on a Los Angeles–style project to cement the river channel to the sea. Mexico carried out its side of the agreement, but then the environmental ethic arose in California, and the cement-lined channel project was blocked in the United States. The result is the large concrete channel in Tijuana sends high-velocity floods charging into the farms and subdivisions of southernmost San Diego.

The Uncoordinated Approach: San Diego

The channelization style of the lower San Diego River has changed with the political winds. In the late 1940s, the U.S. Army Corps of Engineers was left alone to design a flood-control channel for the river mouth. Its channel is 245 m (800 ft) wide, has a natural bottom over which ocean tides roll in and out, and has walls of large boulders (**riprap**). The

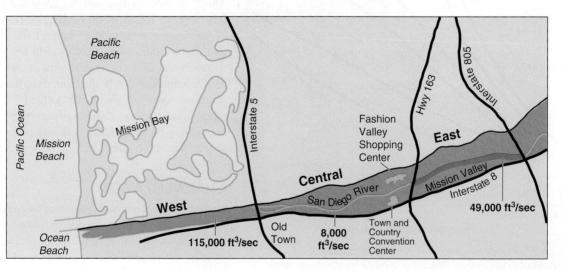

Figure 13.39 Flood channelization styles in the lower San Diego River have varied through time. The west section was built in the late 1940s; it is 245 m (800 ft) wide. The central section was left natural in the late 1950s; it is 7.5 m (25 ft) wide. The east section was built in the 1980s; it is 110 m (360 ft) wide. All channel segments feed directly into one another.

channel will handle a flow of 115,000 ft³/sec, which is estimated to recur about every 400 years (figure 13.39).

San Diego's population swelled after World War II, and the lower river valley, called Mission Valley, started to boom. Hotel builders leaned on a weak city government in the late 1950s to gain permission to build on the floodplain next to the edge of the natural channel. The river channel is about 7.5 m (25 ft) wide, is naturally vegetated, and has a capacity of about 8,000 ft³/sec (figure 13.39).

Mission Valley kept growing to become almost a second downtown. Other developers were attracted to the area. In the 1980s, the developers, interested in maximizing buildable acreage, were confronted by citizen groups that envisioned a vegetated, meandering flood-control channel that would serve as a community park. The resulting compromise favored the developers, who built a 110 m (360 ft) wide concrete-lined channel able to hold 49,000 ft³/sec.

The 20th century's biggest flood in the San Diego River discharged 72,000 ft³/sec in January 1916. Awaiting the next major flood is a concrete-lined channel of 49,000 ft³/sec capacity, which empties into a natural channel that will hold 8,000 ft³/sec, which in turn feeds into a 115,000 ft³/sec capacity channel (see figure 13.39). The planning process does have its flaws.

The Hit-and-Miss Approach: Tucson

In the northern reaches of the great Sonoran Desert lies Tucson, Arizona, where annual rainfall averages only 28.3 cm (11.14 in). September of 1983 was the second rainiest in Tucson history; it was capped off by the arrival of Tropical Storm Octave, which dropped 17.1 cm (6.71 in) of rain.

Rillito Creek cuts across Tucson, collecting much of the urban runoff and feeding the Santa Cruz River. On 1 October 1983, the Santa Cruz River at Congress Street was discharging 1,490 m³/sec (52,700 ft³/sec) of floodwater, causing major damage in the urban area. When the waters abated, 13 people were dead. Based on the entire flood history of Tucson, the Federal Emergency Management Agency had estimated the 100-year flood to be 30,000 ft³/sec. But the 1983 flood was 1.76 times bigger. Was this size flood really that rare an event? Looking at the 20th-century flood record in Tucson shows that six of the seven largest floods occurred between 1960 and 1983. This coincides with growth in population from 265,700 in 1960 to 603,300 in 1984. All the paving and roofing that comes with urbanization may be increasing flood sizes. Computations of 100-year flood size based only on the post-1960 flows yields an estimate of 2,700 m³/sec (96,000 ft³/sec); this tripling of expectations accounts for the effects of urbanization on runoff.

Desert floods are different. Most of the damage is due to bank erosion, not inundation. Stream channels are cut into loose, sandy sediments, forming weak banks that crumble easily. The critical floods here are not those with an urban-style, high-water peak of short duration (see figure 13.36), but longer-duration floods (flat-topped flood hydrograph) that have time to soak the dry streambed and

Figure 13.40 Bank erosion in Pantano Wash in Tucson, Arizona, removed 30 mobile homes and their spaces, 1 October 1983. View to northwest.

Tad Nichols, courtesy of Geophoto Publishing Company.

banks, thus freeing later floodwaters to concentrate their energies on erosion. Flood erosion changes channel location both by cutting new channel segments via overbank flow and by meander migration where the banks erode rapidly. Since the 1940s, some stream banks have eroded laterally more than 300 m (1,000 ft). One big storm and erosive flood can change stream-channel positions dramatically (figure 13.40). This means the defined 100-year floodplain moves with it. The usual static definition of a 100-year floodplain does not mean much when the stream and floodplain migrate widely.

The 1983 Tucson flood showed the problems that come with building protective walls in hit-and-miss fashion. When the stream reaches behind the end of a protective wall, damages are concentrated (figure 13.41). A stream is a system with delicately triggered feedback mechanisms; it must be treated as a whole or not at all. Hit-and-miss bank protection creates problems.

Floods are not inviting to tourists. Evaluate the truth in the message sent after the flood by the Metropolitan Tucson Convention and Visitors Bureau to the national media:

> The 100-year flood has come and gone, so, by all rights, Tucsonans should enjoy another century of great southwestern weather.

The Biggest Floods

In almost every part of the world, tales are told of ancient deluges far greater than any seen in modern times. In India, it is said that Vishnu, the god of protection, used one of his 10 lives to save Mother Earth from a great flood. In China, they celebrate Yu the Great, who helped protect the people from the overwhelming floods of the Yellow River. American

Figure 13.41 View west of the destruction of the Riverview Condominiums by bank erosion along Rillito Creek in Tucson, Arizona, on 1 October 1983. Note the concentrated erosion at the end of the "protective" retaining wall in upper left of photo.

Peter L. Kresan, courtesy of Geophoto Publishing Company.

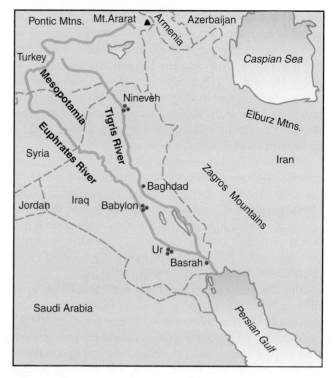

Figure 13.42 Map of the Euphrates and Tigris River plains, the site of the Babylonian (Gilgamesh Epic) and possibly the Hebrew (Noah's ark) flood tales. The lower floodplain receives long-lasting inundation floods from extended heavy rains in the mountains of Turkey and Iran.

Indian origin mythologies begin with an Earth completely flooded. In Babylonia, clay tablets record the Gilgamesh Epic telling of the great flood for which Utnapishtim built an ark that sailed for seven days and seven nights and saved his kin and cattle. The Genesis book of the Bible describes how Noah built an ark to save his family and pairs of all the animals from 40 days and 40 nights of rain that covered the world with water. These sagas recount events that occurred around 6000 BCE to 1000 BCE. Were floods larger in those times? Or were these rare events with very long recurrence intervals—the "1,000-year" inundation floods?

The flood of the Gilgamesh Epic, and possibly the flood described in Genesis, may have occurred within the Tigris and Euphrates Rivers (in modern-day Iraq), which flow across an extensive, low-lying plain to enter the Persian Gulf (figure 13.42). The setting is similar to that of the Mississippi, Huang (Yellow), and other long rivers with their inundation floods. Long-duration rains pouring onto the mountains of Iran and Turkey shed runoff, creating massive flood crests that inundate the floodplains of the lower Tigris and Euphrates Rivers. It is on these fertile floodplains that people congregate to reap agricultural rewards. A long-lasting flood would submerge the world of their existence.

ICE-DAM FAILURE FLOODS

Some of the biggest floods known on Earth occurred during the melting of the continental ice sheets. Glacial melt-water tends to pond in front of glaciers due to downwarped land (isostatic adjustment). Locally, a glacier may flow across a valley, forming an ice dam that causes an enormous lake to fill. Thousands of lakes formed along the glacial front in different locations at different times. The sudden failure of ice dams still occurs today and has been observed and photographed—for example, at ice-dammed Strandline Lake in Alaska.

When ice dams blocking the largest glacial lakes failed, stupendous floods resulted. Their passage is still recorded in lake sediments; by countryside stripped of all soil and sediment cover; by high-elevation flood gravels; by an integrated system of braided channels (a mega-braided stream); by abandoned waterfalls; by high-level erosion; and by large-scale sediment deposits. The most famous of the ice-dam failure floods are preserved in the "channeled scablands"

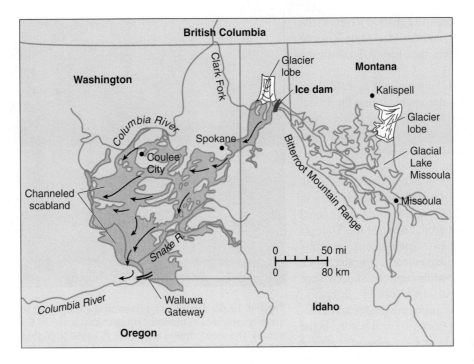

Figure 13.43 The "channeled scabland" is a landscape intensely scoured by stupendous floods that ran southwesterly across Washington. Floodwaters were unleashed when ice dams across Clark Fork in Idaho failed catastrophically and glacial Lake Missoula drained rapidly.

topography in southeastern Washington. Here, the underlying Columbia River flood basalts were swept clean of overlying sediments, while river valleys were cut into the plateau, creating a maze of gigantic channels between scoured flatlands.

Some of the biggest floods burst forth from the widespread glacial Lake Missoula, which was impounded by a glacier that flowed across Clark Fork and acted as a dam (figure 13.43). Ancient wave-cut shorelines testify to a 610 m (2,000 ft) deep lake. When the ice dam failed, a water volume of 2,500 km³ (600 mi³) was released in a flood that lasted up to 11 days. For comparison, the flood volume was more than five times greater than the volume of Lake Erie. Discharge is estimated to have been greater than 13.7 million m³/sec (484 million ft³/sec). Floods moved at velocities greater than 30 m/sec (67 mph).

The immense floods flowed southwestward across the Columbia River plateau, eroding channels and stripping the overlying sediments off an area of 7,250 km² (2,800 mi²) to create the channeled scablands topography. The scablands landscape is marked by scoured bedrock and immense former waterfalls, such as Dry Falls in Grand Coulee, Washington (figure 13.44). The massive floods carried boulders more than 10 m (33 ft) in diameter for miles. Floodwaters moved gravelly sediments in giant ripples up to 15 m (50 ft) high with distances between their crests of 150 m (500 ft) (figure 13.45).

Figure 13.44 The Dry Falls in Washington are a 125 m (400 ft) high cataract that operated when glacial Lake Missoula was rapidly emptying. The floodwaters flowed from the top of the photo toward the bottom. The high cliffs running from left to right across the top of the photo were the waterfalls. The lakes in the left center and center of the photo are giant plunge pools formed where waterfalls eroded big holes into the rock. In the background lies Coulee City.

Keith Dunbar/National Park Service.

The failure of an ice-dammed lake can send so great a volume of water running over and eroding the land that it can change the paths of rivers. In North America, meltwater floods flowed along the Mississippi River to the Gulf of

Mexico, the St. Lawrence and Hudson Rivers to the North Atlantic Ocean, and the Mackenzie River to the Arctic Ocean or via the Hudson Strait to the Labrador Sea.

The discharge of a huge volume of cold, low-salinity glacial meltwater could change the global circulation of deep water through the world ocean (see figure 9.29). The deep-ocean circulation is driven by regional differences in the heat and salinity of ocean water, and this could be changed by a huge influx of cold, fresh meltwater. A change in ocean circulation would in turn make changes in global climate. At 12,900 years ago, climate cooled about 5°C (9°F) in the event known as the Younger Dryas (see figure 12.16). Scientists attribute this dramatic plunge back into colder temperatures to a gigantic meltwater flood through the St. Lawrence River into the North Atlantic Ocean.

The outbursts from glacial meltwater lakes created the largest known floods in Earth history. In the span of a few thousand years, the massive continental ice sheets melted in gigantic quantities, raising sea level by some 130 m (425 ft) (see figure 12.15).

Figure 13.45 Floodwaters crossing the channeled scablands moved gravelly sediment in massive "ripples" that traveled toward the west (top of photo). The mega-ripples are about 15 m (50 ft) high with wavelengths of 150 m (500 ft). This view is down the modern Columbia River near Trinidad, Washington.

University of Washington Libraries, Special Collections, John Shelton Collection, Shelton 5596.

Summary

Floods usually are produced by heavy rains. An intense thundercloud may unleash a fast-moving but short-lasting flash flood in a local stream; they cause many deaths. Widespread rainfall over weeks may produce regional floods on many rivers that last for weeks; they cause extensive economic loss.

Streams are equilibrium systems. Streams convey a volume of water (discharge) and carry sediment (load) toward the ocean. To balance relative differences of discharge versus load, streams vary the slope of their bottom (gradient) and their channel pattern (e.g., meandering or braided).

If a stream has excess discharge, and thus energy, it erodes its bottom and banks, producing a meandering course that decreases its gradient and slows the water flow. If a stream valley has excess sediment, it deposits load on its bottom and straightens its valley, thus increasing its gradient and speeding the water flow. Both mechanisms are examples of negative-feedback, whereby a change in a system provokes other changes that tend to cancel the first change and restore equilibrium.

Floodplains are flattish areas used as stream floors during floods. Small floods happen frequently; large floods happen uncommonly. Flood-frequency curves are produced by plotting flood volumes versus recurrence interval (average number of years between floods of same size). A common standard for design near streams is protection from the 100-year flood. But 100-year floods are only statistical approximations; two 100-year floods can happen on the same stream in the same year.

Streams routinely adopt new courses in a process called avulsion. Humans build levees to try to hold streams in place. For example, the Mississippi River would have changed its course in the 1930s if the U.S. Army Corps of Engineers had not dredged the channel bottom and built bigger levees.

In a rural setting, a heavy rain supplies water to a stream quickly by overland runoff and slowly by underground seepage and flow. After urbanization, pavement and roofs seal off most of the ground, increasing the amount of surface runoff. The result is more high-water levels (peak floods) than a region's preurbanization history predicts.

Channels are constructed to try to control floods. Channelization involves debris clearing, deep digging, widening, and straightening. All these actions cause faster and more powerful water flow.

Some of the biggest floods known occurred thousands of years ago during the retreat of continental glaciers. Dams of ice occasionally failed catastrophically, releasing gigantic volumes of lake water that overwhelmed landscapes and probably changed deep-ocean circulation, which in turn triggered climate changes.

Terms to Remember

Questions for Review

1. Explain what is meant by negative-feedback. Is negative bad?
2. Explain what is meant by positive feedback. Is positive good?
3. Draw a map showing the channel pattern of a stream having excess discharge. Explain the feedback processes involved.
4. When a rainy season causes discharge to increase, what happens to stream gradient? To stream length? To sediment load?
5. How is stream sinuosity related to discharge?
6. When drought causes discharge to decrease significantly, what happens to stream gradient? To sediment load?
7. What stream pattern results from excess sediment load?
8. Geologically speaking, lakes are temporary. Draw a cross-section of a lake and use stream feedback processes to explain how lakes are removed.
9. Explain graded-stream theory.
10. Draw a cross-section and explain floodplains. Label where it is safe to build.
11. Will a 100-year flood occur once every 100 years? Can two 100-year floods occur in the same year?
12. Draw a cross-section through a levee and explain the processes pushing it toward failure.
13. Describe the history of levee control of the Huang (Yellow) River in China. Has it been a success?
14. Draw a flood-frequency curve and explain its use in planning.
15. Draw a hydrograph record for a two-week-long flood in a rural setting. What controls the shape of the hydrograph— that is, what is happening on the rising and falling limbs? Now draw the hydrograph resulting from the same-size rainstorm after the land has been urbanized.
16. What activities are typically involved when humans modify stream channels to provide flood "control"? Use equilibrium stream processes (graded-stream theory) to explain the changes in stream flow characteristics after channelization.
17. What human activities help cause record flood heights?
18. Some of the largest floods of the past 20,000 years resulted from failure of ice dams. Explain this story.
19. What actions could be taken to reduce rainwater runoff in cities?

Questions for Further Thought

1. Make a list of appropriate uses for a floodplain. Make a list of inappropriate uses.
2. Should the federal government (i.e., taxpayers) provide disaster-relief funds to help home and business owners rebuild their flood-ravaged properties on the floodplain?
3. For the stream nearest you, obtain the annual records of largest floods throughout its recorded history and draw a flood-frequency curve. How large a flood is likely to occur every 10, 50, and 100 years?
4. Draw a cross-section and explain why avulsion occurs.
5. Can the United States control the course of the Mississippi River indefinitely?
6. Why are some urban areas receiving more frequent floods and higher peak flows than at any time in their history?

Disaster Simulation Game

Your challenge is to provide flood protection to a European city built on a valley floodplain. The city is near the mountains and the livelihoods of the people are mostly in service industries and energy production. You must protect as many people, buildings, and livelihoods as you can.

The city you must protect has a specified population of people. You are provided with a map of the city. You are given a budget and must make choices about different housing and upgrades plus flood defenses; you must also build a hospital and a school, in addition to covering the wells

Are you ready for the challenge? Go to http://www.stopdisastersgame.org. Click on Play Game. On the next page, click on Play Game again. Select the Flood scenario. Then choose your preferred level of difficulty: Easy (small map); Medium (medium-size map); or Hard (large map).

Hurry up! Some areas will flood and then it will be too late.

CHAPTER 14

Fire

The moving finger writes; and having writ,
Moves on; nor all your piety nor wit
Shall lure it back to cancel half a line,
Nor all your tears wash out a word of it.

—OMAR KHAYYAM, C. 1100, *RUBAIYAT*

LEARNING OUTCOMES

Fire is familiar to all of us in our daily lives. Today, monster wildfires that burn down hundreds of homes are becoming more familiar and more frequent. After studying this chapter, you should:

- understand the relationship between photosynthesis and fire.
- be able to explain the fire triangle from the perspectives of starting a fire, and of extinguishing a fire.
- know the difference between a fuel-driven fire and a wind-driven fire.
- know the relationship between atmospheric high-pressure systems and the spread of fire.
- appreciate the significance of flying embers.
- be able to explain how to protect your house from wildfire.

OUTLINE

Firefighter pauses to look at wall of fire lit as a backfire by a northern California fire crew, 25 October 2007.

Andrea Booher/FEMA.

On Sunday of the 2011 Labor Day weekend, celebrations in Bastrop County, Texas, were halted by raging winds from the remnants of Tropical Storm Lee. The winds knocked trees down across powerlines, which fell on dry grasses and ignited fires in three places. Winds pushed the fires through dead grasses, junipers, and Loblolly pines, combining into one huge wildfire that burned uncontrollably for three days (figure 14.1). Not until the winds subsided could firefighters even begin to gain control. The wildfire killed two people, destroyed 1,691 homes, and burned 139 km² (34,356 acres). This was the most catastrophic wildfire in Texas history.

The Bastrop firestorm did not occur in isolation. More than 180 fires broke out in Texas that weekend. Preceding the fire was an *exceptional drought* that killed grasses and reduced moisture in trees during the hottest summer for any state in U.S. history. In Texas, the June through August high temperature average was 30.4°C (86.8°F); this beat the old record of 29.6°C (85.2°F) set in 1934 Oklahoma.

Fire

Fire is familiar; it is both friend and foe, slave and master. Fire is a natural force, yet it has been used by humans for many thousands of years. Humans developed the ability to artificially generate fire, and the control and use of fire has been a major factor in the development of the human race. The control of fire for warmth allowed humans to migrate into cold climates and build diverse and successful civilizations. Fire for cooking greatly increased the number of foods available and improved their taste, ease of digestion, nutrition, sanitation, and preservation. Fire has long been used to drive game out of hiding during hunting and to scare away predatory animals in the night. Fire aided agriculture

by clearing the land of brush and creating fertilizer with its ashy residue. With the help of fire, humans have been able to expand farmland and pasture against both climatic and vegetational gradients.

The possibilities of fire have stimulated creative thinking, which in turn has spurred human development. One invention has followed another. The use of fire to harden materials led to the creation of pottery, cookware, weapons, and more. The ability to produce ever-higher temperatures led to the discovery of smelting and the use of metals. Fire provided the benefits of sterilization, which advanced public health. Fire inside machinery supplies the energy that underlies our civilization. The heat from burning oil, coal, and natural gas is converted into the electrical and mechanical energy that powers our industries, lights and heats our homes, and enables us to travel quickly to any point in the world. As a rough estimate, the benefits of fire provide each American today with the energy equivalent of employing 100 servants in the past.

Fire has also been put to destructive use, from trying to control wild animals to trying to dominate other humans. Fire evolved from the individual torch to a method of destroying whole cities. For example, Troy was so obliterated by fire that its very site remained unknown to the world for nearly 3,000 years.

Another destructive use of fire has been to deny enemies their prize. For example, on 12 September 1812, Napoleon and the French army reached the hills outside Moscow and looked down on its green, blue, and gold domes. Upon entering the great city, the French found it to be largely deserted, with fires burning throughout. For six days, the fires raged until 90% of the city had been incinerated. The Russians chose to destroy their own heritage rather than let it aid their French conquerors. Denied the support of a conquered populace, Napoleon began his disastrous retreat from Russia during winter's harsh conditions. Napoleon won the military battles but lost the war.

New destructive uses of fire have been conceived with time and technological innovation. In the 20th century, "fire" was packaged into bombs that could be dropped from airplanes. During World War II, entire cities were ignited as thousands of tons of bombs created massive **firestorms,** killing tens of thousands of people in such cities as Hamburg and Dresden.

Fire has always been an important part of the natural world. For example, about 1,800 thunderstorms are in action around the Earth each hour, and their lightning bolts start many fires. Humans have brought fire from the natural world into the cultural world, expanding its role. We control fire for our benefit, but we pay the price when this control is lost. In the United States today, the abundant strikes of lightning start about 15% of all wildfires (figure 14.2). Humans are the main source of fire ignition in the United States (table 14.1). More of the price paid for the cultural use of fire is presented in tables 14.2 and 14.3.

Figure 14.1 The Bastrop wildfire was the most catastrophic wildfire in Texas history, 9 September 2011.
© Mike Stone/Reuters/Corbis.

Figure 14.2 Fires that burned between 1 January and 31 October 2012. Fire intensity ranges from yellow and orange (most intense) to dark red (less). Fires detected by MODIS instruments of NASA.

NASA.

TABLE 14.1
U.S. Wildland Fire Causes in 2000

		Number of Fires	Acres Burned
Natural Causes			
Lightning		18,417 **(15%)**	4,826,643
Human Causes		104,410 **(85%)**	3,595,594
Arson	(26%)		
Equipment	(10%)		
Juveniles	(4%)		
Smoking	(4%)		
Campfires	(3%)		
Railroads	(3%)		
Other & unknown	(50%)		

Source of data: National Interagency Fire Center.

TABLE 14.2
Civilian Fire Deaths and Damages in the United States (1981–2009)

Years	Fire Deaths (annual average)	Damage to Structures (in billions of dollars—annual average)
2006–09	3,251	$24.386
2001–05	4,215	18.777
1996–2000	4,138	12.576
1991–95	4,521	12.177
1986–90	5,696	9.125
1981–85	6,013	6.639

Source of data: U.S. Census Bureau, *Statistical Abstracts of the United States—1996, 2001, 2003, 2008, and 2012.*

What Is Fire?

Fire is the rapid combination of oxygen with carbon, hydrogen, and other elements of organic material in a reaction that produces flame, **heat,** and light. In effect, fire is the **photosynthesis** reaction in reverse.

In the photosynthesis reaction, plants take in water (H_2O) and carbon dioxide (CO_2) and use solar energy to build organic material, their tissue. Oxygen is given off as a by-product of the reaction. The molecules of plants are tied together by chemical bonds between atoms. These bonds store some of the Sun's heat as chemical potential energy. An equation for the photosynthesis reaction is:

$$6\ CO_2 + 6\ H_2O + \text{heat from the Sun} \rightarrow C_6H_{12}O_6 + 6\ O_2$$

The organic molecule ($C_6H_{12}O_6$) in this equation is glucose, which approximates the composition of cellulose, the main component of wood.

TABLE 14.3	
Where People Died in U.S. Fires in 1982	
Residences	82%
One- and two-family homes (66%)	
Apartments (14%)	
Hotels or motels (12.5%)	
Other (7.5%)	
Nonresidential buildings	4.3%
Highway vehicles	9.6%
Other vehicles	2%
Other	2.1%
	Total 100%

Source of data: U.S. Census Bureau, *Statistical Abstract of the United States—1982.*

In the fire reaction, plant material is heated above its point of **combustion,** and oxygen begins combining rapidly with the organic material. The old chemical bonds between carbon and hydrogen are broken; new bonds form between carbon and oxygen and between hydrogen and oxygen; and the stored energy is given off as heat during the fire. The fire equation is identical to the photosynthesis equation, except that it runs in the opposite direction:

$$C_6H_{12}O_6 + 6\ O_2 \rightarrow 6\ CO_2 + 6\ H_2O + \text{released heat}$$

In effect, the solar energy stored by plants during their growth is returned to the atmosphere during fire.

The Need for Fire

Through photosynthesis, plants grow and collectively build large volumes of trunks, branches, leaves, needles, grasses, and such. The mass of organic material is recycled by the combined effects of slow decomposition through rotting and digestion plus rapid burning through wildfire. Decomposition requires warmth and moisture to operate efficiently. In a tropical rain forest with abundant warmth and moisture, rotting can decompose the dead vegetation and recycle the nutrients for the production of new plant material via photosynthesis. There is no need for wildfire to recycle materials (except in cases such as after a hurricane has destroyed large numbers of trees). In the deserts, there is little moisture for either plant growth or decomposition, so fire does not have to be frequent.

In many environments, fire is necessary to recycle nutrients and regenerate plant communities. Fire-dependent ecosystems include grasslands, seasonal tropical forests, some temperate-climate forests, and the Mediterranean-climate shrublands.

Mediterranean climates occur in areas around the Mediterranean Sea and in parts of the Californias, Australia, Chile, and South Africa. There, the wet winters are too cold and the warm summers are too dry for rotting to recycle the products of photosynthesis. The Mediterranean climate areas sit under the downwelling limbs of Hadley cells around 30°N and S latitude, where descending air warms adiabatically and dries out these regions (see figure 9.19). In these regions, fire is necessary for the health of plant communities. Many of the plant species must have the smoke and/or heat of fire to germinate their seeds, to control parasites, and to influence insect behavior. Thus, fire in Mediterranean-type climatic zones is nature's way of cleaning house. But this built-in need for fire should serve as a warning to anyone planning to build a house in this fire-adapted plant community.

We earlier viewed plate tectonics as causing the buildup of strain energy that is released during earthquakes; by analogy, photosynthesis causes the buildup and storage of chemical potential energy in plants that is released during fires.

The Fire Triangle

A fire may begin only when **fuel,** oxygen, and heat are present in the right combination. These three critical components are referred to as the *fire triangle* (figure 14.3). Oxygen makes up 21% of the atmosphere, so as long as a steady supply of air is available, heat and fuel are the most important factors. During summers and/or droughts, heat both warms up and dries out vegetation, making it easier for a lightning strike to ignite a fire. Given the common presence of oxygen and heat, the occurrence of fires is mainly limited by the amount of fuel available.

Any combustible material is fuel. Common categories of fuel include grasses, shrubs, trees, and **slash,** the organic debris left on the ground after logging or windstorms. As the human population spreads into wildlands, houses have become a fifth category of fuel.

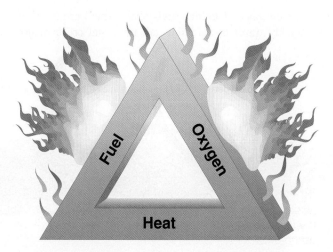

Figure 14.3 The fire triangle. When all three sides are present, a fire results; eliminate one side, and the fire cannot burn.

A Classic Disaster

The Burning of Rome, 64 CE

A famous fire of the ancient world consumed Rome, the capital of the most powerful empire of its time. The popular myth about the fire tells of a vain and bored Emperor Nero, who fiddled while Rome burned. As usual, the truth, as best as we can know it, is a bit different from the myth. On 19 July 64, fire broke out in the Circus Maximus and was swept by strong winds through the small, closely packed buildings along the narrow, winding streets of the city. Upon hearing news of the fire, Nero left Antium and returned to Rome. For six days and seven nights, the fires moved up and down the hills of Rome until only four of its 14 districts were unscathed. Gone were mansions, shrines, and temples built over the centuries; also destroyed was Nero's palace on the Palatine Hill.

Although no eyewitness accounts remain, both Tacitus and Suetonius, writing two and three generations later, describe the actions of Nero that provoked the rage of the populace. Nero is possibly the only ruler of a major empire who considered himself to be primarily a singer and stage performer. During the emotional time of the long-lasting fire, Nero donned his singer's robes, played his lyre, and sang a lengthy song of his own composition called "The Fall of Troy." Expressing his emotions musically during the tragic fire was perceived as callousness, and stories spread until Nero was even rumored to have started the fire. In the absence of contemporary manuscripts, the truth shall probably never be known.

After the fire, Nero quickly rebuilt the Circus Maximus, religious shrines, and other public facilities. He personally paid to have debris from the fire cleared and provided bonuses to building-site owners who quickly completed houses and blocks. For the reconstruction, Nero ordered measures to reduce the probability of future city-destroying fires. Rome was rebuilt with broad streets, spacious and detached houses, and heavy usage of massive stone; households were required to have firefighting apparatus in an accessible place. This story from 2,000 years ago is amazingly similar to events in the news today.

The fire triangle also is useful for visualizing how to fight a fire. Remove one of the sides of the fire triangle, and the fire collapses:

- Firefighters spray water on a fire to reduce its *heat*.
- Air tankers drop reddish-orange viscous fluids in front of a fire to coat the unburned vegetation and block *oxygen* from contacting the plant. Dry chemical extinguishers using "Purple K" (potassium bicarbonate) kill fires by disrupting the chemical reaction.
- Firefighters commonly bulldoze and remove vegetation to eliminate *fuel*. Fuel also may be reduced by setting **backfires** (figure 14.4). A backfire is lit by firefighters in front of the advancing wildfire. As the wildfire draws in oxygen, it also draws in the backfire, thus eliminating fuel in front of the advancing wildfire.

The presence of **ladder fuels** allows small ground fires to quickly travel upward into tall trees and create major wildfires. Vegetation of varying heights makes it easy for fire to climb from grasses to shrubs to trees, similar to climbing up the rungs of a ladder (figure 14.6). How can the ladder fuel threat be reduced? By cutting and pruning vegetation to create vertical separations between layers.

With the emergence of humans and our ability to both make fires and fight the flames, the old natural balance has changed. No longer does fuel supply simply accumulate until ignited by a lightning strike. Now humans extinguish many small fires, causing the buildup of tremendous volumes of dead-plant fuel that can ignite and cause fires bigger than we can handle. We humans interfere with the natural cycle of plant growth and fire, but we don't control it.

The Stages of Fire

Before a fire breaks out, a *preheating* phase occurs during which water is expelled from plants, wood, or fossil fuels by nearby flames, drought, or even a long summer day. The water in wet wood has such a high capacity to absorb heat that the wood becomes extremely difficult to ignite. To burn, wood not only must be dry, but its temperature must also be raised considerably. For example, the cellulose in wood remains stable even at temperatures of 250°C (480°F); however, by 325°C (615°F), it breaks down quickly, giving off large amounts of flammable gases.

The thermal degradation of wood involves the process of **pyrolysis.** During pyrolysis, the chemical structure of solid wood breaks apart and yields flammable hydrocarbon vapors along with water

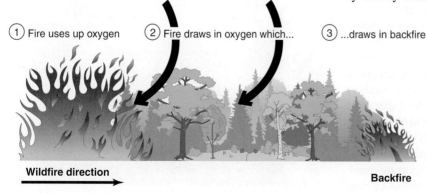

① Fire uses up oxygen   ② Fire draws in oxygen which...   ③ ...draws in backfire

Wildfire direction

Backfire

Figure 14.4 How to fight a fire with fire. Light a backfire in front of an advancing wildfire. As the wildfire draws in oxygen to continue burning, it also draws in the backfire—thus eliminating fuel in front of the wildfire.

An Ancient View of Fire

The Greeks developed influential theories concerning fire. In the fourth century BCE, Aristotle proposed a theory that was taken seriously for nearly 2,000 years. In his view, all matter on Earth was made of varying proportions of four elements: air, earth, fire, and water. He thought the behavior of matter varied according to the relative abundances of two opposing qualities: hotness versus coldness and wetness versus dryness. The four elements combined these qualities as follows:

- air—hotness and wetness
- earth—coldness and dryness
- fire—hotness and dryness
- water—coldness and wetness

The Greeks did not think these elements were identical to everyday materials, but rather that they were essences or purer forms. Their relative qualities may be plotted against temperature and humidity (figure 14.5).

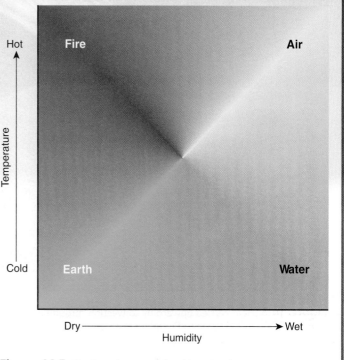

Figure 14.5 The four elements defined by Aristotle.

Figure 14.6 Ladder fuels. Vegetation of varying heights allows fire to climb from grasses to treetops, similar to moving up the rungs of a ladder.

vapor, tar, and mineral residues. If oxygen is present when the temperature is raised, the pyrolized gases can ignite and combustion begins.

In *flaming combustion,* the pyrolized surface of the wood burns fast and hot; this is the stage of greatest energy release in any fire. There are four ways heat can be transferred in a fire. The released heat is carried via **convection** through air flowing up and away from the fire; heat is transferred in **radiation** as electromagnetic waves or particle waves; in **conduction,** the heat energy is transmitted downward or inward through physical contact; and in **diffusion,** heat is in particles that move from hotter to cooler areas (figures 14.7 and 14.8).

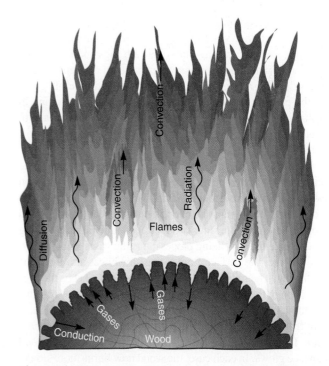

Figure 14.7 Schematic cross-section of a burning log. Heat moves *inward* by conduction, decomposing the cellulose and lignin of wood into gases (pyrolysis) that move through cracks to fuel flames at the log surface. Heat flows *outward* by (1) diffusion of particles from hotter to cooler areas, (2) radiation from flames and from hot surfaces, and (3) convection of hot, lightweight buoyant gases that rise upward.

Figure 14.8 Heat carried upward by convection through air is made visible by smoke and ash in the Hayman fire near Lake George, Colorado, 11 June 2002.
Michael Rieger/FEMA

The Spread of Fire

Wildfires occur in different styles: (1) They may move slowly along the ground, with glowing combustion playing an important role; (2) they may advance as a wall of fire along a flaming combustion front; or (3) they may race through the treetops as a *crown fire* (figure 14.10).

The spread of fire depends on four factors: fuel, the types of plants or other material burned; weather, especially the strength of winds; topography, the shape of the land; and behavior within the fire itself.

FUEL

The energy release in a fire strongly depends on the chemical composition of the plants and organic debris being burned. For example, the eucalyptus family of trees and shrubs has a high oil content that favors easy ignition and intense

The phases of a fire are well known to anyone who starts logs burning in a campfire or fireplace. The preheating phase is usually accomplished by lighting newspaper or kindling with a match to dry out the logs and raise their surface temperature. When kindling burns beneath logs and is supplied with an efficient airflow carrying abundant oxygen, the cellulose and lignin in the logs decompose and give off gases in the process of pyrolysis. As the logs heat, they expand in volume and form cracks that release gases to the surface to feed the flames. Where cracks do not form, the wood will "pop" and throw out sparks and **embers.**

Because of the poor conductance of heat in wood, the interior of a log remains below the combustion point even when the exterior is engulfed in flames. For a log to burn completely, there must be enough outside heat (from embers below, adjoining burning wood, etc.) conducting into the log to continue the pyrolizing process; it is the gases from pyrolysis that fuel the surface flames that slowly eat into the heart of the log. The radiant and convective heat that keeps the logs hot enough to continue burning is also the heat that warms us.

In nature, the flaming combustion stage involves a flaming front that passes by and leaves behind *glowing combustion.* In a fireplace, the active flames disappear, but the wood surface glows. In each case, the wood now burns more slowly and at a lower temperature as the fire consumes the solid wood instead of pyrolyzed gases. The process is a slower **oxidation** of the charred remainders left by the flames.

All the stages of a fire occur simultaneously in different areas of a wildfire (figure 14.9). The character of a wildfire and the area it covers depend on several factors, described next.

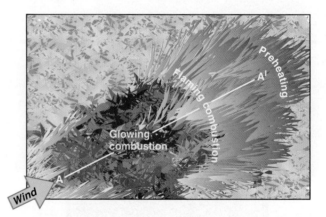

(a)

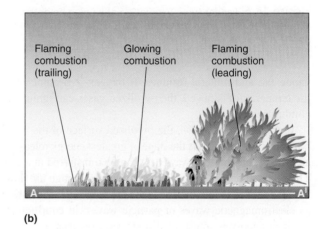

(b)

Figure 14.9 The stages of fire. (a) Horizontal or map view. (b) Vertical slice or cross-sectional view of line A–A^1 in (a).

Figure 14.10 A crown fire races through the treetops, No Name fire, Colorado, 16 June 2002.

Andrea Booher/FEMA

oxygen, distributes heat, pushes the flames forward, and bends them toward preheated plants and other fuel. If winds are absent, a vertical column of convected heat dominates, and the fire may move very slowly (figure 14.11a, type I). If winds are fast, they push the fire front rapidly ahead and prevent a vertical convection column from forming (figure 14.11f, type VI). Strong, gusty winds pick up flaming debris and burning embers called **firebrands** and drop them onto unburned areas, starting new blazes.

TOPOGRAPHY

The topography of the land has numerous effects on fire behavior. Before a fire, the topography sets up microclimates that result in different plant communities that will burn with different intensities. Wind blowing over rugged topography develops turbulence. Canyons with steep slopes and dense vegetation cause high levels of radiant heat that consume virtually all the organic matter in the canyon (figure 14.11g, type VII). Fire burns faster up a slope because the convective heat rising from and above the fire front dries out the upslope vegetation in an intense preheating phase; in effect, a chimney is created up the slope, allowing fire to move quickly (figure 14.12).

FIRE BEHAVIOR

The strength of a fire is partly created by its own actions. The vast quantities of heat given off by a fire create unstable air. Heat-expanded air is less dense and more buoyant; thus, it rises upward in billowing convection columns (figure 14.11a–e, types I–V). The rising columns of hot, unstable air may spin off *fire tornadoes* or *fire whirls* (figure 14.13). Winds sucked into the base of a fire tornado bring oxygen to feed the flames, while huge quantities of heat race up the column, venting above as in a megachimney. Fire tornadoes are commonly 10 to 50 times taller than their diameters, and they may spin at speeds of 250 km/hr (more than 150 mph). Fire tornadoes may carry fiery debris and drop it miles away, starting new fires.

A fatal example showing the strength of a fire tornado occurred on 24 April 2000 near Winkler in the province of Manitoba, Canada. A fire burning 90,000 bales of flax straw spun off a fire tornado that lifted a man and his pickup truck and carried them 50 m (165 ft) away before dumping their remains onto a field.

heat of burning. In the world before global travel and trade, eucalyptus was confined to Australia, but it has been exported in abundance to many areas where it now thrives like a native plant. The eucalyptus fire hazard is a significant threat in Southern California, North Africa, India, and the Middle East.

WIND

Many of the worst fires in history were accompanied by strong winds. The wind brings a continuous supply of fresh

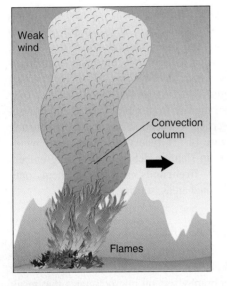

(a) Type I

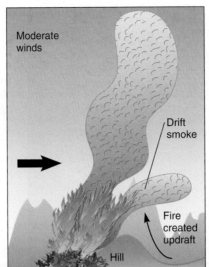

(b) Type II

Figure 14.11 Some fire types.
(a) Type I has a tall convection column and weak winds that push the column at a moderate rate of spread. (b) Type II has a tall convection column that moves rapidly upslope. (c) Type III has a powerful convection column pushed by surface winds; fire tornadoes are spun off. (d) Type IV has a convection column distorted above the ground surface by strong winds; glowing embers are dropped. (e) Type V has a convection column bent by winds of differing speeds. (f) Type VI is a wind-driven fire that spreads rapidly; the winds are too strong for a convection column to function. (g) Type VII shows the effects of topography.

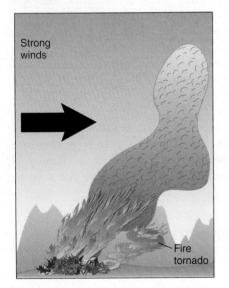

(c) Type III

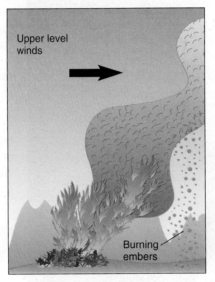

(d) Type IV

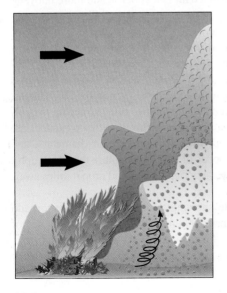

(e) Type V

(f) Type VI

(g) Type VII

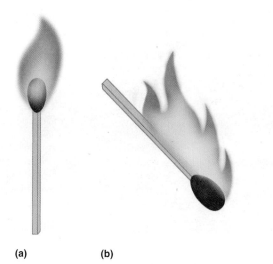

(a)   **(b)**

Figure 14.12 How fires burn on slopes. (a) An upright match burns slowly downward, similar to fire on top of a hill. (b) A sloping match burns rapidly upward, similar to fire moving up a hill.

Figure 14.14 A grassfire moves through the hills of eastern Washington in 2000.

Gary Wilson/USDA Natural Resources Conservation Service.

Figure 14.13 Fire tornadoes form as hot, rising air spins and carries flames upward. Fire tornado along the flaming front of the Jackson Canyon fire near Casper, Wyoming, 16 August 2006.

Dan Borsum, NOAA/NWS/WR/WFO/Billings Montana/Department of Commerce.

The Fuels of Fire

GRASSES

Grasses flourish on broad surfaces of low topographic relief. Grasses cover much of the prairie of the central United States and Canada. Other examples include the fountain grass of Hawaii, sawgrass of Florida, and annual grass understories in forests. Fire is an important part of grassland ecology; grasses are lightweight fuels that ignite easily. When the grass is green, the fire potential is not great. But when grass dries in late summer and early fall, and especially during droughts, then grass fires occur every day. If a grassy area has not had a fire for many years, a thick mass of old dead grass on the ground provides an additional power-laden fuel added to the younger grass above it.

Grassland fires do not get the respect they deserve. A grassy plain being burned by a 1 m (3 ft) high flame front moving 6 km/hr (4 mph) does not instill the same fear that a raging forest fire earns (figure 14.14). But if the winds speed up over the grassland and the flame front jumps to 4 m (15 ft) high moving 15 km/hr (10 mph), communities can be destroyed and people killed. In recent years, the daily news has commonly told us about grass fires, such as the one on 13 March 2006 in the Texas panhandle that burned more than 2,590 km² (1,000 mi²) and killed seven people.

SHRUBS

Shrubs are about 0.6 to 3.5 m (2 to 12 ft) high, but their loose layering allows easy burning. Shrub fires commonly move about 13 km/hr (8 mph) with flames about 15 m (50 ft) high. The most intense fires occur in shrubs with high contents of natural oils. Examples include the palmetto understory in Florida, snowberry beneath Douglas fir forests, Mediterranean-climate chaparral in California, and shrublands regenerated after a fire or grading of the land.

Florida

The 1998 Florida wildfires were set up by natural but extreme conditions. The winter of 1997–98 had unusually warm weather and heavy rainfall, resulting in luxuriant plant growth. In spring 1998, record-breaking drought reduced the water content of plants. In May 1998, clouds brought much lightning but not the usual rains. The result was more than 1,700 destructive wildfires that rank as one of Florida's worst natural disasters (figure 14.15).

Figure 14.15 Fire in the palmetto understory climbs to the tops of the trees in a Florida forest, 1 June 1998.
Liz Roll/FEMA.

Figure 14.16 Dense, impenetrable thickets of highly flammable chaparral cover the hillsides along the dry Santa Margarita River in northern San Diego County.
Photo by Pat Abbott.

California

Much of California is covered by a group of perennial evergreen shrubs known as **chaparral** (figure 14.16). The scarcity of rain makes plant growth difficult. Leaves are "expensive" to make and are protected by containing aromatic oils that discourage insects and other animals from eating them. But these same oils are flammable. Chaparral is one of the most flammable plant groups in the world, and it is a plant community whose generations are in equilibrium with fire.

The many hot and dry months each year do not allow large populations of decomposer organisms to flourish. So, the chaparral community has evolved to let fire take the place of decomposer organisms; it is fire that consumes organic debris and recycles nutrients.

Chaparral plants respond to fire in different ways. Some are "sprouters"; after fire destroys their surface vegetation, shoots spring forth from fire-resistant stems or roots, beginning the next crop of plants (figure 14.17). Some plants are "seeders"; they encase their seeds in shells that open only after the intense heat of fire destroys the outer coat, thus freeing the seed to later absorb water and germinate. Some other plant seeds respond to the smoke of a fire. For example, nitrogen dioxide (one of the gases in smoke) induces complete germination of some seeds in as short a time as 30 seconds. Thus, the very fire that burns away a dense plant cover releases and germinates seeds on ground made bare by the same fire. The brush removal by the fire also allows the seedlings to get the sunlight they need to grow.

FORESTS

Trees and forests are affected by the amount of organic litter or slash on the ground beneath them. If litter is scarce, fires pass through quickly and do little harm to trees. If litter is thick and dry, fire burns hot and slow, and kills the trees (figure 14.18). If the forest has an understory of litter and

Figure 14.17 Some chaparral plants are destroyed by fire above ground, but they resprout after a fire.
Photo by Pat Abbott.

shrubs, they may act as ladders leading the fire from the ground to the treetops, where the fire then moves as a crown fire.

HOUSES

Increasingly important sources of fuel for fires are the houses we build of combustible materials and supply with electricity and natural gas. The live energy supplied via electrical wires and pilot lights inside our houses can turn against us and destroy our homes.

When the northeastern United States was placed under hurricane watch as Sandy approached, fire was not high on the list of worries. However, high winds and storm surges caused damages that disrupted energy systems and let fire loose. For example, the storm-surge water caused something electrical, such as a socket or breaker panel, to short and ignite a fire in an unoccupied house in the Breezy Point neighborhood in the New York City borough of Queens. The fire was seen by firefighters but their access to the fire was hampered by the high levels of flood water. The fire spread and burned down 126 houses in Breezy Point (figure 14.19). The New York City Fire Department tallied 94 fires related to Sandy.

Figure 14.18 A forest fire takes off in Yukon, Canada.
© Digital Vision/PunchStock RF.

major fires commonly break out in bunches. Major fires are of great significance; more than 95% of all burned area is caused by only 2–3% of fires.

COLD-FRONT WINDS

Fire problems increase when a cold front blows in at its typical speeds of 30 to 50 km/hr (20 to 30 mph). As the cold

Fire Weather and Winds

Fire hazards are greatest in those regions with the biggest differences between their wet and dry seasons. Problems occur when a rainy season promotes rapid plant growth, which builds a large biomass, only to be followed by prolonged dry weather. A drought dehydrates plants, both living and dead, making them easier to burn. During a drought, if high winds with low humidity start blowing, small fires may spread into major blazes. And big fires rarely occur alone. A dry, windy weather pattern affects a large region, and

Figure 14.19 Houses burned down in the Breezy Point neighborhood of New York City, 29 October 2012. Fire was triggered by storm-surge damage during hurricane Sandy.
Andrea Booher/FEMA.

front approaches, wind speeds increase, and gusty wind conditions usually persist for hours after it passes. In winter, the cold fronts commonly bring rain, but in summer, the cold fronts are usually dry.

DOWNSLOPE WINDS

Fierce winds occur when a high-pressure air mass flows downslope from a mountain range or high plateau and descends as warm, dry wind toward a low-pressure zone. In the western United States, downslope winds commonly occur in September through April, when a strong, high-pressure air system stagnates over the high-elevation Great Basin and Rocky Mountains (figure 14.20). Winds blow when a low-pressure zone sits on the other side of the mountains and a pronounced pressure gradient exists between the atmospheric air masses. This is dry air flowing down from high elevations; it warms adiabatically as it descends, and it flows fast, commonly at 65 to 100 km/hr (40 to 60 mph) with intervals up to 160 km/hr (100 mph).

If a low-pressure zone sits east of the Rocky Mountains, the downslope winds flow east and are called Chinooks (derived from an American Indian word meaning "snow eater"). Chinook winds can raise surface temperatures by 20°C (36°F) and drop humidities to 5% within a few minutes (figure 14.20). Along the Pacific Coast, the downslope winds are known by several names, including Diablo (Spanish for "devil") in the San Francisco Bay region and Santa Ana in Southern California. A strong airflow from a high-pressure area over the Great Basin may push to the southwest along the ground surface and extend over the Pacific Ocean (figure 14.21). The strong airflow pushes moisture-laden Pacific air away and dries out the vegetation. The

rough topography in California creates turbulence in the airflow, spawning eddies of variable sizes. Hot, dry winds racing over dehydrated California vegetation produce some of the worst fire-weather situations in the world.

LOCAL WINDS

Several local wind patterns are factors in fire behavior: sea breezes, land breezes, slope winds, and valley winds.

Sea breezes are a daily condition in which warm air over the land rises, and cooler air over the ocean flows in to replace the risen air. Sea breezes usually begin in the mid-morning to early afternoon and then cease after sunset. Their onshore wind speeds of 16 to 48 km/hr (10 to 30 mph) reach a maximum during the hottest part of the day (figure 14.22a).

Land breezes are the opposite of sea breezes. After sunset, the land cools more rapidly than ocean and lake surfaces. During the night, the cooler air over the land flows seaward at typical velocities of 5 to 16 km/hr (3 to 10 mph) until shortly after sunrise (figure 14.22b).

Slope winds flow upslope and downslope during an entire day. As the ground heats, the warmed and buoyant air flows upslope with some turbulence. As the ground cools at night, the cooled nighttime air flows downslope but slower and more smoothly than the upslope winds of the daytime.

Valley winds are similar to slope winds. During the day, the warmer air in canyons and valleys flows up-valley. During the night, the cooler air flows down to valley bottoms and then flows down the valley axis.

Regional and local wind behaviors strongly influence the behavior of fires, as is shown in examples from the Great Lakes region and California.

WIND AND FIRE IN THE GREAT LAKES REGION

In the late 1800s, the Great Lakes region was heavily forested, dominantly by pine trees. The settlers in the region logged the forests for timber and clear-cut the native vegetation to expose ground for farming. The widespread removal of forests left abundant woody debris (slash) on the ground, and farmers routinely used fire to clear land for grazing and plowing. Every day, numerous small fires burned. Many individual fires burned for weeks at a time. In the late 1800s, the fire practices of rural America were similar to the slash-and-burn clearing of forests occurring in Brazil today. Nearly half of the U.S. forest fires in 1880 were started by people clearing land (table 14.4). Compare the causes of fire in 1880 (table 14.4) to those in 2000 (see table 14.1).

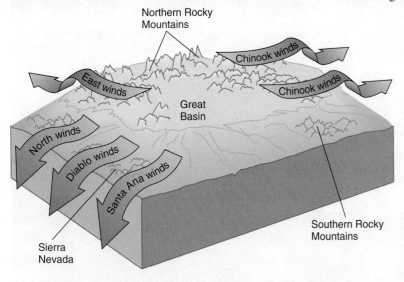

Figure 14.20 Downslope winds have local names. In the western United States, they form when high-pressure air over the Great Basin flows down from the mountains as warm, dry, high-speed winds.

Source: M. J. Schroeder and C. C. Buck, "Fire Weather," 1970. *U.S. Department of Agriculture Handbook 360.*

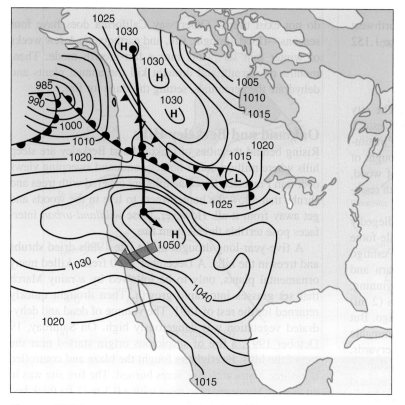

Figure 14.21 A high-pressure center over the Great Basin creates a marked gradient in air pressure to the southwest, producing strong Santa Ana winds in Southern California.

Source: M. J. Schroeder and C. C. Buck, "Fire Weather," 1970. *U.S. Department of Agriculture Handbook 360.*

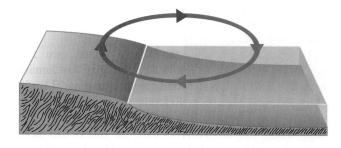

(a) Day

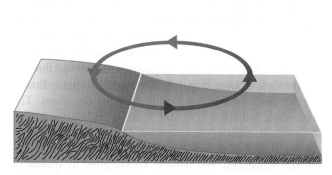

(b) Night

Figure 14.22 Sea and land breezes. (a) During the day, land heats up faster than water. Warm air rises above the land and is replaced by cool breezes off the water. (b) At night, the land cools faster than the water. Warm air rises above the water and is replaced by cool air flowing off the land.

The summer and early autumn of 1871 was a time of drought and excessive dryness. Persistent high-pressure zones over the region blocked moisture-bearing winds from the Gulf of Mexico. The long spell of dry weather sucked the moisture out of logging slash and reduced the water content in forest trees. On the evening of 8 October 1871, gale-force winds blew in from the southwest, and the region's abundant small fires blew out of control from the Ohio Valley to the Great Lakes states and the High Plains. Farms burned in Indiana; towns were incinerated in Michigan, Wisconsin, North Dakota, and South Dakota; and the great city of Chicago was gutted by flames.

Peshtigo, Wisconsin

Small fires were burning most days in late 19th-century Wisconsin. When strong winds arose, they could whip small fires into large blazes, including the deadliest forest fire in U.S. history. Around 9 p.m. on 8 October 1871, the townspeople of Peshtigo heard an approaching roar and saw the southwestern sky glowing red with flames and smoke. The fire did not advance slowly, but raced forward through the dried-out treetops as a crown fire. Tornadoes of fire in the treetops dropped enveloping flames to the ground below. The fire front was 24 km (15 mi) wide and moved at high speed; this did not leave many options for people living in wooden houses and lacking fast transportation to escape. The wide fire

TABLE 14.4	
Causes of U.S. Forest Fires in 1880	
Total Fires = 2,983	
Clearing of land	45.2%
Hunters	21.1%
Locomotives	17.0%
Malice	8.8%
Campfires	2.4%
Indians	1.9%
Smokers	1.2%
Lightning	1.1%
Others at less than 1% each include prospectors, coal pits, woodcutters, carelessness, travelers, spontaneous combustion	

Source of data: Sargent, C. S. *Forests of North America,* U.S. Government Printing Office, 1884.

Southern California

Southern California has a long dry season, chaparral vegetation, and Santa Ana winds—in other words, Southern California is a land that was born to burn. When conditions are ripe for Santa Ana winds, the area all too commonly is ablaze with numerous fires (figure 14.24). In 1993, fires in three counties burned clear to the Pacific Ocean despite the best efforts of firefighters; the ocean was the only firebreak that could stop the advance of the flames. Surfers rode waves in front of bushes burning on the beach.

In California, the destruction of hundreds of homes by wildfire is a frequent event. The Witch Creek fire in October 2007 was one of those fires with staggering statistics: 197,900 acres burned; 1,125 homes destroyed; 239 vehicles burned; 45 firefighters injured; and 2 people dead. But after this catastrophe, one area was studied in unprecedented detail. The study focused on The Trails, a suburban development of 274 houses on a hill in Rancho Bernardo surrounded on three sides by wildland canyons. More than 11,000 pictures were used to document the minute-by-minute advance of the fire front and a house-by-house record of fire initiation and burning. This study provides new insights into the fire threat to anyone's home (table 14.5)

> ### TABLE 14.5
> **Lessons from The Trails in the 2007 Wildfires**
>
> - 274 houses existed; 74 were destroyed and 16 damaged
> - 36 destroyed houses were in the *interior* of the development (figure 14.26)
> - Flying embers started 3 house fires in the interior when the fire front was 9 km (6.5 mi) away
> - Houses were on fire when the fire front was 80 minutes away
> - Flying embers may have started *all* house fires; 55 of the 74 destroyed houses were documented to have ignited from flying embers
> - Some houses took more than 2 hours to burn and were sending off flying embers during that time
> - Streets where many houses had pine trees with pine needle litter on the ground suffered higher percentages of destruction

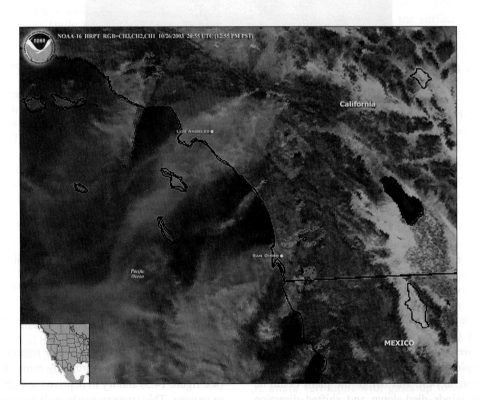

Figure 14.24 Santa Ana winds push fires rapidly westward from Ventura, California, in the north (top left) to Ensenada, Baja California, in the south (bottom right), on 26 October 2003. See the masses of smoke blown over the ocean by these winds. The fires burned 800,000 acres, killing 20 people and destroying 4,500 buildings.

NOAA.

Figure 14.25 Flying embers near The Trails, October 2007.
Courtesy of Captain Steve La Corte.

Figure 14.26 Burned houses in Rancho Bernardo, October 2007. Look in the top center to see two isolated houses burned to the ground thanks to flying embers.
Andrea Booher/FEMA.

and points out the significance of flying embers (figures 14.25 and 14.26).

Flying embers are an underappreciated aspect of fires. We fear the moving wall of flame, but sometimes the flying embers are the greater threat. When a wall of flame enters an area of houses, it is fragmented by streets, driveways, lawns, and such, allowing firefighters to successfully battle the fires. Even after a fire front has moved through a neighborhood with the firefighters following it, flying embers have been documented to start a second round of fires in the neighborhood left behind. Remember that a wildfire is a series of ignitions.

A study by Oren Patashnik of the Scripps Ranch community after the 2003 Cedar fire in San Diego County showed that flying embers were a big problem. It is not surprising that houses with wood-shingle roofs were set afire by flying embers. But it is surprising that many houses with fire-resistant, curved tile roofs caught on fire; flying embers blew through gaps below the tiles and ignited material beneath, such as tarpaper. Flying embers also started fires in houses by passing through mesh screens into attics, by igniting wooden planter boxes, and by entering houses through louvered vents. Oren's mantra is:

- Defensible space
- Remove combustibles
- Remember the embers

The fire peril is evident enough that a few people design their houses to withstand fires. The choice is yours—either pay higher construction costs to fireproof your home, or pay the entire cost to rebuild your home—and your life—after a fire.

Home Design and Fire

Not all of the death and destruction from fires can be called a "natural disaster." Poor decisions on landscaping, home design, and construction materials may be partly to blame (figure 14.27). Many homes are made of wood or roofed with wood shingles. Wooden decks extend out over steep slopes and help fire to concentrate heat, igniting the houses. Natural and planted vegetation commonly continue from the wildland right up to houses or drape over the roofs. All of these flammable materials act to convey fire into and through houses.

Your house can catch on fire in several different ways: (1) Flames can travel to your house by burning through vegetation or along a wood fence. (2) Flames do not have to reach your house; they can generate enough radiant heat to ignite the exterior of your house or even the curtains hanging inside the windows. (3) Firebrands and embers carried by wind can be dropped on or next to your house. Compare the combustion potential of the two houses pictured in figure 14.28.

Stand outside, look at your house and ask yourself: what could catch on fire? Then eliminate the hazard. All it takes to ignite a house is one vulnerable point where combustion can occur. It is possible to build a house and landscape the property so that a fire will pass by (figure 14.29).

The houses that flames pass by commonly have clay- or concrete-tile roofs, stucco or brick exterior walls, double-paned windows, few overhanging roofs or decks, and firebreaks of cleared vegetation extending at least 10 m (more than 30 ft) from the house. Houses built on slopes need even larger areas of cleared space (figure 14.30).

Figure 14.27 Twelve mistakes, or how to sacrifice your house to the fire gods. (1) House is located on a slope. (2) House is made of wood. (3) Wooden deck extends over the slope. (4) Firewood is stored next to the house. (5) Roof is made of flammable wood shingles. (6) Tree limbs hang over roof. (7) Shrubs continue up to house. (8) Large, single-pane windows face the slope. (9) Unprotected louvers face the slope. (10) No spark arrester is on top of chimney. (11) Narrow road or driveway prevents access of fire trucks. (12) Wooden eaves extend beyond walls.

Source: *The Oakland/Berkeley Hills Fire.* Modified from the National Fire Protection Association, Quincy, Mass.

(a)

(b)

Figure 14.28 (a) Bad house—easy to burn. Note the wood exterior walls, wood eaves, wood shingle roof, overhanging eucalyptus tree, and tree litter on the roof. (b) Good house—difficult to burn. Note the stucco walls, boxed eaves of stucco, tile roof, no trees or shrubs next to the house, and no litter.

Photos by Pat Abbott.

Side Note

The Winds of Madness

The Santa Ana winds of Southern California not only push firestorms, they also affect people's moods and behaviors. In his fictional story *Red Winds,* Raymond Chandler wrote:

> There was a desert wind blowing that night. It was one of those hot, dry Santa Anas that come down through the mountain passes and curl your hair, and make your nerves jump and your skin itch. On nights like that, every booze party ends in a fight, meek little wives feel the edge of the

carving knife and study their husbands' necks. Anything can happen.

What is it about Santa Ana winds that affect some people? Apparently it is the high-speed winds, extra-low humidity, and electrically charged air carrying seven to nine times the normal level of positive ions. During "the winds of madness," records of public agencies show increases in domestic violence and household mishaps. Allergic reactions and migraine headaches are common complaints, and the suicide rate is said to double.

Figure 14.29 The fire-safe house. If you use fire-resistant materials to build your house and then landscape it without providing fuel for flames, a wildfire can pass by without harm.

Courtesy CAL FIRE.

HOW WELL HAVE WE LEARNED?

The Oakland/Berkeley Hills firestorm of 1991 was not the first to occur there. In 1923, a firestorm destroyed 584 houses in the same area. After the 1923 fire, a committee was formed to identify the factors that led to building loss. Six factors were listed in order of significance (table 14.6).

In 2003, a task force was formed to identify the factors that led to building losses and deaths during the frequent wildfires in San Diego County. Again, six factors were identified (table 14.6). Compare the lists and evaluate how much progress was made in the 80 years from 1923 to 2003.

If you plan to build a house in the woods or along the urban-wildland interface, your decisions about construction materials and landscaping may well determine whether your house will end up as fuel for flames or not (table 14.7). Even

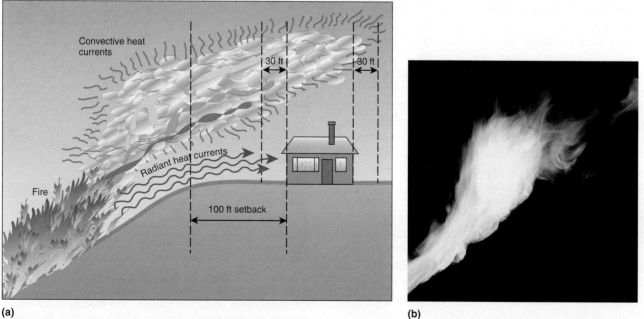

(a)   **(b)**

Figure 14.30 (a) Buildings at the top of a slope require setbacks to avoid the increased heat flow by convection and radiation. The defensive space includes a vertical zone around the building that has no tall trees. (b) A wildfire racing upslope.

(b) © Don Farrall/Getty Images RF.

TABLE 14.6	
Factors That Led to Building Loss	
1923 Oakland/Berkeley Hills	**2003 San Diego County**
1. Flammable roofs	1. Flammable roofs
2. Flammable vegetation near buildings	2. Flammable vegetation near buildings
3. Strong winds	3. Strong winds
4. Narrow, winding roads	4. Narrow, winding roads
5. Inadequate water supplies	5. Inadequate water supplies
6. Lack of modern firefighting equipment	6. Improper design of buildings

little decisions about landscaping can determine whether your house burns or endures (figure 14.31).

Fire Suppression

In August 1910, more than 3 million acres of forest in the Bitterroot Mountains of Idaho and Montana burned in a firestorm known as the Big Blowup. The fire destroyed towns, killed 85 people, and provoked the U.S. Congress into authorizing expenditure of federal money to fight forest fires. And so began the policy of aggressively suppressing forest fires with well-trained armies of professional firefighters.

During the 20th century, people were taught to hate fire and to stamp it out quickly. Fire-suppression tactics and equipment improved during the century, resulting in dramatic reductions in the number of acres burned (table 14.8). Over the decades, forests were transformed by fire-suppression practices.

Under natural conditions, a forest has some large trees with lots of grassy open space between them, as in the Bitterroot National Forest in 1909 (figure 14.32). Forests once held about 30 big trees per acre, and natural ground fires moved quickly through the grasses and thin litter on the forest floor without harming most of the big trees. As fire-suppression efforts increased in number and success, more young trees got the opportunity to grow, which had occurred by the year 1948 (figure 14.32b). With continued suppression of fires, abundant young trees had grown into dense forests by the year 1989 (figure 14.32c). Some of these forests now support 300 to 3,000 trees per acre plus an understory of shrubs. When lightning ignites fires in these dense forests, the flames burn slow and hot, killing the trees. These forest fires sometimes are beyond human ability to extinguish. Our good intentions have backfired. We have created an unnatural environment capable of causing catastrophes.

Our life experience misleads us. What we hike through and see as forest today is not natural. Our hatred of fire coupled with our ability to put out small blazes has created an unnaturally dense forest. The widespread recognition of the fire problem for dense and crowded forests has led to a dilemma. Should the dense forests be thinned? Should natural fires be allowed to burn? Affirmative answers to these questions resulted in the following fires.

TABLE 14.7
How to Protect Your House from Wildfire: A Must-Do List

Roof. The roof is the most vulnerable part of your house because of wind-blown burning embers. You must build or reroof with lightweight materials that will not burn. Remember to remove branches hanging over your roof and sweep off accumulated leaves, needles, and other plant debris. Place screens over all vents.

House. Build or remodel the exterior with fire-resistant materials. It is foolish to build a wood house among abundant trees and shrubs that can burn.

Decks. Decks, balconies, and eaves should be made of or covered by fire-resistant materials, including their undersides.

Windows. Use only double- or triple-paned windows, to reduce the potential of breakage during a fire. Eliminate wood shutters. Beware of the new-style plastic window frames; some have melted and allowed fire into homes.

Flammables. Be sure to place natural-gas tanks and firewood piles at least 10 m (33 ft) from your house.

Trees. Hold back fire from your house by (1) reducing the number of trees in densely wooded areas and (2) cutting off branches within 2 m (6 ft) of the ground so grass fires cannot climb up to the treetops.

Yard. Create a defensible space by (1) replacing fire-loving plants with fire-resistant plants and (2) removing all dry grass, dead brush, and leaves at least 10 m (33 ft) from your home, or 30 m (100 ft) if you are on a slope (figure 14.30). Replace wooden fences with concrete block or metal fences.

Community planning. Place golf courses, grass ball fields, and open-space parks between houses and the wildland so they can serve as firebreaks or buffers.

Figure 14.31 Right versus wrong is illustrated above a busy street in San Diego. The unaware homeowner on the left has dead grasses on a slope leading up to dead trees and an overhanging wood patio. The aware homeowner on the right has planted a nonburning succulent ice plant on the slope leading up to a concrete-block wall.

Photo by Pat Abbott.

(a) 1909 **(b) 1948** **(c) 1989**

Figure 14.32 Photos taken from the same place in the Bitterroot National Forest, Montana, in 1909, 1948, and 1989. (a) In 1909, the natural forest had few large trees and lots of grassy area that burned fast without harming big trees. (b) By 1948, fire suppression had allowed the growth of many more trees and shrubs. (c) By 1989, continued fire suppression had created a dense forest filled with trees and shrubs that will burn slow and hot.

USDA.

TABLE 14.8
U.S. Wildland Fires by Decade

Decade	Number of Acres Burned (annual average)
1920–29	26,004,567
1930–39	39,143,195
1940–49	22,919,898
1950–59	9,415,796
1960–69	4,571,255
1970–79	3,194,421
1980–89	4,236,229
1990–99	3,647,597
2000–09	6,894,478

Source of data: National Interagency Fire Center.

YELLOWSTONE NATIONAL PARK WILDFIRE

Yellowstone is the oldest national park in the United States; it was authorized on 1 March 1872 under President Ulysses S. Grant. The park averages about 15 fires per year started by lightning. This is not an unusual number of fires, considering that Earth as a whole is estimated to have about 8 million lightning discharges per day (see figure 10.36).

Not all lightning strokes have the same fire-igniting potential. The most capable are "hot strokes," which are cloud-to-ground discharges of high amperage and longer duration. About 1 in 25 lightning strokes matches these characteristics. Whether a fire is ignited depends on what the lightning hits. A hot stroke may start a fire if it strikes kindling, such as dry grass, rotten wood, or "organic dust" mechanically blasted into the air by the force of the lightning strike. Lightning has started fires in settings ranging from the tundra of Alaska, the forests of North America, and

the marshes of the Great Lakes to the swamplands of the South, the chaparral of California, and the grassy deserts of Arizona.

The policy of Yellowstone Park from the 1880s to the 1970s was to extinguish all fires as soon as possible. But fire is a natural process. So the question arose, "Who can better manage these wildlands—humans or nature?" Nature became the answer in the 1970s, so the policy was changed to one of putting out human-caused fires but letting natural fires alone to run their course. Lightning-caused fires burn the forest irregularly; they cause formation of a mosaic of meadows, burned-over ground, and forests ranging from young to mature to old growth. The different areas provide a diversity of environments, supporting various communities of plants and animals.

Following the policy change, between 1976 and 1987 there were 235 lightning fires in Yellowstone Park. A typical fire burned about 100 acres, only eight fires burned more than 1,000 acres, and the largest burned 7,400 acres. The policy change was judged to be a great success. Fires were occurring in moderate amounts and were promoting ecological diversity. Then came 1988.

The winter of 1987–88 was a dry one. Lightning started fires as usual in June 1988, but this year, no rains followed as in the previous six June–July intervals. By late July, more than 17,000 acres had burned, and officials decided to extinguish the blazes and suppress any new fires regardless of their origin. But the situation had added complexities: (1) Many forest stands had been killed by infestations of mountain pine beetles, (2) nine decades of fire suppression had created an extensive buildup of dead wood on the ground, and (3) the moisture levels in dead wood had dropped from the usual 15–20% down to 2–7%.

As firefighters struggled to control the situation, the weather worsened; a wave of high temperatures arrived with sustained high winds. In the 24 hours of 20 August 1988, more square miles of Yellowstone Park burned than the total area burned in any preceding decade.

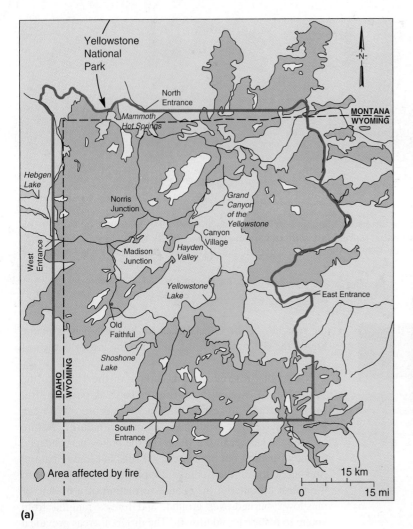

(b)

Figure 14.33 Areas burned by the Yellowstone fires of 1988. (a) Burned areas are orange on map; (b) photo of Yellowstone National Park in late 1988. Burned areas appear maroon; whitish areas near top are smoke.

(b) © InterNetwork Media/Getty Images RF.

Despite the continuing efforts of firefighters, the blazes did not weaken until the arrival of mid-September snows, and the flames did not quit until winter conditions prevailed in November. The weather controlled the flames that humans could not. At the fire's conclusion, 1.4 million acres lay burned, including almost half of Yellowstone Park (figure 14.33). In the previous 116 years, a total of only 146,000 acres had burned.

Ten Years Later

The fires of 1988 drastically changed the look of Yellowstone National Park. Thousands of acres of charred trees held the bodies of 269 elk, nine bison, six black bears, four deer, and two moose. The fires killed trees but opened land to increased sunlight and nutrients that brought forth grasses, wildflowers, and shrubs. Trees are now recolonizing the burnt ground with seedlings of lodgepole pine, Engelmann spruce, subalpine fir, and Douglas fir. Standing dead trees continue to fall and thus enrich the soil as fungi, bacteria, beetles, ants, and other organisms decompose the tree remains. The pre-1988 Yellowstone is gone, but the living and functioning ecosystem is going through many interesting changes, and time will allow a return to the conditions of earlier centuries.

CALIFORNIA VERSUS BAJA CALIFORNIA: PAY NOW OR PAY LATER

Many people hold the view that humanity is separate from the environment, that humans are supposed to "be fruitful, multiply, and subdue the Earth." Many seek to control nature for their own benefit. This includes building houses wherever it strikes their fancy. Fires? No problem, we will just extinguish them before they cause any damage. But what is the long-term effect of short-term suppression of fires? An interesting study by fire ecologist Richard Minnich worked on this question using Landsat imagery ("photos" from satellites) to determine the number and size of all chaparral fires in Southern California and neighboring northern Baja California from 1972 to 1980. The life cycle of the chaparral plant community takes it through a sequence of fire susceptibility. Younger plants do not burn easily, but after 40 years of growth, an increased proportion of dead-plant material acts as fuel, helping fire burn readily and intensely.

In the United States, wildfires are fought energetically and expensively. The goal is to not let fire interfere with human activities, no matter how much money it costs to suppress it.

TABLE 14.9
Chaparral Areas Burned, 1972–1980

	Total Area (thousands of hectares)	Area Burned (thousands of hectares)	% Area Burned	% Burned after September 1	Number of Fires
Southern California	2,019	166	8.2	72	203
Baja California	1,202	95	7.9	20	488

Source of data: R. A. Minnich (1983).

In Mexico, wildfires are simply allowed to burn with little or no human interference. The fire histories of the chaparral areas in the United States versus those in Baja California, the northwestern-most state of Mexico, show interesting differences (table 14.9).

In the United States, fires that break out during the cooler, wetter months are quickly extinguished. Thus, most of the chaparral is allowed to grow older and more flammable. Then, when the hot, dry Santa Ana winds come blowing (commonly in September, October, and November), firestorms are unleashed that firefighters are powerless to stop, and great numbers of acres burn. Southern California has fewer fires than Baja California, but more large ones (figure 14.34).

In Baja California, the wildfires are smaller because older chaparral is commonly surrounded by younger, less-flammable plants. This distribution creates an age mosaic that mixes volatile, older patches with younger, tougher-to-burn growths. Fires are more numerous in Baja California, but they are smaller, and more of them occur during the cooler, wetter months (figure 14.34).

For the 1972–80 period, the percentage of chaparral acreage burned in California and Baja California was about the same. This is despite the enormous expenses and valiant efforts made in the United States to suppress fire. The U.S. fire-control efforts seem to have reduced the number of fires but not the amount of acreage burned. In Southern California, the monster firestorms pushed by Santa Ana winds burn tremendous numbers of acres, killing people and destroying thousands of buildings.

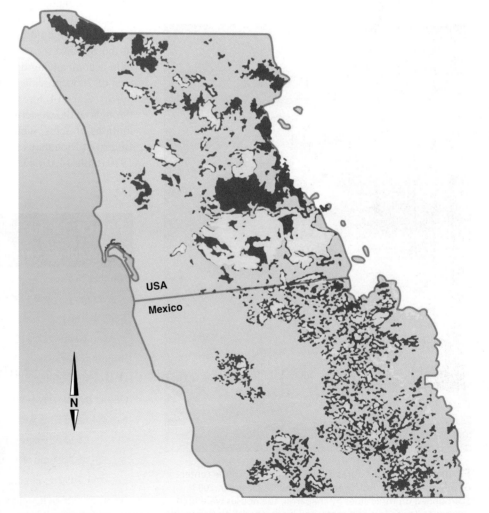

Figure 14.34 Fire burn areas in California and Baja California between 1938 and 1972. Some California wildfires burned huge areas, while the burn areas in Baja California are larger in number but smaller in size. Red and yellow colors distinguish burn areas of different fires.

Figure from Richard A. Minnich, University of California–Riverside.

The Cedar Fire in San Diego County, October 2003: Value of young fuel patches

Fire potential in San Diego County was grim in the fall of 2003. Huge areas of chaparral had grown old and increased their volume of fuel; 48% of the chaparral was more than 50 years old, and another 31% was more than 20 years old. Plant flammability had been further increased by five years of drought and the low moisture conditions following a long, hot summer. When stressed by lack of water, the branches of many chaparral plants die, strips of bark fall, and leaves drop. Half of a mature plant's tissue can be dead, oil-rich, and ready to burn. The buildup of fuel is made even more critical by the thousands of homes located in isolated spots within the sea of chaparral; houses built of fuel and placed within a sea of fuel set up an impending tragedy (figure 14.35).

And then it started. A lost and confused hunter lit a signal fire to guide rescuers to his location. Santa Ana winds blew his flames; the resulting fire and flying embers burned 282,000 acres, destroyed 2,232 buildings, and killed 15 people.

Figure 14.35 The houses on the promontory in the center of this photo were destroyed during the Cedar fire in 2003. Surrounded on three sides by chaparral, the houses were built using wood exteriors and roofs, and landscaped with flammable plants.

USDA.

But it could have been worse if not for some young fuel patches. For example, the fire front moving southwest ran into a young fuel area, the 10,353 acres burned by the Viejas fire in January 2001. These plants, less than three years old, caused the fire to stop in some areas, and led to reduced flame height and heat in other areas, allowing the firefighters to keep the fire from moving into the city of Chula Vista. Also, after the winds reversed, the fire front moving east encountered the Tragedy prescribed burn areas of 2001, 2002, and 2003. The resulting young fuel mosaics there stopped the flames from entering the community of Pine Valley. It is estimated that the Cedar fire would have burned an additional 120,000 acres if it had not run into these recently burned areas with their young, low-fuel chaparral plants.

THE WESTERN AND SOUTHERN UNITED STATES IN 2000

The year 2000 fire season began early in Florida with a big wildfire on January 1. When summer arrived, the conditions for fire in the West and South were ominous: hot weather, low relative humidities, wind, absence of monsoon rains in the Southwest, and a potent source of ignition in lightning accompanying dry thunderstorms. The long, dry summer of 2000 provided a reminder of the problems created by 90 years of forest-fire suppression. Lightning strikes throughout the western states ignited major fires that burned millions of acres, including many of the same acres in the Bitterroot Mountains that had burned in the Big Blowup of 1910 (table 14.10). Despite the valiant efforts of firefighters,

TABLE 14.10	
U.S. Wildland Fires in 2000	
Number of fires	122,827
Average per year (1990–1999)	106,393
Acres burned	8,422,237
Average per year (1990–1999)	3,786,411
Buildings burned	861
Cost of firefighting	$1.36 billion
On the busiest day—29 August 2000	
84 large fires were burning on	
1,642,579 acres in 16 states	
28,462 people were fighting fire using	
1,249 fire engines	
226 helicopters	
42 air tankers	

Source of data: National Interagency Fire Center.

the fires were not extinguished until weather fronts brought welcome rains. The debate about the proper role for humans in the natural forest and fire system will continue for many years. What is the best approach? Logging and mechanical thinning, or controlled burns—or just let nature take its course?

Influence of La Niña on Wildfires

A La Niña condition (see chapter 12) existed in the Eastern Pacific Ocean during 1999 and 2000, and it helped set the stage for the fires of 2000. La Niña brings a pool of cool ocean water to the Americas, leading to drier than average weather in the southern tier of states from Florida to California. The Pacific Northwest receives a great deal of rain, but the other western states receive thunderstorms that commonly are dry. The dry thunderstorms bring lightning that starts fires, but there are no accompanying rains to extinguish the blazes. In 2000, the fires began early in spring and did not quit until rain and snow began falling in autumn.

PRESCRIBED FIRES

One solution to the problem of dense forests and shrublands is for trained people to deliberately set fires—prescribed fires—at times carefully selected for low wind speeds, low temperatures, humid air, good soil moisture, approaching rain, and other factors that limit the size and power of fires (figure 14.36). Prescribed fires have become a widely implemented program. From 1995 to 2000, more than 31,000 prescribed fires were set by federal agencies (table 14.11). The hope is that a prescribed fire will be a controlled fire that burns only the desired area. But with so many fires being set by so many people, mistakes will happen.

Figure 14.36 Prescribed fire ignited from a boat in the Imperial National Wildlife Refuge, Arizona, 18 April 2008.
Photo from U.S. Fish and Wildlife Service.

TABLE 14.11
U.S. Government–Prescribed Fires

Year	Acres Burned	Costs
2000	1,192,220	>$455,000,000
1999	2,240,105	$99,104,000
1998	1,889,564	$70,793,000
1997	1,601,158	$36,146,000
1996	915,163	$29,550,000
1995	918,300	$20,448,000

Source of data: Bureau of Land Management, Bureau of Indian Affairs, Fish and Wildlife Service, National Park Service, U.S. Forest Service.

Los Alamos, New Mexico, May 2000

On 4 May 2000, National Park Service personnel set what they hoped would be a controlled fire at Bandelier National Monument to clear an understory of shrubs that could provide fuel for a wildfire. The next day, high winds whipped the prescribed fire into a wildfire that consumed 50,000 acres of national forest and 235 houses in Los Alamos, displaced 25,000 people, damaged or destroyed 115 buildings in Los Alamos National Laboratory, and came frighteningly close to the hazardous materials sites at the nuclear weapons research facility.

The fire was made worse by some houses in Los Alamos. The initial fire front swept through town quickly; many green trees were left. However, some badly prepared, non-fire-resistant homes caught fire and burned so long and so hot that they ignited many trees that had resisted the original fire front. Trees ignited by burning houses caused the worst damage to the forest and standing houses as the second and hotter wave of fire passed through town.

The prescribed fire caused the very disaster it was supposed to prevent. The federal government has paid $455 million in compensation to victims of the fire. Critics say that the fire-setting personnel were careless and incompetent. Proponents say a lightning strike would have started the same fire anyway. And so the debate rages on. Shall we practice firefighting or fire lighting?

U.S. government–prescribed fires from 1995 through 2000 burned an annual average of 1,459,418 acres (table 14.11). But the pace has quickened. From 2001 through 2011, prescribed fires burned an annual average of 2,437,736 acres—a 70% increase.

WILDFIRES IN AUSTRALIA

Australia is bedeviled by bushfires, major conflagrations swept by high winds through eucalyptus forests that are fire-adapted and fire-maintained. Most eucalyptus trees and bushes have highly flammable oil in their branches and bark. Some trees invite fire by shedding their bark in long, thin strips. The dry strips of bark are kindling that ignites easily and burns fast. Eucalyptus makes good wood for your fireplace, but in a natural setting, the trees may be heated so hot that their sap boils and whole trees explode in flame. This presents obvious problems for houses and towns in the path of a eucalyptus-fueled bushfire. The bluegum eucalyptus is described by poet Robert Sward:

Deadly beauty,
Much to admire.
Until it falls
Or catches fire.

Fire is not a problem for the plant species. Most eucalyptus varieties are designed to burn hot and fast; it is then that their fruit opens and releases unburned seeds to germinate in the fire-cleared ground. The fast-moving fires may burn down trees, but many of the resilient plant species simply send up shoots from their stumps or roots and grow to full-size trees again in several years.

El Niño's Effect

The worst bushfires in history correlate with drought and wind. Many of these weather episodes occur when an El Niño–Southern Oscillation circulation system is operating in the oceans (see figures 12.18 to 12.22). When the ocean waters off Australia are cooler, evaporation and precipitation are reduced. The lessened rainfall creates drought that can extend through an entire Southern Hemisphere winter. Bushfire conditions may start with a high-pressure system off southern Australia. As the high-pressure zone moves slowly eastward, it induces air masses over the central desert to blow southeastward from the interior toward the heavily populated southeastern region of Australia. As the winds descend from the high desert, they warm in downslope-wind fashion, producing temperatures around 40°C (104°F), humidities of less than 20%, and speeds of at least 60 km/hr (40 mph).

The El Niño–Southern Oscillation of 1982–83 was particularly strong, and the Australian summer was the driest in recorded history. On Ash Wednesday, 16 February 1983, strong downslope-type winds from the interior were reinforced by the jet stream in the upper atmosphere. Adelaide, the capital of South Australia, and Melbourne, the capital of Victoria, were ringed by fires pushed by the southeastward-blowing winds.

In midafternoon, a cold front swept through the area, dropping temperatures 10°C (18°F). However, the cooler winds did not cool off the fires, but instead changed their direction to the northeast and increased their speed. The winds were blowing 70 km/hr (45 mph) with gusts up to 170 km/hr (105 mph), and the fire front advanced at speeds up to 20 km/hr (more than 12 mph). The firestorms were so strong that their winds snapped tree trunks.

When the fires were finished, 76 people lay dead, 3,500 had suffered injuries, and 2,500 houses and 20 towns were gone. The livestock toll was also great, as more than 300,000 sheep and 18,000 cattle perished along with uncounted numbers of native animals.

The bushfire hazard is everpresent. The hot, dry, fast winds return almost every year.

Leave Early or Stay and Defend

After the 1994 bushfires, Australia adopted *leave-early-or-stay-and-defend* strategies, also called *shelter-in-place* strategies, that have reduced property losses in some major fires. The strategy utilizes construction and landscaping standards. Buildings must be designed for fire and built with fire-resistant materials; landscaping must have low-fuel content. Additionally, there are requirements for modified vegetation zones surrounding communities, for maintaining an adequate water supply, and for building roads wide enough for firefighting equipment. Finally, able-bodied adults remain in the community during fires and help professional firefighters battle the flames.

The shelter-in-place strategy is now being implemented in some U.S. communities that were built to be fire safe. Opponents of shelter-in-place say the only sure way to survive a wildfire is to evacuate, and if residents stay home, firefighters have to waste time caring for them instead of fighting fires. Proponents say there are never enough professional firefighters and that in a fire-prepared setting, some trained and able-bodied adults can remain to assist in fire suppression. They point out that many of the wildfire deaths occur during evacuation. This pro and con debate will continue for many years.

The leave-early-or-stay-and-defend strategy was tested in the state of Victoria in Australia in 2009. A prolonged drought included temperatures greater than 40°C (104°F) for days. On Saturday, 7 February, the temperature reached 47°C (117°F) with humidity at 7% and wind speeds up to 100 km/hr (60 mph). Then, dozens of bushfires broke out. Winds carried firebrands up to 30 km (19 mi) to start new fires. Temperatures were so intense that some eucalyptus and gum trees with their flammable and aromatic oils literally exploded into flames. Bushfires burned 2,840 km² (1,100 mi²), killing 173 people and about 1 million animals. A resident of Marysville described it as "the inside of hell." Many of those who died waited too long to evacuate, and then tried to outrun flames driven by 60 mph winds, only to die inside their cars.

And the debate continues—leave early or stay and defend? To stay requires preparation and steady nerves. You will hear the wind; you will hear firebrands hitting your house; the power will go out, and you will be plunged into darkness as smoke blocks out the sunshine; your house will rise in temperature. But you cannot panic and flee, or death may catch you.

The Similarities of Fire and Flood

Fire historian Stephen J. Pyne has pointed out that floods can serve as a metaphor for fire. Fires and floods seem so different, yet they have some general characteristics in common:

- Both fire and flood are closely related to weather, plant cover, and topography.
- Both fire and flood are at their strongest when atmospheric conditions are extreme. Fast, dry winds push flames, while heavy rains feed floods.
- Both fire and flood move across the landscape and through human developments as waves of energy. A fire front is a wave of chemical energy released from temporary storage in organic matter. A flood crest is a wave of mechanical energy unleashed when the potential energy of high topographic position is converted to the kinetic energy of motion.
- Both fire and flood become more turbulent the faster they move and the bigger they grow.
- Both fire and flood can be described by their size and frequency. As a first approximation, the bigger the fire or flood, the longer the return time until the next big one.
- Both fire and flood are effectively understood as 50-year, 100-year, or other recurrence-time events.
- Both fire and flood are aggravated by human activity. Fires are made more intense by buildings placed in dense growths of plants and by our habit of quickly suppressing small fires, thus allowing organic debris to build into large masses that provide fuel for gigantic firestorms, such as in Yellowstone National Park in 1988. Floods are made more destructive by buildings placed on floodplains and by levees built to protect against floods that inadvertently help cause record-high flood levels, such as in the Upper Mississippi River Valley in 1993.

Summary

Fire is a natural force used by humans for purposes both good and bad. Controlled use of fire provides energy for industry and travel, as well as lighting, heating, and cooking in our homes. But fire also kills and destroys; in the United States, from 1981 to 2005, fire caused a yearly average of 4,917 deaths (82% in residences) and more than $13 billion in property damage.

Fire is the rapid combination of oxygen with carbon, hydrogen, and other elements of organic material in a reaction that produces flame, heat, and light. In effect, fire is photosynthesis in reverse.

Before a fire starts, a preheating stage occurs in which water is expelled from wood, plants, or fossil fuels by flames, drought, or hot weather. When temperatures exceed 300°C (615°F), wood breaks down and gives off flammable gases in the process of pyrolysis. If oxygen is present, pyrolized gases can ignite, and combustion begins.

In the flaming combustion stage, the pyrolizing wood burns hot. Released heat keeps the wood surface hot through conduction, diffusion, radiation, and convection. After the active flames pass, a glowing combustion stage exists during which fire slowly consumes the solid wood.

The spread of fire depends on several factors: (1) The types of plants or fuel burned make a difference; some plants, such as eucalyptus and chaparral, ignite easily and burn intensely. (2) Strong winds bring oxygen and push flames forward. (3) Topography has numerous effects. For example, before a fire, the terrain helps control plant distribution, and during a fire, a steep slope acts like a chimney that fire races up. (4) Fires heat air, which rises buoyantly and creates its own winds, including fire tornadoes.

Fire threats are greatest in areas with big contrasts between wet and dry seasons. A wet season triggers voluminous plant growth. Then dry conditions dehydrate plants, making it easier for ignition to occur. Add some strong winds, and major fires break out, usually in numerous places.

Buildings can be constructed to withstand fire. Traditional structures can be made safer by eliminating flammable vegetation and woodpiles near them, avoiding wood-shingle roofs and overhanging wood balconies or decks, and using double-paned glass in windows and doors and spark arresters on chimneys. If there is a vulnerable spot in a house, flying embers may enter and set it afire.

Fires cannot be prevented, only deferred. We pay a price for fire suppression. Allowing natural fires to burn helps prevent buildup of extensive debris that can fuel a firestorm during heavy winds. Large firestorms are essentially unstoppable. We must choose between lots of little fires or a few monstrous blazes.

Terms to Remember

backfire   382
chaparral   388
combustion   381
conduction   383
convection   383
diffusion   383
duff   392
ember   384
fire   380
firebrand   385
firestorm   379

fuel   381
fuel-driven fire   393
heat   380
ladder fuel   382
oxidation   384
photosynthesis   380
pyrolysis   382
radiation   383
slash   381
wind-driven fire   394

Questions for Review

1. How many people die on average each year in fires in the United States?
2. Write an equation that describes fire.
3. Compare fire to photosynthesis.
4. What is the fire triangle? What are its components?
5. Explain the process of pyrolysis.
6. Explain the differences between conduction, diffusion, radiation, and convection.
7. Compare the conduction of heat in wood versus metal.
8. What are ladder fuels?
9. What is the difference between flaming combustion and glowing combustion?
10. How does setting a backfire help fight a major wildfire?
11. Will a typical wildfire burn faster upslope or downslope? Why?
12. Explain how a wildfire can create its own winds
13. How does a fire tornado form?
14. How does the daily cycle work for sea breezes and land breezes?
15. Draw a cross-section and explain how downslope winds form. Why are they typically hot and dry?
16. Explain the difference between a fuel-driven fire and a wind-driven fire.
17. Explain the threat that flying embers present to houses.
18. What difficulties are presented when duff catches fire?
19. What is the relationship between a La Niña condition in the Pacific Ocean and wildfires in the southern states from California to Florida?
20. What is the relationship between El Niño and large wildfires in Australia?

Questions for Further Thought

1. Explain how to build a campfire. Explain the fire processes occurring at each stage of your campfire.
2. Make a detailed list of actions you could take to make your current residence safer from fire.
3. If you were designing your dream house on your dream lot, what features could you incorporate into the house and landscape design to better protect your house from destruction by fire?
4. Evaluate the wisdom of quickly suppressing a local wildfire during a time of cold weather with low wind speeds.
5. Should homeowners in houses surrounded by flammable vegetation in wildlands pay the same fire insurance premiums as homeowners in the city?
6. Is the leave-early-or-stay-and-defend concept worthwhile for you? Or do the risks outweigh the benefits?

Disaster Simulation Game

Your challenge is to protect the outer suburbs of an Australian city that is prone to wildfires. The city relies on farming and services. You must save as many people, buildings, and livelihoods as possible. You are given a budget and a set amount of time. Then you have real choices to make.

The city you must protect has a specified population of people. You are provided with a map of the town and choices of different housing and upgrades plus defenses. You also must build a hospital and a school, and you must keep the city water towers from burning. Are you ready for the challenge? Go to http://www.stopdisastersgame.org. Click on Play Game. On the next page, click on Play Game again. Select the Wildfire scenario. Then choose your preferred difficulty level: Easy (small map); Medium (medium-size map); or Hard (large map).

Good luck! A wildfire could start at any time.

CHAPTER 15

Mass Movements

The distinctive features of unintelligent humans are the hastiness and absoluteness of their opinions; scientists are slow to believe, and never speak without modification. The use of science and the scientific method creates an intellectual conscience which believes only up to the evidence in hand, and is always ready to concede that it may be wrong.

—BERTRAND RUSSELL, 1928, *AN OUTLINE OF PHILOSOPHY*

Heavy rains in the mountains set off debris flows carrying huge boulders through Caraballeda, Venezuela, on 15 December 1999, killing thousands of people.

M.C. Larsen and H.T. Sierra/U.S. Geological Survey.

LEARNING OUTCOMES

Gravity pulls unceasingly on the land, bringing down large masses of rock that kill people and destroy structures. After studying this chapter, you should:

- understand the role of gravity in mass movements.
- know the external and internal causes of slope failures.
- be able to explain how hillsides fail in slides.
- understand snow avalanches by applying the same principles that underlie rock movements.
- be familiar with mitigation structures designed to prevent mass movements.
- comprehend how earth masses fail and flow as fluids.
- understand ground subsidence causes in slowly subsiding loose sediments and in catastrophically collapsing limestone caverns.

OUTLINE

Gravity

About 5:30 p.m. on Monday, 21 February 2000, an isolated thundercloud poured rain and dropped hail east of El Cajon, California. Rain ran off the steep hillsides as it has for thousands of years. But this time, the rain run-off washed away enough coarse, sandy soil to destabilize a huge plutonic rock; eight hours later, gravity pulled it free and sent it rolling and bouncing downhill before crashing into a single-story house at 1:27 a.m. (figure 15.1). Neighbors described the noises made by the bouncing boulder and its forceful entry into the house as louder than an earthquake. Luckily, the homeowners were on a ski trip at Lake Tahoe and missed the destruction of their home. The mega-boulder was larger than two cars; it weighed 200 tons (400,000 lbs) and filled the interior of the house, destroying 60% of it. This singular event introduces us to the world of gravity-caused disasters.

Large volumes of material move downslope under the pull of gravity, and some do so catastrophically. These **mass movements** occur throughout the world. In the United States, no state is exempt (figure 15.2). Many mass movements are not reported separately but are included with descriptions of the events that triggered their movements, such as earthquakes, volcanic eruptions, and major rainstorms. Mass movements in the United States annually cause about $1.5 billion in damages and 25 deaths. There are about 2 million mass movements each year in the United States.

A globle data set has been compiled on fatal landslides from 2004 to 2010, excluding earthquake–triggered events. There were 2,620 landslides that killed 32,322 people mostly in China and the Himalaya Mountains.

The Role of Gravity in Mass Movements

The existence of gravity was first discussed scientifically by Isaac Newton (1642–1727), one of the true geniuses in history. Newton's accomplishments were many, including inventing calculus and determining the laws of motion and the universal law of gravitation. The importance of fundamental laws was underscored by Ralph Waldo Emerson in 1841, when he wrote: "Nature is an endless combination and repetition of a very few laws. She hums the old well-known air through innumerable variations."

Gravity is an attraction between objects. It is a force that humans are unable to modify; it cannot be increased, decreased, reversed, or reflected. The law of gravity states that two bodies attract each other with a force directly proportional to the product of their masses and inversely proportional to the square of the distance between them:

$$\text{gravity (g)} = \frac{G \times \text{Mass 1} \times \text{Mass 2}}{\text{distance} \times \text{distance}}$$

where G = a universal constant.

Gravity is relentless—it operates 24 hours a day, every day of the year, and the law of gravity is strictly enforced. The constant pull of gravity is the immediate power behind the agents of erosion. Rain falls, water flows, ice glides, wind blows, and waves break under the influence of gravity. But gravity can also accomplish major changes working largely by itself, without the help of any erosive agent. It is the solo work of gravity that is the subject of this chapter.

Gravity causes the downward and outward movements of landslides and the downward collapse of subsiding ground. Given enough time, gravity would pull all the land into the seas. Over the great lengths of geologic time, all slopes can fail; all slopes should therefore be viewed as inherently unstable. Slope failures may be overpowering, catastrophic events, as when the side of a mountain breaks loose and roars downhill. Or hill slopes may just quietly deform and yield to

Figure 15.1 On 22 February 2000, a 200-ton boulder of plutonic rock rolled downhill under the pull of gravity and crashed into this house east of El Cajon, California.

Photo by Pat Abbott.

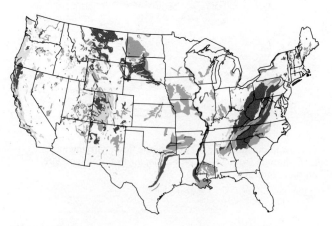

Figure 15.2 Landslide potential of the conterminous United States. Reddish-brown areas have very high potential, yellow areas have high potential, and green areas have moderate potential.

Source: *U.S. Geological Survey.*

the unrelenting tug of gravity in the very slow-moving process known as **creep.**

Gravity pulls materials downslope with a measurable force. For example, consider a 1 lb boulder resting upon a 30° slope (figure 15.3). Gravity exerts 1 lb of pull on the boulder toward the center of the Earth, but the ground is too solid to allow the boulder to move down vertically. The trigonometric relations of a triangle allow the magnitude of the horizontal and vertical forces on the boulder to be determined. The downhill force is calculated as

$$1 \text{ lb} \times \text{sine } 30° = 1/2 \text{ lb}$$

This 1/2 lb of force is directed downslope, toward open space.

Before the boulder moves downhill, or before the whole hillside begins to move, these masses must overcome inertia and **friction.** Inertia is the tendency of a body to remain at rest until an external force is applied. Friction is the resistance to motion of a body that keeps it from moving over another body. Friction comes in large part from the roughness of surfaces that make sliding, flowing, or rolling difficult. The surface could be the ground or some weak rock layer at depth. All that is needed is some initial energy to overcome inertia and friction to begin the boulder's movement or the hillside's failure. Initial energy could come from an earthquake, a heavy rain, a bulldozer, the footstep of a sheep, or . . . .

CREEP

Gravity induces materials to move in many ways, including creep, the slowest but most widespread form of slope failure. Creep is an almost imperceptible downslope movement of the **soil** and uppermost **bedrock** zones. Creep is most commonly seen by its effects on objects, such as telephone poles and fences, that lean downslope or by trees whose trunks have deformed due to growing upward while rooted in material that is slipping downhill (figure 15.4a). The soil zone slips in ultraslow movements as individual particles shift and move in response to gravity; the upper bedrock zone yields to the pull by curving downslope (figure 15.4b).

(a)

(b)

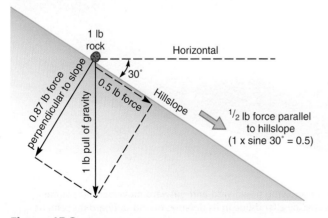

Figure 15.3 Gravitational forces acting on a 1 lb boulder sitting on a 30° slope.

Figure 15.4 (a) Block diagram of a slope showing the effects of creep. Soil moves slowly, and bedrock deforms downhill. (b) Creep has deformed rock layers near Marathon, Texas.

(b) NOAA.

The Role of Gravity in Mass Movements **409**

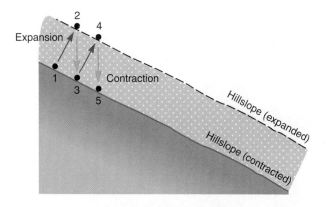

Figure 15.5 How creep works. Surface materials expand perpendicular (e.g., from point 1 to point 2) upon freezing of pore water, swelling when wetted, and heating by the Sun. Surface materials shrink (e.g., from point 2 to point 3) upon thawing, drying, and cooling. Contraction occurs under the pull of gravity and is toward the center of the Earth. The result is a net downslope movement of materials—that is, creep.

We see that ground surfaces do not sit still; they move up and down. By what mechanisms do soil and rock actually move? They move because the volume of soil does not stay constant but swells and shrinks. Several processes cause swelling: (1) Soil has a high percentage of void space, or **porosity.** When water filling these pores freezes, it expands in volume by 9%, swelling the soil volume and lifting the ground surface upward. (2) When soil rich in expandable materials, such as some types of **clay minerals,** is wetted, it absorbs water and expands. (3) Heating by the Sun causes an increase in volume. Soil expands perpendicular to the ground surface (figure 15.5).

Several processes shrink masses of soil, rock, and water, causing the ground surface to lower. Shrinkage occurs when soil (1) thaws, (2) dries, or (3) cools. Lowering of the surface is influenced by the downward pull of gravity, causing a net downslope movement of particles in the soil zone (figure 15.5). Creep is a slow downhill flow, and we will revisit it when discussing types of flow.

External Causes of Slope Failures

Most mass-movement fatalities are caused by the fast-moving varieties, as recorded on a tombstone in Westland, New Zealand:

> Patrick O'Brien
> Who was killed by a landslip
> on the 8th of March 1888
> Aged 37 years
> Death to him short warning gave

> Therefore be careful how you live
> Repent in time and don't delay
> For he was quickly called away

A typical landslide is a mass whose center of gravity has moved downward and outward (figure 15.6). There is a tearaway zone upslope where material has pulled away and a pileup zone downslope where material has accumulated.

(a)

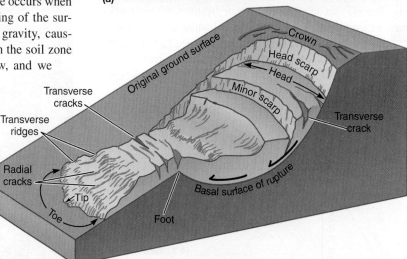

(b)

Figure 15.6 (a) Landslide showing downward-and-outward movement. (b) Some topographic features created by a landslide in its downward-and-outward movement.

(a) U.S. Geological Survey.

In Greater Depth

Energy, Force, Work, Power, And Heat

The effectiveness of agents and events is measured using the related terms of energy: **force, work, power,** and **heat.** Energy is the capacity to do work; it may be potential or kinetic. **Potential energy** is poised and ready to go to work. For example, a house-size boulder resting precariously high on a steep slope has the potential to roll and bounce downhill and do a lot of damage (figure 15.7a). The potential energy (PE) of the huge boulder is equal to its mass (m) times the force of gravity (g) times its height (h) above a certain level, which in this case is the elevation of the valley floor below it:

$$PE = mgh$$

If the boulder starts to roll, its potential energy now becomes kinetic—the energy of motion (figure 15.7b). **Kinetic energy** (KE) is determined by half the product of mass (m) times the velocity (v) squared:

$$KE = 1/2 \, mv^2$$

The kinetic energy of the bouncing boulder adds to its work if its downslope collisions cause other boulders to move and the resultant moving mass brings soil, trees, and other debris downhill with it. The work (w) done on the sliding mass is determined as force (F) times distance (d), where force equals mass (m) times acceleration (a):

$$Force \; (F) = ma$$
$$Work \; (w) = Fd = mad$$

Force can be measured in newtons (N); 1 newton is the force required to accelerate a mass of 1 kilogram at 1 meter per second squared:

$$newton \; (N) = 1 kg \, m/sec^2$$

Force may also be measured using a unit called a dyne. Each dyne equals a mass of 1 gram (gm) accelerated 1 centimeter (cm) per second squared:

$$dyne = 1 \, gm \, cm/sec^2$$

Energy, as well as work, may be expressed in dyne-centimeters (= dyne × cm). Work is defined as force acting over a distance and is dimensionally equivalent to energy. A force of 1 dyne acting over a distance of 1 centimeter is called an erg:

$$erg = dyne \times cm = gm \, cm^2/sec^2$$

A dyne is a small unit of force, and an erg is a small unit of work. Thus, for large-scale phenomena, a larger unit of measurement called a joule is used. A joule is 10 million ergs:

$$joule = 10^7 \, erg$$

Figure 15.7 (a) When the boulder is poised and ready to move, it has potential energy. (b) When the boulder is rolling, its energy is kinetic.

Drawings by Jacobe Washburn.

The landslide triggered by the bouncing mega-boulder may move rapidly (faster than a human can run) or slowly. Whether fast or slow, if the same amount of material ends up at the bottom of the slope, then the amount of work is the same. However, the power is different. Power is defined as the rate at which work occurs and is measured in watts:

$$power = work/time = joule/second = watt$$

After the boulder caused the landslide to move, what happened to slow it down and stop it? Friction—friction with the underlying ground and friction among the boulders, sand grains, trees, and other debris inside the moving mass. As you know from sliding into second base or across a dance floor, friction generates heat. Thus, kinetic energy is related to heat. The definition of heat is the capacity to raise the temperature of a mass; it is expressed in calories. The relationship of work to heat is:

$$4.185 \; joules = 1 \; \textbf{calorie}$$

Calories are units of energy. One calorie is defined as the amount of heat needed to raise the temperature of 1 gram of water by 1 degree centigrade under a pressure of 1 atmosphere. One gram of water is equivalent to about 10 drops. The calorie commonly discussed with food is actually a kilocalorie, or 1,000 of the calories defined here.

External processes that increase the odds of a slope failure include: (1) adding mass high on a slope, as in sediment deposition; (2) steepening the slope, as by fault movements; and (3) removing support from low on a slope, as by stream or ocean-wave erosion.

WATER IN ITS EXTERNAL ROLES

Water plays many important roles in mass movements, both externally and internally. Rainfall is an external factor; rain falls from the atmosphere. Rain runoff causes external erosion that sets masses moving on slopes, and it undercuts the bases of slopes, causing hillsides to fail and move.

Internal Causes of Slope Failures

Beneath the surface, inside the materials underlying a slope, long-term processes are weakening earth and preparing it for failure. Internal causes of slope failure include (1) inherently weak materials, (2) water in different roles, (3) decreasing cohesion, and (4) adverse geologic structures.

INHERENTLY WEAK MATERIALS

The materials most commonly associated with earth failures are the clay minerals. Clays are the most abundant of all sediments. They form during **chemical weathering** as rocks exposed at the surface decompose and form new minerals under conditions of low temperature and pressure. The weathering occurs when acidic fluids, such as water, CO_2-charged water, and organic acids, decompose minerals. These minerals are likely to transform into new atomic structures to achieve equilibrium. In the decomposition process, the relatively simple atomic structure of minerals such as feldspar will transform into the wildly variable structures and compositions of the clay minerals.

Although the bulk chemical compositions of feldspars and clays are similar, their internal structures are radically different. Clay crystals are very small—too small to be seen with a typical microscope. Clay minerals are built like submicroscopic books (figure 15.8). From a top view, they are nearly equidimensional. But a side view shows a much thinner dimension that also is split into even thinner subparallel sheets, like the pages in a book. The booklike structure typically forms in the soil zone where water strips away elements, leaving many unfilled atomic positions in crystal structures. This is like building a Tinker Toy or Lego structure and then removing a tremendous number of pieces.

As clay minerals take in different elements and other elements are removed, they increase and decrease in strength; they expand and contract; and they may absorb water and later have it removed. The constantly changing conditions cause variations in the strength of clay minerals from month

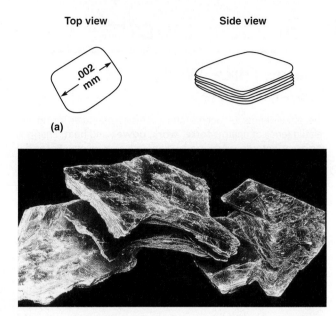

Figure 15.8 (a) Schematic views of the exceedingly small size and structure of a clay mineral. (b) The mineral mica (shown here) provides a mega-scale view of clay minerals. Ultrathin individual sheets have relatively large lengths and widths.

(b) © McGraw-Hill Education/Bob Coyle, photographer.

to month and year to year. Thus, there are certain times when a hill containing clay minerals is weaker, and then gravity has a better chance of provoking a slope failure.

The mechanical or strength characteristics of a rock are usually governed by the 10–15% of the rock having the finest grain size. The strength of clay minerals may be lessened by water (1) adsorbed to the exterior of clays, thus spreading the grains apart, or (2) absorbed between the interlayer sheets, with resultant expansion.

Spectacular examples of slope failures are presented by **quick clays,** the most mobile of all deposits. Quick clays are abundant in Scandinavia and eastern Canada and occur in the northeastern United States. Quick clays begin as fine rock flour scoured off the landscape by massive glaciers and later deposited in nearby seas. The clay and silt sediments sit in a loosely packed "house of cards" structure filled with water and some sea salts that help hold it together as a weak solid (figure 15.9). When glaciers retreat, isostatic rebound lifts these clay-sediment areas above sea level. Freshwater passing through the uplifted sediments dissolves and removes much of the sea salt "glue," leaving quick clay with (1) weak structure, (2) grains mostly less than 0.002 mm diameter (3) water content commonly in excess of 50%, and (4) low salt content. In short, the house-of-cards structure can be collapsed by a jarring event, such as a dynamite blast or vibrations from construction equipment.

What has been solid earth can literally turn to liquid and flow away. How does land that has been plowed and farmed, built on, and occupied for many generations all of a sudden

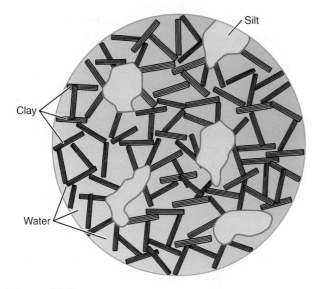

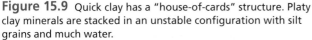

Figure 15.9 Quick clay has a "house-of-cards" structure. Platy clay minerals are stacked in an unstable configuration with silt grains and much water.

Figure 15.10 A 3.5 million m³ mass of quick clay and sand flowed into the South Nation River near Lemieux in Ontario, Canada, on 20 June 1993. In the upper right is County Road 16.

Reproduced with the permission of Natural Resources Canada, 2015
Photo by G.R. Brooks.

turn to fluid and flow away? The collapse of the house of cards with its high water content creates a muddy fluid. To really believe this, you need to see it. Check out the video entitled *Rissa Landslide, Quick Clay in Norway;* it is mind-boggling to watch solid earth suddenly turn to fluid and flow off, carrying houses with it. Quick-clay slope failures have caused the destruction of numerous buildings and the loss of many lives. These events will recur more than once somewhere on Earth during your lifetime.

Canadian Quick-Clay Slope Failures

Quick-clay hazards are common in eastern Canada. Quick-clay disasters (table 15.1) have led to recognizing problem areas and taking preventive actions, such as moving towns. The wisdom of moving people out of harm's way was shown during the high rainfall year of 1993 near the former town-site of Lemieux in Ontario, Canada. During one hour on the afternoon of 20 June, a 3.5 million m³ mass liquefied and flowed, carrying forest and rangeland into the South Nation River (figure 15.10). Eyewitnesses reported flowing waves

of earth carrying trees into the river and sending 3 m (10 ft) high waves both up and down river. Human development of the area has been hindered. How can you live your life on ground that may turn to liquid?

WATER IN ITS INTERNAL ROLES

Water weakens earth materials in several different ways:

1. **Weight of water.** Water is heavier than air. Sedimentary rocks commonly have porosities of 10–30%. If these void spaces are filled with water, the weights of materials are dramatically increased; thus, the driving masses of slope materials are also increased and mass movements may begin.

2. **Absorption and adsorption.** Water is both absorbed (internally) and adsorbed (externally) by clay minerals, with resultant decreases in strength. How does water attach so easily to clay minerals? Because of its unique interplay of charges. Water is a molecule formed by two hydrogen (H^+) atoms linking up with one oxygen (O^-). The two positive charges from the hydrogen atoms should cancel the two negative charges from the oxygen atom

TABLE 15.1
Some Canadian Quick-Clay Slope Failures

Date	Site	Deaths	Damages
June 1993	Lemieux, Ontario	—	County Road 16 severed
May 1971	St. Jean Vianney, Quebec	31	40 houses destroyed
December 1962	Rivière Toulnustouc, Quebec	8	Failure initiated by workers setting off blasts
April 1908	Notre Dame de la Salette, Quebec	33	Houses destroyed
April 1895	Saint Luc de Vincennes, Quebec	5	—
1877	Sainte Geneviève de Batiscan, Quebec	5	—

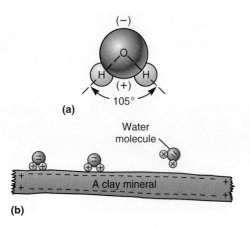

(a)

Water
molecule

A clay mineral

(b)

Figure 15.11 (a) Water is a bipolar molecule. Two hydrogen atoms (each H⁺) attach to one oxygen atom (O⁻), creating an electrically neutral molecule. However, the asymmetry of the molecule makes one side slightly negative and the other side positive. (b) The positive side of water molecules attaches to the negatively charged surface of clay minerals.

to create an uncharged or neutral molecule. But water is a bipolar molecule (figure 15.11a) with its hydrogen atoms on one side (a positive side) and its oxygen on the other side (a negative side). Thus, water molecules can attach their positive sides against clay minerals because clay surfaces are negatively charged (figure 15.11b).

3. **Ability to dissolve cement.** Water flowing through rocks can dissolve some of the minerals that bind the rock together. The removal of cementing material decreases the cohesion of rocks and saps some of a slope's strength, preparing it for failure by mass movement.

A death-dealing example occurred with the failure of the St. Francis dam about 73 km (45 mi) north of Los Angeles, California. The dam was a heavy concrete mass built in 1926 across San Francisquitos Canyon. Unfortunately, part of the foundation of the dam was a poorly bedded mass of gravels, sands, and muds held together by clay coatings and cemented by the mineral gypsum. In two years, lake water dissolved enough gypsum cement and weakened the clays, causing a catastrophic failure of the dam. At about midnight on 12 March 1928, the base of the dam failed, unleashing a 56 m (185 ft)

high wall of water moving 29 km/hr (18 mph). The undammed water took 5 hours and 27 minutes to travel 87 km (54 mi) down the Santa Clara River to the Pacific Ocean. Traveling through the night, the raging water made permanent the sleep of about 420 people.

4. **Piping.** Water flowing underground can not only chemically dissolve minerals but also physically erode loose material. Subsurface erosion (**piping**) can create extensive systems of caverns (figure 15.12). A network of caves obviously makes a hill weaker.

5. **Pore-water pressure.** Pressure builds up in water trapped in pores of rocks being buried deeper and deeper. As sediments pile up on the surface, their weight puts ever more pressure on sediments and pore water at depth. Sediment grains of sand and mud pack into smaller and smaller volumes, while water, which is nearly incompressible, simply stores built-up pressure. When a pile of sediments sits on top of overpressurized pore water, the entire mass becomes less stable. The buildup of pressure within pore water has been referred to as a "hydraulic jack" that progressively "lifts up" sediments until the pull of gravity can start a massive failure. Many mass movements and slope failures are due to abnormally high **pore-water pressures.**

Quicksand occurs where sand grains are supersaturated with pressurized water. For example, if water flows upward through sands, helping lift up the sand grains, the pull of gravity on the sand grains can be effectively canceled, leaving the sand with no strength or ability to carry a load (figure 15.13a). In effect, the pore-water pressure (h_w) equals the weight of the sands (p), leaving the cohesionless sand with no shear strength.

If water-pressurized sand is on a slope, it will flow away; but if it sits on a flat surface, it will be quicksand. Despite what some old movies show, quicksand does not suck people or other objects down. It is rather like stepping in a high-viscosity liquid. Stand there long enough and you will sink below the surface (figure 15.13b). What should you do if you get caught? If the water is not too deep, fall backward and spread your arms out; this distributes your weight broadly, like a boat on water, and you will not sink. If you can float on water, you surely will float on the denser quicksand. Then call

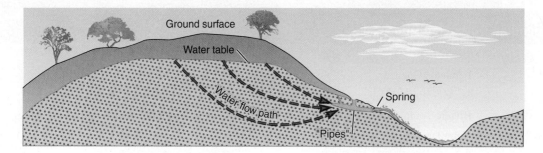

Figure 15.12 Schematic cross-section of groundwater flowing through poorly consolidated rocks. Water will carry sediments to the surface at springs. This erosion creates a network of caverns, or "pipes," that seriously weakens a hill.

In Greater Depth

Analysis of Slope Stability

The analysis of a slope for its resistance to failure is commonly addressed by use of the Coulomb-Terzaghi equation:

$$s = c + (p - h_w)\tan phi$$

where s is **shear** resistance, the sum of characteristics that hold a mass in place; c is cohesion, a measure of how strongly a material is held together by interactions between particles (e.g., there is little or no cohesion between quartz sand grains, but there is much cohesion between flakes of clay); p is the weight per unit area of solids and water above a potential slide surface; h_w is the height of the water column times the unit weight of water; and $tan\ phi$ is the tangent of the angle of internal friction—the slide surface's angle measured in degrees from horizontal.

What does this all mean? The strength of a hillside or mass of material comes from (1) its cohesion—how well it sticks together—plus (2) the weight of all its materials under the pull of gravity. These factors of stability are offset by (1) the pore-water pressure, plus (2) the angle at which a rock mass fractures, the angle of its slide surface. The failure angle is low or near horizontal for weak materials, such as clay-rich sedimentary rock, and is steep or near vertical for strong rocks, such as granites.

When pore-water pressure (h_w) equals the weight per unit area of overlying materials (p), then the only shear resistance the materials possess comes from cohesion. Materials that are low in cohesion, such as quartz sands, effectively have no strength and will fail. An example is quicksand.

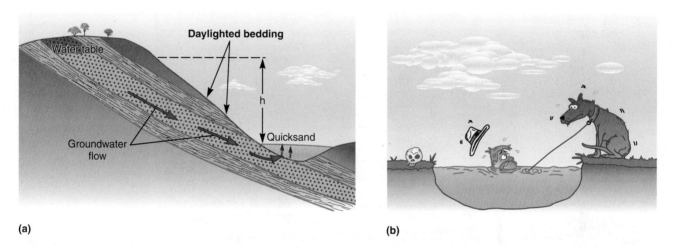

(a)

(b)

Figure 15.13 (a) Schematic cross-section showing groundwater pushed by a high column of water (h) and reaching the surface through loose sands. The uplifting force of the escaping water equals the weight of the sand grains, thus making quicksand. (b) Quicksand!

(b) Drawing by Jacobe Washburn.

to your companions to pull you out, or if alone, slowly slide your way backward, keeping your weight broadly distributed as you pull your legs from the quicksand's grasp. Quicksand holds tightly, so do not panic and flail wildly to get yourself out. Slow and easy, with broad spreading of your body weight, is the answer.

6. **Water table.** At the ground surface, rain, streams, and lakes all supply water to the subsurface. Gravity pulls surface water down to saturate open spaces in subsurface rocks with **groundwater.** A groundwater body is not a lake- or pond-like volume of water. Instead, it is water that saturates a zone of subsurface rock by filling fractures in rocks, pores between sand grains, voids left after shells are dissolved, and other holes. The top of the groundwater body is the **water table.** Below the water table is the saturated zone, and above is the unsaturated zone where groundwater is sparse.

During droughts, water tables may drop hundreds of feet below the ground. During rainy intervals, the water table may reach the ground surface. As water tables rise higher, they trigger increasing numbers of mass movements.

DECREASES IN COHESION

When rocks are buried to depths of hundreds, thousands, and tens of thousands of feet, they are compressed into smaller volumes by the weight of the overlying materials. But this process also works in reverse. When deeply buried rocks are uncovered by erosion and exposed at the surface, the removal of great weight allows the compressed rocks to relax and expand. The expansion in volume produces fractures and increases in porosity. The stress-relaxation process reduces the strength of rocks, opening up passageways and storage places for water to attack and further weaken the rocks.

A Classic Disaster

Vaiont Landslide, Italy, 1963

A shocking example of catastrophe on inclined surfaces occurred in the Vaiont area. A dam was built in 1960 across a deep mountain valley in northeastern Italy. The reservoir impounded 150 million m³ (316,000 acre ft) of water. Several dangerous factors exist at the site: (1) Sedimentary rock layers are folded into a troughlike configuration (**syncline**) with beds on each side dipping toward the river valley. (2) Two sets of fractures split the rocks apart. Fractures formed as mountains rebounded upward after removal of the weight of overlying glaciers. When the ice was gone, the river

downcut a V-shaped canyon. (3) Ancient landslides left old slide surfaces in the rocks. (4) Some rock layers contain thin seams of weak clays. (5) Limestone in the hills has numerous **caverns**. (6) After dam construction, water filling the reservoir saturated the rocks at the toe of the slopes, causing elevated pore-water pressures. These adverse conditions at the reservoir were joined by another: the heavy rains of August-September-October 1963 that raised the water table and saturated the rocks. The rains added a tremendous weight of water into unstable slopes. Forecasters anticipated that the rains would trigger a landslide, but its size was a deadly surprise (figure 15.14).

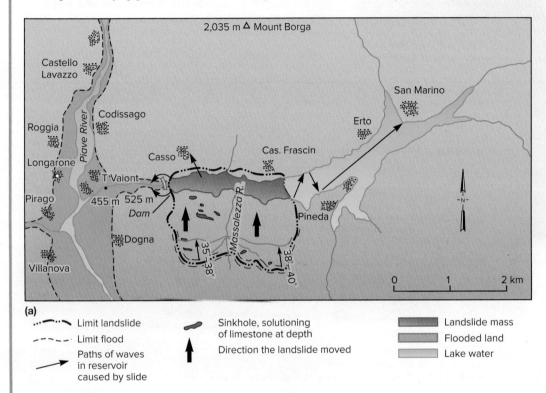

(a)

- ·▪·▪▬ Limit landslide
- ▬ ▬ ▬ Limit flood
- → Paths of waves in reservoir caused by slide
- ◗ Sinkhole, solutioning of limestone at depth
- ⬆ Direction the landslide moved
- ▬ Landslide mass
- ▬ Flooded land
- ▬ Lake water

(b)

Figure 15.14 Vaiont reservoir area. (a) Map showing positions of the landslide and towns affected by the catastrophic flood. (b) Photo after the landslide. Long red arrow shows path traveled by landslide. Note the tall, narrow dam in lower left, the landslide mass in center, and the scarred mountainside in upper right.

© AP Photo.

A Classic Disaster (Continued)

During the heavy rains on the evening of 9 October 1963, most villagers were at home, many still buzzing with the events of the soccer cup matches. At 10:41 p.m., the south wall of the Vaiont reservoir failed massively and rapidly, beginning the events that would kill nearly 3,000 people in the next 7 minutes. The slide mass was 1.8 km (1.1 mi) long and 1.6 km (1 mi) wide, had a volume exceeding 240 million m³, and filled part of the reservoir with rocky debris at heights up to 150 m (500 ft) above the water level. The landslide hit bottom hard, creating an earthquake felt over a wide area of Europe. The mass movements took less than 30 seconds from start to finish, moving at rates of 30 m/sec (68 mph). The slide mass displaced a huge volume of air, blowing outward with enough force to shatter windows and lift roofs from houses.

Most of the death and destruction was caused by displaced reservoir water. A wall of water washed over the dam at 240 m (780 ft) above lake level. Subsequent waves poured out of the reservoir, 100 m (328 ft) above the top of the dam. The combined waves emerged from Vaiont Valley as a wall of water more than 70 m (230 ft) high, slamming head-on into the town of Longarone (figure 15.14). For 6 minutes, Longarone was overwhelmed by raging water. When the waters passed, more than 2,000 residents of Longarone lay dead.

At the upper end of the reservoir, a wave first hit the north wall of Vaiont Valley, demolishing houses. The wave bounced back to the south, hitting the Pineda peninsula, where it again rebounded. Then the wave went northeast across the full length of the reservoir, bypassing without harm the village of Erto, before slamming with full force into San Marino.

What killed the 3,000 people? A tremendous mass of rock slid down an inclined surface and pushed a killing wall of water out of the reservoir. The dam still stands (an engineering success), but the landslide emptied the water (a geologic failure). The event has been called the world's worst dam disaster.

ADVERSE GEOLOGIC STRUCTURES

Many hill-slope masses are weak due to preexisting geologic conditions.

Ancient Slide Surfaces

Ancient slide surfaces are weaknesses that tend to be reused over time. When a mass first breaks loose and slides downslope, it tends to create a smooth, slick layer of ground-up materials beneath it (figure 15.15). The slick layers become especially slippery when wet. It is wise to avoid building on these sites, but if building is necessary, then these slick slide surfaces must be recognized, dug up, and destroyed. Otherwise, they commonly are reactivated when wetted, causing major financial losses and much heartache for building owners. Replays of this scenario are found in the news with saddening frequency.

Daylighted Bedding

The orientation of rock layering within a hill may create either a strong or a weak condition. Where the rock layers dip at angles less than that of the hill slope, the stage is set for slippage and mass movement (see the right-side slope in figure 15.13a). This condition is known as **daylighted bedding** because the ends of shallow-dipping rock layers are exposed to daylight on a steeper slope. On the other side of the same hill, the same rock layers dip into the hill at a steep angle, making it difficult for a massive slide to initiate and break free. This relationship is a good one to keep in mind when you are selecting a home or building site.

Structures Within Rocks

Rocks have weaknesses that set up slope failure. There are several possible causes of failure: (1) rocks that are not cemented together, (2) a clay layer that provides a basal slip surface, (3) soft rock layers sliding off strong materials, (4) joints splitting and separating rock, or (5) an ancient fault acting as a slide surface.

Figure 15.15 A rock surface exposed by bulldozing through a landslide. This surface was smoothed and grooved by a landslide mass that slid across it. A slick surface at depth may be utilized by many generations of landslides. Sample is about 25 cm (10 in) wide.

Courtesy Michael W. Hart.

TRIGGERS OF MASS MOVEMENTS

Slopes usually do not fail for just one reason. Most failures have complex causes. Over the long intervals of time a slope exists, gravity is constantly tugging, and water keeps soaking in and sapping its strength. On numerous occasions, a slope almost fails. Then, along comes another stress for the slope, such as heavy rains, and the slope finally fails in a massive event. Did the last stress, the saturation by heavy rains, cause the slide? Or was it just the trigger for the movement—the proverbial straw that broke the camel's back? Clearly the rains were simply the trigger, or *immediate* cause, for the mass movement.

It is useful to distinguish between immediate and *underlying* causes. The sum of all the underlying causes pushes the slope to the brink of failure, and then an immediate cause triggers the movement. Common triggers for mass movements include heavy rains, earthquakes, thawing of frozen ground, and more and more frequently, the construction projects of humans.

Classification of Mass Movements

Speed of movement and water content vary markedly in different types of mass movements (figure 15.16 and table 15.2). Slow-moving masses cause tremendous amounts

TABLE 15.2	
Rates of Travel for Mass Movements	
Extremely rapid	3 m/sec (10 ft/sec) (6 mph)
⇓	⇓
Very rapid	0.3 m/min (1 ft/min)
⇓	⇓
Rapid	1.5 m/day (5 ft/day)
⇓	⇓
Moderate	1.5 m/month (5 ft/month)
⇓	⇓
Slow	1.5 m/year (5 ft/year)
⇓	⇓
Very slow	0.3 m/5 years (1 ft/5 years)
⇓	
Extremely slow	

Source: D. J. Varnes (1978).

of destruction and property damage, but rapidly moving masses not only destroy, they also kill. Rapid-moving mass movements have been big-time killers all around the world.

The main types of mass movement are downward, as in **falls** or **subsides,** or downward and outward, as in **slides** and **flows** (figure 15.17). Falling is downward from a topographic high place, such as a cliff or mountain, whereas subsiding is downward via collapse of the surface. Sliding occurs when a semicoherent mass slips down and out on top of an underlying failure surface. Flowing occurs when a moving mass behaves like a viscous fluid flowing down and out over the countryside. Next we examine each type of movement in detail.

Falls

Falls occur when elevated rock masses separate along joints, rock layers, or other weaknesses (figure 15.18). When a mass detaches, it mostly falls downward through the air via free-fall and then, after hitting the ground, moves by bounding and rolling. The triggers for falls may be heavy rainfall, frost wedging, earthquakes, or other factors. Perhaps you have been the triggering agent for small falls at one time or another?

Rock masses commonly are fractured into three nearly perpendicular joint directions. Each fracture or joint is a weakness that separates a block of rock. Watch out for rock blocks that are ready to fall.

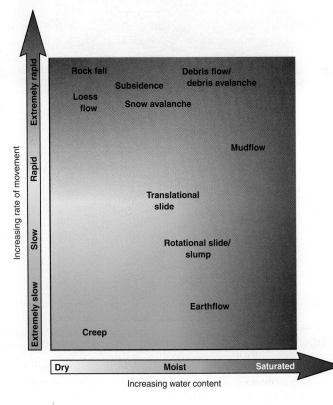

Figure 15.16 Mass-movement speed versus moisture content.

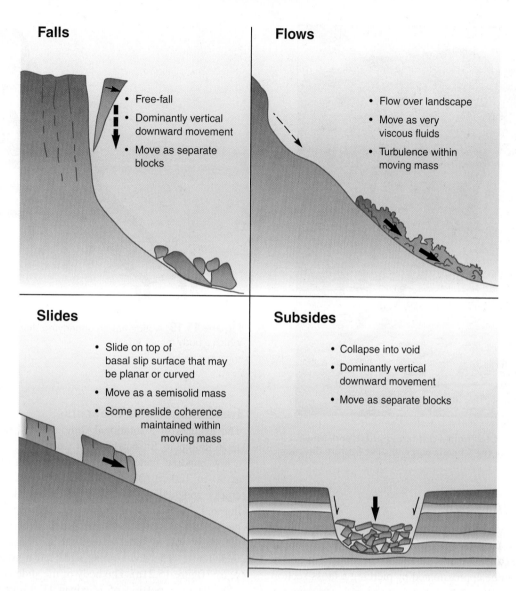

Falls

- Free-fall
- Dominantly vertical downward movement
- Move as separate blocks

Flows

- Flow over landscape
- Move as very viscous fluids
- Turbulence within moving mass

Slides

- Slide on top of basal slip surface that may be planar or curved
- Move as a semisolid mass
- Some preslide coherence maintained within moving mass

Subsides

- Collapse into void
- Dominantly vertical downward movement
- Move as separate blocks

Figure 15.17 Classification of mass movements.

YOSEMITE NATIONAL PARK, CALIFORNIA

During the summertime, Yosemite Valley fills with visitors awed by the steep-walled valley and its waterfalls. But steep walls are also the setting for gravity-powered rock falls. Below Glacier Point, at 6:52 p.m. on 10 July 1996, a 162,000-ton mass of granitic rock pulled away from the canyon wall in two separate masses, 14 seconds apart. Each slid down 165 m (540 ft) and then launched into the air for a 500 m (1,640 ft) drop in an arcing trajectory, reaching a speed of 270 mph before hitting the valley floor and shattering into a roiling cloud of pulverized rock that rolled across the valley and partway up the opposite wall before turning and flowing down the canyon. The falling granitic rock mass pushed air ahead of it in a gale-force blast that knocked down more than 1,000 mature trees and covered 50 acres with an inch-thick blanket of pulverized rock. The fallen rock masses hit the ground, creating a magnitude 3+ earthquake and a vertical column of dust 1 km high that blotted out the Sun. Despite the enormous amount of energy involved in this fall-shatter-flow event, only one person died—a 20-year-old man who was crushed by a tree blown down by the air blast in front of the fallen rock mass.

Slides

Slides, or landslides, are movements above one or more failure surfaces (see figure 15.17). Basal failure surfaces typically are either (1) curved in a concave-upward sense, as in **rotational slides** or **slumps,** or (2) nearly planar, as in **translational slides.** The slide mass maintains some degree of coherence as it slides down the underlying slide surface (figure 15.19).

Figure 15.18 A fall from a cliff face, South Island, New Zealand. Look for the 3 nearly perpendicular fracture directions.

Photo by Pat Abbott.

ROTATIONAL SLIDES

Rotational slides move downward and outward on top of curved slip surfaces (see figure 15.6). Movement is more or less rotational about an axis parallel to the slope. The center of rotation can be approximated by piercing a cross-section with the point of a compass and then swinging the compass in an arc to draw a basal failure surface (figure 15.20a). This is the Swedish circle analysis of slope stability used to calculate driving versus resisting forces, giving a quantitative understanding of how close a slope is to failing. Material above the failure surface can be divided into a driving mass resting on a base inclined out of the hill and a resisting mass sitting on a base inclined back into the hill. In effect, the slope is an equilibrium situation in which a driving mass seeks to break free and move downward and outward, but is blocked by a resisting mass that acts as a wedge holding the driving mass and slope in place.

When a slump occurs, its head moves downward and typically rotates backward (figure 15.20b). Water falling or flowing onto the head of the slide mass ponds in the basin formed by the backward tilt. The trapped water sinks down into the slump mass, causing more instability and movement. Because the scarp at the head of a slump is nearly vertical, it is unstable, thus setting the stage for further mass movement. The toe of a slump moves upward, riding out on top of the landscape (figure 15.20b).

Figure 15.19 A slide has an underlying slippery surface above which a body moves intact.

Ensenada, Baja California

The features of a rotational slide are well illustrated by a 1976 slump in Ensenada, Baja California. Arcuate cracks a few hundred meters long began forming on a hill slope at 90 to 120 m (295 to 400 ft) above sea level. As the cracks widened, most residents heeded this natural warning and evacuated their homes. But not everyone left; two people were asphyxiated during their sleep one night when the slow-moving slide severed the natural gas lines inside their house. The down-going slide left a pronounced head scarp (figure 15.21a). The body of the slide carried a 275 m (900 ft) long portion of Mexican Highway 1 toward the Pacific Ocean (figure 15.21b). The toe exhibited the classic upward bulge as it uplifted the seafloor to 9 m (30 ft) above sea level (figure 15.21c). Local residents lost little time in picking the abalone off the uplifted and exposed seafloor.

Slumps commonly move so slowly that people build and live on them. However, prolonged heavy rains, an earthquake, or some other trigger may increase the movement rate of the slump mass, resulting in significant damages. Can you recognize the head, body, and toe of the slump in figure 15.22?

Although the basal failure surface of a rotational slide is concave in the direction of movement, it rarely is as perfect an arc as is drawn by the Swedish circle analysis method (see figure 15.20a). Why? Because the failure surface usually cuts through heterogeneous materials that influence the rupture path. Discontinuities such as faults, joints, rock layering, and the presence of clay-rich seams can alter the shape of the failure surface. Slope failures beginning as rotational slides may transform downslope into translational slides.

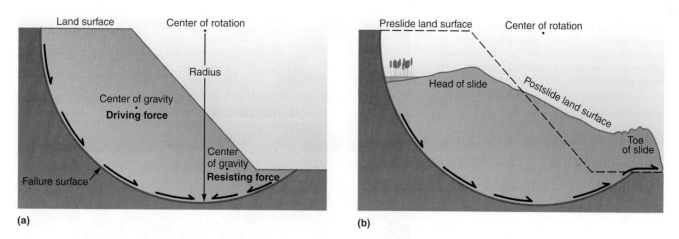

(a)

(b)

Figure 15.20 (a) Swedish circle analysis of slope stability. A compass is set at the center of rotation, and then an arc is swung, approximating the basal failure surface. Computations of driving and resisting forces help determine the stability of the slope. (b) A rotational slide with movement around a center of rotation. Notice the backward-tilted head and bulged toe.

(a)

(b)

(c)

Figure 15.21 (a) A portion of the head scarp of a 1976 rotational slide north of Ensenada, Baja California. The offset shown is the total of weeks of movement. Two people died in this house when the shifting land ruptured their natural gas lines, asphyxiating them during the night. (b) View north across the side of the Ensenada slide. Note the seaward shift of Mexican Highway 1. (c) Toe of the Ensenada slide. The ocean floor was lifted above sea level by the up-bulging toe.

Courtesy Michael W. Hart.

Figure 15.22 View of the international border between Tijuana, Mexico (left), and San Diego, California (right). Can you find the landslide? Look for the arcuate headscarp, slide body with fractures, and bulged-out toe.

Courtesy Michael W. Hart.

TRANSLATIONAL SLIDES

In translational slides, masses move down and out by sliding on surfaces of weakness, such as faults, joints, a clay-rich layer, soft rocks slipping off hard rocks, and hard rocks being spread apart by movements within underlying soft rocks. A translational slide may move as long as it sits on the downward-inclined surface and its driving mass still exists. In contrast, rotational slides move only short distances; their arcuate movements tend to restore equilibrium soon because the driving mass decreases and the resisting mass increases.

Translational slide masses behave in different fashions: (1) They may remain basically coherent as *block slides*. (2) The sliding mass may deform and disintegrate to form a *debris slide*. (3) *Lateral spreading* may occur where the underlying material fails and flows, thus causing the overlying coherent material to break apart and move. The various styles of translational slides may be best understood using specific examples.

Point Fermin, California, Block Slide

An excellent example of a block slide lies on the Palos Verdes Peninsula in Los Angeles. Just east of Point Fermin, the rocks include layers of sandstone and clay-rich mudstone inclined 10° to 22° seaward (figure 15.23a). Beginning in January 1929, a half-mile-long block slide, with 5 acres of its mass on land, began sliding slowly seaward down the inclined bedding on top of a particularly slippery clay layer. The slide surface daylights on an unsupported submarine slope, thus no resisting mass exists to hold the block in place. The slide block remains basically whole during movement. From January 1929 to June 1930, the block shifted from 2 to 2.5 m (6.5 to 8 ft) seaward. No one was killed by

this slow movement, but homes sitting atop 30 m (100 ft) high sea cliffs were slowly twisted out of shape and had to be removed (figure 15.23b). Movement apparently is triggered by excess water from yard irrigation seeping down to layers of weak clays, which expand and lose strength, and sliding begins. Slide movement rates are not constant; they accelerate after earthquakes.

Interstate 40, North Carolina, Debris Slide

Interstate 40 is a lifeline through the Great Smoky Mountains for residents of western North Carolina and Tennessee. But on 25 October 2009, the lifeline was blocked. A major rockslide occurred at mile marker 2.6 in Haywood County (figure 15.24). Layers of hard sedimentary rock slid downslope on top of softer sedimentary rock. While sliding, the hard-rock layers broke along fractures, transforming into a debris slide.

No one saw the slide in action, but several vehicles ran into the debris, causing injuries. Interstate 40 was closed for six months, forcing drivers to make a 53-mile-long detour. It took 7,000 truckloads to remove all the rock debris. Several similar settings along I-40 are now being stabilized before they too break loose and slide.

Turnagain Heights, Anchorage, Alaska, 1964, Lateral Spread

A magnitude 9.2 earthquake triggered numerous mass movements in Alaska. The most destructive slide occurred in the Turnagain Heights residential section of Anchorage as a translational slide of the lateral-spreading type. Part of Anchorage is built above a rock mass called the Bootlegger Cove Clay; it is composed of sediments that were carried

Figure 15.23 Point Fermin translational slide. (a) Cross section showing the block slide on top of an inclined slippery clay layer. Movement is toward the unsupported slope lying offshore below sea level. (b) View of the head of the Point Fermin slide. Note the foundation slabs of destroyed homes. San Pedro is in the background.

(a) Data from W.J. Miller, "The Landslide at Point Fermin, California," in Scientific Monthly 32:464-69, 1931. (b) University of Washington Libraries, Special Collections, John Shelton Collection, Shelton 3172.

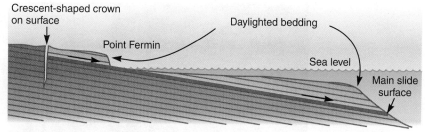

(a)

(b)

away from glaciers and deposited in estuaries and nearshore marine environments. The Bootlegger Cove Clay has a zone at depth rich in clays of low strength, high water content, and high sensitivity to disturbance. Above and below this weak zone are much stiffer clay units.

The Turnagain Heights homes sat above sea cliffs 21 m (70 ft) high. The slide moved toward the sea as a 2.6 km (8,500 ft) long mass extending 365 m (1,200 ft) inland. The ground surface on top of the slide broke into an irregular pattern, dropped an average of 11 m (35 ft), and moved 610 m (2,000 ft) toward the bay (figure 15.25a). Sliding did not begin until 90 seconds of violent earth shaking had liquefied some weak clays at depth. Mass movement began with rotational slides at the sea cliff and extrusion of weak clays from the toe. Rotational slides cut off the weak clay layer, thus preventing further escape of liquefied clays. After the weak clay layer was sealed off and trapped at depth, the violent shaking caused it to deform internally. A large ridge sitting on top of the clay began moving coastward, causing extension of rock layers behind it (figure 15.25b). Extensional pull-apart set the stage for the widespread dropping and sliding of blocks, and oozing upward of clay-rich masses.

Figure 15.24 A debris slide in western North Carolina blocks Interstate 40. Note: (1) tilted rock layers; (2) massive hard rock lying on softer rock; (3) daylighted bedding; and (4) fractures in hard rock.

North Carolina Department of Transportation.

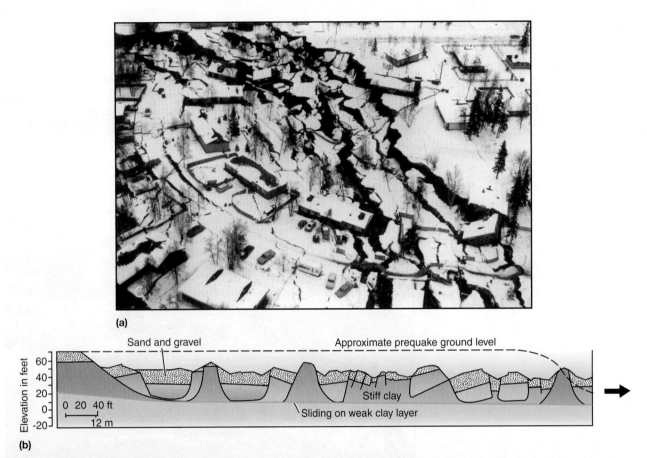

Figure 15.25 Turnagain Heights lateral spread. (a) View of head of Turnagain Heights landslide in Anchorage, Alaska, triggered by the 1964 earthquake. (b) Cross section through the slide. Lateral spreading occurred on top of weak clay layer, causing overlying rock masses to break, drop down, and slide.

U.S. Geological Survey.

During the last three minutes of earthquake shaking and for many seconds thereafter, the weak clay layer deformed and flowed, causing lateral-spreading sliding. Failure at depth caused rupturing of the firm layer at the surface. When movements stopped, some 75 homes lay ruined.

Flows

Flows are mass movements that behave like fluids. The materials within flows range from massive boulders to sand to clay to snow and ice to mixtures of them all. Water contents vary from dry to sloppy wet. Their velocities range from barely moving creep (figure 15.26), to observed speeds of 175 mph, to calculated speeds in excess of 200 mph. Within the moving masses, internal movements dominate, and slip surfaces are absent to short-lived.

There is a complete gradation from debris slides that move on top of slip surfaces to debris flows that do not require a basal slip surface. Many names have been used to describe flows—for example, earthflow, mudflow, **debris flow,** and debris avalanche. Examples will help clarify the different types of flows, but all are characterized by fluidlike behavior.

PORTUGUESE BEND, CALIFORNIA, EARTHFLOW

An expensive example of an ancient earthflow that is still moving exists in the Portuguese Bend area of Los Angeles. Some underlying rock layers are rich in volcanic ashes that have altered to **bentonitic** clays especially susceptible to absorbing water and swelling, with resultant loss of strength.

In the Portuguese Bend area, the rocks are folded into a broad concave warp (syncline) where the rock layers dip seaward (figure 15.27a). Ocean waves eat away at the toe of the slope, thus removing support and helping keep the slope moving slowly downhill. The site is an ancient earthflow, as shown by the rolling and hummocky topography that drops about 300 m (1,000 ft) to the sea, looking like a rumpled carpet with depressions, arcuate scarps, undrained depressions ("lakes"), uptilted and down-dropped blocks, highly fractured material, and crushed plastic, clay-rich materials exhibiting flow structures (figure 15.27b). The area had been used for truck farming—a good usage for unstable ground. There is no great harm in having a plowed field slowly deform.

In the 1950s, a development of 160 fairly expensive houses was built on the former farmland on top of the active

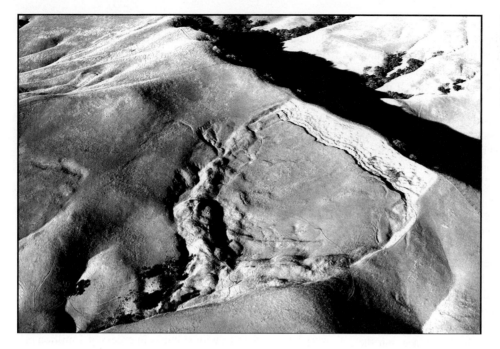

Figure 15.26 Large-scale creep high on a hill (upper right) causes soil and rock to flow downslope (lower left), central California.

University of Washington Libraries, Special Collections, John Shelton Collection, Shelton [no number].

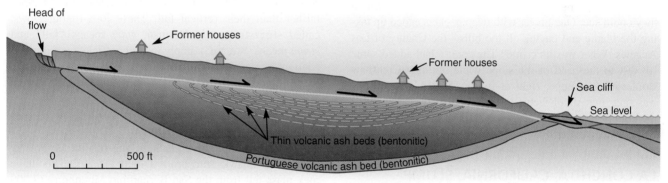

(a)

Head of flow

Former houses

Former houses

Sea cliff

Sea level

Thin volcanic ash beds (bentonitic)

Portuguese volcanic ash bed (bentonitic)

0   500 ft

(b)

(c)

Figure 15.27 Portugese Bend earthflow. (a) Cross section through Portugese Bend showing seaward-dipping rocks, bentonitic clay–rich rock layers, ocean waves eroding the base of the slope, and the pre–existing slide surface (green line). (b) View on 5 March 1958. Note the kink in the pier. Crenshaw Boulevard is the narrow band winding across the upper right of the photo. (c) View of the bulged-up toe of the earthflow, 24 November 1959. Note the remains of houses and roads and the damaged pier.

(b) University of Washington Libraries, Special Collections, John Shelton Collection, Shelton KC13935, (c) University of Washington Libraries, Special Collections, John Shelton Collection, Shelton 3165 slide. The homes were not served by sewers; all fluid wastes were poured below ground via septic tanks, where they reduced the stability of the clay-rich rocks. In 1956, about 25% of the area (400 acres) began moving again. It moved faster in the winter, up to an inch a day, and slower in the summer. Movement rates correlated with rainfall, with a one- to two-month lag time necessary for rainwater to sink below the surface and be absorbed by clay minerals. In three years, parts of the area moved more than 20 m (70 ft) seaward (figure 15.27c).

The homeowners were angry. Their beautiful homes with ocean views were breaking apart. With economic loss comes the urge to sue. Who, or what, gets the blame: (1) the seaward tilt of the rock layers; (2) the bentonitic clays that absorb so much water and then weaken; (3) the ocean waves that erode the toe of the slope, keeping the ocean waters muddy with material eroded from the resisting mass; or (4) the existence of ancient and still-active slide surfaces with water-catching depressions? Each of these features was a good reason not to build there in the first place, but the homeowners were looking for someone with money whom they could sue. One player with deep pockets ended up taking the blame and paying all the bills—the County of Los Angeles. What was its crime? The county had added some fill dirt to the head of the slope while building Crenshaw Boulevard (see upper right of figure 15.27b). The fill dirt was only one small factor in an overall bad situation, but the county had money, so it was found liable and paid the homeowners' expenses.

LA CONCHITA, CALIFORNIA, SLUMP AND DEBRIS FLOWS, 1995 AND 2005

The southern California coastline between Ventura and Santa Barbara features small communities nestled between steep hills and some world-famous surfing locations. La Conchita is one of the communities in this seemingly charmed region. Unfortunately, the stretch of 180 m (600 ft) high cliffs along this coastline has been affected by landslides since historical records began in 1865. The entire cliff behind La Conchita is part of a large ancient landslide. In March 1995, two events brought misery. A deep-seated, coherent slump evolved into an earthflow that moved slowly enough for people to get out of its path, but it destroyed nine houses (figure 15.28a). A few days later, a debris flow damaged five more houses. What triggered these mass movements? Months of extra-high rainfall raised the water table, triggering a failure at depth. The upper parts of the moving masses were dry and gave off clouds of dust.

Despite these destructive events, the lure of the physical setting has kept the community fully populated. Almost 10 years later, on 10 January 2005, about 15% of the 1995 slide mass remobilized into a highly fluid debris flow traveling about 10 m/sec (22 mph). The flow overran a retaining wall built to stop landslides, and then flowed into La Conchita, moving about 10 mph, destroying 13 houses, seriously

damaging 23 others, and killing 10 people (figure 15.28b). This event followed a 15-day period of near-record rainfalls.

Will destructive slumps and deadly debris flows continue to penetrate into La Conchita? Yes.

LONG-RUNOUT DEBRIS FLOWS

Many mass movements involve combinations of fall, slide, and flow at different times and places along their travel route. Common examples include slumps that change into earthflows, debris slides that become rock falls, and slumps that end as topples. But another complex movement is the most spectacular of them all. These are massive rock falls that convert into highly fluidized, rapidly moving debris flows that travel far and kill in great numbers.

Rock falls and small-volume avalanches tend to flow horizontally for distances less than twice their vertical distance of fall; these short distances of transport are due to the slowing effects of internal and external friction. However, very large rock falls, with volumes in excess of 1 million m^3, commonly travel long distances; some travel up to 25 times farther than their vertical fall. These long-runout flows, called **sturzstroms** (in German, *sturz* means "fall" and *strom* means "stream"), imply lower coefficients of internal friction. Sturzstroms have been observed moving at rates up to 280 km/hr (175 mph), even running up and over sizable hills and ridges lying in their paths.

Blackhawk Event, California

A beautiful example of an ancient long-runout debris flow is the Blackhawk sturzstrom (figure 15.29). This huge rock mass fell from the San Bernardino Mountains 17,000 years ago, flowing out onto the Mojave Desert east of Los Angeles. The age is known from fossils in pond sediments on top of the sturzstrom mass. The Blackhawk sturzstrom has a volume of 300 million m^3 that dropped 1.2 km (0.75 mi) vertically and steeply from Blackhawk Mountain and then flowed 9 km (5.6 mi) to form a lobate tongue of rock debris from 10 to 30 m thick, 2 km wide, and 7 km long. The Blackhawk flow traveled 7.5 times farther than it fell and is estimated to have moved at about 120 km/hr (75 mph). This type of event has occurred in historic times.

Elm Event, Switzerland, 1881

With the advent of compulsory education in Europe in the 19th century, there arose a demand for **slate** boards to write on in classrooms. To help satisfy this need, some Swiss farmers near Elm became amateur miners, quarrying slate from the base of a nearby mountain. By 1876, an arcuate fissure had formed about 360 m (1,180 ft) above their quarry, opening about 1.5 m (5 ft) wide. By early September 1881, the quarry had become a V-shaped notch about 180 m (600 ft) long and dug 60 m (200 ft) into the slope. At this time, the upslope fissure had opened to 30 m (100 ft) wide, and falling rocks were frequent, coupled with ominous noises coming from the large overhanging rock mass. These

(a)

(b)

Figure 15.28 La Conchita slump and debris flows. (a) 1995: In March, a slump and relatively slow debris flow moved into town, destroying 12 homes. (b) 2005: On 10 January, following many days of heavy rain, a highly fluid debris flow broke loose from the 1995 mass and moved rapidly through 15 homes, killing 10 people.

(a) R.L. Schuster/U.S. Geological Survey, (b) Mark Reid/U.S. Geological Survey.

Figure 15.29 View of the Blackhawk sturzstrom (debris flow) deposit. It fell from the San Bernardino Mountains 17,000 years ago and flowed out onto the Mojave Desert floor.

University of Washington Libraries, Special Collections, John Shelton Collection, Shelton KC2182.

signs caused the miners to halt work. The inhabitants assumed that the rock mass would fall down; they did not think it would also flow up a steep slope and down their mountain valley to bury 115 of them. But it did.

On 11 September 1881, the Elm event unfolded as a drama in three acts: the fall, the jump, and the surges up a slope and down the nearly flat valley floor (figure 15.30). Act 1, the fall, was described by Mr. Wyss, the Elm village teacher, from the window of his home.

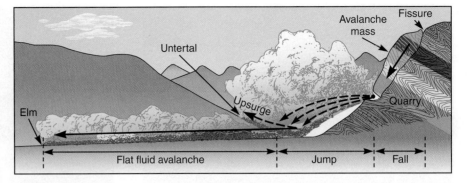

Figure 15.30 Cross-section of the 1881 Elm debris avalanche in Switzerland. A drama in three acts: the fall, the jump, and the surges. The rock debris deposited on the valley floor maintained the same relative positions it had in the bedrock layers of the mountain.

> When the rock began to fall, the forest on the falling block moved like a herd of galloping sheep; the pines swirled in confusion. Then the whole mass suddenly sank.

Apparently, the formerly rigid mass of rock had already begun to disintegrate during its free fall.

In act 2, the jump, the fallen mass hit the flat floor of the slate quarry, completely disintegrated, and then rebounded with a big jump forward, also described by teacher Wyss:

> Then I saw the rock mass jump away from the ledge. The lower part of the block was squeezed by the pressure of the rapidly falling upper part, disintegrated, and burst forth into the air. . . . The debris mass shot with unbelievable speed northward toward the hamlet of Untertal and over and above the creek, for I could see the alder forest along the creek under the stream of shooting debris.

The bottom surface of the jumping mass was sharply defined. Eyewitnesses could see trees, houses, cattle, and fleeing people under the flying debris. The upper surface was not so sharply defined; it was a cloud of rocks and dust. Residents of Untertal who saw the jumping mass coming toward them ran uphill to save themselves. This turned out to be the wrong choice, as part of the flying debris hit their hill slope, fluidized, and flowed upslope 100 m, overtaking and burying them. At the same time, some dogs and cows instinctively moved sideways, thus avoiding the debris flow.

In act 3, the surges, the disintegrated mass of rock was fully in contact with the ground and flowed rapidly down the valley floor. Its motion was described by Kaspar Zentner, who barely eluded the flow:

> "The debris mass did not jump, did not skip, and it did not fly in the air, but was pushed rapidly along the bottom like a torrential flood. The flow was a little higher at the front than in the rear, having a round and bulgy head, and the mass moved in a wave motion. All the debris within the stream rolled confusedly as if it was boiling, and the whole mass reminded me of boiling corn stew. The smoke and rumble was terrifying. I now ran breathlessly over the bridge and bent around the corner of Rudolf Rhyner's house. Then I turned back and held myself

firmly against the house. Just as I went past the corner the whole mass shot right past me at a distance less than 1 meter away. The debris flow must have been at least 4 meters high. A single step had saved me. During the last jump I noticed that small stones were whirling around my legs like leaves in the wind. The house crunched, moved and seemed to be breaking apart. I fled on hands and knees through the garden until I got to the street. I was then safe. I had no pain anywhere and no stones had hit me. I did not feel any particular air pressure."

Although the moving mass at Elm looked and behaved like a "torrential flood" and a "boiling stew," it was not a watery mass but a dry one whose internal fluid was dust and air. Visitors who later viewed the mass of deposited debris remarked on how similar it looked to a "lava flow." The facts are these: a mass of broken rock with a volume of 10 million m^3 dropped 600 m (1,970 ft) and then flowed 2.23 km (1.4 mi) as a dry mass moving at 180 km/hr (110 mph).

Turtle Mountain, Alberta, Canada, 1903

The coal-mining town of Frank, Alberta, occupied a beautiful site in the Oldman River valley and Canadian Rocky Mountains. Its place in history changed on 29 April 1903 when residents were startled at 4:10 a.m. by noise coming down from Turtle Mountain. A 90-million-ton mass of dipping limestone layers slid down their basal surface (daylighted bedding), dropped 1 km (3,000 ft) into the river valley, shattered, and then flowed 3 km (2 mi) across the valley and climbed 130 m (400 ft) up and over the terraced valley wall on the opposite side (figure 15.31). The whole event lasted only 100 seconds, but it pulverized and buried the southern end of town, killing about 70 people. Eyewitnesses included an engineer who had been backing up his train to the coal mine when he heard the rock breaking high on Turtle Mountain. He quickly switched to full-speed-ahead and chugged to safety; as he watched, miners at the loading dock sprinting for their lives were overrun and killed by the rock flow. The site is now an organized tourist attraction with the Frank Slide Interpretive Center, audiovisual program, and self-guided hiking trails.

Figure 15.31 A 90-million-ton mass of rock fell, shattered, and flowed down Turtle Mountain, across the Oldman River valley, and up the opposite wall of the valley on 29 April 1903. About 70 people were killed in the town of Frank, Alberta, Canada.

B. Bradley, University of Colorado/NOAA/NGDC.

Nevados Huascarán Events, Peru

Nevados Huascarán is the highest peak in the Peruvian Andes. The west face of the north peak is granitic rock cut by nearly vertical joints roughly parallel to the face. At 6:13 p.m. on 10 January 1962, with no perceptible triggering event, a huge mass of rock and glacial ice fell, initiating a debris flow. The debris flowed down river valleys like bobsleds, rising higher on the outsides of valley bends and lower on the inside bends. Some debris flowed up and out of the valley at bends, but most stayed between the valley walls and issued forth as a 10 to 15 m (30 to 50 ft) high mass, spreading out as a lobe covering 3.5 km² (1.4 mi²), including part of the town of Ranrahirca (meaning "hill of many stones"); 4,000 people died. The debris flow had a volume of 13 million m³ and traveled at speeds of 170 km/hr (105 mph). The slide left a scar on Nevados Huascarán, including a 1 km high overhanging cliff. A final report on the tragedy stated:

> The people are adjusting to this huge scar that lies across their land and their lives. . . . But they say that Huascarán is a villain who may yet have more to say.

Eight years later, on 31 May 1970, a subduction-zone earthquake of magnitude 7.7 occurred beneath the Pacific Ocean 135 km (84 mi) away, with a hypocenter at 54 km (33 mi) depth. Shaking lasted 45 seconds, but before that time was up, a gigantic portion of the same west-facing slope of the north peak of Nevados Huascarán failed with a sound like a dynamite blast or sonic boom, and a cloud of dark dust obscured the mountain from view. The side of the mountain between 5,500 and 6,400 m (18,000 and 21,000 ft) elevation fell away, including a 30 m (100 ft) thick glacier. The mass was composed of nearly 100 million m³ of granitic rock, ice blocks, glacial sediments, and water. It moved at speeds of 280 to 335 km/hr (175 to 210 mph), devastating an area of 22.5 km² (8.7 mi²) and killing 18,000 people (figure 15.32).

Figure 15.32 Aerial view of Nevados Huascarán and the 1970 debris avalanche that buried 18,000 people. Vertical drop from the summit to Rio Santa is 4,144 m (13,592 ft), and horizontal distance is 16 km (10 mi).

U.S. Geological Survey.

The event began as a fall, transformed into a debris avalanche with an airborne segment, and then moved down the Rio Santa as a debris flow for more than 50 km (30 mi). The sequence was:

1. A vertical fall for 400 to 900 m (1,300 to 3,000 ft).
2. The fallen mass landed on a glacier and slid along its surface, scooping up snow.
3. The debris avalanche raced up the side of a glacial-sediment hill, launching much debris into the air.
4. For the next 4 km (2.5 mi) downslope, boulders weighing up to several tons each rained from the sky, pulverizing houses, people, and other animals. The deadly rain of mega-boulders left the ground pock-marked with craters like a heavily bombed battlefield.
5. The mass recombined as a flow, reaching the 230 m (750 ft) high Cerro de Aira (*cerro* is Spanish for

Flows **429**

"hill"), which had protected the city of Yungay during the 1962 event. No such luck this time. A lobe of debris overflowed Cerro de Aira, burying Yungay and 18,000 people beneath more than 30 m (100 ft) of debris. It was especially bad timing because this was a Sunday afternoon and the population was swollen with visitors from the surrounding region. The scene was well described by survivors who witnessed the event from Yungay's highest point and the safest place in town, the cemetery.

6. Meanwhile, the main mass of material continued racing toward the Rio Santa, preceded by a strong wind pushed in front of it. The main lobe of debris swept across the Rio Santa and ran 83 m (275 ft) up the far slope, killing 60 people in the town of Matacoto, before it fell back like an ocean wave from the shore.

What does the future hold for this site? Today, the glacier above the 1962 and 1970 breakaway sites has large fissures, suggesting that the mountainside beneath is still fractured and unstable. The odds are high that the repopulated areas downslope will again experience another killer debris flow.

Movement of Sturzstroms

In a long run-out flow, how do such large masses of debris move so far and so fast and behave so much like a fluid? Numerous hypotheses have been proposed: (1) Some hypotheses rely on water to provide lubrication and fluid-like flow, but some observed flows are masses of dry debris, as at Elm. (2) Other hypotheses invoke the generation of steam to liquefy and fluidize the moving mass, or (3) call for frictional melting of material within the moving mass; both ideas fail because some long-runout deposits contain blocks of ice or lichen-encrusted boulders, showing that no great amount of internal heat was generated, nor was internal friction ever at a very high level. (4) A popular hypothesis suggests that the falling mass traps a volume of air beneath it and then rides partially supported on a carpet of trapped air that enables it to travel far and fast. This idea has never been verified and seems most unlikely. For example, after its early airborne jump, the Elm sturzstrom was described as being in contact with the ground, and in fact, it dug up pipes buried a meter below the surface. Additional lack of support for the air-cushioned flow hypothesis is the presence of deposits with identical flow features on the ocean floor, and on the Moon and Mars, where no or very little atmosphere is available.

It appears that neither water, heat, nor a trapped cushion of air is necessary for a long-runout debris flow. So how do these large masses do it? A remarkable fact helps guide the formulation of another hypothesis. After the Elm mass fell, disintegrated, jumped, and flowed 2.23 km (1.4 mi) and the rubble had come to rest, the original layering in the bedrock of the mountainside was still recognizable. Even though the debris had flowed, according to all eyewitness accounts, the

hunks of debris stayed in their same relative positions (see figure 15.30). In the words of the German geologist Albert Heim, who studied the scene in 1881:

> When a large mass, broken into thousands of pieces, falls at the same time along the same course, the debris has to flow as a single stream. The uppermost block, at the very rear of the stream, would attempt to get ahead. It hurries but strikes the block, which is in the way, slightly ahead. The kinetic energy, of which the first block has more than the second, is thus transmitted through impact. In this way the uppermost block cannot overtake the lower block and thus has to stay behind. This process is repeated a thousandfold, resulting eventually in the preservation of the original order in the debris stream. This does not mean that the energy of falling blocks from originally higher positions is lost; rather the energy is transmitted through impact. The whole body of the mass is full of kinetic energy, to which each single stone contributes its part. No stone is free to work in any other way.

Who would guess that the pieces of rubble in a rapidly moving debris stream would keep their relative positions next to their neighbors? How can this relationship be explained? A provocative hypothesis involving acoustic energy within the moving mass has been proposed by the U.S. geophysicist Jay Melosh. Apparently, an immense volume of falling debris produces much vibrational or acoustical energy within its mass. The jostling and bumping back and forth of fragments produces acoustic (sound) energy that propagates as internal waves. The trapped acoustic waves may act to fluidize the rock debris, allowing the rapid fall velocity to continue as rapid-flow velocity in a process called **acoustic fluidization.**

Long-runout Debris Flows Strike Again—Washington, 2014

Oso, Washington, is a forested community of 180 people living along the north fork of the Stillaguamish River. March 2014 had been a wet month with 22 inches of rain saturating the area. Then on Saturday the 22nd at 10:37 a.m., the saturated earth of the hill collapsed and a mass of mud up to 21 m (70 ft) thick flowed outward at 100 km/hr (60 mph), demolishing 49 homes on Steelhead Drive, engulfing cars traveling along State Highway 530, and breaking into pieces the house timbers, vehicles, and human bodies. Rescuers could hear voices of trapped people on Saturday night but the mud was still slowly moving and emitting eerie sounds of breaking wood and moving debris. By the time rescuers could get through the oozing mud the voices had stilled, and 43 people were dead.

What happened to turn an idyllic area into a zone of unimaginable horrors? The Steelhead Drive environment of forest-covered hills and river hides the facts that the hills are made of loosely consolidated sediments washed out from retreating glaciers during the past few thousand years; the river is eroding the base of the nearby hill; and landslides have been occurring along the Stillaguamish River for thousands of years (figure 15.33a). The immediate area experienced the huge Hazel landslide in 1937 and several later landslides reactivated from the Hazel mass, including a major landslide in 2006. Some mitigation work was done but the 22 March mudflow went through it as if it were tissue paper.

The river eroding into the resisting mass at the base of the hill sets up inevitable landslides. The unique aspect this time was the water-saturated, clay-rich sediment layers at the base of the hill; these clays, deposited in a former glacial lake, are exceptionally weak. The first landslide broke loose at 185 m (600 ft) above the river; as it fell it greatly increased the pore-water pressure in the clayey sediments at the base of the hill. The overpressurized clay sediments squirted out from below the landslide as a "wave" of mud traveling 100 km/hr (60 mph) that shattered trees and houses in its path. Eyewitness survivors described a "tsunami" of mud 8 m (25 ft) high racing across the valley with sounds like freight trains, plane crashes, and collisions. Even after it stopped the mud was so unstable that objects dropped on top of it quickly sank out of sight (figure 15.33b).

In all there were three major events: (1) At 10:37 a.m. a landslide mass dropped 185 m (600 ft), squeezing out (2) a wall of mud that traveled for about 138 seconds, overwhelming everything in its path; the runout was six times the vertical drop. (3) At 10:42 a.m. the unstable hill failed in a second major landslide that ran into the trailing edge of the first landslide and stopped (figure 15.33c). In the following hour, another dozen smaller landslides occurred from the unstable hill.

Even after all these earth movements the area is not any safer than it was before. More landslides are coming.

Snow Avalanches

Heavy snowfalls on steep slopes yield to the pull of gravity and fail as snow avalanches. They may be understood using the same mass-movement principles that apply to earth and rock. Just like earthen mass movements, snow avalanches creep, fall, slide, and flow. In effect, the snowfall

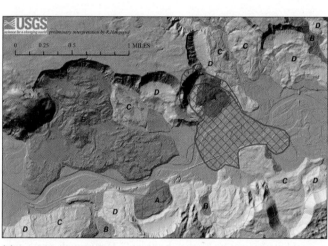

(a)

(b)

(c)

Figure 15.33 Oso landslides and debris flows.
(a) Shaded relief map of numerous past landslides in the Oso area. The March 2014 event is upper A. The red hachured line is the 2014 head scarp and the red cross-hatched pattern is the long runout area buried in mud and debris. Map by Ralph A. Haugerud, USGS. (b) View southward over the toe of the mud- and debris-covered long runout area, 1.13 km (0.7 mi) from the headscarp. (c) Aerial overview shows head scarp amphitheater in upper right. Moving to the left in the direction of movement: in right center lies well-defined mass of landslide number 2; in center is disintegrated mass of landslide number 1; in left center is mud- and debris-covered toe of clayey debris flow (see figure 15.33b).

(b) Photo from U.S. Navy, (c) Photo from Washington State Patrol.

and wind-blown accumulations high on slopes act to "load the head" of the slope, thus setting the stage for failure. Avalanche volumes vary from small to very large. They move at rates ranging from barely advancing to measured speeds of 370 km/hr (230 mph) in Japan. Their travel distance varies from only a few meters to several kilometers.

Avalanches commonly initiate on steep slopes when snowfall builds to 0.5 to 1.5 m (1.5 to 5 ft) thick. But snow thicknesses can reach 2 to 5 m (6 to 16 ft) before failing in big avalanches that can be devastating. Why does snow sometimes build into thick masses on steep slopes? It depends on the internal structure of the snow. The delicate snowflakes are crystals of ice with air in their pore spaces, but they can quickly change. A snowmass is commonly warmer than the surrounding air, so snowflakes within the mass transform. Snowflake exteriors melt and refreeze, becoming cemented together, rounded and packed, allowing the snow mass to build to a greater thickness.

The styles of avalanches vary. Two main types are *loose-powder avalanches* (flows) and *slab avalanches* (slides). Loose, powdery snow has a low amount of cohesion. This weak material may have 95% of its volume as pore space. A loose powder avalanche typically fails at a point source, and then triggers more and more snow into moving during its downhill run and spreading out laterally while moving downslope (figure 15.34). Avalanches begin where slopes are steepest (usually 30° to 45°) and then move down slopes of 20° to 30°, commonly guided by topography, and finally come to a halt in the runout zone (slopes usually less than 20°).

Slab avalanches involve slabs of cohesive snow breaking free from their poorly anchored base (figure 15.35). The failure is analogous to a translational slide, where an upper mass breaks free and slides down and out on top of a layer beneath it. The sliding slabs typically break up and turn into flows during their downslope movement. The key factor in understanding slab avalanches is recognizing that the snow mass on a mountain is made of many different layers of snow that formed at different times. Snow layers develop one by one, storm by storm. Each layer of snow has its own characteristics of thickness, strength, hardness, and density. Slab conditions can develop when a layer of snow deposited by strong winds is added on top of snow deposited by light winds, or a layer of heavily cemented snow sits on a layer of loose snow, or a warm, dense layer of snow sits on a cold, low-density layer. Within the pile of snow layers are melt-freeze crusts that form during times of surface exposure between snowfalls. The result is an inhomogeneous pile of separate and distinct layers that contain numerous potential failure surfaces within them.

A "typical" avalanche may be 0.6 to 0.9 m (2 to 3 ft) deep at its head scarp and 30 to 60 m (100 to 200 ft) wide, and it may descend 90 to 150 m (300 to 500 ft). If the snow is dry, avalanche speeds will be about 65 to 100 km/hr (40 to 60 mph), and if the snow is wet, speeds will be about 30 to 65 km/hr (20 to 40 mph).

Figure 15.34 A loose-powder avalanche flows toward three hikers.
© Royalty-Free/Corbis.

How far can an avalanche travel? How much protection does topography provide? In one example, an avalanche of about 765,000 m³ was unleashed down the slopes of Mount Sanford in Alaska on 12 April 1981; it dropped more than 3 km (2 mi) and flowed for 13 km (8 mi), including running up and over a 900 m (3,000 ft) high ridge.

Are you safe inside a ski resort? The Alpine Meadows ski resort in California was hit by an avalanche at 3:45 p.m. on 31 March 1982. High on a slope, cracks developed about 3 m (10 ft) deep and 900 m (3,000 ft) long. The resulting avalanche of 65,000 tons of snow moved as a 10 m (30 ft) high mass traveling 130 km/hr (80 mph). The avalanche rammed into the ski resort, crushing the ski lift and buildings and killing seven people. Being in the snow-covered mountains is great sport, but while there, thinking about mass-movement principles could be worthwhile.

In today's world, the major effect of avalanches is felt by people during recreational skiing, climbing, snowmobiling,

Figure 15.35 A slab avalanche is triggered by a downhill skier. Notice the vertical cracks at the head, the broad area of rupture, and the moving slabs.

Photograph by Ludwig/U.S. Forest Service/USGS Photo.

and hiking. Average annual death figures for some countries are 35 in Austria, 30 in France, 30 in Japan, 26 in Switzerland, 25 in Italy, 28 in the United States, and 14 in Canada. In the United States, 96% of avalanche deaths occur in eight western states: Colorado (6/yr), Alaska (4/yr), Utah (3.5/yr), Montana (3.5/yr), Washington, Wyoming, Idaho, and California. In Canada, almost all the avalanche deaths are in British Columbia and Alberta. Canadian doctors examined 204 victims of avalanches between 1984 to 2005; the deaths were 75% due to suffocation, 24% to trauma, and 1% to hypothermia (table 15.3). A profile of a typical avalanche victim in the Canadian study is a male in his 20s, a backcountry skier/snowboarder who triggered an avalanche between noon and 2 p.m. in January, February, or March. In an update since 2005, the doctors state that snowmobiling is the activity that now results in the most fatalities.

The chances of surviving an avalanche are good if you can remain near the surface of the avalanche (wear an inflatable airbag to help) and you are found by companions who are prepared to dig you out; wear a beeper and a beacon light to help them find you. If you are alone, or *if your companions have to go find a rescue party, you will probably die*. Remember that suffocation is the leading cause of avalanche deaths.

Almost all avalanche deaths are *not* associated with ski-lift served and maintained slopes. Please be prepared before entering the backcountry and recognize that a sense of "it won't happen to me" can be fatal.

Submarine Mass Movements

A whole range of familiar mass movements takes place below the sea. Rotational slumps occur within the sediment masses deposited at the mouths of large rivers; complex failures occur within the distorted rock masses at subduction zones; and debris avalanches slide down the slopes of submarine volcanoes. These large movements can cause killer tsunami, so they are covered in chapter 8.

TABLE 15.3

Avalanche Deaths in Western Canada, 1984–2005

Activity	Deaths	Males	Suffocation	Trauma
Skiing/snowboarding	123	104	88	35
Snowmobiling	44	42	40	4
Ice climbing	13	13	7	5
Mountaineering	11	8	8	2
Snowshoeing/hiking	8	7	8	—
Working on avalanche control	4	4	3	1
Total	203	178 (88%)	154[*] (77%)	47[*] (23%)

[*] Out of 201 victims

Source: Canadian Medical Association Journal.

Mitigation

Mass movements occur in so many places around the world that almost everyone can feel or see their effects at some time. Gravity pulls on rock, snow, and water everywhere. We try to mitigate the effects of their movements using many types of engineering actions.

RESHAPING TOPOGRAPHY

Earth-moving equipment is used to lessen the likelihood of landslides. The tops of hills can be unloaded by excavation; this lessens the driving mass (figure 15.36). The loads of removed earth can be added as buttresses to the base of the hill; this increases the resisting mass. Equipment is used to remove earth to lessen the steepness of hillslopes and thus lessen the pull of gravity.

STRENGTHENING SLOPES

Rock masses can be strengthened by emplacing rock bolts (figure 15.37). Reinforcing is also accomplished by inserting tie backs and shear pins (see figure 15.36). Hillsides can be held in place by building retaining walls (figure 15.38). The flow of water or snow can be controlled using **gabions** (figure 15.39). Gabions also are used frequently as buttresses or retaining walls; heavy, but loose, rocks held in place by steel mesh have the advantage of shifting about internally to reduce stress; this ability to adjust can be advantageous compared to a rigid concrete barrier, which cannot adjust so it cracks instead.

Figure 15.37 Rock bolts attempt to stabilize a slide mass by extending through the landslide and anchoring into solid bedrock. Wire mesh stops smaller rockfalls.

U.S. Geological Survey.

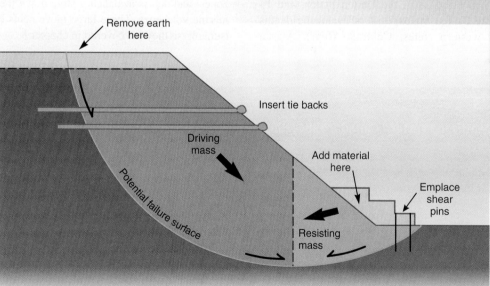

How to prevent a landslide: (1) Unload the head, (2) Strengthen the body, (3) Support the toe

Figure 15.36 A hillslope of homogeneous materials may fail along an arcuate basal surface. The slope is in equilibrium when a driving-mass portion is kept from moving by a resisting-mass portion. Removing material from the driving mass, strengthening the body, or adding to the resisting mass can prevent a landslide. Does this situation bring to mind any construction practices in your area?

Figure 15.38 Retaining wall helps keep steep hillside of sandstone from sliding onto road. La Mesa, California.
Photo by Pat Abbott.

Figure 15.39 A rock wall (gabion) held in place by steel cables directs the flow of water in Alabama. The same type of structure works well for steering snow avalanches or debris flows, or acting as a retaining wall holding back a hillside.
USDA.

Figure 15.40 Concrete-lined drainage ditches carry away rain runoff, thus preventing slope saturation. Note the succulent ice plants on the left that reduce erosion and make a fire break.
Photo by Pat Abbott.

Figure 15.41 Subsurface drainage may be controlled using a lined, gravel-filled trench containing pipes with perforated tops (French drains) that allow water to seep inside the pipes where it then flows away.
© tfoxfoto/Getty Images RF.

DRAINING WATER

Mass movements are more likely to occur in wet earth. There commonly is an incentive to reduce the amount of water that sinks into a slope. On the ground surface, concrete-lined ditches are simple structures that divert water (figure 15.40). If drainage ditches exist near your home remember to periodically clean out the collected rocks, soil, and plants so the rainwater runoff can efficiently flow away. If the ground is too saturated, a simple aid is to dig a trench and fill it with gravel and pipes with holes in the top (French drain); the perforations let water into the pipes, which then flows away (figure 15.41). If the slope is unstable then more drastic measures are needed to reduce the amount of underground water such as eliminating grass lawns, water-loving plants, and sprinkler systems.

Figure 15.42 After a fire on Kwaay Paay, netting was placed on the slope to control erosion and hold soil, allowing new plants to grow.

Photo by Pat Abbott.

Figure 15.43 Water runoff on bare ground can be slowed by using straw wattles.

Photo by Pat Abbott.

CONTROLLING EROSION

Heavy rains can wash down bothersome volumes of soil and earth. An effective way to slow erosion is by placing mats or netting on the slope; nets break up water flow and help hold soil in place while new plants grow their soil-holding roots (figure 15.42). Straw wattles are rolls of snake-like, net-encased straw that can be laid on the ground and staked down to help control erosion (figure 15.43).

How can you protect your property from mass movements? Mitigation of mass movements first involves you stopping, looking, and recognizing hazards in advance. Then visualize how to stop, control, or divert a potential moving mass. Finally, build the appropriate structures to control the hazard.

Subsidence

In **subsidence,** the ground moves downward (see figure 15.17). The surface either sags gently over long lengths of time or drops catastrophically in hours or days as open spaces in rocks collapse. This is not the down-dropping associated with fault movements or volcanism, but either the slow compaction of loose, water-saturated sediments or the rapid collapse of overlying earth into caves.

CATASTROPHIC SUBSIDENCE

There was a loud crash, screams, and then silence. At 11 p.m. on 28 February 2013, in a matter of seconds, a bedroom floor collapsed into a cavern, taking a 37-year-old man down into a 20 ft by 20 ft sinkhole. From the outside, the four-bedroom, sky-blue house in Seffner, Florida, appeared undisturbed; the five other family members inside the house at the time were not physically affected—but a beloved family member had disappeared, and his body was never found. Another Florida sinkhole had formed without warning.

Limestone Sinkholes, Southeastern United States

Limestone is a different kind of rock. Most of it is formed in warm, shallow seas by the accumulation and disintegration of the shells and skeletons of organisms. These organisms remove calcium (Ca), carbon (C), and oxygen (O) from seawater to make their skeletal material of calcium carbonate ($CaCO_3$). Some organisms build limestone directly as reefs. Other organisms die, and then their shells, spines, and other mineralized remains are bound together by $CaCO_3$ precipitated as cement to fill the void spaces.

Limestone formed extensively in the geologic past when the central and southeastern United States were flooded by warm, shallow ocean waters. Today, the limestones in these regions have naturally acidic freshwater flowing through them, dissolving them and forming extensive networks of caverns. When the levels of underground water drop during a drought or due to pumping of groundwater, the removal of the water lessens the internal support that helps hold up the roofs of caves. The loss of buoyant support that occurs when groundwater is drained from caves weakens some so much that their roofs collapse suddenly and catastrophically to form **sinkholes.** Every year, the print and electronic media show us examples of these sudden collapses. In Alabama alone, an estimated 4,000 sinkholes have formed since 1900.

Much of central Florida is underlain by limestone that is covered in most areas by 15 to 30 m (50 to 100 ft) of muddy sands. If the underground water body is lowered because drought has reduced the water supply or humans have overpumped wells, the caverns in limestone may be drained of water. When the water is removed, the weakened cavern may collapse. On 8 May 1981, a small depression on the ground in Winter Park, Florida, grew to a 45 m (150 ft) diameter collapsed pit (sinkhole) within 15 hours. Before a week was up, the sinkhole was 100 m (325 ft) across and 34 m (110 ft) deep (figure 15.44).

Figure 15.44 Aerial view of Winter Park, Florida, in May 1981. The sinkhole is 100 m (325 ft) wide and 34 m (110 ft) deep. Note the failed municipal swimming pool at bottom of photo and the four-lane road on the left.
U.S. Geological Survey.

SLOW SUBSIDENCE

In many areas of the world, the ground surface is slowly sinking as fluids are removed from below the surface (table 15.4). When water or oil are squeezed out or pumped up to the surface, the removal of fluid volume and the decrease in pore-fluid pressure cause rock grains to be crowded closer together; this results in subsidence of the ground surface (figure 15.45).

The fluids within rocks help support the weight of overlying rock layers. The effect is similar to that of carrying a friend in a swimming pool, where the water helps support your friend's body weight. Some examples will help illustrate the subsidence problem.

Groundwater Withdrawal, Mexico City

The Valley of Mexico has been a major population center for many centuries. People need water, so the Aztecs, and later the Spanish, built aqueduct systems to bring water in from the surrounding mountains. In 1846, residents realized that a large volume of groundwater lay beneath the city, so they

In Greater Depth

TABLE 15.4

Some Subsiding Coastal Cities

City	Maximum Subsidence (meters)	Area Affected (km²)
Bangkok, Thailand	1	800
Houston, Texas	2.7	12,000
London, England	0.3	300
Long Beach, California	9	50
Mexico City, Mexico	10	3,000
Nagoya, Japan	2.4	1,300
New Orleans, Louisiana	3	175
Niigata, Japan	2.5	8,300
Osaka, Japan	3	500
San Jose, California	3.9	800
Savannah, Georgia	0.2	35
Shanghai, China	2.7	120
Taipei, Taiwan	1.9	130
Tokyo, Japan	4.5	3,000
Venice, Italy	0.3	150

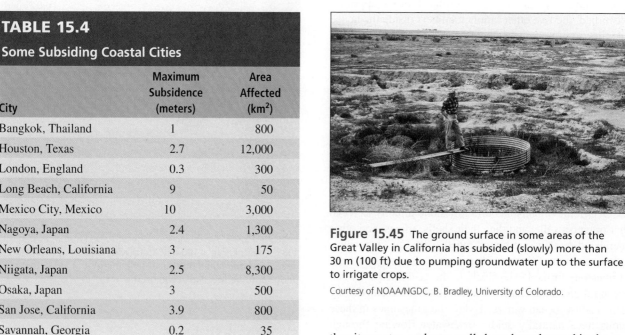

Figure 15.45 The ground surface in some areas of the Great Valley in California has subsided (slowly) more than 30 m (100 ft) due to pumping groundwater up to the surface to irrigate crops.
Courtesy of NOAA/NGDC, B. Bradley, University of Colorado.

the city center, and new wells have been located in the north and south of the valley. Now, land subsidence of 2 to 8 cm (1 to 3 in) a year is occurring in the new areas of groundwater withdrawal. Supplying the water needs of 20 million people is not easy in an area where the evaporation rate is greater than the precipitation rate.

Oil Withdrawal, Houston-Galveston Region, Texas

One of the earliest cases of fluid withdrawal causing ground subsidence occurred around the Goose Creek oil field between Houston and Galveston. Beginning in 1917, pumping brought large volumes of oil, natural gas, and associated water to the surface. As a result, the ground surface subsided enough for seawater in San Jacinto Bay to submerge the area. (Because states own the mineral rights on submerged lands, the state of Texas sued to take over the oil field that now lay below seawater. Ultimately, the state lost its case.)

began drilling wells and making large withdrawals from this convenient underground supply. Through the years, the water has been withdrawn faster than the natural rate of replenishment. People are using more of the underground water each year than rainfall can resupply, so the land subsides. In the center of the city, the land sank about 10 m (30 ft) between 1846 to 1954. The city center now lies lower than the level of nearby Lake Texcoco.

How can land subsidence be stopped? By not pumping out groundwater. Can the land subsidence in Mexico City be reversed? No. Groundwater withdrawal is now banned in

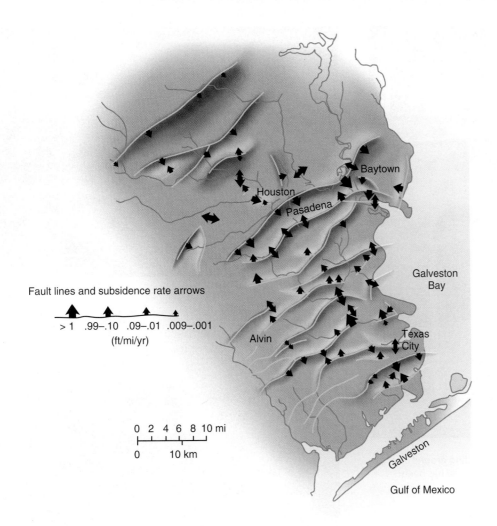

Fatal lines and subsidence rate arrows

> 1   .99–.10   .09–.01   .009–.001
(ft/mi/yr)

0 2 4 6 8 10 mi

0   10 km

Figure 15.46 Map of some active faults in the Houston, Texas, area and relative amounts of subsidence.

Houston-Galveston is one of the largest metropolitan areas in the country. It has relied heavily on groundwater withdrawals to supply its needs. At present, an area greater than 12,000 km² has sunk up to 2.7 m (9 ft). Another significant result of the ground settling is renewed movements on old faults (figure 15.46). These faults act as mega-landslide surfaces upon which the thick pile of coastal sediments can slowly slide toward the Gulf of Mexico. The faults are active but do not produce earthquakes. However, they do break up houses, commercial buildings, roads, and other features.

Long-Term Subsidence, Venice, Italy

Venice is one of the most improbable cities in the world. It was established during the collapse of the Roman Empire in the fifth century CE when local people moved into marshes and islands in a malaria-ridden lagoon to gain some protection from invading armies from the north. The city is built on soft water-saturated sediments that compact and sink under the weight of its buildings at the same time that global sea level is rising. From about 400 to 1900 CE,

Venetians struggled to stay above water as sinking land helped the sea rise about 13 cm (5 in) per century (figure 15.47). To stay above water, Venetians built up the islands using boatloads of imported sand and slowed the sinking of buildings by driving wood poles down into the sediment to make more stable foundations. In the 20th century, the rate of rise of the sea doubled to about 25 cm (10 in) per century due in part to pumping up groundwater during the 1930s through the 1970s. Projections for the 21st century suggest that the sea will rise about 80 cm (30 in). What can Venetians do to save their sinking city? One proposal is to spend billions of dollars to install movable floodgates across the three entrances to the lagoon to stop Adriatic seawater from flooding the lagoon during high tides and storms. However, these same floodgates would harm the economy by disrupting shipping and cause health problems by blocking the outflow of pollutants. Should the floodgates be built? Or should Venetians do as their ancestors did and keep bringing in sediment to raise the ground level? A third proposal is to pump seawater or carbon dioxide into a sand mass lying 600 to 800 m (2,000 to 2,600 ft) below the city in an attempt to "pump up" the region by about 30 cm (12 in) in 10 years. There is no easy and permanent solution for this sinking city.

Delta Compaction, Mississippi River, Louisiana

Deltas are new land formed where rivers drop their loads of sand and mud at the shoreline while building out into seas or lakes. Deltas are popular sites for cities. Many resources exist side by side at deltas: fertile soils, freshwater for drinking and farming, marine foods, tempered climate, and trade and transportation possibilities where rivers from continental interiors meet the world ocean. The problem is that deltas are only loose piles of water-saturated sand and mud, and thus they compact and sink. Each new layer of sediment adds weight on top of older layers, forcing the grains of sediment closer together and squeezing out more of their contained water. The Mississippi River delta at the shoreline is underlain by a 6 km

Figure 15.47 Venice is built on the subsiding ground of a river delta. Notice how the former doorway has been bricked shut to keep out the sea.

© MedioImages/Getty Images RF.

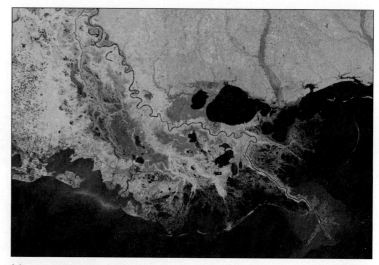

(a)

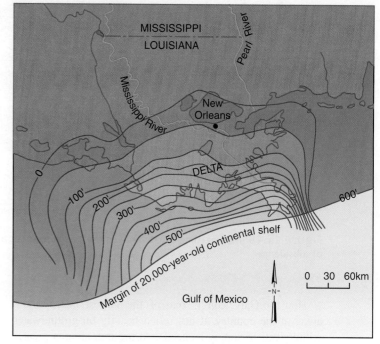

(b)

Figure 15.48 (a) The Mississippi River delta is made of loose sand, mud, and water. The sediments compact, and ground sinks below the sea. (b) Subsidence in the Mississippi River delta area during the last 20,000 years. Contour lines are in feet.

(a) NASA Earth Observatory image by Michael Taylor and Adam Voiland, using Landsat data from the U.S. Geological Survey. (b) Source: After H. Gould, 1970.

(3.7 mi) thick pile of sediments deposited during the last 20 million years (figure 15.48a). In the last 20,000 years, the region has sunk markedly due to sediment compaction, dewatering, and isostatic adjustment (figure 15.48b). Today, the mouth of the Mississippi River is near the eastern edge of the delta it has built in the last 20,000 years. But the river shifts the positions of its course and mouth frequently. At present, it is held in its position only by the ongoing human actions of river-bottom dredging (deepening) and levee building (raising) along its sides.

New Orleans slowly subsides with the delta. The sinking of the city is accentuated in places by pumping up underground water. Parts of New Orleans have sunk about 3 m (10 ft) during the past 50 years. Today, about 45% of New Orleans lies below sea level, making the city ever-more prone to damages from the high-water surges associated with hurricanes. Many other coastal cities also are subsiding due to delta compaction and fluid withdrawal.

Summary

Gravity tugs incessantly at all landforms on Earth, commonly causing failures called mass movements. These movements range from the barely perceptible surface creep of hill slopes to debris flows moving in excess of 200 mph. The movement of surface materials sets the stage for creep. Soil and rock swell upward perpendicular to the surface due to freezing of pore water, wetting of clays, and volume expansion upon solar heating. They shrink straight downward under the pull of gravity upon thawing, drying, and cooling, resulting in a net movement downslope.

Slope failure may be due to external factors, such as steepening of the slope, adding mass upslope, or removing mass low on a slope. Internal factors that make a slope weak include inherently weak materials such as clay minerals; decreasing cohesion through solution or internal erosion; weight of pore water and elevated pore-water pressure; and the presence of adverse geologic structures, such as inclined bedding, fault surfaces, and clay seams. Many factors push a slope toward failure. Movement usually is initiated by a triggering event, such as heavy rains, an earthquake, or human activities.

Major types of mass movement are downward, as in falling and subsiding, or downward and outward, as in sliding and flowing. Falls are rock masses dislodged from elevated slopes; if big enough, they may transform downslope into slides or flows.

Slides are mass movements on top of failure surfaces. Concave-upward, curved failure surfaces produce rotational slides. Here, the slide head tilts backward, the toe bulges upward, and little distance is traveled. Rotational slides are a common destroyer of property, but their slow movements rarely kill anyone. Slides on top of inclined, planar surfaces are called translational; they can travel far and fast. Translational masses may remain essentially coherent as block slides that destroy property, as at Point Fermin, California; they may disintegrate into debris slides, killing thousands of people, as at Vaiont, Italy; or they may spread laterally at the surface when an underlying weak layer of rock deforms, as at Anchorage, Alaska.

Flows are mass movements that behave like fluids, even when they are dry. Flows may be made of gravels, sands, muds, snow, ice, or mixtures of materials. Many names describe them, including earthflow, mudflow, debris flow, avalanche, and sturzstrom. Earthflows destroy many buildings, as in the Portuguese Bend area of Los Angeles. Sturzstroms are flows that run out especially long distances, up to 25 times longer than their vertical drop. The high fluidity of sturzstroms may be due to acoustic or vibrational energy trapped within the fallen mass that keeps grains jostling but apart. Sturzstroms can travel in excess of 325 km/hr (200 mph) and have killed thousands of people.

People are not defenseless against earth movements. Mitigation is achieved through an amazing variety of engineering structures that stop, steer, or scale down mass movements, such as retaining walls, diversion channels, and rock bolts.

Snow avalanches are similar to mass movements of earth. Loose-powder avalanches are flows, and slab avalanches are slides.

Subsidence occurs when the surface drops down either slowly in response to removal of subsurface water or oil or catastrophically, as in the collapse of limestone cavern roofs to form sinkholes. Extensive cavern systems in limestone occur throughout much of the central and southeastern United States.

Terms to Remember

acoustic fluidization 430	mass movement 408
bedrock 409	piping 414
bentonite 424	pore-water pressure 414
calorie 411	porosity 410
carbonic acid 438	potential energy 411
cavern 416	power 411
chemical weathering 412	quick clay 412
clay minerals 410	rotational slide 419
creep 409	shear 415
daylighted bedding 417	sinkhole 437
debris flow 424	slate 426
delta 439	slide 418
fall 418	slump 419
flow 418	soil 409
force 411	sturzstrom 426
friction 409	subside 418
gabion 434	subsidence 437
groundwater 415	syncline 416
heat 411	translational slide 419
kinetic energy 411	water table 415
limestone 437	work 411

Questions for Review

1. What is the role of gravity in causing mass movements?
2. Explain creep. What mechanisms cause soil and surface rock to swell and shrink?
3. Draw a cross-section through a slope and explain external actions that are likely to cause failure as mass movements.
4. Draw a cross-section through a clay mineral. Explain the physical properties that promote swelling and shrinking.
5. Draw a water molecule and explain how it links up so readily with some clay minerals.
6. Draw a cross-section and explain how quicksand forms.
7. How do pore waters become pressurized? What is their role in mass movements?

8. List some adverse geologic structures inside hills that facilitate mass movements.

9. Draw a cross-section and explain the Swedish circle method of analyzing slope stability for rotational slides.

10. Draw a cross-section and explain how translational slides work.

11. What triggering events set off rapid mass movements at Elm, Switzerland? At Vaiont, Italy? At Nevados Huascarán, Peru?

12. Compare snow avalanches to mass movements of soil and rock.

13. Why do coastal cities on river deltas, such as New Orleans, slowly subside? Why does slow subsidence occur in some cities near giant oil fields-—for example, Houston-Galveston, Texas?

14. Why do large cavern systems form in limestone? Why do they sometimes collapse and form sinkholes?

15. What two common substances combine to form carbonic acid?

16. What processes are involved in chemical weathering? How do they relate to clay minerals?

17. Compare potential energy to kinetic energy.

18. In what ways does underground water weaken rocks, i.e., what internal roles does water play?

19. What are the four major categories of mass movements?

20. Compare slides to flows. Which one mainly just affects property? Which one is the bigger killer?

21. Compare loose powder and slab snow avalanches to landslides and debris flows. What is the main way avalanches kill people? What can you wear while skiing, snowboarding, and snowmobiling that will significantly improve your odds of surviving an avalanche?

22. What are several ways that landslide risks can be mitigated?

Questions for Further Thought

1. Roadways are commonly cut into the bases of slopes. When mass movements block the road, the debris is quickly removed. What is wrong with this whole process?

2. Draw a cross-section of a hill made of inclined rock layers. On which side of the hill would you build your house? Consider daylighted bedding.

3. Acoustic fluidization has been proposed as a mechanism for fast-moving, long-distance flows, but it is not widely accepted. What tests or analyses might be conducted to evaluate its reality?

4. Are there dry fluids? Must earth be wet to flow?

5. In your home area, how many engineering structures can you visualize that were built to stop or control mass movements? Look for them as you travel about.

6. Can pumping up huge volumes of groundwater or oil trigger a movement on an active fault?

7. How can you protect yourself against the cavern collapses that form sinkholes?

CHAPTER 16

Coastal Processes and Hazards

Time and tide wait for no man.
—Geoffrey Chaucer, 1386

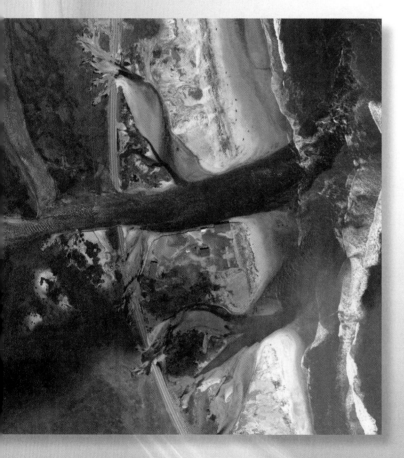

Hatteras Island is a sandy barrier bar located off the North Carolina coast; it is part of the Outer Banks. Flood water from Hurricane Irene cut through Highway 12 and across the island, 28 August 2011.

NOAA Hurricane Irene Project.

LEARNING OUTCOMES

Coasts are hit with waves and tides, sand is pushed around, and we build structures hoping to control it all. After studying this chapter, you should:

- understand how Earth–Moon–Sun gravitational attraction creates tides.
- be able to describe how water moves coastal sand.
- know what to do when caught in a rip current.
- be able to explain the multiple effects of coastal-control structures.
- truly understand the meaning of the term *side effects*.

OUTLINE

- Sand
- Waves
- Tides
- Coastal-Control Structures

oastlines are hit hard by wind-blown waves and tsunami. We need to understand how water moves on, along, and off the shoreline. As greater numbers of people move to the coasts and build communities and cities, they must learn that the sandy coast is a natural defense system helping protect them against wave attack. Nearly the entire U.S. coastline is losing beach sand along both ocean and lakeshores (figure 16.1). To protect our lives and property, we must understand how sand comes and goes from the shore and what effects our human-built structures have on water flow, beaches, and the coastal zone.

Sand

Despite the immense volume of ocean water, there commonly is not enough water in the coastal zone for people to drink. Thus, many dams have been built across rivers leading to the coast. The reservoirs impound freshwater for home and farm use, and the dams allow falling water to create kinetic energy that is converted to electricity.

But dams not only trap a stream's water, they also catch its sand (figure 16.2). The sand being transported by streams is on its way to the coast to become the next generation of beach sand. But when beaches are cut off from their sand supply, they shrink, thus depriving the coastlines of effective protection against waves.

Dams and their reservoirs are not the only human threat to beach sand supply. Sand is needed to make concrete and glass. Without sand, we would not have the houses, office buildings, roads, bridges, and other structures that mark our civilization. Many sand-mining operations have moved into

Figure 16.2 A dam traps sand as well as water. Malibu Canyon dam was built in 1925 in the Santa Monica Mountains, California. Within 13 years, the reservoir was filled with sand that would have been added to the beach otherwise.

University of Washington Libraries, Special Collections, John Shelton Collection, Shelton KC12885.

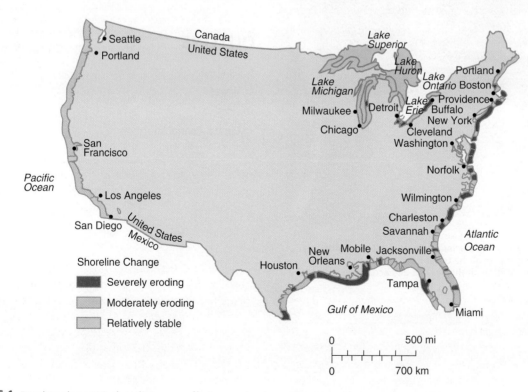

Figure 16.1 Beaches along U.S. shorelines are suffering erosion in varying amounts.

Source: S.K. Williams, et al, in U.S. Geological Survey Circular 1075.

valleys to dig up river sands; these sands were waiting in storage for a big flood to carry them to the beach. Without them, how will the beaches be supplied with their next generation of sand?

SUMMER VERSUS WINTER BEACHES

The Northern Hemisphere summer is a time of warmth and vacations. People flock to the beaches to lie on the sand, bask in the sunshine, and play in the ocean waves (figure 16.3a). But at the same beach during the winter, the conditions are quite different. Not only are the air and water colder and less inviting, but the beach sand may be gone (figure 16.3b). What happened to the sand? It moved offshore, but it will return during the next summer.

During the summer, the Northern Hemisphere is tilted toward the Sun, and weather conditions in the North Pacific and North Atlantic Oceans are relatively calm. With reduced temperature contrasts between the north polar region and the equator, the weather systems have lower wind speeds, which produce ocean waves of shorter wavelength and height (see tables 8.1 and 8.3). Summer waves at the beach are smaller, but they are more numerous. The abundant waves separated by short wavelengths act like bulldozers that push the offshore sand up across the shoreline to build a sandy beach (figure 16.3a).

During the winter, the Southern Hemisphere is tilted toward the Sun, and the north polar region gets very cold. The increased temperature contrast between the North Pole and the equator causes winds to blow faster and stronger. Winter storms are more energetic, and the ocean waves they create are taller and spread farther apart. What happens to the beach when it is attacked by breakers with longer wavelengths? Large breakers rush onto the beach, erode the beach sand, and carry it seaward with the backwash, leaving the beach with less sand (figure 16.3b). Because of the longer wavelengths in winter, much of the beach sand is dragged offshore to sit in submarine sandbars until the shorter-length waves of the next summer push the sand back onto the beaches.

A relationship exists between sandy beaches and wave lengths. Short-length waves push sand onto the beaches, whereas long-length waves pull sand offshore.

Beach Sand as Coastal Protection

A sandy beach is not only good for recreation, it is also a tremendous natural barrier that absorbs the energy of breaking waves. Visualize waves packed full of energy by an intense oceanic storm. The energy-packed waves may carry their intensity for thousands of miles before rearing up and dumping it on the shoreline. What better system to absorb this wave energy than a sandy beach? The waves kick around the sand grains, knocking them every which way, but what is the harm? Sand provides an innocuous way to absorb the ocean's fury. But in winter, the sand is reduced or gone from many beaches. What absorbs the energetic attack of the bigger winter waves? Sea cliffs, houses, roads, or whatever is on the coastline.

(a)

(b)

Figure 16.3 Beach-sand volume varies by season. (a) In summer, small, short-length waves have piled the beach high with sand. View north. (b) In winter, bigger, longer-length waves have removed beach sand, exposing underlying boulders. Boomer Beach, La Jolla, California.

Photos by Pat Abbott.

Waves

Waves in water were discussed in detail in chapter 8 by comparing wind-blown waves with tsunami. Please turn back to the chapter 8 text on "Wind-Caused Waves"; look carefully at figures 8.5 through 8.9 and understand tables 8.1 and 8.3.

RIP CURRENTS

Rip currents are powerful flows of water away from the shoreline. *Rips* typically extend through the surf zone, flattening the incoming waves and carrying brownish sediment offshore (figure 16.4). Rip currents can be narrow or as wide as hundreds of meters (yards). The out-flowing current moves from 0.5 m/sec (1–2 ft/sec) up to 2.5 m/sec (8 ft/sec = 5.5 mph). Rip currents can be killers.

Rip currents have long been famous locally, but they have only been receiving widespread attention in recent years. The National Weather Service has 12 years of records,

Figure 16.4 Rip current flowing out to sea, knocking down incoming waves and carrying brown sediment offshore, July 2001.

Courtesy of Lifeguard Captain Nick Steers, County of Los Angeles Fire Department.

which show an average of 48 rip-current drowning deaths per year, including the 64 deaths in 2013. However, the U.S. Lifesaving Association estimates U.S. rip-current deaths at more than 100 per year, and that rips account for more than 80% of rescues made by surf-beach lifeguards. Rip-current drowning deaths occur when people pulled offshore cannot stay afloat or swim out of the current, due to lack of swimming skills or exhaustion magnified by fear and panic.

Rip currents are most likely to occur during high-surf conditions when bigger, more powerful waves bring water ashore. When too much water builds up near the shoreline, it will create a "channel" so it can flow back offshore (figure 16.5). Rip currents are also common near human-built structures such as jetties or piers where water flow may be directed offshore.

Figure 16.5 A rip current forms as water pushed ashore by Hurricane Jeanne cuts through a beach sand bar and flows back to sea, September 2004.

Dennis Decker/WCM/NWS Melbourne, FL.

How to Recognize and Survive Rip Currents

You can recognize rip currents while standing on the beach by looking for:

- A break through the incoming waves
- A zone with smoother or with choppy water surface cutting through the waves
- A zone of different water color or muddy water
- A zone where water plants or foam are moving seaward

If you are caught in a rip current, do this to survive:

- Remain calm to save your energy and think clearly.
- Do not fight the current. Do not try to swim to shore.
- Do swim parallel to the shoreline to get out of the rip current (figure 16.6).
- If you are not able to swim out of the rip, then float or tread water.
- If you are unable to reach shore, make yourself known by waving your arms and yelling to the shore for help.
- If you see someone in trouble, think before acting. Many people drown while trying to save someone else. Involve a professional lifeguard if possible.

Great Lakes Rip Currents

Rip currents frequently form along seashores, but they also form in the Great Lakes. During the seven-year period from 2002 through 2008, NOAA documented 61 deaths (nearly 8 per year) in the Great Lakes. Deaths occurred in Lakes Michigan (36), Erie (11), Superior (10), Huron (3), and Ontario (1).

The most common conditions for formation of deadly rip currents in the Great Lakes are:

- Waves >3 ft high during swimming season
- Longer-period waves (>6 seconds)
- Winds blowing onshore; the greater their strength, the greater the risk
- Waves coming directly onshore
- Jumping into dangerous water areas such as near docks

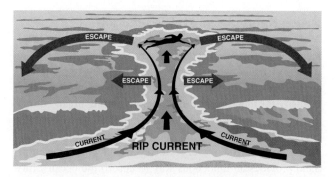

Figure 16.6 Swim parallel to shore to get out of a rip current.

Drawing from NOAA.

Side Note

WAVE REFRACTION

When waves reach shallow water, their velocity becomes controlled by water depth. Waves slow markedly as water depth decreases and bottom friction increases, irrespective of their lengths or heights. Many shallow coastal regions vary in depth, causing incoming waves to vary in speed along different portions of their crests. Because of varying topography and angles of wave approach, waves bend as they approach the shore. The bending process is known as **wave refraction** (figure 16.7a).

The wave segments that first reach shallow water slow down and bunch up (converge); the wave portions still in deeper water continue to roll ahead faster, thus stretching the waves as they spread out to cover a wider area (figure 16.7b). As a result of this refraction of waves, headlands are hit by a concentrated attack of higher-energy waves, while the embayed portions of coastline receive smaller waves.

LONGSHORE DRIFT

Even where the coastline is straight, waves typically approach it at an angle (figure 16.8). The effects of bottom friction and shallow water slow the near-shore end of a wave, thus giving the deeper part of the wave a chance to catch up, which it rarely is able to do.

Figure 16.8 View of groins built out from the shoreline to trap and hold beach sand, Ship Bottom, New Jersey. What is the dominant direction of longshore drift? What would a hurricane do to this coastline?

University of Washington Libraries, Special Collections, John Shelton Collection, Shelton 67-478.

(a)

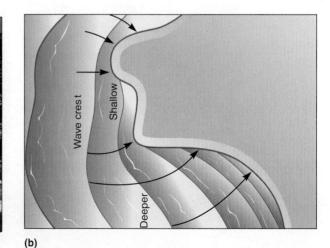

(b)

Figure 16.7 (a) View north of wave refraction around False Point in San Diego, California. (b) Wave refraction around a headland. The portion of a wave that first enters shallow water slows down and converges on the headland. The wave portion still in deeper water races ahead and stretches, thus reducing its impact on the shore.

(a) Photo by Pat Abbott.

When waves arrive at an angle, they run up the beach at an angle and drag the sand grains with them. But the return flow (backwash) is essentially perpendicular to the shoreline (figure 16.9). The process creates a **longshore current** of water, moving along the coast and dragging along the beach sand in longshore transport.

You have probably experienced **longshore drift** if you have gone to the beach, laid down your towel and belongings, and then gone into the waves to play. When you exited the water to return to your towel, was it in front of you? Probably not; it was likely lying some distance away, either up or down the beach. What happened? Your body was moved along the coastline by longshore drift in the same process as the sand grains. Longshore drift is well illustrated in the movie *The Beach: A River of Sand*. Beach sand flows with longshore drift analogous to a river; the river banks are the offshore edge of the breaker zone and the upper limit of water run-up (swash) on the beach.

Tides

Tides are the rise and fall of the surface of the Earth caused by gravitational forces and the rotation of the Earth. The rise and fall are most easily seen with changes in sea level, but they are also visible in water levels in lakes and in magma levels in volcano lava lakes. Although not easily seen, the land surface also rises and falls under gravitational attraction.

In the ocean, it is common to experience two high tides and two low tides each day, but some places have only one high tide and one low tide a day. The times and amplitudes of the tides are influenced by the alignment of the Moon and Sun and by the shape of the coastline and offshore topography. The topography in some places causes the rising tide to arrive with incoming waves known as **tidal bores;** they can be deadly.

TIDAL BORES

A tidal bore is a true *tidal wave*. These bores occur at sites around the world, especially where there is a large range in water level between the high and low tides, usually more than 6 m (20 ft). Tidal bores form where topography funnels the incoming tidal water into a narrow river, lake, or bay (figure 16.11); tidal bores flow *upriver*. The funnel-shape topography forces wide masses of incoming water into progressively narrower space, thus causing the water level to rise. Tidal bores commonly have turbulent water that produces rumbling noises as air mixes with the water. The low-frequency sound may be heard for some distance and can serve as a warning.

Tidal bores can be dangerous. Rivers flowing into the Bay of Fundy between Nova Scotia and New Brunswick, Canada, have tidal bores at times exceeding 2 m (6.6 ft) height. Tourists in the riverbeds have lost their lives to these bores.

The deadliest tidal bores occur along the Qiantang River in China. The extreme bottleneck shape of Hangzhou Bay funnels tidal bores up the river that can be 9 m (30 ft) high and moving up to 40 km/hr (25 mph). The tidal bore is such a spectacular sight that thousands of tourists line up along the river banks to watch, especially during the extra-high tides in September and October. Some of the viewers do not appreciate the power of flowing water, get too close, and are swept away. For example, on 3 October 1993, 86 people were washed into the bore resulting in 19 dead, 27 injured, and 40 missing.

Coastal-Control Structures

The coastal zone is active; conditions are constantly changing. Beaches are moving. Waves are attacking. Sea cliffs are collapsing. Sea level is rising. Yet living at the coastline is highly prized. Many millions of people move to the coastline, then try to control it, or at least coexist with it.

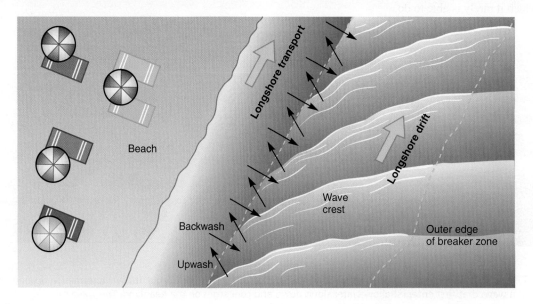

Figure 16.9 Schematic map of longshore drift. Waves run up the beach at an angle but return to sea as perpendicular backwash. The net effect is longshore transport of water and sand along the coast.

In Greater Depth

Gravity and Tides

Newton's law of gravity states that two bodies attract each other with a force directly proportional to the product of their masses and inversely proportional to the square of the distance between them:

$$\text{gravity (g)} = \frac{G \times \text{Mass 1} \times \text{Mass 2}}{\text{distance} \times \text{distance}}$$

where G = a universal constant.

Using Newton's equation to assess the gravitational effects of the Sun and the Moon on Earth requires knowledge of masses and distances. The Moon has a diameter of about 3,500 km (2,160 mi), whereas Earth has an average diameter of about 12,800 km (7,926 mi). For comparison, if Earth were reduced to the size of a basketball, the Moon would be slightly smaller than a tennis ball.

The volume of the Moon is only about 1/49 that of Earth, and the Moon's lower average density of 3.34 means its mass is only about 1/80 that of Earth. By comparison, the Sun's diameter is about 1,395,000 km (864,000 mi), and even though its density is only about 1/4 that of Earth, its mass is still about 332,000 times greater. Because gravitational attraction is directly proportional to mass, the gravitational pull of the massive Sun absolutely dwarfs that of the Moon. But their gravitational attractions are reduced by dividing by distance times distance, and the Sun is 150 million km (93 million mi) away while the Moon is about 386,000 km (239,000 mi) away. Although the Sun's massive gravitational pull is diminished by distance, it still exerts a gravitational pull on Earth more than 170 times stronger than that of the Moon.

The gravitational system of Earth, Moon, and Sun, and their interactions, generates tidal energy. The tidal force is caused by the differences in gravitational forces on the Moon-facing side of Earth compared to the back-side. Isaac Newton was the first to correctly calculate tidal forces as the inverse cube of the distance, meaning in the preceding equation to divide G times mass 1 times mass 2 by distance times distance times distance. These calculations show that the gravitational effect of the Sun on Earth is only 45% as strong as the pull from the Moon; that is, the Moon's role in causing tides on Earth is more than double that of the Sun.

Earth has rather unique tidal effects because (1) 71% of its surface is covered by oceans; (2) it has a long period of rotation compared to many other planets; and (3) its relatively large Moon is nearby. The gravitationally attracted bulges we call tides affect the land, water, and air but are most visible in the daily rises and falls of the ocean surface. The Sun appears overhead once every 24 hours, while the Moon takes about 24 hours and 52 minutes to return to an overhead position. Thus, the Moon appears to move in the sky relative to the Sun. So, too, do the tidal bulges attracted by the Moon move in relation to the tidal bulges caused by the Sun. The two sets of tidal bulges coincide twice a month, at the new and full moons, when the Sun and Moon align with Earth (figure 16.10). These highest tides of the month are called *spring tides*. In the first and third quarters of the Moon, the Sun and Moon are at right angles to Earth, thus producing the lowest tidal ranges, called *neap tides*.

The tidal bulges moving across the face of Earth and within mobile intervals in Earth's interior cause a frictional braking of Earth's rotation. Following Newton's laws of motion, as the rotations of Earth and Moon slow down, they move farther apart, days become longer, and the years have fewer days. At present, Earth and Moon are separating an additional 3.8 cm (1.5 in) per year. Substantiation of the lengthening days is evident in the fossil record. For example, careful counting of growth ridges in the skeletons of corals (broadly similar to tree rings) shows daily additions that vary in size according to the season of the year. A study of 370-million-year-old corals has shown that each day on Earth during their life was about 22 hours long, and a year had 400 days.

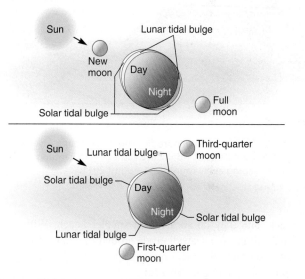

Figure 16.10 Earth tides are caused by the gravitational attractions of the Sun and the Moon. Top: The greatest daily range of tides occurs at the new and full moons. Bottom: The lowest daily range occurs at the first- and third-quarter moons.

SEAWALLS

Beaches are fun. How great it is to live at the beach! But beach sand volumes shrink, or even disappear. What can be done? A common response is to construct a seawall for protection. But every action has multiple reactions—and a beach is an equilibrium system with feedback mechanisms that work to maintain its equilibrium. A natural beach is flattened by tall ocean waves, spreading water over a broad area analogous to a stream overflowing its banks (figure 16.12a). Building a seawall disturbs the equilibrium (figure 16.12b).

After a wall is built, the winds can no longer build sand dunes behind the beach. In a natural setting, sand dunes are redistributed by storm waves, thus dissipating some wave energy. But with a seawall, there is less sand to absorb wave energy. Instead, ocean wave energy is bounced back from the seawall, eroding the beach and steepening the offshore sea bottom. The now-deeper water allows bigger waves to

(a)

(b)

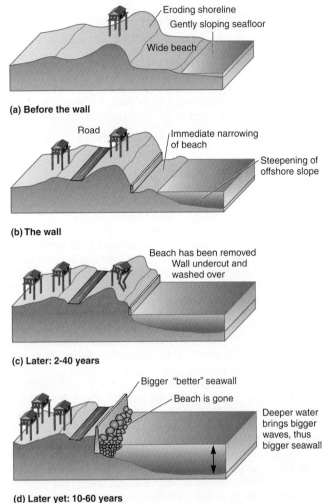

- (a) **Before the wall**
  - Eroding shoreline
  - Gently sloping seafloor
  - Wide beach

- (b) **The wall**
  - Road
  - Immediate narrowing of beach
  - Steepening of offshore slope

- (c) **Later: 2-40 years**
  - Beach has been removed
  - Wall undercut and washed over

- (d) **Later yet: 10-60 years**
  - Bigger "better" seawall
  - Beach is gone
  - Deeper water brings bigger waves, thus bigger seawall

Figure 16.11 (a) Tidal bore flowing *up* Qiantang River, China on 31 August 2011 (b) Tidal wave rushes over Qiantang river bank sweeping away tourists, 31 August 2011.

(a) © Wang Chaoying/Xinhua Press/Corbis (b) © AP Photo.

Figure 16.12 The shoreline and the seawall. (a) Before a wall is built, people attracted to the coastal zone ignore erosion and build houses and businesses. (b) Behind the "protection" of the wall, development increases. But waves rebounding off the wall erode the beach and deepen the coastal waters. (c) Later, beach sand is removed by waves, which then work to undermine the wall. (d) Later yet, citizens band together and buy more "protection" with a bigger seawall. But a bigger wall reflects even bigger waves, and the beach disappears again. And on it goes. . . .

attack, and they do so across a smaller beach. With time, waves hitting the seawall erode around the ends of the wall and undermine the main mass; larger waves break over the top of the wall (figure 16.12c).

What is the typical response to wall failure? Build a bigger wall (figure 16.12d). The seawall and the buildings it protects will then stick farther into the sea each year as waves striking around the ends of the wall erode farther landward. The beach in front of the seawall is steeper and narrower than the beaches at either end. We need to learn how to adapt human activities to the natural processes of the beach.

CLIFF ARMORING

Sea cliffs collapse (figure 16.13). The views from coastal cliffs are among the most enticing in the world. Sea-cliff properties are in such heavy demand that they sell at premium prices (figure 16.14). But sea cliffs retreat under the attack of large waves and dump the cliff-edge buildings onto the beach. What can be done? Keep buildings away from the cliffs. This, however, is easier said than done.

Most homeowners who can afford beach-cliff property also have the money to buy protection. Typically, they erect a barrier of heavy rocks and concrete (figure 16.15). But these barriers are no match for the power of storm waves working over time. If the waves do not destroy a protective wall or mass of riprap by frontal assault, they may erode underneath and around the sides, undermining the barrier and helping break it apart. Barriers also reflect waves hitting them, and the rebounding water surges powerfully erode the beach, thus removing protective sand.

The long-term picture is an unstoppable retreat of sea cliffs under the attack of ocean waves. Humans can delay the inevitable, but they cannot stop it. Anything built on the sea cliffs must be thought of as a temporary structure; it will be destroyed.

Figure 16.13 A block of sea-cliff sandstone gets ready to fall on Moonlight State Beach in Encinitas, California. Do you see the three mutually perpendicular joints (fractures) that allow this block to separate from the cliff?

Photo by Pat Abbott.

Figure 16.14 Condominiums built to the edge of a retreating sea cliff, Solana Beach, California.

Photo by Pat Abbott.

Figure 16.15 Along Sunset Cliffs, the city of San Diego built cement walls and dumped riprap to "protect" the sea cliffs, and then billed the homeowners. Meanwhile, a charming cove and sandy beach have been obliterated.

Photo by Pat Abbott.

GROINS AND JETTIES

Because beach sands are disappearing in so many places, various structures are built to trap and retain sand. One of the most popular techniques is to place sets of short, elongated masses perpendicular to the coastline, the so-called **groins** (see figure 16.8). Groins interfere with longshore drift, causing sand to be deposited on the up-drift side; however, erosion takes place on the down-drift side. Emplacement of groins usually must be accompanied by artificial replenishment of sands.

Other problems with groins include the erosion of sand by storms coming from other directions and their tendency to direct rip currents offshore, carrying sand (and swimmers) beyond the breaker zone and thus out of the longshore drift system.

Even larger masses than groins are **jetties,** which are built perpendicular to the coastline to create inlets to harbors and channels for boat passage. The trick is to design jetties of the right length to create a large enough tidal prism (in-and-out volume of seawater) to naturally scour the channel and keep it open. Of course, this means the jetties extend beyond the

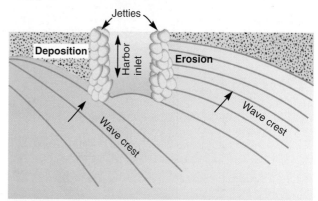

Figure 16.16 Jetties of riprap are extended beyond the surf zone, allowing boats to get in and out of a harbor. Longshore-transported sand accumulates on the up-drift side, and erosion occurs on the down-drift side.

breaker zone and interfere with the longshore-transport system (figure 16.16). Beach sand builds up in large volumes on the up-drift side, while significant amounts of erosion occur on the down-drift side.

BREAKWATERS

Another class of structures, known as **breakwaters,** are built to protect shorelines or harbors from wave attack. Breakwaters may be either attached to or detached from the shoreline. The goal of providing boats a safe haven from heavy waves is reasonable, but preventing waves from hitting the shoreline also cuts off the energy that drives the longshore-transport system (figure 16.17). The sand, deprived of the energy that moves it, just stops its travels and fills in the area behind the breakwater. To keep the harbor or sheltered area open, a permanent dredging operation must be set up to move the sand back into the longshore-transport system down-drift from the breakwater.

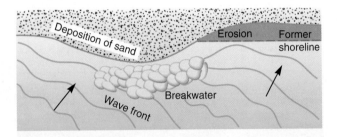

Figure 16.17 A breakwater is built parallel to shore to provide protection from powerful waves. But it also stops waves from hitting the beach, thus shutting down the longshore-transport system.

MASSIVE STRUCTURES IN FUTURE

In 2012, Hurricane Sandy showed how vulnerable some urban areas are to rising water levels. Even ignoring hurricanes, rising sea level presents problems. For example, New York City has seen sea level rise by about 3 cm (1.2 in) per decade since 1900; this is nearly double the global rate. A 2013 study by the NYC Panel on Climate Change suggests that the coastal area affected by floods could double by 2100 (figure 16.18). New York City has 400,000 people, 270,000 jobs, and 68,000 buildings inside the 100-year floodplain; that is a lot to protect.

The post-Sandy plan for New York City asks for levees and floodwalls to protect vital infrastructure in urban areas; repaired natural wetlands and sand beaches around the outer boroughs; and new building codes that require new structures to be elevated. These are some of the suggestions for coping until 2040, but they may not be enough for later in the century. Post-2040 suggestions include massive storm barriers and huge floodgates holding back Atlantic Ocean water; an offshore chain of artificial barrier islands; underwater "breakwaters" around Staten Island; and floodgates across rivers. Global climate changes are requiring local adaptations.

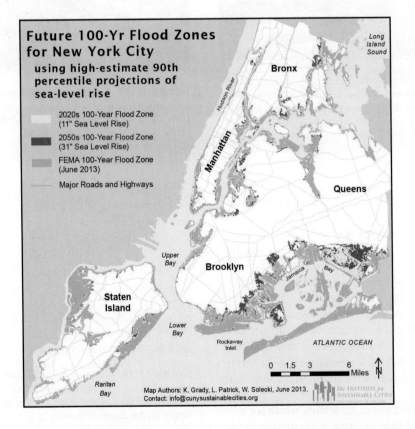

Figure 16.18 Future flood zones in New York City. As sea level rises, the risk grows for increasing numbers of people, the economy, and the infrastructure.

Source: NYC Panel on Climate Change 2013.

Side Note

You Can Never Do Just One Thing

Many human actions along the coastline cause unintended problems. For example, building jetties to allow boats to use the ocean and a bay is fine in itself, but the total effects of a construction project must be evaluated in advance. The jetties are great for boat users, but what happens to the longshore drift of beach sand? Beach sand movement is blocked and deposited on the up-drift side of jetties, building a large beach (see figure 16.16). However, the down-drift side of the jetties is deprived of new beach sand, so waves erode the beach, triggering construction to slow the erosion. In the example in figure 16.15, the city emplaced riprap and built cement walls, and then assessed property owners for the cost of the construction. But the walls and riprap were largely needed because the jetties had stopped the longshore drift, thus reducing the protective strip of beach sand. Boaters gained, homeowners lost.

The planning process ignored a law of nature—you can never do just one thing. You cannot build a jetty and achieve only one effect. Every action has multiple reactions. When planning an action, you must remember that it will have several reactions and then evaluate the pluses and minuses in advance to see if going ahead with the project makes sense. The positive effects should dominate.

Garrett Hardin has stated this principle another way: there is no such thing as side effects. We all know the term **side effects,** but the English language is playing a trick on us here. We learn a term and then incorporate it into our thought processes and use it as if it describes reality. Consider the example of taking a prescription drug: the drug is prescribed for a reason, and you are supposed to be aware of the side effects. But this is misleading; there are no side effects—every action has multiple reactions. Take a drug (action) and you will experience several effects (reactions). There are no side effects, just a group of reactions to evaluate. Do the benefits of taking the drug outweigh the negative consequences? If so, then proceed with the action.

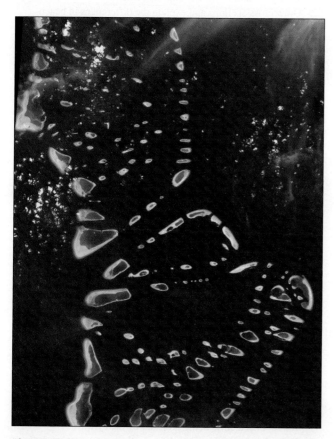

Figure 16.19 Satellite view of some coral-atoll islands in the Maldives.

NASA/GSFC/METI/ERSDAC/JAROS, and U.S./Japan ASTER Science Team.

MOTHER NATURE AT WORK

As global sea level rises, coastal zones are flooding worldwide. Consider the situation of the Maldives, a nation on 1,192 coral-atoll islands in the Indian Ocean (figure 16.19). The country is the smallest in Asia, both in population and in land area, 330,000 people living on 298 km^2 (115 mi^2). The average ground elevation is 1.5 m (4.9 ft); the highest elevation land is 2.4 m (7.9 ft).

The Maldives have beautiful white beaches and a nearly constant year-round temperature where the all-time recorded high was 35°C (95°F) and the record low was 25°C (77°F). Not surprisingly, tourism brings in >60% of their foreign exchange.

What will rising sea levels bring to the Maldives? The problem may be solved by Mother Nature. Recent studies suggest that coral reefs can grow and rise at the same rate as the sea level. Coral reefs on some atolls have been shown to grow 10 to 15 mm per year, which is 1 to 1.5 m per century. As long as a coral reef is healthy and generating a lot of sand, an atoll island should be okay.

To date, where problems on atolls have occurred they are usually due to poor shoreline management by humans cutting passageways through reefs, removing sand, building causeways and sea walls, and discharging waste water into the sea—all of which interfere with coral-reef growth and their ability to rise at the same rate as sea level.

The global rise in sea level may not be a problem for the Maldives, but it will be a problem for New York City.

Summary

As mountains wear down to sand, heavy rains and rivers carry the sand to the shoreline to form beaches. Our dam building and use of sand in concrete have reduced the supply of sand to the beach.

In summer, waves have shorter heights and lesser wavelengths, causing sand to be pushed onto beaches. During winter, the backwash from taller waves, separated by greater wavelengths, drags beach sand offshore to be stored as submarine sandbars. Beach sand provides excellent protection for coastlines under heavy wave attack. Loose sand grains absorb tremendous amounts of wave energy and then quickly fall back into place.

A wave approaching a coastline that has different water depths will refract or bend as the wave portion in shallower water slows more than the portion in deeper water. When waves strike the beach at an angle, they create a longshore current that carries beach sand along the coastline as a "river of sand."

Rip currents are powerful flows of water away from the seashore or lakeshore. Rip currents can carry bathers and swimmers offshore past the surf line. Rips drown more than 100 people each year in the United States. Stay calm and swim parallel to the beach to escape.

Tides are caused primarily by the gravitational attraction of the Moon and Sun. The surface of the Earth, both water and land, rises and falls in tides that sweep around our planet. In areas with funnel-shaped bays and river mouths, the rising tide may flow upriver as a tidal bore (tidal wave).

Humans have major effects on the coast. We reduce the volumes of new beach sand, which allows greater wave attack on sea cliffs, where expensive buildings commonly are constructed. Humans build groins, jetties, and breakwaters to try to control waves and sand movement, and they place riprap and concrete walls to try to stop wave attack.

Each human-built structure triggers multiple reactions, including some negative effects. It is a law of nature that every action has multiple reactions. But many beach construction projects are done with only one purpose in mind, and then people are surprised at the "side effects" that occur. For example, concrete seawalls built to protect against wave attack also reflect waves seaward, eroding the beach sand as they return to the sea. Then, with beach sand gone, waves attack the ends and bottom of the wall, causing it to fail.

Terms to Remember

breakwater 452	rip current 445
groin 451	side effect 453
jetty 451	tidal bore 448
longshore current 448	tides 448
longshore drift 448	wave refraction 447

Questions for Review

1. Why does our relatively small Moon have a greater tidal effect on Earth than the gigantic Sun?
2. What is a tidal bore? Is it the same phenomenon as a tidal wave?
3. How does a rip current form?
4. When you are standing on the beach, how can you recognize a rip current?
5. What should you do if you are caught in a rip current?
6. Where does beach sand come from? What can stop a new supply of sand from reaching the beach?
7. Draw cross-sections and explain the differences in beach-sand volume between summer and winter.
8. Discuss the role of beach sand in providing protection from strong waves.
9. Draw a map of a peninsula jutting into the ocean. Diagram the refraction of ocean waves hitting the shoreline.
10. Draw a map and explain how the longshore current works.
11. Draw a map showing the effects of a jetty on longshore-drifted beach sand. Explain what happens on both the up-drift and down-drift sides of the jetty.
12. Draw a map showing a breakwater and its effect on longshore-drifted beach sand.
13. Draw cross-sections and explain the effects of a seawall on beach sand.
14. Is the term *side effects* misleading?
15. Which U.S. states have the most shark-bite fatalities? Which shark species are killing the most people in U.S. coastal waters?
16. How does the rate of growth of coral reefs on atolls compare to the rate of sea-level rise?

Questions for Further Thought

1. In your home area, how many engineering structures can you visualize that were built to control coastal processes? Look for them as you travel about.
2. Can placing riprap and building cement walls prevent the retreat of sea cliffs being attacked by heavy waves?
3. Does the term *side effects* describe the way nature works?

CHAPTER 17

Impacts with Space Objects

Comets are members of the solar system, these dazzling and perplexing rangers, the fascination of all astronomers, rendered themselves still more fascinating by the sinister suspicion attaching to them of being possibly the ultimate destroyers of the human race.

—THOMAS HARDY, 1882, *Two on a Tower*

Imagine the effects of a large asteroid slamming into the Earth at high velocity.

© Brand X Pictures/PunchStock RF.

LEARNING OUTCOMES

Asteroids and comets are zooming through space; sometimes they hit Earth and cause huge explosions. After studying this chapter, you should:

- be familiar with the travel speeds of Earth, asteroids, and comets, and the amount of energy released upon impacts.
- be familiar with the number and volume of meteoroids that hit Earth each day.
- know the sequence of events starting when a meteoroid first impacts Earth until the final crater has formed.
- understand what happened when an asteroid hit 65 million years ago in what is now Yucatan, Mexico.
- be familiar with the biggest impact on 20th-century Earth.
- know how we could prevent future impacts.

OUTLINE

- Energy and Impacts
- Impact Scars
- Sources of Extraterrestrial Debris
- Rates of Meteoroid Influx
- The Crater-Forming Process
- Crater-Forming Impacts
- Impact Origin of Chesapeake Bay
- The End Cretaceous Impact
- Biggest Events of the 20th and 21st Centuries
- Frequency of Large Impacts

Impacts

Soon after the Sun had risen on 15 February 2013, a fiery mass came out of the sunny eastern sky racing over Siberian Russia before exploding with a blast of light brighter than the sunlight. The dazzling light sent people running to their windows to see what happened, only to be met with atmospheric shock waves that shattered their windows, cutting about 1,500 people so severely that they needed medical attention. The experience was a reminder that light travels faster than shock waves, or sound. If you see a bright flash in the sky, get away from windows and seek protection.

What happened? An undetected meteor entered the atmosphere and was slowed and heated by friction, before exploding in the atmosphere above Chelyabinsk, Russia, an industrial city of 1 million people. The meteor was house size, about 20 m (65 ft) diameter, and weighed about 10,000 tons. It became visible in the atmosphere about 92 km (57 mi) above ground traveling 19 km/sec (42,000 mph) while descending at a shallow angle along a 254 km (158 mi) flight path for about 30 seconds before exploding at 30 km (19 mi) elevation with a bright flash, a hot cloud of dust and gas, many fragments of asteroid, and a large shock wave (figure 17.1). The meteor explosion damaged walls and roofs of more than 7,000 buildings and left many thousands of windowless buildings to face the below freezing temperatures of the Siberian winter. It could have been worse; if the meteor had traveled through the atmosphere to Chelyabinsk at a steep angle, it could have remained whole and struck the ground, killing many people.

Earth moves rapidly through the same space occupied by **comets** and **asteroids** traveling on different paths (figure 17.2). Sometimes our paths cross, and life on Earth feels the effects. Could a comet or asteroid impact do serious damage to the human race? The statistics calculated in 1992 suggest that it is possible (table 17.1). But the situation has improved. We have within reach the technological ability to deflect asteroids from Earth-colliding paths and to send them harmlessly away. The technologically more difficult task of landing humans on the Moon and then returning them to Earth was achieved by the United States in July 1969. We will consider a defense plan for Earth at the end of the chapter, but first, let's understand space objects and impacts.

Energy and Impacts

Earth, asteroids, and comets move through space at high rates of speed. When their paths intersect, there are explosive impacts. Earth travels more than 950 million km (590 million mi) around the Sun each year—an orbital speed in excess of 108,000 km/hr (67,000 mph) (figure 17.3). The kinetic energy of this orbital motion is about 2.7×10^{33} joules. When this tremendous amount of energy is involved in a head-on collision with a large asteroid moving 65,000 km/hr (40,000 mph) or a comet traveling 150,000 km/hr (93,000 mph), the effects on life are catastrophic and worldwide.

Additional sources of energy lie in the rotational motions of Earth—the daily rotation of Earth about an axis that pierces its center, and the monthly rotation of the

Figure 17.2 The Earth presents a small target in a cosmic shooting gallery.

NASA.

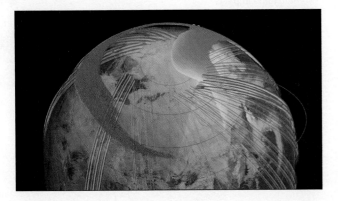

Figure 17.1 Space dust weighing hundreds of tons reached an altitude of 40 km (25 mi) within hours of the Chelyabinsk meteor explosion; it then flowed around the Northern Hemisphere for over 3 months. The pictured band shows the extent of the debris 4 days after the explosion as modeled from satellite data.

NASA Goddard's Scientific Visualization Studio.

TABLE 17.1	
Frequency of Globally Catastrophic Impacts	
Average interval between impacts	**500,000 years**
Annual probability a person will be killed	**1/500,000**
Assumed fatalities from impact	**1/4 of human race**
Total annual probability of death	**1/2,000,000**

Source: D. Morrison (1992).

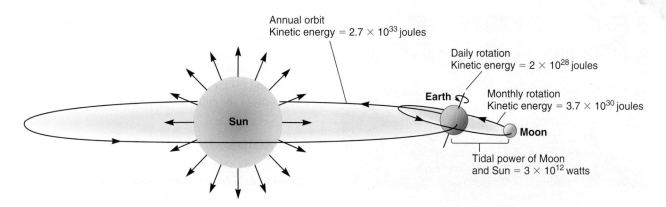

Annual orbit
Kinetic energy = 2.7×10^{33} joules

Daily rotation
Kinetic energy = 2×10^{28} joules

Earth

Monthly rotation
Kinetic energy = 3.7×10^{30} joules

Sun

Moon

Tidal power of Moon
and Sun = 3×10^{12} watts

Figure 17.3 The rotations and orbits of the Earth-Moon-Sun system result in tremendous amounts of energy.

Earth-Moon system about its common center of gravity lying about 4,680 km (2,900 mi) away from the center of Earth toward the Moon.

Impact Scars

A good place to see impact scars made by collisions with space debris is the surface of the Moon (figure 17.4). In its first few hundred million years of existence, the Moon was a violent place as millions of objects slammed into it. The intense bombardment apparently occurred as a sweeping-up of debris left over from the formation of the planets. The Moon's surface still displays tens of millions of ancient impact craters, some with diameters of hundreds of miles. By 3.9 billion years ago, the heavy barrage of space debris had died down.

Flood basalts poured forth on the Moon from about 3.8 to 3.2 billion years ago. They created the dark-colored **maria** (*mare* is Latin for "sea") so prominent on the Moon's surface today. The maria have relatively few impact scars on them, thus providing evidence that the period of intense bombardment was over before 3.8 billion years ago. Impact craters on the lunar maria were made by collisions with asteroids and comets in the same process that continues today.

For more than 3 billion years, the Moon has been essentially "dead"; it is an orbiting museum showing the scars of its ancient past. The Moon has no plate tectonics, no liquid water or agents of erosion, and no life. About the only event that disturbs the cemetery calm of the Moon is the occasional impact of an asteroid or comet.

Why are impact craters so common on the Moon but so rare on Earth? The Moon is geologically dead, so impact scars remain. But the Earth is dynamic: it destroys most of the record of its past. Plate-tectonic movements consume impact scars during subduction (see figure 2.26) and crumple them during continent collisions. The hydrologic cycle puts the agents of erosion working to erase all impact craters on Earth (see figure 9.6). Nevertheless, some impact

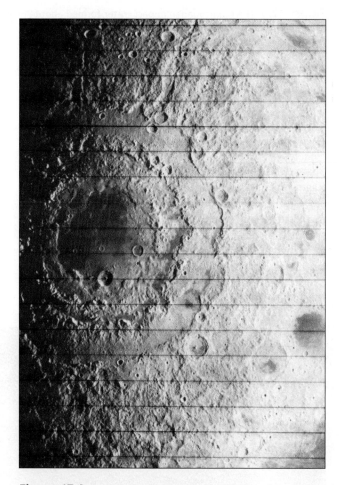

Figure 17.4 The Moon's surface is ancient and pockmarked by numerous impact craters. Notice the large Orientale multiring basin. Its outer ring is 1,300 km (800 mi) across; on Earth, it could stretch from the Great Lakes to the Gulf of Mexico.
NASA.

scars have avoided destruction, especially the geologically younger ones (figure 17.5). U.S. geologist Robert Dietz calls impact scars **astroblemes,** or literally, star wounds (the Greek word *astron* means "star" and *blema* means "wound" from a thrown object).

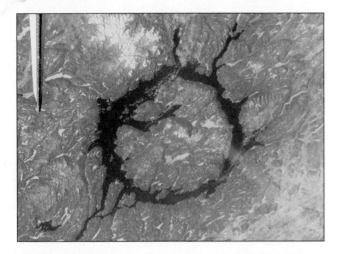

Figure 17.5 The Manicouagan impact crater formed about 214 million years ago in Late Triassic time, in northern Quebec, Canada. It is 75 km (45 mi) across, but probably exceeded 100 km before glacial erosion stripped away its upper levels.

NASA/JSC.

Sources of Extraterrestrial Debris

Space debris that collides with Earth comes primarily from fragmented asteroids and secondarily from comets. The pieces of asteroids and comets that orbit the Sun are called **meteoroids.** When meteoroids blaze through Earth's atmosphere as a streak of light or **shooting star,** they are referred to as **meteors.** The objects that actually hit Earth's surface are called **meteorites.** The main types of meteorites are either "irons" (metallic) or "stones" (rocky). Although most space objects that reach Earth's atmosphere are "stones,"

they are not very abundant on the ground. Stony meteorites are less commonly collected because (1) they break up more readily while passing through Earth's atmosphere, (2) those that reach the ground are weathered and destroyed more rapidly, and (3) they are not as easily recognized as "irons." Thus, most of the collected meteorites are "irons."

ASTEROIDS

The Solar System has eight planets in orbit around the Sun (figure 17.6). The four inner planets are small, close together, near the Sun, and rocky. The outer planets are larger, spaced far apart, located at great distances from the Sun, and composed mainly of hydrogen and helium gas surrounding rocky cores; they commonly are orbited by icy moons and rings of icy debris with compositions of water (H_2O), ammonia (NH_3), carbon dioxide (CO_2), and methane (CH_4). Between the inner and outer planets lie the asteroids, a swarm of small (less than 1,000 km diameter) rocky, metallic, and icy masses (table 17.2).

Meteorites appear to come from the inner part of the Solar System and especially from the asteroid belt (figure 17.7). Asteroids are small bodies orbiting the Sun. The three largest asteroids make up about half the combined total mass of all asteroids; they are Ceres, Vesta, and Pallas, with respective diameters of 950, 525, and 513 km. There are more than 200 asteroids with diameters greater than 100 km, about 1,000 with diameters greater than 30 km, and another million with diameters of more than 1 km. If all the asteroids were brought together, they would make a planet about 1,500 km (932 mi) in diameter; this body would have less than half the diameter of our Moon (table 17.2).

The asteroids lie mostly between Mars and Jupiter in a zone where a ninth planet might have been expected to form. The ingredients composing many asteroids are similar to those from which the planets were assembled via low-velocity collisions. However, the asteroids were apparently too strongly influenced by the gravitational pull of Jupiter and thus have been unable to combine, or recombine, and form a planet. The gravitational acceleration caused by the massive planet Jupiter creates asteroid velocities that are too fast and individual collisions that are too energetic to allow the asteroids to collide, unite, and stick together to form a planet.

Notice on figure 17.7 that the asteroids are concentrated in belts and that there are gaps between the belts. The gaps occur at distances related to the

Figure 17.6 The Solar System bodies. Sun, eight planets, and a new and growing category of dwarf planets.

International Astronomical Union.

Side Note

Dwarf Planets

Dwarf planets are a category of Solar System objects smaller than planets yet larger than asteroids. The eight planets are each large enough to have cleared their orbit region of other objects. Asteroids are smaller bodies with irregular shapes. Dwarf planets fit in between: (1) they are massive enough to be rounded by their own gravity, (2) they have not cleared their orbit region of smaller objects, and (3) they are not moons of a planet.

The term *dwarf planet* was adopted by the International Astronomical Union in 2006. The recognition of dwarf planets is still in the early stages. Pluto was downgraded from a planet to a dwarf planet. The asteroid Ceres is now raised to dwarf planet status. With continuing study, dozens of objects probably will be labeled as dwarf planets.

TABLE 17.2

Bodies in the Solar System

	Diameter (miles)	Specific Gravity (water = 1)	Gravity (Earth = 1)
Sun	864,886	1.4	27.9
Inner Rocky Planets			
Mercury	3,024	5.4	0.4
Venus	7,522	5.3	0.9
Earth	7,918	5.5	1.0
Moon	2,160	3.3	0.2
Mars	4,200	3.9	0.4
Asteroid belt			
Outer Ice and Gas Planets			
Jupiter	86,692	1.3	2.6
Saturn	72,352	0.7	1.1
Uranus	29,168	1.3	0.9
Neptune	28,230	1.6	1.2
Dwarf Planet			
Pluto	1,472	1.9	0.06
Ceres	590	2.1	0.03

Source: R. T. Dodd (1986).

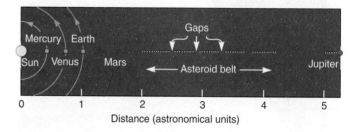

Figure 17.7 The Solar System from the Sun to Jupiter. The asteroid belt is composed of millions of rocky and metallic objects that did not combine to form a ninth planet. One astronomical unit equals 150 million km (93 million mi), the distance between the Sun and Earth.

orbit time of Jupiter. The asteroids rarely collide, but when they do, their collisions may be spectacular impacts at 16,000 km/hr (10,000 mph). The force of these smashups may bump an asteroid into one of the gaps in the asteroid belt. An asteroid nudged into a gap experiences an extra gravitational acceleration from Jupiter that makes its orbital path more eccentric, and thus, it becomes more likely to collide with a planet.

A photo of the asteroid Ida shows it has impact craters and its own moon, now named Dactyl (figure 17.8). When Ida and Dactyl collide with a planet, craters will form simultaneously in two different areas. This observation explains some of the double-impact sites found on Earth (figure 17.9).

Recent radar data show that some asteroids are not solitary, solid masses but rather are made of two or more similar-size bodies bound together by gravitational attraction. Calculations indicate that the amount of energy needed to break up an asteroid is much less than the amount of energy required to scatter all its fragments. Thus, an asteroid may be broken into pieces during a collision, and then the pieces may be held together by gravity, creating a loose collection of rocky debris—a rubble pile.

The realization that collisions have made some asteroids into rubble piles raises interesting questions. Did some of the impact craters on Earth form within hours of each other when a multichunk asteroid hit the surface? One line of impact structures being investigated to see if they are the same age includes Manicouagan (see figure 17.5). Removing the effects of 214 million years of plate tectonics places three impact sites along a 4,462 km (2,766 mi) long line parallel to the ancient 22.8°N latitude. Saint Martin in Manitoba, Canada (40 km diameter), lines up with Manicouagan in Ontario, Canada (100 km diameter), and both line up with Rochechouart in France (25 km diameter). If these three impacts occurred hours apart, life on Earth must have suffered a terrible blow.

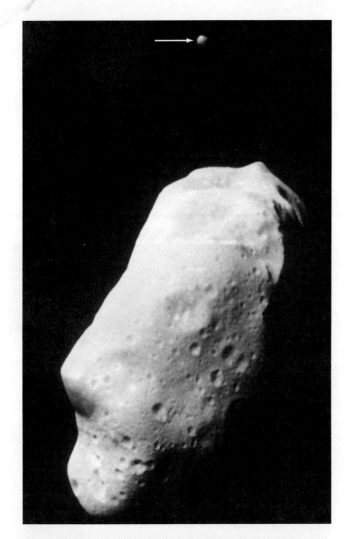

Figure 17.8 The asteroid Ida is 56 km (35 mi) long and pockmarked with impact craters. Traveling with Ida is its near-spherical moon Dactyl (arrow) with dimensions of 1.2 × 1.4 × 1.6 km.

NASA/JPL.

Figure 17.9 Space-shuttle view of the Clearwater Lakes double impact sites, Quebec, Canada. About 290 million years ago, a two-part asteroid hit the ground. The western crater is 32 km (20 mi) across and has a central uplift. The eastern crater is 22 km (less than 14 mi) across; its central uplift is below water level.

NASA.

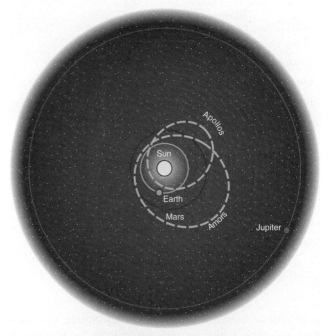

Figure 17.10 Earth's orbit (green) around the Sun is intersected by the Apollo asteroids. The Amor asteroids cross the orbit of Mars (red) and pass near the Earth.

Clusters of asteroids such as the Apollos have orbits that intersect Earth's orbit (figure 17.10). There are more than a thousand mountain-size Apollo asteroids that could hit the Earth. Other groups of asteroids, such as the Amors, pass near Earth but intercept the orbit of Mars. The Apollo and Amor asteroid groups are sources for large asteroids and meteorites that occasionally slam into Earth's surface. With so many asteroids whizzing about, we are lucky that space has such an immense volume and that Earth is such a small target.

COMETS

Comets are icy bodies moving through space that release gas and dust (figure 17.11). Comets are commonly divided into short and long-period classes. A short-period comet makes a complete orbit in less than 200 years. Some of them are found in the **Kuiper belt,** a flattened disk of comets with orbits ranging from near Neptune out to about 50 astronomical units (an astronomical unit is 93 million mi—i.e., the distance between Earth and Sun). The Kuiper-belt comets

Figure 17.11 Computer-enhanced view of a comet.
© StockTrek/Getty Images RF.

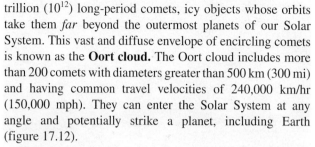

probably are icy debris left over from the formation of the outer icy planets. They are the icy bodies that never collided and accreted onto a larger planet, analogous to the asteroid belt of rocky bodies that never accreted onto the inner rocky planets. There may be a billion Kuiper-belt comets greater than 5 km (3 mi) in diameter.

Most of the short-period comets were captured by the gravitational pulls within the Solar System. These comets have had their orbits changed and now can spend thousands of years until they meet their fate by (1) erosional destruction or (2) colliding with a planet or the Sun.

The long-period comets have orbits lasting longer than 200 years. The Solar System is surrounded by about a trillion (10^{12}) long-period comets, icy objects whose orbits take them *far* beyond the outermost planets of our Solar System. This vast and diffuse envelope of encircling comets is known as the **Oort cloud.** The Oort cloud includes more than 200 comets with diameters greater than 500 km (300 mi) and having common travel velocities of 240,000 km/hr (150,000 mph). They can enter the Solar System at any angle and potentially strike a planet, including Earth (figure 17.12).

The planets, the asteroid belt, and the Kuiper belt orbit the Sun in planes, whereas the Oort cloud has a spherical distribution. The comets pulled into the inner parts of the Solar System are only a miniscule fraction of the Oort cloud. Most of the comets we see have wildly eccentric orbits that bring them in near the Sun at one end of their orbit (**perihelion**), but they swing out beyond the outermost planet at the other end of their journey (**aphelion**). A comet may travel 100,000 astronomical units away during its orbit.

Comets are called "dirty snowballs" to describe their composition of ice and rocky debris. When an incoming comet passes Saturn on its journey toward the Sun, it begins to be affected by sunlight and the **solar wind** (the stream of subatomic particles flying outward from the Sun). Material from the frozen outer portion of a comet **sublimates** directly to vapor, thus liberating gases and trapped dust to form the distinctive luminous "tail" of a comet (figure 17.13a). The term *comet* is derived from a Greek word for "long-haired." A comet's tail is produced by the charged particles of the solar wind acting on comet ices. The nearer an icy comet approaches the Sun, the larger its tail becomes. As the comet curves around the Sun, its tail rotates also, always pointing away from the Sun. The tail lines up with the solar wind moving outward from the Sun; the tail is not simply a vapor trail behind a comet. A comet that has lost most of its ices over time has a dim and small tail. Despite their visibility in the sky, comets are surprisingly small; most have heads less than 15 km (10 mi) in diameter.

Some comets operate more directly under the gravitational influence of the Sun and Jupiter. As a comet's orbit loses eccentricity and begins passing by the Sun more often, its ice volume declines and its tail shrinks. There are at least 800 of these short-period comets with orbits of less than 200 years' duration. Some short-period comets may be from orbits about the outer gas-giant planets and are thus distinct from the Oort-cloud comets. If their ices are completely sublimated or melted, the remaining rocky body is quite similar to an asteroid.

Figure 17.12 A comet enters the Solar System heading toward the Sun.
© Digital Vision/PunchStock RF.

(a)

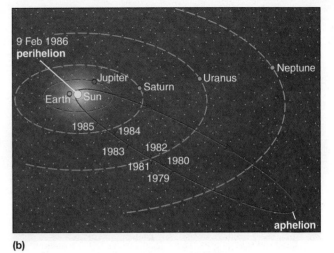

9 Feb 1986
perihelion

Jupiter

Saturn

Uranus

Neptune

Earth Sun

1985   1984

1983   1982

1981   1980

1979

aphelion

(b)

Figure 17.13 (a) Halley's comet viewed from Easter Island, 8 March 1986. (b) Orbit of Halley's comet during its 76-year round trip. Halley's elongate elliptical orbit is steeply inclined to Earth's orbit.

(a) © James Balog/Getty Images.

The most famous of the comets is the one named for Edmund Halley, the man who calculated its orbit in 1682 and predicted its return to the inner Solar System. Halley's comet travels from near the Sun (its perihelion) to beyond Neptune (its aphelion). The orbit of Halley's comet takes from 74 to 79 years, averaging 76 years (figure 17.13b). Near Neptune, the comet travels 1.5 km/sec (more than 3,300 mph), but it speeds up to 55 km/sec (more than 120,000 mph) as it nears the Sun due to the Sun's immense gravitational acceleration. On its round-trip journey, Halley's comet spends only about 15 months inside the orbital region of Jupiter, but this is where its size is reduced most and its tail develops and glows bright before it returns to the deep freeze of its outer orbit. Halley's comet glowed brightly during the Norman conquest of England in 1066; it passed Earth in 1835 as Samuel Clemens (Mark Twain) was born and then returned again in 1910, the year he died. On 20 May 1910, Earth passed through the tail of Halley's comet. The latest visit of Halley's comet was in 1986, and with a little luck, you will get to see it on its next visit.

The ices of comets contain carbon compounds, some of which are important building blocks of life on Earth. For example, Halley's comet contains carbon (C), hydrogen (H), oxygen (O), and nitrogen (N) in ratios similar to those in the human body. Many scientists think the compounds used to build life were brought to Earth by comets. Are we the offspring of comets?

Rates of Meteoroid Influx

An estimated 100,000 million or more meteoroids enter the Earth's atmosphere every 24 hours. The numbers of incoming objects are directly related to their size; the smaller the meteoroids, the greater their abundance (figure 17.14). Earth is largely protected from this bombardment by its atmosphere. At about 115 km (70 mi) above the ground, the

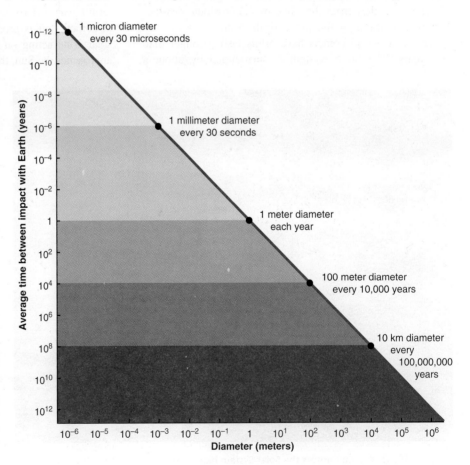

Figure 17.14 Frequency versus size of impacting space debris.

In Greater Depth

Shoemaker-Levy 9 Comet Impacts on Jupiter

A once-in-a-lifetime event occurred during the week of 16–22 July 1994, as a series of comet fragments plunged into Jupiter. The comet was named after its discoverers: geologists Eugene and Carolyn Shoemaker and comet hunter David Levy. In 1992, the comet had flown too close to Jupiter, and the planet's immense gravitational attraction pulled in the comet and broke it into pieces. In 1994, the broken-up comet was stretched out like a string of beads as it again approached Jupiter. In succession, 21 large fragments plunged into Jupiter's dense atmosphere at speeds up to 60 km/sec (134,000 mph). Each impact caused (1) an initial flash as a fragment collided with the heavy atmosphere, (2) a superheated fireball of hot gas rising upward as a plume thousands of kilometers above Jupiter's clouds, and (3) radiation as the plume crashed back down at high velocity.

The largest fragment (G) apparently was only 1 km (0.6 mi) across, yet it left an impact scar larger than the diameter of the Earth (figure 17.15). Each impact caused a rising plume of hot gas that expanded and cooled as it rose (figure 17.16). Although the impacting fragments were small and penetrated only into Jupiter's upper atmosphere, the impact energy released by fragment G was equivalent to about 315 million World War II atomic bombs.

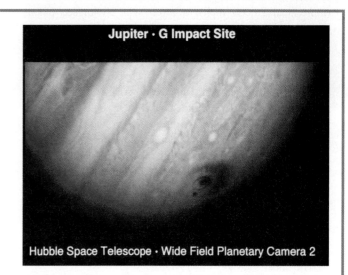

Figure 17.15 Impact scar of Shoemaker-Levy 9 comet fragment G on Jupiter (in lower right of photo), 17 July 1994. To the lower left of G is the smaller impact scar of fragment D. NASA.

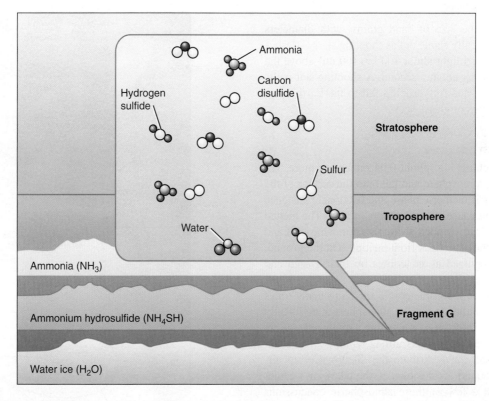

Figure 17.16 Path and impact plume of Shoemaker-Levy 9 comet fragment G plunging through cloud layers thought to make up Jupiter's upper atmosphere.

atmosphere is dense enough to cause many meteoroids to begin to glow. A typical meteor is seen about 100 km (60 mi) above the ground and has largely or entirely vaporized before reaching 60 km (35 mi) above the surface. All this incoming debris adds from 100 to 1,000 tons of material to Earth's surface each day.

Earth is also protected from meteoroids by the very great speeds with which they hit the atmosphere—from 11 km/sec (more than 25,000 mph) to more than 30 km/sec (about 70,000 mph). At high impact speeds, the low-viscosity atmosphere behaves more like a solid. Remember how hard the water in a pool or lake feels when you hit it doing a belly flop? Meteoroids hitting the atmosphere at incredible velocities experience a similar effect; they may be destroyed on impact with the atmosphere, deflected into space if their angle of approach is low enough, or slowed down due to friction. Incoming objects weighing more than about 350 tons are big enough to be largely unaffected by the atmosphere. Both the smallest and the largest meteoroids pass through the atmosphere with little change. However, the intermediate sizes may suffer significant alterations.

COSMIC DUST

The littlest meteoroids are so small that they pass downward through the atmosphere effectively unchanged and settle onto the surface as a gentle rain. Particles with diameters around 0.001 mm have so much surface area compared to their volume that their frictional heat of passage is radiated as quickly as it develops, and they escape melting.

SHOOTING STARS

Incoming debris the size of sand grains, with diameters around 1 mm, typically flame out as shooting stars—flashes of friction-generated light about 100 km (60 mi) above the ground that blaze for about a second. A shooting star melts in the atmosphere, and tiny droplets fall to the Earth's surface as little spheres of glassy rock.

METEORITES

Meteoroids weighing 1 g (about 0.04 ounce) or more will pass through the atmosphere and fall onto the surface of the Earth. During their meteoric phase, the frictional resistance of the atmosphere causes their exteriors to melt. The melted surface materials are stripped away, also removing heat and thus protecting their interiors from melting. On the ground, they can be recognized as meteorites by their glazed and blackened outer crusts.

Consider a multipound meteoroid traveling many times faster than a rifle bullet. It hits the atmosphere and violently compresses the air in front of it, creating a mini sonic boom. We all experience loud booms, rattling windows, and shaking houses; usually, we attribute the disturbances to airplanes (sonic booms), explosions, earthquakes, or weather phenomena. But a few of these atmospheric concussions may be caused by meteors hitting the top of the atmosphere (figure 17.17). To be heard on the ground, an incoming meteor must be at least as big as a basketball.

On 3 October 1996, the sky had two brilliant light shows. First, a meteorite hit the atmosphere over New Mexico, lighting up the New Mexico and Texas sky before bouncing back into space. Second, after a 100-minute orbit, the meteorite

was pulled back into the atmosphere over the Pacific Ocean, crossing over the California coast at Point Conception and then exploding in midair at least twice, sending off sound waves recorded by seismographs. The meteorite fragments apparently fell in the eastern Sierra Nevada foothills, but no pieces have been found—yet.

The heat of atmospheric friction may raise the surface temperature of an incoming object to 3,000°C (about one-half the temperature of the Sun's surface). Melted surface material is stripped off to feed the glowing tail of a fireball, lighting up thousands of cubic miles of sky.

Friction with the atmosphere also slows a meteorite down; it may hit the ground at only 320 to 640 km/hr (200 to 400 mph). In 1954, an 8.5 lb stony meteorite crashed through the roof of a woman's home in Sylacauga, Alabama, bounced off several walls, and then hit her and severely bruised her hip.

On 28 March 2003, the midnight sky over Illinois, Indiana, Ohio, and Wisconsin flashed an eerie blue as a meteorite sped overhead and then broke apart. Four houses, the fire department, and one car were damaged in Park Forest, Illinois, and two more houses were hit in nearby Matteson. Residents found more than 60 pieces of meteorite, ranging from gravel to softball-size.

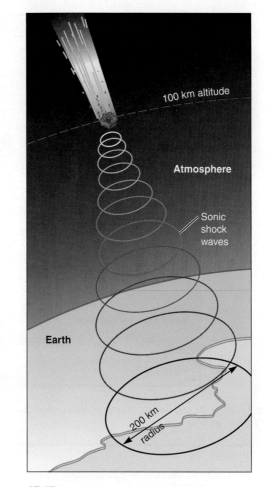

Figure 17.17 A meteor hitting the top of the atmosphere sends off sonic shock waves that might be heard on the ground.

Around 6:30 p.m. on 27 September 2003, a large meteorite racing through the sky lit up like a giant greenish-blue floodlight over the coastal state of Orissa in eastern India. The meteorite broke apart with a thunderous cracking sound and then rained debris. A thatched-roof house in Mayurbhanj was set afire by meteorite debris, injuring three people. One 55-year-old man viewing the fireball collapsed from shock and died.

It has been suggested that some mysterious deaths may be attributable to meteorite impact. This claim has been argued by a defense lawyer in a murder case.

Antarctica

The best place on Earth to find meteorites is Antarctica. Meteorites slam into its great ice sheet and are stored and preserved as if they were in your freezer. Meteorites encased in ice are protected from rainwater, carbonic acid, and plant chemicals that would decompose them elsewhere. The flow of ice can concentrate meteorites at bends in a flowing glacier, thus making it easier for us discover and collect them. In the past 30 years, more than 16,000 meteorites have been collected, including many unique samples.

The Crater-Forming Process

The amount of energy released by an asteroid or comet impact depends on its speed and mass. Asteroids may impact at 14 km/sec (more than 30,000 mph), and a long-period comet could hit at 70 km/sec (more than 150,000 mph).

It is all too easy to visualize an impact by throwing a rock as hard as you can into a pile of loose sand, but this is a poor analogy. Even at a relatively low impact velocity of 12 km/sec (27,000 mph), the kinetic energy released at impact of a meteoroid is 25 times as much as the chemical energy freed by the blast from an equivalent amount of dynamite. An impact almost instantaneously releases energy, creating an explosion that converts much of the meteorite/asteroid/comet and impact-site rock into a superhot vapor. The impact and blast leave a crater with a diameter about 20 times greater than the diameter of the space object.

The impact of smaller meteorites creates *simple craters* with raised rims and concave bottoms lacking central uplifts, such as Meteor Crater, Arizona (figure 17.18).

Figure 17.18 View northwest of Meteor Crater, Arizona. Notice the upturned rock layers in the crater rim, the little hills of ejected debris surrounding the crater, and the individual blocks of resistant rock (e.g., limestone) strewn about the plateau.

The impact of larger bodies forms *complex craters* with central uplifts and collapsed outer rims. As the asteroid or comet hits the surface, a shock wave passes radially into the ground and compresses the rocks (figure 17.19a). Large impacts generate so much heat and pressure that much of the asteroid and crater rock are melted and vaporized (figure 17.19b). In an instant, temperatures may reach thousands of degrees centigrade, and pressure may exceed 100 gigapascals (more than the weight of a million atmospheres). The shock wave pushes rocks at the impact site downward and outward in rapid acceleration of a few kilometers per second. The rocks in the crater and the debris thrown out of the crater are irreversibly changed by the short-lived high temperature and pressure. Rocks are broken, melted, and vaporized; new minerals, such as diamond, are created; and a common mineral such as quartz has its atomic structure transformed by the high-pressure impact into its high-density form as the mineral **stishovite.**

Still within the initial second, a release or dilatation wave follows into the Earth and catches up with the accelerating rocks, deflecting material upward and outward and forming a central uplift on the crater floor (figure 17.19c). The crater that exists in this split second is transient and soon to be enlarged.

As the crater is emptied of vaporized and pulverized asteroid and rock, the fractured walls of the transient crater fail and slide in toward the center of the crater (figure 17.19d). Some rocks and debris fall back into the crater. This is the final enlarged crater, with an upraised center surrounded by a circular trough and then by an outermost fractured rim. The outer circle of the final crater may be 100 times wider in diameter than the crater is deep.

The Manicouagan impact crater shows an upraised central area surrounded by a circular trough (see figure 17.5). An outer circle of a final crater may have existed at higher elevations at Manicouagan and then been eroded away by post-impact continental glaciation. Similar features are seen on other planets as well. The Yuty crater on Mars (figure 17.20) has a well-developed central peak surrounded by a circular trough. Apparently, subsurface ice deposits were melted at Yuty, yielding muddy, liquefied ejecta that flowed over the adjacent area. Similar features would form on Earth if impact occurred on the frozen ground of Siberia, northern Canada, or Alaska.

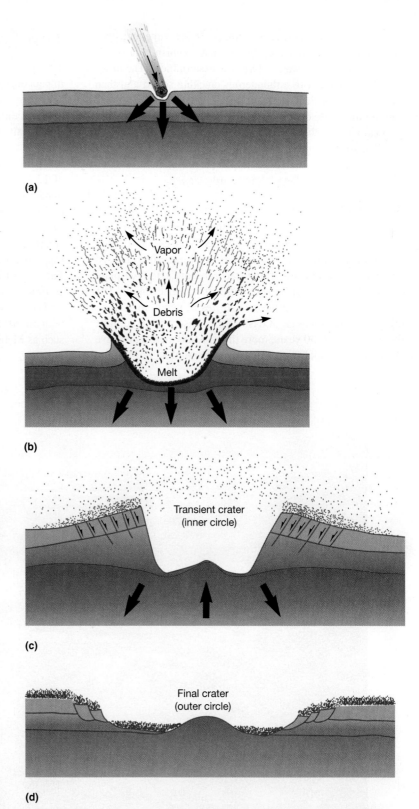

Figure 17.19 Impact of a meteoroid. (a) An incoming meteoroid heavier than 350 tons may be moving faster than 50,000 km/hr (30,000 mph). The impact sends a shock wave into the ground. (b) The impact shock causes such high temperatures and pressures that most of the meteoroid and crater rock are vaporized and melted. (c) The release wave following the shock wave causes the center of the floor in the transient crater to rise. (d) The fractured walls fail and slide into the crater, creating a wider and shallower final crater littered with some fall-back debris.

Figure 17.20 Yuty crater on Mars has a well-developed central peak and surrounding circular trough.

NASA/JPL.

evidence of impact, rocks altered by impact remain in their broken, melted, and shocked states for the rest of their existence.

The impact process has been studied by firing bullets into soft sand. When a bullet hits sand, it fragments and explodes, creating a circular crater whether the bullet path is vertical or at a shallow angle to the surface. After the event, what evidence remains? There is a circular crater with small fragments of metal that are difficult to find.

Crater-Forming Impacts

Meteoroids with weights greater than 350 tons are not slowed down much by the atmosphere. The big ones hit the ground at nearly their original speed, explode, and excavate craters. The record of crater-forming impacts on Earth is sparse. Craters are erased by erosion, consumed by subduction, mangled by continent collisions, and buried beneath younger sediments. So far, there are 166 known impact craters, including 58 in the United States and Canada (figure 17.22 and table 17.3).

The impact process may be visualized in miniature using a falling drop of water hitting a still body of water (figure 17.21). At the point of impact, water springs upward, ripples and troughs surround the impact, and a spray of fine water shoots upward and outward. Although the water quickly returns to its normal still condition and retains no

Figure 17.21 Drop of water hitting a body of water. Note the central rebound.

© Stockbyte/PunchStock RF.

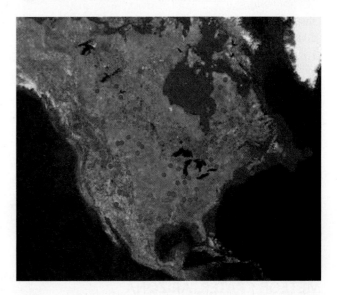

Figure 17.22 Impact crater locations, Canada and the United States. See table 17.3 for names, sizes, and ages.

World Impact Crater map produced by the Planetary and Space Science Centre, University of New Brunswick, Fredericton, Canada. For further details see: http://www.passc.net/EarthImpactDatabase

TABLE 17.3
Names, Sizes, and Ages of Canada and United States Impact Craters

Crater Name	Location	Diameter (km)	Age (millions of years)	Crater Name	Location	Diameter (km)	Age (millions of years)
Ames	Oklahoma	16	470 ± 30	Manson	Iowa	35	73.8 ± 0.3
Avak	Alaska	12	3 − 95	Maple Creek	Saskatchewan	6	<75
Barringer	Arizona	1.2	0.049 ±	Marquez	Texas	12.7	58 ± 2
Beaverhead	Montana	60	~600	Middlesboro	Kentucky	6	<300
Brent	Ontario	3.8	396 ± 20	Mistastin	Labrador	28	36.4 ± 4
Calvin	Michigan	8.5	450 ± 10	Montagnais	Nova Scotia	45	50.5 ± 0.8
Carswell	Saskatchewan	39	115 ± 10	New Quebec	Quebec	3.44	1.4 ± 0.1
Charlevoix	Quebec	54	342 ± 15	Newporte	North Dakota	3.2	<500
Chesapeake Bay	Virginia	90	35.5 ± 0.3	Nicholson	NWT Canada	12.5	<400
Clearwater East	Quebec	26	290 ± 20	Odessa	Texas	0.17	<0.05
Clearwater West	Quebec	36	290 ± 20	Pilot	NWT Canada	6	445 ± 2
Cloud Creek	Wyoming	7	190 ± 30	Presqu'ile	Quebec	24	<500
Couture	Quebec	8	430 ± 25	Red Wing	North Dakota	9	200 ± 25
Crooked Creek	Missouri	7	320 ± 80	Rock Elm	Wisconsin	6	<505
Decaturville	Missouri	6	<300	Saint Martin	Manitoba	40	220 ± 32
Deep Bay	Saskatchewan	13	99 ± 4	Serpent Mound	Ohio	8	<320
Des Plaines	Illinois	8	<280	Sierra Madera	Texas	13	<100
Eagle Butte	Alberta	10	<65	Slate Islands	Ontario	30	~450
Elbow	Saskatchewan	8	395 ± 25	Steen River	Alberta	25	91 ± 7
Flynn Creek	Tennessee	3.8	360 ± 20	Sudbury	Ontario	250	1850 ± 3
Glasford	Illinois	4	<430	Upheaval Dome	Utah	10	<170
Glover Bluff	Wisconsin	8	<500	Viewfield	Saskatchewan	2.5	190 ± 20
Gow	Saskatchewan	5	<250	Wanapitei	Ontario	7.5	37.2 ± 1.2
Haughton	Nunavut	23	39	Wells Creek	Tennessee	12	200 ± 100
Haviland	Kansas	0.015	<0.001	West Hawk	Manitoba	2.44	351 ± 20
Holleford	Ontario	2.35	550 ± 100	Wetumpka	Alabama	6.5	81 ± 1.5
Ile Rouleau	Quebec	4	<300	**Three new discoveries:**			
Kentland	Indiana	13	<97	Decorah	Iowa	5.5	470
La Moinerie	Quebec	8	400 ± 50	Santa Fe	New Mexico	6–13	<1200
Manicouagan	Quebec	100	214 ± 1	Whitecourt	Alberta	0.04	<0.0011

METEOR CRATER, ARIZONA

The world's classic meteorite crater lies on an arid portion of the Colorado Plateau in north-central Arizona about 26 km (16 mi) west of the town of Winslow. Meteor Crater, also known as Barringer Crater, is more than 1.2 km (4,100 ft) wide, excavated about 175 m (575 ft) below the plateau, and surrounded by a rock rim rising 35 to 60 m (115 to 200 ft) above the countryside (see figure 17.18).

Several pieces of evidence demonstrate that the crater formed by meteorite impact: (1) The crater is steep-sided and closed; (2) the rim of surrounding rock was created by uplifting the horizontal sedimentary-rock layers of the region and tilting them away from the crater (figure 17.23); (3) little hills of rock outside the crater rim are inverted piles of the rock sequence exposed in the crater walls; (4) huge blocks of limestone are strewn around outside the crater; (5) the crater floor holds a 265 m (870 ft) thickness of shattered rock; (6) numerous pieces of nickel-iron metallic meteorite with a combined weight of nearly 30 tons have been collected in the area; and (7) several features indicate the occurrence of high temperature and pressure, such as unusual varieties of quartz (the minerals **coesite** and stishovite), cooled droplets of once-melted metal, fused masses of sand grains, and **shatter cones** (a structure of cones inside of cones) that form under intense pressure.

There also is negative evidence that argues against other processes being responsible for the crater: (1) No volcanic material has been found within or nearby, and thus the crater is not a volcanic-explosion pit. (2) No solutional features are present to argue for a solution-collapse process of subsidence similar to that forming sinkholes.

When all the evidence is considered, the words of Sherlock Holmes in "The Adventure of the Bruce-Partington Plans" apply: "Each fact is suggestive in itself. Together they have a cumulative force." In sum, the evidence at Meteor Crater, Arizona, is overwhelming; the site has become the most photographed meteorite crater in the world.

Meteor Crater formed about 50,000 years ago when a nickel-iron metallic meteorite came blazing through the atmosphere. The meteorite had a diameter of about 40 m (130 ft) and hit the ground traveling about 12 km/sec (27,000 mph). The enormous energy of impact was largely converted into heat, which liquefied about 80% of the meteorite and the enveloping ground in less than a second. Impact craters are almost always circular because the meteorite explosion upon impact creates shock waves that excavate

Figure 17.23 Close-up view of Meteor Crater showing peeled-back rock layers and ejected debris.

University of Washington Libraries, Special Collections, John Shelton Collection, Shelton KC12507.

the crater. About 100 million tons of rock were pulverized in this Arizona event, generating about double the energy released by the Mount St. Helens volcanic eruption. The shock wave leveled all trees in the region, wildfires broke out, and dust darkened the sky.

When the surface features are as clearly evident as they are at Meteor Crater, Arizona, there is little debate about the origin of a crater. But how are geologically ancient meteorite-impact craters recognized? What clues remain after the obvious topographic features have been removed by erosion or buried beneath later sediments? Some ideas have been generated during the worldwide search for evidence to explain the great dying of organisms that occurred 66 million years ago at the close of Cretaceous time (see geologic time scale in Epilogue). These new approaches have increased our understanding of Chesapeake Bay.

Impact Origin of Chesapeake Bay

By drilling bore holes and analyzing seismic lines, it has recently been shown that Chesapeake Bay is the site of a 90 km (56 mi) diameter crater formed by asteroid impact 35.5 million years ago (figure 17.24). The crater has a 25 km (16 mi) diameter central peak surrounded by a circular trough or moat up to 400 m (more than 1,300 ft) deep and is 30 km (19 mi) across. The shape of the impact structure is similar to that of Yuty crater (see figure 17.20). The impact crater was a topographic low spot that rivers flowed toward, thus forming the network of river valleys that are today drowned by the Atlantic Ocean to form Chesapeake Bay.

Geologists have long puzzled over the source of **tektites,** glassy spherules formed by in-air cooling of impact-melted rock. The North American tektite field consists of glassy spherules spread over the southeastern United States, Gulf of Mexico, and Caribbean Sea area. Now we know that an asteroid heading south 35.5 million years ago slammed into Virginia and showered tektites over a 9 million km^2 area south of the impact site.

The End Cretaceous Impact

To learn what happened at the close of Cretaceous time, we should follow the advice spoken by Sherlock Holmes in "The Problem of Thor Bridge": "If you will find the facts, perhaps others may find the explanation." To find the facts of 66-million-year-old events means to examine rocks of that age.

The modern search began near Gubbio, Italy, where the latest Cretaceous rocks are limestone loaded with fossil foraminifera, microscopic animal protists with diameters up to 1 mm. The younger limestone lying above it contains an impoverished and markedly changed fossil assemblage. Between the limestones lies a 1 cm thick clay layer that

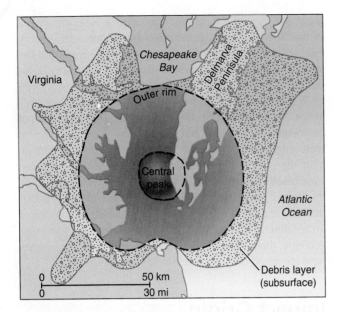

Figure 17.24 Outline of the buried impact crater in lower Chesapeake Bay, Virginia.

marks the boundary between the Cretaceous and the overlying Paleogene (Tertiary) rocks. Does the boundary clay layer hold facts that might explain the events of its days? The late Luis Alvarez, a Nobel Prize–winning physicist, and his geologist son Walter focused their research group's efforts on this topic. Their investigation determined that the clay layer contains a high percentage of the element iridium, an enrichment about 300 times greater than the normal abundance. Here is a fact begging for an explanation.

Iridium is a siderophile, or iron-loving element, whose concentration on Earth parallels that of iron. Most of Earth's iridium lies deep in its iron-rich core; it migrated with iron to the core during the time of early heating when the Earth separated into layers of different density (see figure 2.3). But the boundary clay layer, which is found in many places around the world, holds an estimated one-half million tons of iridium. How did this layer become so enriched with iridium? Luis Alvarez reasoned that because meteorites are rich in iridium, a 10 km (more than 6 mi) diameter asteroid could have supplied the volume of iridium estimated to be present in the boundary clay.

In a condensed version, here is a popular theory of our times: an asteroid with a diameter equal to the height of Hawaii (from seafloor to the peak of Mauna Kea) slammed into the Earth 66 million years ago (figure 17.25). The impact, called the Chicxulub impact, caused a great dying among life worldwide, including the extinction of dinosaurs (excluding birds), and left its incriminating fingerprint as iridium in a global clay layer. The theory is intriguing, easy to grasp, and beguilingly simple to accept. But for a theory to gain widespread approval in the scientific community, it must explain all relevant facts and allow predictions to be made. If these predictions later become supported by facts, then a theory gains wider acceptance. Thus, before the end Cretaceous impact theory could gain wide acceptance in the scientific community, many more facts had to be discovered to verify predictions made by the theory.

Figure 17.25 Artist's view of the Chicxulub impact 65 million years ago.
© Brand X Pictures/PunchStock RF.

EVIDENCE OF THE END CRETACEOUS IMPACT

Once the impact theory was proposed, it excited scientists worldwide, and the search for facts shifted into high gear. Researchers around the world began searching for other end Cretaceous boundary clay layers to look for other evidence of Earth-like versus meteorite-like components. Here are their findings: (1) The clay layer was found on the continents, thus ruling out the possibility that the iridium enrichment was due simply to a change in ocean composition. (2) The boundary clay minerals have a different composition from clays in the limestone layers above and below it; they might be explained by a mixture of one part asteroid to 10 parts Earth crust. (3) Quartz grains with shocked crystal structures are present, indicating a short and violent impact. Shocked quartz, with its planar deformation features, has only been found in association with impacts, so its discovery at the end Cretaceous boundary is strong evidence of impact. (4) Sand-size spherules of minerals are present, suggesting melting and resolidification. (5) Ratios of the radioactive element rhenium to its decay product osmium are similar to those in meteorites and are quite different from the ratios in Earth surface rocks. (6) Abundant microscopic diamonds, found in some meteorites, occur in the boundary clay layer. (7) Carbon-rich grains with "fluffy" structures indicative of fire are abundant in the boundary clay layer.

SITE OF THE END CRETACEOUS IMPACT

The facts from the boundary clay layer compelled more scientists to agree that a massive impact had occurred, but even more evidence was needed. If an asteroid slammed into Earth some 66 million years ago, where was the impact site? Could it be found? Or had it been (1) subducted and destroyed, (2) buried beneath a continental glacier, (3) hidden under piles of sediment on land or seafloor, (4) covered by flood basalt, (5) eroded and erased from the face of the Earth, or (6) crunched into oblivion by a continent collision? Geologists searched for impact scars worldwide, but some were too small, while others were too old or too young.

Then related evidence from latest Cretaceous rocks began to focus the search. On Haiti was found a sedimentary rock layer containing shocked quartz grains and 1 cm diameter glassy spherules formed from melted rock. In Cuba, a thick, chaotic sedimentary deposit with huge angular blocks was discovered. In northeastern Mexico, similar particles were found in a thick bed of sediment containing land debris, ripple marks, and other features interpreted as a tsunami deposit. Similar but thinner deposits were found in the end Cretaceous boundary position in the banks of the Brazos River in Texas and in coastal deposits in New Jersey and the Carolinas. The sedimentary features suggested a Caribbean region impact, but where?

The excitement of the search led to the Yucatan Peninsula of Mexico, where the Mexican national petroleum company (PEMEX) had drilled exploratory wells in the region of Merida. At depths of 2 km, the PEMEX well bores had encountered a 90 m (300 ft) thick zone of shattered rock containing shocked quartz grains and glassy blobs of once-melted rock. On the ground surface lie solution pits (sinkholes) aligned in a circular pattern. Geophysical measurements show circular patterns of gravity and magnetic anomalies suggesting a circular disturbance at depth. Seismic surveys reveal a raised inner ring 80 km (50 mi) in diameter and an outer ring about 195 km (120 mi) in diameter.

These data all help define the Chicxulub structure formed by a massive asteroid slamming into a shallow, tropical sea around 66 million years ago.

SIZE AND VELOCITY OF IMPACTOR

Pieces of asteroid debris define the composition of the impacting body. The size of the impact crater helps define an asteroid diameter of 9–14 km (5.6–8.7 mi), most commonly estimated at about 10 km (6 mi). Measurements of the sizes and volume of impact debris lead to estimates of impact velocity. Purdue University geophysicists B. C. Johnson and H. Jay Melosh estimate the velocity of the end Cretaceous asteroid at about 21 km/sec (47,000 mph).

ANGLE OF IMPACT

An asteroid can hit the Earth at angles ranging from 90° (vertical impact) to 0° (grazing blow). At Chicxulub, the subsurface features measured by gravity and magnetism show some opening to the northwest, like a horseshoe (figure 17.26a). This asymmetry could be the result of the asteroid coming in from the southeast (figure 17.26b), hitting and excavating a deep crater that shallows to the northwest. An oblique impact of 20° to 30° would concentrate its energy into vaporizing surface rocks and making a mammoth vapor/dust cloud. This impact scenario would cause great grief to life in North America by spraying the continent with a high-speed vapor cloud hot enough to ignite plants.

The impact was so great that its effects were felt not just regionally but worldwide. They probably played a significant role in the mass extinction of plants and animals that marked the end of Cretaceous time (see Epilogue).

PROBLEMS FOR LIFE FROM THE END CRETACEOUS IMPACT

What did living things have to tolerate when the massive asteroid slammed into shallow sea and the land? Many difficult conditions resulted, on both regional and global scales: (1) At Chicxulub, the impact certainly created an earthquake of monumental magnitude along with numerous gigantic aftershocks. Seismologist Steven M. Day has estimated the magnitude of the earthquake by scaling up from the energy released in a nuclear explosion. Extrapolating upward from an atomic bomb blast of magnitude 4 leads to an end Cretaceous impact earthquake of magnitude 11.3

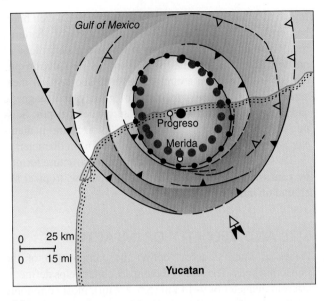

(a)

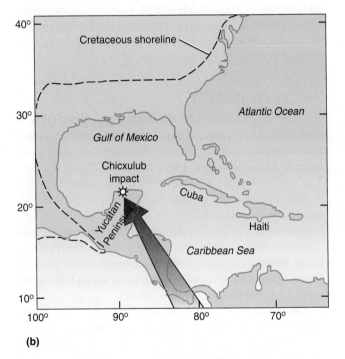

(b)

Figure 17.26 (a) A buried impact crater is shown on the tip of the Yucatan Peninsula. Notice how the gravity data appear to open to the northwest, suggesting that the asteroid came from the southeast. (b) An approximate path of the end Cretaceous asteroid. Its impact would have sent a superhot vapor cloud over North America.

(see figure 3.26). In addition, several other events are probable: (2) Wildfires raged regionally, or even globally. One study suggests the impact ejected so much hot debris into the atmosphere that massive wildfires resulted, and consumed much of the vegetation in North America, the Indian subcontinent, and the equatorial region of the world. (3) Huge amounts of nitrogen oxides in the atmosphere probably fell as acid rain and acidified surface waters. (4) Dust

and soot in the atmosphere blocked sunlight and turned day into night, thus making photosynthesis difficult and plunging much of the world into dark wintry conditions for weeks to several months. (5) After the atmospheric dust settled, the water vapor and CO_2 remaining in the atmosphere would have led to global warming for years (figure 17.27).

What additional insults must life survive after a 10 km diameter asteroid splashes into the ocean? (1) Tsunami up to 300 m (1,000 ft) tall will likely occur. (2) A bubble of steam up to 500 km^3 (120 mi^3) volume will probably blow into the upper atmosphere, carrying Earth rock and asteroid debris. (3) At Chicxulub, where the asteroid landed in shallow, tropical marine water, the impact must have vaporized enormous quantities of underlying limestone ($CaCO_3$), thus increasing atmospheric CO_2, maybe by an order of magnitude. After the winter-causing asteroidal dust settled, the added CO_2 in the atmosphere may have elevated Earth's climate into global-warming conditions. Average temperatures may have risen 10°C (18°F), forcing life on Earth to endure the shift from an "extra cold winter" to an "overly hot and long summer."

It must be noted that the Chicxulub impact was not the only global disaster for life around the close of Cretaceous time. The massive Deccan flood basalt (India) erupted more than 1.1 million km^3 (265,000 mi^3) of lava in about 750,000 years. The basalt lavas devastated the regional land but its volcanic gases changed the climate and ocean composition. The Deccan flood basalt eruption began about 250,000 years before the Chicxulub impact and continued for about 500,000 years after the impact. The one-two punch of asteroid impact and volcanism caused global mass extinctions on land and in the oceans (see Epilogue).

Biggest Events of the 20th and 21st Centuries

The biggest event, so far, in the 21st century was the meteor explosion over Chelyabinsk, Russia, in 2013 (see figure 17.1). But an even bigger meteor explosion occurred over Russia in 1908.

TUNGUSKA, SIBERIA, 1908

The morning was sunny in central Siberia on 30 June 1908. Then, just after 7 a.m., a massive fireball came streaking in from the east. It exploded about 8 km (5 mi) above ground in a monstrous blast heard 1,000 km (more than 600 mi) away. No humans lived immediately under the blast point, but many reindeer did, and they died. A man 60 km (37 mi) away was enveloped in such a mass of heat that he felt his shirt almost catch fire before an air blast threw him 2 m (7 ft). People and horses 480 km (300 mi) from the explosion site were knocked off their feet. At the site, a huge column of fire rose 20 km (12 mi) high and was visible 650 km

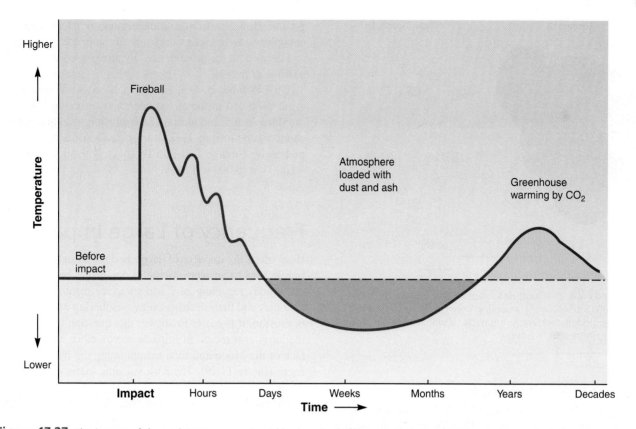

Figure 17.27 The impact of the end Cretaceous asteroid had marked effects on Earth's surface temperatures. First, there was a fireball, followed by hot gases that lasted for many hours. Second, temperatures dropped to wintry conditions as airborne dust and soot blocked much incoming sunlight for several months. Third, after the dust settled, CO_2 remained aloft, creating a greenhouse effect that lasted for years.

After David A. Kring, "Impact Events and Their Effect on the Origin, Evolution and Distribution of Life." *GSA Today,* v. 10, no. 8 (2000), p. 4.

(400 mi) away. The ground shook enough to be registered on seismometers in Russia and Germany. Barometric anomalies were recorded as the air blast traveled twice around the world. In Sweden and Scotland, an extraordinarily strong light appeared in the sky about an hour after sunset; it was possible to read by this light until after 2 a.m.

Scientists around the world speculated on what had happened, but it was years before an expedition went to the remote area to search for evidence. Near the Tunguska River, the forest in an area greater than 1,000 km² (more than 400 mi²) was found to have been knocked down and destroyed; many trunks were charred on one side. Over a broader area exceeding 5,000 km² (2,000 mi²), 80 million trees were down, and many others had tilted trunks, broken branches, and other signs of disturbance. But there was no impact crater or even broken ground. The relative lack of facts led unrestrained minds to invent all sorts of wild stories. It was not until 1958 that scientists returned to the site and collected little globules of once-melted metal and silicon-rich rock.

Several important facts must be explained. Witnesses had reported an intense, bluish-white streak in the sky, a horrendous explosion, a searing blast of heat, blasts of air that encircled the globe, a brilliant sunset, and a bright night. Yet, there was no impact crater—only little globules of melted material. What happened? Evidence indicates that a meteoroid racing through the atmosphere broke up and exploded about 8 km (5 mi) above the ground. It was either a fragment of an icy comet about 50 m (165 ft) in diameter, or a large, stony meteorite about 30 m (100 ft) in diameter. The object was traveling about 15 km/sec (33,000 mph) when it disintegrated in a spectacular midair explosion. If it had been a metallic body, it would almost certainly have slammed into the ground. But comets and stony meteorites are weak bodies traveling at outrageous speeds, and the resistance of Earth's atmosphere is so strong that they typically break apart. At the end of June 1908, the comet Encke was passing by Earth; one of its fragments is the likely culprit for the Tunguska event.

The Tunguska comet explosion rocked a sparsely inhabited area and devastated a forest. Imagine if it had exploded over a city like Washington, DC, obliterating its buildings and people (figure 17.28). The broader area of knocked-down trees was big enough to include Baltimore. How common are these Tunguska-like events? Are such events frequent enough for humans to be concerned about?

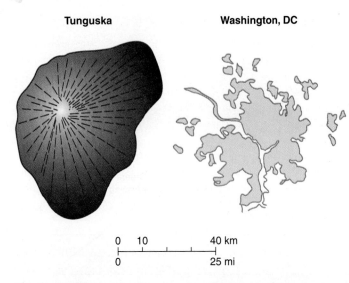

Tunguska   **Washington, DC**

0   10   40 km

0   25 mi

Figure 17.28 The Tunguska event in 1908 devastated more than 1,000 km² (400 mi²), knocking down most trees. Had this comet exploded over Washington, DC, it would have been one of the major events in history.

BIGGEST "NEAR EVENTS"

On 22 March 1989, the asteroid 1989 FC with a diameter of about 500 m (0.3 mi) crossed Earth's orbit at almost the wrong moment; Earth had been at the spot only six hours earlier. The asteroid missed us by less than 700,000 km (400,000 mi). Had a collision occurred on land, the impact would have created a crater about 7 km (4.4 mi) across. Was this close call a freak occurrence? Apparently not: several thousand such bodies are in Earth-approaching orbits.

On 19 May 1996, a 150 m (500 ft) wide asteroid missed Earth by only 453,000 km (less than 281,000 mi). This asteroid is almost four times wider than the meteorite that excavated Meteor Crater, Arizona (see figure 17.18).

In March 1998, the media widely and excitedly reported that an asteroid labeled 1997 XF11 might hit Earth in the year 2028. In the same year, two big-budget Hollywood movies, *Deep Impact* and *Armageddon,* sensationalized the effects of collision with a comet and an asteroid, respectively. To communicate calmly about the threat of comet and asteroid impacts, Richard Binzel developed the Torino scale (table 17.4), which assesses the threat on a scale of 0 to 10.

For smaller objects, the number of near misses is surprisingly high. Detailed telescopic examination has shown that up to 50 house-size bodies pass between Earth and the Moon each day. Should we worry about the consequences of impacts from speeding bodies of 20 m (65 ft) diameter? Apparently not, because Earth has its own defense system against small bodies—its atmosphere. When comets and stony meteorites traveling at 50,000 km/hr (more than 30,000 mph) hit the atmosphere, the impact breaks most of them into smaller pieces that burn up explosively. Most of the flameouts occur 10 to 40 km (6 to 25 mi) above the surface, which is too high to cause significant damage on the

ground. However, most iron meteorites are internally strong enough to stay intact as they pass through the atmosphere, and they do hit the ground. Luckily, iron meteorites are relatively uncommon.

On 8 October 2009 at 11 a.m., a 10 m (33 ft) wide meteoroid exploded in the atmosphere above the town of Bone on Muna Island, Indonesia. The explosion was equivalent to about 50,000 tons of TNT; it triggered sensors of the Comprehensive Nuclear-Test-Ban Treaty Organization more than 10,000 km (6,000 mi) away.

Frequency of Large Impacts

How often do impacts of large bodies occur? This question is hard to answer looking at our planet because of the continuous recycling of Earth's surface materials by plate tectonics and their destruction by weathering and the agents of erosion. It is easier to answer this question by looking at the long-term record of impacts preserved on the dead surface of the Moon and then extrapolating the results back to Earth (figure 17.29). The dark volcanic maria on the Moon formed after a few-hundred-million-year period of intense asteroidal bombardment more than 3.9 billion years ago.

TABLE 17.4

The Torino Scale

Assessing Comet and Asteroid Impact Hazards

Events with No Likely Consequences (white zone)

0   No collision hazard, or object is small

Events Meriting Careful Monitoring (green zone)

1   Collision is extremely unlikely

Events of Concern (yellow zone)

2   Collision is very unlikely

3   Close encounter with >1% chance of local destruction

4   Close encounter with >1% chance of regional devastation

Threatening Events (orange zone)

5   Significant threat of regional devastation

6   Significant threat of global catastrophe

7   Extremely significant threat of global catastrophe

Certain Collisions (red zone)

8   Collision will cause localized destruction (one event each 50 to 1,000 years)

9   Collision will cause regional devastation (one event each 1,000 to 100,000 years)

10   Collision will cause global catastrophe (one event each 100,000 years)

Figure 17.29 Close-up view of craters on the Moon.
NASA.

The basalt-flooded maria cover 16% (6 million km²) of the Moon's surface; they formed by about 3,200 million years ago. The maria are scarred by five craters with diameters greater than 50 km (more than 30 mi) and by another 24 craters with diameters between 25 and 50 km (15 and 30 mi). This averages to one major impact occurring somewhere on the maria every 110 million years.

Applying these impact rates to Earth generates the following numbers: Earth's surface area is more than 80 times the area of the lunar maria, so it would have had more than 80 times as many impacts—that is, about 2,400 impacts leaving craters greater than 25 km (15 mi) in diameter. Land comprises about 30% of the Earth's surface, so about 720 of these craters should have formed on land. More than 160 craters have been discovered so far, but most of them are less than 25 km in diameter. Most of the missing craters have probably been destroyed or buried.

The odds are extremely small that a large asteroid will hit the Earth during your lifetime. However, so many people will be killed when a big space object does hit that it skews the probabilities. Statistically speaking, every individual has a greater chance of being killed by a comet or an asteroid than of winning a big jackpot in a lottery! The probabilities of death by meteoroid impact were indirectly assessed in the words of paleontologist George Gaylord Simpson: "Given enough time, anything that is possible is probable." Because the risks from large meteoroid impact are high, they should be of concern to humans. U.S. astronomer David Morrison has described the Earth as a target in a cosmic shooting gallery of high-speed asteroids and comets. The situation has been evaluated for defensive actions we humans might take.

LIFETIME RISKS OF IMPACT

The problems presented by meteoroid impacts were addressed at National Aeronautics and Space Administration (NASA) international workshops in 1991–92, as requested by NASA and the U.S. Congress. The conclusions are presented in a report entitled *The Spaceguard Survey.* The risks were analyzed for space objects with diameters greater than 1 km (0.62 mi) (table 17.5; see table 17.1). About 90% of the potential impactors are **near-Earth asteroids (NEAs),** or short-period comets; collectively, they are known as **near-Earth objects (NEOs).** The other 10% are intermediate or long-period comets (those having greater than 20-year return times).

In 1992, more than 200 NEOs were known to exist, and about 25–50% of them were predicted to eventually hit the Earth. The risk of being killed in the United States via the impact of a large asteroid in the next 50 years was thought to be about one chance in 20,000. The surprisingly high risk was an artifact of the calculations—a tremendous number of people could be killed in a big impact (e.g., 1.5 billion). This is a classic case of frequency versus magnitude; big impacts don't happen often, but their effects can be global (table 17.6).

Our lack of knowledge about future impacts received a lot of attention after David Morrison remarked that there are fewer people searching for asteroids and comets that might impact the Earth than there are people working a shift in a small McDonald's restaurant. In 1998, the U.S. Congress authorized $40 million for NASA to find 90% of the near-Earth asteroids greater than 1 km (0.62 mi) in diameter by 2008. The first step in protecting ourselves is to locate the near-Earth objects, determine their orbits, and learn which ones present immediate threats. The search was conducted by six international observatories. By the end of 2014, there were 864 large NEAs discovered; this is estimated to be more than 90% of the total population. None of the located large NEAs are on a near-future collision course with Earth.

Early success led to further action. In 2005, Congress instructed NASA to extend the search through 2020 to locate and track 90% of NEOs with diameters greater than 140 m (460 ft). These NEOs are the so-called *city killers*—objects

TABLE 17.5

Frequency of Impacts and Annual Risks of Death

For Tunguska-size Events:	
Average interval between impacts	300 years
Average interval for populated areas only	3,000 years
Average interval for urban areas	100,000 years
Average interval for U.S. urban areas only	1,000,000 years
Total annual probability of death	1/30,000,000

Source: D. Morrison (1992).

TABLE 17.6

Odds of Dying in the United States from Selected Causes (2007)

Cause of Death	Odds of Happening
Motor vehicle accident	1 in 90
Murder	1 in 185
Falls	1 in 250
Fire	1 in 1,100
Firearms accident	1 in 2,500
Drowning	1 in 9,000
Flood	1 in 27,000
Airplane crash	1 in 30,000
Tornado	1 in 60,000
Asteroid/Comet Impact (global)	**1 in 75,000**
Earthquake	1 in 130,000
Lightning	1 in 135,000
Asteroid/Comet Impact (regional)	**1 in 1,600,000**
Food poisoning by botulism	1 in 3,000,000
Shark attack	1 in 8,000,000

Source: C. Chapman (2007).

at least the size of a 45-story building, or much bigger. By March 2015, 12,250 NEOs had been located, including 91 near-Earth comets. The NEAs are dominated by Apollo (62%) and Amor (32%) asteroids (see figure 17.10).

New information from NASA's Wide-field Infrared Survey Explorer (WISE) indicates there are more **potentially hazardous asteroids (PHAs)** than previously thought. PHAs are a subset of the larger population of near-Earth asteroids (figure 17.30). The low inclination of PHA orbits is more aligned with the plane of Earth's orbit and brings them closer than 8 million km (5 million mi) from Earth. PHAs are big enough to cause damage on a regional, or larger, scale (figure 17.31). The WISE sample of 107 PHAs indicates that the entire population is 4,700 PHAs +/− 1,500. By March 2015, there were 1,557 known PHAs.

As the numbers of known NEOs increase and their orbits are documented, we find that almost all of them are harmless in the near future. The largest asteroids, such as the end Cretaceous asteroid of 66 million years ago, will impact Earth about once every 10^8 years. Thus, the risk of a civilization-changing impact in the 21st century is significantly reduced. Objects in the Tunguska-size range will impact about once every 300 to 500 years.

PREVENTION OF IMPACTS

What can be done about a large PHA that is on course to impact Earth? Taking appropriate engineering actions could make this a preventable natural disaster. Early suggestions

Figure 17.30 Edge-on view of near-Earth asteroids (NEAs). Many potentially hazardous asteroids (PHAs) are closely aligned with the plane of our Solar System. Earth's orbit = green line; simulated NEAs (blue dots) and PHAs (orange dots) are for a typical day.

Image credit: NASA/JPL-Caltech.

include (1) attaching a rocket engine to the PHA to drive it away from us, (2) using a big mirror to focus sunlight on it and vaporize the rock, and (3) scooping rock (mass) from it and tossing it away.

A gentler form of protective action is to launch a spacecraft, a heavy *gravity tractor,* that would hover near an asteroid, allowing the resultant gravitational attraction to pull the

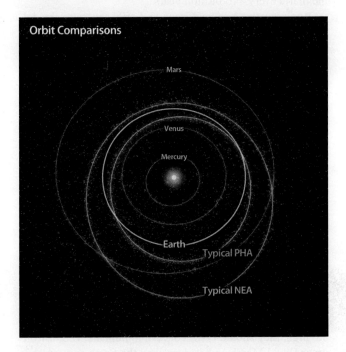

Figure 17.31 Orbits of a typical near-Earth asteroid (NEA) in blue, a potentially hazardous asteroid (PHA) in orange, and Earth in green.

Image credit: NASA/JPL-Caltech.

TABLE 17.7 First Landings of Spacecraft on Space Objects

- Moon: Russia landed their Luna 2 on 13 September 1959; then the United States landed Apollo 11 and placed the first men on the Moon on 20 July 1969.
- Venus: Russia landed their Venera 3 on 1 March 1966.
- Mars: Russia landed their Mars 3 on 2 December 1971.
- Near-Earth asteroid Eros (Amor group): NASA landed its spacecraft NEAR Shoemaker on the 34 km long asteroid traveling 24.36 km/sec (54,000 mph) on 12 February 2001. Eros is a potential Earth impactor in about 2 million years.
- Titan, moon of Saturn: European Space Agency aided by NASA/JPL landed spacecraft Huygens on 14 January 2005.

- Near-Earth asteroid Itokawa (Apollo group): Japanese Space Agency (JAXA) landed their spacecraft Hayabusa on the 535 m long body traveling 25.37 km/sec (56,000 mph) on 20 November 2005.
- Comet 67P Churyumov/Gerasimenko: European Space Agency launched spacecraft Rosetta, which dropped the lander Philae onto the 4 km (2.5 mi) long comet traveling 37,000 mph.
- Asteroid Vesta and dwarf planet Ceres: NASA spacecraft Dawn orbited around Vesta for 14 months in 2011–2012, traveled to Ceres and began orbiting it in March 2015.

asteroid into an Earth-avoiding path. If the gravity tractor effect would be too small, the PHA could be rammed with a *kinetic impactor,* a heavy, high-speed projectile that would transfer its momentum to alter the PHA's velocity or course. If a PHA or comet is discovered too late for these actions to be implemented, the only option may be to hit it with a nuclear explosion that would blast some of its rocky/icy material away and alter its path.

We talk of plans to intercept moving objects in space and redirect them, but does our current technology allow us to do this? Russians, Americans, Europeans, and Japanese have made real progress in sending spacecraft to Solar System objects, not only to big planets but also to fast-moving, small asteroids and comets (table 17.7).

We can send spacecraft to asteroids, but can we change their orbits? We missed an opportunity to try on 15 February 2015, the same day the meteor exploded over Chelyabinsk (see figure 17.1). The asteroid called 2012 DA14 passed over the Indian Ocean only 27,000 km (17,200 mi) above Sumatra. It traveled *inside* the orbits of our communications satellites. This asteroid has a diameter of 30 m (100 ft) and weighs 20 times more than the Chelyabinsk meteor. If 2012 DA14 had hit the ground directly it would have wiped out life in an area of about 1,940 km^2 (750 mi^2).

If an asteroid passes through a 365 km (227 mi) wide region in space known as a **keyhole,** then Earth's gravitational pull could change its orbit enough to bring it back for a collision. Is a large impact too unlikely an event to take seriously? Arthur C. Clarke said:

> We tend to remember only the extraordinary events, such as the odd coincidences; but we forget that almost every event is an odd coincidence. The asteroid that misses the Earth is on a course every bit as improbable as the one that strikes it.

Summary

Space debris colliding with Earth is primarily rocky (stones), metallic (irons), or icy. Stony or metallic bodies are called meteorites if they land on the surface of Earth; icy masses are comets. Irons are the most common form of space debris found by collectors because they pass through the atmosphere more readily, are more resistant to weathering, and are more easily recognized. Impact scars are hard to find on Earth because they are erased by plate-tectonic actions and the agents of erosion. Conversely, the geologically dead Moon is a museum of impact scars.

Asteroids abound in a Sun-orbiting belt between Jupiter and Mars. The strong gravitational pull of the massive planet Jupiter sends collided fragments outbound on collision courses with nearby planets, including Earth.

Comets surround the Solar System in a vast envelope known as the Oort cloud. Some comets are pulled inward close to the Sun and may collide with the Sun or a planet, as occurred in July 1994, when the Shoemaker-Levy 9 comet collided with Jupiter.

About 100,000 million meteoroids enter Earth's atmosphere every 24 hours. Most are small and burn up due to atmospheric friction before reaching 55 km (35 mi) aboveground. Still, meteorites add up to 1,000 tons of material to Earth's surface each day.

Meteoroids hit Earth's atmosphere at speeds ranging from 25,000 to 70,000 mph; some are destroyed on impact, some are deflected into space, and others are slowed by friction as they pass through the atmosphere. Meteoroids weighing more than 350 tons are slowed little by the atmosphere and may hit the ground at high speeds, explode, and excavate craters.

The impact of a large asteroid generates such tremendous heat and pressure that much of the asteroid and crater rock is vaporized. Rocks in the crater and debris thrown out of the crater are broken, melted, and vaporized; minerals take on new high-pressure atomic structures and include forms such as diamonds. As the initial or transient crater is emptied, the crater bottom rebounds upward, and the fractured walls slide inward toward the crater center, forming a final, enlarged crater. The crater may be 100 times wider than it is deep.

Life on Earth is subjected to great stress by an impact. A large asteroid impact can generate an earthquake of greater than magnitude 11, cause widespread wildfires, create nitrous oxides in the atmosphere that fall as acid rain, and place dust in the atmosphere that blocks incoming sunlight to create "winter." After the dust settles, water vapor and CO_2 remain in the atmosphere, causing a global-warming "summer." Additionally, if the impact occurs in the ocean, tsunami of 1 to 3 km height can occur.

About 50 house-size bodies pass between Earth and the Moon each day, but Earth is protected by its atmosphere. It is difficult to estimate how often large bodies impact Earth because plate tectonics and erosion destroy the evidence. However, looking at the Moon's surface indicates that Earth sustains a crater greater than 24 km (15 mi) in diameter every 1.33 million years.

Statistically, your risk of being killed by a meteoroid impact is about 1 in 720,000. We have the ability to locate incoming meteoroids and could change their paths by sending explosives or a rocket out to redirect them.

Terms to Remember

aphelion 461	near-Earth object
asteroid 456	(NEO) 475
astrobleme 457	Oort cloud 461
coesite 469	perihelion 461
comet 456	potentially hazardous asteroid
keyhole 477	(PHA) 476
Kuiper belt 460	shatter cone 469
maria 457	shooting star 458
meteor 458	solar wind 461
meteorite 458	stishovite 466
meteoroid 458	sublimates 461
near-Earth asteroid	tektite 469
(NEA) 475	

Questions for Review

1. At what speed does Earth travel around the Sun?
2. Why does the Moon display impact scars so clearly?
3. Why are impact scars relatively rare on Earth?

4. Distinguish between a meteor, a meteoroid, a meteorite, an asteroid, and a comet.
5. Why are metallic meteorites so commonly collected?
6. Why did the asteroids of the asteroid belt not assemble into a ninth planet?
7. Why does a comet's tail glow brighter as it nears the Sun? Why does a comet's tail point away from the Sun?
8. How much space debris is added to Earth each day? How big must a meteoroid be to pass through the atmosphere with little slowing?
9. Draw a series of cross-sections showing what happens when a 10 km (6 mi) diameter asteroid hits the Earth at 50,000 km/hr (30,000 mph).
10. List the evidence you could collect to demonstrate that a specific area was the site of an ancient asteroid impact.
11. Describe the sequence of life-threatening events that occur when a 10 km (6 mi) diameter asteroid slams into Earth.
12. Explain the Torino scale of impact threats.
13. What is a keyhole? How could one help cause an asteroid impact on Earth?
14. The Chelyabinsk meteor entered Earth's atmosphere at an 18° angle. What would have happened if it had entered at 90°?
15. Rank in order the speed of light, shock waves, and sound. What should you do when an intensely bright light flashes through the sky?
16. How does a dwarf planet differ from a planet? How does it differ from an asteroid?
17. What is the Oort cloud?
18. How many space objects have Earthlings landed on with space crafts? Which countries have the technologic ability to accomplish this?

Questions for Further Thought

1. Extrapolating impact rates from the Moon, about 720 craters with diameters greater than 24 km (15 mi) should have formed on land on Earth; only about 160 have been found so far. Might some of the missing sites have been created by big impacts that caused mass extinctions? How could you proceed scientifically to investigate this possibility?
2. Should the United States spend the money and effort to develop engineering devices capable of landing on large asteroids and comets and diverting their courses away from Earth?
3. It is proposed that we send rockets or explosives to divert incoming large asteroids or comets. Might this action just shatter the incoming object into many devastating impactors? Or cause the object to hit Earth on a more direct path?
4. An In Greater Depth early in chapter 15 showed that kinetic energy $= 1/2 \ mv^2$ where m is mass and v is velocity. Think through this equation, and assess the impact energy of an asteroid and a comet both having diameters of 10 km (6 mi). Asteroids may have four times as much mass, but comets can easily travel twice as fast. Do they bring equivalent amounts of kinetic energy? Upon impact, does it matter that the asteroid is metal and/or rock, whereas the comet is mostly ice?

Epilogue:
Mass Extinctions

Figure 1 Petrified tree logs of the genus *Araucaria*, Late Triassic Period, Petrified Forest National Park, Arizona. The genus still lives today, more than 200 million years later.

© Dr. Parvender Sethi.

LEARNING OUTCOMES

The deadliest natural disasters kill many people, but the deadliest conditions on Earth bring on mass extinctions during which time most species on Earth go extinct. After studying the Epilogue, you should:

- understand fossils.
- understand how extinctions are documented.
- be familiar with the geologic time scale.
- recognize the mass extinctions that have occurred.
- be familiar with the recent and ongoing mass extinction of large-bodied animals.

We live with horrifying natural disasters. Individual earthquakes and floods have each killed in excess of three-quarters of a million people; single cyclones have drowned more than one-half million humans; and individual volcanic eruptions have killed tens of thousands. Yet these disasters etched into the historic record seem small when compared to the mass **extinctions** documented in the fossil record. The human tragedies wrought by natural disasters involve individuals of a **species,** not the entire species. Several times in the past 541 million years, many, or even most, of the species on Earth became extinct in a geologically short time. These great dyings or mass extinctions are the biggest natural disasters known to have occurred on Earth.

How do we know about the lives of extinct species? We read the fossil record.

The Fossil Record

Fossils are evidence of former life. Two common requisites for organisms to become fossilized are (1) possession of hard parts, such as teeth, shells, or bones, and (2) rapid burial, which protects deceased organisms from being scavenged and disintegrated. Fossilization occurs in numerous ways.

For example, tree trunks can be buried by mud and sand dropped by floods. After burial, the wood may be slowly replaced by minerals that grow out of underground water to make petrified wood (figure 1).

Dinosaurs provide another example. They are known to us primarily by their original hard parts—their fossil bones, skulls, and teeth. Their fossils are carefully removed from the rock and then assembled like a jigsaw puzzle to show us extinct dinosaurs (figure 2).

Profound changes are seen in fossil species within the sedimentary rock record. Fossil species make abrupt first appearances, are found in abundance in overlying rock layers, but then are not found in higher rock layers (figure 3). Species go extinct. Sometimes entire communities of plants and animals die out in geologically short lengths of time.

Figure 2 An *Allosaurus* skeleton from 150 million years ago. University of Utah Museum.

Photo by Pat Abbott.

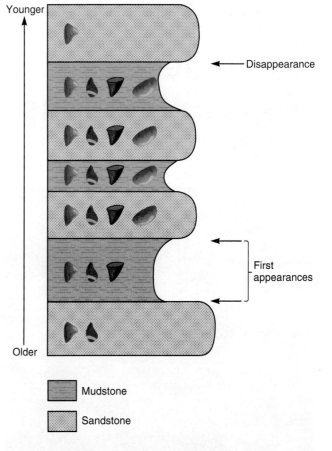

Figure 3 A vertical sequence of sedimentary rock layers illustrates the laws of superposition (the lower the layer, the older) and faunal succession (older forms of life die out and new forms develop).

In the late 1700s it was further recognized that fossils collected from older rock layers (lower by superposition) differ more from present-day organisms than fossils collected from younger rock layers (higher by superposition). This relationship is known as the law of faunal succession. Geologists around the world reacted to this law by going out into the field and recording the order of sedimentary strata and the vertical sequence of their contained fossils. The sequences from around the world were compiled into a standard geologic column in the early 1800s (figure 4). In the early 20th century, quantification was added to the fossil record using dates measured on radioactive isotopes in igneous rocks associated with fossil-bearing sedimentary rocks.

The fossil record documents the appearances and disappearances of millions of species. There have been remarkable changes in life during the billions of years of Earth history.

Mass Extinctions

Extinction occurs on the species level. When so many individuals of a species die that reproduction fails, the continuity of their kind is stopped, and extinction occurs. The average "life span" of a species is about 4 million years. Most of the species that have ever lived on Earth are now extinct; in fact, more than 99.9% of all plant and animal species that have ever lived are extinct. At present, species diversity on Earth is estimated to range from 40 to 80 million, but consider that these large numbers are less than 0.1% of the species in Earth's history.

There are so many millions of fossil species that plotting all of their extinctions through geologic time is difficult. However, a major effort was carried out by the U.S. paleontologist John Sepkoski at the next hierarchial level—the genus. He plotted the times of extinction for about 25,000 genera of marine invertebrates and protozoa (unicellular "animals"). The results are fascinating (figure 5). The plot of generic extinctions versus time produces a highly variable,

Eon	Era	Period		Millions of years ago	Major appearances
Phanerozoic	Cenozoic	Quaternary			Humans
				2.6	
		Tertiary	Neogene		Direct human ancestors
				23	
			Paleogene		
				66	
	Mesozoic	Cretaceous			Flowering plants in abundance
				145	
		Jurassic			
				201	
		Triassic			
				252	Birds
	Paleozoic	Permian			Mammals and Dinosaurs
				299	
		Carbon-iferous	Pennsylvanian		
				323	Reptiles
			Mississippian		
				359	Amphibians
		Devonian			(vertebrates on land)
				419	
		Silurian			
				444	Land plants
		Ordovician			
				485	
		Cambrian			Fishes
				541	Great diversification and
Pre-Cambrian		Ediacaran		635	abundance of life in the sea
		Proterozoic			Sexual reproduction
				2,500	
		Archean			
				4,000	Oldest fossils Oldest Earth rocks
Hadean					
				4,570	Origin of Earth

Figure 4 Geologic timescale based on superposition of sedimentary rock layers and the irreversible succession of fossils. Numerical ages were measured on igneous rocks found in association with fossil-bearing sedimentary rocks.

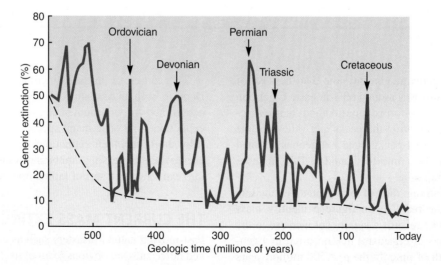

Figure 5 Extinction patterns of genera of marine invertebrates and protozoans versus time. The percentage of extinctions was calculated by dividing the number of extinctions by the number of genera alive at that time. The dashed line represents background extinctions.

From J.J. Sepkowski, "Phanerozoic Overview of Mass Extinction," in *Patterns and Processess in The History of Life*, Springer/Verlag, New York, 1986.

In Greater Depth

Causes of Mass Extinctions

PLATE TECTONICS

- Changes in sea-floor spreading rates change sea level. For example, when spreading rates increase, the greater volumes of risen magma build larger volcanic mountain ranges on the sea floor. This greater bulk causes sea level to rise with seawater spilling over onto the continents, decreasing the amount of land available for life.
- Sea-level changes due to sea-floor spreading can cause sea level to rise and fall up to 200 m (660 ft). The buildup and melting of huge glaciers also can cause sea level to rise and fall up to 200 m (660 ft). The amount of land can change back and forth between being 40% to 17% of Earth's surface (for example, see figure 12.14).
- Number and sizes of continents can have dramatic effects on life. Consider the number of species that could live in a one-continent world (see figure 2.24, Pangaea) versus the number of species that could live at a time of multiple continents and large islands such as today. For example, think of how different the animals are today on Africa vs. Australia vs. North America.

VOLCANISM

- Changes in atmosphere and ocean composition occur when flood basalts pour out on the Earth's surface (see Flood Basalts in Chapter 6).

CLIMATE CHANGE

- Climate changes from extra-warm times during increased Greenhouse conditions (see Late Paleocene Torrid Age in Chapter 12) into chilly times during Ice Ages when glaciers cover much of the land (see figure 12.13 and Late Cenozoic Ice Age in Chapter 12).

OCEAN COMPOSITION CHANGES

- Today the oceans are well stirred thanks to winds and to density differences (see figure 9.29). However, during some warmer climates, as polar seawaters warm they reduce in density, thus reducing the density differences that cause deep-water ocean circulation; when this happens deep-ocean waters can become depleted in oxygen, causing organisms to die. At other times, during Ice Ages, catastrophic glacial melt-water floods can cover the ocean with cold fresh water, causing problems for life (see figure 13.43). A less stirred ocean can kill much life.

EXTRATERRESTRIAL IMPACTS

- Large asteroids and comets traveling more than 15 km/sec (33,000 mph) have hit the Earth's surface and caused die offs around the world (for example, see figure 17.27).

BIOLOGIC CHANGES

- Species can be driven into extinction by excessive predation, pandemics, and habitat loss. And when times are tough during environmental change, population sizes may decrease so extensively that a point of no return is reached; there simply are not enough individuals left to breed and continue the species.

MULTIPLE CAUSES

- Life adapts to existing conditions, and when existing conditions change, life must change as well, or else extinction can occur. All of the above processes bring change, and they all can cause extinction. However, to cause a global *mass extinction* probably requires two or more of these processes of change occurring simultaneously or closely spaced in time.

---

jagged line that obscures trends. But note that the level of background extinctions was around 50% in early Cambrian time when many new life-forms appeared and their extinctions were common. Ignoring the peaks and valleys of the red line, we can see that background extinctions (dashed line) have declined with time to around 5–10% in more recent (Late Cenozoic) time.

Against this declining rate of background extinction, marked spikes in the red line on figure 5 identify mass extinctions; each spike records a significant increase in the number of extinctions of organisms that occurred in a geologically short length of time. In the past 500 million years of Earth history there have been a *Big 5* mass extinctions that have received the most attention. The Big 5 occurred at the end of Ordovician time, during the late Devonian, and at the ends of Permian, Triassic, and Cretaceous time.

The causes of mass extinctions vary (see In Greater Depth). Each mass extinction has a unique blend of co-occurrences of environmental changes that lead to species extinction. And once many species go extinct in a geologically short time, their extinctions set off a positive feedback event where increasing numbers of other species are drawn into extinction in a sort of falling dominoes effect.

THE CURRENT MASS EXTINCTION

Fast-moving natural disasters such as earthquakes, tsunami, and hurricanes are obvious to all of us. Slow-moving natural disasters such as climate change and mass extinctions can be difficult to recognize. It appears that we are living during an ongoing sixth mass extinction (*Big 6*), a wave of extinctions hitting large-bodied animals (>45 kg or 100 lbs) extra hard.

What are the causes of this 6th major mass extinction? One cause of environmental stresses for life are the changes in climate that occur in the current Ice Age. As continental glaciers advance, cover land, and chill the climate, life must adapt. And then change occurs again as the ice sheets melt and retreat. These glacial advances and retreats have been occurring for millions of years, but the extinctions of large-bodied animals are concentrated within the last advance and retreat cycle.

What other major stress for life on Earth has arrived during the past 100,000 years? The growth and worldwide spread of the human population with their increasingly sophisticated ways of hunting. Humans hunt large-bodied animals for food, and emigrating humans bring with them plant and animal species that devastate many native species in the newly occupied lands.

Australia

Humans migrated to Australia about 56,000 years ago. The new arrivals found 24 genera of large-bodied animals. By 40,000 years ago, 23 of the genera (marsupials, reptiles, bird) were extinct; only a genus of large kangaroos survived. The extinctions occurred in all climate zones and habitats. The die-offs took place thousands of years before the extinctions in the Americas, Madagascar, or New Zealand, suggesting that the Australian extinctions were a regional problem, rather than part of a global event. The most likely cause for the extinctions was humans overhunting naïve, large-bodied herbivores (plant eaters), which in turn helped cause the carnivores (meat eaters) who preyed upon them to die out.

Madagascar and New Zealand

Humans have been documented as the exterminating agent in the extinctions on Madagascar, New Zealand, Hawaii, and Chatham Island. The first humans to reach Madagascar and New Zealand found that the largest animals living there were flightless birds. On Madagascar, the largest elephant bird species stood 3.4 m (11 ft) tall and weighed up to 500 kg (1,100 lbs). On New Zealand, the largest moa reached 4 m (13 ft) tall but was not as heavy. Humans killed the big birds and stole their eggs, driving them into extinction.

It appears that the heavy effects of humans on the environment did *not* just arrive with the Industrial Revolution but rather have accompanied every human advance from toolmaking to control of fire, to agriculture, to the taming of companion animals. The rate of human-induced or related extinction has increased during the past 12,000 years. The past two centuries have seen even faster rates of extinction that, if continued for another two centuries, could equal the great mass extinctions of the geologic past.

The fossil record of large-bodied animal extinctions shows variations during the last 100,000 years (table 1). African animals fared best; this may be because humans originated and evolved in Africa where the animals had already coexisted with humans for many thousands of generations. For the other continents and islands, the extinctions appear to follow the arrival of human immigrants—wherever humans went, extinctions followed (table 2).

TABLE 1
Extinctions of Large-Bodied Animals

	Genera Extinct in Last 100,000 Years	Genera Still Living	% Genera Gone Extinct
Africa	7	42	14%
North America	33	12	73
South America	46	12	79
Australia	19	3	86

Source: Martin and Klein (1989)

TABLE 2
Human Migrations and Large-Animal Extinctions

	Appearance of Humans (years ago)	Concentrated Extinctions (years ago)
Africa	~200,000	—
Europe	Over 100,000	12,000–10,000
Australia	56,000	40,000
North America	15,000	12,000–10,000
South America	15,000	12,000–8,000
Madagascar	1,500	By 500
New Zealand	1,000	900–600

Source: Martin and Klein (1989)

In Greater Depth

La Brea Tar Pits, Metropolitan Los Angeles

A stunning example of this mass extinction lies in the heart of downtown Los Angeles at the La Brea Tar Pits. The tar pits formed where oil from underground reservoirs seeped upward through fractures in overlying rocks to reach the surface. The natural gas and lighter-weight oils evaporated, leaving behind sticky, high-viscosity **asphalt** in pools. In the last 40,000 years, individual organisms from more than 660 species have become stuck and entombed in this asphalt.

If an animal steps into the asphalt, it usually cannot get out (figure 6). Why would sinking into a few inches of asphalt be enough to trap a large mammal? Visualize this scenario: you step into an asphalt pool and sink above your ankles. When you lift one foot to escape, what does your other foot do? It pushes down into the asphalt. You are trapped. Escape is even more difficult for a four-legged animal such as a mammoth.

What would a trapped animal do? Probably scream as loud as it could. Who would answer the distress call? Mostly carnivores and scavengers. More than 85% of the larger-bodied mammal fossils found at La Brea are carnivores (table 3). This is a curious fact, since ecosystems have herbivores in abundance and a lesser number of carnivores that feed on them. Apparently each herbivore trapped in the asphalt attracted hungry carnivores. This interpretation is supported by the bird fossils, which are mostly predators and scavengers, such as vultures, condors, and eagles.

So far, 59 species of mammals have been found, but much more work remains to be done. In addition to the animals listed in table 3, there are mammal fossils of bears, mammoths, mastodons, deer, tapir, and peccaries. The list of larger-bodied mammals is incomplete without mentioning the partial skeleton of a 9,000-year-old human. This is significant. Human beings were living in the area while the extinction of large-bodied mammals was occurring.

When you are in Los Angeles, please take the time to visit the still-active La Brea Tar Pits and see the artistic displays inside the adjoining George C. Page Museum; you will be glad you did. The site is easy to reach; it is at 5801 Wilshire Boulevard.

Figure 6 Statue of an extinct mammoth trapped in an asphalt deposit at the La Brea Tar Pits in metropolitan Los Angeles, California.

Photo by Pat Abbott.

TABLE 3

Proportions of Larger Mammals at La Brea[*]

Dire wolf	48%
Saber-toothed cat	30
Coyote	7
Bison	5
Horse	4
Sloth	3
Large cats	2
Camel	1
Antelope	1

[*]Based on more than 3,400 specimens.

Glossary

A

aa Lava flow with a rough, blocky surface.

acceleration (1) To cause to move faster. (2) The rate of change of motion.

acceleration due to gravity The acceleration of a body due to Earth's gravitational attraction, expressed as the rate of increase of velocity per unit of time (32.17 ft/sec/sec at sea level and 45° latitude).

Accumulated Cyclone Energy (ACE) Measure of energy expended by a tropical cyclone. Values are totaled to measure energy released during an entire season.

acoustic fluidization A theorized process whereby sound waves trapped inside a dry, fallen mass lessen internal friction to enable fluidlike flow.

acre foot A measure of water volume in which 1 acre of surface is covered 1 ft deep. An acre is about 90% of the area between the goal lines on a football field.

actualism Using the actual processes operating on Earth today to interpret the past; not inventing unrecognized processes to explain the past.

adiabatic process The change in temperature of a mass without adding or subtracting heat. Examples are cooling with expansion and warming upon compression.

aerosol A suspension of fine solid or liquid particles in air.

aftershock A smaller earthquake following a mainshock on the same section of a fault. Aftershocks can continue for years following a large mainshock.

air-mass thunderstorm Local, short-lived thunderstorm that may form and dissipate in several hours.

albedo The reflectivity of a body; for Earth, how much solar radiation is reflected back to space.

alluvial fan A gently sloping cone of sediment formed where a stream leaves the hills or mountains and flows out onto a plain or valley.

amplitude The maximum displacement or height of a wave crest or depth of a trough.

andesite A volcanic rock named for the Andes Mountains in South America. It is intermediate in composition between basalt and rhyolite and commonly results from melting of continental rock in basaltic magma.

anoxic Depleted of oxygen.

anticline A fold where rock layers are compressed into a convex-upward pattern.

anticyclone A region of high atmospheric pressure and outflowing air that rotates clockwise in the Northern Hemisphere.

aphelion The point in the orbit of a body that is farthest from the Sun.

archaea An ancient branch of life whose species can thrive under high pressures and temperatures; they derive their energy by breaking the chemical bonds of inorganic molecules.

ash Fine pyroclastic material less than 4 mm in diameter.

asphalt Dark-colored, high-viscosity petroleum residue remaining after natural gas and lightweight oil evaporate.

asteroid Stony or metallic mass that orbits the Sun.

asthenosphere The layer of the Earth below the lithosphere in which isostatic adjustments take place. The rocks here deform readily and flow slowly.

astrobleme An ancient impact site on Earth, usually recognized by a circular outline and the presence of highly disturbed, shocked rocks.

Atlantic Multidecadal Oscillation (AMO) Decades-long intervals of colder or warmer sea-surface temperatures in the North Atlantic Ocean. May relate to hurricane formation.

atmosphere The gaseous envelope around Earth, composed chiefly of nitrogen and oxygen. The average weight of the atmosphere on Earth's surface is about 14.7 lb/in^2.

avalanche A large mass of snow, ice, soil, or rock that moves rapidly downslope under the pull of gravity.

avulsion An abrupt change in the course of a stream and the adoption of a new channel.

B

backfire A fire deliberately set to consume fuel in front of an advancing wildfire in order to stop it.

barometer An instrument for measuring atmospheric pressure.

barrier island An elongated sandbar parallel to the land but separated by a lagoon or marsh.

basalt A dark, finely crystalline volcanic rock typical of low-viscosity oceanic lavas.

base isolation Protecting buildings from earthquakes by isolating the base of the building from the shaking ground via rollers, shock absorbers, etc.

base level The level below which a stream cannot erode; usually sea level.

base surge A cloud of volcanic gas and suspended debris that flows rapidly over the ground.

bauxite Aluminum ore formed as a residual soil in the tropics where warm, heavy rainfalls dissolve and carry off most other materials.

BCE Before the common era. Equivalent to BC.

bedding The layering of rocks, especially sedimentary rocks.

bedrock Solid rock lying beneath loose soil or unconsolidated sediment.

bentonite A general term for clays derived from volcanic ash; these clays have extreme swell-shrink properties.

blind thrust A reverse fault of shallow inclination that does not break the surface.

blizzard Strong, cold winds filled with snow.

body wave Seismic wave that travels through the body of the Earth—for example, primary and secondary waves.

Bowen's reaction series The order of crystallization of common minerals from a cooling magma.

braided stream An overloaded stream so full of sediment that water flow is forced to divide and recombine in a braided pattern.

breakwater An offshore structure built parallel to the coastline to provide shelter from wave attack.

brittle Behavior of material whereby stress causes abrupt fracture.

buffer A mixture of substances in solution that neutralize changes, thus acting to maintain an equilibrium composition. For example, the world ocean has several buffer systems.

buoyancy The quality of being able to float, usually on water or rock.

C

caldera A large (more than 2 km diameter), basin-shaped volcanic depression, roughly circular in map view, that forms by a pistonlike collapse of a cylinder of overlying rock into an underlying, partially evacuated magma chamber.

calorie The amount of heat required to raise the temperature of 1 gram of water 1 degree centigrade at a pressure of 1 atmosphere.

carbonic acid A common but weak acid (H_2CO_3) formed by carbon dioxide (CO_2) dissolving in water (H_2O).

carnivore A flesh-eating animal.

carrying capacity The maximum population size that can be supported under a given set of environmental conditions.

cavern A large cave.

CE Common era. Equivalent to AD.

centigrade A temperature scale that divides the interval between the freezing and boiling points of water into 100°. Conversion from the Fahrenheit scale is by $C = 5/9(F - 32)$.

chaparral A dense, impenetrable thicket of stiff shrubs especially adapted to a dry season about six months long; abundant in California and Baja California. Fire is part of the life cycle of these plants.

chemical weathering The decomposition of rocks under attack of base- or acid-laden waters.

chondrite A stony meteorite characterized by the presence of small rounded grains or spherules.

cinder cone Steep volcanic hill made of loose pyroclastic debris.

clay minerals Very small (less then 1/256 mm) minerals with a sheet- or booklike internal crystal structure. Many varieties absorb water or ions into their layering, causing swelling or shrinking.

climate The average weather conditions at a place over many years.

coal A sedimentary rock made largely of plant remains. Coal is readily combustible.

coesite A very dense mineral of SiO_2; a polymorph of quartz created under a pressure of about 20,000 atmospheres.

cohesion A mass property of sediments whereby particles cohere or stick together.

combustion Act of burning.

comet An icy body moving through space that releases gas and dust. They commonly are referred to as dirty snowballs, but some are icy dirtballs.

compaction The decrease in volume and porosity of a sediment via burial.

composite volcano A volcano constructed of alternating layers of pyroclastic debris and lava flows. Also known as a *stratovolcano*.

compression A state of stress that causes a pushing together or contraction.

condensation The change of state of a substance from vapor to liquid.

conduction Transfer of heat downward or inward through material by communication of kinetic energy from particle to particle.

conglomerate A sedimentary rock dominated by gravel (pebbles, cobbles, boulders).

continent Lower-density masses of rock, exposed as about 40% of the Earth's surface: 29% as land and 11% as the floor of shallow seas.

continental drift The movement of continents across the face of the Earth, including their splitting apart and recombining into new continents.

convection A process of heat transfer whereby hot material at depth rises upward due to its lower density while cooler material above sinks because of its higher density.

convergence zone A linear area where plates collide and move closer together. This is a zone of earthquakes, volcanoes, mountain ranges, and deep-ocean trenches.

core The central zone or nucleus of Earth about 2,900 km (1,800 mi) below the surface. The core is made mostly of iron and nickel and exists as a solid inner zone surrounded by a liquid outer shell. Earth's magnetic field originates within the core.

Coriolis effect Moving objects experience Earth move out from beneath them; in the Northern Hemisphere, bodies move toward their right-hand sides, while in the Southern Hemisphere, they move toward their left.

crater An abrupt basin commonly rimmed by ejected material. In volcanoes, craters form by outward explosion, are commonly less than 2 km in diameter, and occur at the summit of a volcanic cone. Similar rimmed basins form by impacts with meteorites, asteroids, and comets.

creep The slow, gradual, more-or-less continuous movement of ice, soil, and faults under stress.

cross-section A two-dimensional drawing showing features in the vertical plane, as in a canyon wall or road cut.

crust The outermost layer of the lithosphere, composed of relatively low-density materials. The continental crust has lower density than oceanic crust.

crystallization The growth of minerals in a fluid such as magma.

cumulonimbus A dense, to wering, vertical cloud up to 20 km (12 mi) high. It is the final stage of growth of cumulus clouds.

cumulus A tall, puffy cloud made of rounded masses stacked upon each other, generally with a flat base and height <1 km (0.6 mi).

curie A measure of radioactivity equal to 3.7×10^{10} disintegrations per second.

Curie point The temperature above which a mineral will not be magnetic.

cyclone A region of low atmospheric pressure and converging air that rotates counterclockwise in the Northern Hemisphere.

D

daylighted bedding Rock layers that dip at a lower angle than the slope of a hill, thus the ends of the beds "see daylight." These layers are prone to mass movement due to lack of support of their ends.

debris Any accumulation of rock fragments; detritus.

debris flow Loose sediment plus water that is pulled downslope directly by gravity.

decompression melting The most common process creating magma; achieved by reducing pressure on hot rock, not by adding more heat.

delta The mass of sediment brought by a river and built outward into a standing body of water.

demographic divide Most wealthy countries have low birth rates and long life expectancies. Mostly poor countries have high birth rates and shorter life expectancies.

demographic transition The change from a human population with high birth rates and high death rates to one with low birth rates and low death rates.

demography The statistical study of populations.

density The mass per unit volume of a substance.

derecho Winds that blow straight ahead.

dew point temperature The air temperature when the relative humidity of an air mass reaches 100% and excess water vapor condenses to liquid water.

dielectric constant A measure of the displacement currents occurring after applying an electric field.

diffusion Intermingling movement caused by thermal agitation with flow of particles from hotter to cooler zones.

dike Magma injected as tabular masses into older rocks.

dip The angle of inclination measured in degrees from the horizontal.

dip-slip fault Fault where most of the movement is either up or down in response to pushing or pulling.

directivity Tendency for a rupturing fault to direct more energy in the direction it is moving.

discharge The volume of water flowing in a stream per unit of time.

divergence zone A linear zone formed where plates pull apart, as at a spreading center.

downburst A localized, severe downdraft of air that includes an outburst of strong winds on the ground.

drought A prolonged interval of dryness causing damage to plants and animals.

ductile Behavior of material whereby stress causes permanent flow or strain.

ductility The ability to change shape markedly without breaking; "plastic" behavior.

duff A mat of organic debris in which fire can smolder for days.

dust devil Tornado-like vortex that forms when air heated by hot ground rises in a spinning column.

E

earthquake The shaking of the Earth by seismic waves radiating away from a disturbance, most commonly a fault movement.

earthquake cluster A sequence of large earthquakes that occur closely in time and space.

earthquake weather A common misconception; there is no connection between weather and earthquakes. Weather occurs at the Earth's surface, and earthquakes occur deep below the surface.

easterly wave A wave disturbance oriented north–south within the trade winds. Easterly waves have converging winds with clouds and rain east of their axis, and diverging winds with

clear skies west of their axis. Some easterly waves host the growth of hurricanes.

eddy A circular moving water current; a whirlpool.

elastic Behavior of material whereby stress causes deformation that is recoverable; when stress stops, the material returns to its original state.

element Distinct varieties of matter; an atom is the smallest particle of an element.

El Niño A climate pattern that occurs every two to seven years when the trade winds relax and warm ocean water in the equatorial Pacific Ocean flows to the west coast of North America.

embayment An indentation of the shoreline; depressed land near the mouth of a river.

ember A small, glowing piece of burning material.

energy Capacity for performing work.

epicenter The point on the surface of the Earth directly above a fault movement (i.e., earthquake location).

epidemic An outbreak of disease.

equilibrium A state of balance in a system; a condition in which opposing processes are so balanced that changes cause compensating actions.

erosion The processes that loosen, dissolve, and wear away Earth materials. Active agents include gravity, streams, glaciers, winds, and ocean waves.

escape tectonics Collision of continents may cause large areas to move away from the pressure, such as Turkey moving west away from Arabia, and Indo-China squeezing eastward away from India.

evolution The change in life-forms (species) over time.

exponential growth Growth in a compound fashion that, given time, leads to incredible numbers.

extinction The die-off or elimination of a species.

eye Forms in the center of a hurricane when wind speeds exceed 119 km/hr (74 mph). Descending air in the eye warms, causing clear air.

eyewall Area of a hurricane surrounding the eye where winds spiral upward. Wind speeds and precipitation are greatest here.

F

Fahrenheit Temperature scale in which the boiling point of water is 212° and the freezing point is 32°. Conversion from the centigrade scale is by $F = 9/5\ C + 32$.

failed rift Site of a spreading center that did not open far enough to create an ocean basin.

fall A mass moving nearly vertically and downward under the influence of gravity.

fault A fracture or belt of fractures where the two sides move past each other.

feedback A change in a system that provokes further changes. See *positive feedback* and *negative feedback*.

Ferrel cell Surface air flows toward the poles from about 30° latitude until rising at the subpolar lows around 60° latitude; a secondary air cell formed between the Hadley and Polar cells.

fertility The ability to produce offspring; the proportion of births to population.

fetch The length of water surface the wind blows across to create waves or lake-effect snow.

fire The rapid combination of oxygen with organic material to produce flame, heat, and light.

firebrand Burning debris such as branches and embers that are lifted above a fire and carried away to possibly start new fires.

firestorm A fire large enough to disturb the atmosphere with excess heat, thus creating its own winds.

fissure A narrow parting or crack in rock.

flank collapse A catastrophic event whereby the side of an oceanic volcano falls into the sea.

flash flood A local and sudden flood of relatively great volume and high velocity that lasts for a short time following a few hours of intense rainfall. Flash floods cause many deaths.

flood basalt Tremendous outpourings of basaltic lava that form thick, extensive plateaus.

floodplain The nearly flat lowlands that border a stream and act as the stream bed during floods.

flood stage Water level at which a river overflows its banks or levees.

floodway River channels and adjacent lowlands reserved to handle flood water diverted from a larger river.

flow A mass movement whereby the moving body of material behaves like a liquid.

fluid Implies a flowing of any material, from gases to highly viscous magma, to something that usually is solid but may be mobilized by heating.

foehn wind A warm, dry wind on the leeside of a mountain range, pronounced *fewn*.

fold Beds of rock that have been bent and warped. See *anticline* and *syncline*.

footwall The underlying side or block of a fault.

force Mass times acceleration.

foreshock A smaller earthquake that precedes a mainshock on the same section of a fault.

fossil Evidence of former life, including bones, shells, teeth, leaves, and footprints.

fracking *See hydraulic fracturing.*

fracture A general term for any breaks in rock. Fractures include faults, joints, and cracks.

fracture zone Major lines of weakness in oceanic crust; former transform faults.

freezing rain Supercooled rain that turns to ice when it touches objects such as trees and powerlines.

frequency Number of events in a given time interval. For earthquakes, it is the number of cycles of seismic waves that pass in a second; frequency = 1/period.

friction The resistance to motion of two bodies in contact.

front A boundary separating air masses of different temperature or moisture content.

fuel Any substance that produces heat by combustion.

fuel-driven fire Fire burning on calm weather days that advances slowly through the fuel, giving firefighters opportunities to stop the fire.

fusion Act of melting.

G

gabion Literally means large cage. A mass of stones held in place by steel mesh or cables.

gene The fundamental unit of inheritance; carries the characteristics of parents to their offspring.

genome The common pool of genetic material shared by members of a species.

geothermal energy Energy derived from subsurface water heated to high temperatures by nearby magma.

geyser A hot spring that gushes magma-heated water and steam.

glacier A large mass of ice that flows downslope or outward due to the internal stresses caused by its own weight.

glass Matter created when magma cools too quickly for atoms to arrange themselves into the ordered atomic structures of minerals. Most glasses are supercooled liquids.

global climate model (GCM) A three- and four-dimensional computer model of Earth's atmosphere that simulates global climates produced by varying temperature, rainfall, atmospheric pressure, winds, and ocean currents.

global positioning system (GPS) Accurate measurement by satellite of monitored ground sites.

global warming potential (GWP) The ability of a greenhouse gas to trap heat in the atmosphere as compared to CO_2.

Gondwanaland A southern super-continent that included South America, Africa, Antarctica, Australia, New Zealand, and India during Jurassic time.

graded stream An equilibrium stream with an evenly sloping bottom adjusted to efficiently handle water flow (discharge) and sediment (load) transport.

gradient The slope of a stream channel bottom; change in elevation divided by distance.

granite A quartz-rich plutonic rock.

gravel Sediment pieces coarser than 2 mm in diameter, including granules, pebbles, cobbles, and boulders.

gravity The attraction between bodies of matter.

great natural disaster A disaster so overwhelming that outside assistance is needed to handle the rescue and recovery for the region.

greenhouse effect The buildup of heat beneath substances such as glass, water vapor, and carbon dioxide that allow incoming, short-wavelength solar radiation to pass through, but block the return of long-wavelength radiation.

groin A low, narrow barrier built perpendicular to shore to slow the longshore transport of sand.

groundwater The volume of water that has soaked underground to fill fractures and other pores; it flows slowly down the slope of the underground water body.

H

haboob Strong duststorms powered by thunderstorm downbursts in desert regions of the world.

Hadley cell A thermally driven atmospheric circulation pattern whereby hot air rises at the equator, divides and flows toward both poles, and then descends to the surface at about 30°N and S latitude.

hail Precipitation of hard, semispherical pellets of ice.

half-life The length of time needed for half of a radioactive sample to lose its radioactivity via decay.

halite Table salt; a mineral made of sodium chloride (NaCl) formed from evaporating seawater.

hangingwall The overlying side or block of a fault.

harmonic tremors Nearly continuous, small earthquakes created by underground magma on the move.

haze Fine dust, smoke, water, and salt particles that reduce the clarity of the atmosphere.

heat The capacity to raise the temperature of a mass, expressed in calories.

heat capacity The amount of heat required to raise the temperature of 1 gram of a substance by 1°C.

herbivore A plant-eating animal.

hertz One hertz (Hz) equals one cycle per second.

hook echo A band of heavy rain and hail in rotation around a supercell thunderstorm. Usually recognized by radar.

hot spot A place on Earth where a plume of magma has risen upward from the mantle and through a plate to reach the surface.

humidity A measure of the amount of water vapor in an air mass.

hurricane A large, tropical cyclonic storm with wind speeds exceeding 119 km/hr (74 mph); called a *typhoon* in the western Pacific Ocean and a *cyclone* in the Indian Ocean.

hydraulic fracturing Creation of cracks in subterranean rocks by pumping liquids underground under high pressure. Process especially used to extract natural gas or oil.

hydrograph A plot of water volume, height, flow rate, etc., with respect to time.

hydrologic cycle The solar-powered cycle in which water is evaporated from the oceans, dumped on the land as rain and snow, and pulled by gravity back to the oceans as glaciers, streams, and groundwater.

hydrosphere The waters of the Earth, including the oceans, lakes, and rivers.

hyperthermia Overheating of the body, commonly due to extreme weather. Body temperature greater than 40°C (104°F) can be life threatening.

hypocenter The initial portion of a fault that moved to generate an earthquake. Hypocenters are below the ground surface; epicenters are above them on the surface.

hypothermia Occurs when body temperature drops below 35°C (95°F) and the heart, respiratory, and other systems cannot function normally.

I

igneous rock Rock formed by the solidification (crystallization) of magma.

impermeable Impervious; the condition of rock that does not allow fluids to flow through it.

inertia The property of matter by which it will remain at rest unless acted on by an external force.

influenza Highly contagious virus-caused diseases. The word *influenza* is commonly shortened to *flu*.

insolation Amount of solar radiation received at any area on Earth.

intertropical convergence zone (ITCZ) The zone where collision occurs between the trade winds of the Northern and Southern Hemispheres.

inversion layer An atmospheric layer in which the upper portion is warmer or less humid than the lower.

ion An electrically charged atom or group of atoms.

isobar A map line connecting points of equal pressure.

isostasy The condition of flotational equilibrium wherein the Earth's crust floats upward or downward as loads are removed or added.

isotope Any of two or more forms of the same element. The number of protons is fixed for any element, but the number of neutrons in the nucleus can vary, thus producing isotopes.

J

jet stream Fast-moving belts of air in the upper troposphere that flow toward the east.

jetty A structure built perpendicular to the shore, usually to protect the entrance to a harbor.

joint A fracture or parting in rock.

jokulhlaup Glacial outburst flood.

K

keyhole A small region of space near Earth where gravitational attraction can change the course of a space object.

kilopascal A unit of pressure equal to 0.1 millibar.

kinetic energy Energy due to motion. *See potential energy.*

Kuiper belt A flattened disk of comets with orbital periods of less than 200 years traveling in an orbital plane similar to the planets of the Solar System but extending out 50 astronomical units.

L

ladder fuel Vegetation of varying heights in an area that allows fire to move easily from the ground to the treetops.

lahar A volcanic mudflow composed of unconsolidated volcanic debris and water.

lake-effect snow Snow produced after cold air blows over large areas of warmer water, picking up water vapor that forms clouds that precipitate snow just past the downwind shoreline.

La Nada A climate pattern that occurs when seawater temperatures in the tropical eastern Pacific Ocean are neither excessively warm nor cool but neutral.

La Niña A climate pattern that occurs when cooler than normal seawater exists in the tropical eastern Pacific Ocean.

lapse rate The rate at which Earth's atmosphere cools with increasing altitude, or warms with decreasing altitude.

latent heat The energy absorbed or released during a change of state.

latent heat of condensation The heat released when vapor condenses to liquid. For water, the heat release is about 600 calories per gram.

latent heat of fusion Water releases about 80 calories per gram when it freezes. In reverse, ice absorbs about 80 calories per gram when it melts.

latent heat of vaporization Water absorbs about 600 calories per gram when it evaporates. This stored heat is released during condensation.

latitude Reference lines that encircle Earth parallel to the equator. The equator is 0° latitude, and other lines are proportioned up to 90°N (North Pole) or 90°S (South Pole). Perpendicular to longitude.

Laurasia A northern supercontinent that included most of North America, Greenland, Europe, and Asia (excluding India) from about 180 to 75 million years ago.

lava Magma that flows on the Earth's surface.

lava dome A mountain or hill made from highly viscous lava, which plugs the central conduit of volcanoes.

law of cross-cutting relationships A feature (rock body, fault, erosion surface) is younger than any rock body it cuts across.

law of faunal assemblages Similar assemblages of fossil organisms indicate similar ages for the rocks that contain them.

law of faunal succession Fossil organisms succeed one another in a definite and recognizable order.

law of original continuity A water-laid sediment body continues laterally in

all directions until it thins out due to nondeposition or butts against the edge of the basin of deposition.

law of original horizontality Sediments are deposited in nearly horizontal layers.

law of superposition In a sequence of sedimentary rock layers, the oldest layer is at the base, and ages are progressively younger toward the top.

left-lateral fault A strike-slip fault where most of the displacement is toward the left hand of a person straddling the fault.

levee A natural or human-built embankment along the sides of a stream channel.

lifting condensation level The altitude in the atmosphere where rising air cools to saturation (100% humidity) and condensation begins.

lightning A flashing of light as atmospheric electricity flows between clouds or between clouds and the ground.

limestone A sedimentary rock composed mostly of calcium carbonate ($CaCO_3$), usually precipitated from warm saline water. Limestones on continents may later be dissolved by acidic groundwater to form caves.

liquefaction The temporary transformation of water-saturated, loose sediment into a fluid; typically caused by seismic waves.

liquid Implies a flow characteristic of water. It has a definite volume but no definite shape.

lithosphere The outer rigid shell of the Earth that lies above the asthenosphere and below the atmosphere and hydrosphere.

Little Ice Age A colder interval between about 1400 and 1900 CE marked by renewed glaciation in the Northern Hemisphere.

load The amount of material moved and carried by a stream.

loess Extensive deposits of wind-blown fine sediment (silt, very fine sand, clay) commonly winnowed from glacially dumped debris.

logarithm The exponent of that power of a fixed number that equals a given number.

longitude Reference lines connecting the North and South Poles. The line running through Greenwich, England,

is the 0° line; all other lines are counted away toward either 180°W or 180°E.

longshore current Waves in the breaker zone hit the beach at an angle, causing a "current" to flow down the beach.

longshore drift Beach sediment (and bathers) are carried along the shoreline if waves are striking the beach at an angle.

M

magma Molten or liquid rock material. It crystallizes (solidifies) on the Earth's surface as volcanic rock and at depth as plutonic rock.

magnetic field A region where magnetic forces affect any magnetized bodies or electric currents. Earth is surrounded by a magnetic field.

magnetic pole The point where Earth's magnetic field flows back into the ground. Currently, this point is near the North Pole.

magnetism A group of physical phenomena associated with moving electricity.

magnitude An assessment of the size of an event. Magnitude scales exist for earthquakes, volcanic eruptions, hurricanes, and tornadoes. For earthquakes, different magnitudes are calculated for the same earthquake when different types of seismic waves are used.

mainshock The largest earthquake in a sequence.

major hurricane The most damaging of hurricanes. They are category 3 to 5 strength with winds exceeding 179 km/hr (111 mph).

mantle The largest zone of the Earth, comprising 83% by volume and 67% by mass.

map A two-dimensional representation showing the features in a near-horizontal surface, such as the ground.

marble Metamorphosed limestone.

maria Dark, low-lying areas of the Moon filled with dark volcanic rocks.

mass A quantity of material.

mass movement The large-scale transfer of material downslope under the pull of gravity.

Maunder Minimum A cooler interval that occurred between 1645 and 1715 CE when astronomers noted minimal sunspots on the Sun's surface.

meander The curves, bends, loops, and turns in the course of an underloaded stream that shifts its bank erosion from side to side of its channel.

Medieval Maximum A relatively warm interval in the Northern Hemisphere between about 1000 and 1300 CE.

mesocyclone A rotating mid-level region within a supercell thunderstorm.

mesosphere (1) The mantle from the base of the asthenosphere to the top of the core. (2) The atmospheric layer above the stratosphere and below the thermosphere.

metamorphic rock A former igneous or sedimentary rock whose mineralogy, chemistry, and texture have been changed due to high temperature and pressure.

metamorphism The changes in minerals and rock textures that occur with the elevated temperatures and pressures below Earth's surface.

meteor The light phenomenon that occurs when a meteoroid enters Earth's atmosphere and vaporizes; commonly called a *shooting star*.

meteorite A stony or metallic body from space that passed through the atmosphere and landed on the surface of Earth.

meteoroid A general term for space objects made of metal, rock, dust, or ice.

methane A gaseous hydrocarbon (CH_4).

methane hydrate An icelike deposit in deep-sea sediments of methane combined with near-freezing water.

Milankovitch theory Glacial advances and retreats on Earth are controlled by variations in the amount of solar radiation received at high latitudes during summer due to changes in Earth's orbit around the Sun and in the tilt angle and direction of Earth's axis of rotation.

millibar A unit of atmospheric pressure. At sea level, the pressure is about 1,013 millibars.

mineral A naturally formed, solid inorganic material with a characteristic chemical composition and physical properties that reflect an internally ordered atomic structure.

mitigation Actions taken by humans to minimize the possible effects of a natural hazard.

monsoon Winds that reverse direction seasonally. In summer, warm air rises above hot land, drawing in rain-bearing winds from over the ocean. In winter, the flow reverses.

mortality Death rate; the proportion of deaths to population.

N

natural disaster An event or process that destroys life and/or property.

natural hazard A source of danger to life, property, and the environment. The probability that a dangerous event will occur.

near-Earth asteroid (NEA) An asteroid that comes within 0.3 astronomical unit (Au) of Earth's orbit. Au is the mean distance from Earth to the Sun.

near-Earth object (NEO) A comet or asteroid whose orbit makes it potentially hazardous to Earth.

negative feedback Occurs in equilibrium systems where one change triggers another change that tends to negate the initial change and restores equilibrium.

neotectonics The study of the youngest faults and tectonic movements.

niche A site in the environment where an organism or species can successfully exist.

nor'easter A winter weather condition involving a low-pressure system with its center offshore of the Atlantic coasts of the United States and Canada. The counterclockwise rotation brings warm, moist offshore air into contact with cold Arctic air.

normal fault A dip-slip fault in which the upper fault block has moved downward in response to tensional stresses.

North Atlantic Oscillation (NAO) A shifting of atmospheric pressures over the North Atlantic Ocean occurring on a multiyear timescale.

nuclear fission Splitting the nucleus of an atom, with the resultant release of energy, neutrons, and large daughter products.

nuclear fusion Combining smaller atoms to make larger atoms, with a resultant release of energy.

nuée ardente A turbulent "glowing cloud" of hot, fast-moving volcanic ash, dust, and gas; a pyroclastic flow.

O

obsidian Dark volcanic glass.

Oort cloud A vast and diffuse envelope of comets surrounding the Solar System.

order A unit in taxonomy that contains one or more families; its rank is below class.

oxidation Combination with oxygen. In fire, oxygen combines with organic matter; in rust, oxygen combines with iron.

ozone A gaseous molecule composed of three atoms of oxygen.

P

Pacific Decadal Oscillation (PDO) A 20- to 30-year-long climate pattern in North America influenced by sea-surface temperatures in the midlatitudes of the north Pacific Ocean. Warm phases bring increased storminess.

pahoehoe Lava flow with a smooth, ropy surface.

paleontologist Scientist who studies the fossils of animals, plants, and other life-forms.

paleontology The study of fossils and the evolution of life through time.

paleoseismology The study of prehistoric earthquakes.

pandemic A disease occurring over a wide area and affecting many people.

Pangaea A supercontinent that existed during Late Paleozoic time when all the continents were unified into a single landmass.

Panthalassa A massive, single ocean that occupied 60% of Earth's surface in Late Paleozoic time.

perihelion The point in the path of an orbiting body that is closest to the Sun.

period The length of time for a complete cycle of seismic waves to pass; equals 1/frequency.

permeability The capacity of a porous material to transmit fluids.

photosynthesis The process whereby plants produce organic compounds from water and carbon dioxide using the energy of the Sun.

pillow lava Lava cooled underwater into pillow-shaped masses.

piping Formation of conduits due to erosion by water moving underground.

plastic The behavior of a material that flows as a fluid (liquid) over time, but is strong (solid) at a moment in time.

plate A piece of lithosphere that moves atop the asthenosphere. There are a dozen large plates and many smaller ones.

plateau An elevated, extensive tract of land; a tableland.

plate tectonics The description of the movements of plates and the effects of plate formation, collision, subduction, and slide-past.

Plinian eruption The eruptive phase of a volcano during which an immense column of pyroclastic debris and gases is blown vertically to great heights.

plume An arm of magma rising upward from the mantle.

plutonic rock Rock formed by the solidification of magma deep below the surface. Named for Pluto, the Roman god of the underworld.

Polar cell Cold, dense air over the poles flows along the surface to about 60° latitude where it is warm and moist enough to rise at the subpolar lows and flow back toward the poles at higher altitudes.

polymorph The characteristic of a chemical substance to crystallize in more than one mineral structure.

pore An opening or void space in soil or rock.

pore-water pressure Pressure buildup in underground water that offsets part of the weight (pressure) of overlying rock masses.

porosity The percentage of void space in a rock or sediment.

positive feedback Occurs in nonequilibrium systems where one change triggers more changes in the same direction.

post-tropical cyclone A hurricane may transition into a post-tropical cyclone as it travels into the mid-latitudes and moves over cold water, collides with a front, and encounters increasing vertical wind shear.

potential energy The energy a body possesses because of its position; for example, a large rock sitting high on a steep slope. See *kinetic energy*.

potentially hazardous asteroid (PHA) Asteroid with an orbit close to Earth's orbit; they come within 8 million km (5 million mi) of Earth

and are big enough to survive passing through Earth's atmosphere and cause damage on a regional or global scale.

power The rate of work.

predation The biologic activity in which a predator organism feeds on a prey organism.

pressure gradient force Perpendicular to the air pressure isobars on a map is a gradient or slope of pressure change. Air flows from high to low pressure.

primary (P) wave The first seismic wave to reach a seismometer. Movement is by alternating push-pull pulses that travel through solid, liquid, and gas.

processes of construction Land-building processes of volcanism, seafloor formation, earthquakes, and continent collisions fueled by Earth's internal energy.

processes of destruction Land-destroying processes such as erosion and landsliding that are fueled by Earth's external energy sources, the Sun and gravity.

pumice Volcanic glass so full of holes that it commonly floats on water.

pyroclastic Pertaining to magma and volcanic rock blasted up into the air.

pyroclastic fall Volcanic material that falls back to earth after an explosive eruption. Pyroclasts range in size from microscopic to bigger than a bus.

pyroclastic flow A high-temperature, fast-moving cloud of fine volcanic debris, steam, and other gases.

pyroclastic surge A type of pyroclastic flow having higher steam content and less pyroclastic material. Surges are lower density, more dilute, high velocity, and may flow outward in a radial pattern.

pyrolysis Chemical decomposition by the action of heat.

Q

quarry An openly mined area on the surface.

quartz A resistant, rock-forming mineral made of silicon and oxygen (SiO_2). The most common mineral found as a sand grain.

quick clay A clay that loses nearly all its shear strength after being disturbed.

quicksand Loose sand partially supported by upward-flowing water to create a semiliquid mass into which heavy objects sink.

R

radiation Heat emitted as rays.

radiative forcing Changes in energy flow in and out of Earth's atmosphere measured in watts per meter squared.

radioactive isotope Unstable element containing excess subatomic particles that are emitted to achieve a smaller, stable atom.

radioactivity The breakdown of unstable atomic nuclei by emission of particles or radiation. The decay process produces smaller atoms and gives off heat.

radiometric dating The determination of ages using measured laboratory amounts of (1) radioactive parent atoms and (2) decay-product atoms, and then (3) using half-lives to compute the lengths of time of radioactive decay.

rain bands Spiraling bands of thunderstorms in a hurricane that yield intense rainfall.

recurrence interval The average time interval between floods or earthquakes of a given size.

reef An organism-built structure or current-deposited mound of $CaCO_3$ material (limestone).

regional floods Huge, long-lasting floods in large river valleys with low topography resulting from prolonged heavy rains over an extensive region. They cause heavy economic losses.

resonance The act of resounding, ringing. A vibrating body moves with maximum amplitude when the frequency of seismic waves is the same as the natural frequency of the body.

resurgent caldera A large topographic depression formed by pistonlike collapse of the overlying roof into a magma chamber, with a later central uplift of the caldera floor.

resurgent dome The uplifted floor and mass of magma in the center of a large volcanic caldera.

retrofit Reinforcement or strengthening of an existing building or other structure.

return period Amount of time between an event of a given size.

reverse fault A dip-slip fault where the upper fault block has moved upward in response to compressional stresses.

rhyolite A volcanic rock typical of continents. Typically forms from high-viscosity magma.

ridge A volcanic mountain range that lies along the spreading centers on the floors of the oceans.

rift The valley created at a pull-apart zone. Term commonly describes the valley that occurs along the axis of the volcanic mountain ranges of seafloor spreading centers.

right-lateral fault A strike-slip fault where most of the displacement is toward the right hand of a person straddling the fault.

rip current A strong current flowing seaward from the shore; erroneously called riptide.

riprap Large, irregular boulders placed to slow the attack of erosion.

rock A solid aggregate of minerals.

rogue wave An unusually tall wave created when several wave systems briefly and locally combine their energies.

rotational slide Downward-and-outward movement of a mass on top of a concave-upward failure surface.

S

salinity The total quantity of dissolved salts. For seawater, salinity is about 35 parts per thousand.

salt dome An upward-risen plug of salt that behaves like a viscous fluid when the high pressures of deep burial cause the salt to deform and flow.

sand Sediment grains with diameters between 1/16 and 2 mm.

scarp A steep, clifflike face or slope.

scoria Basaltic rocks with numerous holes formed by gases escaping from magma.

scoria cone A small cone or horseshoe-shaped hill made of pyroclastic debris from Hawaiian or Strombolian-type eruptions. They commonly occur in groups.

seafloor spreading Where tectonic plates pull apart, magma wells up and solidifies to create volcanic mountains, which in turn are pulled apart as new ocean floor.

seamount Submarine hill or mountain, typically a former volcano.

secondary (S) wave Second seismic wave to arrive at the seismometer. Movement occurs by shearing particles at right angles to the travel path. S waves move through solids only.

sector collapse When a volcano becomes structurally unable to support its own mass, a large portion of the volcano can fail catastrophically.

sediment Fragments of material of either inorganic or organic origin. Sizes are gravel (more than 2 mm), sand (2 to 0.0625 mm), silt (0.0625 to 0.0039 mm), and clay (less than 0.0039 mm). A mixture of silt and clay equals mud.

seiche An oscillating wave on a lake or landlocked sea that varies in period from a few minutes to several hours. Pronounced *saysh*.

seism Earthquake.

seismic-gap method Earthquakes are expected next along those fault segments that have not moved for the longest time.

seismic moment A measure of earthquake size that involves amount of movement on the fault, the shear strength of the rocks, and the area of fault rupture.

seismic wave A general term for all waves generated by earthquakes.

seismogram The record made by a seismograph.

seismograph An instrument that records vibrations of the Earth.

seismology The study of seismic waves generated by earthquakes.

seismometer An instrument that detects Earth motions.

sensible heat Heat carried by conduction or convection in air, water, or rock.

severe thunderstorm Thunderstorm with winds faster than 58 mph; it can produce large hail and tornadoes.

shatter cone Distinctively grooved and fractured cone-shape structure in rock.

shear The failure of a body where the mass on one side slides past the portion on the other side.

shield volcano A very wide volcano built of low-viscosity lavas.

shooting star Tiny space particles (about 1 mm in diameter) that burn up with a flash of friction-generated light in Earth's atmosphere.

side effect A misleading term that directs attention away from the concept that every action has multiple reactions.

sill A tabular body of igneous rock intruded between and parallel to rock layers.

silt Sediment grains with diameters between 1/16 and 1/256 mm.

single-cell thunderstorm A single thunderstorm with one main updraft, They pass through their life cycle of development, maturity, and decay with a few hours. See *air-mass thunderstorm*.

sinkhole A circular depression on the surface created where acidic water has dissolved limestone.

sinuosity The length of a stream channel divided by the straight-line distance between its ends.

slash Debris such as logs, branches, and needles left on the ground by logging or high winds.

slate Mud changed to hard rock by the high temperatures and pressures of metamorphism.

sleet Precipitation of fine icy particles formed as frozen rain.

slide A gravity-pulled mass movement on top of a failure or slide surface; a landslide.

slip The actual displacement along a fault surface of formerly continuous points.

slump A landslide above a curved failure surface.

slurry A highly mobile, low-viscosity mixture of water and fine sediment.

soil The surface layers of sediment, organic matter, and decomposing bedrock.

solar radiation Energy emitted from the Sun mostly in the infrared, visible light, and ultraviolet wavelengths.

solar wind The outflow of charged particles (ions) from the Sun.

species Organisms similar enough in life functions to breed freely together.

specific gravity The ratio of the density of a material to the density of water.

specific heat The amount of heat required to increase the temperature of 1 gram of a substance by 1°C.

spreading center The site where plates pull apart and magma flows upward to fill the gap and then solidifies as new lithosphere/ocean floor.

spring A place where groundwater flows out onto the surface.

stishovite A high-pressure, extremely dense mineral made of SiO_2. It is a polymorph of quartz produced by shock metamorphism.

strain A change in the form or size of a body due to external forces.

strata Sedimentary or volcanic rock layers with distinct physical or paleontological characteristics. Singular form is *stratum*.

stratigraphic sequence A stack of rock layers accumulated over time.

stratosphere The stable atmospheric layer above the troposphere.

stratovolcano A composite cone built of layers of both lava and pyroclastics.

stress External forces acting on masses or along surfaces; forces include shear, tension, and compression.

strike The compass bearing of the trend of a rock layer as viewed in the horizontal plane.

strike-slip fault Fault where most of the movement is horizontal or slide-past in character.

sturzstrom Long-run-out movements of huge masses at great speeds.

subduction The process of one lithospheric plate descending beneath another one.

sublimation Changing from solid to gas without passing through a liquid phase.

subside To sink or settle downward; to become lower.

subsidence Downward movement of the ground either slowly or catastrophically.

sunspot A dark area on the Sun's surface with below-average temperature.

supercell thunderstorm A complex thunderstorm formed from a huge updraft. Commonly is tilted, allowing rain, hail, and tornadoes to exist side-by-side.

supernova The cataclysmic eruption of a star that releases tremendous quantities of energy.

super position Law of The principle that successively younger rock layers are deposited on top of lower older rock layer.

surface tension The attractive force between molecules at a surface.

surface wave A class of seismic waves that travel along the surface only—for example, Love and Rayleigh waves.

surge A large mound of seawater that builds up within the eye of a hurricane and then spills onto the land.

swell One of a series of regular, long-period, somewhat flat-crested waves that travel outward from their origin.

syncline A fold where rock layers are compressed into a concave-upward position.

T

tectonic cycle New lithosphere forms at oceanic volcanic ridges, the lithospheric plates spread apart to open ocean basins, and then the oceanic plates are reabsorbed into the mantle at subduction zones.

tectonics The deformation and movement within the Earth's outer layers.

tektites Rounded pieces of silicate glass formed by impact melting.

tension A state of stress that tends to pull a body apart.

thermohaline flow The flow of deep-ocean waters made denser by coldness (thermo) and saltiness (haline).

thrust fault A reverse fault where the upper fault block is pushed up a shallow-dipping fault surface.

thunder The sound given off by rapidly expanding gases along the path of a lightning discharge.

thunderstorm A tall, buoyant cloud of moist air that generates lightning and thunder, usually accompanied by rain, gusty winds, and sometimes hail.

tidal bore During exceptionally high, rising tides a body of sea water may be funneled through a river mouth and then flow upriver; a *tidal wave*.

tidal friction Gravitational attraction between Earth, Moon, and Sun stretches the solid mass of the Earth and converts some energy from Earth's rotation into heat.

tides The alternate rising and falling of land and water surfaces due to the gravitational pulls of the Moon and Sun.

tipping point The point at which long-term small changes suddenly produce large effects.

topography The shape of Earth's surface both above and below sea level.

topple A large rock mass that has fallen over.

tornado Spinning funnels of wind whose rotating wind speeds can exceed 480 km/hr (300 mph).

tornado outbreak Six or more tornadoes within 24 hours from a single weather system.

tornado warning An alert from the National Weather Service that a tornado is occurring or is imminent.

trade winds Drying winds that blow toward both sides of the equator; from 30°N latitude, they move to the southwest, and from 30°S latitude, they blow to the northwest.

transform fault A strike-slip fault that connects the ends of two offset segments of plate edges, such as spreading centers or subduction zones.

translational slide A mass that slides downward and outward on top of an inclined planar surface.

trench The elongate and narrow troughs where ocean water can be more than twice as deep as usual. Trenches mark the down-going edges of subducting plates.

triple junction A place where three plate edges meet.

tropical cyclone Any weather system formed over tropical waters that rotates counterclockwise in the Northern Hemisphere.

tropical depression A tropical cyclone with wind speeds less than 63 km/hr (39 mph).

tropical disturbance A low-pressure system in the tropics characterized by thunderstorms and weak surface wind circulation.

tropical storm A tropical cyclone with wind speeds between 63 and 119 km/hr (39 and 74 mph).

tropical wave Surface low-pressure system over northwest Africa that moves westward within the trade winds. Above warm Atlantic Ocean water, such systems may grow into tropical storms or hurricanes.

tropopause The top of the troposphere.

troposphere The lowest layer of the atmosphere, ranging from 18 km (11 mi) thick at the equator to 8 km (5 mi) thick at the poles.

tsunami Giant, long-period sea waves caused by oceanic disturbances, such as fault movements, volcanic eruptions, meteorite impacts, and landslides.

turbidite A rock layer that grades from coarse sediment at the bottom to fine at the top. It is deposited from underwater density (turbidity) current events.

typhoon A large, tropical cyclonic storm with wind speeds exceeding 119 km/hr (74 mph); called a hurricane in the Western Hemisphere.

U

ultra-Plinian Exceptionally large outpourings of pyroclastic material in high eruption columns and voluminous ash-flow sheets that cover wide areas.

uniformitarianism The concept that the same laws and processes operating on and within Earth throughout geologic time are the same laws and processes operating today.

V

vaporization Act of conversion to gas.

virus Submicroscopic agents of many infectious diseases. Viruses replicate inside living cells of organisms.

viscosity The property of a material that offers internal resistance to flow, its internal friction. The lower the viscosity, the more fluid the behavior.

viscous Ease of flow. The more viscous a substance, the less readily it flows.

volatile Describes substances that readily become gases when pressure is decreased or temperature increased.

volcanic plug Vertical mass of igneous rock that cooled inside the central conduit of a volcano.

volcanic rock Rock formed by solidification of magma at Earth's surface.

volcano An opening of the Earth's surface where magma has poured or blown forth, typically creating hills or mountains.

vortex A whirling body of water or air with circular rotation around an axis that may be oriented vertically, horizontally, or somewhere in between.

W

wall cloud A markedly lower cloud that may form in the area of strongest updraft in a supercell thunderstorm. Many strong tornadoes form within wall clouds.

waterspout Tornado-like vortex that forms above warm water when rising water vapor condenses into a column.

water table The upper surface of the groundwater body. It is nearer the surface during rainy intervals and deeper below the surface during droughts.

wavelength The distance between two successive wave peaks, or troughs, in seismic waves or ocean waves.

wave refraction The bending of waves. A segment of ocean wave in shallower water slows while the segment in deeper water continues to race ahead, thus causing a bend in the wave. A seismic wave will bend when it passes into rocks with different physical properties.

weather The state of the air at a locale with respect to hot or cold, wet or dry, calm or storm.

weathering The surface processes that physically disintegrate and chemically decompose rock to produce soil and sediment.

wind-driven fire Wind-driven fire fronts that move fast. The wind carries firebrands forward, starting spot fires several miles ahead. Firefighters scramble to put out spot fires but can do little against the flame front.

work Distance times force, where force equals mass times acceleration.

Y

yield stress The stress difference at which permanent deformation first occurs.

Text Credits

Index

M

macrobursts, 258
magma
    andestic, 146–148, 146f, 148t
    basaltic, 143f, 147, 148t
    chemical composition of, 145
    crystallized form, 32
    decompression melting, 150
    falling, as volcanic bomb, 153f
    rapid assembly and rise, 183, 183f
    rhyolitic, 146–149, 146f, 148t
    temperature of, 146–150
    viscosity of, 146–150, 152, 155t
    volatiles of, 146, 152, 155t
    volcanic eruptions and, 142
    volume of, 152, 155t
    water content of, 150–151
magnetic field, 36, 36f
magnetic polarity, 38, 38f
magnetic pole, 36
magnetism, 36
magnetization of volcanic rocks
    Atlantic Ocean floor magnetic map, 38f
    continents, 41–42
    deep oceanic trenches, 41
    earthquake evidence, 39–40, 40f
    lava and, 37–38
    ocean basin ages, 40–41
    oceanic mountain ranges, 41
    seafloor depth, systematic increases, 41, 42f
    seafloor magnetization patterns, 38–39
magnitude, 10–11, 61, 64
mainshock, 63–64
major flood stage, 364
major hurricanes, 289
Maldives coral-atoll islands, 453, 453f
Malibu Canyon dam, California, 444f
Mammoth Lakes, 167f
Mammoth Mountain, California, 195–197
Madagascar, extinctions, 483
Manicouagan impact crater, Quebec, Canada, 458f, 459, 466
Mann, Michael, 330
mantle, 25, 26
manufactured homes, 306
map, 51
maria, 457
Marina district, San Francisco, 100–101
Marques (ship), 205
Mars, 313, 313t, 458, 466, 467f
mass extinctions. *See* extinctions
mass movements
    adverse geologic structures, 417
    ancient slide surfaces, 417
    catastrophic subsidence, 437
    classification of, 418, 419f
    creep, 409–410
    daylighted bedding, 417
    defined, 408
    delta compaction, Mississippi River, 436fX, 439–440, 440f
    external causes of slope failures, 410–412
    falls, 418–419
    flows, 418 (*See also* flows)
    gravity, and, 408–410
    groundwater withdrawal, Mexico City, 437–438
    internal causes of slope failures, 412–418
    mitigation efforts, 434–436f
    oil withdrawal, Houston-Galveston, 439f
    rates of travel, 418t
    rotational slides, 419–421
    slides, 419–424
    slow subsidence, 437–439
    speed vs. moisture content, 418f
    submarine, 433
    subsidence, 437–440
    translational slides, 422–424

    triggers of, 418
    Venice, Italy, 439, 440f
    water, and, 412–415
    weak materials, 412–413, 412f
    Yosemite National Park, 419
Matthes, Francois, 330
mature stage of thunderstorm, 255f, 256f
Mauna Loa, Hawaii, 155, 155f, 161
Maunder Minimum, 330, 332
Mayan civilization, 329, 329f
McCormick, John, 361–362
meandering, 351, 351–352f
Medieval Maximum, 330
Mediterranean climate, 381
Meers fault, 133
mega-killer earthquakes (2003-2015), 78t
Melosh, Jay, 430, 471
melting, 150, 150f
Mercalli, Giuseppi, 66
Mercalli Intensity Scale, 66–71, 67–68t
mesocyclone, 257, 265f, 266
mesosphere, 226, 236
Meteor Crater, Arizona, 465, 465f, 468–469, 469f
meteorites, 24, 40, 458, 461–465, 464f
meteoroids, 458, 462–465, 462f
meteors, 458
methane (CH_4), 317, 332, 334, 334t, 346
methane hydrates, 317, 318f
Mexico
    Baja California fires and, 400–402, 401t
    Cocos plate, 86
    El Chichón Volcano, 188t, 327, 327f
    La Gloria, Veracruz, flu epidemic, 19
    Paricutin Volcano, 149f, 158f
    Popocatépetl Volcano, 10, 10f
    Veracruz, flu epidemic (2009), 19
Mexico City, groundwater withdrawal, 437–438
Mexico City earthquake (1985), 85–87, 87–88f
Mexico-type hurricanes, 293–294
Michener, James A., 5
Michoacan seismic gap, 86
microbursts, 258
middle latitudes, 241–244
Milankovitch, Milutin, 319
Milankovitch theory, 320
millibars, 236
Milne, A. A., 248
minerals, 147–148, 147f
Minnich, Richard A., 400
minor flood stage, 364
Mission Valley, California, 373
Mississippi River
    delta compaction, 439–440, 440f
    floods, and, 363–365f, 363–366, 364t
    intraplate earthquakes, 130, 130f
    New Orleans, 304, 304–305f
mitigation, 8
moderate flood stage, 364
moist adiabatic lapse rate, 233
moment magnitude scale, 63, 64f
money, compound interest on, 14
monsoon, 234
Montaigne, Michael de, 281
Mont Pelée, Martinique, eruption (1902), 186–187
Moon
    asteroidal bombardment of, 474–475, 475f
    gravity, and tides, 449, 449f
    impact craters, space objects, 457f
    impact origin of, 25
Moon rocks, 40
Morrison, David, 475
mortality rate, 13
Mount Etna, Italy, 171
Mount Fuji, Japan, 159f
Mount Kilimanjaro, 80
Mount Mayon, Philippines, 187, 187f
Mount Mazama, Oregon, 162–164, 164f
Mount Nyiragongo, Zaire, 194
Mount Ontake, Japan, 172

Mount Pinatubo, Philippines, 144, 144f, 151f, 153f, 157, 188, 197, 197f, 327, 328f
Mount Rainier, Washington, 155, 155f, 189–190
Mount Sanford, Alaska, 432
Mount Shasta, California, 181–182, 182f
Mount St. Helens, Washington, 165, 175–180f, 176–178
Mount Tambora, Indonesia, eruption (1815), 192, 328
Mount Unzen, Japan, 185
Mount Vesuvius eruption (79 CE), 144–145, 145f, 159–160, 159f
M_0 (seismic moment), 63
mud volcanoes, 334
Muir, John, 124
Mullineaux, Donal R., 181
Munch, Edvard, 164
Myanmar cyclone in (2008), 6, 7t, 8

N

Nashville, Tennessee, tornadoes, 271
National Oceanic and Atmospheric Administration (NOAA), 222–223
natural disasters
    deadliest, list of (1970-2013), 7t
    economic losses, 8, 9t
    fatality rates, 6
    frequency, 11
    government, role in, 6–8, 308
    human responses to, 8
    insurance, 8, 9t
    insured losses (1970-2013), 9t
    magnitudes, 10–11
    Popocatépetl Volcano (Mexico), 10, 10f
    population growth, and, 12–16
    return period, 11
    statistics (2013), 6
natural gas production, 117–118
natural hazards, 8–10
natural storage areas, 367–368
Nazca plate, Chile, 87–88
neap tides, 449
near-death stage, 338
near-Earth asteroids (NEAs), 475–476, 476f
near-Earth objects (NEOs), 475–476
negative feedback, 354
neotectonics, 113–115
Neptune, 462
Nero, Emperor, 382
Nevada
    Lake Mead, 29–30, 30f
    Lake Tahoe, 220, 220f
    Pleasant Valley, earthquake (1915), 125
    Reno, earthquake (2008), 127
Nevado del Ruiz eruption, 188–189, 189f
Nevados Huascarán events, Peru, 429–430
New England, earthquakes, 134
Newfoundland, Canada, tsunami (1929), 217–218, 218f
New Madrid, Missouri, 116, 130–132, 133f
New Orleans, 304–305f, 363, 440. *See also* Hurricane Katrina (2005)
Newton, Isaac, 65, 408, 449
New York
    Breezy Point, Queens, 389, 389f
    Oakwood Beach, Staten Island, 308
New Zealand
    extinctions, 483
    North Island, 79
    South Island, 50, 50–51f
Nicaragua tsunami (1992), 221–222
nitrous oxide (N_2O), 332, 335
NOAA (National Oceanic and Atmospheric Administration), 222–223
noncondensing greenhouse gases, 333, 334t
nonexplosive eruptions, 151
N_2O (nitrous oxide), 332, 335